Human Development

ACROSS THE LIFESPAN

second edition

John Dacey
Boston College

John Travers
Boston College

Book Team

Editor *Michael Lange*
Developmental Editor *Sheralee Connors*
Production Editor *Marlys Nekola*
Designer *Jeff Storm*
Art Editor *Kathleen M. Huinker-Timp*
Photo Editor *Robin Storm*
Permissions Coordinator *Vicki Krug*
Art Processor *Joyce E. Watters*
Visuals/Design Developmental Consultant *Marilyn A. Phelps*
Visuals/Design Freelance Specialist *Mary L. Christianson*
Publishing Services Specialist *Sherry Padden*
Marketing Manager *Elizabeth Haefele*
Advertising Manager *Nancy Milling*

WCB Brown & Benchmark
A Division of Wm. C. Brown Communications, Inc.

Executive Vice President/General Manager *Thomas E. Doran*
Vice President/Editor in Chief *Edgar J. Laube*
Vice President/Sales and Marketing *Eric Ziegler*
Director of Production *Vickie Putman Caughron*
Director of Custom and Electronic Publishing *Chris Rogers*

Wm. C. Brown Communications, Inc.

President and Chief Executive Officer *G. Franklin Lewis*
Corporate Senior Vice President and Chief Financial Officer *Robert Chesterman*
Corporate Senior VicePresident, and President of Manufacturing *Roger Meyer*

Cover Photo © Curt Maas/Tony Stone Images

Copyedited by Christianne Thillen

Freelance Permissions Editor Karen Dorman

The credits section for this book begins on page 645 and is considered an extension of the copyright page.

A Times Mirror Company

Library of Congress Catalog Card Number: 92–76068

ISBN 0–697–12732-X

Printed in the United States of America by Wm. C. Brown Communications, Inc., 2460 Kerper Boulevard, Dubuque, IA 52001

10 9 8 7 6 5 4 3 2 1

This book is dedicated with deep affection to the two people who have helped us the most—Linda Schulman and Barbara Travers, our wives.

Brief Contents

Part One Introduction

1 Lifespan Psychology: An Introduction 2

2 Theoretical Viewpoints 30

Part Two Beginnings

3 The Biological Basis of Development 56

4 Pregnancy and Birth 81

Part Three Infancy

5 Physical and Cognitive Development in Infancy 112

6 Psychosocial Development in Infancy 142

Part Four Early Childhood

7 Physical and Cognitive Development in Early Childhood 170

8 Psychosocial Development in Early Childhood 199

Part Five Middle Childhood

9 Physical and Cognitive Development in Middle Childhood 230

10 Psychosocial Development in Middle Childhood 262

Part Six Adolescence

11 Adolescence: Background and Context 298

12 Adolescence: Rapidly Changing Bodies and Minds 322

13 The Troubled Adolescent 346

14 Adolescence: Gender Roles and Sexuality 378

Part Seven Early Adulthood

15 Passages to Maturity 408

16 Early Adulthood: Physical and Cognitive Development 431

17 Early Adulthood: Psychosocial Development 453

Part Eight Middle Adulthood

18 Middle Adulthood: Physical and Cognitive Development 478

19 Middle Adulthood: Psychosocial Development 498

Part Nine Late Adulthood

20 Late Adulthood: Physical and Cognitive Development 524

21 Late Adulthood: Psychosocial Development 549

22 Dying and Spirituality 568

Contents

List of Boxes xiii
Preface xvii

Part One INTRODUCTION

Chapter 1
Lifespan Psychology: An Introduction **2**

A Biopsychosocial Model of Development **5**
Why Study Lifespan Development? **6**
The Meaning of Lifespan **6**
The Developmental Task Concept **9**
Needs across the Lifespan **10**
History of Lifespan Studies **14**
Child Studies **14**
Great Names in the History of Child Development **15**
Adolescence **15**
Adulthood **17**
Issues in Lifespan Development **17**
Culture and Development **18**
Continuity versus Discontinuity **18**
Stability versus Change **19**
Sensitive Periods vs. Equal Potential **20**
Developmental Research **20**
Data Collection Techniques **21**
Time Variable Designs **22**
Conclusion **26**
Chapter Highlights **27**
Key Terms **29**
What Do You Think? **29**
Suggested Readings **29**

Chapter 2
Theoretical Viewpoints **30**

The Importance of Theories **30**
Critical Issues **31**
Biopsychosocial Causes of Change **32**
The Psychoanalytic Approach: *Sigmund Freud's Theory [BIOPSYCHOsocial]* **32**
Defending the Unconscious Mind **33**
Structures of the Psyche **33**
The Developing Personality **34**
The Cognitive Structures Approach **36**
Jean Piaget's Theory [BIOPSYCHOsocial] **36**
Beginning the Study of Intellectual Development **36**
The Role of Intelligence **36**
Stages of Cognitive Development **38**
Lev Vygotsky's Theory [bioPSYCHOSOCIAL] **40**
The Behavioral Approach **41**
B. F. Skinner's Theory [biopsychoSOCIAL] **41**
Albert Bandura's Theory [biopsychoSOCIAL] **43**
The Psychosocial Crises Approach **45**
Erik Erikson's Theory [BIOPSYCHOSOCIAL] **45**
Conclusion **51**
Chapter Highlights **51**
Key Terms **53**
What Do You Think? **54**
Suggested Readings **54**

Part Two BEGINNINGS

Chapter 3
The Biological Basis of Development **56**

The Fertilization Process **58**
The Beginnings **58**
The Menstrual Cycle **60**
Infertility **64**
In Vitro Fertilization **67**
Chromosomes and Genes **69**
Heredity at Work **70**
DNA: Structure and Function **71**
How Traits Are Transmitted **73**
What Is Transmitted **73**
Examples of Chromosomal Disorders **75**
Examples of Genetic Disorders **76**
Conclusion **77**
Chapter Highlights **79**
Key Terms **80**
What Do You Think? **80**
Suggested Readings **80**

Chapter 4
Pregnancy and Birth **81**

The Prenatal World **82**
The Germinal Period **83**
The Embryonic Period **84**
The Fetal Period **85**
Development of the Senses **86**

Influences on Prenatal Development 89
Developmental Risk 89
Teratogens 89
Maternal Influences 96
Fetal Problems: Diagnosis and Counseling 99
The Birth Process 100
Stages in the Birth Process 100
Myths and Facts of Birth 101
Birth Complications 103
Childbirth Strategies 104
The Special Case of Prematurity 104
Conclusion 107
Chapter Highlights 107
Key Terms 108
What Do You Think? 109
Suggested Readings 109

Part Three INFANCY
Chapter 5
Physical and Cognitive Development in Infancy 112

Physical and Motor Development 114
Neonatal Reflexes 114
Newborn Abilities 115
Neonatal Assessment Techniques 117
Brain Development 118
Motor Development 118
Neonatal Problems 121
Perceptual Development 123
The Meaning of Perception 124
Visual Perception 126
Cognitive Development 128
Piaget's Sensorimotor Period 129
Criticisms of Piaget 130
Infants and Memory 133
Language Development 134
Key Signs of Language Development 135
Language Acquisition: The Theories 136
Conclusion 139
Chapter Highlights 140
Key Terms 141
What Do You Think? 141
Suggested Readings 141

Chapter 6
Psychosocial Development in Infancy 142

Relationships and Development 144
Background of the Relationship Research 144
The Importance of Reciprocal Interactions 145
The Meaning of Relationships 148
The Characteristics of a Relationship 149
The Givens in a Relationship 150
First Relationships 153
Relationships: Beginnings and Direction 153
The Developmental Sequence 154
Attachment 156
Bowlby's Work 156
Attachment Research 158
The Klaus and Kennell Studies 160
Fathers and Attachment 162
Current Research into Attachment 163
Early Emotional Development 164
Signs of Emotional Development 164
Theories of Emotional Development A Summary 165
First Feelings 165
Conclusion 166
Chapter Highlights 167
Key Terms 168
What Do You Think? 168
Suggested Reading 168

Part Four EARLY CHILDHOOD
Chapter 7
Physical and Cognitive Development in Early Childhood 170

Features of the Early Childhood Years 172
Physical and Motor Development 174
Characteristics of Physical Development 174
Continuing Brain Development 176
Physical Development: An Overview 177
Growing Motor Skills 179
The Special Case of Drawing 181
Cognitive Development 182
Piaget's Preoperational Period 182
Intellectual Differences 188
Language Development 192
Language as Rule Learning 193
The Pattern of Language Development 194
Speech Irregularities 194
Conclusion 196
Chapter Highlights 196
Key Terms 197
What Do You Think? 197
Suggested Readings 198

Chapter 8
Psychosocial Development in Early Childhood 199

The Family in Development 201
The Changing Family 202
Parenting Behavior 203
Homeless Children 204

Children of Divorce 206
Day Care 209
Early Childhood Education 213
Early Childhood Education: A Positive Outlook 213
Miseducation 214
The Self Emerges 215
The Development of Self 215
The Special Case of Self-Esteem 215
Gender Identity 217
The Importance of Play 222
The Meaning of Play 222
The Development of Play 224
Conclusion 225
Chapter Highlights 226
Key Terms 226
What Do You Think? 227
Suggested Readings 227

Part Five **MIDDLE CHILDHOOD**

Chapter 9
Physical and Cognitive Development in Middle Childhood **230**

Physical Development 232
Physical Changes in Middle Childhood 232
Motor Skills 233
Cognitive Development 235
Piaget and Concrete Operations 235
The Psychometric View of Intelligence 237
New Ways of Looking at Intelligence 239
Thinking and Problem Solving 243
Thinking Skills 243
Decision Making and Reasoning 245
Problem-Solving Skills 247
Moral Development 248
Piaget's Explanation 248
Kohlberg's Theory 250
In a Different Voice 252
Language Development 253
Using Language 254
Bilingualism 255
The Importance of Reading 256
Conclusion 259
Chapter Highlights 259
Key Terms 260
What Do You Think? 261
Suggested Readings 261

Chapter 10
Psychosocial Development in Middle Childhood **262**

Siblings and Development 264
The Developing Sibling Bond 264
How Siblings Help Each Other 265
How Siblings Affect Development 267
The Influence of Peers 268
Children's Friendships 269
Peers in Middle Childhood 270
Schools and Middle Childhood 272
Schools and Social Development 273
Middle Childhood Educational Change 274
Our Multicultural Schools 276
Bilingual Education 277
Television and Development 278
Television and Cognitive Development 278
Television and Violence 280
Television and Prosocial Behavior 281
Stress in Childhood 282
Abused Children 283
The Nature of the Problem 284
Types of Stress 285
Developmental Effects of Stress 287
Children and Violence 287
The Invulnerable Child 290
Conclusion 293
Chapter Highlights 295
Key Terms 296
What Do You Think? 296
Suggested Readings 296

Part Six **ADOLESCENCE**

Chapter 11
Adolescence: Background and Context **298**

How Should We Define Adolescence? 299
When Does It Start? 299
Ancient Times 301
The Middle Ages 301
The "Age of Englightenment" 302
The Twentieth Century 302
G. S. Hall and the Theory of Recapitulation 303
Theories of Adolescence: Anna Freud 305
Theories of Adolescence: Robert Havighurst 306
Theories of Adolescence: Erik Erikson 307
The Search for Identity 308
The Moratorium of Youth 308
Negative Identity 309
Identity Status 310

Changing American Families and Their Roles in Adolescent Life 312
The Loss of Functions 312
The Increase in Age-Related Activites Among Older Adolescents 313
The Effects of Divorce 313
The Effects of Gender 315
The Nurturing Parent 315
Interactions with Peers 316
Adult Anxieties about Adolescent Subculture 316
The Origins of Subcultures 317
Conclusion 319
Chapter Highlights 319
Key Terms 321
What Do You Think? 321
Suggested Readings 321

Chapter 12
Adolescence: Rapidly Changing Bodies and Minds 322

Puberty 322
Early Studies 322
Your Reproductive System 323
When Does Puberty Start? 326
The Effects of Timing on Puberty 327
Cognitive Development 332
Variables in Cognitive Development: Piaget 332
Variables in Cognitive Development: Flavell 333
Adolescent Egocentrism 336
Critical Thinking 337
Creative Thinking 338
The Use of Metaphor 339
Creativity, Giftedness, and the IQ 340
Conclusion 342
Chapter Highlights 343
Key Terms 344
What Do You Think? 344
Suggested Readings 345

Chapter 13
The Troubled Adolescent 346

Substance Abuse 347
Prevalence of Use 347
Ethnic Group and Abuse 349
Crime and Abuse 352
Substance Use and Personal Relationships 352
Combating Substance Abuse 353
Mental Health Issues 355
Types of Mental Disorders 356
Eating Disorders 357
Depression 358
The Crisis of Death 361
Suicide 363
Delinquent Behavior 369
The Nonaggressive Offender 369
The Juvenile Delinquent 371
Learning and Delinquency 372
Gangs 372
Conclusion 375
Chapter Highlights 375
Key Terms 377
What Do You Think? 377
Suggested Readings 377

Chapter 14
Adolescence: Gender Roles and Sexuality 378

Sexual Identity and Gender Roles 379
Aspects of Gender Role 379
Erik Erikson's Studies 380
Androgyny 381
Sexual Behavior 382
The Sexual Revolution 382
Stages of Sexuality 383
Autosexual Behavior 383
Homosexual Behavior 384
Causes of Homosexuality 385
The Onset of Homosexuality 385
Heterosexual Behavior 386
First Coitus 387
The Many Nonsexual Motives for Teenage Sex 389
The Janus Report 390
Sexual Abuse 391
Sexually Transmitted Diseases 392
AIDS 392
Other Sexually Transmitted Diseases 394
The Teenage Parent 396
Trends in Behavior 397
The Role of Family in Teen Pregnancies 398
Causes of Teenage Pregnancy 400
Conclusion 402
Chapter Highlights 402
Key Terms 404
What Do You Think? 404
Suggested Readings 405

Part Seven EARLY ADULTHOOD
Chapter 15
Passages to Maturity 408

Initiation Rites 409
Analysis of an Initiation Rite 410
The Passage to Adulthood in Western Countries 412
The Transition to Adulthood in the United States 412
Types of Initiation Activities in the United States 413

The Adolescent Moratorium 413
Implications of the Lack of an Initiation Ceremony 414
Two Proposals 416

Dealing with the Stresses of Adulthood 418
Change as a Source of Stress 419
Stimulus Reduction versus Optimum Drive Level 421
The General Adaptation Syndrome 422
Disease as a Result of Stress 424
Measuring the Relationship between Stress and Physical Illness 426
Risk and Resilience 426

Conclusion 427
Chapter Highlights 428
Key Terms 429
What Do You Think? 429
Suggested Readings 430

Chapter 16
Early Adulthood: Physical and Cognitive Development 431

Physical Development 432
The Peak Is Reached 432
Organ Reserve 433
The Effect of Life-Style on Health 433

Sexuality 439
Freud's Theory 439
Sexual Scripts: Gagnon and Simon 439
Sexual Motivations: Mitchell 440
The Sociobiological View: Wilson 441
Practices of Young Adults 442

Cognitive Development 444
Intellectual/Ethical Development 444
"Women's Ways of Knowing" 446

Physical and Cognitive Aspects of Love 448
The Seven Forms of Love: Sternberg 448
Validation: Fromm 449

Conclusion 450
Chapter Highlights 450
Key Terms 451
What Do You Think? 451
Suggested Readings 452

Chapter 17
Early Adulthood: Psychosocial Development 453

Marriage and the Family 454
Changing American Marriages and Families 454
Types of Marriage 457

Patterns of Work 458
Employment Patterns 459
How People Choose Their Careers 459
The Phenomenon of the Dual-Career Family 460

Stages of Personal Development 462
Transformations: Gould 462
The Adult Life Cycle: Levinson 466
Seasons of a Man's Life: Levinson 468
Intimacy versus Isolation: Erikson 470
Male versus Female Individuation 471

Conclusion 474
Chapter Highlights 474
Key Terms 475
What Do You Think? 475
Suggested Readings 475

Part Eight MIDDLE ADULTHOOD

Chapter 18
Middle Adulthood: Physical and Cognitive Development 478

Physical Development 478
Health 479
Muscular Ability 480
Sensory Abilities 480
The Climacteric 482

Sex in Middle Adulthood 484

Cognitive Development 485
Theories about Intelligence 485
New Views of Intelligence 487
The Development of Creativity 488
Information Processing 492
Learning Ability 493

Conclusion 495
Chapter Highlights 495
Key Terms 496
What Do You Think? 497
Suggested Readings 497

Chapter 19
Middle Adulthood: Psychosocial Development 498

Marriage and Family Relations 499
Marriage at Middle Age 499
Relationships with Aging Parents 500
Relationships with Siblings 500
The Middle-Aged Divorcée 502

Personality Development: Continuous or Changing 504
Continuity versus Change 504
Transformations: Gould 505
Seasons of a Man's Life: Levinson 506
Seasons of a Woman's Life: Levinson 511
Adaptations of Life: Vaillant 512
Generativity versus Stagnation: Erikson 514

Continuous Traits Theory 515
Patterns of Work 516
Special Problems of the Working Woman 517
The Mid-Career Crisis 518
Some Suggestions for Dealing with the Mid-Life Crisis 520
Conclusion 520
Chapter Highlights 520
Key Terms 521
What Do You Think? 522
Suggested Readings 522

Part Nine LATE ADULTHOOD

Chapter 20
Late Adulthood: Physical and Cognitive Development 524

Must We Age and Die? 526
Physiological Aspects of Aging 526
Genetic Aspects of Aging 529
Effects of the Natural Environment on Aging 529
Other Modifiers of Ability 530
Physical Development 531
Reaction Time 531
Sensory Abilities 534
Other Body Systems 535
Health 536
Appearance 539
Cognitive Development 541
Cognitive Decline in the Elderly: Tests versus Observations 541
Terminal Drop 542
Creativity 543
Conclusion 546
Chapter Highlights 547
Key Terms 548
What Do You Think? 548
Suggested Readings 548

Chapter 21
Late Adulthood: Psychosocial Development 549

Social Development 550
Gender Roles 550
Sexuality 550
The Elderly and Their Families 555
The Older Worker 558
Retirement 559
Personal Development 562
Personal Development: Committee on Human Development 562
Personal Development: Eriikson 565
Conclusion 566
Chapter Highlights 566
Key Terms 567
What Do You Think? 567
Suggested Readings 567

Chapter 22
Dying and Spirituality 568

The Role of Death in Life 568
What Is Death? 569
Four Types of Death 569
The Legal Definition of Death 570
Dealing Successfully with the Death of Others 571
Grief Work 571
Pathological Grieving 572
The Role of Grief 573
The Role of the Funeral 574
Dealing Successfully with One's Own Death 576
Kübler-Ross: The Stages of Dying 576
Suicide: The Saddest Death 579
The Overall Picture 579
The Influence of Race and Place of Residence on Suicide 580
The Influence of Gender on Suicide 580
Successful Dying 581
The "Death with Dignity" Law 581
The Hospice: "A Better Way of Dying" 582
Spirituality 583
Religious Participation 584
Frankl's Theory of Spirituality 585
Jung's Theory of Spirituality 586
Wilson's Theory of Spirituality 587
Fowler's Theory of Spirituality 588
Conclusion 590
Chapter Highlights 590
Key Terms 591
What Do You Think? 591
Suggested Readings 592

Glossary 593
References 610
Credits 645
Name Index 648
Subject Index 654

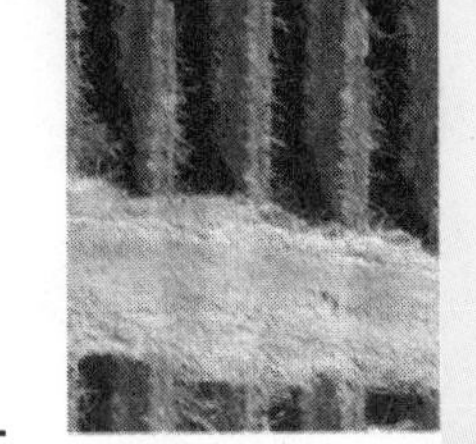

An Applied View

Chapter 1

Chart Your Own Lifespan 9
Maslow and Street Gangs 18
Understanding the Research Article 27

Chapter 2

Conducting Your Own Cognitive Experiment 39
Who Can Find the Raisin? 43

Chapter 3

The Human Genome Project 69
The DNA Alphabet 72

Chapter 4

Guidelines for Parents 106

Chapter 5

Toilet Training—Easy or Difficult 120
What Do Infants Think About? 132

Chapter 6

Do Infants Prefer Attractive Faces? 150
The Significance of the Mother–Infant Interaction 155

Chapter 7

Height and Weight in the Early Childhood Years 176
Children and Their Humor 187
The Case of Genie 196

Chapter 8

Desirable Qualities of Day-Care Centers 212

Chapter 9

How Would You Answer These Questions? 246
Try This Problem 247
The Role of Memory 249
Schools and Character Development 252

Chapter 10

The Extent of Television Watching 279
A Life of Crime 289

Chapter 11

An Average Day in the Life of Some North American Teens 299
What Were You Like? 305
The Erikson Psychosocial Stage Inventory 307
Identity Rating 311

Chapter 12

How Well Do You Know Your Reproductive System? 323
Dealing with Early or Late Development 330
Defining Democracy 333
Tommy's Case 341
Guidelines for Improving Your Own Creativity 342

Chapter 13

Do You Have a Drinking Problem? 349
School Climate Survey 354
Warning Signs of a Potential Suicide Attempt 369

Chapter 14

The Young Adolescent's Rigid View of Gender 381
How to Talk to Teens About Sex (or Anything Else, for that Matter) 388
Talking to Teens About Pregnancy 400

Chapter 15

The Components of Maturity 414
How Young Cancer Patients Feel When They Learn They Have Cancer 427

Chapter 16

Dealing with Adult Children of Alcoholics (ACoA) 435

Chapter 17

Why Marry? 457
Getting a Job 459
Responsibility for Self 466
How's Your Individuation Index? 473

Chapter 18

Obstacles and Aids to Creativity 491
The Lifelong Learning Resource System 494

Chapter 19

The Mid-Life Crisis Scale 509

Chapter 20

The Facts on Aging Quiz 532
Facts About Alzheimer's 538
A Comparison of Physical Abilities 540

Chapter 21

Decisions Most Older Couples Must Make 555
Do You Know Your Grandparents? 558
Pet Ownership by Elderly People 562
National Retirement Programs 564

Chapter 22

A Personal Experience of One of the Authors 573
My Attempts at Suicide (Anonymous) 581
If You Had Your Life to Live Over Again, What Would You Do Differently? 584

A Multicultural View

Chapter 2

Modeling Among Latino Youth 44

Enrichment through Diversity 51

Chapter 6

Different Cultures, Different Interactions 146

Chapter 8

Self and Others 216

Chapter 9

Immigrant Children and Tests 239

Language, Thought, and Immigrants 255

Chapter 11

Is Adolescence a Cultural Phenomenon? 302

Racial Influences on Peer Groups 318

Chapter 12

Following Vygotsky's Model in a Spanish Speaking Community 337

Chapter 13

No to Drugs, Yes to Helping Others 349

Chapter 14

The Role of Race in Teen Pregnancy 399

Chapter 15

Yudia and Mateya Come of Age 411

Initiation Rites in the United States 412

Membership in a Minority Group 425

Chapter 19

Work in the Lives of African-American and White Women 517

Chapter 20

Gateball and the Japanese Elderly 539

Chapter 21

Caregiving for Elderly in Swarthmore, PA and Botswana, Africa 557

Chapter 22

The Funeral in Other Times and Countries 575

What's Your View

Chapter 1
15

Chapter 2
34

Chapter 3
60
63
66
68

Chapter 4
88
99

Chapter 5
115

Chapter 6
148

Chapter 7
172

Chapter 8
223

Chapter 9
234
244

Chapter 10
281
282

Chapter 11
304

Chapter 13
372

Chapter 14
380
386

Chapter 15
418

Chapter 16
437
442
447

Chapter 17
456
465

Chapter 18
490

Chapter 19
516

Chapter 20
534

Chapter 21
550
565

Chapter 22
582
585

Preface

Studying the human lifespan (from conception to death) is both informative and exciting. Your work in this field is important to you personally because knowledge of the lifespan helps you to understand your own behavior. The better you are able to knit your personal experiences into a meaningful pattern, the more content you will be with your life. A famous psychologist, Abraham Maslow, once said that human beings are integrated, organized wholes. The greater your organization, the greater your happiness will be. Recognizing the significant features and crises of the human lifespan can also give you insights into the behavior of others and help you to achieve better relationships with those around you.

The opportunity to fulfill your potential is not confined to any one time of your life. *Development is a lifelong process.* Psychologists now know that development, once thought to end at childhood or possibly adolescence, is a process that continues from conception to death. Today we realize that the changes of adulthood—maturity and aging—matter as much as those of any other period. Consequently, studies of human development focus on the lifespan *within* the cultural context in which development occurs. By analyzing the various developmental periods, researchers are discovering the features of each period and uncovering the mechanisms by which we move from one stage to the next.

In this book, we'll trace the significant changes that occur at various times in the lifespan, emphasizing that the potential for development is constant throughout our lives—from youth to old age. Change is a constant in our lives, and we all attempt to make change as positive as possible. For example, today we are convinced that aging isn't necessarily accompanied by decline and deterioration.

You have come to the study of the human lifespan at a particularly exciting time. Recent biological breakthroughs offer the promise of a brighter and healthier future for an unlimited number of individuals who otherwise might have been condemned to shorter, pain-filled lives. Biological discoveries have also been accompanied by heightened sensitivity to the influence of culture on development. All humans proceed through the lifespan exposed to the opportunities and limitations of a particular culture. At no other time in the history of psychology has greater emphasis been placed on the formula $B = P \times E$; that is, behavior results from the interaction of personal endowment and experience.

Basic Themes of the Text

In deciding how best to present the complex data relating to the human lifespan, we had to choose between topical and chronological organization. In a topical approach, you study specific topics—physical development, cognitive development, language development—as they occur at each stage of life. In a chronological approach, you study the lifespan as you live it: infancy, early childhood, middle childhood, adolescence, and adulthood.

One approach is not necessarily better than the other, so we have decided to incorporate the best features of both. The basic framework of the book is chronological, but within each of the developmental stages we present many topics that incorporate the latest theory, research, and applications to real situations. In this way, we hope you acquire a developmental perspective on the significant and sensitive topics you will soon be reading about. To achieve this objective, we present the lifespan as follows:

- Prenatal beginnings: Conception to birth
- Infancy
- Early childhood
- Middle childhood
- Adolescence
- Early adulthood
- Middle adulthood
- Late adulthood

To help you understand the important developmental ideas presented in this book, we include the most recent research currently available (as well as classic studies that have stood the test of time). You'll find major terms in boldface and important ideas presented in italics to help you focus on critical issues. Items in boldface also appear at the end of each chapter and in a glossary of definitions at the end of the book.

We have tried to write this book as simply and clearly as possible, with a minimum of psychological jargon. We have carefully reviewed each sentence in the first edition and ruthlessly changed, eliminated, and updated each topic discussed. As a result, this edition is the product of lengthy discussions, data searches that persisted until the manuscript went into production, intensive reviews by respected colleagues, and, finally, extensive revisions.

Major Changes in the Second Edition

Thanks to suggestions from students who used the first edition of our text and the insightful comments of reviewers, we have made many substantial changes in the second edition. Here are some of the most important changes:

- We have made greater use of the biopsychosocial model. Because the data relating to developmental psychology are so extensive, we believe that this model helps you to better organize developmental data, and it also serves as an effective memory aid. We are likewise aware that it is impossible to understand development from a single perspective. That is, biological development has both psychological and social consequences; social development has both biological and psychological effects; and psychological development has both biological and social influences.

- Due to the continued advances in the biological sciences, we have expanded our discussion of such biologically based topics as external fertilization techniques, genetics, AIDS, and menopause. Faithful to our model, we have interpreted these data from a biopsychosocial viewpoint.

- We have substantially increased our analysis of the influence of context on development. The multicultural emphasis of the first edition of *Human Development Across the Lifespan* has been expanded in the second edition. Boxes entitled "A Multicultural View" appear in each chapter. The content of each chapter also reflects growing research into the contributions each culture makes to the development of its members (and vice versa). For example, we have expanded our discussion of language to include positive aspects of bilingualism and the need for bilingual education. Our intent is to indicate how knowledge of cultural diversity enriches our understanding of human development.

- We decided that the effect of schooling on development needed more discussion. Consequently, we have explored in considerable detail the influence that the America 2000 educational strategies can have on development. We have linked changes in the teaching of such subjects as language and mathematics to development and also examined how homework and grade retention practices affect a child's growth.
- Earlier in this preface, we mentioned that development is a lifelong process, which strongly implies that developmental psychology textbooks must constantly change to accommodate fresh insights into the developmental process. To meet this challenge, we have incorporated such new topics as recent results of the Human Genome Project, the effect of homelessness and violence on children. At the adolescent and adult levels, we have included new information on such subjects as sexuality (including the 1993 Janus report), sexually transmitted diseases, gang behavior, cognitive changes, menopause, and Alzheimer's disease.

Teaching-Learning Features of the Second Edition

We firmly believe that this textbook's success (that is, how meaningful it is to our readers) depends on content and organization. We have already discussed the content of *Human Development Across the Lifespan;* now we will describe its organization. Helping you to master the book's contents in as uncomplicated and meaningful a manner as possible has been the most important pedagological goal of our work. To accomplish this task, we have built a number of features into each chapter:

- *Chapter outlines.* The major topics of each chapter are presented initially so that you can quickly find the subject you need. An outline helps you to retain material (a memory aid) and is an efficient method for reviewing content.
- *Opening vignette.* Each chapter opens with a vignette that illustrates the chapter content. These vignettes are intended to demonstrate how the topics described in the chapter actually "work" in the daily lives of human beings, young and old.
- *List of objectives.* Following the introductory section of each chapter, we present a carefully formulated list of objectives to guide your reading. When you finish reading the chapter, return to the objectives and test yourself to see if you can respond to their intent; that is, can you analyze, can you apply, can you identify, can you define, can you describe?
- *"View" boxes.* We have designed our boxes to expand on the material under discussion and to do so in a manner calculated to aid your retention. The view boxes are of three types:

  –*What's Your View?* Here we present controversial issues and ask for your opinion after you have studied the facts.

–*An Applied View*. Here you will see how the topics under discussion apply to an actual situation, in settings such as a classroom, a residential center, or a medical facility.

–*A Multicultural View*. Here we analyze the contributions of different cultures individual development.

- *Conclusion*. At the end of each chapter, you will find a brief concluding statement that summarizes the main themes of the chapter. This statement gives you a quick check of the purpose of the chapter and the content covered.
- *Chapter highlights*. Following this brief concluding section is a more detailed number of summary statements, grouped according to the major topics of the chapter. This section should help you to review the chapter quickly and thoroughly. Turn to the chapter table of contents and then check against the chapter highlights to determine how successful you are in recalling the pertinent material of the chapter.
- *Key terms*. You will find at the end of each chapter a list of terms essential to understanding the ideas and suggestions of that chapter. These terms were highlighted and explained in the context of the chapter. They also appear in the glossary. We urge you to spend time mastering the meanings of these terms and relating them to the context in which they appear.
- *What do you think?* Following the key terms, you will find a series of questions intended to help you demonstrate your knowledge of the chapter content, not only by applying the material to different situations but also by being creative in answering the question or solving the problem.
- *Suggested readings*. At the end of each chapter, you will find an annotated list of four or five books or journal articles that we think are particularly well suited as supplements to the contents of the chapter. These references are not necessarily textbooks; they may not deal specifically with either education or psychology. We believe, however, that they shed an illuminating light on the chapter material.

Supplementary Materials

We have worked with the publisher and a group of very talented individuals to put together a high-quality set of supplementary materials to assist instructors and students who use this text.

The *Student Study Guide* was created by Richard Morehouse of Viterbo College and Lynne Blesz Vestal. The study guide begins with a section designed to help students become better learners. Then, for each chapter, the study guide provides a chapter overview, learning objectives, key terms and guided review exercises, a set of application-type questions for review and discussion, and a multiple-choice practice test.

An *Instructor's Manual* has also been prepared by Richard Morehouse and Lynne Blesz Vestal. Each chapter of the manual includes a chapter overview, learning objectives, key terms (page referenced to the text), lecture suggestions, classroom/student activities, and questions for review and discussion. The instructor's manual is conveniently housed within an attractive 11″ × 13″ × 9″ carrying case. This case is designed to accommodate the complete ancillary package. It contains the material for each chapter within a separate handling file, allowing you to keep all your class materials organized and at your fingertips.

The instructor's manual includes a comprehensive *Test Item File* consisting of over 1,000 items. Each test item is referenced to its related learning objective and text page and is classified as factual, conceptual, or applied, based on the first three levels of Benjamin Bloom's taxonomy.

TestPak 3.0 is an integrated program designed to print test masters, permit online computerized testing, help students review text material through an interactive self-testing, self-scoring quiz program, and provide instructors with a gradebook program for classroom management. Test questions can be found in the Test Item File, or instructors can create their own. Instructors may choose to use *Testbank A* for exam questions and *Testbank B* in conjunction with the quiz program. Printing the exam yourself requires access to a personal computer—an IBM that uses 5.25- or 3.5-inch diskettes, an Apple IIe or IIc, or a Macintosh. TestPak requires two disk drives and works with any printer. Diskettes are available to instructors through their local Brown & Benchmark sales representative, or by phoning Educational Services at (319) 588–1451. The package contains complete instructions for making up an exam.

The *Brown & Benchmark Developmental Psychology Transparency/Slide Set* consists of 100 newly developed acetate transparencies or slides. These full-color illustrations include graphics from various outside sources. Created by Lynne Blesz-Vestal, these transparencies were expressly designed to provide comprehensive coverage of all major topic areas generally covered in developmental psychology. A comprehensive annotated guide provides a brief description for each transparency and helpful suggestions for use in the classroom.

Finally, we believe the most important feature of the text is its usability. We have attempted to write a book that you will turn to, even after you have finished the course—when you have a question about your work or your family, for example—as a source of ideas, support, and reassurance.

Acknowledgments

We express particular thanks to Michael Lange, our editor and friend, for his insights and support during the writing of both editions of this book. We also thank Sheralee Connors, our development editor, for her management skills and ability to smooth troubled waters. Chris Thillen was a superb copy editor whose ability considerably sharpened the focus of our book. Marlys Nekola, our production editor, cheerfully helped us to keep our book on its production schedule.

Our graduate research assistants also contributed through researching topics and making suggestions for the text. We are especially grateful to Tracy Hurd, Kathy Lennon, and Jacquie Scarbrough. We also thank Gretchen Cahaly, Daniel Harrell, Deborah Margolis, Sue Pratt, and Jane Siblin.

As we mentioned, several colleagues contributed to our work. We are grateful to the following individuals for their helpful suggestions:

- John B. Benson, Texarkana College
- Andrea Chen, State University of New York at Binghamton
- Frances M. Droddy, Lamar University
- Nolan Embry, Lexington Community College
- Cecile Hundley Fitt, Indian River Community College
- Linda E. Flickenger, St. Clair County Community College
- Drusilla Glascoe, Salt Lake City Community College
- Gayla Preisser, Eastern Montana
- Judith M. Sgarzi, Salem State College
- Alice M. Tate, Kankakee Community
- Barbara D. Lyles, Howard University
- Pamela Manners, Troy State University
- Mary Mindess, Lesley College
- Richard E. Morehouse, Viterbo College
- Kim L. Morrisey, Rutgers University
- Edward R. Mosely, Passaic County Community College
- Jeffrey L. Naglebush, Ferris State University
- Janine A. Watts, University of Minnesota–Duluth
- Tony L. Williams, Marshall University

PART I

Introduction

CHAPTER 1

Lifespan Psychology: An Introduction

A Biopsychosocial Model of Development 5

Why Study Lifespan Development? 6

The Meaning of Lifespan 6

The Developmental Task Concept 9

Needs across the Lifespan 10

History of Lifespan Studies 14

Child Studies 14

Great Names in the History of Child Development 15

Adolescence 15

Adulthood 17

Issues in Lifespan Development 17

Culture and Development 18

Continuity vs. Discontinuity 18

Stability vs. Change 19

Sensitive Periods vs. Equal Potential 20

Developmental Research 20

Data Collection Techniques 21

Time Variable Designs 22

Conclusion 26

Chapter Highlights 27

Key Terms 29

What Do You Think? 29

Suggested Readings 29

This is a book about lifespan psychology, that is, those changes in life that occur from conception to death. If you think about changes in your own life—beginning school, perhaps going off to college, beginning a job, marriage, childbirth—you begin to appreciate the complexity of development. It's usually difficult, however, to look at ourselves objectively, so let's examine the life of an outstanding individual whose rise to fame and power offers an insightful view into what is meant by lifespan development—Colin Powell.

Powell, whom America came to know quite well as chairman of the Joint Chiefs of Staff, was born in Harlem on April 5, 1937, the son of Jamaican immigrant parents. He gained confidence and self-esteem from the care, love, and attention that he received from his family. His parents valued education and he quickly realized the importance of learning. As a young African-American, Powell decided that opportunities were greater in the army than in the corporate world.

A product of the ROTC program at the City College of New York, Powell soon demonstrated those leadership qualities that would mark him as an outstanding soldier. In 1963, as a second lieutenant and newly married, Powell was sent to Vietnam. Wounded in action, he showed the bravery and compassion for his colleagues that has been a constant theme in his career. Returning to the United States, he obtained a master's degree in business administration, and in 1972 was selected as a White House Fellow.

During the following years, Powell served in various parts of the world, gradually attaining the rank of Major General. He was recalled to the White House to be military advisor to the Secretary of Defense in 1982, then in 1986 returned to the army. In 1987, he was asked to be National Security Advisor to the president, and in 1989 he became chairman of the Joint Chiefs of Staff. During these years, he became the first African-American to attain the rank of a four-star general.

His rise to the top military post in the United States confounded many who thought that only West Point graduates could ascend to such heights. His courage in Vietnam, however, resulted in six medals (including the Purple Heart and a Bronze Star) and earned him the respect and admiration of President George Bush, who recommended his appointment. During the Gulf War, Powell played a pivotal role in determining strategy and in advising the president.

As you examine the paths that Colin Powell followed in his life, you can begin to identify those themes that make studying the human lifespan so fascinating. You can see how his family was a powerful and positive influence throughout his early years. Later, his marriage to Alma Johnson and their three children provided additional support. His decision to make the army a career led to international experiences and a network of friends who would help shape his future. These developmental milestones in Colin Powell's life served to make him the person he is today, but at the same time it is possible to discern behaviors that have been remarkably persistent throughout his lifespan: courage, compassion for others, and devotion to family and country. ■

In our work, we'll trace the significant changes that occur at various times in the lifespan, emphasizing that the potential for development is constant throughout our lives—from youth to old age. For example, today we are convinced that aging isn't necessarily accompanied by decline and deterioration. Change is a constant in our lives, and we all attempt to make that change as positive as possible. While these changes may not be as dramatic as those we have just traced in Colin Powell's life, they are important to us as individuals making our own progress throughout our lifespans.

We all attend to those people and events that we think are significant; we all engage in different experiences; and we all trace our own unique routes as we move through the lifespan. Development, however, does not always proceed smoothly. For example, under positive conditions developmentally delayed infants can catch up (they usually do) and become robust 10-year-olds (Bukatko & Daehler, 1992; Rutter, 1981; Tronick, 1989). In high school, an adolescent girl may do well in her classwork but physical development may proceed more slowly and prevent her from participating in athletic events. We all know people who were shy adolescents but who became outgoing, socially attractive adults.

These examples help to explain the appeal of studying lifespan development. Before beginning our work on analyzing the lifespan in this chapter, however, we'll first present a model—the **biopsychosocial model**—that is the framework of this book. This model will help you to integrate the wealth of information that you'll find in the pages to come. If you think of lifespan development as the product of the interaction of biological, psychological, and social forces, you can better understand and appreciate the complexity of development. Once you apply this model to the various chapters, you will find that you have a useful means of recalling the information in any chapter.

We'll then explore the meaning of lifespan development in this chapter and attempt to indicate its importance to you by illustrating how peaks and valleys come into all of our lives. Although we all chart our own course, we can still identify considerable similarities in our lives. We walk, we talk, we attend school, and we search for a satisfying career. Yet within this sameness, we all have different experiences that cast a unique shadow over our journey through the lifespan.

Following this discussion, we'll briefly trace the role that different disciplines played in the emergence of lifespan studies. In any subject as complex as the study of human development, several issues have arisen that affect the manner in which you identify and interpret developmental data. These deserve our attention. Finally, we'll examine those research techniques that will help you to assess and interpret the theories and studies you will meet in the coming pages.

When you complete your reading of this chapter, you should be able to:

- Use the biopsychosocial model as a means of interpreting and recalling developmental data.
- Define lifespan development.
- Identify major developmental issues.
- Describe the antecedents of lifespan developmental studies.
- Apply your knowledge of lifespan techniques to the studies you read.

These photos (a–e) show Bill Clinton at various stages of his life span.

(b)

(a)

(c)

(d)

(e)

A Biopsychosocial Model of Development

As you read through the pages of this book, you'll find the material organized around the concept of a biopsychosocial model, which illustrates how biological, psychological, and social-cultural forces influence development. Biological elements range from the role of our genes in development to adult health issues; psychological elements include all aspects of cognitive and personality development; social elements refer to such influences as family, school, and peers. We believe that the biopsychosocial model encompasses the concept of heredity and environment to explain development since it is the interaction of these forces that explains the complexity of development. (See Figure 1.1.)

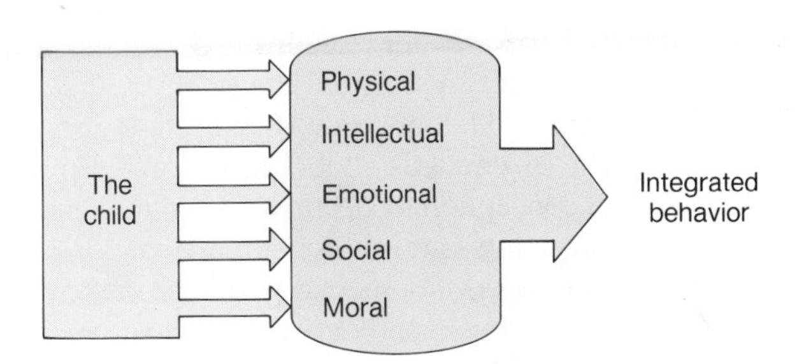

Figure 1.1
The interactive nature of development

For example, let's assume that Hilary, a 5-year-old who entered first grade in September, is unhappy. Most mornings, she pleads that she is too sick to go to school. She is having difficulty with her schoolwork, and her teacher reports that she disrupts the class by interrupting others and wandering around the room. Because of her classroom behavior, she is unpopular with the other children in the class. After complaining of headaches, Hilary is examined by the school nurse, who finds that she needs glasses. Here is an example of how a physical problem can affect psychological development (learning and reading) as well as social development (popularity with classmates). Table 1.1 summarizes several important developmental topics that help to make up the biopsychosocial model.

In chapters 5 through 10, we preview the topics of the chapter by presenting them according to the biopsychosocial model and frequently by identifying the theorist(s) who have most extensively studied that developmental period. (You will read about these theorists in chapter 2.) These previews also include tables (like

Table 1.1 The Biopsychosocial Model

Bio	Psycho	Social
Genetics	Cognitive development	Attachment
Fertilization	Information processing	Relationships
Pregnancy	Problem-solving	Reciprocal interactions
Birth	Thinking skills	School
Physical development	Perceptual development	Peers
Motor development	Language development	Television
Health issues	Moral development	Stress
Puberty	Formation of the self	Marriage
Menstruation	Body image	Family

Table 1.1) identifying the characteristics of each developmental period and defining them as biological, psychological, or social factors. When we leave the middle childhood years, it becomes more difficult to categorize these factors. Consequently, we'll no longer use tables, but will continue to interpret development from the biopsychosocial perspective. We believe that by analyzing lifespan development with the biopsychosocial model, you will have a tool that better enables you to understand and remember the material of any chapter.

Why Study Lifespan Development?

Lifespan development is the study of human development from conception to death. As such, it is important to you personally because it helps you to understand your own behavior once you have grasped its main ideas. Studying lifespan development should also provide you with insights into the behavior of others and, because of this, you should achieve better relationships with those around you.

We are aware today, as you can tell from the brief discussion at the chapter's opening, that development is a lifelong process. Psychologists now realize that development, once thought to end at childhood, or possibly adolescence, is a process that continues from conception to death. Today we realize that the changes of adulthood—maturity and aging—are as developmental as those of any other period. By analyzing the various developmental periods—infancy, early childhood, middle childhood, adolescence, and adulthood—researchers are trying to discover the features of each period and to uncover the mechanisms by which we move from one stage to the next.

The Meaning of Lifespan

The meaning of **lifespan psychology** has been nicely captured by Harre and Lamb (1983) when they note that lifespan psychology focuses on the life course of the individual within the social and political context in which the individual's development occurs. A lifespan perspective recognizes the need to search for a general sequence of change over a life's course. Yet for purposes of research and speculation, we arbitrarily divide the lifespan into segments, while recalling that each segment is part of a whole. In this book, the lifespan is divided as follows:

Prenatal—conception to birth
Infancy—birth to 2 years
Early Childhood—2 to 5 years
Middle Childhood—6 to 11 years
Adolescence—12 to 18 years
Early Adulthood—19 to 34 years
Middle Adulthood—35 to 64 years
Later Adulthood—65+

Development does not proceed randomly. It is tightly linked to what psychologists call context; that is, the circumstances in which an individual develops. Context is such an important feature of development that we will return to it frequently throughout this book. To understand development as fully as possible, you must also understand its context. The changes in various models of development, which are used to explain developmental processes, illustrate the critical role that context plays in development.

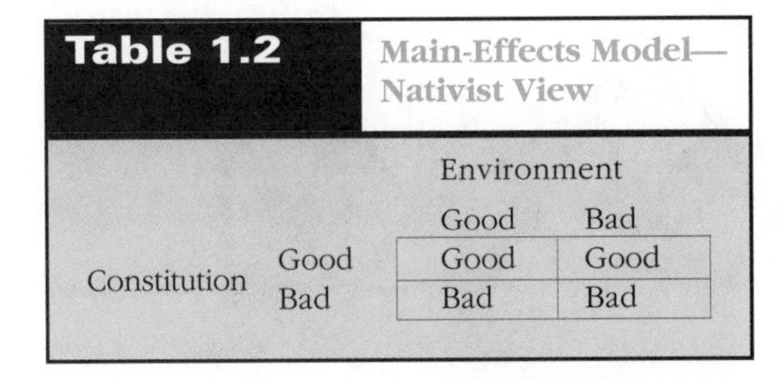

Table 1.2 Main-Effects Model—Nativist View

		Environment	
		Good	Bad
Constitution	Good	Good	Good
	Bad	Bad	Bad

Table 1.3 Main-Effects Model—Nurturist View

		Environment	
		Good	Bad
Constitution	Good	Good	Bad
	Bad	Good	Bad

Table 1.4 An Interaction Model

		Environment	
		Good	Bad
Constitution	Good	Good	Medium
	Bad	Medium	Bad

Models of Development

One of the earliest examples was called the **main-effects** model and was similar to a medical diagnosis: *this* condition produced *this* problem. Little, if any, attention was paid to the possible interaction among causes. There are two possible interpretations of a main-effects model. A **nativist** view means that physical causes dominate developmental outcomes (biology is destiny). For example, poor eyesight causes poor reading; no consideration is given to the possibility of damaged relationships (a child becomes unruly and unpopular with peers). A **nurturist** view implies that the environment is all-important. Tables 1.2 and 1.3 illustrate both interpretations of the main-effects model, that is, the type of individual produced according to the nativist and nurturist view.

A second attempt to explain development was by an **interaction** model, which states that both the individual and the environment interact with each other. Although an improvement over the main-effects model, its weakness was in the assumption of an unchanging body and an unchanging environment. Poor eyesight causes poor reading and results in poor grades, which does not tell us how these changes came about. Table 1.4 illustrates the type of individual produced by the interaction model.

Finally, we know today that heredity and environment produce their results in a complex, interactive manner. To complicate our analysis even more, however, we ourselves are interacting with the environment. Remember, at birth we have had nine months of development. So we help to shape the reactions of those around us; that is, our responses cause parents and siblings to respond in a unique fashion.

Sameroff (1975) used the idea of reciprocal interactions to propose what he called the **transactional model** and in so doing argued against views that attribute development to either heredity or environment. This process never ceases and is also known as **reciprocal interactions** (Brazelton, 1984). We respond to those around us and they change; their responses to us thus change and we in turn change. Figure 1.2 illustrates the transactional and reciprocal interaction models.

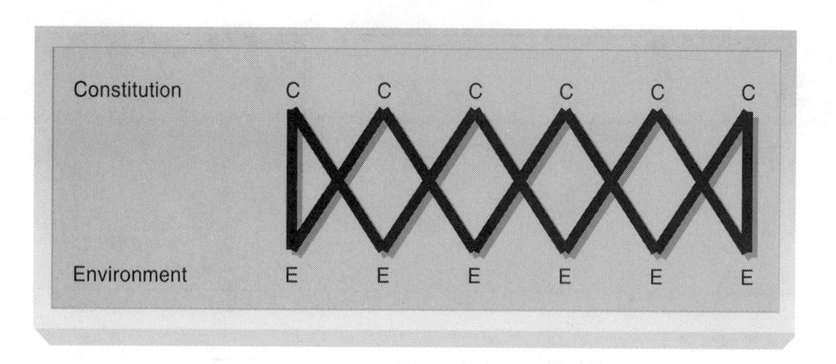

Figure 1.2
Transactional model.
Source: Sameroff: (1975).

The reciprocal interaction and transactional models of development are basically identical; both emphasize the constant state of reorganization we all experience through development. Both stress the complexity of developmental analysis by recognizing

- There are no simple cause-and-effect explanations of development.
- All of our internal growth forces are interacting.
- All relevant external forces are likewise interacting.
- Internal and external forces are interacting with each other.

In examining these various models, you can see the greater role that context is given in development. For example, Bronfenbrenner has suggested a systems analysis of development that highlights the importance of context.

A Systems Analysis

The environment acts in ways that we are only beginning to comprehend. Thanks to the work of Uri Bronfenbrenner (1978, 1989; Bronfenbrenner & Crouter, 1983), we now realize that there are many environments acting on us. Bronfenbrenner conceives of the environment as a set of nested structures, each inside the next. The deepest level is the immediate setting in which the developing person resides, which is called the **microsystem** (for example, the home or school).

The next level moves beyond the immediate setting and necessitates examining the relations between immediate settings in which the person actually participates. This is called the **mesosystem** (the relationship among microsystems) and an example can be seen in a child's school achievement. Those children who are fortunate enough to be in a family that maintains a close and warm relationship with the school can be expected to do well in their classroom work.

The **exosystem** (an environment in which you are not present, but which nevertheless affects you (your parents' jobs, for example) and the **macrosystem** (the blueprint of any society) are environmental forces that affect development. To illustrate the interactive nature of these forces, consider the youngster whose father has just lost his job (changes in the exosystem), which causes the family to move to another location with different friends and schools for the child (changes in the micro- and mesosystems).

A systems analysis helps to emphasize the significance of context to development. Also note the crucial tasks that were identified in Bronfenbrenner's work: getting along with family members, going to school, and careers. The concept of critical tasks in our lives is one that has received considerable attention from developmental psychologists.

An Applied View — Chart Your Own Lifespan

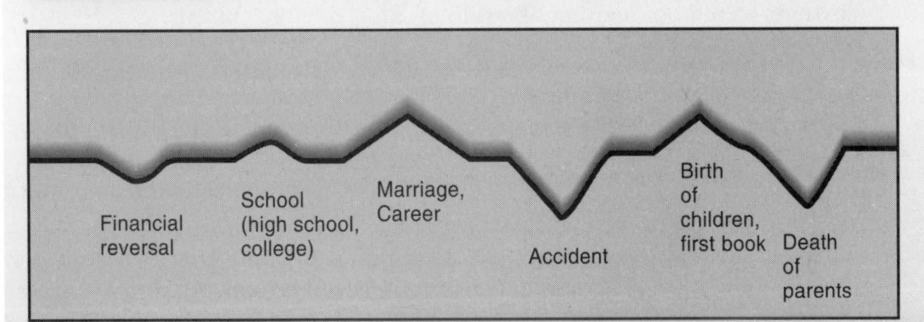

Endeavoring to illustrate how important knowledge of the lifespan is to each of us, Sugarman (1986) has devised a simple exercise that you can do quickly. Using a blank sheet of paper, assume that the left edge of the page represents the beginning of your life and the right edge where you are today. Now draw a line across the page that indicates the peaks and valleys that you have experienced so far.

For example, the chart for one of the authors would be as follows:

In this chart, the first valley was a financial reversal for the author's parents. The first peak represents happy and productive high school years, followed by entry into teaching, and then marriage a few years later. The deep valley was a serious accident followed by years of recuperation and then the birth of children and the publication of a first book. You can see that it looks like a temperature chart. Try it for yourself. Sugarman (1986) suggests that when you finish, you then sit back and ask yourself these questions:

- Are there more peaks than valleys?
- Is there a definite shape to your chart?
- Would you identify your peaks and valleys as major or minor?
- What caused the peaks and valleys?
- Could you have done anything to make the peaks higher and the valleys more shallow?
- What happened during the plateaus?
- What's your view of these highs and lows in your life?

You have drawn a picture of your lifespan, and the questions that you have just answered are actually the subject matter of lifespan development.

The Developmental Task Concept

From examining the changes in your own lifespan and reading about the events in Colin Powell's life, you can see that critical tasks arise at certain times in our lives. Mastery of these tasks is satisfying and encourages us to go on to new challenges. Difficulty with them slows progress toward future accomplishments and goals. As a mechanism for understanding the changes that occur during the lifespan, Robert Havighurst (1952, 1972, 1982) has identified critical developmental tasks that occur throughout the lifespan. Although our interpretations of these tasks naturally change over the years and with new research findings, Havighurst's developmental tasks offer lasting testimony to the belief that we continue to develop throughout our lives.

Havighurst (1972) defines a **developmental task** as one that arises at a certain period in our lives, the successful achievement of which leads to happiness and success with later tasks; while failure leads to unhappiness, social disapproval, and difficulty with later tasks. Havighurst uses slightly different age groupings, but the basic divisions are quite similar to those used in this book. He identifies three sources of developmental tasks (Havighurst, 1972):

- *Tasks that arise from physical maturation.* For example, learning to walk, talk, and behave acceptably with the opposite sex during adolescence; adjusting to menopause during middle age.

- *Tasks that arise from personal sources.* For example, those that emerge from the maturing personality and take the form of personal values and aspirations, such as learning the necessary skills for job success.
- *Tasks that have their source in the pressures of society.* For example, learning to read or learning the role of a responsible citizen.

According to our biopsychosocial model, the first source corresponds to the "bio" part of the model, the second to the "psycho," and the third to the "social" aspect.

Havighurst has identified six major age periods: infancy and early childhood (0–5 years), middle childhood (6–12 years), adolescence (13–18 years), early adulthood (19–29 years), middle adulthood (30–60 years), and later maturity (61+). Table 1.5 presents typical developmental tasks for each of these periods.

The developmental task concept has a long and rich tradition. Its acceptance has been partly due to a recognition of sensitive periods (see pp. 20 of this chapter) in our lives and partly due to the practical nature of Havighurst's tasks. Knowing that a youngster of a certain age (say, 11) is encountering one of the tasks of that period (learning an appropriate sex role) helps adults to understand a child's behavior and establish an environment that helps the child to master the task. Another good example is that of acquiring personal independence, an important task for the middle childhood period. Youngsters test authority during this phase and, if teachers and parents realize that this is a normal, even necessary phase of development, they react differently than if they see it as a personal challenge (Hetherington and Parke, 1986).

For example, note Havighurst's developmental tasks for middle adulthood, one of which is a parent's need to help children become happy and responsible adults. Adults occasionally find it hard to "let go" of their children. They want to keep their children with them far beyond any reasonable time. For their own good, as well as that of their children, they must reconcile their wants and needs with those of their children. Once they do, they can enter a happy time in their own lives if husbands and wives are not only spouses but friends and partners as well.

Havighurst is not alone in the importance he places on the developmental task concept (Cole, 1986; Goetting, 1986; Cristante & Lucca, 1987; Cangemi and Kowalski, 1987). For example, Goetting (1986) has examined the developmental tasks of siblings and identified those that last a lifetime, such as companionship and emotional support. Other tasks seem to be related to a particular stage in the life cycle, such as caretaking during childhood and later the care of elderly parents.

Identifying and mastering developmental tasks help us to understand the way change affects our lives. Another way to understand lifespan changes is to identify those needs that must be satisfied if personal goals are to be achieved. To help you recognize the role that needs play in our lives, let's examine the work of Abraham Maslow and his needs hierarchy.

Needs across the Lifespan

Assume that Amy, a high school senior, is concerned that her English class is not helping her to prepare for college work. In other words, this student believes that her needs are not being met; there is a lack of need satisfaction. One of Maslow's most famous concepts is that of **self-actualization,** which means that we use our abilities to the limit of our potentialities (Maslow & others, 1987). If people are convinced that they should—and can—fulfill their promise, they are then on the path to self-actualization. *Self-actualization is a growth concept,* and individuals move toward this goal (physical and psychological health) as they satisfy their basic needs.

Table 1.5 The Developmental Tasks

Infancy and Early Childhood (0–5)	Middle Childhood (6–12)	Adolescence (13–18)
1. Learning to walk 2. Learning to take solid foods 3. Learning to talk 4. Learning to control the elimination of body wastes 5. Learning sex differences and sexual modesty 6. Acquiring concepts and language to describe social and physical reality 7. Readiness for reading 8. Learning to distinguish right from wrong and developing a conscience	1. Learning physical skills necessary for ordinary games 2. Building a wholesome attitude toward oneself 3. Learning to get along with age-mates 4. Learning an appropriate sex role 5. Developing fundamental skills in reading, writing, and calculating 6. Developing concepts necessary for everyday living 7. Developing conscience, morality, and a scale of values 8. Achieving personal independence 9. Developing acceptable attitudes toward society	1. Achieving mature relations with both sexes 2. Achieving a masculine or feminine social role 3. Accepting one's physique 4. Achieving emotional independence of adults 5. Preparing for marriage and family life 6. Preparing for an economic career 7. Acquiring values and an ethical system to guide behavior 8. Desiring and achieving socially responsible behavior
Early Adulthood (19–29)	**Middle Adulthood (30–60)**	**Later Maturity (61+)**
1. Selecting a mate 2. Learning to live with a partner 3. Starting a family 4. Rearing children 5. Managing a home 6. Starting an occupation 7. Assuming civic responsibility	1. Helping teenage children to become happy and responsible adults 2. Achieving adult social and civic responsibility 3. Satisfactory career achievement 4. Developing adult leisure time activities 5. Relating to one's spouse as a person 6. Accepting the physiological changes of middle age 7. Adjusting to aging parents	1. Adjusting to decreasing strength and health 2. Adjusting to retirement and reduced income 3. Adjusting to death of spouse 4. Establishing relations with one's own age group 5. Meeting social and civic obligations 6. Establishing satisfactory living quarters

Source: From Robert Havighurst, *Developmental Tasks and Education,* 3d ed. Copyright © 1972 David McKay Company, Inc., a division of Random House, New York, NY. Reprinted by permission.

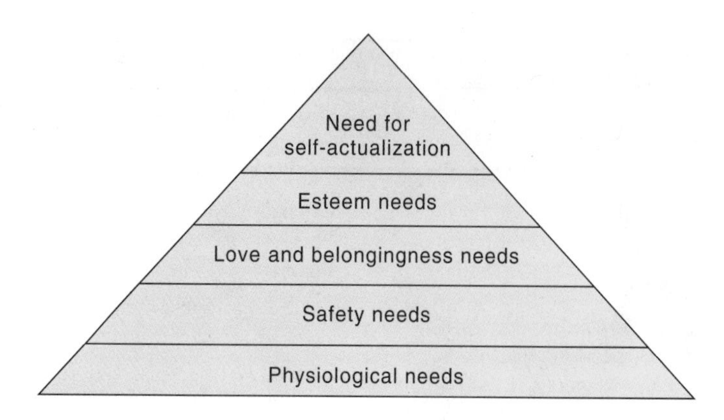

Figure 1.3
Maslow's hierarchy of needs.
Source: Data for diagram based on Hierarchy of Needs, in "A Theory of Human Motivation" in *Motivation and Personality,* 2d ed., 1970, by Abraham H. Maslow.

Need Satisfaction

Growth toward self-actualization requires the satisfaction of a hierarchy of needs. In Maslow's theory there are five basic needs: physiological, safety, love and belonging, esteem, and self-actualization. Figure 1.3 illustrates the hierarchy of needs, with those needs at the base of the hierarchy assumed to be more basic relative to the needs above them in the hierarchy.

- *Physiological needs.* **Physiological needs,** such as hunger and sleep, are dominant and are the basis of motivation. Unless they are satisfied, everything else recedes. For example, students who frequently do not eat breakfast or suffer from poor nutrition generally become lethargic and withdrawn; their learning potential is severely lowered. Note: This is particularly true of adolescents, who can be extremely sensitive to their weight.
- *Safety needs.* **Safety needs** represent the importance of security, protection, stability, freedom from fear and anxiety, and the need for structure and limits. For example, individuals who are afraid of school, of peers, of a superior, or of a parent's reaction have their safety needs threatened and their well-being can be affected.
- *Love and belongingness needs.* **Love and belongingness needs** refers to the need for family and friends. Healthy, motivated people wish to avoid feelings of loneliness and isolation. People who feel alone, not part of the group, or who lack any sense of belongingness usually have poor relationships with others, which can then affect their achievement in life.
- *Esteem needs.* **Esteem needs** refer to the reactions of others to us as individuals and also to our opinion of ourselves. We want a favorable judgment from others, which should be based on honest achievement. Our own sense of competence combines with the reactions of others to produce a sense of self-esteem. Consequently, we must acquire competence and find the opportunities that permit us to achieve and to secure reinforcement, both from others and our own sense of satisfaction in what we have done.
- *Need for self-actualization.* By **self-actualization needs,** Maslow was referring to that tendency, in spite of the lower needs being satisfied, to feel restless unless we are doing what we think we are capable of doing. As Maslow noted (Maslow & others, 1987), musicians must make music, artists must paint, and writers must write. The form that needs take isn't

An Applied View — Maslow and Street Gangs

Maslow's hierarchy of needs has widespread application to all members of society, including those youth who find themselves in a dangerous environment. In his analysis of Chicago's African-American street gangs, Perkins (1987) identifies several features of street gangs that many youths find appealing:

- *A sense of identity.* Gang membership provides recognition for those youths who have not received it elsewhere, thus compensating for low self-esteem.
- *A sense of belonging.* Belongingness gives gang members a feeling of acceptance by those who face similar problems and hardships. These youths find a sharing of common values and needs to be personally rewarding (Perkins, 1987, p. 56).
- *A sense of power.* Many African-American youths think power is crucial for their survival. The gang affords such youths a buffer to feelings of alienation and low self-esteem.
- *A sense of security.* Youths who live in high-risk settings often need protection; for many of them, gang membership is their only means of obtaining security.
- *A sense of discipline.* Gangs demand a strict adherence to structure that appeals to many African-American youths.

Interpreting these five features as needs, Perkins (1987) then relates them to Maslow's need hierarchy in an attempt to explain why a street gang appeals to some youth. For example, the gang member's need for security and power can be found at Maslow's levels of physiological and safety needs; identity, belonging, and discipline are needs found at the love and belongingness level; identity, belonging, and power appear as esteem needs. Perkins believes that satisfaction of self-actualizing needs probably can never be achieved through membership in a street gang.

Examining the needs of gang members in this manner helps to explain how the need satisfaction of many youth can be achieved through gang membership. Perkins (1987) concludes by noting that gangs will continue to attract African-American youths because self-preservation will always be a priority for those who live in an environment that does not fulfill their basic needs.

important: one person may desire to be a great parent; another may desire to be an outstanding athlete. Regardless of professions, *what human beings can be, they must be* (Maslow & others, 1987, p. 22).

Closely allied to these basic needs are cognitive needs (the desire to know and understand) and aesthetic needs. But as Maslow noted, we must be careful not to make too sharp a distinction between these and the basic needs; they are tightly interrelated (Maslow & others, 1987). As you can see, Maslow's remarkably perceptive analysis furnishes us with rich general insights into human behavior, especially those needs that lead to developmental changes during the lifespan.

Using these ideas in the following pages, we'll attempt to identify those factors that lead to success and adjustment (high self-esteem, parental warmth and love, family support). We can't stop there, however. We must also attempt to discover how self-esteem develops and why parental love and support are so significant; that is, what mechanisms are involved. In your reading and in your search for answers to the important questions raised throughout this book, don't let yourself be lulled into an easy acceptance of age as the solution. Remember that individuals mature at quite different rates, and that various aspects of development (physical, cognitive, psychosocial) proceed at different rates within the same individual. Here once again you can see the usefulness of the biopsychosocial model. If you understand it, you'll not be content with merely identifying one cause of a person's behavior. Rather, you'll search for the interaction of factors and thus acquire a deeper and richer explanation of that behavior. Look for the causes, search for the processes, and you'll discover the excitement of studying the lifespan!

History of Lifespan Studies

Now that you understand the meaning of lifespan development, you may be interested in learning about the roots of lifespan studies. These roots have led developmental psychologists to interesting and, in some cases, surprising discoveries. What seems to have happened is that each developmental epoch—infancy, early childhood, middle childhood, adolescence, and early, middle, and late adulthood—has in its turn become a primary focus of research and speculation.

Child Studies

Children and childhood are relatively new concepts; for example, celebrating a child's birthday did not begin until the end of the eighteenth century. In tracing the roots of child development study, Robert Sears (1975) concludes that the stirrings of empathy toward children first began in the eighteenth century. The period of the American Civil War brought concern for the mentally ill and delinquent children—a clear humanitarian appeal—that signalled solicitude for children's welfare.

Children were not always prized as they are by most parents today. This photo shows children in a typical 19th-century sweatshop, reflecting an attitude in which children as young as 5 or 6 were valued for their economic work.

In the twentieth century, these changes in attitude have been accompanied by remarkable advances in the way we view children. From a belief that children passively accept what is done to them, we now think that children actively shape their surroundings: people, events, and situations. They change as a result of what is done to them, and the changes in the children produce changes in those around them, which is the meaning of reciprocal interaction.

In explaining typical development and individual differences in normal development, Scarr (1992) notes that during the past decade European psychologists have argued that children participate in their own development. This belief, called **action theory,** proposes that people influence, in important ways, the path of their development through the lifespan. Children construct reality from the opportunities that their environments present. This constructed reality emerges from the interaction among children's heredity, the behavior of others, and the environments children select because of their interests, talent, and personality (Scarr, 1992).

Thus, from seeing children as the objects of adult actions, we now view them as active, creative individuals who help to shape their environments. Fussy babies, for example, elicit a much different response from those around them than happy, cheerful babies. Both types of children have shaped their environments and tend to see their worlds differently. Once you realize that children process and interpret information from their environments, and then change their behavior because of it, you can understand how their reactions affect those around them.

As the twentieth century comes to a close, the agenda for children has taken on a decidedly different look. Although developmental psychologists still wrestle with the relative importance of heredity and environment, new problems, especially those involving the context of development, have captured their attention. For example, how can we best understand the rich multicultural backgrounds of our changing population (Konner, 1991)? What will be the long-term impact of daycare on children who are placed in centers soon after birth (Belsky & Braungart, 1991)? How does homelessness affect children's development (Rafferty & Shinn, 1991)? We address these questions in the pages to come.

What's Your View?

?

As an example of a youth who got off to a rocky start but selected his own environments, consider the early school problems of Winston Churchill. Coming from a family environment in which he was emotionally neglected, Churchill constructed a reality that enabled him to overcome these early disadvantages. Beginning school at seven years of age, he immediately set his own agenda; he refused to learn anything he wasn't interested in; he was rebellious and troublesome; he waged a constant battle with authority.

At the bottom of his class, Churchill maintained this pattern through his years at Harrow, the famous British prep school. As one of his biographers, William Manchester, writes:

> Churchillian stubbornness, which would become the bane of Britain's enemies, was the despair of his teachers. He refused to learn unless it suited him. He was placed in what today would be called a remedial reading class, where slow boys were taught English. He stared out the window. Math, Greek, and French were beneath his contempt. (Manchester, 1983, p. 157)

These words describe one of the world's greatest leaders, a statesman who led the English people through the darkest days of the Second World War and ultimately triumphed. Yet, by today's standards, he probably would be labelled a difficult child "with a behavior disorder." Do you think other children can overcome the effects of an early negative environment? What's your view? (Remember your answer to this question, because we'll return to it later in this chapter.)

Great Names in the History of Child Development

You may well recognize some of these famous figures, who have been associated with the study of child development over the years. They are but a few of the well-known individuals who have contributed so much to the child development movement.

- *G. Stanley Hall* is often referred to as the founder of the child development movement in the United States. He became famous for using questionnaires to discover what children were thinking about.
- *Sigmund Freud* startled the world with his psychoanalytic theory (see chap. 2) and stressed the importance of the early years in a child's life, a belief that remains strong today.
- *John Watson's* views on child rearing were extremely popular in the 1920s and the 1930s, especially his belief that children should be conditioned early in life.
- *Arnold Gesell* pioneered the use of creative methods in child development research, such as one-way mirrors and motion pictures to record development.
- *Jean Piaget* has dominated our thinking about the cognitive development of children since the 1930s. His views are presented in detail throughout this book.

Adolescence

After recounting the picture of adolescence as described by an amused psychiatrist—"think of your children as poisoned by hormones for five to ten years"—Konner (1991) states that we can do better than this definition. For most parents of adolescents, tears, trouble, and turbulence never materialize. In fact, most adolescents display noteworthy responsibility and even idealism.

Adolescent girls come to their growth spurt earlier than boys, which creates some interesting social situations.

It was G. Stanley Hall's book, *Adolescence,* published in 1904, that signalled the beginning of intense and continuing study of adolescence as a distinct developmental period. Hall's views of adolescence (including that of storm and stress) greatly influenced those writing about adolescence for the next fifty years. But by the 1950s, a more positive view of adolescence was beginning to emerge (Petersen, 1988). Researchers began to concentrate less on the problems of adolescence and more on the developmental characteristics of the period, which produced a great deal of basic data about adolescents.

Today research on this period has increased dramatically. The number of articles and journals devoted to adolescence has grown sharply, and a professional society—The Society for Research on Adolescence—has come into existence. Current research on adolescence emphasizes social development, family relationships, peers, and the wider social environment. Three topics have received considerable attention (Petersen, 1988).

Adolescent Adjustment or Adolescent Turmoil

Research has seriously questioned the early belief of Hall and others that for adolescents turmoil is natural, a phenomenon they outgrow with continued development. For example, Petersen (1988) states that those adolescents demonstrating severe turmoil frequently exhibit later serious problems. In fact, it is becoming clear that those adolescents who have stormy lives belong to the relatively small group who suffer from two or more of three linked problems: substance abuse, mental disorders, and juvenile delinquency (Elliot, Huizinga, & Menard, 1989).

Changes in adolescence–physical, social, cognitive–require teenagers to develop different methods of coping. Still searching for identity and susceptible to peer pressure, adolescents can engage in thrill-seeking and dangerous behavior. It is a time when the adults around teenagers need to respond sensitively and treat adolescents as "almost adults."

If not all adolescents experience problems, how can we identify those at risk? One marker seems to be gender, which offers the following clues:

- Adolescent boys with psychological difficulties usually have a history of childhood problems.
- More girls first show signs of psychological problems in adolescence (greater tendency toward depression, for example).
- Anxiety seems to play a key role in the appearance of adolescent difficulties (achievement situations are more likely to arouse anxiety in males, while interpersonal relations are more likely to cause anxiety in females).

Puberty and Its Effects

The mechanisms that control pubertal change develop prenatally. For most children, pubertal hormonal levels begin increasing at about seven years, with bodily changes appearing about four or five years later. Girls begin these changes about a year or two earlier than boys.

You probably remember how intensely aware you were of these changes within yourself. Also, the timing of puberty is crucial in relation to others because of the psychological and social processes involved. The well-developed, physically mature girl can only wonder about male classmates still running around in baseball caps and chewing bubble gum.

Here are some of the major findings that have resulted from studies of the relationship between puberty and psychological functioning:

- There is little evidence that puberty is linked to psychological problems; its effects may be positive as well as negative.
- While puberty may affect psychological variables, its impact is specific—on depression, aggression, or self-concept, for example.
- Puberty affects boys and girls differently.

Adolescent-Family Interactions

Recent research into adolescent-family relations has focused on the reciprocal effects of the family on the adolescent. (This research focus reflects the current interpretation of development as one of reciprocal interactions—see "Models of Development" earlier in this chapter.)

Is there a generational gap between adolescents and their families? As you can see, this question is an extension of the belief that "adolescence is an age of turmoil" for all adolescents. Again, research has shown that such beliefs are incorrect; parents and their adolescents have more similar attitudes and values than adolescents and their friends. (Adolescent peer similarities are more common in the adolescent culture, such as dress and music).

Adolescent-parent conflict, when it exists, results not only from adolescent difficulties but also from parental factors, such as divorce, excessive parental control, or a parental midlife crisis. Today's research into adolescent-family interactions looks at the family as a system, particularly the developmental status of its members. If an adolescent does indeed face a problem, are the parents sufficiently mature to adjust (Dacey & Kenny, in press)?

Adulthood

Adulthood is that time of our lives when we begin a career, form long-lasting relationships, assume personal and civic responsibilities, care for aging parents, and adjust to the aging process. Is it any wonder that this period has attracted the interest of scholars and become a discipline in itself, much as we have seen happen in childhood and adolescence?

Former president George Bush, Mother Teresa, and Jonas Salk (who at age 83 is searching for an AIDS vaccine), remind us of the growing number and contributions of senior citizens. Given the accomplishments of many individuals in their later years and the crucial developmental tasks of adulthood, we can understand why these years have attracted so much scholarly attention. Evidence of the growing interest in adulthood may be seen in the rapidly growing number of studies of this period of life (U.S. Department of Health and Social Services, 1991). Among the topics most heavily researched is that of intellectual ability: What is the reality and extent of intellectual decline with age? This is an important question when we consider America's aging population—"the graying of America."

As each of these age periods assumed its place in the developmental mainstream, psychologists felt a growing need to integrate findings, to devise some mechanism that would renew focus on the totality of human development. Consequently lifespan development was born, a study committed to the view that development is not confined to any period or periods, but rather is a lifelong process encompassing the time from conception to death.

As with any discipline, there are several controversial issues that lifespan studies must address. These topics engage the time and energy of developmental psychologists and help to give direction to the field.

Issues in Lifespan Development

In lifespan psychology, certain basic issues demand attention because they affect the manner in which we interpret behavior and development. We now discuss several of these issues, which appear repeatedly throughout the book. In chapter 2 we present the views of the leading developmental theorists, each of whom has taken a position on these issues.

Culture and Development

We pride ourselves on being a culturally diverse nation that encourages newcomers to share our way of life. Even under the best of conditions, however, immigrants can experience difficulties: language, customs, acceptance, job opportunities. The receiving country must also adapt. Schools, churches, markets, and politics all must change accordingly. We explore these ideas in later chapters, but first, what do we know about immigration today?

We pride ourselves on being a nation of immigrants, a country that welcomes newcomers with the promise of unrestricted opportunity. To achieve this objective during a time of increasing immigration demands consideration and tolerance for those of different color, nationality, and beliefs.

The United States is now experiencing its *third* great immigration wave. The federal government began keeping immigration records in 1820, which was the peak of the first significant immigration movement. Most of these immigrants arrived from Great Britain, Ireland, Germany, and Scandinavia, with a smaller number of Chinese entering California (Kellogg, 1988).

The second major wave occurred between 1900 and 1920, mainly immigrants from Italy, Hungary, Poland, and Russia. In 1907, more than 1.3 million immigrants entered the United States; during the decade 1900–1910, 8.8 million newcomers arrived, representing more than 40 countries. It wasn't until the economy collapsed in the late 1920s that immigration slowed.

The third significant wave began in the late 1960s and still continues. If we count the three categories of immigrants—legal immigrants, refugees, and undocumented immigrants—today's migration is the largest in our history. More than two-thirds of this latest migration are Asian and Hispanic (Kellogg, 1988).

Children from different cultures bring their differences with them; thus, their different customs may influence the relationships they form. Certainly their diet is different. Learning styles vary, which can affect their classroom achievement. Frequently fleeing war and poverty, they carry emotional scars. As you can well imagine, all of these conditions have developmental consequences.

For example, a white female elementary school teacher in the United States recently gave her students a math problem: If there are four blackbirds sitting in a tree and you shoot one of them with a slingshot, how many are left? A white student quickly answered, "three." An African immigrant student answered, with equal confidence, "zero." The teacher was puzzled, thinking that the new student either had misunderstood the problem or had a math problem. Actually the immigrant student reasoned that if you shoot one bird, the others will fly away (Wing Sue, 1992). Here is a good example of the need to understand the backgrounds of our changing population.

Continuity versus Discontinuity

We can summarize the issue of **continuity** versus **discontinuity** as follows: Do developmental changes appear abruptly, or as the result of a slow but steady progression (Rutter, 1989)? As a rather dramatic illustration, consider the phenomenon known as attachment in infancy. Sometime after the age of 6 months, babies begin to show a decided preference for a particular adult, usually the mother. We then say that the infant has *attached* to the mother. During any time of stress—anxiety, illness, appearance of strangers—the child moves to the preferred adult. With regard to continuity or discontinuity, does attachment appear suddenly as completely new and different behavior, or do subtle clues signal its arrival? (For an excellent overview of this topic, see Robins & Rutter, 1990.)

Continuities and discontinuities appear in all our lives, because the term *development* implies change. Puberty, leaving home, marriage, and career all serve to shape psychological functioning. Continuities occur, however, because our initial experiences, our early learning, our temperaments remain with us. The form of the behavior may change over the years, but the underlying processes may remain the same. For example, the conduct disorders of childhood (stealing, trouble in school, truancy) may become the violence of adulthood (theft, wife abuse, child abuse, murder, personality disorders). There may be surface dissimilarities in the types of behavior, but the processes that cause both kinds of behavior may be identical, which has caused some developmental psychologists to argue for continuity in development (Rutter, 1989).

We must, however, explain those periods in our lives that seem to be quite different from those that preceded them; for example, walking and talking. We also negotiate *transitions* at appropriate times in our lives, such as leaving home, beginning a career, getting married, adjusting to the birth of children. It's not just a matter of "doing these things," because the circumstances surrounding them also have important developmental effects. What is the "right" age to get married? Is it a positive occasion and not something that "should" be done? The reality of these events has caused other developmental psychologists to see development as mainly discontinuous. Most developmental psychologists now believe that both continuity and discontinuity characterize development.

Stability versus Change

Whether children's early experiences (either positive or negative) affect them throughout their lifespans is a question that today intrigues most developmental psychologists. If a child suffers emotional or physical neglect, abuse, or malnutrition, does it mean the child is scarred for life? If you answer yes, you believe in **stability;** if your answer is no, then you accept the likelihood of change. Although it seems to be a fairly simple question, firm answers remain elusive.

We know, for example, that human beings show amazing resiliency; recovery from damage (both physical and psychological) has been a well-documented fact of biological and psychological research (Robins & Rutter, 1990). Yet flexibility (or the more technical term, **plasticity** has its limits, which brings us to a critical question: Under what conditions do children and adults recover from damage (Goldman-Rakic, 1983)?

As Gardner (1983) notes, there seems to be considerable flexibility in human growth, especially during the early months. For example, the human infant who is swaddled during the first year of life still walks normally in the second year. What stand you take on this issue has serious implications. Consider, for example, those who believe that whatever happens to a child in the early years leaves an indelible mark; that the damage remains with the child for life. Adherents of flexibility take exactly the opposite view: since human beings demonstrate tremendous recuperative powers, refusal to offer help borders on the criminal in their minds. (For an excellent discussion of this topic, see Gardner, 1983, chap. 3; and Shatz, 1992.)

The child-rearing, educational, economic, and political consequences of either of these positions spread throughout the entire fabric of our society. As one example, processes such as psychotherapy are encouraged because they are believed to help overcome earlier traumatic experiences. At the other extreme are those who believe that since early damage is irreversible, intervention is useless. Although advocates of both views have moderated their ideas somewhat, the lines are clearly drawn; and identical evidence has been interpreted differently by both sides. We analyze this important topic at greater length in several chapters.

Sensitive Periods versus Equal Potential

Our earlier discussion of developmental tasks should have alerted you to the significance of the *timing* of experiences and individual reactions to them. At different times in our lives, certain experiences are more meaningful than similar events at other times. Going off to school for the first time was more intense than walking into this lifespan class; a pregnant teenager reacts differently to thoughts of a family than a pregnant young adult woman with considerable support or an older woman who has successfully conceived after years of unsuccessful attempts.

Many developmental psychologists believe there are certain times in the lifespan when a particular experience has a greater and more lasting impact than at other times (Sroufe, Cooper, & DeHart, 1992; Rutter, 1989). These are called **sensitive periods.** For example, the care and attention of a loving adult during the first months of life seems obviously linked to the appearance of attachment. If such an experience is lacking during these early days, but comes at a later age (say five years), does it produce the same results? Or was the sensitive period missed? Here we have another explosive question that cuts across many of society's activities—socially, politically, economically.

Let's assume for the sake of argument (and we don't agree with this conclusion) that once a sensitive period is missed, the damage can never be repaired. Why, then, should we bother to pour hundreds of millions of dollars into intervention programs (such as Head Start) in the hope of "making up" for lost experiences? Politicians who vote on the use of public monies must take a stand on this question; voters either agree or disagree with their decisions. Most developmental psychologists don't agree that "once missed—never catch up." But they also believe that nobody, child or adult, can withstand severe deprivation, either physical or psychological, indefinitely. We all have our limits.

These issues help to identify lifespan psychology as a dynamic discipline, one that has great theoretical and practical implications. But, as fascinating as these issues are, we must never forget the integrated nature of development. With these ideas in mind to help you interpret developmental data, we turn now to those research techniques that developmental psychologists use in resolving these questions.

Developmental Research

Having identified several key developmental issues and explored a promising model of development, we must now ask: How can we obtain reliable data about these topics so that we can better understand them?

Today we use many approaches to understanding human behavior. Each has its strengths and weaknesses; none is completely reliable. Most developmental psychologists employ one of three data collection methods: **descriptive studies, manipulative experiments,** and **naturalistic experiments.** In the first type, information is gathered on subjects without manipulating them in any way. In the second two, an experiment is performed before the information is gathered.

Developmental psychologists also use one of four time-variable designs: **one-time, one-group studies; longitudinal studies; cross-sectional studies;** and a combination of the last two, called **sequential studies.** Each type of study varies according to the effect of time on the results. We now describe each of these aspects of research.

Data Collection Techniques

The three data collection techniques are described in the following subsections.

Descriptive Studies

Descriptive studies are quite common. Most are numerically descriptive: For example, how many 12-year-olds think the government is doing a good job, versus 17-year olds who think so? How much money does the average 40-year-old woman have to spend per week? How many pregnant teenage girls were or were not using birth control? How happily or unhappily does the average 66-year-old man view his sex life? Some studies ask people their opinions about themselves (called self-report studies) or other people. These studies may use interviews or questionnaires. Other studies describe people simply by counting the number and types of their behaviors (called observational studies). A third type, *case studies,* presents data on an individual or individuals in great detail, in order to make generalizations about a particular age group.

An example of the case study approach is Mack and Hickler's *Vivienne: The Life and Suicide of an Adolescent Girl* (1982). After Vivienne's death, the researchers obtained the family's permission to read her diary, poems, and letters. They also interviewed her relatives, friends, and teachers in order to shed light on her thinking as she came closer and closer to committing this tragic act. Although their findings may explain the suicide of only this one person, the researchers' hope was to discover the variables that caused such a decision. Many of the best theories about development have been based on detailed case studies of small numbers of individuals.

Descriptive studies have the advantage of generating a great deal of data. Because the sequence of events is not under the observer's control, however, causes and effects cannot be determined.

Manipulative Experiments

In the quest for the causes of behavior, psychologists have designed many manipulative experiments. In these, they attempt to keep all variables (all the factors that can affect a particular outcome) constant except one, which they carefully manipulate. It is called a treatment. If there are differences in the results of the experiment, they can be attributed to the variable that was manipulated in the treatment. The experimental subjects must respond to some test selected by the investigator in order to determine the effect of the treatment. Figure 1.4 illustrates this procedure.

In the figure, E is the experimental group and C is the control group, which receives no special treatment; *x* stands for the treatment; and the lowercase *b* and *a* refer to measurements done before and after the experiment. There must be no differences between the two groups, either before or during the experiment (except the treatment). Otherwise, the results remain questionable.

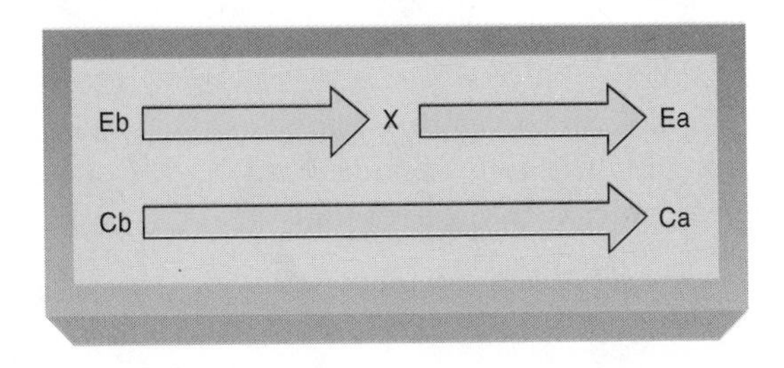

Figure 1.4
The classic experiment

An example would be Dacey's study (1993) in which two similar groups of eighth-graders were randomly selected from all those in two inner-city middle schools. The experimental group was given the treatment (a series of fourteen lessons in self-control) while the control group studied the traditional curriculum. At the end of this experiment, researchers observed both groups to see whether there was any decrease in drug use or dropping out of school. Since the experimental group did significantly better on these criteria, we can assume that the lessons in self-control were effective.

Though manipulative experiments often can lead us to discover what causes what in life, there are problems with them. How do you know your results are reliable? Was the treatment similar to normal conditions? Do subjects see themselves as special because you picked them, and thus react atypically? For these reasons, researchers may turn to naturalistic experiments.

Naturalistic Experiments

In naturalistic experiments, the researcher acts solely as an observer and does as little as possible to disturb the environment. "Nature" performs the experiment, and the researcher acts as a recorder of the results. (Note: Don't confuse these experiments with descriptive studies that are done in a natural setting—such as a park—those are not experiments). An example is the study of the effects of the Northeast blizzard of 1978 by Nuttall and associates, (1980). These researchers compared the reactions of people whose homes were destroyed to the reactions of people whose homes suffered only minor damage.

Only with a naturalistic experiment do we have any chance of discovering causes and effects in real-life settings. The main problems with this technique are that it requires great patience and objectivity, and it is impossible to meet the strict requirements of a true scientific experiment.

Time Variable Designs

In the following subsections, we describe the four time variable designs.

One-Time, One-Group Studies

As the name implies, one-time, one-group studies are those that are carried out only once on one group of subjects. Thus it is impossible to investigate causes and effects, because the sequence of events cannot be known.

Longitudinal Studies

The longitudinal study, which makes several observations of the same individuals at two or more times in their lives, can answer important questions. Examples are determining the long-term effects of learning on behavior, the stability of habits and intelligence, and the factors involved in memory.

Although much of childhood behavior disappears by adulthood, there has long been a suspicion that some adult traits develop steadily from childhood and remain for life. In his search for such stable characteristics, Benjamin Bloom, in his classic work *Stability and Change in Human Characteristics* (1964), notes that the development of some human characteristics appears visible and obvious, while that of others remains obscure. The following are three growth studies in which more than 300 persons have participated for over thirty years:

- *The Berkeley Growth Study,* begun in 1928, was designed to study the mental, motor, and physical development of a sample of full-term, healthy babies.

- *The Guidance Study* took youngsters born in 1928 and 1929 and began to study them at 21 months of age. The aim was to study physical, mental, and personality development in a normal group.
- *The Oakland Growth Study* of 200 fifth- and sixth-graders, begun in the early 1930s, was designed to study many interrelations between developmental changes and behavior. The investigators tried to discover whether developmental changes affect a child's potential.

One of the longitudinal growth studies most often quoted is that of the Fels Research Institute (Kagan & Moss, 1962). The subjects were forty-five girls and forty-four boys, all white, whose personality development was traced from birth through early adulthood. The investigators conducted extensive interviews with both the children and their parents. Among the particular techniques used were

- *Personality tests* given at regular intervals. The child was asked to react to a picture (the Thematic Apperception Test) or to a design such as a Rorschach inkblot. Trained persons analyzed responses for clues to personality, including motives, attitudes, and problems.
- *Observation of the mother* in the home with the child present, and also annual interviews with the mother.
- *Measurement of the intelligence* of both the mother and the father, using the Otis IQ test.
- *Regular observation of the child's behavior* in the home, in school, and at day camp. The child was also interviewed by researchers.

Kagan and Moss summarize the obvious advantages of the longitudinal method when they note that it permits the discovery of lasting habits and of the periods in which they appear. A second advantage is the possibility of tracing those adult behaviors that have changed since early childhood.

There are, however, many problems with longitudinal research. It is expensive and often hard to maintain because of changes in availability of researchers and subjects. Changes in the environment can also distort the results. For example, if you began in 1960 to study changes in political attitudes of youth from 10 to 20 years of age, you would probably have concluded that adolescents become more and more radical as they grow older. But the war in Vietnam would surely have had much to do with this finding. The results of the same study done between 1970 and 1980 would probably not show this trend toward the radical.

Cross-Sectional Studies

The cross-sectional study method compares groups of individuals of various ages at the same time, in order to investigate the effects of aging. For example, if you want to know how creative thinking changes or grows during adolescence, you could administer creativity tests to groups of 10-, 12-, 14-, 16-, and 18-year-olds, and check on the differences of the average scores of the five groups. Jaquish and Ripple (1980) did just this, but their subjects ranged in age from 10 to 84!

As with each of the others, there is a problem with this method. Although the effects of cultural change can be minimized by careful selection, it is possible that the differences you may find may be due to differences in **age cohort,** rather than maturation. Age cohorts are groups of people born at about the same time. Each cohort has had different experiences throughout its history, and this can affect the results as well as the actual differences in age. Figure 1.5 compares these two approaches.

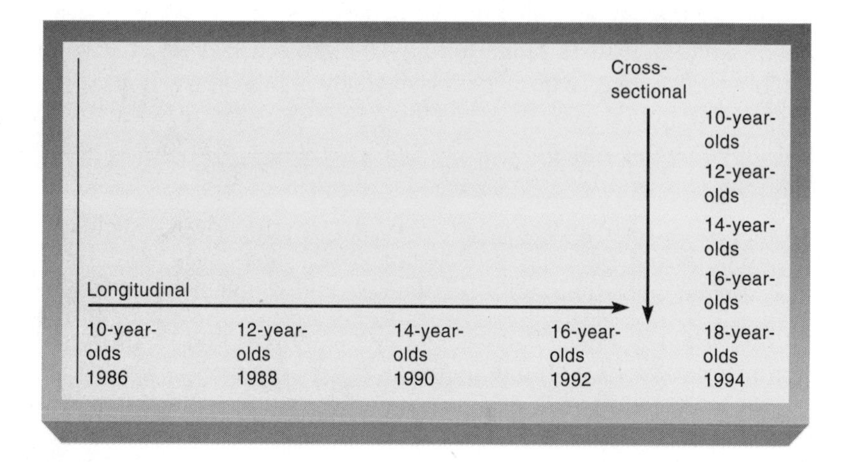

Figure 1.5 Comparison of the longitudinal and cross-sectional approaches

Sequential (Longitudinal/Cross-Sectional) Studies

When a cross-sectional study is done at several times with the same groups of individuals (such as administering creativity tests to the same three groups of youth, but at three different points in their lives), the problems mentioned before can be alleviated. Table 1.6 illustrates such a study. Although this type of research is complicated and expensive, it may be the only type that is capable of answering important questions in the complex and fast-changing times in which we live.

Table 1.6 Illustration of a Longitudinal/Cross-Sectional Study

Creativity Test			
Test I March 4, 1988	Test II March 4, 1990	Test III March 4, 1992	
Group A (12 years old)	Group A (14 years old)	Group A (16 years old)	Mean Score Group A
Group B (14 years old)	Group B (16 years old)	Group B (18 years old)	Mean Score Group B
Group C (16 years old)	Group C (18 years old)	Group C (20 years old)	Mean Score Group C
Mean Score 1988	Mean Score 1990	Mean Score 1992	

Table 1.7 shows how each of the data collection methods may be combined with each of the time variable designs. For each of the cells in this table, a number of actual studies could serve as examples. Can you see where each study mentioned in this section would go?

Table 1.7 Relationships of Data Collection Techniques and Time Variable Designs

Time Variable Designs	Data Collection Techniques: Descriptive	Manipulated	Naturalistic
One-time, one-group			
Longitudinal			
Cross-sectional			
Sequential			

To conclude this section, Figure 1.6 compares the various research techniques. By *controlled,* we mean the degree to which the investigator can control the relevant variables. By *inclusive,* we mean the degree to which all relevant information is included in the data.

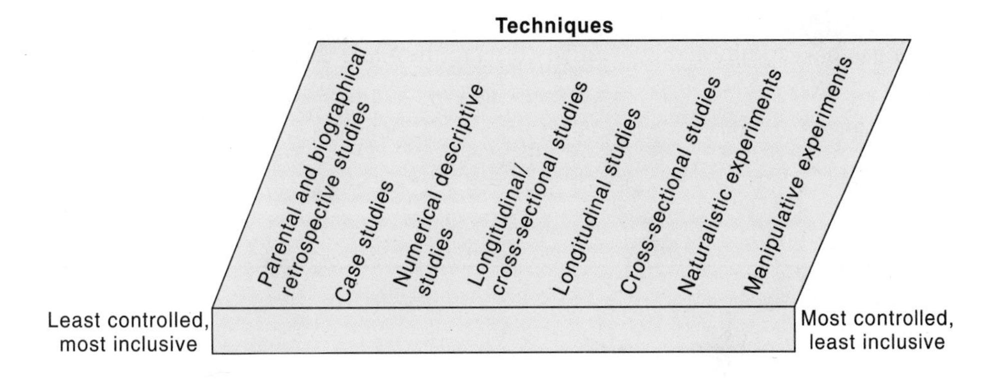

Figure 1.6
A comparison of research techniques

As you now realize, lifespan studies focus on successive periods of development. Table 1.8 illustrates how age periods are reflected in this text's coverage.

Table 1.8 Age Periods and Chapter Coverage

Age Period	Chapters
Beginnings	3. The Biological Basis of Development 4. Pregnancy and Birth
Infancy (0–2)	5. Physical and Cognitive Development in Infancy 6. Psychosocial Development in Infancy
Early Childhood (2–6)	7. Physical and Cognitive Development in Early Childhood 8. Psychosocial Development in Early Childhood
Middle Childhood (7–11)	9. Physical and Cognitive Development in Middle Childhood 10. Psychosocial Development in Middle Childhood
Adolescence (12–18)	11. Background and Context 12. Rapidly Changing Bodies and Minds 13. The Troubled Adolescent 14. Sexuality and Gender Roles
Early Adulthood (19–34)	15. Passages to Maturity 16. Physical and Cognitive Development 17. Psychosocial Development
Middle Adulthood (35–64)	18. Physical and Cognitive Development 19. Psychosocial Development
Late Adulthood (65+)	20. Physical and Cognitive Development 21. Psychosocial Development 22. Dying and Spirituality

Conclusion

In this chapter we have presented a model of development—the biopsychosocial model—that forms the structure of the book. We urge you to use this model to help you grasp and retain the material and meaning of the various chapters. We have also identified the various age groups that constitute the lifespan and that are the focus of this book. Lifespan study can aid us in adjusting to a society in which rapid change seems to be an inevitable process. By acquiring insights into your own development and recognizing the developmental characteristics of people of differing ages, you can hope to have more harmonious relationships with others. Also, as a result of reading about the strengths and weaknesses of different research methods, you should be more analytical and critical of the studies that are presented.

With these basic ideas firmly in place, we now turn to a closer look at those theorists who have made lasting contributions to the field of developmental psychology. As you read their explanations of development, keep in mind the purposes they were trying to achieve: Freud's search for the causes of behavior in the unconscious, Piaget's endeavors to identify the cognitive structures that underlie cognitive

An Applied View — Understanding the Research Article

As you continue your reading and work in lifespan development, your instructor will undoubtedly ask you to review pertinent articles that shed light on the topic you're studying. Many of these articles present the results of an experiment that reflects the scientific method.

The typical research article contains four sections: the **Introduction,** the **Method** section, the **Results** section, and **Discussion** (Moore, 1983). We'll review each of these sections using a well-designed study—*The Effects of Early Education on Children's Competence in Elementary School,* published in *Evaluation Review* (Bronson, Pierson, & Tivnan, 1984)—to illustrate each of the four parts.

1. **The Introduction**
   The introductory section states the purpose of the article (usually as an attempt to solve a problem) and predicts the outcome of the study (usually in the form of hypotheses). The introduction section also contains a review of the literature. In the introductory section of the article by Bronson and associates, the researchers state that their intent is to coordinate the effects of early education programs on the performance of pupils in elementary school. They concisely review the pertinent research and suggest a means of evaluating competence.
2. **The Method Section**
   The method section informs the reader about the subjects in the experiment (Who were they? How many? How were they chosen?), describes any tests that were used, and summarizes the steps taken to carry out the study. In the study by Bronson and her associates, the subjects were 169 second-grade children who had been in an early education program and 169 other children who had not been in the preschool program. The outcome measure was a classroom observation instrument. The authors then explained in considerable detail how they observed the pupils.
3. **The Results Section**
   In the results section, the information gathered on the subjects is presented, together with the statistics that help us to interpret the data. In the article we are using, the authors present their data in several clear tables and present differences between the two groups using appropriate statistics.
4. **Discussion**
   Finally, the authors of any research article discuss the importance of what they found (or didn't find) and relate their findings to theory and previous research. In the Bronson article, the authors report that the pupils who had experienced any early education program showed significantly greater competence in the second grade. The authors conclude by noting the value of these programs in reducing classroom behavior problems and improving pupils' competence.

Don't be intimidated by research articles. Look for the important features and determine how the results could help you to understand people's behavior at a particular age.

development, Vygotsky's emphasis on the context of development, Erikson's effort to describe the psychosocial crises of life, and the behaviorists' attempt to explain the role that the environment plays in development.

Chapter Highlights

A Biopsychosocial Model of Development

- The interaction of biological, psychological, and social factors explains development.
- Changes in one domain (for example, the biological) can affect all aspects of development.

Why Study Lifespan Development?

- As psychologists realized that development did not cease at adolescence but continued into adulthood and old age, lifespan psychology assumed an important place in developmental psychology.
- Several models of development have been proposed to describe the interaction of heredity and environment.
- Reciprocal interactions is a valuable concept in understanding development.
- The developmental task concept is a good reminder of the critical tasks we face at different ages.
- The timing of experiences as well as the transitions during the lifespan help us to gain insights into developmental processes.
- The satisfaction of needs is important for harmonious progress through the lifespan.
- Today we are aware of the significance of context in development.

History of Lifespan Studies

- The concept of childhood is relatively new.
- Children are now seen as active participants in their development.
- Most adolescents are not as troubled as the individuals we read about so frequently.
- Today's adolescent research places greater emphasis on the family's role.
- Studies of adulthood have indicated that individuals continue to grow and change during these years.

Issues in Lifespan Development

- Any analysis of lifespan development must address key developmental issues such as the importance of culture and development if it is to present a complete picture of development.
- Many psychologists believe that development occurs as a steady progression of small accomplishments (for example, most infants begin to move on the floor by pulling themselves on their stomachs; they then move to a position on their hands and knees and move much more quickly, which is an example of continuous development); other psychologists believe that development occurs in spurts or stages, such as the marked difference between crawling and walking, which is an example of discontinuity.
- The controversy over stability or change continues to divide developmental psychologists; does the aggressive toddler become the aggressive adult, or does this behavior change over time?
- Sensitive periods seem to exist; that is, there are times in our lives when we acquire new behaviors more easily than at other times.

Developmental Research

- To explain the various ages and stages of development, we must use the best data available to enrich our insights and to provide a thoughtful perspective on the lifespan.
- Good data demand careful research methods; otherwise, we would be constantly suspicious of our conclusions.
- The most widely used research techniques include descriptive studies, manipulative experiments, and naturalistic experiments.
- Developmental psychologists also use four time variable designs: one-time, one-group; longitudinal studies; cross-sectional studies; and sequential studies.

Key Terms

Action theory
Age cohort
Biopsychosocial model
Continuity
Cross-sectional studies
Descriptive studies
Developmental task
Discontinuity
Esteem needs
Exosystem
Interaction
Lifespan psychology
Longitudinal studies
Love and belonging needs
Macrosystem
Main-effects model
Manipulative experiments
Mesosystem
Microsystem
Nativist
Naturalistic experiments
Nurturist
One-time, one-group studies
Physiological needs
Plasticity
Reciprocal interactions
Safety needs
Self-actualization needs
Sensitive period
Sequential studies
Stability
Transactional model

What Do You Think?

1. We urged you to refer to the biopsychosocial model as you continue your reading of the text. Can you explain its potential value? Now think of an example in your own life, or in the life of a family member, and describe how biological, psychological, and social factors interacted to produce a particular effect. Do you think the model helped you to explain that person's behavior?
2. We presented several issues that thread their way through lifespan studies: for example, the role of sensitive periods, continuity versus discontinuity, stability versus change. Why do you think these are issues? Examine each one separately and defend your reasons for stating that they have strong developmental implications.
3. Throughout the chapter, we have stressed the important role that context plays in development. What do you think of this emphasis? Although this is only the first chapter of your reading, recall your own life and think about how those around you have influenced you (in both positive and negative ways). Use these personal experiences in your answer.

Suggested Readings

Havighurst, R. (1972). *Developmental tasks and education*. New York: David McKay. Your school library probably has a copy of this little classic. As you read through it, note Havighurst's perceptive explanation of developmental tasks and what they mean for development.

Maslow, A. (1987). *Motivation and personality*. New York: Harper & Row. Maslow's description of the road to self-actualization. This book is easy to read, thoughtful, and a helpful guide to analyzing human behavior. You will find Maslow's ideas about a hierarchy of needs both insightful and provocative.

Woodward, B. (1991). *The commanders*. New York: Simon & Schuster. A fascinating look at the Gulf War, in which Colin Powell played such a key role. Woodward's analysis of Powell's behavior during these tense days provides telling insights into Powell's development and character.

CHAPTER 2

Theoretical Viewpoints

The Importance of Theories 30
Critical Issues 31
The Biopsychosocial Causes of Change 32
The Psychoanalytic Approach: *Sigmund Freud's Theory* [BIOPSYCHOsocial] 32
Defending the Unconscious Mind 33
Structures of the Psyche 33
The Developing Personality 34
The Cognitive Structures Approach 36
Jean Piaget's Theory [BIOPSYCHOsocial] 36
Beginning the Study of Intellectual Development 36
The Role of Intelligence 36
Stages of Cognitive Development 38
Lev Vygotsky's Theory [bioPSYCHOSOCIAL] 40
The Behavioral Approach 41
B. F. Skinner's Theory [biopsychoSOCIAL] 41
Albert Bandura's Theory [biopsychoSOCIAL] 43
The Psychosocial Crises Approach 45
Erik Erikson's Theory [bioPSYCHOSOCIAL] 45
Conclusion 51
Chapter Highlights 51
Key Terms 53
What Do You Think? 54
Suggested Readings 54

In this chapter, you will begin to understand how theories help to explain human development. We'll look at the critical issues in the field, and then we'll start our exploration of four of the most important theories—those of Sigmund Freud, Jean Piaget, B. F. Skinner, and Erik Erikson.

As a result of having read this chapter, you should be able to:

- Describe the purposes of theory-making and the relationships of theory to three other aspects of science.
- List and differentiate the four critical issues that are of concern when we examine theories of human development.
- Explain biopsychosocial causes.
- Itemize Freud's stages of development, together with his concepts of the functions and constructs of the human psyche.
- Define Piaget's four stages of cognitive development, as well as explain the roles played by the two central processes of organization and adaptation.
- Explain the role of reinforcement, punishment, and extinction in Skinner's behavioral theory, as well as detail the contributions of Ivan Pavlov, John Watson, and Albert Bandura to behavioral theory.
- Compare and contrast Erik Erikson's eight stages of psychosocial crisis.
- Discuss these issues from an applied, a multicultural, and your own point of view. ■

The Importance of Theories

True, my theory is no longer accepted, but is was good enough to get us to the next one!

Donald Hebb (whose theory of intelligence was popular in the 1940s)

For many people, the word *theory* means someone's guess about why something happens the way it does. For example, Bob might say, "It's my theory that Joe quit the team because he thinks we don't like him." Used in a textbook, the word *theory* often makes readers think of complicated arguments between experts—"Highbean's theory disagrees with Numbskull's, in that . . ."

In this book, we use the word differently. We believe that theories are essential in psychology, serving several vital functions. Good theories

- Are helpful tools for organizing a huge body of information. The published studies on human development number in the tens of thousands. The results of these findings would be incomprehensible unless they were organized in some meaningful way. A theory is a shorthand description of this complexity. It forms a framework of "pegs" on which we can hang similar kinds of research findings.
- Help us to focus our search for new understandings. They offer guideposts in our quest for the truth about the complicated human body and its development.
- Do not just describe; they try to explain how findings may be interpreted. They offer building blocks that can help us to understand *which* facts are important, and *what* conclusions to draw.

- Draw attention to major disagreements among scholars and scientists. By making these differences clearer, they offer testable ideas that can be confirmed or refuted by research. There can be different interpretations of the same facts. This can be confusing, but it is important that you learn to recognize these differences and draw your own conclusions.

Theories do not stand alone; they are related to other aspects of science. At the basic level of all the sciences is *empirical research*. For social science, of which lifespan psychology is a part, empirical research means studying real people under carefully arranged conditions. From this research, scientists form *constructs*. A construct is an idea about some particular aspect of the human being. The ego and intelligence are two examples. Constructs are used to build theories. A theory is a system of ideas that attempts to explain research findings by showing how the constructs are related. The ultimate goal of this process is to produce greater *understanding* of how and why humans think, feel, and behave as they do. Figure 2.1 illustrates how these four factors of science affect each other.

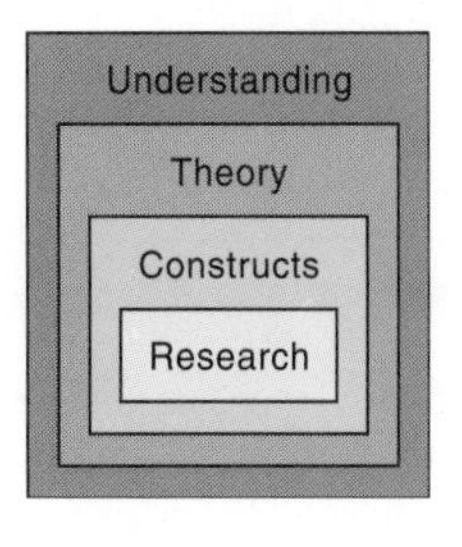

Figure 2.1
The relationships of theory to other aspects of science

We cannot give you the one final answer on how humans develop—no one can do that. We can, however, introduce you to the best current thinking in the study of human development. The three aspects of the biopsychosocial model that was introduced in chapter 1 serve as overriding themes.

The chapters that follow are organized by age level. As each age is considered, the relevant part of these three aspects is discussed in greater detail. By the end of the book, we believe you will have a clear understanding of the biopsychosocial model. This in turn will provide you with the best comprehension of the splendid journey that is known as the human lifespan.

Critical Issues

Let us first remind you of the four major issues to which you were also introduced in chapter 1. These four questions are central to the study of human development, and thus arise again and again:

- *Biopsychosocial forces.* Of these forces, which is more important to human development, the genes we inherit from our parents, the psychological development of our minds and personalities, or the specific social events in our lives since we were conceived? (See section below.)

- *Continuity versus discontinuity.* Do the changes that happen to us as we grow up occur slowly and imperceptibly, or are they abrupt and discrete?

- *Stability versus change.* What happens when a child suffers a traumatic accident? Does this experience have to scar the child's personality forever? Those who take the *stability* position say yes, it is likely the child will be greatly affected throughout life. Those who believe in *plasticity* say no, there is a chance that the experience can be overcome so that it has little later effect.

- *Sensitive periods versus equal potential.* Are there certain times in the lifespan when a particular experience has a greater and more lasting impact than at any other time, or do all periods have an equal potential for affecting development?

As you will see, each of the theorists we have chosen to highlight in this chapter and in the book takes a position on each of these issues. We will compare and contrast them so that you will be better able to decide where you stand on these critical issues.

The Biopsychosocial Causes of Change

All theorists agree that humans change over time, and that there are biological, psychological, and social causes of this development. They differ, however, on the emphasis they put on each of these three factors. Depending on how they think about human nature, they view change as being caused mainly by one of the following factors:

- *Biological.* The genes are the major factor in our development from one discrete stage of life to the next. Change tends to be rather abrupt.
- *Psychological.* Changes in personality and intellect are behind our development through life. Change depends on the way these internal states mature.
- *Social.* Development depends greatly on what is happening in our environment. Because so many things happen in the course of our lives, change tends to be a gradual, ever-present process.
- *Some combination* of two or three of these factors.

We have designated each of the theories that follow as being in one of these four camps by capitalizing some of the letters in the word *biopsychosocial.* As we said in chapter 1, all developmental theorists agree that these factors—biology, psychology, and social environment—play a role in human development. Therefore, by capitalizing the part of *biopsychosocial* emphasized by the theorist, we show that each theory acknowledges the whole person, but gives more importance to one or more parts than another.

In the sections that follow, we describe theories that have been considered the most comprehensive in the field of lifespan psychology. We might have included several other highly regarded theories, but they have mainly to do with specific aspects of the mind. We consider those theories in detail later in this book.

The Psychoanalytic Approach: Sigmund Freud's Theory [BIOPSYCHOsocial]

Dr. Sigmund Freud was a medical doctor before he invented psychiatry.

In more than one hundred years of psychological research, it is impossible to think of anyone who has played a larger role than Sigmund Freud. Even his most severe critics admit that his theory on the development of personality is a milestone in the social sciences. In fact, many people mistakenly think **psychoanalytic theory,** the name he gave to his theory, is the same as psychology. Because Freud felt that the personality develops as a result of biological and psychological forces, with social influences playing a smaller role, we have labelled this a BIOPSYCHOsocial theory.

Probably because of his experiences as a medical doctor, Freud doubted the reliability of people's testimony about themselves. He also distrusted behavior as a source of the truth. For him, the *unconscious* was the key to understanding the human being. It is there that the most important motives and values reside.

Because many of the ideas in the unconscious are primitive, they are not acceptable to the conscious mind. For example, if a child is furious with her mother, she may not be able to acknowledge it because she is not supposed to hate her mother. Only bad people do that.

Defending the Unconscious Mind

Freud believed that important information in the unconscious is kept from awareness by an array of **defense mechanisms** (Gay, 1988). There are unconscious attempts to prevent awareness of unpleasant or unacceptable ideas. The *psychic censor,* a function of the mind, stands guard over these unconscious thoughts by using defense mechanisms to block awareness. Table 2.1 describes some of the most common defense mechanisms.

Table 2.1 Some Common Defense Mechanisms

Repression	Unconsciously forgetting experiences that are painful to remember. Example: sexual abuse
Compensation	Attempting to make up for an unconsciously perceived inadequacy by excelling at something else. Example: learning to play the guitar if unable to make the soccer team
Rationalization	Coming to believe that a condition that was contrary to your desires is actually what you had wanted all along. Example: being glad a trip was cancelled, because "it would have been boring anyway"
Introjection	Adopting the standards and values of someone you are afraid to disagree with. Example: joining a gang
Regression	Reverting to behaviors that were previously successful when current behavior is unsuccessful. Example: crying about getting a low grade in school with the unconscious hope that the teacher will change the grade
Displacement	When afraid to express your feelings toward one person (e.g., anger at your teacher), expressing them to someone less powerful, such as your sister

From *Adolescents Today,* 3/e, by John S. Dacey. Copyright © 1986, 1982 by Scott, Foresman and Company. Reprinted by permission of HarperCollins Publishers.

Anna Freud, Sigmund Freud's daughter, discovered two exclusively adolescent defense mechanisms: *asceticism,* in which, as a defense against the sexual, "sinful" drives of youth, teenagers frequently become extremely puritanical; and *intellectualization,* in which adolescents defend against emotionality of all kinds by becoming extremely intellectual and logical about life.

Structures of the Psyche

Freud divided the mind into three structures: the **id,** the **ego,** and the **superego.** These structures appear at different stages of the young child's development. They are empowered by the *libido,* Freud's term for psychic energy. It is similar to the physical energy that fuels bodily functions. He argued that the libido is motivated mainly by sexual and aggressive *instincts.* The characteristics of the three structures of the mind are as follows:

- *The id.* This structure is the only one present at birth. It contains all of our basic instincts, such as our need for food, drink, dry clothes, and nurturance. It is the simplest of the structures, operating only in the pursuit of bodily pleasures. It is not realistic; it can be satisfied merely by

What's Your View?

Defense mechanisms function to protect the conscious mind from the truth. They distort the realities that we find too painful to face. Because they mislead us, and because their maintenance requires spending a lot of energy that could better be spent elsewhere, many psychologists feel we should try to eliminate them if possible.

Others argue that some truths are just too painful to face, and that therefore, at least in the short run, defense mechanisms are useful. They say that most people go through difficult periods (e.g., breaking up with your first love), and that during these periods, defense mechanisms can provide a beneficial buffer for a vulnerable self-image. There will be time enough later to "face the music" of reality. What do you think?

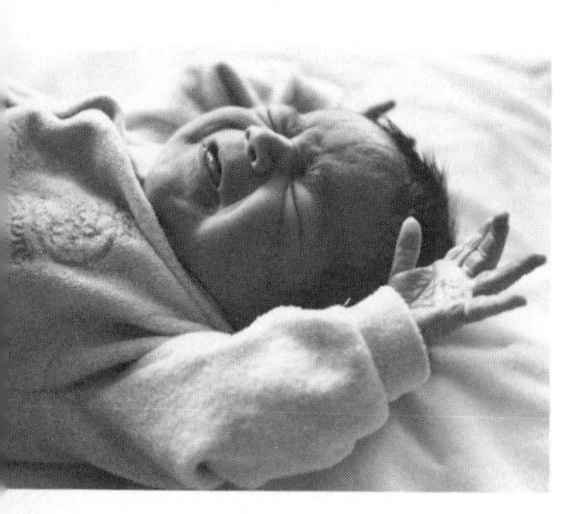

Sigmund Freud suggested that babies react to needful feelings such as hunger in several steps. First they become aware of the need, then they cry; next they imagine that the need has been met, and then they fall back to sleep. Slowly they learn that imagination is no substitute for real satisfaction of a need.

imagining that we have gotten what we wanted. It has no clear notion of time, space, or any other aspect of reality. Think about your dreams; they are a good example of the id in operation.

- *The ego.* The ego is the central part of our personality, the (usually) rational part that does all the planning. It keeps us in touch with reality. It begins to develop from the moment of birth.

  Imagining that you get what you want does not work very well, nor for very long. Consider a baby that awakens from a nap and starts to cry. Soon it stops and drifts back into a doze. Before long it awakens again and cries longer this time. If the baby is left unattended, the dozing periods get shorter and the crying periods longer.

  Freud suggested that the awakening baby probably is hungry (or thirsty or wet). At first, it imagines that the problem is solved and is comforted, but this works less and less well as it becomes hungrier. Over time, the baby learns to cry loudly if it wants its mother, and not to stop until she comes. This is the beginning of an ego.

  The ego is necessary so that we can learn to live in the real world. The stronger the ego becomes, the more realistic, and usually the more successful, the person is likely to be.

- *The superego.* Throughout infancy, we gain a clearer and clearer conception of what the world is like. Then, toward the end of the first year, our parents and others begin to teach us what they believe it *should* be like. They instruct us in right and wrong, and expect us to begin to behave according to the principles they espouse. This is the beginning of the superego. Freud disagreed with the religious idea of an inborn conscience. He argued that all morality is learned, as a function of the superego.

Now starts the never-ending battle between the desires of the id and the demands of the superego. The main job of the ego is to strive unceasingly for compromises between these two "bullies."

The Developing Personality

Now we get to the discrete stages. For Freud, development means moving through five instinctive stages of life, each of which he assigns to a specific age range. Each stage is discrete from the others. Each stage has a major function. Each function is based on a pleasure center, and unless this pleasure center is stimulated appropriately (not too much, not too little), the person becomes *fixated* (stuck at that stage) and is unable to become a fully mature person. The five stages are

- The *oral stage* (0 to 1 1/2 years old). The oral cavity (mouth, lips, tongue, gums) is the pleasure center. Its function is to obtain an appropriate amount of sucking, eating, biting, and talking.
- The *anal stage* (1 1/2 to 3 years old). The anus is the pleasure center. The function here is successful toilet training.
- The *phallic stage* (3 to 5 years old). The glans of the penis and the clitoris are the pleasure centers in this stage and in the two remaining stages. (That he named this stage the *phallic* stage is one of several reasons Freud is not popular among feminists! [e.g., Horney, 1967]) The major function of this stage is the healthy development of sexual interest. This is achieved through masturbation and unconscious sexual desire for the parent of the opposite sex. Resolution of the conflicts caused by this desire (called the *Oedipal conflict* in males and the *Electra conflict* in females) is the goal.
- The *latency stage* (5 to 12 years old). During this stage, sexual desire becomes dormant. This is especially true for males, through the defense mechanism of introjection (see Table 2.1). They refuse to kiss or hug their mothers, and treat female age-mates with disdain. Because our society is more tolerant of the daughter's attraction to her father, the Electra complex is less resolved and girls' sexual feelings are less repressed during this stage.
- The *genital stage* (12 years old and older). Now there is a surge of sexual hormones in both genders, which brings about an unconscious recurrence of stage 3. Normally, however, youths have learned that desire for one's parents is taboo, and so they set about establishing relationships (bumblingly at first) with members of the opposite sex who are their own age. Freud believed that if these five stages are not negotiated successfully, homosexuality or an aversion to sexuality itself results. (It should be noted that this concept is not popular among homosexuals, many of whom believe that their sexual orientation goes much deeper than this—see chap. 14.) If fixation occurs at any stage, anxiety results, and defense mechanisms will be used to deal with it.

Table 2.2 summarizes the psychoanalytic position from the standpoint of our four major issues.

Table 2.2 Freud's Position

Issue	Position
Biopsychosocial forces	Genetic inheritance and psychological growth are more important than social forces.
Continuity vs. discontinuity	Development is continuous.
Stability vs. change	Development is stable, because fixation at some stage tends to be permanent (unless successful psychotherapy can be achieved).
Sensitive periods vs. equal potential	There are five sensitive periods in life: the oral, anal, phallic, latency, and genital stages.

The Cognitive Structures Approach

The two most highly regarded "fathers" of the study of cognitive development are the Swiss biologist Jean Piaget and the Russian psychologist Lev Vygotsky. We now briefly examine their views, which are the foundation of **cognitive structures** theory.

Jean Piaget's Theory [BIOPSYCHOsocial]

While Freud was concerned with the structures of the personality, the Swiss biologist/psychologist Jean Piaget (1896–1980) sought to understand the cognitive structures of the intellect (1953, 1966; Flavell, 1963).

Beginning the Study of Intellectual Development

Jean Piaget was among the first to study normal intellectual development.

Piaget is the foremost contributor to the study of cognitive development. Beginning his scholarly career at the age of 11(!), Piaget published numerous papers on birds, shellfish, and other topics of natural history. As a result, the diligent Swiss was offered the curator's position at the Geneva Natural History Museum. He was only 15 at the time, and turned the offer dow to finish high school. He received his PhD in biology at the age of 21, and wrote more than fifty books during his lifetime.

Piaget began his research on cognitive development in 1920, when he took a position in the Binet Laboratory in Paris. He was given the task of standardizing a French version of an English-language test of reasoning ability, which enabled him to observe how children responded to the questions. He discovered that there were similarities in the wrong answers given by each age group. For example, 5-year-olds would give a wrong answer for one reason, whereas older age groups gave the same wrong answer for other reasons. This discovery led Piaget to the idea that children of different age groups have different thinking patterns. Prior to 1920, little research had been done on the nature of intelligence. Most scientists viewed children as miniature adults who used adult thinking methods but used them poorly. The scientists felt that as information was poured into the child, mental maturity gradually developed. Piaget discovered that specific abilities must be acquired before the child's intellect could fully mature; information alone is not enough. Furthermore, he observed that these mental abilities develop in stages, each one preparing the way for the development of the next.

The Role of Intelligence

Piaget's background in biology formed the basis for his view of intelligence. If you have ever seen an episode of "The Undersea World of Jacques Cousteau," you have probably been amazed by the way ocean dwellers adapt themselves to their particular environments. The puffer fish swells to twice its size when threatened by an enemy. The anglerfish uses its dorsal fin as bait to lure smaller fish to its mouth. Nature is filled with examples of superb adaptations of animals to their environments. But nature has not fitted humans to any specific environment. It has equipped us to adapt to most environments through the gift of intelligence. Like the swelling of the puffer fish, human intellect seemed to Piaget to be another example of biological adaptation.

Intelligence matures through the growth of increasingly effective *mental structures*. These structures can best be defined as the blueprints that equip us to affect our environment. They are the tools of adaptation. At birth, all babies simply reflect the environment. When a specific event (a stimulus) occurs, infants react automatically to it. Sucking and crying are examples of these reflexes.

Soon after birth, reflexes are transformed into *schemata*. Schemata are patterns of behavior that infants use to interact with the environment. Figure 2.2 depicts this relationship. Infants develop schemata for looking, for grasping, for placing objects in their mouths. As infants grow older and begin to encounter more elements of their world, schemata are combined and rearranged into more efficient structures. By the time children reach age 7, higher forms of psychological structures are developing. These structures are called *operations*.

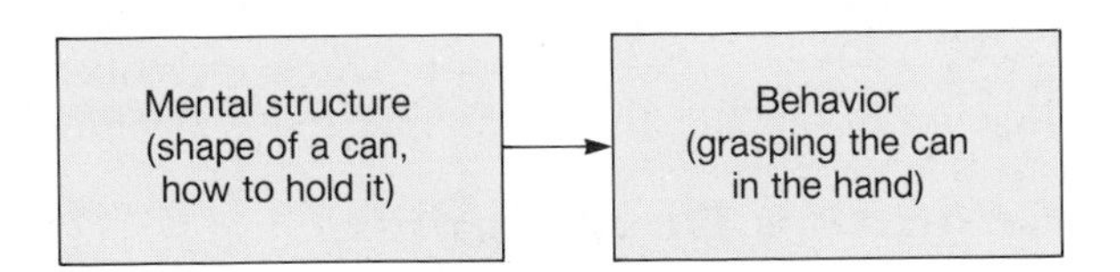

Figure 2.2
Piaget's concept of the schema. Mental structures held in the mind direct and control our behaviors.

According to Piaget, adult thinking is composed of numerous operations that enable the individual to manipulate the environment. Operations are mental, internalized actions. They are similar to programs in a computer. Programs enable the computer to manipulate the data fed into it in various ways. Mental operations do the same. In fact, mental operations are able to take things apart and reassemble them without actually touching the object. To continue the metaphor, schemata are like files used in computer programs.

In addition to reflexes, the infant has at birth two basic tendencies or drives that affect intellectual functioning throughout his or her lifetime. They are **organization** and **adaptation,** and they govern the way we use our schemata to adjust to the demands of the environment.

Organization

Our innate tendency to organize causes us to combine our schemata more efficiently. The schemata of the infant are continuously reorganized to produce a coordinated system of higher-order structures. Each time a new schema is acquired, it is integrated into existing schemata. Consider children learning to throw a ball. They may understand the various parts of a good throw, but they cannot throw the ball well until they have integrated these parts into a smooth and efficient movement.

Adaptation

The second tendency in all human beings is to adapt to the environment. Adaptation consists of two complementary processes: **assimilation** and **accommodation.** We assimilate when we perceive the environment in a way that fits our existing schemata. That is, we make reality fit our minds. We accommodate when we modify our schemata to meet the demands of the environment. That is, we make our minds fit reality.

We try to assimilate as much as possible because it is easier than accommodation. It is easier to see situations, events, and objects as something we understand and can work with. When asked to describe an unfamiliar object, we say: "It looks like an orange," or "It's hard like a rock." To make these comparisons, we perceive the object in a way that fits what we know. We look for similarities. For example,

You are able to read these words because you altered your perception of them mentally to suit the structures in your mind.

You may have physically assimilated this idea by turning this book upside down. When we mentally alter what we see, we *recognize* or rethink it and thus assimilate it.

Does _ . . . _ . . _ . _ . . 88 * * * 8 _'"? To answer this question written in code, you would have to learn the key. In doing so, you would be acquiring a new schema. This is accommodation. All mental activity uses assimilation and accommodation.

It is necessary to balance these processes. A person who is incapable of assimilation, such as a mentally retarded person, does not have the capacity to take advantage of his previous experiences. A person who is incapable of accommodation, such as the rigid schizophrenic, is unresponsive to his environment.

Stages of Cognitive Development

Mental growth takes place in four discrete stages, each one laying the foundation for the next. They are the *sensorimotor stage* (birth to 2 years old); the *preoperational stage* (2 to 7 years old); the *concrete operational stage* (7 to 11 years old); and the *formal operational stage* (11 years old and up). The change from one stage is quite clear, and regression from later to earlier stages does not happen. Let us look at each stage more closely.

Sensorimotor Stage (Birth to 2 Years Old)

Infants are egocentric at birth: they do not experience the external world as separate from themselves. The existence of objects in the environment of infants is entirely dependent on their sensory perception of them. When an object is removed from their sight or touch, they do not search for it. They act as though it no longer exists. As the sensorimotor stage proceeds, infants develop object permanence, the awareness that objects exist independent of their perception. An example is the child who looks for the rattle he has dropped from his high chair.

Preoperational Stage (2 to 7 Years Old)

Until this stage, children's intellectual tools consist of the ability to use their senses and to interact with the world through movement. When children learn to talk, they acquire the ability to represent objects symbolically in their minds. Consider the statement, "I want a cookie." To make this demand, children must be aware that cookies exist as a class of objects. They must also have acquired the verbal symbol *cookie* and have connected feelings of hunger with the symbol for an object that will satisfy it. Now that they have symbols to represent objects in their world, they can manipulate these objects mentally. Now they can think.

Concrete Operational Stage (7 to 11 Years Old)

Children now become concerned with *why* things happen. The intuitive thinking style of the preoperational stage is replaced by elementary logic, as operations begin to develop that enable children to form more complex mental actions on concrete elements of their world.

Logical thinking requires that one have an understanding of the physical properties of the world. Knowing the correct answer to the question "Which is heavier, a pound of feathers or a pound of lead?" depends on understanding the concept of density in relation to measurement. Preoperational children cannot

An Applied View — Conducting Your Own Cognitive Experiment

Most readers of this text are in Piaget's formal operational stage. This means that they already have the ability to make reasonable hypotheses. If you would like to test yourself on this ability, try the following experiment (you may want to do it with several friends/classmates). To do it, you need two sheets of paper 10×10 inches in size, both ruled off in 1–inch squares, with each of the hundred squares numbered consecutively; and three crayons: one yellow, one orange, and one red. Now follow these instructions:

1. Hypothesize which part or parts of the squared sheet most people will point to, and say why.
2. Make a second hypothesis, explaining why you think that part or parts will be chosen second most often.
3. Show one of the ruled sheets to at least fifty people (the more, the better), and ask them to point to one square, any square, at random.
4. Make a note of how many times each of the squares is chosen, and whether the chooser is left- or right-handed.
5. Color the squares on one of the sheets, using your data for right-handers only. Leave those squares with no choices white. Color those with one choice yellow, those with two choices orange, and those with three or more choices red.
6. Do the same thing on the second sheet for the data for left-handers.

Do the data support your hypotheses? If not, can you think of reasons why? Can you think of better ways to go about hypothesizing in general?

understand that the weight is the same regardless of the density of the objects. The operations necessary for this understanding are developed in the concrete operational stage.

When children acquire these operations, they become able to solve problems logically by the use of elementary deductive reasoning. However, their thinking style still needs refinement. Ask the 9-year-old, "How would things be different if we had no thumbs?" and he is likely to respond, "But we *do* have thumbs!" The concrete operational child does not consider possibilities that are not real. The tools of thought are assembled but still need the refinement that takes place in the formal operational stage.

Formal Operational Stage (11 Years Old and Up)

In adolescence, one's thinking style takes wing. Formal operations expand thought to the abstract. Reality is represented by symbols that can be manipulated mentally, just as data are represented by electromagnetic code that can be programmed in the computer.

Only when we reach adolescence are we capable of answering questions such as "What would happen if it always rained up?" and "What if we had no thumbs?"

The adolescent is capable of forming conclusions based on hypothetical possibilities. Answering the question, "What would things be like if it rained up?" involves mentally picturing rain rising from the ground. This mental picture is contrasted with reality and various conclusions are produced.

Thinking in the formal stage becomes much more orderly and systematic. Most 8-year-olds would be unable to answer the question, "If Jane's hair is darker than Susan's, and Susan's hair is darker than Mary's, whose hair is darkest?" Although an 8-year-old is capable of "ranking" the children by darkness of hair, she is unable to manipulate facts concerning imaginary people.

Lev Vygotsky's Theory [bioPSYCHOSOCIAL]

A major criticism of Piaget's theory is that he has not adequately considered the role of social interactions in cognitive development. The work of Russian psychologist Lev Vygotsky (1896–1934) has attracted considerable attention because he did emphasize social processes.

As Wertsch and Tulviste (1992) explain it,

> ***In Vygotsky's view, mental functioning in the individual can be understood only by examining the social and cultural processes from which it derives. [His view] . . . differs from that which is typically assumed in contemporary Western psychology. Instead of beginning with the assumption that mental functioning occurs first and foremost, if not only, within the individual, it assumes that one can speak equally appropriately of mental processes*** **between** ***people.*** **(pp. 548–49)**

That is, cognitive growth depends on children's direct interactions with those around them. Adults interact with children in a way that emphasizes the things a culture values. For example, a child points to a book. Her parent says the word *book,* and then proceeds to demonstrate that books are fun to read.

Vygotsky's emphasis on the role of culture and society in cognitive development (1978, 1981) is in sharp contrast with Piaget's child struggling to understand a problem on his own. Development of higher mental processes such as memory, attention, and reasoning often occur as children learn to use the inventions of society. Examples of such inventions are language, mathematical systems, and memory devices. Young learners are frequently guided by others who are already skilled in using these tools.

Therefore, we cannot be content with the results of measurements such as ability or intelligence tests, which provide only a static measure of a student's present ability. Vygotsky was concerned with how a child grows intellectually through the help of others who are better informed. He introduces the notion of the **zone of proximal development.** This zone refers to the distance between a child's actual developmental level (problem solving she can do independently), and a higher level of *potential* development (problem solving done with adult guidance or in collaboration with more capable peers). In other words, it is the difference between what pupils can do by themselves and what they can do with help. The zone of proximal development describes a range of learning from the actual to the potential.

As Vygotsky noted (1978), instruction is useful only when it precedes development. That is, teaching awakens those functions that are ready to mature and that are in the zone of proximal development. Instruction is an interactive process in which the instructor constantly adjusts task goals to meet a child's abilities, as well as increasing or decreasing the amount of assistance in direct response to a child's success or failure. We will have more to say about Vygotsky's research in several later chapters.

Table 2.3 summarizes the cognitive structures position as viewed by Piaget and Vygotsky, in terms of our four major issues.

Table 2.3 The Cognitive Structures Position

Issue	Theory: Piaget	Theory: Vygotsky
Biopsychosocial forces	All change is the result of biologically inherited schemata and psychological forces. Piaget has little to say about social causes.	Change results less from biological forces; it is seen as being primarily from psychological and social forces.
Continuity vs. discontinuity	Development is discontinuous. It depends on the emergence of mental states that are the direct result of the maturation of the brain. Transition between the several age-stages is clearly detectable, especially when the person is presented with tasks designed to detect it.	Development is primarily continuous. It depends on the interaction with adult and peer teachers.
Stability vs. change	Development is marked by stability.	Development is marked by plasticity.
Sensitive periods vs. equal potential	Sensitive periods play a vital role in development.	Sensitive periods are mentioned, but less clearly defined, by Vygotsky.

The Behavioral Approach

Harvard psychologist B. F. Skinner is known as the father of American behaviorism.

We turn now to a third approach to human development, **behavioral theory.** Two of its best-known proponents are B. F. Skinner and Albert Bandura.

B. F. Skinner's Theory [biopsychoSOCIAL]

When Skinner (1938, 1948, 1953, 1971, 1983) began to study learning in the late 1920s, he wanted only to satisfy his curiosity about how organisms are changed by their environment. He believed that scientists should start studying simpler organisms and work up to investigating the complexity of humans. Pigeons and rats were chosen for study because their nervous systems are reasonably similar to those of humans. Also, manipulating their lives does not pose the ethical problems that arise with human experimentation. Skinner felt that knowledge gained about these animals would provide clues to the study of humans. If so, this would save the researcher time, effort, and costly mistakes.

Skinner made two basic assumptions about the learning of all organisms, whether an amoeba or a human being. First he assumed that all learning is ultimately determined by forces outside the control of the organism. What we learn and how we learn it is determined completely by genetic inheritance and by the influences of our environment, past and present. He agreed that inside forces such as values and beliefs are involved, but thought that these are formed as the result of the environment. Thus he did away with the concept of **free will.**

He further argued that the study of learning is analogous to a training technique used in engineering schools called the "black box method." In this technique, engineering students are given a black box with two terminals. They vary the voltage, amperage, and wattage of the electricity put in one side, and then meter the results coming out the other side. They are not allowed to know what is inside the black box; it is their job to infer what is in there by watching how the black box responds to the electrical stimuli. Likewise, in the study of learning there is no way for us to know what is going on inside the human head. It is presumptuous to make guesses about the mechanisms of the mind at this time, he felt.

Having made these assumptions, Skinner studied the relationship between three variables: stimuli, responses, and the reactions of the environment to responses. His formula for learning is simple. It follows these three steps:

1. A *stimulus occurs in the environment.* For instance, this might be a question from a teacher to a student. Often, it is impossible to tell what the stimulus is.

2. *A response is made in the presence of that stimulus.* An example is a correct or an incorrect answer on the part of the student.

3. *One of three things happens as a reaction to this response.* The response is reinforced, the response is punished, or the response is ignored.

Hence, learning follows this formula:

Stimulus (?) →Response → Reaction of the environment to that response.

Skinner argued that if the response is reinforced, it is more likely to result the next time that stimulus occurs. Thus, if the teacher asks that question again, the student is more likely to give the right answer, because that response was reinforced. If the response is punished, the response is less likely in the future. If the response is simply ignored, it also becomes less likely in the future. Each of these three concepts—**reinforcement, punishment,** and no response (which Skinner calls **extinction)**—needs further explanation.

Notice that Skinner did not use the word *reward.* He felt that reward is distinct from reinforcement. A reward is something you give a person because *you* think it will make that person feel good for having made a response. There is a problem with this: It assumes that we know what is going on in the mind of another person. We know what is rewarding to us, but we can only guess what is rewarding to someone else. Skinner preferred to define reinforcement as anything that makes a response more likely to happen in the future, without making any reference to how it makes the individual feel inside.

There are two kinds of reinforcement, and the difference between them is important. **Positive reinforcement** refers to any event that, when it occurs after a response, makes that response more likely to happen in the future. **Negative reinforcement** is any event that, when it ceases to occur after a response, makes that response more likely to happen in the future. Notice that both types of reinforcement make a response *more* likely to happen. Giving your daughter candy for doing the right thing would be positive reinforcement. Ceasing to twist your brother's arm when he gives you back your pen would be negative reinforcement.

To reduce the likelihood of inappropriate responses, Skinner suggested the use of a fourth technique, extinction. Extinction simply means that when an undesirable response occurs, it is disregarded. When a response is unreinforced for a long enough time, it is discontinued.

Skinner's theory is also known as **operant conditioning.** This is because the basis of his theory is that when operants (actions that people or animals take of their own accord) are reinforced, they become conditioned (more likely to be repeated in the future).

An Applied View — Who Can Find the Raisin?

Probably the best way to understand Skinner's view of behaviorism is to apply its principles to teaching someone. For this suggested activity, you need two cups, a raisin, and two squares of paper, of different colors, big enough to cover the cups.

For 1-to 2-year-olds, you must start very simply. Follow these steps:

1. Place a raisin in one of the cups and cover that cup with a large square of either color.
2. Cover the empty cup with the large square of the other color. Do this so that the child cannot see what you are doing.
3. Ask the child to guess which cup the raisin is in by pointing to one of the squares. If the child gets the right one, he or she wins the prize. If not, say, "Too bad. Maybe you'll get it next time. Let's try again."
4. Now, out of the child's sight, switch the positions of the cups, but cover the cup with the raisin with the same square. Do this until the child regularly guesses correctly. This is called *continuous reinforcement.* Generally speaking, the older the child, the quicker the learning.
5. Now cut out two large circles, each a different color, and repeat steps 1 through 4 using circles instead of squares. Does the child learn more quickly this time?

For older children, the game can be made more and more complex. Use a greater variety of shape combinations, and use three or even four cups. Finally, try to get the child to verbalize what he or she thinks is the principle behind this activity. Perhaps the child could even invent another way of doing it! As you perform this experiment, you will gain insights into the behaviorist ideas of reinforcement and extinction.

What does operant conditioning have to do with development? In Skinner's opinion, human development is the result of the continuous flow of learning that comes about from the operant conditioning we receive from the environment every day.

These concepts of change in human behavior have been enhanced by the ideas of Albert Bandura and his associates. They extend the behaviorist view to cover social behavior.

Albert Bandura's Theory [biopsychoSOCIAL]

Albert Bandura, one of the chief architects of social learning theory, has stressed the potent influence of modeling on personality development. He calls this **observational learning.** In a famous statement on social learning theory, Bandura and Walters (1963) cite evidence to show that learning occurs through observing others, even when the observers do not imitate the model's responses at that time and get no reinforcement. For Bandura, observational learning means that the information we get from observing other people, things, and events influences the way we act.

Social learning theory has particular relevance for development. As Bandura and Walters note, children often do not do what adults tell them to do but rather what they see adults do. If Bandura's assumptions are correct, adults can be a potent force in shaping the behavior of children because of what they do.

The importance of models is seen in Bandura's interpretation of what happens as a result of observing others:

- The observer may acquire new responses, including socially appropriate behaviors.
- Observation of models may strengthen or weaken existing responses.

- Observation of a model may cause the reappearance of responses that were apparently forgotten.
- If children witness undesirable behavior that is either rewarded or goes unpunished, undesirable pupil behavior may result. The reverse is also true.

Bandura, Ross, and Ross (1963) studied the relative effects of live models, filmed human aggression, and filmed cartoon aggression on preschool children's aggressive behavior. The filmed human adult models displayed aggression toward an inflated doll; in the filmed cartoon aggression, a cartoon character displayed the same aggression.

Later, all the children who observed the aggression were more aggressive than youngsters in a control group. Filmed models were as effective as live models in transmitting aggression. The research suggests that powerful, competent models are more readily imitated than models who lack these qualities (Bandura & Walters, 1963).

A Multicultural View 

Modeling Among Latino Youth

For most teens the family plays an important role, but for Latino adolescents this is especially true. This quotation shows how a Latino teenage girl chose her mother as her model of behavior and values:

As far as I can remember, my mother was strong and independent. She loved us so much that she protected us from the dangers of the barrio. She kept the family together as long as she could, and the traditions were a big part of her life. She is a very pretty woman with strong Mexican Indian features: high cheek-bones and a tired clear face. She is short, heavyset, and has a physically tired body. She always wore a little makeup and red lipstick. . . . "Mi madre" is the pride and joy of my life, and she is not only my mother but my closest friend (Kunjifu, 1985, p. 29).

In summary, the behaviorist position holds that human development does not happen in predictable stages but as the result of stimuli from the environment. Since there are millions of stimuli in a person's lifetime, behaviorists see development as continuous, something that usually happens in small steps every day. Table 2.4 reviews the behaviorist position, as viewed by Skinner and Bandura.

Table 2.4 The Behaviorist Position

Issue	Position
Biopsychosocial forces	Social forces are by far the most important influences on development.
Continuity vs. discontinuity	Development is continuous, and depends on the four forces that control learning: positive and negative reinforcement, punishment, and extinction.
Stability vs. change	Because learning is so important, human growth is highly subject to change. It has great plasticity.
Sensitive periods vs. equal potential	All times of life have equal potential for change.

The Psychosocial Crises Approach

The chief proponent of **psychosocial theory** approach is Erik H. Erikson. We now briefly review his theory.

Erik Erikson's Theory [bioPSYCHOSOCIAL]

Among other important books, Erik Erikson wrote *Childhood and Society* (1963). It is an amazingly perceptive and at times poetically beautiful description of human life. Erikson's view of human development derives from his extensive study of people living in an impressive variety of cultures: Germans, East Indians, the Sioux of South Dakota, the Yuroks of California, and wealthy adolescents in the northeastern United States (1959, 1968). His ideas also stem from intensive studies of historical figures such as Martin Luther (1959) and Mahatma Gandhi (1969). He sees human development as the interaction between your genes and the environment in which you live.

Psychologist Erik Erikson is best known for his theory of psychosocial crises.

According to Erikson, human life progresses through a series of eight stages. Each of these stages is marked by a crisis that needs to be resolved so that the individual can move on. Erikson uses the term *crisis* in a medical sense. It is like an acute period during illness, at the end of which the patient takes a turn for the worse or better. At each life stage, the individual is pressured, by internal needs and the external demands of society, to make a major change in a new direction.

The ages at which people go through each of the stages vary somewhat, but the sequence of the stages is fixed. The ages of the first five stages are exactly the same as in Freud's theory (Erikson is an ardent student of Freud). Unlike Freud and Piaget, however, Erikson believes that the stages overlap.

Each of the crises involves a conflict between two opposing characteristics. Erikson suggests that successful resolution of each crisis should favor the first of the two characteristics, although its opposite must also exist to some degree. Table 2.5 gives an overview of his psychosocial theory.

Table 2.5 Erik Erikson's Psychosocial Theory of Development

Stage	Age	Psychosocial Crisis
1	Infancy (0 to 1 1/2)	Trust vs. mistrust
2	Early childhood (1 1/2 to 3)	Autonomy vs. shame and doubt
3	Play age (3 to 5)	Initiative vs. guilt
4	School age (5 to 12)	Industry vs. inferiority
5	Adolescence (12 to 18)	Identity and repudiation vs. identity confusion
6	Young adult (18 to 25)	Intimacy and solidarity vs. isolation
7	Adulthood (25 to 65)	Generativity vs. stagnation
8	Maturity (65+)	Integrity vs. despair

It is necessary to have experienced each crisis before proceeding to the next. Inadequate resolution of the crisis at any stage hinders development at all succeeding stages, unless special help is received. When a person is unable to resolve a crisis at one of the stages, Erikson suggests that "a deep rage is aroused comparable to that of an animal driven into a corner" (1963, p. 68). This is not to say that anyone ever resolves a crisis completely. It is important to note that Erikson's description of the eight stages of life is a picture of the ideal, and that no one ever completes the stages perfectly. However, the better a person does at any one stage, the more progress. Let us look at each stage more closely.

1. *Basic trust versus mistrust (birth to 1 1/2 years old).* In the first stage, which is by far the most important, a sense of basic trust should develop. For Erikson, trust has an unusually broad meaning. To the trusting infant, it is not so much that the world is a safe and happy place, but rather that it is an orderly, *predictable* place. There are causes and effects that one can learn to anticipate. For Erikson, then, trust flourishes with warmth and care, but it might well include knowledge that one will be spanked regularly for disobeying rules.

   If the infant is to grow into a person who is trusting and trustworthy, it is essential that a great deal of regularity exist in its early environment. The child needs variation, but this variation should occur in a regular order that the child can learn to anticipate. For example, the soft music of an FM radio can provide regular changes in sound level. So does the movement of the colorful toy birds hanging over a child's crib.

   Some children begin life with irregular and inadequate care. Anxiety and insecurity have a negative effect on family and other relationships so important to the development of trust. When a child's world is so unreliable, we can expect mistrust and hostility, which under certain circumstances can develop into antisocial, even criminal, behavior. Of course, not all such people become criminals. It is also possible to gain basic trust in infancy and then lose it later. Sometimes people who have not suffered an injurious childhood can lose their basic sense of trust because of damaging experiences later in life.

2. *Autonomy versus shame and doubt (1 1/2 to 3 years old).* When children are about 1 1/2 years old, they should move into the second stage, characterized by the crisis of autonomy versus shame and doubt. This is the time when they begin to gain control over their bodies and is the usual age at which toilet training is begun.

   Erikson agrees with other psychoanalysts that toilet training has far more important consequences in one's life than control of one's bowels. The sources of generosity and creativity lie in this experience. If children are encouraged to explore their bodies and environment, a level of self-confidence develops. If they are regularly reprimanded for their inability to control excretion, they come to doubt themselves. They become ashamed and afraid to test themselves.

Of course, excretion is not the only target for regulation in this stage. Children of this age usually start learning to be self-governing in all of their behaviors. Although some self-doubt is appropriate, general self-control should be fostered at this stage.

3. *Initiative versus guilt (3 to 5 years old).* The third crisis, initiative versus guilt, begins when children are about 3 years old. Building on the ability to control themselves, children now learn to have some influence over others in the family and to successfully manipulate their surroundings. They should not merely react, they should initiate.

   If their parents and others make them feel incompetent, however, they develop a generalized feeling of guilt about themselves. In the autonomy stage they can be made to feel ashamed by others; in this stage, they learn to make themselves feel ashamed.

4. *Industry versus inferiority (5 to 12 years old).* The fourth stage corresponds closely to the child's elementary school years. Now the task is to go beyond imitating ideal models and to learn the elementary technology of the culture. Children expand their horizons beyond the family and begin to explore the neighborhood.

   Their play becomes more purposeful, and they seek knowledge in order to complete the tasks that they set for themselves. A sense of accomplishment in making and building should prevail. If it does not, children may develop a lasting sense of inferiority. Here we begin to see clearly the effects of inadequate resolution of earlier crises.

   As Erikson puts it, the child may not be able to be industrious because he may "still want his mother more than he wants knowledge." He suggests that the typical American elementary school, staffed almost entirely by women, can make it difficult for children (especially boys) to make the break from home and mother. Under these circumstances, children may learn to view their productivity merely as a way to please their teacher (the mother substitute), and not as something good for its own sake. Children may perform in order to be "good little workers" and "good little helpers" and fail to develop the satisfaction of pleasing themselves with their own industry.

5. *Identity and repudiation versus identity confusion (12 to 18 years old).* The main task of the adolescent is to achieve a state of identity. Erikson, who originated the term **identity crisis,** uses the word in a special way. In addition to thinking of identity as the general picture one has of oneself, Erikson refers to it as *a state toward which one strives.* If one were in a state of identity, the various aspects of one's self images would be in agreement with each other; they would be identical. Ideally, a person in the state of identity has no internal conflicts whatsoever.

   Repudiation of choices is another essential aspect of reaching personal identity. In any choice of identity, the selection we make means that we have repudiated (turned down) all the other possibilities, at least for the present. When youths cannot achieve identity, when *identity confusion* ensues, it is usually because they are unable to make choices.

According to Erikson, a sense of intimacy with another person of the opposite gender should develop between the ages of 18 and 25. If it does not, a sense of isolation results.

As Biff, the son in Arthur Miller's *Death of a Salesman,* says, "I just can't take hold, Mom, I can't take hold of some kind of life!" Biff sees himself as many different people; he acts one way in one situation and the opposite way in another—a hypocrite. Because he refuses to make choices and shies away from commitments, there is no cohesiveness in his personality. He is aware of this lack, but is unable to do anything about it. (We will have much more to say about this stage in chap. 12.)

6. *Intimacy and solidarity versus isolation (18 to 25 years old).* In the sixth stage, intimacy with others should develop. Erikson is speaking here of far more than sexual intimacy. He is talking about the essential ability to relate one's deepest hopes and fears to another person, and to accept another person's need for intimacy in turn.

   Each of us is entirely alone, in the sense that no one else can ever experience life exactly the way we do. We are imprisoned in our own bodies, and can never be certain that our senses experience the same events in the same way as another person's senses. Only if we become intimate with another are we able to understand and have confidence in ourselves. During this time of life, our identity may be fulfilled through the loving validation of the person with whom we have dared to be intimate.

7. *Generativity versus stagnation (25 to 65 years old).* Generativity means the ability to be useful to ourselves and to society. As in the industry stage, the goal here is to be productive and creative. However, productivity in the industry stage is a means of obtaining recognition and material reward. In the generativity stage, one's productivity is aimed at generating a sense of personal fulfillment. Thus, the act of being productive is itself rewarding, regardless of whether recognition or reward results.

   Furthermore, there is a sense of trying to make the world a better place for the young in general, and for one's own children in particular. In this stage many people become mentors to younger individuals, sharing their knowledge and philosophy of life.

   When people fail in generativity, they begin to stagnate, to become bored and self-indulgent, unable to contribute to society's welfare. Such adults often act as if they are their own child.

8. *Integrity versus despair (65 years old and older).* When people look back over their lives and feel they have made the wrong decisions, or more commonly, that they have too frequently failed to make any decision at all, they see life as lacking integration. They feel despair at the impossibility of "having just one more chance to make things right." They often hide their terror of death by appearing contemptuous of humanity in general, and those of their own religion or race in particular. They feel disgust for themselves.

   To the extent that they have been successful in resolving the first seven crises, they achieve a sense of personal integrity. Adults who have a sense of integrity accept their lives as having been well spent. They feel a kinship with people of other cultures and of previous and future generations. They have a sense of having helped to create a more dignified life for humankind. They have gained wisdom. Table 2.6 reviews the psychosocial position in terms of our four major issues.

Table 2.6 The Psychosocial Position

Issue	Position
Biopsychosocial forces	Psychological and social forces predominate.
Continuity vs. discontinuity	There are eight universal periods of life that proceed in a set order, making development discontinuous. Because it is possible to regress in any of the stages, however, and progress can be uneven, Erikson's theory can also be seen as continuous.
Stability vs. change	To the extent that the individual is able to resolve each of the eight psychosocial crises, there are many possibilities for change.
Sensitive periods vs. equal potential	Each of Erikson's eight stages is definitely a sensitive period.

Table 2.7 summarizes the three theories that take a stage view of human development. (Reminder: The behaviorists do not take such a view.)

Table 2.7 Comparison of Three Developmental Stage Theories

Ages	Psychosexual Stages (Freud)	Psychosocial Stages (Erikson)	Cognitive Stages (Piaget)
0 to 1 1/2	Oral	Basic trust vs. mistrust	Sensorimotor
1 1/2 to 3	Anal	Autonomy vs. doubt and shame	2 years—Preoperational
4 to 5	Phallic	Initiative vs. guilt	
6 to 11	Latency	Industry vs. inferiority	7 years—Concrete operational
12 to 17	Genital	Identity vs. confusion	11 years—Formal operational
Young adulthood	*	Intimacy vs. isolation	*
Middle adulthood	*	Generativity vs. stagnation	*
Late adulthood	*	Ego integrity vs. despair	*

Note: An asterisk (*) means the theorist suggests no stage for this age group.

Table 2.8 summarizes the positions that each of the four theories takes on each of the critical issues we first described in chapter 1. Please remember that summaries always leave out vital details. The letters in the table stand for the issues. For example, *C* stands for continuous development. Thus Skinner's theory is continuous, Freud's and Piaget's are discontinuous, and Erikson's combines the two.

Table 2.8 Comparison of Theories Presented in This Chapter, According to the Four Major Issues

Issue	Freud	Piaget	Vygotsky	Skinner	Bandura	Erikson
Biopsychosocial	BP	BP	PS	S	S	PS
Continuity vs. discontinuity	D	D	C	C	C	C/D
Stability vs. change	S	S	C	C	C	C
Sensitive periods vs. equal potential	SP	SP	EP	EP	EP	SP

A Multicultural View 

Enrichment through Diversity

One final point: As research on human beings includes more and more diverse groups, our understanding of how different we are from each other grows. For instance, as more women conduct research, our view of what it means to be human is changing. Psychology now takes a lifespan perspective, which means we must examine not only each stage of life, but the relationships between them. More and more research is looking at diverse ethnic, cultural, and racial groups. It is increasingly important that we learn about all these differences as our society becomes more diversified.

However, this new emphasis on difference can lead to the conclusion that one group must be somehow inferior to another. In the past, assumptions that non-whites are more likely to be unmarried parents or to commit criminal acts were accepted stereotypes. Women were either considered as imperfect men or assumed to develop exactly like men. Latinos or Native Americans were totally missing from our research. In treating the white middle-class male point of view as the norm, the rest of society is either ignored or becomes abnormal by comparison. Rather than treating women or other cultural or racial groups as needing more appreciation or help, recent research is seeking to understand diversity as a valuable contribution to expanding knowledge.

Psychological research has become increasingly attuned to the importance of context. No one lives in a social vacuum. Each theory is imbedded in a time and place, just as each person is. Over the course of this book, we attempt to explore the diversity of human development, in part through the regular sections of the book, and in part through separate "A Multicultural View" boxes like this one. In addition, you should ask yourself, as you read various research reports, "Does this study reflect nearly all people of this group, or is it only relevant to the kinds of people under study?"

Conclusion

In this chapter, we covered a great deal of important ground. We introduced you to the framework on which, in many ways, the rest of the book depends. Although there are many theories we might have chosen to include in this chapter (some of which we will discuss later), the ones we have presented here have played or are playing major roles in our understanding of human development.

There is a lot to remember, but we will be coming back again and again to these seminal ideas to help you to gain a firm understanding of them. In these first two chapters, you have received an overview of the fascinating study of human development. Now we begin to take a much closer look at each of the major aspects of life, using the powerful insights of theory and research.

Chapter Highlights

What Theories Do

- Good theories help to organize information, to focus our search for understanding, to describe and explain interpretations and conclusions, and to recognize major disagreements among scholars and scientists.
- A theory attempts to explain empirical research by showing how constructs are related.

Critical Issues

- We will be discussing theories from the standpoint of four critical issues: the relative influence of biopsychosocial forces, continuity versus discontinuity, stability versus change, and sensitive periods versus equal potential.

The Biopsychosocial Causes of Change

- While all theorists believe that there are biological, psychological, and social causes of human development, they differ on the emphasis they give to each of the three factors.
- The biological view is that genes are the major factor in our development.
- The psychological view is that changes in personality and intellect guide our development.
- The social view is that the environment plays the major role in human development.
- Most theorists believe that a combination of two or three of these factors guides human development.

Sigmund Freud—Psychoanalysis [BIOPSYCHOsocial]

- The unconscious mind is the key to understanding human beings.
- Important information in the unconscious mind is kept hidden through an array of defense mechanisms.
- The mind is divided into three constructs: the id, the ego, and the superego, each of which appears at different stages of a child's development.
- Personality development is divided into five instinctive stages of life, each stage serving a major function: oral, anal, phallic, latency, and genital.
- Failure to pass through a stage of development results in fixation, which halts a person from becoming fully mature.
- Anna Freud believed that puberty disrupts the delicate balance between the superego and the id, causing the adolescent to regress to earlier stages of development.

Jean Piaget—Cognitive Structures [BIOPSYCHOsocial]

- Piaget focused on the development of the cognitive structures of the intellect during childhood and adolescence.
- Organization and adaptation play key roles in the formation of structures.
- The infant and child pass through Piaget's first three stages: sensorimotor, preoperational, and concrete operational.
- Piaget's highest stage of cognitive development, formal operations, begins to develop in early adolescence.

Lev Vygotsky—Cognitive Structures [bioPSYCHOSOCIAL]

- The capacity to learn depends on abilities of the child's teachers as well as on the child's abilities.
- The difference between the child's ability to learn independently and to learn with help is called the zone of proximal development.

B. F. Skinner—Behaviorism [biopsychoSOCIAL]

- Classical conditioning forms the basis of behavorism, and was founded by Ivan Pavlov and John Watson.
- Skinner's paradigm involves three steps: a stimulus occurs in the environment; a response is made in the presence of that stimulus; and the response is reinforced, punished, or extinguished.

Albert Bandura—Behaviorism [biopsychoSOCIAL]

- Bandura has extended Skinner's work to the area of social learning, which he calls observational learning.

Erik Erikson—Psychosocial Crises [bioPSYCHOSOCIAL]

- Human life progresses through eight "psychosocial" stages, each one marked by a crisis and its resolution.
- While the ages at which one goes through each stage vary, the sequence of stages is fixed. Stages may overlap, however.
- A human being must experience each crisis before proceeding to the next stage. Inadequate resolution of the crisis at any stage hinders development.

Key Terms

Accommodation
Adaptation
Assimilation
Behavioral theory
Cognitive structures theory
Defense mechanisms
Ego
Extinction
Free will
Id
Identity crisis
Negative reinforcement
Observational learning
Operant conditioning
Organization
Positive reinforcement
Psychoanalysis
Psychoanalytic theory
Psychosocial theory
Punishment
Reinforcement
Superego
Zone of proximal development

What Do You Think?

1. What is your reaction to the statement, "The truth or falseness of a theory has little to do with its usefulness"?
2. Some people say that in his concept of human development, Freud emphasizes sexuality too much. What do you think?
3. Which is better, assimilation or accommodation? If you could do only one, which would it be? Why?
4. Skinner criticized the other theorists in this chapter for believing they can describe what goes on in the human mind. After all, he said, no one has ever looked inside one. What's your position?
5. Is it possible for a person to be deeply intimate with another person and still be in a state of identity confusion?

Suggested Readings

Clark, Ronald W. (1980). *Freud: The man and the cause.* New York: Random House. This is one of the most judicious and even-handed books written about the father of psychoanalysis.

Erikson, Erik. (1958). *Young man Luther.* New York: Norton. Martin Luther was the main force behind the Protestant Reformation. In Erikson's penetrating analysis of the causes behind Luther's actions, we have a wonderfully clear example of his ideas about adolescence in general and negative identity in particular.

Skinner, B. F. (1948). *Walden two.* New York: Macmillan. Many students find it hard to see how Skinner's behaviorism would function in everyday life. In this novel, we see how a community based on his principles would operate. In fact, for a while at least, several such communities really existed. This is a good way to understand this theory.

Tyler, Anne. (1986). *The accidental tourist.* New York: Knopf. This story of a man who tries desperately to avoid the bumps of life offers a fine example for you to use in analyzing each of the theories presented in this chapter.

PART II

Beginnings

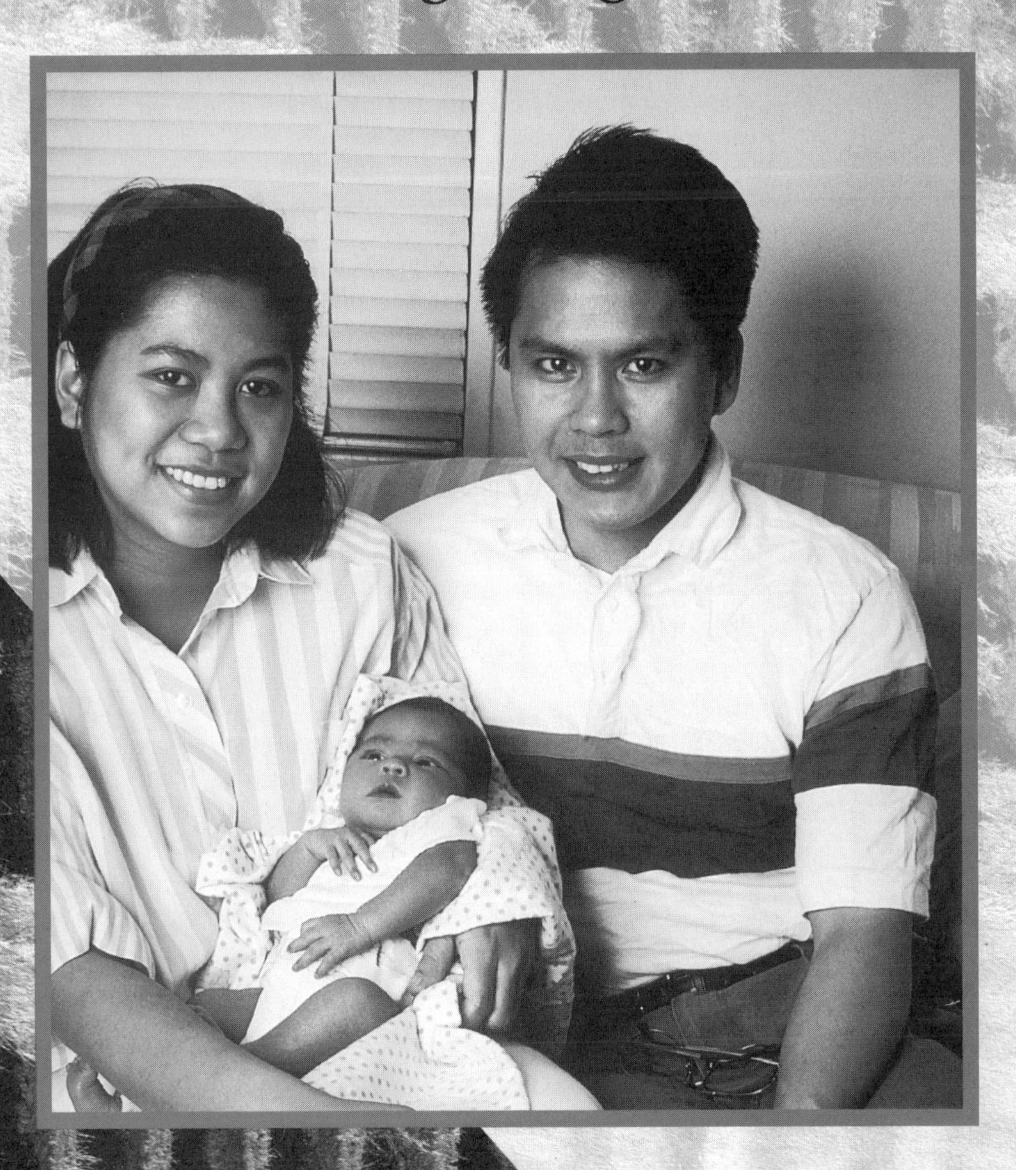

CHAPTER 3

The Biological Basis of Development

The Fertilization Process 58
The Beginnings 58
The Menstrual Cycle 60
Infertility 64
In Vitro Fertilization 67
Chromosomes and Genes 69
Heredity at Work 70
DNA: Structure and Function 71
How Traits Are Transmitted 73
What Is Transmitted 73
Examples of Chromosomal Disorders 75
Examples of Genetic Disorders 76
Conclusion 77
Chapter Highlights 79
Key Terms 80
What Do You Think? 80
Suggested Readings 80

On October 6, 1992, Frank and Ellen Smith arrived at Dr. James Otis' office a few minutes early for their appointment. Married for seven years, they had been unable to conceive and had found that their problem was caused by male infertility, in this case a low sperm count. As they attempted to learn more about their problem, they discovered that they were not alone; estimates are that one in six heterosexual couples has a fertility problem.

Other couples whom they had met in their search for a solution had recommended Dr. Otis as a physician who was sympathetic, knowledgeable, and successful in using DI—donor insemination. (Close to 100,000 women in the United States will use DI this year.) Frank and Ellen chose this procedure because the screening process in accepting sperm for freezing lowers the risk of sexually transmitted diseases (STD) and also provides detailed information about the donor: race, ethnic background, blood type, hair and eye color, physical characteristics, and personal background information.

The Smiths found that Dr. Otis was exactly as he had been described. Unhurried and calm, he told the couple just what would be expected of them. He also stressed that he received sperm from a large, nationally known, respected sperm bank. He explained that freezing techniques had improved in the last few years. Cryobiology, the study of how best to freeze living tissue, had made great strides by adding chemicals such as glycerol to liquid nitrogen at a temperature of −190°C. This technique provides protection to cells and tissues by preventing any damage that could be caused by the formation of ice crystals.

Frank and Ellen felt reassured by their discussion with Dr. Otis. They couldn't afford the time and money required for adoption, but from their reading had decided their chances of conceiving through donor insemination were high. Healthy couples with no other problems have a 70 to 80 percent success rate within one year. This figure compares with a natural conception rate of 85 to 90 percent for fertile couples.

The Smiths were also pleased that the procedure was relatively simple. After carefully tracking a woman's menstrual cycle and searching for accurate clues to the time of ovulation by examining her vaginal mucus and urine, the physician injects the semen through the cervix with a needleless syringe.

Their hopes and expectations were justified. Ellen conceived on the couple's fourth attempt, and later gave birth to a healthy baby boy. ■

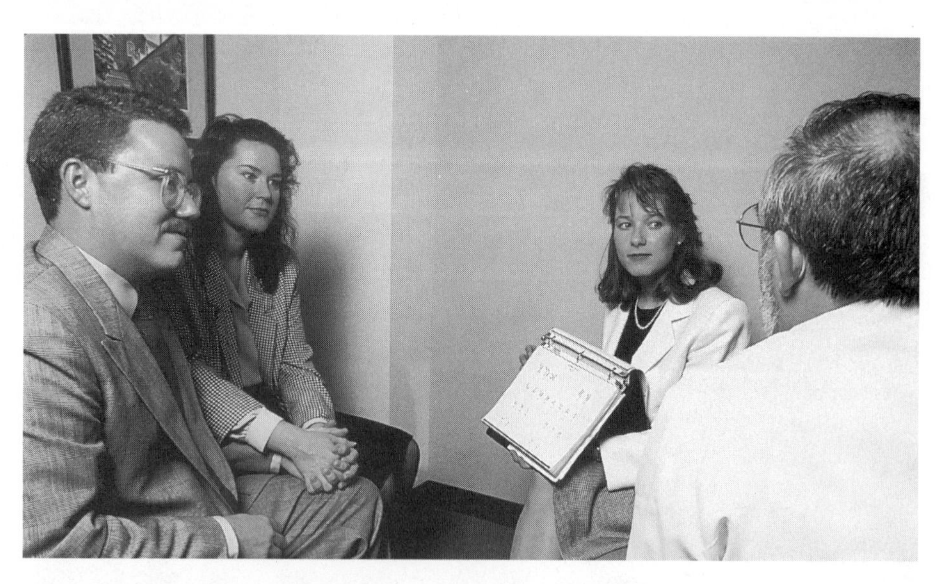

Advances in reproductive technology provide many options for those requiring alternative methods of conception.

Any journey begins with the first step. For students of the lifespan, this means the time that development begins: when sperm joins with egg, uniting the mother's genetic endowment with that of the father. As you can tell from the chapter's opening, today's technology gives couples a bewildering array of reproductive choices. In this chapter, you will explore a world so tiny that it is almost impossible to imagine. You will read about some great biological discoveries of the past century, and about recent breakthroughs that have led scientists to the origins of life itself.

Think for a moment about yourself. You probably think of yourself as an individual—and you are right! Thanks to your genetic inheritance, there's no one quite like you in the world. You may not be impressed by this fact, but you must occasionally have wondered what made you what you are. How much of what you are came from your parents, and how much came from your contacts with your environment? Remember, however, that heredity and environment are so interrelated that we really can't divide them; they must be considered together. We'll remind you of this at key points in the chapter.

In this chapter you'll read about the fertilization process, during which the sperm and egg unite. Today, however, we can no longer refer to "the union of sperm and egg." We must ask additional questions: Whose sperm? Whose egg? Where did the union occur? Was it in the woman's body? Was it in a test tube? Which woman will carry the fertilized egg? You can see, then, that it's a process filled with the potential for conflict and controversy because of these new techniques that enable fertilization to occur outside of a woman's body.

Sometime in your life you probably have said, "Oh, I've inherited that trait." That's the easy answer. *How* did you inherit that characteristic? Do you remember that DNA you heard about in your science classes? Here you'll learn that the discovery of DNA involved some of the world's greatest biologists in a race to be the first to find the "secret of life." Today, scientists are engaged in a struggle to "map the human genome," that is, to identify our genetic endowment—the 50,000 to 100,000 genes that lie in the nucleus of every cell (Bishop & Waldholz, 1990). Once this accomplishment is within our grasp, we will then be able to eliminate over three thousand genetic diseases! Perhaps equally as exciting is the attempt to identify "susceptibility" genes, which do not of themselves cause disease but make certain individuals susceptible to such diseases as breast cancer, colon cancer, and Alzheimer's, among others.

The strong resemblance of these family members to each other is testimony to the role of heredity in our makeup.

These characteristics, however, don't just appear; they're passed on from generation to generation. So following our discussion of genes, we'll trace the manner in which hereditary traits are transmitted. Unfortunately, we are all too well aware that occasionally the transmission of traits produces abnormalities, which we also discuss. Finally, no discussion of human heredity is complete without acknowledging the ethical issues that have arisen because of these new developments. Answers still elude us, but at least we can ask several critical questions, such as: Should scientists be allowed to implant specific genes to satisfy a couple's preferences? Should society determine that certain types of genes be implanted to ensure "desirable" products? No easy questions, these, but you can be sure that they will arise in the future.

Any discussion of the biological basis of development must place considerable emphasis on our genetic makeup. To avoid any misunderstanding, however, remember that development results from the interaction of heredity *and* development. It is the process by which our **genotype** (the genetic contribution of our parents) is expressed as a **phenotype** (our observable characteristics). Genes always function in an environment, and environmental circumstances affect the way that genes express themselves. For example, a child may be born with great intellectual potential, but because of a lack of opportunity and education the child's intellectual capacity may never reach fulfillment. These ideas help to explain why we have turned to a biopsychosocial model of development that encompasses both heredity and environment.

When you finish reading this chapter, you should be able to:

- Recall that development results from the interaction of heredity and environment.
- Identify the essential elements in the reproductive process.
- Distinguish between internal and external fertilization.
- List the steps that lead to ovulation.
- Describe the manner in which the genetic process functions.
- Formulate questions relating to the sensitive nature of the new reproductive technology.

THE FERTILIZATION PROCESS

The fusion of two specialized cells, the **sperm** and the egg or **ovum,** mark the beginning of development and the zygote (the fertilized ovum) immediately begins to divide. This fertilized ovum contains all of the genetic material that the organism will ever possess. During the initial phase of development following fertilization, it is almost impossible to distinguish the male from the female (Tapley & Todd, 1988).

The Beginnings

Any discussion of fertilization today must account for the advances that both research and technology have made available. Consequently, our discussion will be broken into two parts:

- **Internal,** or natural fertilization
- **External techniques,** such as **in vitro fertilization** (the famous "test-tube" babies).

Table 3.1 contains a glossary of many of the terms you will find in this discussion. Be sure to refer to it when you meet an unfamiliar term. Otherwise, the amazing richness of the genetic world can escape you.

In our analysis of genetic material and its impact on our lives, we should attempt to follow the manner in which we receive genes from our parents. (Even this fundamental fact today requires further explanation. Genes may come from surprising sources, thanks to our technology (Grobstein, 1988). More about this later.) But our story begins with the male's sperm and the female's egg.

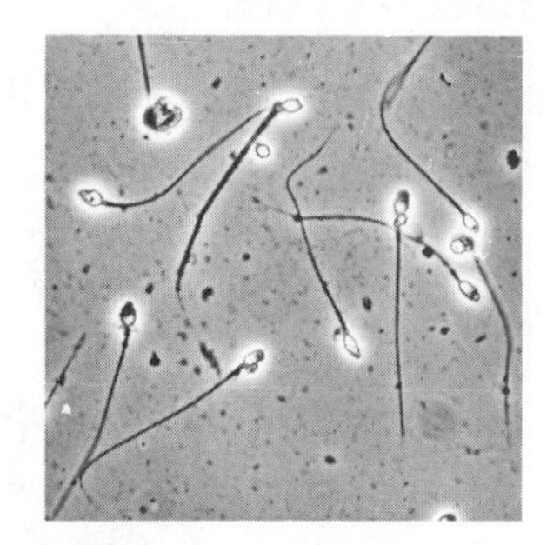

(a)

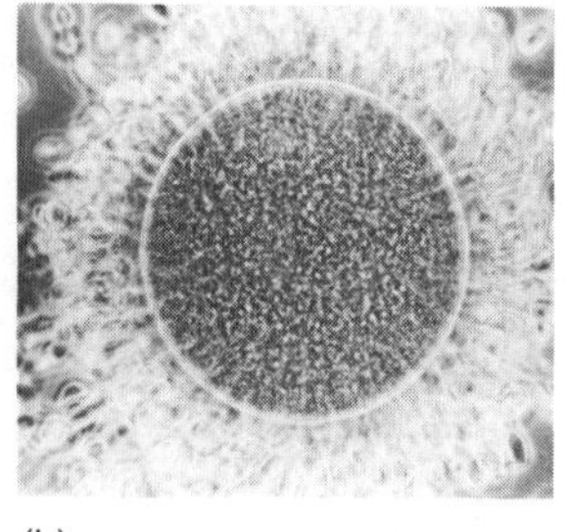

(b)

The sperm (a), in its search for the egg (b), carries the 23 chromosomes from the male.

The Sperm

Certain cells are destined to become the sperms and eggs. The chief characteristics of the sperm are its tightly packed tip (the *acrosome*), containing twenty-three chromosomes, a short neck region, and a tail to propel it in its search for the egg. Sperm are so tiny that estimates are that the number of sperm equal to the world's population could fit in a thimble (Travers, 1982).

As Tapley and Todd (1988) note, a major purpose of a male's reproductive organs is to manufacture, store, and deliver sperm. Males, at birth, have in their testes those cells that will eventually produce sperm. At puberty, a *meiotic* division occurs in which the number of chromosomes is halved and actual sperm are

Table 3.1 A Genetic Glossary

Acrosome: Area at the tip of the sperm.

Allele: Alternate forms of a specific gene; there are genes for blue eyes and brown eyes.

Autosome: Chromosomes other than the sex chromosomes.

Chromosome: Stringlike bodies that carry the genes; they are present in all of the body's cells.

DNA: Deoxyribonucleic acid, the chemical structure of the gene.

Dominance: The tendency of a gene to be expressed in a trait, even when joined with a gene whose expression differs; brown eyes will appear when genes for blue and brown eyes are paired.

Fertilization: The union of sperm and egg to form the fertilized ovum or zygote.

Gametes: The mature sex cells, either sperms or eggs.

Genes: The ultimate hereditary determiners; they are composed of the chemical molecule deoxyribonucleic acid (DNA).

Gene locus: The specific location of a gene on the chromosome.

Genotype: The genetic composition of an individual.

Heterozygous: The gene pairs for a trait differ; a person who is heterozygous for eye color has a gene for brown eyes and one for blue eyes.

Homozygous: The gene pairs for a trait are similar; the eye color genes are the same.

Meiosis: Cell division in which each daughter cell receives one–half of the chromosomes of the parent cell. For humans this maintains the number of chromosomes (46) at fertilization.

Mitosis: Cell division in which each daughter cell receives the same number of chromosomes as the parent cell.

Mutation: A change in the structure of a gene.

Phenotype: The observable expression of a gene.

Recessive: A gene whose trait is not expressed unless paired with another recessive gene; both parents contribute genes for blue eyes.

Sex chromosome: Those chromosomes that determine sex; in humans they are the 23d pair, with an XX combination producing a female, and an XY combination producing a male.

Sex-linkage: Genes on the sex chromosome that produce traits other than sex.

Trisomy: Three chromosomes are present rather than the customary pair; Down syndrome (mongolism) is caused by three chromosomes at the 21st pairing.

Zygote: The fertilized egg.

formed. Simultaneously, the pituitary gland stimulates the hormonal production that results in the male secondary sex characteristics: pubic hair, a beard, and a deep voice.

The Ovum (Egg)

The egg is larger than the sperm, about the size of the period at the end of this sentence (Travers, 1982). The egg is round and its surface is about the consistency of stiff jelly. You may find it hard to believe, but a whale and a mouse come from eggs of about the same size. In fact, the eggs of all mammals are about the same size and appearance (Guttmacher & Kaiser, 1986). When females are born, they already have primal eggs. From one to two million eggs have been formed in the ovaries. Since only one mature egg is required each month for about thirty-five years, the number present far exceeds the need.

What's Your View?

As you read this chapter, the amazing biological advances in studies of human development will probably fascinate you. You may be tempted to the view that biology holds clues to most of the secrets of development. Keep this question in mind: What's your view as to the relative importance of heredity and environment?

This box is intended to help you to remember the environmental role in development. As a pertinent reminder, consider the longitudinal study conducted by Emmy Werner (1989) that began in 1954. Studying all of the infants born on the Hawaiian island of Kauai in 1955 (698 infants), Werner and her colleagues analyzed their development at 1, 2, 10, 18, and 31 or 32 years of age. Of the 698 infants, 276 were identified as "at risk": reproductive stress, dysfunctional families, mentally disturbed, or alcoholic parents. It was a study involving remarkable cultural diversity since the children were of Hawaiian, Japanese, Philipino, Portuguese, Chinese, and Korean descent. It was also a stable population: 88 percent of the group were available for the 18-year follow-up. (Recall the strengths and weaknesses of longitudinal studies that we described in chapter 1.)

Gathering a vast array of data on the children, Werner was able to sort various high-risk categories:

- Prenatal or perinatal stress (complications during pregnancy, labor, or delivery) was found in 69 children.
- Severe prenatal or perinatal stress was diagnosed in 23 children. Only 14 of these children lived until the age of 2 years.
- One of every 6 children had physical or intellectual complications of perinatal or neonatal origin that required long-term care.
- One of 5 children developed serious learning or behavior problems.
- By the time the children were 10 years of age, twice as many needed mental health services as required medical care.

As they followed these children from birth to 18 years of age, Werner found that the impact of reproductive stress lessened with time, and that the outcome of every biological problem depended on the child's environment. Family dysfunction, parental mental illness, or generally poor home conditions were consistently related to long-term biological complications.

One in three of the high-risk children who grew up in troubled environments turned out to be competent young adults. Using several of these children as case studies, Werner discovered several predictive factors that helped the children to overcome their difficulties: a child's temperament (active, low excitability), a high degree of sociability, alertness, and concentration. There were four or fewer children in the family, with at least two years between each child. The competent children also established a bond with someone who gave them positive attention (a grandparent, uncle, or older sibling). These relationships helped the children to find meaning in their lives and gave them a belief that they could control their destinies. In their thirties, these individuals were leading satisfactory adult lives.

Heredity *or* environment? Heredity *and* environment? What's your view?

Many of these primal eggs succumb before puberty. They simply shrivel up and disappear. At puberty the pituitary gland stimulates the hormonal production that results in the female secondary sex characteristics: pubic hair, breasts, wider hips, and a higher voice.

The Menstrual Cycle

The pituitary gland secretes another hormone that stimulates the ripening of eggs, and after two weeks one egg, which has ripened more than the others, is discharged from the ovary's surface. (Figure 3.1 illustrates the relationship of the ovum to the ovary, the fallopian tubes, and the uterus.) This process, called **ovulation,** triggers a chemical reaction that inhibits the ripening of further eggs. It also prepares the uterine lining for a potential fertilized ovum.

If fertilization does not occur, the prepared uterine lining is shed in menstruation. When the menstrual bleeding ceases, the entire process begins again. During each menstrual cycle many eggs are discarded. As a woman approaches the end of

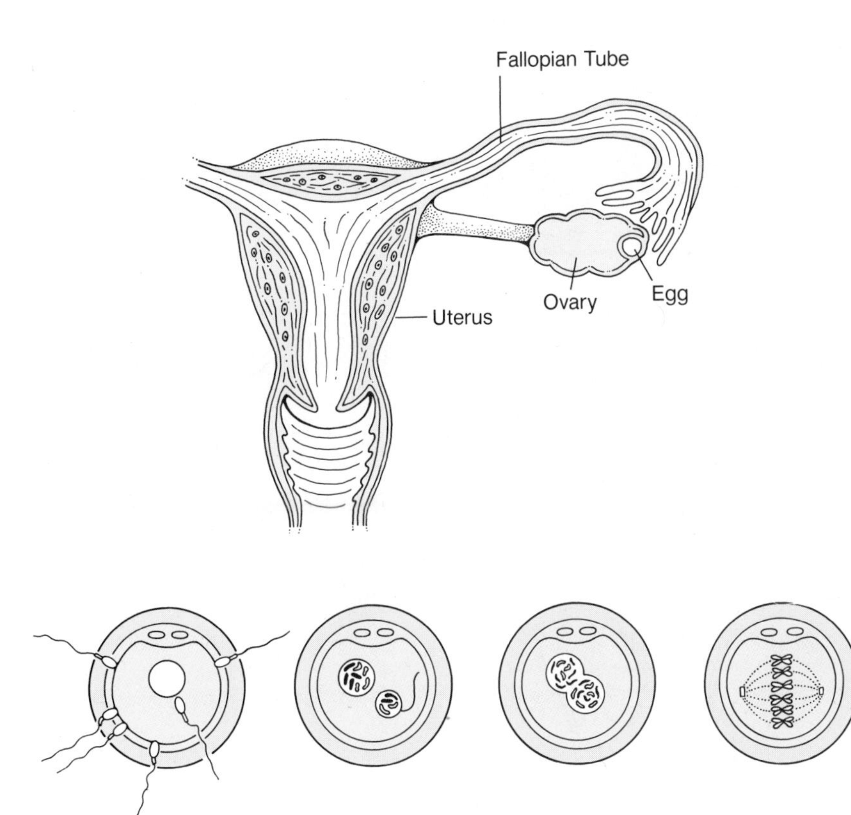

Figure 3.1
The relationship of ovary, egg, fallopian tube, and uterus

Figure 3.2
Fertilization of the egg. The sperm, carrying its 23 chromosomes, penetrates the egg, with its 23 chromosomes. The nuclei of sperm and egg fuse, resulting in 46 chromosomes.

her egg-producing years, these last ova have been present for as many as forty years. This may explain why the children of older women are more susceptible to genetic defects. The eggs have been exposed to environmental hazards (such as radiation) too long to escape damage (Singer, 1985).

Implantation

When the egg is discharged from the ovary's surface, it is enveloped by one of the **fallopian tubes.** The diameter of each fallopian tube is about that of a human hair, but it almost unfailingly ensnares the egg and provides a passageway to the uterus. If fertilization occurs, it takes place soon after the egg enters the fallopian tube. Figure 3.2 illustrates the fertilization process.

Fusion of the two cells is quickly followed by the first cell division. As the zygote (the fertilized egg) travels toward **implantation** within the uterus, cell division continues. The cells multiply rapidly and after about seven days reach the uterine wall. The fertilized egg is now called a **blastocyst.** The journey is pictured in Figure 3.3.

The Zygote

After the sperm and egg unite, the new cell (the potential individual) possesses 23 pairs of chromosomes, or 46 chromosomes. One member of each pair has been contributed by the father and one by the mother. Figure 3.4 illustrates the chromosomal arrangement of a typical male and female. The fertilized egg at this stage is called the **zygote.**

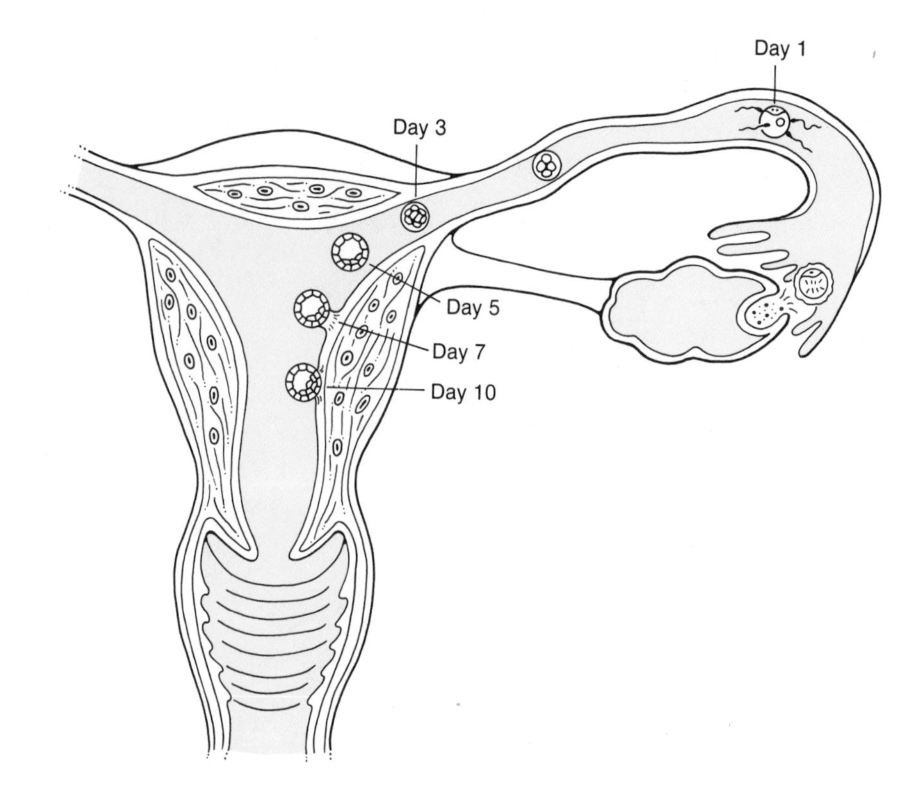

Figure 3.3
From ovulation to implantation
L.P. Wisniewski and K. Kirshhorn, *A Guide to Human Chromosome Defects,* 2d ed. © March of Dimes

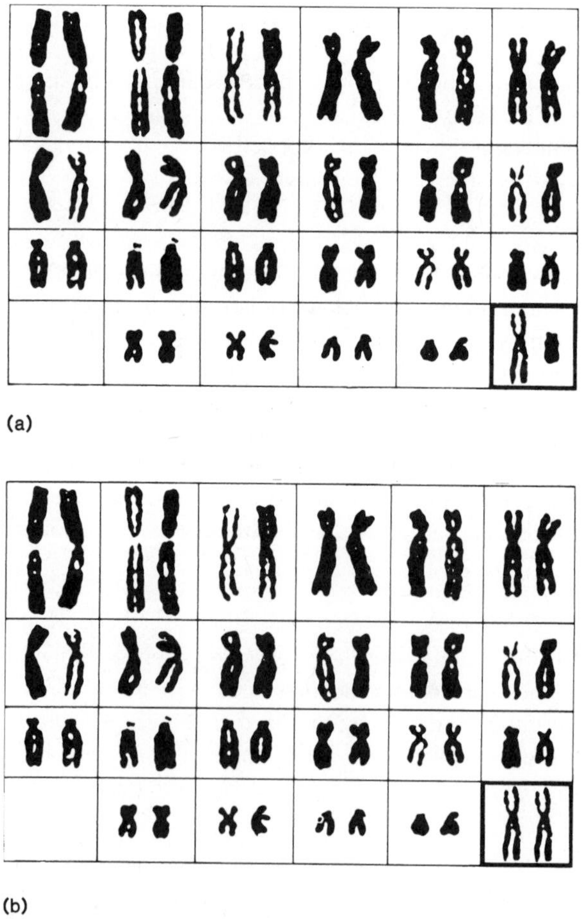

(a)

(b)

Figure 3.4
(a) At the top is the chromosome structure of a male, and (b) at the bottom is the chromosome structure of a female. The 23d pair is shown in the bottom right box of each figure; notice that the Y chromosome of the male is smaller. To obtain this chromosomal picture, a cell is removed from the individual's body, usually from the inside of the mouth. The chromosomes are magnified extensively and then photographed.

What's Your View?

?

Occasionally, and for reasons that still elude us, twins are born. Most twins occur when a woman's ovaries release two ripened eggs (rather than one) and both are fertilized by separate sperm. These twins are called **nonidentical,** or **dizygotic.** Their genes are no more alike than those of siblings born of the same parents but at different times.

Less frequently, twins develop from a single fertilized egg that divided shortly after conception. This process leads to **identical,** or **monozygotic,** twins whose genes are identical; that is, they share the same genotype.

There are an estimated 50 million pairs of twins around the world, with about two million pairs in this country (Powledge, 1983). Regardless of country, the incidence of identical twins is about the same—3 in every 1,000 live births. In the United States, two-thirds of all twin births are dizygotic (Alexander, 1987). The rate of nonidentical (dizygotic) twin births, however, varies considerably from country to country. For example:

- In the United States, about 11 or 12 pairs of dizygotic twins are born per 1,000 live births (once in 90 white births, once in 73 black births). The rate is about the same in Great Britain and other European countries.
- In Japan, about 2 or 3 pairs of dizygotic twins are born per 1,000 live births.
- In Nigeria, 40 pairs of dizygotic twins are born per 1,000 live births (Alexander, 1987).

Although it is difficult to explain these different figures, certain facts about twin births are now clear. For example, a mother's chances of giving birth to dizygotic twins increases with age, peaking at about the age of 37. The chances of a mother of twins giving birth to twins again is one in twenty. Birth control pills also seem to be linked to an increase in twin births. Twins likewise seem to appear more frequently in some families. With regard to sex, almost one-third of all twins are boy-girl. The remaining two-thirds (same-sex twins) are equally divided between boys and girls (Alexander, 1987).

As you can well imagine, twins, especially identical twins that have been separated at birth, have long fascinated psychologists. By comparing these twins who have grown up in different environments, researchers can estimate the influence of heredity and environment on behavior. For example, studies have consistently shown that the IQs of monozygotic (identical) twins are quite similar, while those of dizygotic (nonidentical) are much less so (Berndt, 1992). Yet Riese (1990), while studying differences in temperament between monozygotic and dizygotic twins, found little evidence of genetic influence during the neonatal period. How do you explain these differences? Is it heredity or environment? What's your view?

The 46 chromosomes that the zygote possesses represent the individual's total biological heritage. By a process of division, each cell in the body will have a replica of all 46 chromosomes. The significance of the chromosomes is that they carry the genes, the decisive elements of heredity.

The size of the elements involved is almost bewildering. We have commented on the size of the sperm—so small that it can be seen only microscopically. The head of the sperm, which is about one-twelfth of its total length, contains the 23 chromosomes.

While individuals may change in the course of their lives, their hereditary properties do not change. The zygote, containing all 46 chromosomes, represents the "blueprint" for our physical and mental makeup. Under ordinary circumstances, environmental conditions leave the 46 chromosomes unaltered. (We know today that environmental agents such as drugs, viruses, or radiation may cause genetic damage.) Figure 3.5 illustrates the process by which the number of chromosomes remains the same from generation to generation.

Figure 3.5
The hereditary process

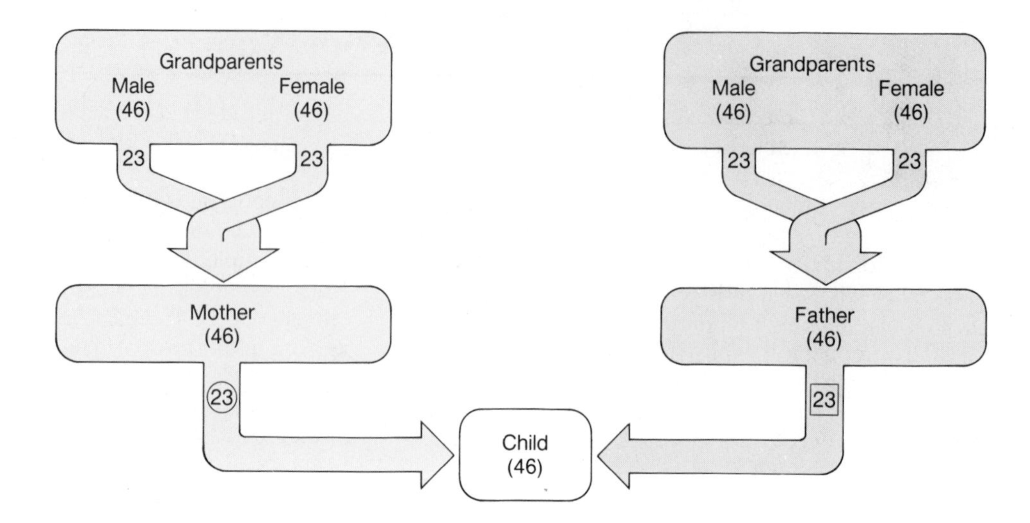

Once fertilization occurs, the zygote is not yet free from risk. Let's use a figure easy to work with and assume that 100 eggs are exposed to sperm. Here is the estimated mortality rate from fertilization to birth:

84 are fertilized
69 are implanted
42 survive one week
37 survive seven weeks
31 survive to birth

Thus nature's toll results in about a 70 percent mortality rate.

Infertility

Although the fertilization process just described is the normal process for most women, there are exceptions. For some couples, **infertility,** for whatever reason, is an inescapable problem. Today there are hundreds of thousands of childless couples who desperately desire children. These couples share a growing problem in our society: infertility.

Causes of Infertility

Estimates are that one in five American couples meet the criteria for infertility: an inability to achieve pregnancy after two years. Here are some reasons for this increased rate:

- **Sexually transmitted diseases (STD)** show a marked increase. These can lead to **pelvic inflammatory disease (PID),** which in turn can cause infertility.
- A growing number of women are delaying childbearing until their thirties or later. Older women are more likely to become infertile because they may have used **intrauterine devices (IUDs),** had more abdominal surgery, and may be subject to **endometriosis,** which increases with age. (Endometriosis is the growth of endometrial tissue outside of the uterus. It is typically found in the lining of the uterus.)
- Males are responsible in 35 percent of the cases.

Table 3.2 Most Common Causes of Infertility

Female	Male	Female	Male
General		**Hormonal**	
Overweight	Overweight	Pituitary failure	Pituitary failure
Underweight	Underweight	Thyroid disease	Thyroid disease
Anemia	Excessive alcohol use	Ovarian failure	Adrenal disorders
Hostile mucus	Excessive smoking	Adrenal disorders	
Turner syndrome	Klinefelter syndrome		
Developmental		**Genital Disease**	
Uterine malformations	Undescended testicles	Infection(PID, chlamydia, gonorrhea, cervicitis, etc.)	Infection(PID, (gonorrhea, other sexually transmitted diseases)
Undeveloped ovaries	Varicocele	Tubal scarring	Injury to testes
Incompetent cervix	Underdeveloped testicles	Other abnormalities	Hydrocele
	Poor sperm count	Endometriosis	Orchitis
	Defective sperm	Cervical polyps	Prostatitis
	Impotence	Uterine Fibroids	
	Ejaculatory disorders		
Both Sexes			
Marital maladjustment			
Poor timing of intercourse			
Immunological incompatibility			
Genetic disorders			

Adapted from Albert Decker, M.D. and Suzanne Loebl, *Why Can't We Have a Baby?* Copyright © 1978 Dial Press; as printed in D. Tapley and W. Todd, *Complete Guide to Pregnancy*. Copyright © 1988 Crown Publishing.

- No cause is identified in 10 percent of the cases.
- In about 30 percent of all cases, both partners have a fertility problem (Tapley & Todd, 1988).

Table 3.2 describes the most common causes of infertility.

Simply because a woman does not become pregnant within a specific time does not necessarily mean that a couple is infertile. The best advice given to couples is to wait for two years before suspecting infertility if a woman is in her twenties, one year if she is between 30 and 35, and six months if she is over 35. Even if a couple finds that they are indeed infertile, they needn't abandon their hopes of parenthood. Many **external fertilization** techniques are available today. The new

What's Your View?

Closely allied to the new external fertilization techniques has been the development of sperm banks. Many couples have also opted for **donor insemination** (DI). In 1987, the Office of Technology Assessment announced that 80,000 American women had attempted artificial insemination using anonymous donor sperm, resulting in 30,000 live births. (Single women make up about 15 percent of this group.) Since 1960, about 400,000 babies have been born through this process.

Today about 135 sperm banks operate in the United States. Sixteen cryobanks store and sell frozen sperm, which are frozen in liquid nitrogen at –196° C and remain potent for years. As we mentioned in the chapter's opening, much of the appeal of this technique lies in the screening processes used, which are intended to reduce the possibility of sexually transmitted diseases.

The donors' traits are carefully recorded, since many couples want to match the male's characteristics as closely as possible. Donors usually receive a fee of about $100 or less and are not told about the use of their sperm. They must sign a form waiving all parental rights to any children conceived with their sperm (Tapley & Todd, 1988).

Although donor insemination has undoubtedly been in practice for most of this century, it came to public notice in the 1950s with the founding of the Repository for Germinal Choice. The repository originally was intended to accept and freeze the sperm of Nobel Prize winners, but due to the few who volunteered and their age (most were in their seventies and had less vigorous sperm), its acceptance standards were relaxed to include scientists and mathematicians.

As we have seen, many people have eagerly accepted the idea of sperm banks. On Madison Avenue in New York City sits the largest sperm bank in the world, holding over 35,000 samples of frozen sperm. Do you think this practice should be encouraged or discouraged? What's your view?

technologies we are about to discuss offer couples new hope. As a response to this need, there are now hundreds of in vitro fertilization centers and sperm banks around the country.

Although you are probably most familiar with in vitro fertilization, another process, **AID** (artificial insemination by donor) is by far the most widely used procedure. Here are some techniques now available (Grobstein, 1988):

- Artificial insemination
- In vitro fertilization (fertilization of egg and sperm occurs in dish)
- Embryo transfer (an embryo from one woman's womb is transferred to that of another woman)
- Women sell their eggs
- Human embryos are frozen for long-term storage
- Surrogate motherhood (a woman agrees to be inseminated by the sperm of a man whose wife is infertile and returns the child to the couple after birth)

Consequently, today a child may have as many as five parents:

- A sperm donor (father or other male)
- An egg donor (mother or other female)
- A surrogate mother
- The couple that raises the child

"Let me get this straight. One bouquet goes to the mother who donated the egg. A second goes to the mother who housed the egg for insemination. A third goes to the mother who hosted the embryo and gave birth to the child. A fourth goes to the mother who raised it and a fifth goes to the mother with legal custody."

In Vitro Fertilization (IVF)

In vitro fertilization is the external fertilization technique you are probably most familiar with. (At this point we urge you to reexamine Figure 3.1 detailing the control system of a woman's menstrual cycle, since the in vitro procedure attempts to imitate both these steps and their timing.) The steps in the procedure are as follows (Grobstein, 1981):

1. The woman is usually treated with hormones to stimulate maturation of eggs in the ovary, and she is observed closely to determine the timing of ovulation (i.e., the time at which the egg leaves the surface of the ovary).
2. The physician makes an incision in the abdomen and inserts a laparoscope (a thin tubular lens through which the physician can see the ovary) to remove mature eggs.
3. The egg is placed in a solution containing blood serum and nutrients.
4. Capacitation takes place—this is a process in which a layer surrounding the sperm is removed so that it may penetrate the egg.
5. Sperm are added to the solution; fertilization occurs.
6. The fertilized egg is transferred to a fresh supporting solution.
7. Fertilized eggs (usually three) are inserted into the uterus.
8. The fertilized egg is implanted in the uterine lining.

What's Your View?

?

For many couples who remain childless in spite of several attempts at external fertilization, adoption offers a viable option. The process of adoption has changed radically in the past few years because of a limited number of children and an increase in the number of couples who wish to adopt. This statement is not quite as simple as it appears. More children are available for adoption than is commonly thought, but they fall into several categories.

- Older children
- Minority children
- Handicapped children

While these children are available for immediate adoption, the waiting period for healthy white infants may run into years.

Adoption procedures were formerly *closed,* that is, the biological parents were completely removed from the life of their child once the child was surrendered for adoption. The bonds between birth parent(s) and child were legally and permanently severed; the child's history was sealed by the court. The child was effectively cut off from its genetic past (Gilman, 1987). Supposedly, this prevented the natural parents and the adoptive couple from emotional upset. Many biological mothers, however, reported in later interviews that they never recovered from the grieving process. This procedure fits the standard definition of adoption: to take a child of other parents voluntarily as one's own.

Today, however, if a pregnant woman approaches an adoption agency, she gets what she wants. She can insist that her child be raised by a couple with specific characteristics: nationality, religion, income, number in family. She can ask to see her child several times a year, perhaps take the youngster on a vacation, and telephone the child frequently. The adoption agency will try to meet these demands. This process is called *open adoption,* and while many adoptive couples dislike the arrangement, they really have no choice. Thus we see a new definition of adoption: the process of accepting the responsibility of raising an individual who has two sets of parents (Gilman, 1987, p. 29).

Open adoption is a radical departure from the days when a woman who had decided to give up her baby for adoption had to wear a blindfold and earplugs during delivery so she wouldn't see or hear her baby. Today the natural mother may actually select the adoptive couple from several profiles that are given to her. (These profiles contain information about the adoptive couple: food preferences, television, politics, how the couple deals with stress, how they would handle a two-year-old with a temper tantrum.) Most officials at adoption agencies agree that biological mothers rarely select a couple solely on the basis of income. Religion, life style, and family stability seem much more important. Face-to-face meetings between the couples are becoming more common and often are decisive in the natural mother's decision.

Although there are problems with open adoption—the adoptive parents frequently resent the continuing presence of the biological mother—most experts agree that a change was needed. Too many adopted children have shown emotional difficulties on learning that they were relinquished, and a large number of biological mothers have prolonged difficulty as a result of relinquishing their babies. Only time can tell how successful this new procedure will be.

What's your view—closed or open adoption?

During this period, the woman is being treated to prepare her body to receive the fertilized egg. For example, the lining of the uterus must be spongy or porous enough to hold the zygote. The fertilized egg must be inserted at the time it would normally reach the uterine cavity. Variations on this procedure include **ZIFT** (zygote intrafallopian transfer), in which the fertilized egg is transferred to the fallopian tube, and **GIFT** (gamete intrafallopian transfer), in which sperm and egg are placed in the fallopian tube with the intent of achieving fertilization in a more natural environment.

A recent interesting development relates to the success rate that in vitro fertilization centers have claimed. Noting that some advertising seemed inflated, the Society for Assisted Reproductive Technology (SART), a division of the American Fertility Society representing most of the IVF centers in this country, reported that it would disclose success rates for its member clinics. This action was in response to

An Applied View — The Human Genome Project

On August 24, 1989, Pete Rose was banished from baseball. For baseball fans, that was the news of the day. In fact, it was headline news around the country. That same day, another announcement far more important for human health appeared in most news outlets, but with far less fanfare. Researchers had found the gene that caused **cystic fibrosis** (CF).

For Margaret Smith, the announcement was momentous. Looking at her three-year-old son, she could only wonder if the breakthrough came in time to help her child, who was afflicted with cystic fibrosis. She knew what the future held for almost all youngsters born with this disease: lungs becoming filled with mucus, chronic coughing, wheezing, repeated bouts with pneumonia, and a life expectancy of about twenty years. To her, discovery of the CF gene was stunning news.

The CF gene discovery came as part of the **Human Genome Project,** an undertaking comparable to the Manhattan Project that resulted in the atomic bomb and the Apollo Project that produced the first moon landings. The Human Genome Project is nothing less than an attempt to identify and map the 50,000 to 100,000 genes that constitute our genetic makeup. It is a project, international in scope, that will take about fifteen years and cost at least three billion dollars!

To give you an idea of the potential uses of a gene map, Bishop and Waldholz (1990) estimate that as many as 2.4 million women each year will undergo prenatal tests for chromosomal abnormalities. Almost 2.8 million women will try to discover if they are carriers of any of the genes of the four major inherited diseases: cystic fibrosis, sickle cell anemia, hemophilia, and muscular dystrophy.

Studies of genetic susceptibility will undoubtedly follow the same pattern. The genes that make people susceptible to certain diseases do not, of themselves, cause disease. It is the combination of a particular environment with a particular gene. For example, a person may be born with a susceptibility to lung cancer and never develop any malignancy. If that same person is a heavy smoker, however, the chances of developing cancer are much greater.

Once the mechanisms that cause a susceptibility gene to spring into action are more fully understood, such preventative measures as screening techniques and drug therapy will save many lives.

congressional criticism of centers that were advertising a success rate of up to 40 percent. Federal investigations had shown that nationally, a woman has a 9 to 10 percent chance of success from any single treatment attempt.

Many centers resisted the disclosure of results because an apparently low success rate would damage their reputation, resulting in a loss of potential patients. This would be unfair to those centers that report their results honestly but work with a wide range of patients. Some centers work with an older population and make no restrictions about the male's sperm count. Consequently, their success rate is lower than that for centers accepting a more limited group of patients.

Chromosomes and Genes

The fertilized egg, the zygote, contains the individual's genetic endowment, represented by the 46 chromosomes in its nucleus. Human chromosomes appear in 23 pairs. Each pair, except the 23d, is remarkably alike. The 23d pair defines the individual's sex: an XX combination indicates a female; XY indicates a male. The sperm actually determines sex, since it alone can carry a Y chromosome. Thus there are two kinds of sperm: the X chromosome carrier and the Y chromosome carrier.

The Y carrier is smaller than the X, which contains more genetic material. The Y carrier is also lighter and speedier and can reach the egg more quickly. But it is also more vulnerable; if ovulation has not occurred, the X carrier survives longer than the Y carrier.

Consequently, the male, from conception, is the more fragile of the two sexes. Estimates are that 160 males are conceived for every 100 females. However, so many males are spontaneously aborted that only 105 males are born for every 100

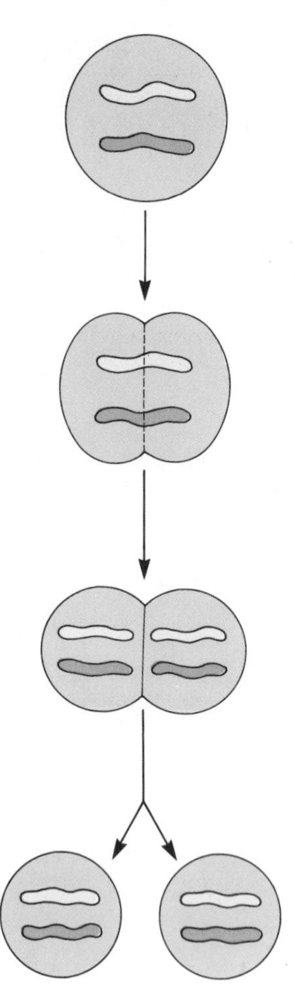

Figure 3.6
A mitotic division is a cell division in which each daughter cell receives the same number of chromosomes as the parent cell—46.

females. A similar pattern appears in neonatal life and continues throughout development, until women finally outnumber men, reversing the original ratio. Certain conclusions follow:

- Structurally and functionally, females resist disease better than males.
- The male is more subject to hereditary disease and defect.
- Environmental elements expose the male to greater hazards.
- Females are born with and retain a biological superiority over males (Singer, 1985).

The significance of the chromosomes lies in the material they contain—the genes. Each gene is located at a particular spot on the chromosome, called the *gene locus.* The genes, whose chemical structure is **DNA** (deoxyribonucleic acid), account for all inherited characteristics, from hair and eye color to skin shade, even the tendency toward baldness.

Aside from performing their cellular duties, the genes also reproduce themselves. Each gene constructs an exact duplicate of itself, so that when a cell divides, the chromosomes and genes also divide and each cell retains identical genetic material. As the cells divide, however, they do not remain identical. Specialization appears and different kinds of cells are formed at different locations.

The genes, then, are continuously active in directing life's processes according to their prescribed genetic codes. The action of the genes is remarkable not only because of complexity but also because of an elegant simplicity.

For example, the genes initially form the body's basic materials. Although this activity continues throughout our lives, more specialized functions gradually appear and begin to form a circulatory system, a skeletal system, and a nervous system. The process continues until a highly complex human being results (Grobstein, 1988). (Remember the distinction that you read about in the genetic glossary. A **genotype** is a person's genetic composition, while a **phenotype** is the observable expression of a gene.)

Heredity at Work

The original cell, the possessor of 46 chromosomes, begins to divide rapidly after fertilization, until at birth there are billions of cells in the infant. The cells soon begin to specialize: some become muscle, some bone, some skin, some blood, and some nerve cells. These are the somatic or body cells. The process of division by which these cells multiply is called **mitosis** (Figure 3.6). In a mitotic division, the number of chromosomes in each cell remains the same.

A second type of cell is also differentiated: the germ or sex cell that ultimately becomes either sperm or egg. These reproductive cells likewise divide by the process of mitosis until the age of puberty. But then a remarkable phenomenon occurs—another type of division called **meiosis,** or reduction division (Figure 3.7). Each sex cell, instead of receiving 46 chromosomes upon division, now receives 23.

Mitosis is basically a division of cells in which each chromosome is duplicated so that each cell receives a copy of each chromosome of the parent cell. What is doubled in mitotic division is the amount of DNA, the chief component of the genes. Meiosis, which occurs only in germ cell reproduction, is responsible both for the shuffling of hereditary characteristics received from each parent and for their random appearance in offspring. During the reduction division, the chromosomes separate longitudinally so that 23 go to one cell and 23 to another.

For the male, reduction division begins to occur just before puberty. For the female, the process differs slightly. Since she is required to produce only one mature egg a month, there is no provision for an indefinitely large number of eggs, as there is for sperm. A woman normally sheds only three to four hundred mature ova in her lifetime, while the normal male in a single ejaculation emits hundreds of millions of sperm.

At birth the female's ovaries contain tiny clusters of all the eggs that will mature in later years. Just before puberty, the final phases of the reduction division occur, and mature eggs are formed. It is as if there is a lengthy waiting period, from birth until about the age of 12 or 13, before the process is finally completed. The 23 chromosomes with their hereditary content are present at birth but must await the passage of time before biological maturity occurs in the female.

Now that we have traced the process by which fertilization occurs and discovered what is passed from parents to child, we still need to know "why I am who I am." This brings us to the discovery of DNA, one of this century's greatest achievements.

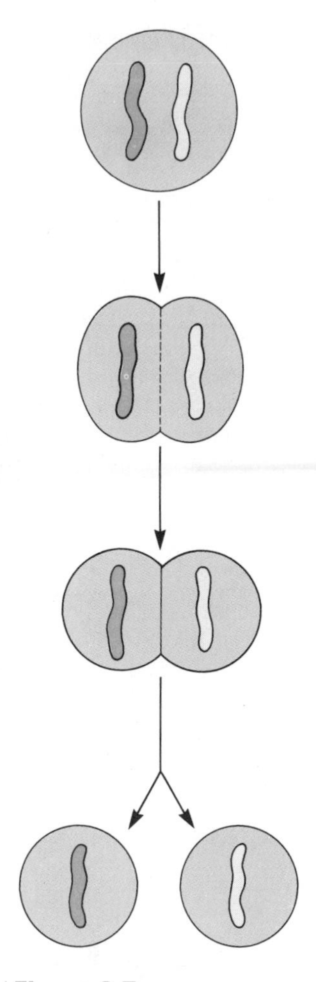

Figure 3.7
A meiotic division is a cell division in which each daughter cell receives one half of the chromosomes of the parent cell—23.

DNA: Structure and Function

We know that each chromosome contains thousands of genes, and that each of these thousands of genes has a role in the growth and development of each human being. The chemical key to the life force in humans, animals, and plants is the amazing chemical compound **DNA.** It constitutes about 40 percent of the chromosomes and is the molecular basis of the genes (Singer, 1985). Genes not only perform certain duties within the cell, but join with other genes to reproduce both themselves and the whole chromosome (see Figure 3.8).

Each gene follows the instructions encoded in its DNA. It sends these instructions, as chemical messages, into the surrounding cell. The cell then produces certain substances or performs certain functions according to instructions. These new products then interact with the genes to form new substances. The process continues to build the millions of cells needed for various bodily structures.

Examine Figure 3.8 and note how the strands intertwine. The strands, similar to the sides of a ladder, are connected by chemical rings: adenine (A), guanine (G), cytosine (C), and thymine (T). The letters are not randomly connected: A joins with T, G with C. If a code were written as AGCTTGA, it must appear as:

A G C T T G A
T C G A A C T

Thus, one sequence determines the other.

A remarkable feature of DNA is its ability to reproduce itself and ensure that each daughter cell receives identical information. During mitosis the DNA splits as readily as a person unzips a jacket. Each single strand grows a new mate, A to T and G to C, until the double-helix model is reproduced in each daughter cell.

The four letter possibilities, A-T, T-A, G-C, and C-G, seem to limit genetic variation. But when we consider that each DNA molecule is quite lengthy, involving thousands, perhaps millions of chemical steps (TA, GC, AT, CG, AT, CG, TA), the possible combinations seem limitless. The differences in the DNA patterns account for the individual genetic differences among humans and for differences between species.

One intriguing question is how the encoded information contained in the DNA is transmitted to the surrounding cell. The process is essentially as follows: RNA (ribonucleic acid) forms within the nucleus of the cell, acts as a messenger for DNA, and moves into the cell body to direct the building of the body's substances.

AN APPLIED VIEW The DNA Alphabet

If you examine Figure 3.8, you will notice that the various rungs and sides of the DNA ladder resemble small blocks. These are called **nucleotides,** and come in four varieties depending on the AT, GC pairings. Carl Sagan (1977, p. 23) states that although the language of heredity is written in an alphabet of only *four letters,* the final book is very rich. The average DNA molecule consists of about 5 billion pairs of nucleotides.

Sagan attempts to estimate the amount of information that our genes contain. Since there are four different kinds of nucleotides, the number of bits of information in a single chromosome is 20 billion. To what can we compare 20 billion bits of information?

They are the equivalent of about 3 billion letters. Since the average word contains six letters, each chromosome incorporates information equal to about 500 million words. At about three hundred words per typical printed page, this translates into the equivalent of about 2 million pages. The average book consists of about five hundred pages. Thus the information in one human chromosome corresponds to that in four thousand books. As Sagan says, the rungs on the DNA ladder represent a library of information.

Sagan's comment about this human uniqueness is pertinent for lifespan development:

> ■ *We have made a kind of bargain with nature: our children will be difficult to raise but their capacity for new learning will greatly enhance the chances of survival of the human species.* (1977, p. 23)

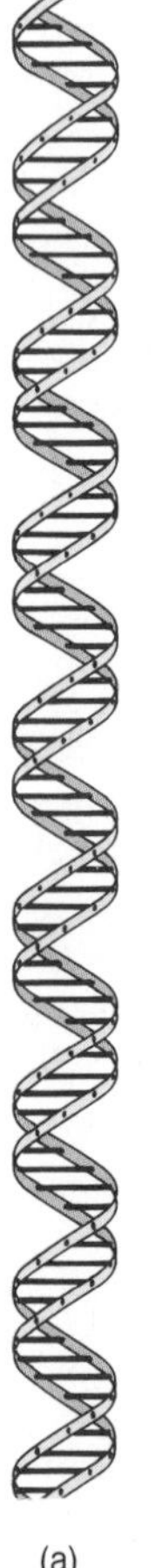

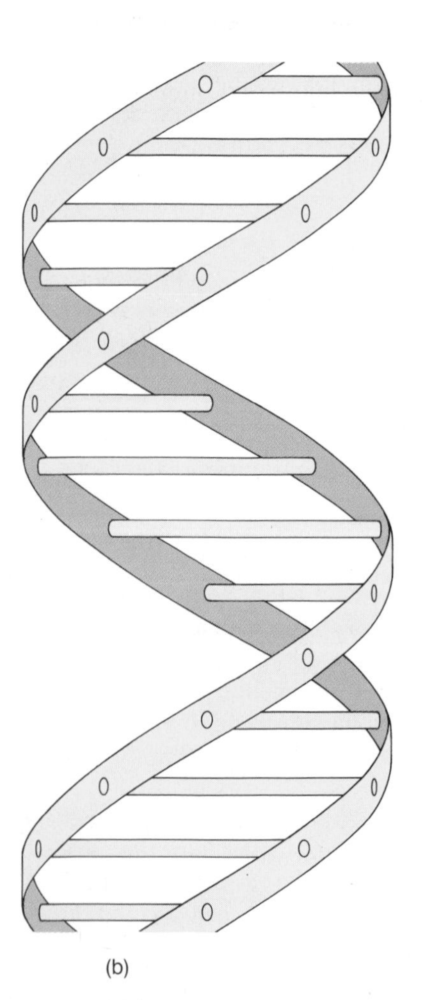

(a) (b)

Figure 3.8
The DNA double helix. (a) The overall structure that gave DNA its famous name. (b) A closer examination reveals that the sides of the spiral are connected by chemicals similar to the rungs of a ladder.

How Traits Are Transmitted

Hemophilia is an example of sex-linked inheritance, which affected several of the royal families of Europe (such as the Romanovs, pictured here)

What color are your eyes? Do your brothers and sisters have the same color eyes? Theirs could be different from yours; the mother and father who produced a brown-eyed child can also have a youngster with blue eyes. For centuries, guesses, myths, and speculation were used to explain the bewildering and mysterious happenings of heredity.

It was not until the end of the nineteenth century that Gregor Mendel offered a scientific explanation. (Note the relatively recent emergence of genetic facts: the transmission of traits by Mendel in the 1860s, the Watson-Crick double-helix model of DNA in 1953, and the number of human chromosomes in 1956.)

Gregor Mendel, an Austrian monk who studied plants as a hobby, attempted to crossbreed pure strains of pea plants. He used pure sets of plants, that is, peas that were either round or wrinkled, yellow or green, tall or dwarf. He discovered that the first generation of offspring all had the same trait: After crossbreeding round and unwrinkled peas, Mendel noted that the offspring were all round. Did the offspring inherit the trait from only one parent?

Mendel quickly eliminated this explanation, since the missing trait (wrinkled) reappeared in the second generation. But the trait that was exclusively expressed in the first generation (roundness) was the majority trait in the second generation. That is, the ratio of round peas to wrinkled peas consistently remained at 3 to 1 in the second generation.

Thus roundness is the **dominant** trait and wrinkled is the **recessive** trait. These two genes, round (R) and wrinkled (W), yield four possible peas: RR, WW, RW, and WR. Peas RR and WW are pure strains and breed true; RW and WR contain both the dominant and recessive trait.

A genetic grid based on Mendel's first pairing (round and wrinkled) is shown in Table 3.3. Note how the 3 to 1 ratio appears in the second generation.

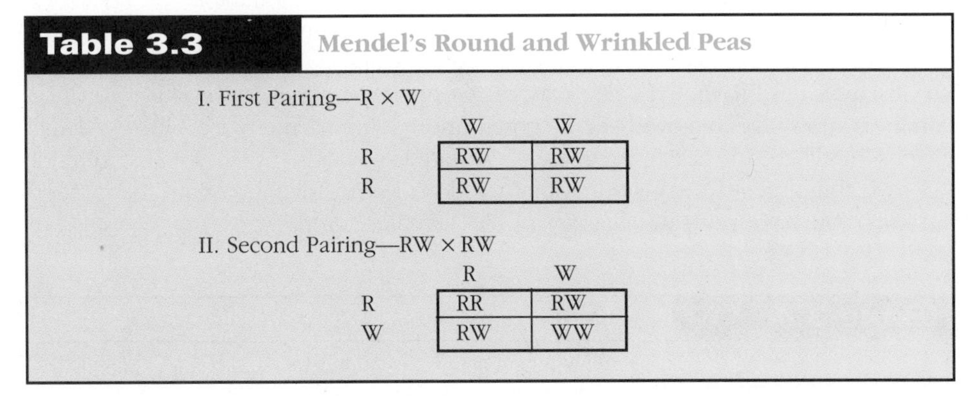

Table 3.3 Mendel's Round and Wrinkled Peas

I. First Pairing—R × W

	W	W
R	RW	RW
R	RW	RW

II. Second Pairing—RW × RW

	R	W
R	RR	RW
W	RW	WW

What Is Transmitted

Is it a boy or a girl? Is it healthy? These are the two questions asked most frequently by parents following the birth of a child. As mentioned earlier, the 23d pair of chromosomes determines the sex of a child: XY = boy; XX = girl. Remember: it is the male who determines the baby's sex, since the egg always carries the X chromosome, while the sperm may contain either an X or a Y.

Sperm X + egg X = female
Sperm Y + egg X = male

There also occurs what is known as **sex-linked inheritance.** If you recall, the X chromosome is substantially larger than the Y (about three times as large). Therefore the female carries more genes on the 23d chromosome than the male. This difference helps to explain sex linkage (Berndt, 1992).

Think back now to the difference between dominant and recessive traits. If a dominant and recessive gene appear together, the dominant trait is expressed. There must be two recessive genes for the recessive trait (say, blue eyes) to appear.

But on the 23d set of chromosomes, there is nothing on the Y chromosome to offset the effects of a gene on the X chromosome. As far as we know today, the Y chromosome almost completely lacks genes except for the one that determines maleness—with the possible exception of a condition producing exceptionally hairy ears (Singer, 1985).

Perhaps the most widely known of these sex-linked characteristics is **hemophilia** (the bleeder's disease). The blood of hemophiliacs does not clot properly. Several of the royal families of Europe were particularly prone to this condition. Another sex-linked trait attributed to the X chromosome is **color blindness.** The X chromosome contains the gene for color vision and, if it is faulty, there is nothing on the Y chromosome to counterbalance the defect.

In 1970, a condition called **fragile X syndrome** was discovered. The end of the X chromosome looks ready to break off. Although fragile X seems to cause no physical problems, about 80 percent of these boys are mentally retarded. Fragile X appears in about 1 in 2,000 live births.

Prospective parents also usually speculate about the color of their child's eyes. What is the genetic explanation for the presence of different eye colors? There is a gene responsible for blue eyes and another responsible for brown eyes. The gene for brown eyes is dominant and the gene for blue eyes is recessive. For example, a child receiving two "brown" genes will have brown eyes; a child receiving two "blue" genes will have blue eyes. Finally, the youngster who receives one of each will have brown eyes, because the gene for brown eyes is dominant.

The basic principle is that genes producing dark eye colors (brown, black) dominate over those producing light eye colors (blue, green, gray). If the mother, for instance, has brown eyes, she may carry only those genes for brown eyes (if her family is mostly brown-eyed) or a brown and a blue (if there are blue-eyed relatives somewhere in the family tree). If the father has blue eyes, he most certainly has two "blue" genes. Let's assume that we don't know anything about the mother's family. Table 3.4 reports the chances of blue-eyed children.

Table 3.4 Predicting Eye Color

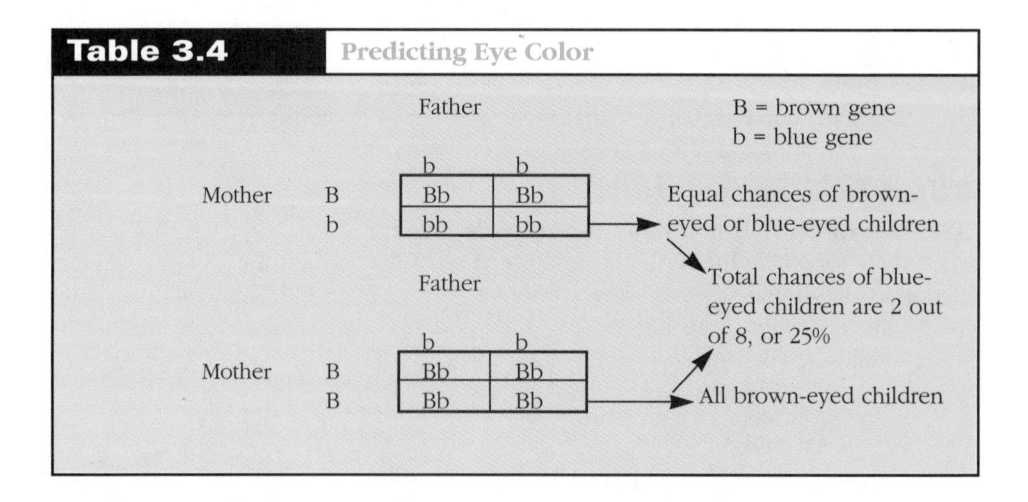

Examples of Chromosomal Disorders

Although the chance of a chromosomal error in pregnancy is small, the risk increases with the age of the woman. Increasing age of the father may also be a factor, though less significant (Tapley & Todd, 1988). (See Table 3.5.)

Table 3.5 The Risk of Abnormalities

The risk of Down syndrome and other chromosomal abnormalities increases with the mother's age.

Risk of Significant Chromosomal Abnormality	Risk of Down Syndrome	Maternal Age
1:455	1:1,000	15 or under
1:526	1:1,429	18
1:526	1:1,429–1:2,000	21
1:476	1:1,111–1:1,429	24
1:455	1:1,000–1:1,250	27
1:385	1:833–1:1,111	30
1:323	1:667–1:909	32
1:244	1:417–1:526	34
1:179	1:256–1:400	35
1:149	1:200–1:313	36
1:123	1:156–1:244	37
1:105	1:123–1:192	38
1:81	1:95–1:152	39
1:63	1:73–1:118	40
1:49	1:56–1:93	41
1:39	1:43–1:72	42
1:31	1:33–1:57	43
1:24	1:25–1:44	44
1:18	1:19–1:35	45
1:15	1:15–1:27	46
1:11	1:11–1:21	47
1:9	1:9–1:17	48
1:7	1:6–1:13	49

Reprinted with permission from The American College of Obstetrics and Gynecologists (*Obstetrics and Gynecology,* 1981, Vol. 58, No. 3, pages 282-285).

Down Syndrome

Down syndrome is caused by a deviation on the 21st pair of chromosomes; the individual may have 47 chromosomes. This defect was discovered in 1866 by a British doctor, Langdon Down, and produces distinctive facial features, small hands, a large tongue, and possible functional difficulties such as mental retardation, heart defects, and an added risk of leukemia.

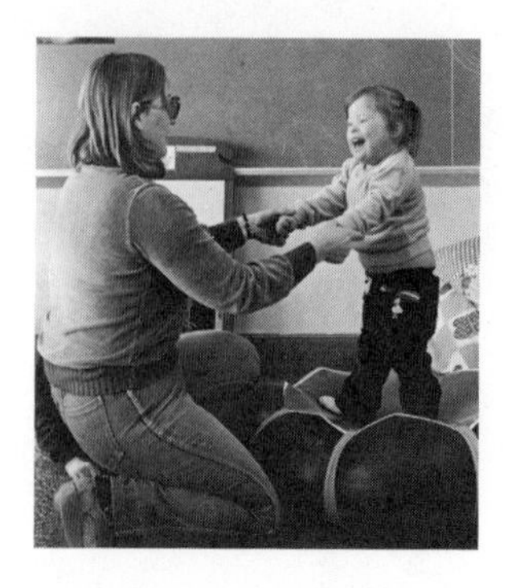

Down syndrome is caused by a deviation on the 21st pair of chromosomes. These children have distinctive facial features and are usually motorically and mentally retarded.

The appearance of Down syndrome is closely related to the mother's age: chances are about 1 in 750 between the ages of 30 and 35; 1 in 300 between 35 and 39; 1 in 80 between 40 and 45; and after 45 years the incidence jumps to 1 in 40 births. Under age 30, the ratio is only 1 in 1,500 births.

Although the exact cause of Down syndrome remains a mystery, the answer may lie in the female egg production mechanism, which results in eggs remaining in the ovary for forty or fifty years. The longer they are in the ovary, the greater the possibility of damage. There is no treatment for Down syndrome other than good medical supervision and special education. These individuals usually are cheerful, perhaps slightly stubborn, with a good sense of mimicry and rhythm. Since the severity of the defect varies, institutionalization of the child is no longer immediately recommended. Some youngsters develop better in the home, especially if the parents believe they can cope successfully.

Other Chromosomal Disorders

If you recall, the 23d pair of chromosomes are the sex chromosomes, XX for females, XY for males. Estimates are that one in every 1,200 females and one in every 400 males has some disorder in the sex chromosomes.

Occasionally a male will possess an XXY pattern rather than the normal XY. This is call Klinefelter syndrome, and eventually it causes small testicles, reduced body hair, and possible infertility (about 1 in 1,000 male births). Another pattern that appears in males is XYY, which may cause larger size and increased aggression (about 1 in 1,000 male births). Heated controversy and inconclusive results have surrounded the study of the "super male."

Females occasionally possess an XO pattern (lack of a chromosome) rather than XX. This is called Turner syndrome and is characterized by short stature, poorly developed secondary sex features (such as breast size), and usually sterility (about 1 in 2,500 female births).

Examples of Genetic Disorders

Jews of Eastern European origin are struck hardest by **Tay-Sachs disease,** which causes death by the age of 4 or 5. At birth the afflicted children appear normal, but development slows by the age of 6 months, and mental and motor deterioration begin. About 1 in every 25 Jews of Eastern European origin carries the defective gene, which is recessive; thus danger arises when two carriers marry. The disease results from a gene failing to produce an enzyme that breaks down fatty materials in the brain and nervous system. The result is that fat accumulates and destroys nerve cells, causing loss of coordination, blindness, and finally death.

Sickle-cell anemia, which mainly afflicts those of African descent, appeared thousands of years ago in equatorial Africa and increased resistance to malaria. Estimates are that 10 percent of the African-American population in the United States carry the sickle-cell trait. Thus, two carriers of the defective gene who marry have a 1 in 4 chance of producing a child with sickle-cell anemia.

The problem is that the red blood cells of the afflicted person are distorted and pointed. Because of their shape, they encounter difficulty in passing through the blood vessels. They tend to pile up and clump, producing oxygen starvation accompanied by considerable pain. The body then acts to eliminate these cells, and anemia results.

In the population of the United States, **cystic fibrosis** is the most severe genetic disease of childhood, affecting about 1 in 1,200 children. About 1 in 30 individuals are carriers. The disease causes a malfunction of the exocrine glands, the glands that secrete tears, sweat, mucus, and saliva. Breathing is difficult because of the thickness of the mucus. The secreted sweat is extremely salty, often producing heat exhaustion. Cystic fibrosis kills more children than any other genetic disease. Today, if diagnosed early and treated properly, those who are afflicted can lead a fairly lengthy life.

Cystic fibrosis is deadly for several reasons: Its causes remain unknown (although recent research has identified a suspect gene), and carriers cannot be detected. It is not until a child manifests the breathing and digestive problems characteristic of the disease that identification is possible. New research, however, offers hope.

Phenylketonuria (PKU) results from the body's failure to break down the amino acid phenylalanine, which then accumulates, affects the nervous system, and causes mental retardation. Most states now require infants to be tested at birth. If PKU is present, the infants are placed on a special diet that has been remarkably successful.

But this success has produced future problems. Women treated successfully as infants may give birth to retarded children because of a toxic uterine environment. Thus at the first signs of pregnancy, these women must return to a special diet. The "cured" phenylketonuric still carries the faulty genes.

Spina bifida (failure of the spinal column to enclose completely) is an example of a genetic defect caused by the interaction of several genes. During the first few weeks following fertilization, the mesoderm sends a chemical signal to the ectoderm that causes the beginnings of the nervous system. The process is as follows:

1. The chemical signal is sent from the mesoderm to the ectoderm.
2. A tube (the neural tube) begins to form, from which the brain and spinal cord develop.
3. Nerve cells are formed within the tube and begin to move to other parts of the developing brain.
4. These neurons now begin to form connections with other neurons.
5. Some of the neurons that don't connect with other neurons die.

If the neural tube does not close, spina bifida results, which can cause mental retardation.

While these are the more frequent chromosomal and genetic diseases, other diseases also have, or are suspected of having, a strong genetic origin: diabetes, epilepsy, heart disorders, cancer, arthritis, and some mental illnesses. (See Table 3.6 for a summary of this discussion.)

Conclusion

In this chapter we explored the biological basis of our uniqueness. We considered not only the power and beauty of nature in establishing our genetic endowment but also the growing influence of technology.

The genes provided by the mother and father unite to produce a new and different human being. Yet this new life still shows many of the characteristics of both parents. We saw how this newness and sameness challenged researchers for decades.

Table 3.6 Chromosomal and Genetic Disorders

Name	Effects	Incidence
Chromosomal Disorders		
Down syndrome	47 chromosomes; distinctive facial features, mental retardation, possible physical problems	Varies with age; older women more susceptible (under 30—1 in 1,500 births, 35–39—1 in 300 births)
Klinefelter syndrome	XXY chromosomal pattern in males; possible infertility, possible psychological problems	About 1 in 1,000 male births
Turner syndrome	One X chromosome missing in female; lack of secondary sex characteristics, infertile	About 1 in 2,500 female births
XYY syndrome	XYY chromosomal pattern in males; tend to be large, normal intelligence and behavior	About 1 in 1,000 male births
Genetic Disorders		
Tay-Sachs disease	Failure to break down fatty material in central nervous system; results in blindness, mental retardation, and death (usually by 4 or 5 years of age)	One in 25 Jews of Eastern European origin is a carrier
Sickle-cell anemia	Blood disorder producing anemia and considerable pain; caused by sickle shape of red blood cells	About 1 in 10 African-Americans is a carrier
Cystic fibrosis	Body produces excessive mucus, causing problems in the lungs and digestive tract; may be fatal, suspect gene recently identified	About 1 in 2,000 births
Phenylketonuria (PKU)	Body fails to break down amino acid (phenylalinine); results in mental retardation, treated by special diet	About 1 in 15,000 births
Spina bifida	Neural tube problem in which the developing spinal column does not close properly; may cause partial paralysis and mental retardation	About 1 in 1,000 births (depending on geographical area)

Beginning with the discoveries of Mendel and continuing today, the secrets of hereditary transmission have intrigued scientists to this day. Today's work, building on our knowledge of DNA, provides hope for the future while simultaneously raising legal and ethical questions that have yet to be resolved.

Following fertilization—either by natural or external methods—there begins a 9-month period (at least for most children) that concludes with expulsion into the waiting world. It is this period that occupies us in chapter 4.

Chapter Highlights

Heredity and Environment

- Any analysis of development must be from the perspective of heredity *and* environment, that is, by using the biopsychosocial model.
- Development is the expression of our genes (genotype) as a form of behavior (phenotype).

The Beginnings

- Studies of fertilization now must include both internal and external techniques.
- Sperm and egg unite to produce a zygote.

The Fertilization Process

- Knowledge of hormonal control of the menstrual cycle is crucial for understanding fertilization.
- The study of twins, especially monozygotic twins, has long fascinated psychologists.
- The increasing number of infertile couples has led to a growing demand for external fertilization.
- The most widely used external fertilization technique is AID (artificial insemination by donor).
- The success rate of external fertilization procedures has improved with increasing knowledge.
- Today's adoption procedures include both closed and open adoption.

Heredity at Work

- Mitosis and meiosis are the means of cell division.
- The Human Genome Project is an endeavor to identify and map all human genes.
- Understanding how traits are transmitted requires a knowledge of the workings of dominant and recessive genes.
- Chromosomal as well as genetic abnormalities can occur.

Key Terms

Artificial insemination by donor (AID)
Amniocentesis
Blastocyst
Color blindness
Cystic fibrosis (CF)
Dizygotic
DNA
Dominant
Donor insemination (DI)
Down syndrome
Endometriosis
External fertilization
Fallopian tubes
Fragile X syndrome
Fraternal
Gamete intrafallopian transfer (GIFT)
Genotype
Hemophilia
Human Genome Project
Identical
Implantation
Infertility
Internal fertilization
Intrauterine device (IUD)
In vitro fertilization (IVF)
Klinefelter syndrome
Laparoscope
Meiosis
Menstrual cycle
Mitosis
Monozygotic
Non-identical
Nucleotides
Ovulation
Ovum
Pelvic inflammatory disease (PID)
Phenotype
Phenylketonuria (PKU)
Recessive
Sex-linked inheritance
Sickle-cell anemia
Sperm
Spina bifida
Tay-Sachs disease
Turner syndrome
Zygote
Zygote intrafallopian transfer (ZIFT)

What Do You Think?

1. You may have begun your work in lifespan psychology with the idea that either heredity or environment was all-important. Do you still think that way? Or do you believe that one may be somewhat more important than the other? Does the biopsychosocial model help you to answer these questions?
2. There have been several controversies lately about surrogate mothers and the children they bear. Do you have any strong feelings about surrogacy? Can you defend it? Regardless of your personal feelings, can you present what you see as the pros and cons of surrogacy?
3. In your reading, perhaps you noticed that the process of in vitro fertilization depended on research findings from studies of the menstrual cycle. Can you explain this, paying particular attention to the administering of hormones and the timing of their administering?
4. James Watson and Francis Crick received a Nobel Prize for their discovery of the double-helix structure of DNA. Why is the discovery of DNA so important in our lives? Can you think of anything you have read about in the newspapers or seen on television that derives from this discovery?
5. Given what you know about the role of heredity in development, how would you evaluate the importance of genetic counseling? If you were thinking of having children, and you were concerned about the genetic background of yourself or your partner, would you seek genetic counseling?

Suggested Readings

Bishop, J. and Waldholz, M. (1990). *Genome.* New York: Simon & Schuster. An excellent, readable account of the Human Genome Project. The personal histories are especially appealing.

Sagan, C. (1977). *The dragons of Eden.* New York: Random House. Sagan's penetrating and well-written account of genes and their function offers an intriguing entry into the world of heredity. (Available in paperback.)

Watson, J. (1968). *The double helix.* Boston: Atheneum. This personal, colorful look at the discovery of DNA is guaranteed to hold your attention. If you are the least bit intimidated by the thought of reading about your genetic heritage, this book should eliminate your concerns. (Available in paperback.)

CHAPTER 4

Pregnancy and Birth

The Prenatal World 82

The Germinal Period 83

The Embryonic Period 84

The Fetal Period 85

Development of the Senses 86

Influences on Prenatal Development 89

Developmental Risk 89

Teratogens 90

Maternal Influences 96

Fetal Problems: Diagnosis and Counseling 99

The Birth Process 100

Stages in the Birth Process 100

Myths and Facts of Birth 101

Birth Complications 103

Childbirth Strategies 104

The Special Case of Prematurity 104

Conclusion 107

Chapter Highlights 107

Key Terms 108

What Do You Think? 108

Suggested Readings 109

Ellen and Kevin were delighted. The parents of a 3-year-old boy, they were now looking forward to their second child. Kevin, having shared in the birth of their first child, was calmer but even more excited as he looked forward to the events of this pregnancy and birth. Ellen, a healthy 31-year-old, was experiencing all the signs of a normal pregnancy. The morning sickness abated at 12 weeks. She felt movement at 17 weeks, and an ultrasound at 20 weeks showed normal development. Weight gain, blood glucose levels, blood pressure, and AFP (**alpha-fetoprotein**) test results were all within acceptable ranges.

Since this was Ellen's second pregnancy, she felt more comfortable with her changing body. Visits to the obstetrician were pleasant and uneventful. At 28 weeks, loosening ligaments caused her pelvis to irritate the sciatic nerve at the base of the spine. Her obstetrician recommended Tylenol, a heating pad, and rest. He also recommended a visit to an orthopedist to confirm the treatment. Ellen refused to take any medication and decided not to bother with a second opinion. Since she would not be X-rayed or take even a mild painkiller, she felt she could tolerate the pain for the 12 weeks until delivery. She rationalized that the back pain was acceptable because it accompanied a normal pregnancy.

At 31 weeks, Ellen noticed episodes of unusual movement and became concerned. A week later, fearing that the baby might be in distress, she called the obstetrician to describe the jabs, pokes, and excessive movements she was experiencing. The doctor immediately scheduled a biophysical profile: an eight-point check of the internal organs and another ultrasound.

Ellen nervously gulped the required 32 ounces of water an hour before the examination. Arriving at a glistening new medical center, she was quickly escorted to an examining room. The technician was both professional and serious as he looked for a problem (increasing Ellen's anxiety). The examination included a thorough check of the baby's heart chambers, a type of EKG (electrocardiogram) measurement, and an assessment of blood flow—all displayed in vibrant colors. The cord, internal organs, position of the baby, and body weight were all evaluated. Ellen was able to listen to the baby's heartbeat while watching the heart chambers function.

Later that day, Ellen's obstetrician called to tell her the results: no apparent medical problems, just an unusually active baby. Ellen and Kevin sagged with relief. They also received a bonus: a reassuring ultrasound image of their unborn child's face—in living color!

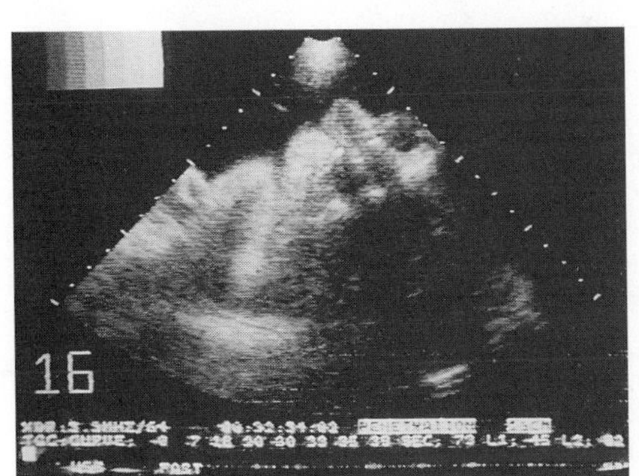

Ultrasound is frequently used when questions arise about a pregnancy. Soundwaves directed onto the uterus bounce off the bones and tissues of the fetus and are formed into an image.

Ellen's journey through the nine months of her pregnancy is quite typical: normal prenatal development accompanied by occasional worrisome moments. As you read about her journey in this chapter, you'll discover that a myth is being destroyed. In fact, several myths are on the road to extinction. The notion that babies at birth know nothing and are passive sponges has been totally disproved. We now know that newborns are amazingly competent. After all, they have been flourishing in a highly protected environment, sometimes humorously called "the prenatal university," for nine months. Researchers, however, are asking serious questions about the potential of prenatal life. The basic issue is this: can fetal enrichment (a fancy name for fetal education) produce a better baby? ■

Is it fact or fancy to think that prenatal learning is a possibility? While we cannot as yet answer this question with certainty, several clues can help us. First, there is little doubt that the senses (hearing, taste, smell, touch, vision) develop during pregnancy; they do not appear instantly at birth. Taste and smell are present by 20 weeks; touch appears at about 25 weeks; fetuses respond to sound by about 27 or 28 weeks; brain life begins by the seventh month. EEGs (electroencephalograms) taken just before birth show brain waves similar to those of infants.

One of the leading centers of research into fetal life is at the University of North Carolina. Here, Anthony DeCasper and his colleagues are attempting to piece together the puzzle of fetal learning. For example, they had infants suck on a nipple attached to a tape recorder. Fast sucking activated a recording of their mothers' voices, while slower sucking produced another woman's voice. The babies showed a preference for their mothers' voices, as indicated by fast sucking.

Taking these results one step further, DeCasper asked a group of pregnant women to read a children's story aloud twice a day for six weeks prior to their expected delivery day. At birth, the infants' patterns showed that they preferred the familiar story to one they had never heard (DeCasper & Fifer, 1980). Obviously, there has been some fetal "learning." Results are sufficiently promising to stimulate continued research.

We'll first explore the prenatal world, that nine-month period that provides nourishment and protection and serves as a springboard for birth. Next, we'll turn to those agents that can influence prenatal development. These are both physical and psychological, and can be either positive or negative. We'll then look at birth itself, the completion of a journey that has involved remarkable development. Finally, for various reasons, some fetuses can't endure this nine-month journey. These early births are called premature. In the past few years, great advances—technological, medical, and psychological—have resulted in more and more of these babies surviving.

When you complete your reading of this chapter you should be able to:

- Describe the periods of prenatal development
- Analyze the major features of each period
- Indicate the times of greatest sensitivity to insult
- Identify dangerous maternal diseases
- Distinguish those drugs that can cause permanent damage during prenatal development
- Assess the potential of fetal surgery
- Discriminate the various stages of the birth process
- Designate the possible causes of prematurity

The Prenatal World

While it may be difficult to imagine, you are the product of one cell, the *zygote,* or fertilized egg. Once the union of sperm and egg took place, it was only a matter of hours (about 24–30) before that one cell began to divide rapidly. The initial phase of the event occurred in a very protected world—the prenatal environment.

The fertilized egg must pass through the fallopian tube to reach the uterus, a journey of about seven days to travel five or six inches. Hormones released by the ovary stimulate the muscles of the fallopian tube wall so that it gently pushes the zygote toward the uterus. During its passage through the fallopian tube, the zygote receives all of its nourishment from the tube. Figure 4.1 illustrates passage into the uterus and implantation.

Implantation seems to occur in three phases:

1. **Apposition,** during which the fertilized egg, now called a blastocyst, comes to rest against the uterine wall.
2. **Adhesion,** during which the prepared surface of the uterus and the outer surface of the fertilized egg, now called the **trophoblast,** touch and actually "stick together."
3. **Invasion,** during which the trophoblast digs in and begins to bury itself in the uterine lining (Beaconsfield, Birdwood, & Beaconsfield, 1983).

Two events in this process are particularly remarkable:

- During the invasion phase, the trophoblast penetrates the uterine lining only so far and then stops. The distance varies depending on the species. Why? As yet, we have no answers to this question.
- Why doesn't the mother reject this invasive tissue? Again, there are no answers to this question.

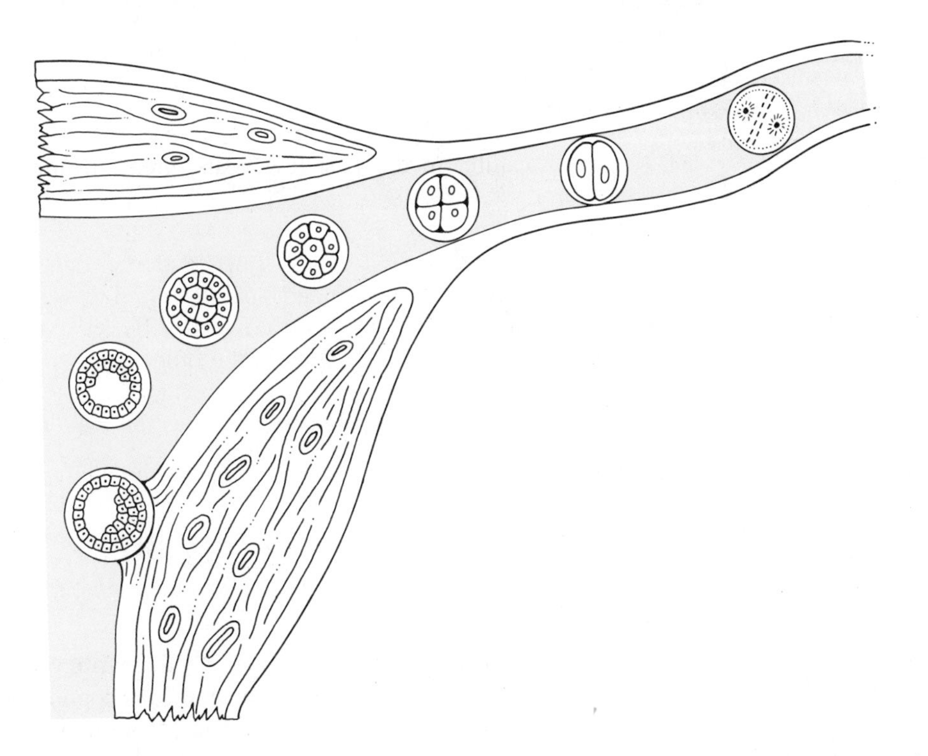

Figure 4.1
Passage of the zygote into the uterus

We can identify three fairly distinct stages of prenatal development: germinal, embryonic, and fetal.

The Germinal Period

The **germinal period** extends through the first two weeks. Since the passage through the fallopian tube takes seven days, the zygote is now one week old and called a blastocyst. During the second week, the blastocyst becomes firmly implanted in the wall of the uterus. From its outer layer of cells, a primitive **placenta,** an

umbilical cord, and the **amniotic sac** begin to develop. The inner cell layer develops into the embryo itself. Figure 4.2 illustrates the developmental significance of the blastocyst.

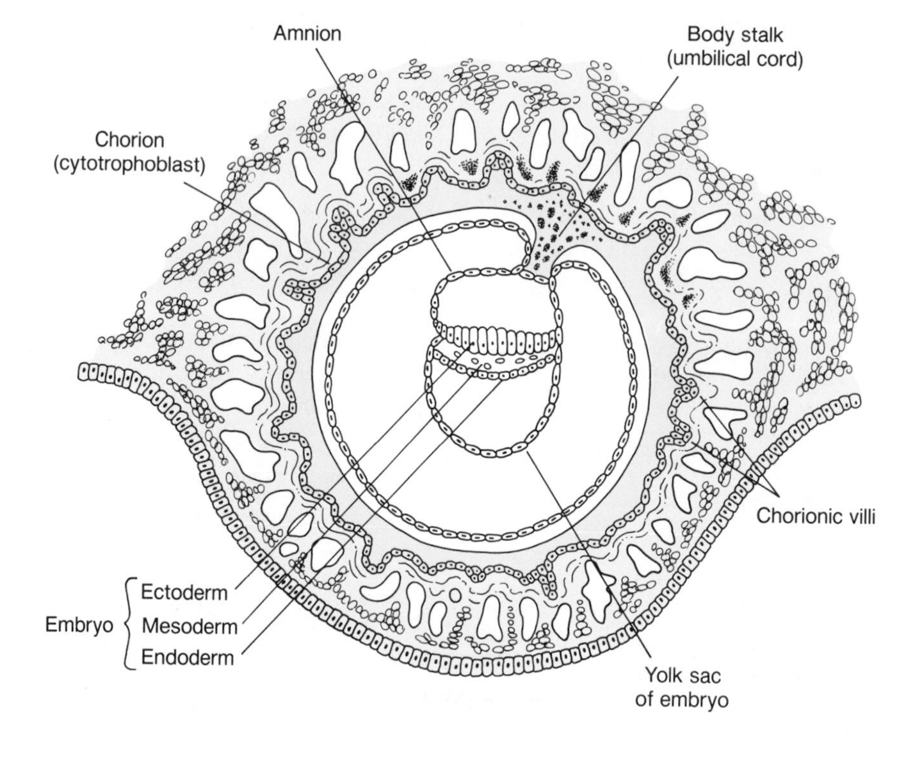

Figure 4.2
During the second week, the blastocyst becomes firmly implanted in the wall of the uterus and the placenta, umbilical cord, and embryo itself begin to form from its outer layer of cells.

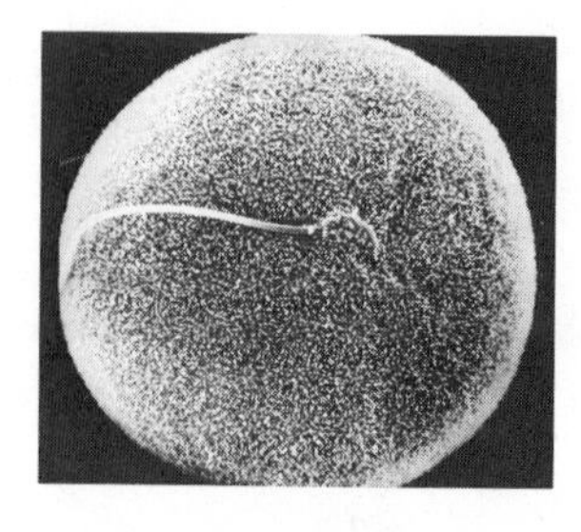

Fertilization through the embryonic period:
(a) The moment of fertilization.

The placenta and the umbilical cord serve a critical function during developmenlt. The placenta supplies the embryo with all its needs, carries off all its wastes, and protects it from danger. The placenta has two separate sets of blood vessels, one going to and from the baby through the umbilical cord, the other going to and from the mother through the arteries and veins supplying the placenta.

We can summarize the first two weeks following conception as follows:

Week 1: Movement of the zygote through the fallopian tube to the uterus; continued cell division

Week 2: Blastocyst adheres to uterine wall; forms placenta, umbilical cord, amniotic sac

The Embryonic Period

When the second week ends, the germinal period is complete. In the **embryonic period,** from the third through the eighth week, development of a recognizable human being begins. The nervous system develops rapidly, which suggests that the embryo at this time is quite sensitive to any obstructions to its growth. Our earlier discussion of sensitive periods as a time when certain experiences have a significant impact on the developing organism is clearly supported during the embryonic period. With the rapid formation of many different organ systems, any negative agent (drugs, disease) can have long-lasting effects.

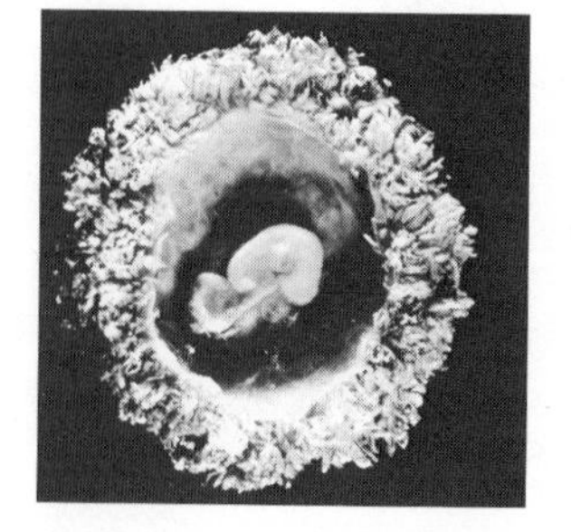

(b) This 4-week-old embryo now has a beating heart, body buds are beginning to emerge, and the eye region is becoming discernible.

Perhaps the most remarkable change in the embryo is cellular differentiation. Three distinct layers are being formed: the **ectoderm,** which will give rise to skin, hair, nails, teeth, and nervous system; the **mesoderm,** which will give rise to muscles, skeleton, and the circulatory and excretory systems; and the **endoderm,** which will give rise to lungs, liver, and pancreas. (See figure 4.3 for details.)

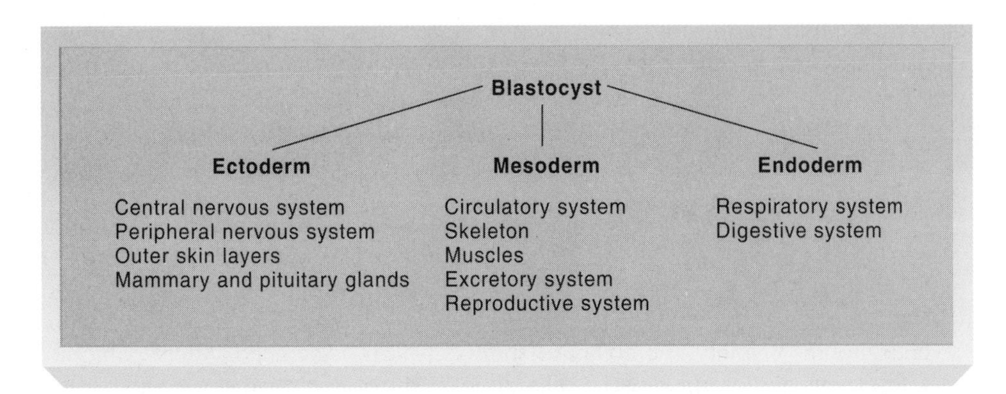

Figure 4.3
Development from the three layers of the blastocyst

Usually by the completion of the fourth week, the heart begins to beat—the embryo's first movement. The accompanying photographs show that during the fifth week eyes and ears begin to emerge, bodily buds give clear evidence of becoming arms and legs, and the head area is the largest part of the rapidly growing embryo.

During the sixth and seventh weeks, fingers begin to appear on the hands, the outline of toes is seen, and the beginnings of the spinal cord are visible. During the germinal period, the number and differentiation of cells rapidly increase; in the embryonic period, the organs are formed **(organogenesis)** (Sadler, 1985).

After eight weeks, 95 percent of the body parts are formed and general body movements are detected. During these weeks embryonic tissue is particularly sensitive to any foreign agents during differentiation, especially beginning at the third or fourth week of the pregnancy.

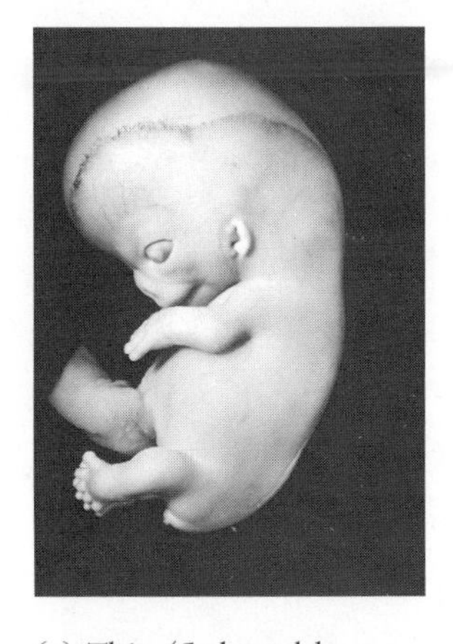

(c) This 45-day-old embryo has visible arms and legs, with the beginnings of toes and fingers. It is possible to detect eyes and ears in the rapidly growing head area.

We can summarize the embryonic period as follows:

Weeks 3+: Rapid development of nervous system
Week 4: Heart beats
Week 5: Eyes and ears begin to emerge, bodily buds for arms and legs
Week 6 and 7: Fingers and toes, beginning of spinal cord
Week 8: About 95 percent of body parts differentiated—arms, legs, beating heart, nervous system

The embryonic period can be hazardous for the newly formed organism. Estimates are that about 30 percent of all embryos are aborted at this time without the mother's knowledge; about 90 percent of all embryos with chromosomal abnormalities are spontaneously aborted (Travers, 1982).

At the end of this period there is a discernible human being with arms, legs, a beating heart, and a nervous system. It is receiving nourishment and discharging waste through the umbilical cord, which leads to the placenta. The placenta itself never actually joins with the uterus, but exchanges nourishment and waste products through the walls of the blood vessels (Guttmacher & Kaiser, 1986). The future mother begins to experience some of the noticeable effects of pregnancy: the need to urinate more frequently, morning sickness, and increasing fullness of the breasts.

The Fetal Period

The **fetal period** extends from the beginning of the third month to birth. During this time, the fetus grows rapidly both in height and weight. The sex organs appear during the third month, and it is possible to determine the baby's sex. Visible sexual differentiation begins, while the nervous system continues to increase in size and complexity.

By the fourth month, the fetus is about one-half its birth size. The fourth to the fifth month is usually the peak growth period. During this time, the mother begins to feel movement. The fetus now swallows, digests, and discharges urine.

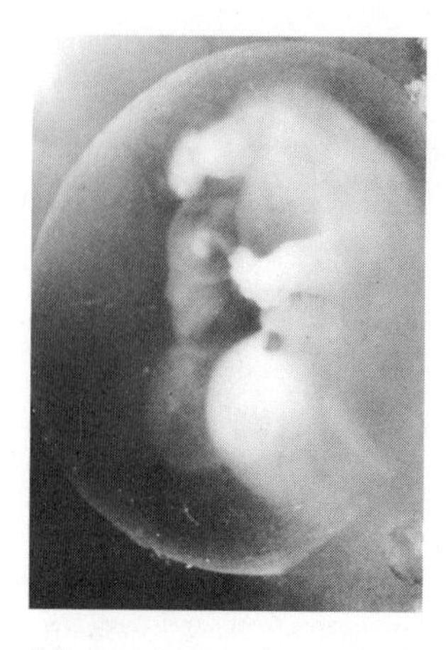

(d) Coming to the end of the embryonic period, this 7-week-old embryo has begun to assume a more human appearance. It is now about 1 inch in length with discernible eyes, ears, nose, mouth, arms, and legs.

Growth is rapid during the fourth month to accommodate an increasing oxygen demand (Nilsson & others, 1987). The fetus produces specialized cells: red blood cells to transport oxygen and white blood cells to combat disease.

The fetus is now active—sucking, turning its head, and pushing with hands and feet—and the mother is acutely aware of the life within her. Figure 4.4 represents the fetus in the fourth month.

By the end of the fifth month, the baby is ten to twelve inches long and weighs about a pound. The fetus sleeps and wakes like the newborn does, even manifesting a favorite sleep position. Rapid growth continues in the sixth month, with the fetus gaining another two inches and one pound, but slows during the seventh month. Viability, the ability to survive if born, is attained. After six months very few new nerve and muscle cells appear, since at birth the nervous system must be fully functioning to ensure automatic breathing.

During the two final fetal months, organ development prepares the fetus for the shock of leaving the sheltered uterine world. The senses are ready to function (some are already functioning). For example, the fetus is able to hear sound—the silent world of the fetus is a myth. The fetus hears many environmental sounds—voices, stomach rumblings, and the pulsing of the mother's blood.

We can summarize these developments as follows:

Third month: Sex organs appear
Fourth month: Rapid growth, red blood cells, white blood cells; active sucking
Fifth month: Hears sound, sleeps, 10–12 inches long, 1 pound
Sixth month: Rapid growth, 12–14 inches, 2 pounds
Seventh month: Growth slows, viability attained
Eighth & ninth months: Preparation for birth; senses ready to function, brain is 25 percent of adult weight

At the end of the ninth month, the fetus (just before birth) is about 20 inches long, weighs about 7 pounds, 6 ounces, and its brain at birth is 20 to 25 percent of its adult weight.

Few babies (about 1 in 20) are born on the day predicted. There are several reasons for this discrepancy. The varying length of the menstrual cycles can affect estimates of the time of ovulation. Also, conception occurs at different times. Finally, fetuses mature at different rates, which affects fetal size and the onset of labor.

Most women begin to experience some discomfort as the time of birth approaches. The extra weight, body changes, and sheer effort of movement all contribute to this discomfort. During this period of preparation, the major influence on the growing child is its mother. If the mother is healthy, happy, and reasonably cautious, both she and her child will be the beneficiaries. Table 4.1 summarizes the course of prenatal development.

Figure 4.4
The fetus at 4½ months–a time of rapid growth and considerable activity

Development of the Senses

There has been considerable speculation about the role of the senses before birth. The developmental path is as follows:

- *Touch.* Touch refers to the reaction to pressure, temperature, and pain, and seems to produce a generalized response. If a stimulus is applied to the fetus, there is a definite reaction. The specialized skin senses are capable of functioning long before birth.

Table 4.1 Prenatal Development

Name	Development
Zygote	Fertilization (union of sperm and egg)
	Cleavage (about 30 hr)
Blastocyst	Rapid cell division (about four days)
	Implantation (about sixth or seventh day)
Embryo	Begins at third week
	Central nervous system grows rapidly
	Heart begins to beat at about 28 days
	Digestive organs form during the second month
	Muscular system appears during the second month
	Sex differentiation at about seven weeks
	95% of body parts differentiated at the end of the eighth week
Fetus	Organs begin to function from 8–12 weeks
	Rapid growth during fourth month
	Viability at about 27 weeks
	Exercises functioning systems until birth

- *Taste.* Taste buds appear as early as the third fetal month and seem to be more widely distributed in fetal life than in adult life. Initially, taste buds appear on the tonsils, palate, and part of the esophagus. In the adult, taste cells are restricted to the tongue. Although the mechanism for taste is present before birth, the presence of the amniotic fluid limits the taste response until after birth.
- *Smell.* Like taste, the neurological basis for smell appears before birth; thus there exists the possibility of response. But since the nasal cavity is filled with amniotic fluid, it is not likely that the sense of smell functions before birth. Premature infants, in the last month, can smell substances when air enters the nasal cavity.
- *Hearing.* As previously noted, most fetuses can hear sound by the fourth month. The auditory mechanism is well developed structurally in later fetal life, but since the middle ear is filled with fluid, the fetus cannot respond to sounds of normal intensity. Strong auditory stimuli, however, produce a response.
- *Vision.* There is general agreement that the absence of adequate retinal stimulation eliminates the possibility of true sight during prenatal life. Muscular development enables the fetus to move its eyes while changing body position, but little is known about what or how much the fetus sees. By the end of pregnancy, the uterus and the mother's abdominal wall may be stretched so thin that light filters through, exposing the fetus to some light and dark contrast. Visual development begins in the second week following fertilization and continues until after birth. At birth, the eye is sufficiently developed to differentiate light and dark.

What's Your View?

The first meeting of the Fetal Medicine and Surgery Society took place in 1982. In slightly more than a decade, we have seen the emergence of several new surgical techniques (Kolata, 1990). As amazing as it may sound, it is now possible to operate on a fetus. Fetal surgery has saved several lives and with continued refinement promises a healthy future for many babies. Three types of fetal surgery have received considerable attention:

- To cure a condition in which the brain ventricles fill with fluid and expand (called **fetal hydrocephalus**), surgeons now operate within the womb. They must pierce the woman's abdomen and uterine wall and penetrate the fetal skull. They then insert a tube (*catheter*) into the brain to drain this region until birth. Unless this condition is treated, fluid presses against the walls (membranes) of the fetal brain ventricles and can cause mental retardation or even death.
- To correct a blocked urinary tract in the fetus (called **fetal hydronephosis**), the surgeon removes the fetus from the womb, operates, and then returns the fetus to the uterus. The surgeon then adds either saline solution (salt water) or amniotic fluid that has been saved and warmed, and finally sews the uterus. This surgery could not be accomplished unless drugs preventing labor were available. If the condition is not corrected, the lungs of the fetus cannot develop. The blocked urinary tract causes urine to stay in the bladder, which can actually burst. Or urine can back up into the kidneys, causing serious damage. The major problem is that the fetus stops producing amniotic fluid, which is mainly fetal urine. The fetus swallows amniotic fluid, which causes the lungs to grow. Without the amniotic fluid, the lungs don't grow; and at birth, the fetus simply can't breathe.
- To correct a condition in which the fetus has a hole in its diaphragm (the muscle separating the abdomen from the chest cavity), surgeons make an incision across the uterus and cut into the fetal chest and abdomen. (This condition is called **diaphragmatic hernia.**) They push back into the abdomen any abdominal organs that might have moved through the hole in the diaphragm into the chest cavity. They then close the hole. Without this surgery, the abdominal organs that have pushed into the chest cavity restrict the growth of the lungs and eventually the fetus cannot breathe.

These three procedures are merely examples of what the future holds: lung and liver transplants using adult tissue. Another intriguing finding of fetal surgery is that cutting fetal tissue leaves no scars. The implications of this discovery for surgery treating such problems as cleft lips and palates are enormous.

These advances, however, have not occurred without giving rise to new ethical problems. Here are a few examples:

- Who decides which fetuses will benefit from surgery and which will not?
- Are the outcomes of fetal surgery sufficiently known to undertake the risk?
- Does this mean that the fetus is an actual patient?
- What are the implications of the growing ability to perform pregnancy reductions? (A woman finds that she has four fetuses, not all of which can survive under these conditions. Who decides which fetuses will be eliminated?)

These are not easy questions. Do you think fetal surgery should continue until these questions are answered? What's your view?

What does all of this mean? In an attempt to put the development of the senses into perspective, Tapley and Todd (1988) note that the foundation of the nervous system forms in the first few weeks following conception; by six weeks the reflexes are active and electrical brain-wave patterns appear. Touching the fetal forehead as early as nine weeks causes the fetus to turn away, and if the soles of its feet are stroked, the toes curl up. The entire body is sensitive to touch, with the exception of the head. By mid-pregnancy, the inner ear is fully developed, and the fetus reacts with movement to external sound.

During amniocentesis, the fetal heart rate changes and movement increases, suggesting that the fetus has sensed tactile stimulation. Muscular development of the eyes enables the fetus to move its eyes during sleep or changing its position; from about the 16th week, the fetus is sensitive to any light that penetrates the uterine wall and the amniotic fluid. Toward the end of pregnancy, a bright light pointed at the mother's abdomen causes the fetus to move. The fetus begins to swallow amniotic fluid early in the pregnancy and demonstrates taste by turning toward and swallowing more of a sweet substance injected into the amniotic fluid.

This summary of fetal life leads to an inevitable conclusion: Given adequate conditions, the fetus at birth is equipped to deal effectively with the transition from its sheltered environment to the extrauterine world.

Influences on Prenatal Development

When we speak of "environmental influences" on children, we usually think of the time beginning at birth. But remember: At birth an infant has already had 9 months of prenatal living, with all of its positive and negative features.

Many women today experience the benefits of the latest research and thinking about prenatal care. Diet, exercise, and rest are all carefully programmed to the needs of the individual woman. (Where women, especially pregnant teenagers, lack such treatment, the rates of prenatal loss, stillbirths, and neonatal (just after birth) mortality are substantially higher.)

Developmental Risk

Before we discuss the possibility of prenatal problems, the concept of **developmental risk** helps to put this topic in perspective. The term obviously is intended to describe children whose well-being is in jeopardy. Although the term has been widely used for only the past 15–20 years, the concept itself has been a matter of concern since the 1920s (Kopp, 1987).

As you can probably guess, developmental risks incorporate a continuum of biological and environmental conditions. These range from the very serious (genetic defects) to the less serious (mild oxygen deprivation at birth). Thus a child may be developmentally at risk.

Kopp (1987) summarizes the findings of decades of research on biologically based risk as follows:

- The earlier the damage (a toxic drug or maternal infection), the greater the chance of negative long-term effects. If you recall our earlier discussion of sensitive periods (see chap. 2), times of rapid growth (especially during the embryonic period with the accelerated development of the central nervous system) are also times of particular susceptibility.
- Risks that arise just before, during, and after birth show the most serious consequences during infancy and early childhood and gradually recede during the school years.

- Recent research has shown that prenatal risk accounts for the largest percentage of individuals with IQs below 50 and with severe neurological and sensory problems.
- A warm and supportive environment helps to lesson any negative long-term effects.
- With the exception of the most severe risks, the basic behaviors of infants remain unaltered but may be delayed. (For example, certain types of brain damage may remain undetected until school entrance.)
- The developmental outcomes of risk are often associated with the support that a child's environment offers.
- Developmental risk may have negative consequences for all behavior, although it is most frequently associated with lack of intellectual achievement.

As you can well imagine, these findings have important social, financial, and political implications. (You may want to discuss these in class.) They also offer hope for the at-risk child: A sensitive environment can help.

Teratogens

With regard to developmental risk, our major concern here is with those substances that exercise their influence in the prenatal environment, a time of increased sensitivity. Teratogenic agents, which are any agents that cause abnormalities, especially demand our attention. **Teratogens** that can cause birth defects are drugs, chemicals, infections, pollutants, or a mother's physical state, such as diabetes. The study of these agents is called teratology (Kelley-Buchanan, 1988). Table 4.2 summarizes several of the more common teratogenic agents and the times of greatest potential risk.

By examining Table 4.2, you can see that these teratogenic agents fall into two classes: infectious diseases and different types of chemicals.

Infectious Diseases

Some diseases that are potentially harmful to the developing fetus and that are acquired either before or during birth are grouped together as the **STORCH** diseases (Blackman, 1984):

Syphilis
Toxoplasmosis
Other infections
Rubella
Cytomegalovirus
Herpes

Syphilis Syphilis is sexually transmitted and, if untreated, may affect the fetus. It makes no difference whether the mother contracted the disease during pregnancy or many years before. If the condition remains untreated, about 50 percent of the infected fetuses will die any time during or after the second trimester. Of those who survive, serious problems such as blindness, mental retardation, and deafness may affect them. Given the advances in antibiotic treatments, the incidence of congenital syphilis has steadily decreased.

Toxoplasmosis Toxoplasmosis is caused by a protozoan (a single-celled microorganism) that is transmitted by many animals, especially cats. Because the infection is usually undetected, the woman may pass the organism to the fetus. The

Table 4.2 Teratogens, Their Effects, and Time of Risk

Agent	Possible Effects	Time of Risk
Alcohol	Fetal alcohol syndrome (FAS), growth retardation, cognitive deficits	Throughout pregnancy
Aspirin	Bleeding problems	Last month, at birth
Cigarettes	Prematurity, lung problems	After 20 weeks
DES	Cancer of female reproductive system	From 3–20 weeks
LSD	Isolated abnormalities	Before conception
Lead	Death, anemia, mental retardation	Throughout pregnancy
Marijuana	Unknown long-term effects, early neurological problems	Throughout pregnancy
Thalidomide	Fetal death, physical and mental abnormalities	The first month
Cocaine	Spontaneous abortion, neurological problems	Throughout pregnancy
AIDS	Growth failure, low birth weight, developmental delay, death from infection	Before conception, throughout pregnancy, during delivery and breast feeding
Rubella	Mental retardation, physical problems, possible death	First three months, may have effects during later months
Syphilis	Death, congenital syphilis, prematurity	From five months on
CMV	Retardation, blindness, deafness	Uncertain, perhaps 4–24 weeks
Herpes simplex	Central nervous system (CNS) damage, prematurity	Potential risk throughout pregnancy and at birth

results include both spontaneous abortions and premature deliveries. Low birth weight, a large liver and spleen, and anemia characterize the disease. Serious long-term consequences include mental retardation, blindness, and cerebral palsy. The incidence of toxoplasmosis is about one or two per thousand live births.

Other Infections This category includes such diseases as influenza, chicken pox, and several rare viruses.

Rubella (German measles) When pregnant women hear the name **German measles** (the technical term is *rubella*), warning signals are raised, and with good reason. German measles is typically a mild childhood disease caused by a virus. Children who become infected develop a slight fever and perhaps swollen glands behind the ears. A rash usually appears at about the second or third day. By the fourth or fifth day, all the symptoms have usually disappeared.

For pregnant women, however, the story is quite different. Women who contract this disease may give birth to a baby with a serious defect: congenital heart disorder, cataracts, deafness, or mental retardation. The risk is especially high if the disease appears early in the pregnancy, when a spontaneous abortion may result. The infection appears in less than one per thousand live births.

If it does occur, estimates of the relationship between damage and timing are as follows (Sadler, 1985):

- 50% chance if the infection occurred during the first four weeks of pregnancy
- 20–25% following infection in the fifth to eighth week
- 10–15% in the ninth to twelfth week
- 5% in the 13th to 16th week
- Damage to the fetus is rare during the 17th week and later

Any woman who has had German measles as a child cannot catch it a second time; she is immune. But it is wise to have a blood test taken to be on the safe side. The American Medical Association recommends that any woman of childbearing age who has not been vaccinated for German measles be immunized. These women should then avoid becoming pregnant for at least three months.

Cytomegalovirus (CMV) This imposing title indicates a disease that can cause damage ranging from mental retardation to blindness, deafness, and even death. It is the most common STORCH infection, with an incidence of ten to twenty per thousand live births. One of the major difficulties in combatting this disease is that it remains unrecognized in pregnant women. Consequently, we do not know the difference in outcome between early and late infection.

Herpes Simplex In the adult, the type I herpes virus usually appears in the mouth, while type II herpes appears in the genital area. If the disease is passed on to the fetus (usually during the passage through the birth canal), a child develops symptoms during the first week following birth. The central nervous system seems to be particularly susceptible to this disease, with serious long-term consequences. The incidence is less than one per thousand live births (Blackman, 1984).

The Special Case of AIDS

The final infectious disease we wish to discuss is **AIDS** (Acquired Immune Deficiency Syndrome). There probably isn't a reader of this text who hasn't heard of AIDS. To give you some idea of the seriousness of the problem, estimates are that for every three children born to mothers with AIDS, one will die and the other two will become orphans. An infected mother can pass the virus to the fetus during pregnancy, during delivery, and after birth, occasionally through breast milk (Kelley-Buchanan, 1988). Through 1991, one million children worldwide have been born with the HIV infection. These figures help to explain the growing and intensifying movement to provide school children with AIDS education as soon as they can grasp the concepts involved.

Effects What is this disease that has caused such worldwide alarm? We know today that AIDS is a venereal disease, and that the virus causing it can lie dormant for years. What triggers full-blown AIDS remains unknown, but remember that AIDS is the end stage of the infection and is not in itself a disease. To help you sort out the various stages, terms, and their meanings, think of AIDS as following a three-stage progression:

1. A person becomes infected with **HIV (Human Immunodeficiency Virus),** commonly called the AIDS virus. This initial stage encompasses by far the largest number of cases. Most of these individuals are sexually active homosexual and bisexual males, plus men and women who abuse intravenous drugs.

   There are four possible ways of sexually transmitting HIV: male to male, male to female, female to female, and female to male (Hochhauser & Rothenberger, 1992). The initial appearance of AIDS was primarily in male homosexuals or bisexuals, but recently the number of cases in this category has dropped due to increased awareness and the use of such drugs as AZT (azidothymidine). While this is encouraging, the number of drug users with AIDS has grown markedly.

A small percentage of HIV-infected individuals at this first stage have received infected blood, are the heterosexual partners of someone with AIDS, or are infants born to infected mothers. Individuals in this stage show laboratory signs of infection. What this means is that they have antibodies to the virus in their blood. They show *no* symptoms, but they can transmit the disease if they engage in high-risk behavior such as anal intercourse, sharing needles, or are the heterosexual partners of non-infected individuals. While no exact time can be given, once individuals are infected, they will probably show positive signs (the antibodies previously mentioned) in about two to eight weeks (Langone, 1991). They are then referred to as HIV positive. A person who is HIV positive does *not* have AIDS.

2. Next, the person develops an AIDS-related complex, often referred to as ARC, which is a more severe, intermediate stage. Currently there seems to be a 20 to 50 percent chance that those who test HIV positive will develop ARC characteristics (although some investigators believe that the chances are almost 100 percent). Defining this stage is difficult, because any associated symptoms are not life-threatening. The same symptoms that appear in HIV-positive individuals may also be present in people who are not infected with HIV (swollen glands, persistent diarrhea, weight loss, night sweats, dry coughing, and lymphadenopathy [swollen lymph nodes]). These same symptoms may also be present in full-blown AIDS victims. Today, many believe that the ARC category should be dropped and these individuals included in stage three AIDS, thus making them eligible for social services and disability allowances.
3. AIDS is the final stage. As the immune system breaks down, infected individuals become subject to opportunistic infections that actually cause death. AID (acquired immune deficiency—note that there is no *S*) does not kill, but it destroys the body's ability to fight off disease. People are identified with AIDS if they test positively to antibodies for the disease, if they show a positive culture for the virus itself, and if they demonstrate a loss of disease-fighting white cells, called T–4 lymphocytes (Langone, 1991). They also have one or several of the following opportunistic diseases:

- Pneumocystis carinii pneumonia (also called PC pneumonia) is probably the most common life-threatening disease affecting AIDS victims.
- Cytomegalovirus (CMV) is a herpes virus that attacks the central nervous system.
- Tuberculosis is an ancient disease that strikes AIDS-infected individuals particularly hard.
- Kaposi's sarcoma is a cancer of the walls of the blood and lymphatic vessels. Homosexual men with AIDS are especially susceptible to this cancer.

Background When an alarming increase in PC pneumonia and Kaposi's sarcoma began appearing in young homosexual men in June 1981, scientists at the Centers for Disease Control immediately speculated that it was caused by a virus.

(Conservative estimates are that viruses cause 60 percent of human diseases.) Viruses are ultramicroscopic parasites, really biochemicals, that require a living cell to flourish and reproduce. Once they invade a cell, they take over the cell's metabolic machinery and reproduce rapidly.

The effects of teratogens: Mothers may pass the AIDS virus to their babies during pregnancy, delivery, and through breast milk. While babies of AIDS-infected mothers may not necessarily receive the virus, those that do are likely to succumb by 5 or 6 years of age.

While all living cells contain both DNA and RNA, viruses may contain *either* DNA or RNA (which makes them so difficult to classify). So viruses have to compensate. If the virus contains DNA, it usurps the cell's metabolic machinery and begins to reproduce the viral DNA. If the virus contains only a strand of RNA (called a retrovirus), it must convert itself into DNA once it infects the host cell—which it does by means of an enzyme called reverse transcriptase (Langone, 1991). The converted viral DNA now moves into the nucleus of the host cell and assumes control of its functioning. As mentioned previously, antibodies will appear in about eight weeks. The person is now HIV positive, and the virus may remain dormant for years. Estimates are that the median incubation period from HIV infection to active AIDS is 9.8 years (Hochhauser & Rothenberger, 1992). During this period, the infected person may pass the virus to someone else.

Once activated, the process that leads to the final stage of AIDS commences. The virus now reproduces rapidly, destroys T–4 cells, and ravages the immune system, thus making the individual susceptible to a wide variety of diseases. When the AIDS virus is fully activated, its victim usually dies within two to three years. With the discovery of new anti-AIDS agents and the innovative use of existing drugs, however (for example, AZT), survival time may be extended. Thus the time from initial infection to death may range from ten to twenty years (Hochhauser & Rothenberger, 1992).

With regard to the fetus, estimates are that an infected mother transmits the virus from 30 to 50 percent of the time. Thus, 50 to 70 percent of these fetuses remain unaffected. When the virus is transmitted, a condition called AIDS embryopathy may develop. This causes growth retardation, small head size (microcephaly), flat nose, and widespread, upward-slanted eyes, among other characteristics. Also associated with this disease are higher rates of preterm disease, low birth weight, and miscarriage.

For those fetuses who become infected, AIDS has a shorter incubation period than for adults. Symptoms may appear as early as six months after birth—and include weight loss, fever, diarrhea, and chronic infections. Once symptoms appear, babies rarely survive more than five to eight months.

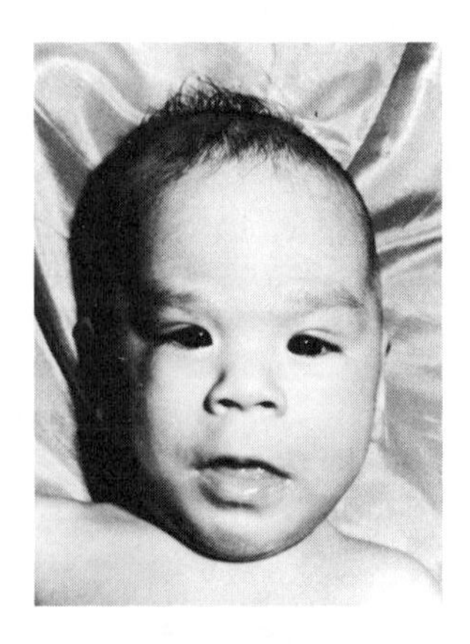

Babies born to women who drank heavily during their pregnancies may manifest distinctive characteristics such as those shown in this photo.

Chemicals

Many women of childbearing age in the United States use one or more of the following drugs: alcohol, cocaine, marijuana, and nicotine. Fifteen percent of these women use drugs with sufficient frequency to cause damage to a fetus during pregnancy. Estimates are that 30 to 40 percent of pregnant women smoke; 60 to 90 percent use analgesics during pregnancy; 20 to 30 percent use sedatives; and an undetermined number continue to use illicit drugs (Stimmel, 1991). Also, a number of women continue to use drugs *before they realize they are pregnant.* Table 4.3 illustrates the potential adverse effects of several classes of drugs.

Prescription drugs such as **thalidomide** have also produced tragic consequences. During the early 1960s, this drug was popular in West Germany as a sleeping pill and an antinausea measure that produced no adverse reactions in women. In 1962, physicians noticed a sizable increase in children born with either partial or no limbs. In some cases, feet and hands were directly attached to the body. Other outcomes were deafness, blindness, and occasionally, mental retardation. In tracing the cause of the outbreak, investigators discovered that the mothers of these children had taken thalidomide early in their pregnancies.

Table 4.3 Adverse Effects of Mood-Altering Drugs on Pregnancy and the Newborn

Drug	Spontaneous Abortion	Premature Delivery	Perinatal Mortality	Neonatal Withdrawal	Fetal Distress	Congenital Abnormality
Alcohol	+		+		+	+
Amphetamines		+	+			
Barbiturates, sedatives, tranquilizers				+	+	
Cannabis					+/-	
Cocaine	+	+	+		+	+
Heroin	+	+	+	+	+	+
Marijuana						+/-
Methadone				+		
Nicotine	+				+	
Phencyclidine		+		+/-	+	

+ Adverse effects
+/- Effects not consistently documented
"The Facts About Drug Use" Copyright 1991 by Consumers Union of U.S., Inc., Yonkers, NY 10703-1057. Reprinted by permission from CONSUMER REPORTS BOOKS, April 1991.

DES (diethylstilbestrol) is another example. In the late 1940s and 1950s, DES (a synthetic hormone) was administered to pregnant women, supposedly to prevent miscarriage. It was later found that the daughters of the women who had received this treatment were more susceptible to vaginal and cervical cancer. These daughters also experienced more miscarriages when pregnant than would be expected. Recent suspicions have arisen about the sons of DES women; they seem to have more abnormalities of their reproductive systems.

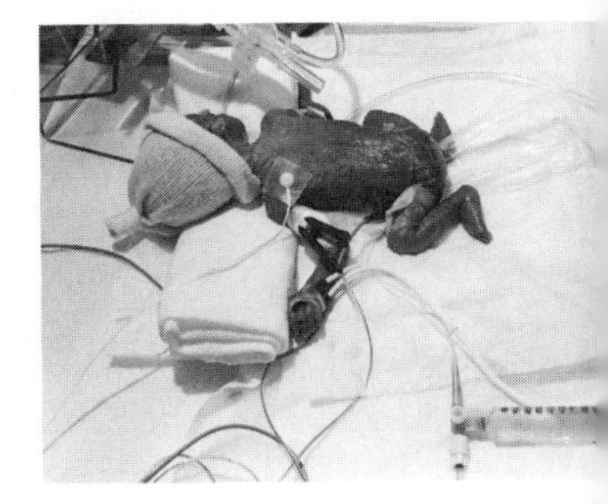

This cocaine-addicted baby was born prematurely and suffers from such behavior disturbances as tremulousness, irritability, and muscular rigidity.

As knowledge of the damaging effect of these agents spreads, women have grown more cautious once they realize they are pregnant. We know now that these agents pass through the placenta and affect the growing embryo and fetus. We also know that certain prenatal periods are more susceptible to damage than others; for example, the embryonic period. Figure 4.5 illustrates times of greater and lesser vulnerability.

To keep a pregnancy as safe as possible, a woman should begin by avoiding the obvious hazards. For example, as Tapley and Todd (1988, pp. 103–104) note, smoking negatively affects everything about the reproduction process: fertility, conception, possible spontaneous abortion, fetal development, labor and delivery, and a child's maturation. Smoking is probably the most common environmental hazard in pregnancy, and it results in a smaller than normal fetus. Babies of smoking mothers may have breathing difficulties and low resistance to infection, and they seem to suffer long-lasting effects after birth.

Other agents—such as alcohol (see pp. 106), almost all drugs (including aspirin), unnecessary medication, and risky chemicals at work or at home—should be avoided (Tapley & Todd, 1988). Most pregnant women today are also cautious about the amount of caffeine and sweeteners they use (Guttmacher & Kaiser, 1987). For example, the FDA has cautioned pregnant women to moderate their consumption of caffeine-containing foods and beverages. These simple precautions will eliminate danger for most women.

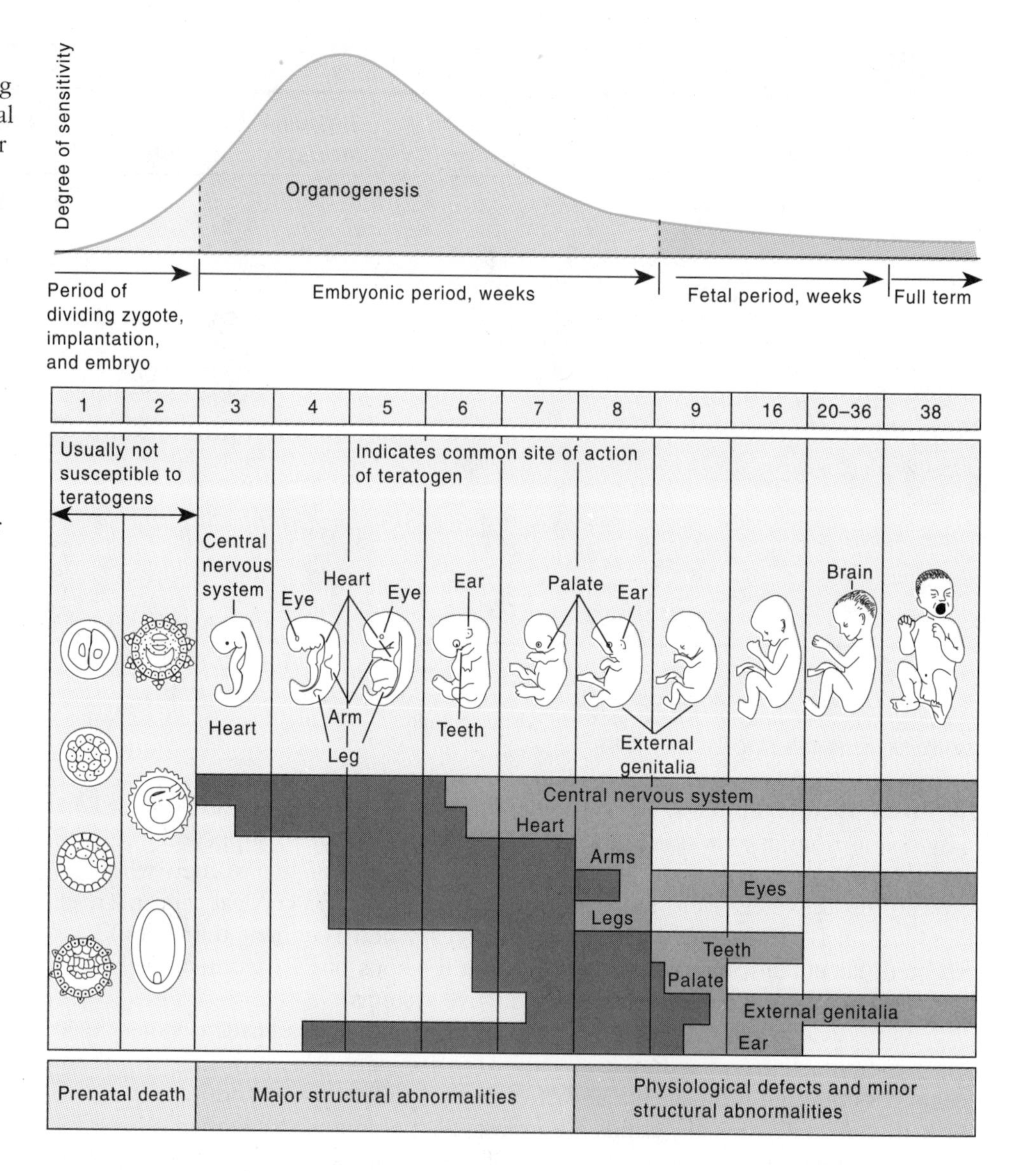

Figure 4.5
Teratogens and the timing of their effects on prenatal development. The danger of structural defects caused by teratogens is greatest early in embryonic development. This is the period of organogenesis, which lasts for several months. Damage caused by teratogens during this period is represented by the dark-colored bars. Later assaults by teratogens typically occur during the fetal period and, instead of structural damage, are more likely to stunt growth or cause problems of organ function.

Maternal Influences

Among the significant influences on prenatal development are maternal nutrition and maternal emotions.

Maternal Nutrition

Because the fetus depends on its mother for nourishment, most women today are keenly aware of the need to have a proper diet that will help them give birth to a healthy, happy baby. When you consider the rapidity of prenatal growth (especially from two to seven months), you can understand the importance of a mother's diet, both for her and the child she is carrying.

Women of childbearing age who wish to have children need to evaluate their weight and nutritional habits well before pregnancy. In this way they can establish good eating habits and attempt to maintain normal weight for their size. Such planning will help them to accommodate the recommended 25 pounds that they will gain during their pregnancy. How much weight to gain is always an important question for pregnant women.

Most doctors offer the following weight gain guide to women:

- If you are of normal weight, try to restrict weight gain to 22–28 pounds.
- If you are underweight, a gain of 25 to 30 pounds is acceptable.
- If you are overweight, do not gain more than 20 to 25 pounds.

Table 4.4 illustrates normal ranges of weight gain.

Table 4.4 Normal Ranges of Weight Gain

Number of Weeks of Pregnancy	Low Side of Normal Weight Gain	High Side of Normal Weight Gain
12	2	4
16	4	8
20	7	13
24	10	18
28	13	23
30	14	25
32	16	27
34	17	30
36	18	32
37	19	33
38	20	34
39	20.5	35
40	21	more than 35

From M.A. Jonaitis, "Nutrition During Pregnancy" in *Complete Guide to Pregnancy*, edited by D. Tapley and W. Todd. Copyright © 1988 Crown, New York.

If the pregnant woman is young, she may need more calories than an adult woman. A weight gain of up to 35 pounds may be necessary for her to produce a baby of normal weight. Here we see one of the problems of teenage pregnancy: A young girl is still growing and needs additional calories for herself. If she resists this need because of a concern for appearance or a desire to shield her pregnancy, the fetus may not receive enough nourishment (Jonaitis, 1988).

Baker and Henry (1987) recommend that pregnant women should plan to add 300 calories to their regular diets, including the following four basic food groups:

- Milk and dairy products
- Meat or protein equivalent
- Fruit and vegetables
- Bread and cereal

The woman's physician will usually recommend supplements such as additional protein, iron, calcium, sodium, fiber, and vitamins. We have previously mentioned the dangers of alcohol use and cigarette smoking, while coffee, tea, and soft drinks should be taken with caution.

The effects of drugs, disease, and diet, while dramatic, are not the only influences on prenatal development. How a woman feels about her pregnancy is also highly significant.

Maternal Emotions

Most women report that delight, anxiety, worry, and irritability are common reactions during pregnancy. Mood swings are characteristic, especially during the first trimester when so much is happening to the woman: hormonal changes, increased fatigue, cravings, and sickness. These feelings usually diminish during the second trimester, when the woman is more accustomed to the changes in her body. But worry and anxiety may increase during the third trimester, as sleep becomes difficult and birth draws near (Bowe, 1988).

These feelings are normal and typically not sufficiently intense to affect the fetus. Seriously stressed mothers often have babies who are restless, irritable, and have feeding problems or bowel difficulties. Maurer and Maurer (1988) conclude that the fetus is not perfectly insulated from the mother's stress. Unless the stress that the mother experiences is unusually severe and prolonged, the effects on the fetus are usually of short duration.

There is no direct evidence that a mother's emotions affect prenatal growth. Nevertheless, data continue to accumulate suggesting that a woman under stress releases hormones that may influence prenatal development. While a definite link between maternal emotions and prenatal growth and even later neonatal behavior is still lacking, it is possible to trace a pattern of events. Stress activates the mother's autonomic nervous system to produce hormones, which enter the mother's blood, cross the placenta, and enter the fetal bloodstream (Travers, 1982).

Women adjust to pregnancy differently. Since it is a condition that affects the total system, there is an immediate biological difference: Some women tire more easily than others and require more sleep and rest. Women differ in their more obvious physical reactions, such as nausea and vomiting: Some react better by eating several small servings than two or three large meals. Some women begin their pregnancies with feelings of depression, while others avoid depression completely. How can we explain these differences?

- The events surrounding the pregnancy are crucial: career status, money, whether the pregnancy was expected or wanted.
- The women's personal experience with the mother-child bond probably reflects the mother's personality.
- Another major influence on the mother's personality and subsequent attitude toward the child is her relationship with the child's father.
- The mother's acceptance of pregnancy also affects her specific attitudes toward the unborn child.
- The psychological journey that women travel usually takes them from a time of intense self-preoccupation to a gradual recognition of the new person with whom they form a complex relationship.
- The mother's expectation for the child is also significant. Does she see the child as an independent human being who will forge his or her own way, or as an extension of herself?

The woman, as an individual, interprets pregnancy either as a crisis and an abnormal state of illness or as a normal occurrence and a state of health. Regardless of pregnancy, everyday life continues and most babies are born normal and healthy

What's Your View?

Some women have a greater chance of developing difficulties during pregnancy or delivering a child with problems. To cope with these conditions, the rapidly expanding field of fetal diagnosis not only identifies problems but also offers means of treatment. Consequently, women with high-risk pregnancies have available today greater access to prenatal testing and genetic counseling, which can raise several ethical and legal issues. As Sroufe, Cooper, and DeHart (1992) note, knowledge entails responsibility.

- **Confidentiality** Issues here include the disclosure of sensitive information to third parties and the possibility of unauthorized individuals attaining access to private information stored in data banks.
- **Autonomy** One of the major questions about genetic and prenatal counseling is: Should such programs remain voluntary for individuals if there is a known history of family genetic problems, or should they be forced to participate? Thus far, opinion is firmly tied to the principle of autonomy—people should have freedom of choice about genetic services.
- **Knowledge** Counseling should help people become informed decision-makers about their own well-being. One way of accomplishing this, of course, is to provide as much information as possible. But consider this scenario for a moment: You are the counselor involved, and in the course of obtaining data about a specific problem, you discover that the supposed father is not., in fact, the biological father (Nightingale & Goodman, 1990). What do you do? There is no easy answer. If you withhold this information, you could cause future problems. If you present it objectively, you could destroy a relationship.

Think about this. What's your view? Prenatal diagnosis and genetic couseling raise issues and questions for all involved.

because their mothers coped with emotional situations without harming themselves or their child. Perhaps Nilsson and his colleagues (1987) offer the best advice when they suggest that the pregnant woman live her usual life but avoid excesses.

Nevertheless, prenatal problems arise, and today's diagnostic techniques often enable early detection of these difficulties.

Fetal Problems: Diagnosis and Counseling

When a prenatal problem is suspected, both diagnostic procedures and counseling services are available. Among the diagnostic tools now available are the following.

Amniocentesis

Amniocentesis is probably the technique you have heard most about. It entails inserting a needle through the mother's abdomen, piercing the amniotic sac, and withdrawing a sample of the amniotic fluid. (Amniocentesis may be done from the 15th week of pregnancy on.) The fluid sample provides information about the child's sex and almost 70 chromosomal abnormalities. For example, spina bifida (see chap. 3) produces a raised level of a protein called alpha-fetoprotein (AFP), which may be detected by amniocentesis (Guttmacher & Kaiser, 1986).

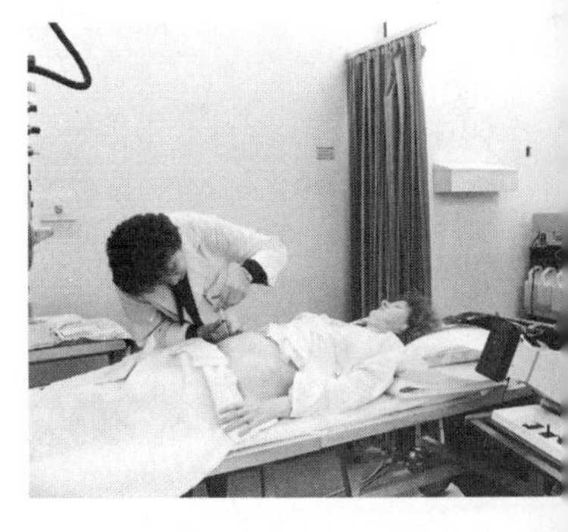

Here a pregnant woman is undergoing amniocentesis. Amniotic fluid is withdrawn and analyzed to determine sex and any chromosomal abnormalities. Amniocentesis may be done from the 15th week onward.

In a **fetoscopy,** a tiny instrument called a fetoscope is inserted into the amniotic cavity, making it possible to see the fetus. If the view is clear, defects of hands and legs are visible. (Fetoscopy is usually performed after the 16th week.)

Chorionic Villi Sampling (CVS)

The outer layer of the embryo is almost covered with chorionic villi, fingerlike projections that reach into the uterine lining. A catheter (small tube) is inserted through the vagina to the villi, and a small section is suctioned into the tube. **Chorionic villi sampling** is an excellent test to determine the fetus' genetic structure and may be performed beginning at 8 weeks, usually between 8 and 12 weeks.

Ultrasound

Ultrasound is a relatively new technique that uses sound waves to produce an image that enables a physician to detect structural abnormalities. Useful pictures can be obtained as early as 7 weeks. Ultrasound is frequently used in conjunction with other techniques such as amniocentesis and fetoscopy (Sadler, 1985).

About 1 percent of infants suffer from some genetic defect, while another 0.5 percent suffer from defective chromosomes. As a result, prenatal testing is steadily becoming more common, especially for older women. Testing and counseling are intended to help couples who are concerned about the possibility of inherited problems.

For example, children born with cystic fibrosis or sickle-cell anemia acquire these diseases from parents who are both carriers. Tests are now available to determine if a person is a carrier of a particular genetic disease. If both potential partners are carriers, the chances of children acquiring the disease can be calculated. The counselor would then explain how severe the problem is, what treatment is available, and what the developmental outcomes would be.

The Birth Process

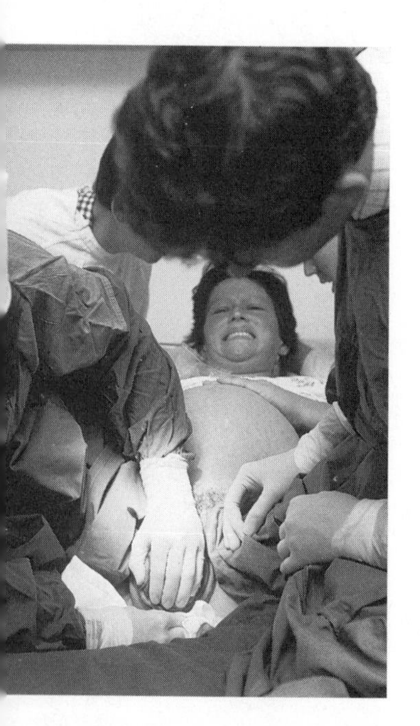

The presence of the father during birth can be a source of physical and psychological support for the mother. Fathers present during birth describe it as an "unforgettable experience."

The odyssey that began approximately nine months earlier reaches its climax at birth. Before this moment arrives, the mother has to make certain decisions. Does she, for example, ask the physician to use an anesthetic, or does she want natural childbirth? Both methods have their advantages and disadvantages.

Natural childbirth provides an unforgettable experience for the mother (and father), but it is hard, painful work that some women prefer to avoid. The use of anesthesia prevents much of the birth pain, but the drug may affect the baby adversely, decreasing alertness and activity for days after birth (Maurer & Maurer, 1988).

Stages in the Birth Process

A woman usually becomes aware of the beginning of labor by one or more of these signs:

- The passage of blood from the vagina
- The passage of amniotic fluid from the ruptured amniotic sac through the vagina
- Uterine contractions and accompanying discomfort

The first two clues are certain signs that labor has begun; other pains (false labor) are occasionally mistaken for signs of true labor.

Three further stages of labor can also be distinguished:

1. *Dilation*. The neck of the uterus (the cervix) dilates to about four inches in diameter. This is the process responsible for labor pains and may last for twelve or thirteen hours, or even longer.

   Think of the baby at this stage as enclosed in a plastic cylinder. It is upside down in the mother's abdomen, with the bottom of the cylinder under the mother's rib and the tip buried deep in her pelvis.

   The cervix is about one-half inch long and almost closed. Before the next stage, expulsion, occurs, the diameter of the cervix must be stretched to a diameter of four inches. (The comedienne Carol Burnett has said that the only way you can imagine this feeling is if you pulled your upper lip over your head!)

2. *The expulsion.* Once the cervix is fully open, the baby passes through the birth canal. This phase typically lasts about 90 minutes for the first child, and about 30 to 45 minutes for subsequent children (although it can last longer). This is the phase when most fathers, if they are present, become exultant. They describe the appearance of the head of the baby (called the crowning) as an unforgettable experience.

   With the cervix fully dilated, the fetus no longer meets resistance and the uterine contractions drive it through the birth canal. Uterine pressure at this stage is estimated to be 60 pounds.

   It is important to note that the times for expulsion (90 minutes and 30–45 minutes) are averages. If this second stage of labor is prolonged—with no evidence of a problem—surgical intervention remains unnecessary. Occasionally, women spend five or six hours (or more) in a normal first birth.

3. *The afterbirth.* In the **afterbirth** stage, the placenta and other membranes are discharged. It is measured from the birth of the baby to the delivery of the placenta and may last only a few minutes. If the spontaneous delivery of the placenta is delayed, it may be removed manually. Figure 4.6 illustrates the birth process.

For most women, the birth process, as painful as it may be, proceeds normally. Occasionally, however, problems arise.

Myths and Facts of Birth

In spite of what you may have heard, no one knows exactly what causes labor to begin or why it begins about 280 days after the first day of the last menstrual period. On the other hand, we do know certain things about birth (Guttmacher & Kaiser, 1986):

- Births are not equally distributed among the days of the week. Deliveries occur much more frequently on Monday through Friday than on weekends. The highest rate is on Tuesdays and the lowest on Sunday.

- Delivery rates seem to peak during August and September, with the lowest rates occurring in December and January.

- When a pregnancy ends spontaneously before the 20th week, a spontaneous abortion, commonly called a **miscarriage** has occurred. After the 20th week, the spontaneous end of a pregnancy is called a **stillbirth** if the baby is born dead, or a *premature birth* if the baby survives.

- Occasionally a pregnancy occurs outside of the uterus. In an **ectopic pregnancy,** the fertilized egg attempts to develop outside the uterus, usually in one of the fallopian tubes. This is sometimes referred to as a *tubal pregnancy*. About one in every two hundred pregnancies is ectopic.

- Many women feel "down" a few days after giving birth. This is fairly common and is now thought to be a normal part of pregnancy and birth for some women. Called **postnatal depression,** this condition may be caused by the sudden change in hormones after birth. Also, a woman may have a sense of anticlimax after completing something she has anticipated for so many months. Women also tire easily and feel some tension about care of the baby, especially after a first birth. Postnatal depression usually leaves quickly.

Figure 4.6
Stages in the birth process

Stage One:
Baby positions itself

Stage Two:
Baby begins to emerge

Stage Three:
Placenta is expelled

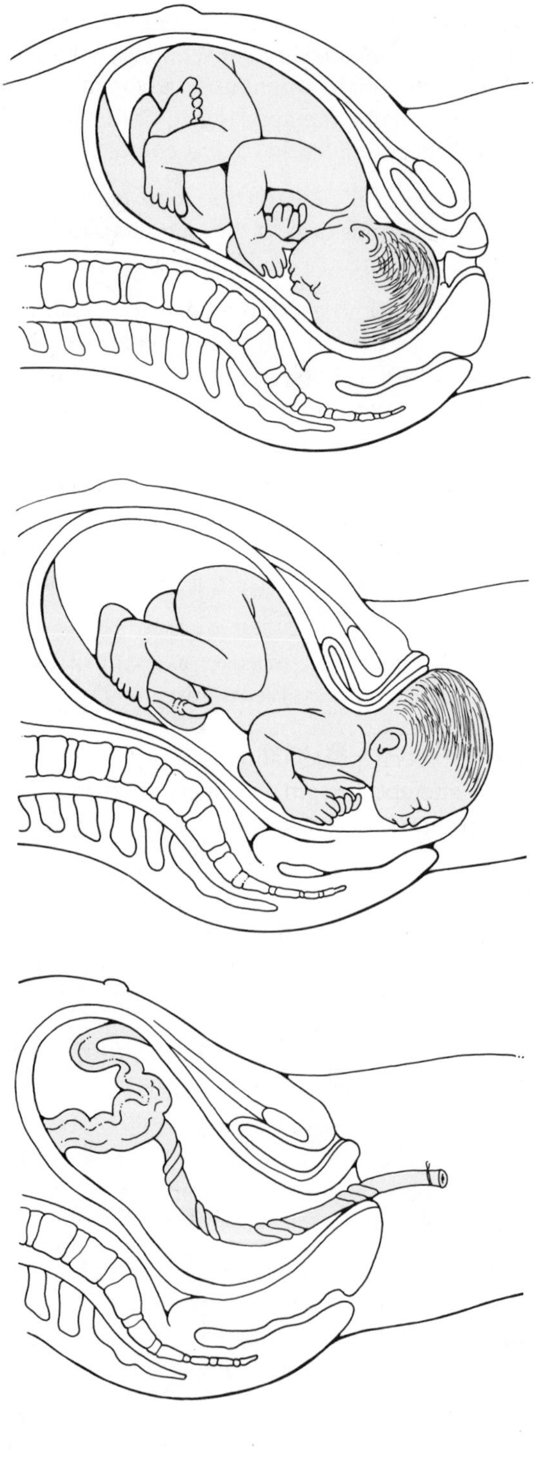

Birth Complications

Birth can sometimes be exceptionally difficult, even dangerous. The following are a few of the more common complications.

Breech Birth

About four out of every hundred babies are born feet first, or buttocks first, while one out of a hundred are in a crosswise position (transverse presentation). These are called **breech births.**

Forceps Delivery

Occasionally, for safety, the physician will withdraw the baby with forceps during the first phase of birth. A **forceps delivery** presents some danger of rupturing blood vessels or causing brain damage.

A decision about a forceps delivery depends on two conditions: those involving the fetus and those related to the mother. Is the fetus in such distress that there is a dangerous slowing of the heart rate? Is the baby in the correct position? Has the mother sufficient strength for the final push?

Cesarean Section

If for some reason the child cannot come through the birth canal, surgery is performed to deliver the baby through the abdomen, in a procedure called **cesarean section.** Although now fairly safe, this operation is considered major surgery and is not recommended unless necessary. More than 20 percent of all live births are cesarean, a figure many consider to be excessive. Among the reasons for this high rate are fetal distress, breech presentation, maternal age, or a previous cesarean. Today many women attempt a vaginal delivery following a cesarean; the success rate is from 60 to 80 percent.

Prematurity

About seven out of every hundred births are premature, occurring less than 37 weeks after conception. Fortunately, today it is possible to simulate womb conditions so that the correct temperature and humidity, bacteria control, and easily digestible food can be provided for the child. Still, prematurity presents real dangers, ranging from mental deficiency to death. (We'll shortly discuss this topic in more detail.)

Anoxia (Lack of Oxygen)

If something during the birth process should cut the flow of oxygen to the fetus, brain damage or death can result. There is a substantial need for oxygen during birth because pressure on the fetal head can cause some rupturing of the blood vessels in the brain. After the umbilical cord is cut, delay in lung breathing can also produce **anoxia.** Failure here can cause death or brain damage.

Controversy surrounds infants who have experienced anoxia, survived, but show evidence of mental dullness. Does anoxia cause long-term developmental impairment? Sameroff (1975) has reviewed the literature concerning delivery and birth complications and the pertinent studies of anoxia. Investigators assumed that early cerebral oxygen deprivation would cause later intellectual difficulty, so youngsters were studied during infancy, at 3 years of age, and finally at 7 years.

A definite pattern emerged: A few days after birth, infants seemed impaired on visual, sensorimotor, and maturational levels. At 3 years, studied with perceptual-motor, cognitive, personality, and neurological tests, they showed lower than normal cognitive functioning and an improved performance on the other items. By

7 years, significant IQ differences had disappeared. Sameroff concludes that anoxia is a poor predictor of later intellectual functioning, and that socioeconomic characteristics still remain the best single predictor of future adjustment.

The Rh Factor

Rh factor refers to a possible incompatibility between the blood types of mother and child. If the mother is Rh-negative and the child Rh-positive, miscarriage or even infant death can result. During birth some of the baby's blood inevitably enters the mother's bloodstream. The mother then develops antibodies to destroy fetal red blood cells. This usually happens after the baby is born, so the first baby may escape unharmed. During later pregnancies, however, these antibodies may pass into the fetus' blood and start to destroy the red blood cells of an Rh-positive baby.

Estimates are that about 10 percent of marriages are between Rh-negative women and Rh-positive men. Today, a protective vaccine (RhoGam) has almost eliminated the possibility of Rh incompatibility when Rh-negative women are identified. In a case where the first baby's blood causes the mother to produce antibodies, exchange blood transfusions may be given to the baby while still in the uterus.

Childbirth Strategies

Most babies escape complications and experience little if any birth difficulty. To help the newly born child adjust to a new environment, Leboyer (1975) believes that we must stop "torturing the innocent." Traditionally, newborns encounter a cold, bright world that turns them upside down and slaps them. Leboyer advocates a calmer environment. He suggests extinguishing all lights in the delivery room except a small night light, and making the room silent at the time of birth. Immediately after birth, the child is placed not on a cold metal scale, but on the mother's abdomen, a natural resting place. After several minutes, the child is transferred to a basin of warm water. Leboyer claims that this process eases the shock of birth and that babies are almost instantly calm and happy.

Another technique, called **prepared childbirth,** or the Lamaze method after French obstetrician Fernand Lamaze, has become extremely popular with the medical profession. For several sessions women are informed about the physiology of childbirth and instructed in breathing exercises. The technique is intended to relieve fear and pain by relaxation procedures.

A range of birth options is now available to couples. Some, for example, are choosing home births under the guidance of midwives, who are trained delivery specialists. About 50 percent of all nonhospital deliveries are by midwives. Some hospitals are providing birthing rooms, which have a more relaxed and homelike atmosphere than the typical delivery room and may provide birthing beds or birthing chairs for greater comfort (Tapley & Todd, 1988). Still, between 95 and 99 percent of all births occur in hospitals.

The Special Case of Prematurity

The average duration of pregnancy is 280 days. Occasionally, however, some babies are born early; they are premature or preterm, often called "preemies." Formerly these babies had high mortality rates, but with today's sophisticated technology their chances of survival are much greater. Before we discuss the condition of these babies and the reasons for their early appearance, let's establish some pertinent facts.

Facts about Prematures

In 1949 the World Health Organization identified premature infants as those weighing 2,500 grams or less (i.e., about 5 1/2 pounds or less). This figure was problematic, however, since some babies are naturally small but perfectly healthy.

In 1961, the WHO redefined **prematurity** to include those infants with a birth weight of 1,500 grams (about 3 pounds) or less, and who were born before 37 weeks gestation. Thus two criteria were suggested: (1) low birth weight and (2) immaturity.

Within this definition, two additional classifications are possible (Spreen & others, 1983):

- Infants born before 37 weeks whose weight is appropriate for their age; these are called *preterm AGA*—Appropriate for Gestational Age.
- Those born before 37 weeks whose weight is low for their age; these are called *preterm SGA*—Small for Gestational Age.

A third classification has recently been proposed—very low birth weight (VLBW), which is defined as below 1,500 grams (about 3 pounds).

Causes of Prematurity

About 250,000 of all infants born in the United States can be classified as premature. Although it is still impossible to predict which women will begin labor prematurely, prematurity has been linked to certain conditions (Avery & Litwack, 1983). *Once a woman has given birth prematurely,* the risk of prematurity in the next pregnancy is about 25 percent. If two pregnancies have ended prematurely, the risk in the third pregnancy rises to 70 percent. Is this tendency to prematurity inherited? To date, there is no evidence pointing to a genetic connection.

Multiple Births *Multiple births*—twins or triplets—usually produce babies whose birth weights are lower than that of a single baby. This condition results in prematurity and accounts for about 10 percent of premature births. There are signs that *multiple abortions* performed in the second trimester increase the risk of prematurity. Age has been identified as a correlate of prematurity. If the mother is under 17 or over 35, the risk is substantially increased.

Low Socioeconomic Status *Low socioeconomic status* is also a frequent accompaniment. In underdeveloped countries as many as one infant in four is born prematurely. In the United States more premature babies are born to poor than affluent women. The reasons remain a mystery, although frequent pregnancies, maternal malnutrition, and poor (if any) prenatal care may be responsible.

Smoking *Smoking* is a significant factor in any discussion of prematurity. Avery and Litwack (1983) state that, regardless of social class, cigarette smoking is associated with infants of low birth weight. The relationship between prematurity and smoking is found with those who smoke one pack per day. For those women who smoke more than one pack per day, the chances of a baby of low birth weight more than doubles.

Alcohol *Alcohol* also increases the likelihood of prematurity. Women who consume alcohol daily during their pregnancy can produce damage in their babies especially a condition called **fetal alcohol syndrome** (FAS), which has four clusters of clinical features (Vorhees & Mollnow, 1987):

- Psychological functioning, which may include mild to moderate retardation, irritability, hyperactivity, and possible learning disabilities
- Growth factors, primarily growth retardation

An Applied View — Guidelines for Parents

Although the new technology designed for premature is marvelous, these babies need to sense parental love. This might seem next to impossible, given the technological nature of the premature nursery.

Here are some guidelines parents of prematures can follow (Harrison, 1983):

- *Try to understand the baby.* Begin by observing carefully, learning what upsets and soothes. How does the baby respond to different types of stimulation? How long does it take to calm the baby after some upset? What kinds of clues are being given to signal discomfort (changes in skin color, muscular reaction, breathing rate)?
- *Use as much body contact as possible.* Since these babies came into the world early, they often seem physically insecure (when compared to the normal infant after birth). Touching the baby with the whole hand on back or chest and stomach often relaxes the premature. Massaging both relaxes and shows affection.
- *Talk to the baby.* Studies have repeatedly shown that prematures, while seemingly unresponsive, show better rates of development when exposed to the mother's voice as often as possible.

When the premature infant can be taken from the incubator and given to the parent, a delicate moment has arrived. Some parents find it difficult to react positively to a premature; they feel guilty, occasionally harbor feelings of rejection, and must fight to accept the situation. They are simply overwhelmed. Usually this reaction passes quickly. On the occasion of this initial contact, parents should have been well prepared for holding a baby that is still entangled in wires and tubes.

- Physical features such as a small head, and possible defects in limbs, joints, face, and heart
- Structural effects, which may include major malformations such as brain damage

About 60 percent of American women drink. For those taking 10 drinks per week while pregnant, the chance of having a low-birth-weight baby doubles.

Other causes of prematurity include maternal infection, cervical problems, high blood pressure, unusual stress, diabetes, and heart disease. As Avery and Litwack (1983) note, even when all of these causes are enumerated, it still is impossible to explain exactly what happened in any given pregnancy. These authors emphasize that having a baby early is not usually anyone's fault.

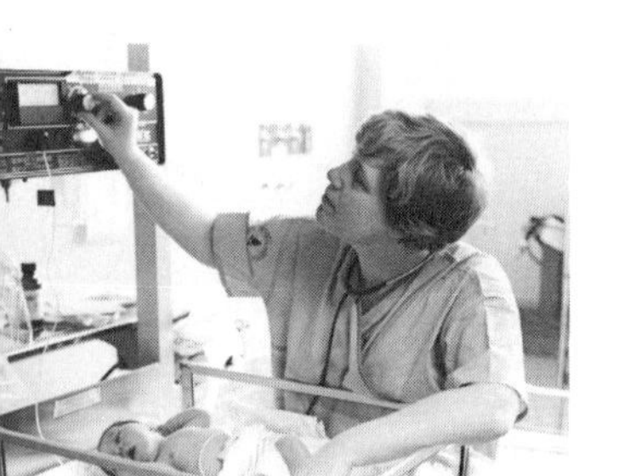

With advances in the treatment of prematures (temperature control, nutrition), the outlook for these babies has greatly improved. Psychological insights into the development of prematures have led to the conclusion that parental support and stimulation are needed during the baby's hospitalization to ensure that attachment proceeds as normally as possible.

A particularly interesting fact about prematures (one that has changed in the past few years) relates to outcome. Early outcome studies indicated that the incidence of major handicaps for prematures ranged from 10 to 40 percent. (These handicaps include mental retardation, cerebral palsy, serious vision defects.) Recent studies in industrialized nations have shown a reduced rate of problems, from 5 to 15 percent. We can conclude, then, that preterm birth and low birth weight do not normally place the baby at severe risk (Field, 1990).

If you think our discussion is overly optimistic, consider the conclusions of Goldberg and DiVitto (1983). Of the 150,000 premature babies born in the United States each year:

- Almost all preterms weighing over 3 1/2 pounds survive.
- Of those weighing between 2 1/2 and 3 1/2 pounds, 80–85 percent survive.
- Of those weighing between 1 1/2 and 2 1/2 pounds, 50–60 percent survive.

Though these children may differ from full-term babies in the early days of their development, most of these differences eventually disappear. Most prematures reach developmental levels similar to those of full-term babies. The only difference is that it takes premature babies a little longer to get there.

What can we say about the future development of these infants? Can we predict which youngsters will have later problems? If we eliminate known hazards—genetic defects and congenital malformations—prediction becomes less certain. One conclusion, however, seems inevitable: The younger and smaller the infant, the greater the risk for later difficulties.

We can best summarize this brief discussion by quoting Avery and Litwack:

> ***The outlook for normal development has improved for all infants, including those born prematurely. Very small infants continue to have risks for some problems that relate to the reasons for their premature birth, or to the difficulties they may encounter during the precarious days in intensive care. Overall, 90% of them will be normal. The remaining 10% will for the most part represent the infants who are the smallest or most premature. Some will have major disabilities, such as cerebral palsy or blindness. Some will have lesser disabilities such as crossed eyes, wheezing, and perhaps some motor incoordinations. Continuing research into the causes of these problems and means of prevention remains a high priority.*** **(1983, pp. 18–19)**

Conclusion

In this chapter, you have seen how a human being begins its journey through the lifespan. Nature's detailed choreography of prenatal development provides a remarkably complex yet elegantly simple means of ensuring the survival of generations. Once conception occurs, uniting the genetic contribution of both mother and father, the developmental process is underway, sheltered for the first nine months in the protective cocoon of the womb.

For some, the process is interrupted and the uterine stay is shortened. Today these prematures have a heightened chance of survival and of normal physical and psychological development, thanks to technological advances.

Chapter Highlights

The Prenatal World

- Once fertilization has occurred, implantation occurs in three stages.
- The germinal period is the time when the fertilized egg passes through the fallopian tube.
- The embryonic period is a time of rapid development and great sensitivity.
- The fetal period is a time of preparation for life outside the womb.
- The senses develop during the prenatal months and are ready to function at birth.

Influences on Prenatal Development

- *Developmental risk* is a term that applies to those children whose welfare is in jeopardy.
- Teratogens are those agents that cause abnormalities.

- Infectious diseases and chemical agents are the two basic classes of teratogens.
- Today AIDS is recognized as a potential danger for newborns.
- Maternal nutrition and emotions are important influences during pregnancy.
- Advancing technology has provided diagnostic tools for the detection of many fetal problems.

The Birth Process

- Birth occurs as a series of stages.
- Complications can develop during the birth process.
- Childbirth strategies are evolving that are designed to ease the transition from womb to world.
- Today the outlook for prematures is much more optimistic than in previous times.

Key Terms

Adhesion
Alpha-fetoprotein (AFP)
Afterbirth
AIDS (acquired immune deficiency syndrome)
Amniocentesis
Amniotic sac
Anoxia
Apposition
Breech birth
Central nervous system (CNS)
Cesarean section
Chorionic villi sampling (CVS)
Cytomegalovirus (CMV)
DES (diethylstilbestrol)
Developmental risk
Diaphragmatic hernia
Ectoderm
Ectopic pregnancy
Embryonic period
Endoderm
Fetal alcohol syndrome (FAS)
Fetal hydrocephalus
Fetal period
Fetoscopy
Forceps delivery
German measles
Germinal period
Herpes simplex
Human immunodeficiency virus (HIV)
Implantation
Invasion
LSD
Mesoderm
Miscarriage
Organogenesis
Placenta
Postnatal depression
Prematurity
Prepared childbirth
Rh factor
Stillbirth
Teratogens
Thalidomide
Toxoplasmosis
Trophoblast
Ultrasound
Umbilical cord

What Do You Think?

1. There has been considerable discussion recently about the possibility of prenatal learning. Where do you stand on this issue? Be sure to support your opinion with facts from this chapter.

2. You probably have heard how careful women must be when they are pregnant. They are worried about such things as smoking and drinking. Do you think we have become too nervous and timid about these dangers? Why?

3. Turn back to Table 4.2. From your own knowledge (relatives and friends, for example), indicate which of these teratogens do you think are most common. Select one and explain why you think it is a common threat and what could be done to help prevent it. (Lead paint is a good example.)

4. There are significant medical and ethical questions surrounding such techniques as fetal surgery. Assume that a physician does not inform a woman that her fetus is a good candidate for fetal surgery. The baby is stillborn. Is the doctor guilty of malpractice or any crime? Can you think of other examples?

Suggested Readings

Avery, M. C. and G. Litwack (1983). *Born early*. New York: Little, Brown. Vital data, good writing, and a positive outlook make this an excellent introduction to the topic of prematurity.

Guttmacher, A. and I. Kaiser (1986). *Pregnancy, birth, and family planning*. New York: Signet. This popular book, available in paperback, has been a best-seller for many years. It carefully presents detailed information about pregnancy and birth.

Kolata, G. (1990). *The baby doctors*. New York: Dell. A riveting account of the types of surgery now being performed on fetuses and newborns.

PART III

Infancy

CHAPTER 5

Physical and Cognitive Development in Infancy

Physical and Motor Development 114

Neonatal Reflexes 114

Newborn Abilities 115

Neonatal Assessment Techniques 117

Brain Development 118

Motor Development 118

Neonatal Problems 121

Perceptual Development 123

The Meaning of Perception 124

Visual Perception 126

Cognitive Development 128

Piaget's Sensorimotor Period 129

Criticisms of Piaget 130

Infants and Memory 133

Language Development 134

Key Signs of Language Development 135

Language Acquisitions: The Theories 136

Conclusion 139

Chapter Highlights 140

Key Terms 141

What Do You Think? 141

Suggested Readings 141

Liz gazed at the infant girl in her arms. Although she had carried this tiny creature within her for nine months, it was still a stranger. Would it be an easy baby? Who would she really look like? Would she do well in school? Of course she would be smart! As these thoughts flashed through her mind, she remembered a course in child psychology she had taken. How had the instructor referred to a newborn? Neonate; yes, that was it. She had read the books on what to expect: feedings, sleep patterns, possible illnesses. But what really was a neonate? What was an infant? Catching herself in these musings, she began to laugh.

Liz certainly was not the first person to speculate about infancy. It is interesting to trace several of these speculations, since they offer insights into the remarkable changes that have occurred in our interpretation of infancy. Do you remember Shakespeare's character Jaques in *As You Like It?* In his speech about the seven ages of our lives, he describes an infant in the first age as "mewling and puking in the nurse's arms."

Freud's description of infancy as richly laden with sexual experiences presents a far different view, one that met with fierce opposition. Noting that it is a serious error to believe that children have no sexual life and that sexuality only begins at puberty, Freud (1966) argues that "from the very first children have a copious sexual life." They direct their first sexual lusts and their curiosity to those who are nearest and for other reasons dearest to them—parents, brothers, and sisters. Freud claimed that infants give clear evidence that they expect to derive pleasure not only from their sexual organs, but that many other parts of their body "lay claim to that same sensitivity." ■

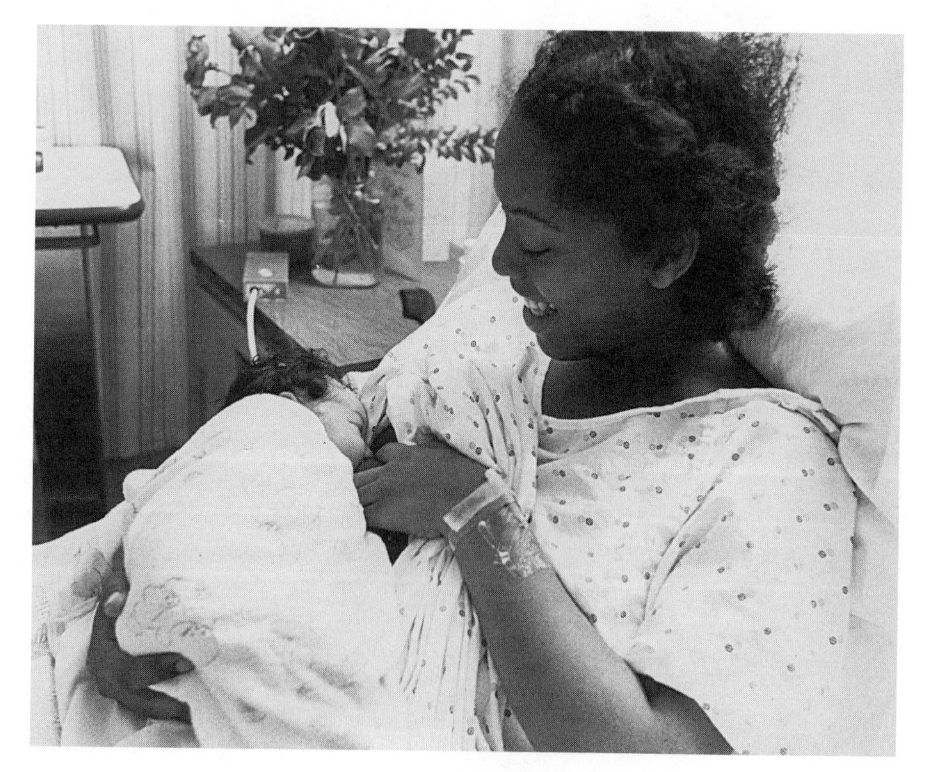

The bond between mother and infant begins to form immediately after birth.

John Watson, one of America's early behaviorists, had discovered the writings of Pavlov on classical conditioning and believed that conditioning answered all of our questions about human behavior. Under Pavlov's influence, Watson viewed infants as a source of potential stimulus-response connections. He believed that if you turn your attention to an infant early enough, you can make that infant into anything you want—it's all a matter of conditioning. His famous statement about infancy is worth repeating here.

> *Give me a dozen healthy infants and my own specified world to bring them up in and I'll guarantee to take any one at random and train him to become any type of specialist I might select—doctor, lawyer, artist, merchant-chief and, yes, even beggar-man and thief, regardless of his talents, penchants, tendencies, abilities, vocations, and race of his ancestors.* (Watson 1924, p. 104)

Piaget's view of infancy differed radically from Watson's. Rather than seeing children as small sponges waiting for something to be poured into them, Piaget believed that children actively construct their own views of the world during these early years. One of the first psychologists to question the apparently passive state of an infant, Piaget argued that infants are much more competent than originally thought.

Infants actually construct their view of the world, a view that changes with age. Although infants don't as yet possess language, Piaget argues for the existence of intelligence before language. Intelligence initially develops from infants using their bodies to explore the environment and then developing perceptions about their experiences. Piaget states that cognitive development during infancy determines the entire course of an individual's mental growth (Piaget, 1967). It is during these years that children build the cognitive structures that are the foundation of their intelligence.

The growing acceptance of an infant's competence has caused some to see a baby as a finely tuned computer, ready to begin such sophisticated activities as reading. That belief is far removed from an earlier concept of an infant as a passive, inert organism. Today's developmental psychologists would most likely recommend to Liz that she think of her infant daughter as an individual with her own unique personality and potential. ■

To help you understand an infant's world, you'll examine the methods used to assess the well-being of infants following birth. Then you'll trace various aspects of infant development: physical, motor, perceptual, cognitive, language, and social/emotional. You'll also begin to discern the issues and themes discussed in chapter 1. For example, the issue of sensitive periods is particularly important in any discussion of infancy. Do the events that occur in infancy leave an indelible mark that lasts a lifetime?

In chapter 1, we mentioned that each chapter, beginning with infancy, would contain a table highlighting the biopsychosocial features of that period. The topics just mentioned—cognition, language, sensitive periods—"fit" our biopsychosocial model as shown in Table 5.1. (Note that we have highlighted the headings "Bio" and "Psycho" to stress their emphasis in the chapter.)

Remember, however, that each of these topics is shaped by both biological and environmental influences. For example, brain development, which fits so neatly into our "Bio" column, depends on such environmental agents as nutrition and stimulation. When you think about development in this way—from a biopsychosocial perspective—it reinforces an important developmental principle: All development is integrated.

Table 5.1 Characteristics of the Period

Bio	Psycho	Social
Neonatal reflexes	Perceptual development	Interaction
Brain development	Cognitive development	
Motor development	Language development	
Neonatal assessment	Nonorganic failure to thrive (FTT)	
Sudden infant death syndrome (SIDS)	Sleeping disorders	
Respiratory distress syndrome (RDS)		

After reading this chapter, you should be able to:

- Describe those abilities that infants begin to demonstrate at birth
- Identify critical neonatal reflexes that enhance survival and enable neonates to adapt to their environment
- Distinguish the typical clues that signal normal physical and cognitive development
- Appraise an infant's cognitive development by observing behavior in a variety of situations
- Evaluate an infant's language development using the key signs of language acquisition

Physical and Motor Development

As Field (1990) has stated, developing motor skills influence cognitive, psychosocial, and emotional development. Growing children experience changes in shape and body composition and in the distribution of tissues. For example, the infant's head at birth is about a quarter of the body's total length, but in the adult it is about one-seventh of body length. Different tissues (muscles, nerves) also grow at different rates, and total growth represents a complex series of changes. At birth, of course, the infant must assume those life-sustaining functions that the mother had provided for nine months. Consequently, any analysis of an infant's physical growth must take into consideration the important role that native reflexes play.

Neonatal Reflexes

When a stimulus repeatedly elicits the same response, that behavior is usually called a **reflex.** Popular examples include the eye blink and the knee jerk. All of the activities needed to sustain life's functions are present at birth (breathing, sucking, swallowing, elimination). These reflexes serve a definite purpose: The gag reflex enables infants to spit up mucus; the eye blink protects the eyes from excessive light; an antismothering reflex facilitates breathing.

In an attempt to rank an infant's reflexes in order of importance, Harris and Liebert (1992) note that the most crucial reflexes are those associated with breathing. Breathing patterns are not fully established at birth, and sometimes infants briefly stop breathing. These periods are called **apnea,** and while there is some concern that they may be associated with sudden infant death, periods of apnea are

What's Your View?

?

Before continuing your reading, try to decide how competent you think a newborn baby is. In one view, infants are empty, unresponsive beings merely waiting for things to be done to and for them. They become active, healthy babies only because of maturation and the actions of those around them. In another view, newborns are seen as amazingly competent and capable of much more than is now expected of them. Consequently, some form of education is needed from the first months of an infant's life. In a third view, infants are seen as neither passive objects nor as superbabies, designed for instant greatness. Newborn babies are seen as bringing abilities with them into the world, a cluster of competencies that enables them to survive but also permits them to engage in a wider range of activities than was previously suspected (Brazelton, 1987).

The expectations that parents have for their babies determine how babies are treated and have important physical, cognitive, and psychosocial consequences. As we know, babies "tune into" their environments and are quite skillful in detecting the moods of those around them. Adherents of this view advise: Let them be babies, using the natural methods that have proven successful: love, attention, and warmth (Brazelton, 1987).

Which view do you think is most realistic? Which will most help infants to fulfill their potential? What's your view? Note: After you have read this chapter, return to this box and see if you would answer the question in the same way.

quite common in all infants. Usually they last for about 2 to 5 seconds; episodes that extend from about 10 to 20 seconds may suggest the possibility of a problem (Berg & Berg, 1987). Sneezing and coughing are both reflexes that help to clear air passages.

Next in importance are those reflexes associated with feeding. Infants suck and swallow during the prenatal period and continue at birth. They also demonstrate the **rooting** reflex, in which they'll turn toward a nipple or finger placed on the cheek and attempt to get it into the mouth. Table 5.2 describes some of the more important neonatal reflexes.

Newborn Abilities

In the days immediately following birth until about two weeks to one month, the infant is called a **neonate.** During this period, babies immediately begin to use their abilities to adapt to their environment. Among the most significant of these are the following:

- Infants display clear signs of **imitative behavior** at 7–10 days. (Try this: Stick out your tongue at a neonate of about 10 days—the baby will stick its tongue out at you!)
- Infants can *see* at birth and, if you capture their attention with an appropriate object (such as a small, red rubber ball held at about 10 inches from the face), they will track it as you move the ball from side to side. Infants react to color at between 2 and 4 months, while depth perception appears at about 4 to 5 months.
- Infants can not only *hear* at birth (and prenatally), but they also can perceive the direction of the sound. In a remarkably perceptive yet simple experiment, Wertheimer (1962) sounded a clicker (similar to those children play with) from different sides of a delivery room only 10 minutes after an infant's birth. The infant not only reacted to the noise, but attempted to turn in the direction of the sound.

Table 5.2 Neonatal Reflexes

Name of Reflex	How Elicited	Description of Response
Plantar grasp	Pressure applied to bottom of foot	Toes tend to curl
Babinski	Gently stroke sole of foot	Toes spread in an outward and upward manner
Babkin	Press palm of hand while infant lies on back	Mouth opens; eyes close
Rooting	Gently touch cheek of infant with light finger pressure	Head turns toward finger in effort to suck
Sucking	Mouth contacts nipple of breast or bottle	Mouth and tongue used to manipulate (suck) nipple
Moro	Loud noise or sudden dropping of infant	Stretches arms and legs and then hugs self; cries
Grasping	Object or finger is placed in infant's palm	Fingers curl around object
Tonic neck reflex	Place infant flat on back	Infant assumes fencer position: turns head and extends arm and leg in same direction as head
Stepping	Support infant in upright position; soles of feet brush surface	Infant attempts regular progression of steps

- Infants are *active seekers* of stimulation. Although their main efforts are devoted to controlling bodily functions such as breathing and heart rate, they occasionally, for brief moments, pay close attention to the environment. These moments signal their search for stimulation; they indicate an ability to process information.
- Infants manifest a willingness, even a need, to *interact* with other human beings (Hinde, 1987). Compelling evidence suggests that interpersonal relationships have a powerful effect on development. (See Suomi, Harlow, & Novak, 1974, for studies of rhesus monkeys and the impact of altered relationships on their development.) What has startled investigators of human relationships is the active role that infants play in controlling their parents' responses, a phenomenon called **reciprocal interactions** (Hinde, 1979).
- Infants, using these abilities, begin their efforts to master the developmental tasks of infancy: learning to take solid foods; learning to talk and walk.

As impressive as these accomplishments are, we must still be rather cautious to avoid overestimating an infant's abilities. Remember: Different parts of our brains mature at different rates. For example, the brain fibers for hearing develop early, but the brain parts for understanding what is heard develop much later. The brain areas controlling the upper body develop much more rapidly than those controlling other bodily areas, in a pattern having strong survival overtones.

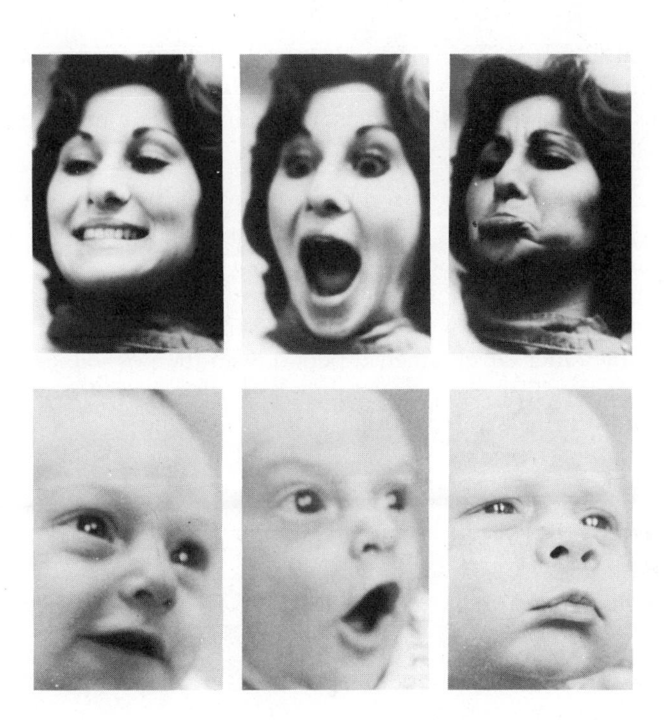

Sample photographs of a model's happy, surprised, and sad expressions, and an infant's corresponding expressions.

Although all infants are born with these reflexes and abilities, not all possess them to the same degree. For example, some neonates demonstrate much weaker reflex action than others, a condition that affects their chances of surviving. Consequently, efforts to develop reliable measures of infant behavior have increased sharply.

Neonatal Assessment Techniques

There are three basic classifications of neonatal tests that assess infant reflexes and behavior.

1. *The Apgar.* In 1953 Virginia Apgar proposed a scale to evaluate a newborn's basic life signs. The **Apgar** is administered one minute after birth, and repeated at three-, five-, and ten-minute intervals. Using five life signs—heart rate, respiratory effort, muscle tone, reflex irritability, and skin color—an observer evaluates the infant by a three-point scale. Each of the five dimensions receives a score of 0, 1, or 2. (0 indicates severe problems, while 2 suggests an absence of major difficulties.)

2. *Neurological Assessment.* The **neurological assessment** is used for three purposes:

   - Identification of any neurological problem
   - Constant monitoring of a neurological problem
   - Prognosis about some neurological problem

   Each of these purposes requires testing the infant's reflexes, which is critical for neurological evaluation and basic for all infant tests (Prechtl, 1977).

3. *Behavioral Assessment.* The Brazelton Neonatal Behavioral Assessment Scale (named after T. Berry Brazelton) has become a significant worldwide tool for infant assessment. While the Brazelton tests the reflexes we have just discussed, its major emphasis is on how the infant

interacts with its environment. In other words, it also permits us to examine the infant's behavior. Brazelton believes that the baby's state of consciousness is the single most important element in the examination (1984, 1990).

All three of these assessment techniques provide clues about the infant's ability to function on its own. Tests such as these, plus careful observation, have given us much greater insight into infant development. These tests have also helped us to realize that infants are much more competent than we previously suspected.

To summarize, then, we can say that infant growth occurs at a rate unequalled in any other developmental period, with the possible exception of adolescence. If you recall, the average weight at birth is about seven pounds and length is about 20 inches. Table 5.3 illustrates height and weight increases during this period.

Table 5.3 Physical Growth—Infancy

Age (months)	Height (in.)	Weight (lb.)
3	24	13–14
6	26	17–18
9	28	20–22
12	29.5	22–24
18	32	25–26
24	34	27–29

Brain Development

Three distinct layers appear during the initial stages of the embryonic period: **ectoderm, mesoderm,** and **endoderm.** As we have seen, the ectoderm acts as the basis for the developing nervous system. Estimates are that the baby's brain, at birth, is about a quarter of its adult size. At 6 months it is about 50 percent of its adult weight, 60 percent at 1 year, 75 percent at 2 1/2 years, 90 percent at 6 years, and 95 percent at 10 years. The developmental pattern is seen in Figure 5.1.

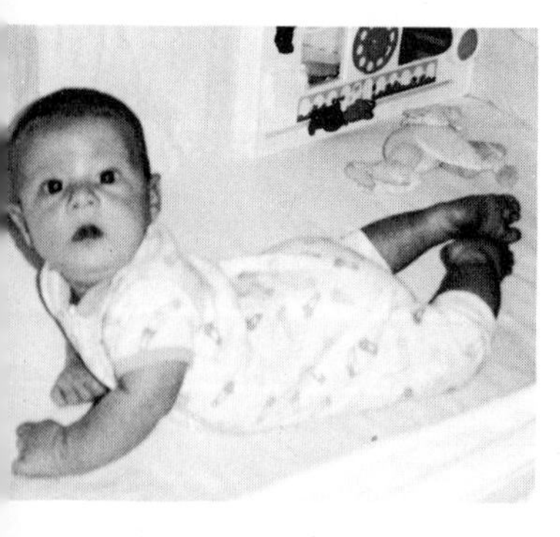

Note the steady development of body control in this picture, especially the head and upper body. Control of the lower body and legs follows by several months.

We know, however, that different parts of the brain develop at different rates. For example, the cerebellum, which controls the fine motor activities, develops more slowly than the brain areas controlling the muscles of the upper body. During the first month after birth, the cortical areas of the brain show an increasing thickness. Brain development seems to follow a definite schedule. First to show signs of development is the motor area, followed by the sensory region, the auditory area, and then the visual region. Consequently, we should expect motor development to proceed rapidly, which is just what happens.

Motor Development

Motor development proceeds at a steady pace, and the rate of motor activity seems to have a genetic component. Studying motor activity level, Saudino and Eaton (1991) used actometers—mechanical motion recorders—to assess activity level. They found that identical twins were significantly more similar in their activity levels than fraternal twins. These findings can also help parents adjust to the temperaments of their children.

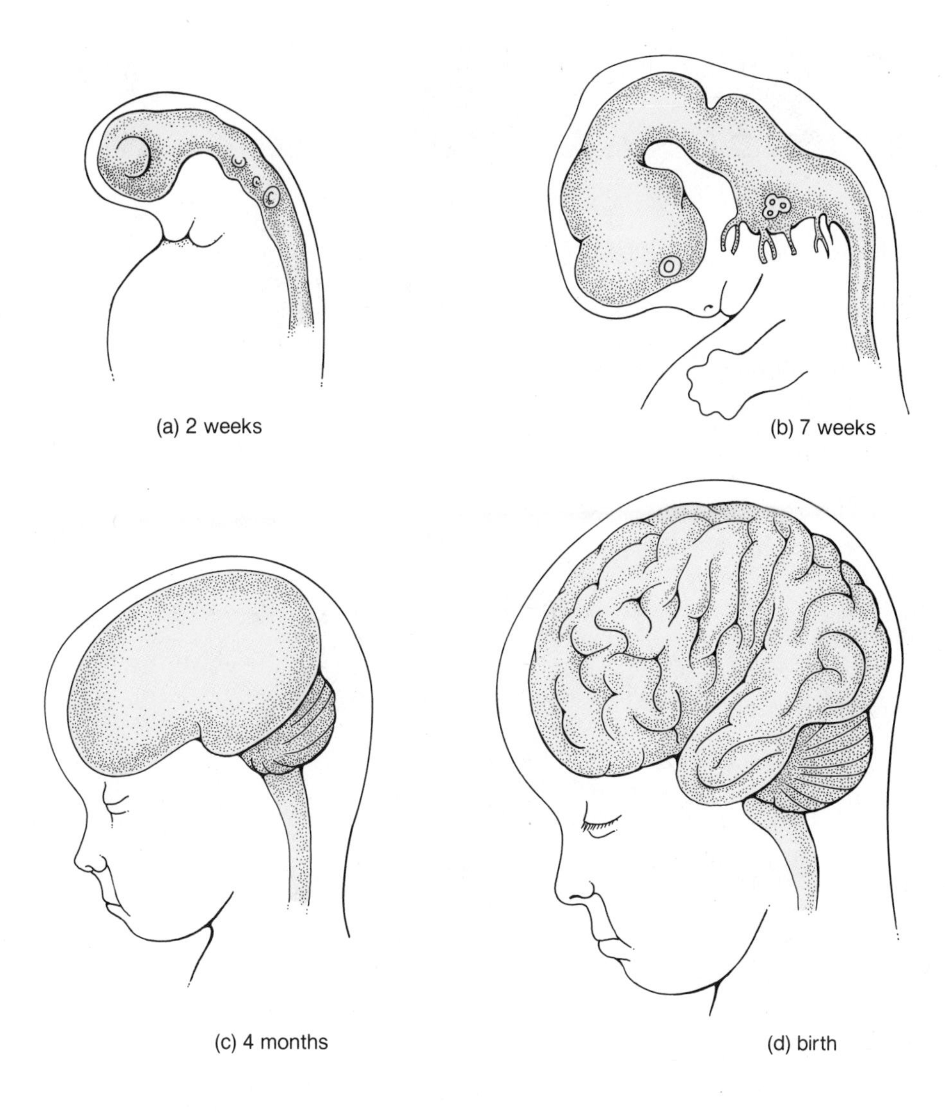

Figure 5.1
Brain development

Since motor development occurs in a head-to-feet direction (called cephalocaudal, as opposed to a proximodistal direction, which is from the center of the body to the extremities), an infant's ability to control its head signals advancing motor development. Particularly interesting in any analysis of motor development is the reported motor precocity of black infants. In a study of black infants at 2 days and then at 1 month, Rosser and Randolph (1989) duplicated this finding, and reported that the black infants in their study also performed well on all standard behavioral scales. This finding, which differs from those of other studies that relied on low-income samples, was attributed to the prenatal and perinatal conditions of the mothers. Of the 80 mothers, 49 were of low to lower-middle socioeconomic status (SES) and 31 were of middle SES (Rosser & Randolph, 1989, p. 138).

Following are several important characteristics of motor control.

Head Control

The most obvious initial head movements are from side to side, although the 1-month-old infant occasionally lifts its head when in a prone position. Four-month-old infants can hold their heads steady while sitting and will lift their head and shoulders to a ninety-degree angle when on their abdomens. By the age of 6 months, most youngsters can balance their heads quite well.

An Applied View — Toilet Training—Easy or Difficult?

By the end of their child's infancy period, most parents begin to think about toilet-training. One important fact to remember is that voluntary control of the sphincter muscles, which are the muscles that control elimination, does not occur until about the eighteenth month. (For some children, control is not possible until about 30 months.) Attempting to train children before they are ready can only cause anxiety and stress for the child and frustration for the parents.

As Caplan and Caplan (1984) note, children really can't be trained; they learn when to use the toilet. It is not something parents do *to* a child; rather, parents *help* the child without feelings of tension and fear. Certain signs of readiness can alert parents that they can initiate the process (Caplan & Caplan, 1983):

- As we noted, necessary muscle control does not occur until well into the second year.
- A child must be able to communicate, either by words or gesture, the need to use the bathroom.
- At about 2 years, almost all children want to use the toilet, to become more "grown-up," and to rid themselves of the discomfort of wet or soiled diapers.

Parents should try to obtain equipment that the child feels most comfortable with—either a chair that sits on the floor or one that fits over the toilet seat. Brazelton (1981), studying the children in his clinical practice, offers the following estimates for ages of control:

- Most children start training at from 24–30 months.
- The average age of daytime control was 28.5 months.
- The average age of nighttime control was 33.3 months. (Girls attain nighttime control about 2 1/2 months before boys.)

Most parents expect their children to be toilet trained by the end of infancy, usually sometime between 2 and 3 years of age (Charlesworth, 1987). Physiologically, most children are ready to learn control; socially it is desirable; psychologically it may be traumatic unless parents are careful. If youngsters are punished for a behavior that they find difficult to master, their perception of the environment is affected, which may produce feelings of insecurity. The common sense of most adults results in a combination of firmness and understanding, thus helping youngsters master a key developmental task (Travers, 1982, p. 57).

Locomotion: Crawling and Creeping

Crawling and creeping are two distinct developmental phases. In **crawling,** the infant's abdomen touches the floor and the weight of the head and shoulders rests on the elbows. Locomotion is mainly by arm action. The legs usually drag, although some youngsters push with their legs. Most youngsters can crawl after age 7 months.

Creeping is more advanced than crawling, since movement is on hands and knees and the trunk does not touch the ground. After age 9 months, most youngsters can creep.

Most descriptions of crawling and creeping are quite uniform. The progression is from propulsion on the abdomen to quick, accurate movements on hands and knees, but the sequence is endlessly varied. Youngsters adopt a bewildering diversity of positions and movements that can only loosely be grouped together.

Locomotion: Standing and Walking

After about age 7 months, infants when held will support most of their weight on their legs. Coordination of arm and leg movements enables babies to pull themselves up and grope toward control of leg movements. The first steps are a propulsive, lunging forward. Gradually a smooth, speedy, and versatile gait emerges. The world now belongs to the infant.

Once babies begin to walk, their attention darts from one thing to another, thus quickening their perceptual development (our next topic). Tremendous energy and mobility, coupled with a growing curiosity, push infants to search for the boundaries of their world. It is an exciting time for youngsters but a watchful time for parents, since they must draw the line between encouraging curiosity and initiative and protecting the child from personal injury. The task is not easy. It is, however, a problem for all aspects of development: What separates unreasonable restraint from reasonable freedom?

Table 5.4 summarizes milestones in motor development.

Table 5.4 Milestones in Motor Development

Age	Head Control	Grasping	Sitting	Crawling-Creeping	Standing-Walking
1–3 months	Can lift head and chest while prone	Grasps objects; briefly holds objects: carries objects to mouth	Sits awkwardly with support		
4–8 months	Holds head steady while sitting; balances head	Develops skillful use of thumb during this period	Transition from sitting with slight support to brief periods without support		
8–12 months	Has established head control	Coordinates hand activities; handedness begins to appear	Good trunk control; sits alone steadily	Crawling movements appear (trunk touches floor); begins about 7 months	
		Handedness pronounced; holds crayon; marks lines	Can sit from standing position	Creeping (trunk raised from floor) begins at 9–10 months and continues until steady walking	Can stand holding onto something; will take steps when held; by 12 months will pull self up
14 months					Stands alone; begins to walk alone
18 months					Begins to run

Neonatal Problems

Not all infants enter the world unscathed. Occasionally the developmental sequence that we have just discussed does not run smoothly. Among the most prominent of possible problems are the following.

Failure to Thrive (FTT)

The weight and height of **failure-to-thrive** infants consistently remain far below normal. They are estimated to be in the bottom 3 percent of height and weight measures. They account for about 3 percent of pediatric hospital admissions.

There are two types of FTT: those with organic causes (the problem is usually some gastrointestinal disease; this category accounts for 30 percent of FTT cases) and those with nonorganic causes (no physical cause for the problem can be found).

Nonorganic FTT is difficult to treat, and the problem may originate in family interactions (see chap. 13). The seriousness of this problem is evident from the outlook for FTT infants: Almost 50 percent of them will continue to experience physical, cognitive, and behavioral problems.

Sudden Infant Death Syndrome (SIDS)

Discussion of the survival value of reflexes introduces one of the most perplexing problems facing both parents and researchers, **sudden infant death syndrome** (SIDS). An estimated 10,000 infants 2–4 months old die each year from SIDS. There is little warning, although almost all cases are preceded by mild cold symptoms and usually occur in late winter or early spring.

SIDS rarely occurs before age 1 month or after age 1 year; most victims are between 2 and 4 months old. Deaths peak between November and March. Once again, boys are more vulnerable, in this case by a 3-to-2 margin.

SIDS is particularly devastating for parents because of the lack of warning. These infants are apparently normal. Parents put them in a crib for a nap or for the night and return later to find them dead. (Hence the common name, "crib death.") You can imagine the effect this has on parents, particularly the feelings of guilt: What did I do wrong? Why didn't I look in earlier? Why didn't I see that something was wrong? Today, special centers have been established to counsel grieving parents.

While no definite answers to the SIDS dilemma have yet been found, current research points to a respiratory problem. Control of breathing resides in the brain stem, and autopsies indicate that the infant may not have received sufficient oxygen while in the womb. (This condition is called fetal hypoxia.)

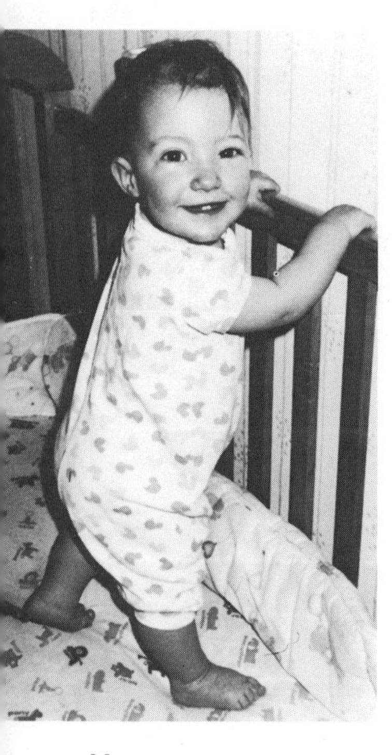

Most youngsters somewhere in the seven-to-nine-month period begin to pull themselves up to a standing position. Their legs are now strong enough to support them while standing.

Sleeping Disorders

Most **sleeping disorders** are less serious than FTT or SIDS. Nevertheless, infant sleeping problems negatively affect growth and trouble parents. As Ferber (1985) states:

> ***The most frequent calls I receive at the Center for Pediatric Sleep Disorders at Children's Hospital in Boston are from a parent or parents whose children are sleeping poorly. When the parent on the phone begins by telling me "I am at the end of my rope" or "We are at our wits' end" I can almost predict what will be said next.***

Ferber goes on to explain that the parent has a child between the ages of 5 months and 4 years who does not sleep readily at night and wakes repeatedly. Parents become tired, frustrated, and often angry. Frequently, the relationship between the parents becomes tense.

Usually a sleep problem has nothing to do with parenting. There is also usually nothing wrong with the child, either physically or mentally. Occasionally problems do exist: a bladder infection, or, with an older child, emotional factors causing night terrors. A sleep problem is not normal and should not be waited out.

Neonates sleep more than they do anything else (usually from 14 to 15 hours per day) and have three sleep patterns: light or restless, periodic, and deep. There is little if any activity during deep sleep (about 25 percent of sleep). Neonates are mostly light sleepers and have the brain wave patterns associated with dreaming (although infants probably do not dream).

Some internal clock seems to regulate sleep patterns, with most deep sleep spells lasting approximately twenty minutes. At the end of the second week, a consistent and predictable pattern emerges. Neonates sleep in short stretches, about seven or eight per day. The pattern soon reverses itself, and infants assume an adult's sleep schedule.

Normal sleep patterns are as follows:

- *During the first month,* infants reduce their seven or eight sleep periods to three or four and combine two of them into one lasting about five hours. (If parents are lucky, this longer period will occur after a late evening feeding.) Infants are thus establishing a night and day routine.
- *During the third month,* sleep patterns are usually regulated. Morning and afternoon naps supplement night stretches ranging from six or seven to as much as ten or eleven hours.
- *At five or six months,* infant sleep patterns change. As part of the night sleep cycles, the infant is usually wide awake at dawn, bursting with excitement, and demanding an audience.
- *By eight months,* naps are shorter and some infants may require only one in the afternoon. A problem that most infants begin to show at this age is a reluctance to go to bed at night. They now begin to sleep most of the night.
- *At twelve months,* napping may be difficult and infants begin to set their schedules; that is, they nap only when tired. There is continued resistance to go to bed at night, but once asleep they usually sleep through the night.

Children suffering from sleep disorders ordinarily have nothing organically wrong with them. Changing their activities before sleep usually cures the problem.

For an idea of normal sleep patterns, see Figure 5.2.

To determine whether a child has a sleeping problem, Ferber (1985) suggests these criteria: (1) the child's sleep patterns cause problems for parents or the infant; (2) obvious problems exist: an inability to sleep or, with older children, sleep-walking and sleep terrors; (3) more subtle symptoms are at work: excessively loud snoring may signal a breathing problem; the child is unable to go back to sleep after waking.

Respiratory Distress Syndrome (RDS)

The last of the disorders to be discussed is **respiratory distress syndrome** (also called *hyaline membrane disease*). While this problem is most common with prematures, it may strike full-term infants whose lungs are particularly immature.

RDS is caused by the lack of a substance called surfactant, which keeps open the air sacs in the lungs. When surfactant is inadequate, the lung can collapse. Since most babies do not produce sufficient surfactant until the 35th prenatal week, you can see why it is a serious problem for prematures. (Only 10 percent of a baby's lung tissue is developed at full-term birth.)

Full-term newborns whose mothers are diabetic seem especially vulnerable to RDS. Babies whose delivery has been particularly difficult also are more susceptible. The good news is that today 90 percent of these youngsters will survive and, given early detection and treatment, the outlook for them is excellent.

Perceptual Development

From what we've said so far, the current picture of an infant is that of an active, vibrant individual vigorously searching for stimuli. How do infants process these stimuli? Answering this question moves us into the perceptual world. For babies not only receive stimuli, they interpret them. Before attempting to chart perceptual development, let's explore the meaning of perception in a little more detail.

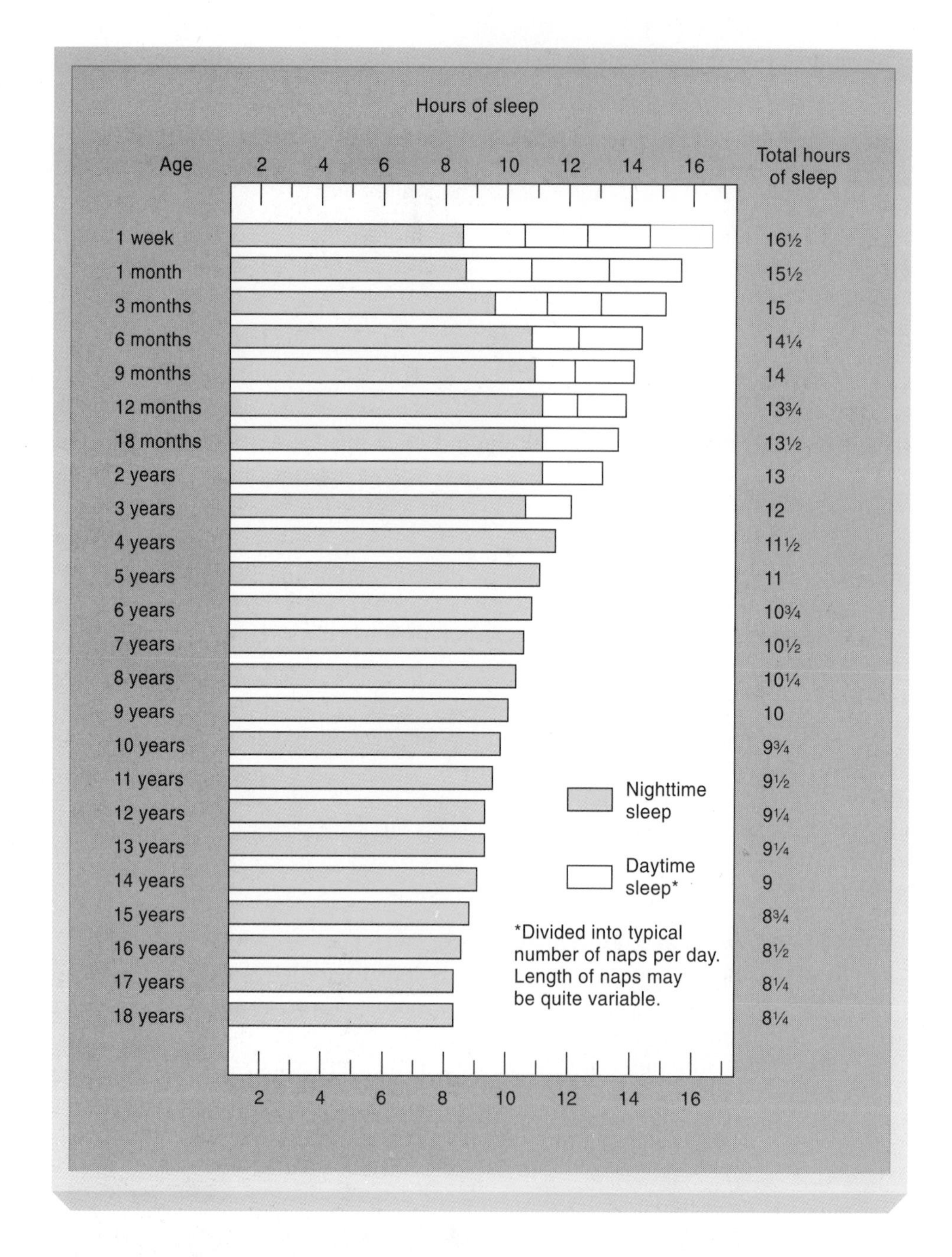

Figure 5.2
Typical sleep requirements in childhood

From Richard Ferber, *Solve Your Child's Sleep Problems*. Copyright © 1985 by Richard Ferber, M.D. Reprinted by permission of Simon & Schuster, Inc.

The Meaning of Perception

Infants acquire information about the world and constantly check the validity of that information. This process defines perception: getting and interpreting information from stimuli (Harre & Lamb, 1983). Infants are particularly ingenious at obtaining information from the stimuli around them. They attend to objects according to the perceptual information the objects contain.

Seeking information leads to meaning. Objects roll, bounce, or squeak—in this way infants learn what objects are and what they do. During infancy, youngsters discover what they can do with objects, which furthers their perceptual development.

Remember: Infants are born ready to attend to changes in physical stimulation. Stimuli presented frequently cause a decrease in an infant's attention (*habituation*). If the stimuli are altered, the infant again attends, indicating awareness of the difference. For example, if you show an infant a picture (flower, birds, anything attractive), the child is first fascinated, then becomes bored; the child has habituated. If you now change the picture, you again capture the child's attention.

Infants, however, encounter a wide variety of objects, people, and events, all of which differ in many dimensions: color, size, shape. They must learn how to react to each, and how they react depends on many factors. For example, Clifton, Perris, and Bullinger (1991) found that when 6- and 7-month-old infants were in a dark setting and were presented with objects that made a sound, they reached accurately for the objects as long as they were within their reach. When the objects were out of reach, they were far less accurate. They thus perceive distance and direction and seem able to define their auditory space as long as an object is within reach of the body.

In a classic study, Brooks and Lewis (1976) studied how infants responded to four different types of strangers, a male and female child, a female adult, and a female midget. In this way, facial configurations and height were varied. The infants reacted to the children by continuous looking and some smiling. They reacted to the midget with considerable puzzlement but no positive response such as smiling or movement toward her. They reacted to the adult by sporadic looking, averting their eyes, frowning, and even crying. Thus the infants used size and facial configuration cues.

We may conclude, then, that perception depends on both learning and maturation. An infant's perceptual system undergoes considerable development following birth, resulting from greater familiarity with objects and events in the world as well as from growth.

Infants quickly begin to attend to the objects in their environments, thus constructing their views about how the world "works."

Visual Perception

Humans are born able to see and quickly exhibit a preference for patterns. Recent research on vision report the following (Aslin, 1987):

- Studies show that **variable accommodation** (focusing on objects at various distances) appears at about age 2 months.
- **Binocular coordination** appears at around 4 months. Studies indicate that there may be a critical period for the attainment of stereopsis (three-dimensional vision), since infants born with congenital esotropia (lack of ability to develop three-dimensional vision) have a greater chance of acquiring stereopsis if surgery is performed before the age of 2.
- Infants see *color* sometimes in the 2–4 month period. (A major difficulty in establishing an exact time for color recognition has been in separating color from brightness; it has only recently been overcome.)
- Individual differences in *visual tracking* ability exist at birth.

Visual Preference

Do infants prefer looking at some objects more than others? In an exciting yet simple series of experiments, Robert Fantz provided dramatic documentation of an infant's perceptual ability. Fantz (1961) states that the best indicator of an infant's visual ability is eye activity. Infants who consistently gaze at some forms more than others show perceptual discrimination; that is, there is something in one form that holds their attention.

Using a "looking chamber" in which an infant lies in a crib at the bottom of the chamber and looks at objects placed on the ceiling, Fantz could determine the amount of time that infants fixated on different objects.

He tested thirty infants from ages 1 week to 15 weeks on four pairs of test patterns: horizontal stripes and a bull's eye, a checkerboard and two sizes of plain squares, a cross and a circle, and two identical triangles. The more complex patterns attracted the infants' attention significantly longer than either the checkerboard and square or the triangles.

The next step involved testing to discover whether infants preferred facial patterns. Three flat objects shaped like a head were used. One had regular features painted in black on a pink background; the second had scrambled features; the third had a solid patch of black at one end. The three forms were shown to 49 infants from 4 days old to 6 months old. Infants of all ages looked longest at the real face. The plain pattern received the least attention.

Fantz next tested pattern perception by using six objects, all flat discs six inches in diameter: face, bull's-eye, newsprint, red disc, yellow disc, and white disc. The face attracted the greatest attention, followed by the newsprint, the bull's-eye, and then the three plain-colored discs (none of which received much attention). Infants, then, show definite preferences based on as much complexity as they can handle (human faces are remarkably complex).

Infants' ability to detect complexity is also seen in studies by Hirshberg (1990), in which he worked with sixty-six 12-month-old infants who were responding to emotional signals given by their parents. Using toys as stimuli, the parents would give an emotional response to the infant when the child looked at them after picking up a toy. Sometimes the parent gave a happy response; at other times, with a different toy, the parent gave a fearful response; occasionally one parent gave a fearful response while the other parent gave a happy response to the infant's playing with the same toy.

The results showed that the infants did not place more emphasis on the mothers' or fathers' reactions, but responded equally to the positive and negative signals of each. Infants showed the greatest decrease in toy exploration when they received conflicting signals from the parents, that is, when one gave a happy response and the other gave a fearful response. This study reinforces the belief that infants are capable of detecting considerable complexity in their surroundings.

Visual Adaptation

Studying visual development spurs speculation about how growing visual skill helps infants to adjust to their environment. Gibson and Walk (1960), in their famous "visual cliff" experiment, reasoned that infants would use visual stimuli to discriminate both depth and distance.

The visual cliff is a board dividing a large sheet of heavy glass. A checkerboard pattern is attached flush to one half of the bottom of the glass, giving the impression of solidity. The investigators then place a similar sheet on the floor under the other half, creating a sense of depth—the visual cliff (see Figure 5.3).

Figure 5.3
A child's depth perception is tested on the visual cliff. The apparatus consists of a board laid across a sheet of heavy glass, with a patterned material directly beneath the glass on one side and several feet below it on the other.

Thirty-six infants from ages 6 to 14 months were tested. After the infant was placed on the center board, the mother called the child from the shallow side and then the cliff side. Twenty-seven of the youngsters moved onto the shallow side toward the mother. When called from the cliff side, only three infants ventured over the depth. The experiment suggests that infants discriminate depth when they begin crawling.

Is this ability present before 6 months? Investigating the value of heart rate changes in analyzing infant behavior, Campos (1976) extended the visual cliff experiments. Noting that cardiac deceleration indicated infant attention while acceleration suggested infant fear, Campos placed infants of 1, 2, 3, and 5 months of age on the cliff. Expecting to find fear responses (cardiac acceleration) even with those who were not yet crawling, Campos was surprised to discover cardiac deceleration (attention) in the 2-, 3-, and 5-month-old infants. The reactions were as follows:

- One month—no change in cardiac rate
- Two months—cardiac deceleration (attention)

- Three months—cardiac deceleration (attention)
- Five months—decreasing cardiac deceleration (decreasing attention)
- Nine months—cardiac acceleration (fear)

The pattern is fascinating and raises several questions. Prelocomotor infants perceive something at 2, 3, and 5 months, but is it depth? If they do indeed perceive depth, why don't they show fear? Thus infants demonstrate visual ability at birth and quickly show signs of increasing visual skill. Infants at 3 months also make primitive attempts to organize their visual surroundings and to integrate vision with other infant activities.

To conclude, we can state that by 2–4 months of age, infant perception is fairly sophisticated. They perceive figures as organized wholes; they react to the relationship among elements rather than single elements; they perceive color; and complex rather than simple patterns fascinate them.

Cognitive Development

Can infants really think? If they can, what is their thinking like? Do they understand what is happening in the world around them? How do we explain the change in thinking from the newborn to the 2-year-old? Analyzing the way that children think led Schulman (1991) to conclude that infants, like the rest of us, are trying to answer four basic questions:

- *What's out there?* The first task infants must master is to distinguish the objects in their world; faces, voices, milk, rattles, and others (Schulman, 1991). To accomplish this task, infants discover similarities and differences and begin to recognize patterns. In a sense, this task remains with us for life; for example, you're trying to discover "what's out there" in lifespan psychology.
- *What leads to what?* Infants quickly learn that some experiences follow others: Crying leads to attention (most of the time). When infants learn about before-after sequences, they are beginning to acquire an understanding between past, present, and future.
- *What makes things happen?* Infants become aware that some events cause another event; it's not just a matter of before-after. For example, pushing a toy makes it move; letting go of a doll causes it to fall. They also begin to realize that it is people who *do* things.
- *What's controllable?* Infants gradually learn that they can cause things to happen. In our discussion of reciprocal interactions, we mentioned that infants exercise control of those around them. By doing certain things such as making pleasant noises or smiling, the baby causes those around them to smile or begin to play with the baby. Thus infants begin to learn that they have the power to cause things to happen, to control their world.

How do infants develop an understanding of the objects around them? In their first year of life, they seem to proceed through several stages. In *the first month,* they have no idea that objects are permanent—out of sight, out of mind. From *1 to 4 months,* infants will continue to stare at the spot where an object disappeared and then turn their attention to something else. From *4 to 8 months,* infants begin to show signs that an object still exists even if they can't see it; they'll look for a toy after they drop it; they love playing peek-a-boo. From *8 to 12 months,* infants develop the notion of **object permanence** and will continue to hunt for a hidden object (Maurer & Maurer, 1988).

Recent studies indicate that object permanence is related to later cognitive development suggesting that roots of cognition lie in infancy. For example, Rose and associates (1991) compared object permanence (among other cognitive indicators) in premature infants at 1 year with IQ measures at 5 years. They found a significant relationship between the two, suggesting developmental continuities in cognition.

To answer Schulman's four questions and to discover how infants develop such concepts as object permanence, we now turn to the work of the great Swiss scholar, Jean Piaget. You previously read about several of his important ideas in Chapter 2; here we'll examine his interpretation of infancy in some detail.

Piaget's Sensorimotor Period

Piaget (1967) states that the period from birth to language acquisition is marked by extraordinary mental growth and influences the entire course of development. **Egocentrism** describes the initial world of children. Everything centers on them; they see the world only from their point of view. Very young children lack social orientation. They speak at and not to each other, and two children in conversation may be discussing utterly unrelated topics. Egocentric adults know that there are other viewpoints, but they disregard them. The egocentric child simply is unaware of any other viewpoint.

The remarkable changes of the **sensorimotor period** (about the first two years of life) occur within a sequence of six stages. Most of Piaget's conclusions about these stages were derived from observation of his own three children. (Jacqueline, Lucianne, and Laurent have become as famous in psychological literature as some of Freud's cases or John Watson's Albert.)

Stage 1 During the first stage, children do little more than *exercise the reflexes* with which they were born. For example, Piaget (1952) states that the sucking reflex is hereditary and functions from birth. At first, infants suck anything that touches their lips; they suck when nothing touches their lips; then they actively search for the nipple. What we see here is the steady development of the coordination of arm, eye, hand, and mouth. Through these activities, the baby is building a foundation for forming cognitive structures.

When infants begin to move things to get what they want, they are "coordinating their secondary schemata." This is a clear signal of advancing cognitive development.

Stage 2 Piaget refers to stage 2 (from about the first to the fourth month) as the stage of first habits. During stage 2, **primary circular reactions** appear, in which infants repeat some act involving their bodies. For example, they continue to suck when nothing is present. They continue to open and close their hands. There seems to be no external goal, no intent in these actions other than the pleasure of self-exploration. But infants are learning something about that primary object in their world: their own bodies.

Stage 3 Secondary circular reactions emerge during the third stage, which extends from about the fourth to the eighth month. During this stage, infants direct their activities toward objects and events outside themselves. Secondary circular reactions thus produce results in the environment, and not, as with the primary circular reactions, on the child's own body.

For example, Piaget's son, Laurent, continued to shake and kick his crib to produce movement and sound. He also discovered that pulling a chain attached to some balls produced an interesting noise, and he kept doing it. In this way, babies learn about the world "out there," and feed this information into their developing cognitive structures.

Stage 4 From about 8 to 12 months of age, infants **coordinate secondary schemes** to form new kinds of behavior. Now more complete acts of intelligence are evident (Piaget & Inhelder, 1969).

The baby first decides on a goal (finding an object that is hidden behind a cushion). The infant attempts to move objects to reach the goal. In stage 4, part of the goal object must be visible behind the obstacle. Here we see the first signs of intentional behavior.

Stage 5 Tertiary circular reactions appear from 12 to 18 months of age. In the tertiary circular reaction there is again repetition, but it is repetition with variation. The infant is exploring the world's possibilities. Piaget thinks that the infant deliberately attempts to provoke new results instead of merely reproducing activities. Tertiary circular reactions indicate an interest in novelty for its own sake.

A continuing interest in novelty produces the curiosity that motivates continuous growth and change in an infant's cognitive processes. For example, how many times have you seen a baby standing in a crib and dropping everything on the floor? But listen to Piaget: Watch how the baby drops things, from different locations and different heights. Does it sound the same when it hits the floor as the rug? Is it as loud dropped from here or higher? Each repetition is actually a chance to learn. Thanks to Piaget, you will be a lot more patient when you see this behavior.

Stage 6 During stage 6, the sensorimotor period ends and children develop a basic kind of *internal representation*. A good example is the behavior of Piaget's daughter, Jacqueline. At age 20 months, she approached a door that she wished to close, but she was carrying some grass in each hand. She put down the grass by the threshold, preparing to close the door. But then she stopped and looked at the grass and the door, realizing that if she closed the door the grass would blow away. She then moved the grass away from the door's movement and then closed it. She had obviously planned and thought carefully about the event before acting. Table 5.5 summarizes the accomplishments of the sensorimotor period.

As infants acquire the ability to form representations of objects, they begin to move through their environments more skillfully.

Progress through the sensorimotor period leads to four major accomplishments:

- *Object permanence:* Children realize that there are permanent objects around them; something out of sight is not gone forever.
- *A sense of space:* Children realize there is a spatial relationship among environmental objects.
- *Causality:* Children realize there is a relationship between actions and their consequences.
- *Time sequences:* Children realize that one thing comes after another.

By the end of the sensorimotor period, children move from purely sensory and motor functioning (hence the name *sensorimotor*) to a more symbolic kind of activity.

Criticisms of Piaget

As we noted in chapter 2, although Piaget has left a monumental legacy, his ideas have not been unchallenged. Piaget was a believer in the stage theory of development; that is, development is seen as a sequence of distinct stages, each of which entails important changes in the way a child thinks, feels, and behaves (Scarr & others, 1986). Rest (1983), however, argues that the acquisition of cognitive structures is gradual rather than abrupt and is not a matter of all or nothing; for example, a child is not completely in the sensorimotor or preoperational stage. A child's level of cognitive development seems to depend more on the nature of the task than on a rigid classification system.

Table 5.5 Outstanding Characteristics of the Sensorimotor Period

The Six Subdivisions of This Period	
Stage 1	During the first month the child exercises the native reflexes, for example, the sucking reflex. Here is the origin of mental development, for states of awareness accompany the reflex mechanisms.
Stage 2	Piaget refers to stage 2 (from 1 to 4 months) as the stage of *primary circular reactions*. Infants repeat some act involving the body, for example, finger sucking. (*Primary* means first, *circular reaction* means repeating the act.)
Stage 3	From 4 to 8 months, *secondary circular reactions* appear; that is, the children repeat acts involving objects outside themselves. For example, infants continue to shake or kick the crib.
Stage 4	From 8 to 12 months, the child "coordinates secondary schemata." Recall the meaning of *schema*—behavior plus mental structure. During stage 4, infants combine several related schemata to achieve some objective. For example, they will remove an obstacle that blocks some desired object.
Stage 5	From 12 to 18 months, *tertiary circular reactions* appear. Now children repeat acts, but not only for repetition's sake; now they search for novelty. For example, children of this age continually drop things. Piaget interprets such behavior as expressing their uncertainty about what will happen to the object when they release it.
Stage 6	At about 18 months or 2 years, a primitive type of representation appears. For example, one of Piaget's daughters wished to open a door but had grass in her hands. She put the grass on the floor and then moved it back from the door's movement so that it would not blow away.

An Applied View — What Do Infants Think About?

Examining cognitive development in infancy may lead you to say: Well, yes, infants are much more competent than I thought, but what do they think about? What do they feel? Answering these questions takes us from the observational and experimental world we have just explored into a psychodynamic view of infancy.

A Path of Sunshine: 7:05 a.m.

> A space glows over there. A gentle magnet pulls to capture. The space is growing warmer and coming to life. Inside it, forces start to turn around one another in a slow dance. (Stern, 1990, p. 55.)

Stern, a sensitive interpreter of an infant's thoughts and feelings, has used these words to describe how a path of brilliant sunshine can attract an infant's attention. Infants react to attractive stimuli for all the reasons we described in this chapter: intensity and complexity of the stimulation, visual ability, need for novelty, etc. As we watch infants react to such stimuli, we may ask what subjective experiences accompany these behaviors.

We can't crawl inside an infant's mind, but speculating about what an infant's experiences may be like can shape our idea of what an infant is. We saw a good example of the power of these ideas in chapter 2. Freud's notion of infancy as a period seething with emotions is in stark contrast to Piaget's belief that the infant is like a little scientist busily constructing a model of its world. Both these theorists have made inferences about an infant's subjective experiences.

Is there a starting point for attempting to explain an infant's subjective experiences of its own social life? An infant's sense of self is probably the best guide for us to follow, because the qualitative changes we see in development testify to new forms of personality integration (Stern, 1985). Infants, as they undergo these changes, seem to portray a new "presence" or "social feel." For example, when an infant smiles into a parent's eyes and coos at about 3 months of age, it is more than a change of behavior. As observers, we recognize something unique and we react differently.

Stern (1985) believes that infants are predesigned to be aware of these self-organizing processes and experience four different senses of self:

- The *emergent self* appears in the time from birth to 2 months. If you observe infants of this age, they show joy, distress, anger, and surprise, clear signs of subjective experience.
- The *core self* emerges between 2 and 4 months. Infants use memory and a growing sense of physical competence to organize their experiences. They slowly realize that their actions have consequences.
- The *subjective self* develops between 7 and 15 months. Infants begin to realize that they can share their experiences with others; for example, the mother knows the infant wants the cookies, and the child knows the mother knows. How many times have you seen a child of this age find something and smilingly show it to the mother?
- The *verbal self* follows after 15 months. From this time on, language and symbolic play are tangible clues about what is occurring in the infant's subjective world.

We should also remember, however, that the subjective world of parents also influences their children's development. Parents tend to see their children in a way that relates to their own needs, values, desires, and experiences. For example, Brazelton and Cramer (1990) have identified three parental fantasies:

- *The infant as ghost,* in which the baby reminds the parents of someone (usually dead), which in turn unleashes a flood of emotional feelings and may affect a parent's relationships with the child; Selma Fraiberg (1980) has referred to this situation as the "ghost in the nursery."
- *A parent's relationship with the infant reenacts a past mode of relationship.* Parents sometimes seek to recapture the relationships of their childhood through interactions with the infant. A mother who teased and fought with her brothers and sisters, or who might have kept a distance from them, adopts that same pattern with her children.
- *The infant as part of the parent.* Some parents attempt to project part of their selves on the infant: lazy, greedy, determined, stubborn.

The fantasy worlds of infants and parents affect children's development, and parents should be aware of such tendencies and avoid them where possible.

Such criticisms have led to a more searching examination of the times during which children acquire certain cognitive abilities. For example, Piaget believed that infants will retrieve an object that is hidden from them in stage 4, from 8–12 months. Before this age, if a blanket is thrown over a toy that the infant is looking at, the child stops reaching for it as if it doesn't exist.

Tracing the ages at which object permanence appears, Baillargeon (1987) devised an experiment in which infants between 3 1/2 and 4 1/2 months old were seated at a table where a cardboard screen could be moved back and forth. It could be moved forward (toward the baby) until it was flat on the table or backward (away from the baby) until the back of the cardboard touched the table.

Baillargeon then placed a painted wooden block behind the cardboard screen so that the infant could see the block when the screen was in a forward, flat position. But when the screen was moved backwards, it came to rest on the block, removing it from the infant's sight. Occasionally Baillargeon secretly removed the block so that the screen continued backwards until it rested flat on the table. The 4 1/2-month-old infants looked surprised at the change (they looked at the screen longer); even some of the 3 1/2-month-olds seemed to notice the "impossible" event. These findings suggest that infants may develop the object permanence concept earlier than Piaget originally thought.

Infants and Memory

As infants progress through these first two years, behavior appears that can only be attributed to memory. Discussing the appearance of memory, Kail (1990, p. 1) states that between 4 and 7 months, a marvelous change occurs in the relationship between infants and their parents. At 4 months, infants can distinguish between human and nonhuman objects in their environment (they smile and babble more to the human figures), but they are just as likely to smile at strangers as at parents. By 7 months, this has changed; infants will not smile at strangers and may appear threatened by them. (We'll discuss the psychosocial reasons for this change in chapter 6. Here we'll concentrate on cognitive explanations.)

As you can well imagine, testing infant memory is difficult because infants can't tell us whether or what they remembered. Consequently, investigators have relied on experiments that measure the time that infants look at familiar and novel objects. Kail (1990) gives the example of presenting two groups of infants with a bull's-eye pattern and a set of stripes. One group saw the bull's-eye first and then the stripes; the other group had the procedure reversed: first the stripes and then the bull's-eye. If both groups looked longer at the second presentation (the stripes for the first group and the bull's-eye for the second), it shows that they preferred the novel stimuli, thus giving evidence of infant memory. That is, they remembered the first and habituated and were then more interested in the novel stimulus.

Newborns seem to recognize events they have heard or seen before—recall the work of DeCasper and Fifer (1980) that we mentioned in chapter 4. By 2 to 3 months, infants will remember an event for several days, perhaps as long as a week. Memory continues to improve as infants "economize," that is, rather than remembering specific events, they begin to integrate their experiences (Shields & Rovee-Collier, 1992). This helps to explain how they find hidden objects (such as in the object permanence experiments); they can now recall things and events (Kail, 1990). The growth of memory ability is a significant developmental accomplishment of the last half of the first year of life (Perris & others, 1990).

Olson and Sherman (1983) summarize infant recognition memory as follows. During the first three months, infants show growing ability to retain what they have experienced for a relatively brief time (one to several days). From age 3 to 6 months, fairly consistent patterns of memory are present—lasting from 3 to 10 days.

From age 6 to 12 months, infants show joy when the mother returns; infants also tend to repeat first words, two practical reminders of their improving memory. From 12 to 24 months, language usage and reactions to family, friends, and strangers testify to an active, competent memory. You will have noticed in discussing the latter stages of the sensorimotor period how language becomes increasingly important.

Language Development

All children learn their native language at about the same time and in a similar manner. Here is the basic sequence of language acquisition:

- At about 3 months, children use sounds in a similar manner to adults.
- At about 1 year, they begin to use recognizable words.
- At about 4 years, they have acquired the complicated structure of their native tongue.
- In another two or three years, they speak and understand sentences that they have never previously used or heard.

The specific sequence of language development appears in Table 5.6.

Table 5.6 The Language Sequence

Language	Age
Crying	From birth
Cooing	5–8 weeks
Babbling	4–5 months
Single words	12 months
Two words	18 months
Phrases	2 years
Short sentences and questions	3 years
Conjunctions and prepositions	4 years

During the first year of life, the sounds that infants make change dramatically, from simple crying in the newborn to complex babbling just before speech begins (Sachs, 1985, p. 45). Tracing these changes in sound production, Sachs (1985) has identified four stages.

- *Stage 1* (0 to 8 weeks) This is a time of reflexive crying and vegetative sounds, when infants are quite limited in the sounds they can produce.
- *Stage 2* (8 to 20 weeks) Infants begin to coo and laugh, seeming more social and interactive with their parents.
- *Stage 3* (16 to 30 weeks) Infants use single syllables; this stage seems to lead into babbling.
- *Stage 4* (25 to 50 weeks) This is a time of advanced babbling.

- *Stage 5* Infants begin what Sachs (1985) calls "jargon babbling," that is, strings of sound uttered with a variety of stress and intonation patterns. These sounds in late babbling are similar to the first words that infants attempt.

Key Signs of Language Development

Table 5.6 describes the timetable of language acquisition until about age 4. Note that children combine words into sentences and use all categories of vocabulary (conjunctions and prepositions as well as nouns and verbs).

Several things signal difficulty in language acquisition. **Babbling** is a good example. When children babble, they make sounds that approximate speech. For example, you may hear an "eee" sound that makes you think that the infant is saying "see." This is to be expected. Deaf children, however, continue to babble past the age when other children begin to use words.

If you have the opportunity, listen to a child's speech when single words begin to appear. You will notice a subtle change before the two-word stage. *Children begin to use one word to convey multiple meanings;* For example, youngsters say "ball" meaning "give me the ball," "throw the ball," or "watch the ball roll." They have now gone far beyond merely labeling this round object as a ball.

When the *two-word stage* appears (at about 18 months), children initially struggle to convey tense and number. They also experience difficulty with grammatical correctness. Children usually employ word order ("me go") for meaning, only gradually mastering inflections (plurals, tense, possessives) as they begin to form three-word sentences. A youngster's efforts to inject grammatical order into language are a good sign of normal language development.

Sounds to Words

At about age 4 months, children make sounds that approximate speech. These increase in frequency until the children are about a year old, when they begin to use single words. After children commence using words, babbling still appears among the simple words.

We do not yet understand the relationship between babbling and word appearance. Babbling probably appears initially because of biological maturation. (Deaf children babble, which would seem to suggest that babbling does not depend on external reinforcement.)

Late in the babbling period, children use consistent sound patterns to refer to objects and events (Devilliers & Devilliers, 1978). These are called **vocables** and suggest children's discovery that meaning is associated with sound. For example, a lingering *l* sound may mean that someone is at the door. The use of vocables is a possible link between babbling and the first intelligible words.

At about age 1, the first words appear. Often called **holophrastic speech** (one word to communicate many meanings and ideas), it is difficult to analyze. These first words, or **holophrases,** are usually nouns, adjectives, or self-inventive words and often contain multiple meanings. "Ball" may mean not only the ball itself but "throw the ball to me."

Infants "tune into" the speech they hear and immediately begin to discriminate distinctive features. They also seem to be sensitive to the context of the language they hear; that is, they identify the emotional nature of speech. So the origins of language appear immediately after birth in infant gazes and vocal exchanges with those around them.

The precursors of language blend with babbling, which then merges into the first words, and is then continuous with the appearance of two words, phrases, and sentences. Remember: The period of single words is more than a time of merely accumulating more and more words. Although vocabularies increase, there are notable changes in both the kinds of words and the ways they are used between ages 1 and 2.

First Words

Children begin to use multiple words to refer to the things that they previously named with single words. Rather than learning rules of word combination to express new ideas, children learn to use new word forms. Later, combining words in phrases and sentences suggests that children are learning the structure of their language.

At about 2 years of age children's vocabularies expand rapidly, and simple two-word sentences appear. They primarily use nouns and verbs (not adverbs, conjunctions, or prepositions), and their sentences demonstrate grammatical structure. Although the nouns, adjectives, and verbs of children's sentences differ from those of adults, the same organizational principles are present. These initial multiple-word utterances (usually two or three words: "Timmy runs fast") are called **telegraphic speech.** Telegraphic speech contains considerably more meaning than superficially appears in the two or three words.

Word order and inflection (changing word form: e.g., "word"/"words") now become increasingly important. During the first stages of language acquisition, word order is paramount. Children combine words without concern for inflections, and it is word order that provides clues as to their level of syntactic (grammatical) development.

Once two-word sentences are used, inflection soon appears, usually with three-word sentences. The appearance of inflections seems to follow a pattern: first the plural of nouns, then tense and person of verbs, and then possessives.

Vocabulary Growth

Vocabulary constantly expands, but it is difficult to estimate the extent of a child's vocabulary since youngsters know more words than they articulate. Table 5.7 presents the approximate number of words that appear at various ages.

Table 5.7 The Pattern of Vocabulary Development—Estimates

Age (years)	Number of Words: Low	Number of Words: High
1	2–3	4–6
2	50	250
3	400	2,000
4	1,200	6,000
5	1,500	8,000
6	2,500	10,000

Language Acquisition: The Theories

Many of the language achievements we take for granted are actually amazing accomplishments that defy easy explanation. Imitation, although a powerful linguistic force, does not seem to be the sole explanation for a youngster's intuitive grasp of grammar, because a child hears so many incorrect utterances. Nor does imitation explain the manner by which thoughts are translated into words. Although we don't have a totally satisfactory explanation of how children acquire their language, four major theories have been proposed.

Biological Theory

A **biological,** or **nativist,** explanation focuses on innate language mechanisms that automatically unfold. Linguistic achievements that cannot be explained by imitation or some other cause have led some to a biological interpretation of language. What else but an innate capacity for language can explain the innovative nature of language? Children do not merely imitate those around them. If you listen carefully to young children, you will distinguish unique comlbinations of words, words that they never heard before. They may have heard the words *man, doll,* and *walk,* but never the combination "man walk doll." Such novel utterances testify to the creative aspects of language.

The work of Eric Lenneberg offers insights into the biological bases of the capacity for language (1967). Although the specific causal elements and the underlying cerebral mechanisms for the language explosion are still unknown, Lenneberg believes that imitation, conditioning, and reinforcement—all external factors—are inadequate explanations for language development, and that anatomical and physiological agents—internal factors—play a major part.

Lenneberg postulates that language development follows a biological schedule, which is activated when a state of "resonance" exits, that is, when children are "excited" in accordance with the environment, the sounds that they have been hearing suddenly assume a new, meaningful pattern. Consequently, if children of an appropriate age are placed in any language community, they will immediately and with little difficulty acquire that language.

Cognitive Theory

Piaget's *cognitive* explanation of language development views language as part of a child's emerging cognitive abilities. Piaget believes that language emerges not from a biological timetable, such as Lenneberg suggests, but from existing cognitive structures and in accordance with the child's needs. Piaget begins his basic work on language, *The Language and Thought of the Child* (1926), by asking, What are the needs that children tend to satisfy when they talk? What is the function of language for a child? Piaget answers this question by linking language to his belief in cognitive structures.

Recording the speech of two 6-year-old children, Piaget identified two major speech categories of the preoperational child: egocentric speech and socialized speech. Children engage in **egocentric speech** when they do not care to whom they speak, or whether anyone is listening to them (Piaget, 1926, p. 32). Children use **socialized speech** when they exchange views with others, criticize one another, ask questions, give answers, and even command or threaten. Piaget estimates that about 50 percent of the 6-year-old's speech is egocentric, and that what is socialized is purely factual. He also warns that although most children begin to communicate thought at between 7 and 8 years of age, their understanding of each other is still limited.

Seven or eight years of age is the beginning of the slow but steady disappearance of egocentrism, except in verbal thought, in which traces of egocentrism remain until about 11 or 12 years of age. Usage and complexity of language increases dramatically as children pass through the four stages of cognitive development. Piaget insists that the striking growth of verbal ability does not occur as a separate developmental phenomenon, but reflects the development of cognitive structures.

A different cognitive interpretation has been proposed by Vygotsky (1962), who argues that the roots of language and thought are separate and only become linked through development (1962, p. 119). Thought isn't just expressed *in* words; it comes into existence *through* words.

Vygotsky also sharply disagreed with Piaget's explanation of egocentric speech. Rather than seeing it as contributing little to cognitive development, Vygotsky believed that children use egocentric speech to "grasp and remedy the situation"; that is, to solve problems.

> ***Where's the pencil? I need a blue pencil. Never mind, I'll draw with the red one and wet it with water; it will become dark and look like blue.*** **(1962, p. 16.)**

Egocentric speech (now more commonly called private speech), which Vygotsky believed is a form of social communication, gradually becomes internalized. Thus the true direction of thought and language is not from the individual to the socialized, but from the social to the individual (Vygotsky, 1962).

Psycholinguistic Theory

A **psycholinguistic** interpretation attempts to explain how native speakers can understand and produce sentences that were never spoken or written. Noam Chomsky, a professor of linguistics at the Massachusetts Institute of Technology, believes that all humans have an innate capacity to acquire language as a result of their biological inheritance. Trained in linguistics, mathematics, and philosophy, Chomsky goes further than Lenneberg in his biological views: Not only do we have a biological predisposition for language, we also have an innate knowledge of language. He calls it our **language acquisition device (LAD).** Chomsky (1965) also states that no one acquires a language by learning billions of sentences of that language. Rather, children acquire a grammar that can generate an infinite number of sentences in their native language.

Chomsky's work is usually referred to as psycholinguistics, a combination of psychology and linguistics. Linguistics is the study of the rules of any language, whereas psychology focuses on behavior. Linguists assume that the rules of language are part of our knowledge; psychologists have attempted to discover how these rules are represented in our minds (especially the capacities that children must have to master the rules of their language). The combination of the two approaches has produced the field of psycholinguistics (Gardner, 1982).

Children possess an innate competence for language acquisition, just as they possess an innate capacity for walking. No one has to tell children *how* to walk—they walk and talk without consciously knowing how they did either. Although all normal children possess approximately the same language competence, their performance, or use of language, varies increasingly as they grow older. This variation is largely due to differences in opportunities to learn how to use language, which includes not only speaking but also listening, writing, and reading.

Behaviorist Theory

A *behavioristic* explanation concentrates on language as a learned skill. Children utter sounds that are reinforced and shaped by the environment, especially parents and teachers, and gradually learn to make distinguishable sounds and to form correct sentences. B. F. Skinner, a leading behaviorist, proposed a detailed behavioristic theory of language acquisition in one of his early books (1957).

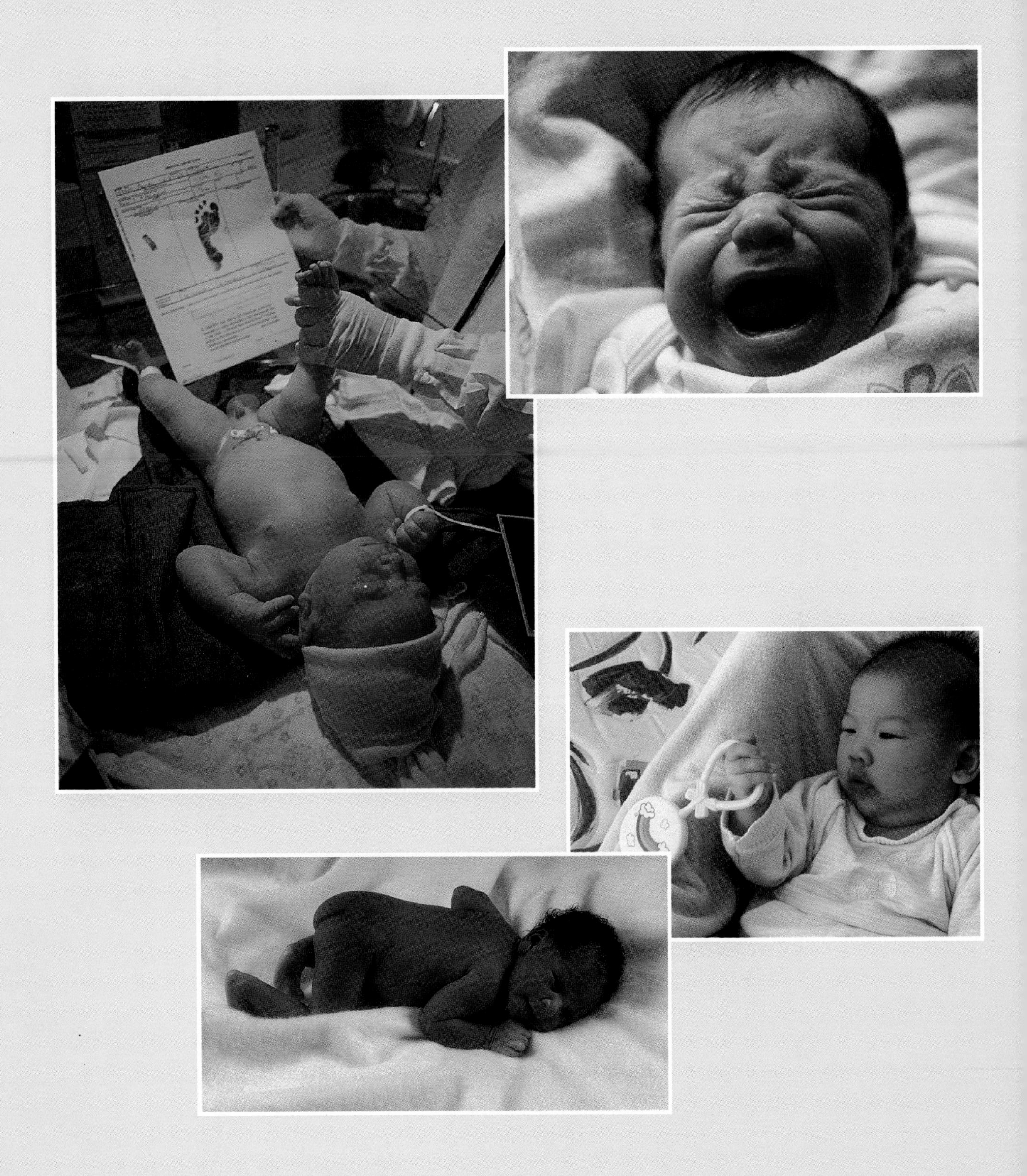

During the infancy years, babies use the native talents with which they were born until, by the end of the infancy period (about two years of age),they typically are walking, talking children with an insatiable thirst for knowledge. Beginning with the reflexes that help them to survive, infants experience one of the most rapid periods of development in the human lifespan. Brain, body, and mind develop at a pace that challenges observers not only to chronicle but also to interpret the many changes that occur. Physical growth is obvious; less so are the exciting changes that take place in the

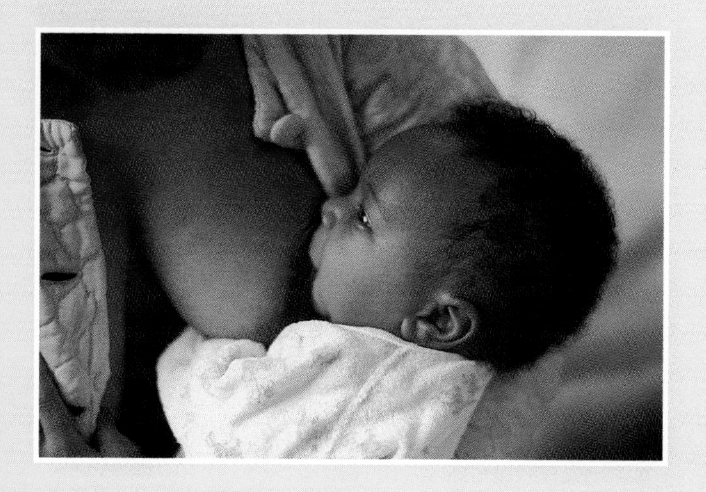

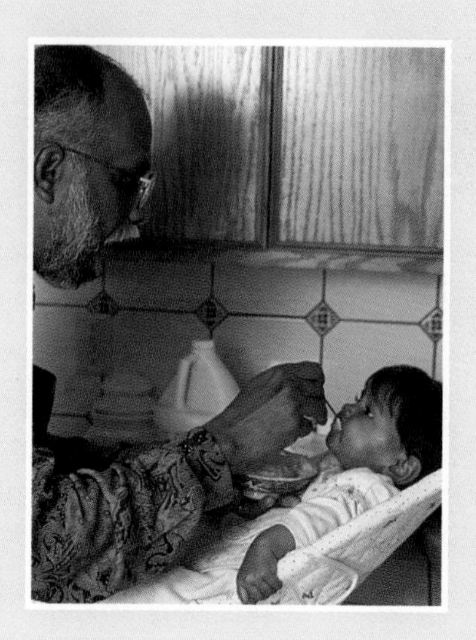

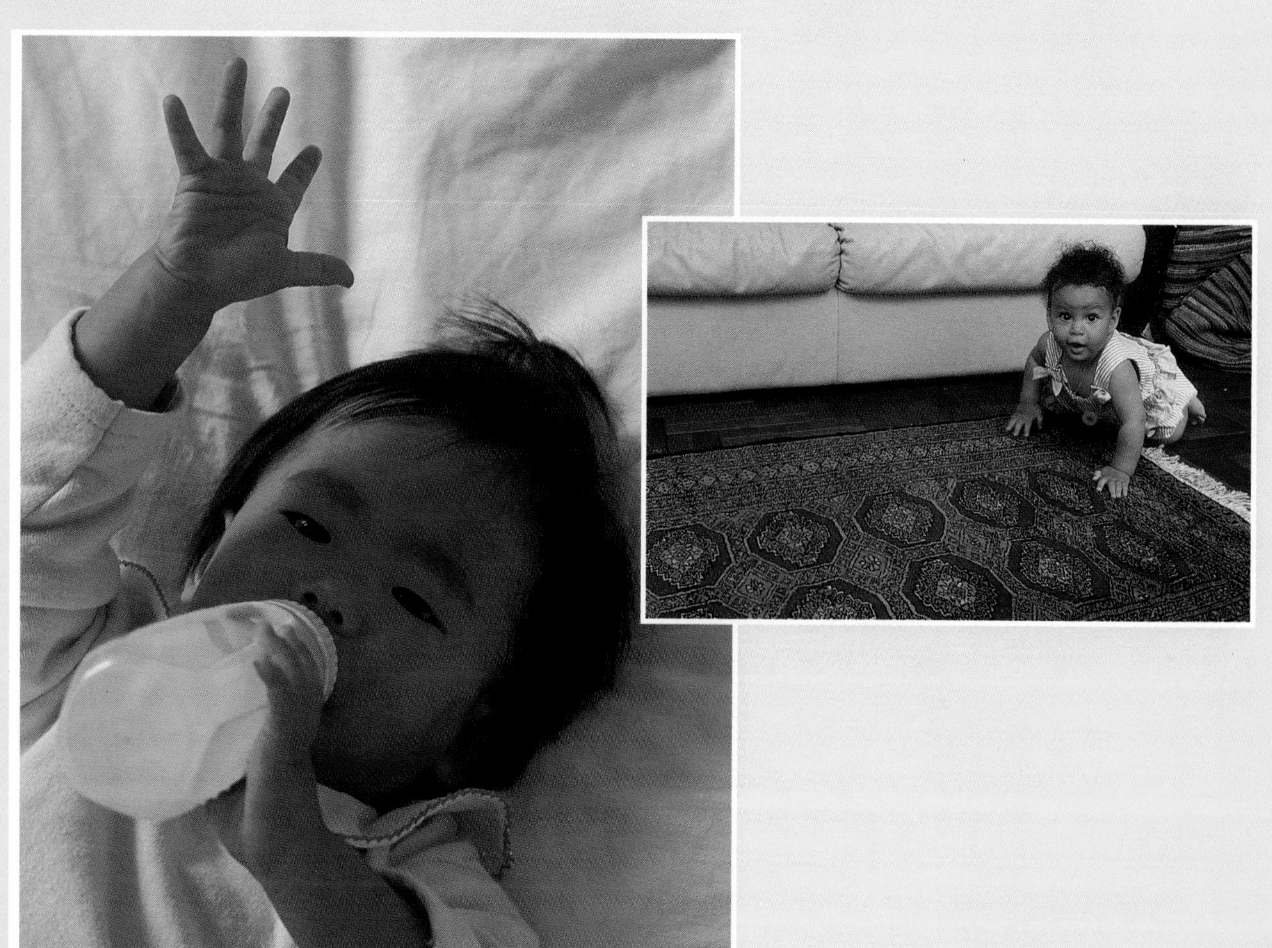

brain, thus enabling infants to acquire the competencies that so dramatically appear. Crawling, standing, walking, and talking highlight motor development. These abilities, coupled with cognitive and language development, help infants to construct a picture of their world. A warm, supportive environment in which infants can attach to loved ones provides the emotional security that leads to needed self-esteem. With our current knowledge that infants are active partners in their own development has come the realization that these years have important implications for a child's future.

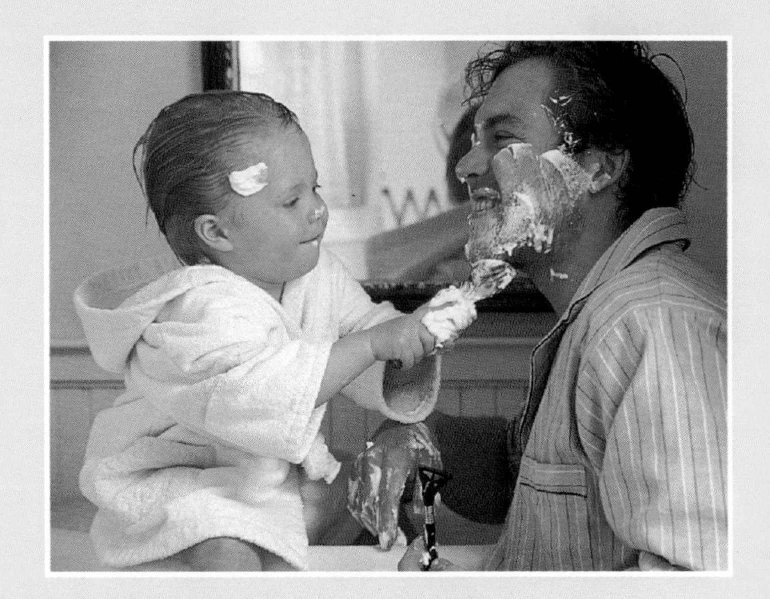

Note how frequently we have said that cognitive development and language development occur in relation to other people. Huttenlocher and associates (1991) have found that parental speech input at 16 months accounts for considerable variation among children in vocabulary growth. This finding points to the critical role that social development plays in an infant's growth.

Table 5.8 summarizes several of the developmental highlights we have discussed in this chapter.

Table 5.8 Developmental Characteristics of Infancy

Age (months)	Height (in.)	Weight (lb.)	Language Development	Motor Development	Cognitive (Piaget)
3	24	13–14	Cooing	Supports head in prone position	Primary circular reactions
6	26	17–18	Babbling—single-syllable sounds	Sits erect when supported	Secondary circular reactions
9	28	20–22	Repetition of sounds signals emotions	Stands with support	Coordinates secondary schemata
12	29.5	22–24	Single words—"mama," "dada"	Walks when held by hand	Same
18	32	25–26	3–50 words	Grasps objects accurately, walks steadily	Tertiary circular reactions
24	34	27–29	50–250 words, 2–3 word sentences	Runs and walks up and down stairs	Representation

As we complete this initial phase of examining infant development, remember that all phases of development come together in an integrated manner. Motor development is involved when a child moves excitedly toward its mother on her return. Language development is involved when infants intensify their relationships with their mothers by words that are now directed toward her. Cognitive development is probably less obvious but just as significant: Children are excited by their mothers' return because they remember their mothers. Consequently, a sound principle of development remains—all development is integrated.

Conclusion

You should now view an infant as an individual of enormous potential, one whose activity and competence is much greater than originally suspected. It is as if a newborn enters the world with all its systems ready to function and eager for growth. What happens during these first two years has important implications for future development. Setbacks—both physical and psychological—will occur, but they need not cause permanent damage. From your reading in chapter 1, you realize that human infants show remarkable resiliency.

How do infants first learn that they can trust those around them? The answer lies in the way their initial needs are satisfied. They have the ability to detect and react to parental signals. In chapter 2 you read about Erikson's stages of development. In light of what you now know about infant potential, it is easier to accept Erikson's great contribution to our understanding of infants' development of trust.

You are also now aware that proper need satisfaction entails psychological as well as physical comfort. These first parent-infant interactions furnish the basis for attachment, the knowledge that others can be trusted. In chapter 6, we turn to an infant's social development and analyze how interactions with others affect development.

Chapter Highlights

Physical and Motor Development

- Newborns display clear signs of their competence: movement, seeing, hearing, interacting.
- Infants' physical and motor abilities influence all aspects of development.
- Techniques to assess infant competence and well-being are widely used today.
- Motor development follows a well-documented schedule.
- Infants can develop problems such as SIDS and FTT for a variety of reasons.

Perceptual Development

- Infants are born with the ability to detect changes in their environment.
- Infants are capable of acquiring and interpreting information from their immediate surroundings.
- Infants from birth show preferences for certain types of stimuli.

Cognitive Development

- Infants, even at this early age, are attempting to answer questions about their world, questions that will continue to occupy them in more complex and sophisticated forms throughout their lives.
- One of the first tasks that infants must master is an understanding of the objects around them.
- Piaget's theory of cognitive development has shed considerable light on the ways that children grow mentally.
- A key element in understanding an infant's cognitive development is the role of memory.

Language Development

- Infants show rapid growth in their language development.
- Language acquisition follows a definite sequence.
- Language behaviors in infancy range from crying to the use of words and phrases.

KEY TERMS

Apgar
Apnea
Babbling
Binocular coordination
Biological, or nativist theory
Coordination of secondary schemes
Crawling
Creeping
Ectoderm
Egocentric speech
Egocentrism
Endoderm
Failure to thrive
Holophrase
Holophrastic speech
Imitative behavior
Language acquisition device (LAD)
Mesoderm
Neonate
Neurological assessment
Object permanence
Primary circular reactions
Psycholinguistic
Reciprocal interactions
Reflex
Respiratory distress syndrome
Rooting
Secondary circular reactions
Sensorimotor period
Sleeping disorders
Socialized speech
Sudden infant death syndrome
Telegraphic speech
Tertiary circular reactions
Variable accommodation
Vocables

WHAT DO YOU THINK?

1. The shift from considering an infant as nothing more than a passive sponge to seeing infants as amazingly competent carries with it certain responsibilities. We can't be overly optimistic about a baby's abilities. Why? What are some of the more common dangers of this viewpoint?
2. Testing infants has grown in popularity these past years. You should consider some cautions, however. Remembering what you have read about infants in this chapter, mention several facts you would be careful about.
3. You have been asked to babysit your sister's 14-month-old baby. When you arrive, the mother is upset because she has been repeatedly picking up things that the baby has thrown out of the crib. With your new knowledge, you calm her down by explaining the baby's behavior. What do you tell her?
4. After reviewing the infancy work, what do you think about this period as "preparation for the future?" Select one phase of development (for example, cognitive development) and show how a stimulating environment can help to lay the foundation for future cognitive growth.

SUGGESTED READINGS

Field, T. (1990). *Infancy.* Cambridge, MA: Harvard University Press. A clear, simple, and carefully written account of infancy by a well-known commentator on these years.

Harris, J. R. and R. Liebert. (1992). *Infant & child.* Englewood Cliffs, NJ: Prentice-Hall. A thorough, well-documented account of the early years; particularly strong in its use of the transactional model.

Maurer, D. and C. Mowrer (1988). *The world of the newborn.* New York: Basic Books. A fascinating explanation of the newborn's experiences—scientifically sound and enjoyable to read.

Osofsky, J. (ed.) (1987). *Handbook of infant development.* New York: Wiley. This book is really the "bible" of infancy. If you have any questions about infant behavior, this is the text to use.

Restak, R. (1986). *The infant mind.* New York: Doubleday. You will find this popular, well-written account of an infant's world not only informative but entertaining.

CHAPTER 6

Psychosocial Development in Infancy

Relationships and Development 144

Background of the Relationship Research 144

The Importance of Reciprocal Interactions 145

The Meaning of Relationships 148

The Characteristics of a Relationship 149

The Givens in a Relationship 150

First Relationships 153

Relationships: Beginnings and Direction 153

The Developmental Sequence 154

Attachment 156

Bowlby's Work 156

Attachment Research 158

The Klaus and Kennell Studies 160

Fathers and Attachment 162

Current Research into Attachment 163

Early Emotional Development 164

Signs of Emotional Development 164

Theories of Emotional Development: A Summary 165

First Feelings 165

Conclusion 166

Chapter Highlights 167

Key Terms 168

What Do You Think? 168

Suggested Readings 168

Janice watched with exasperation as her 1-year-old son, Joseph, sat crying. Her friend Laura had dropped in for a cup of coffee and had started to play with Joseph when he began to cry and pull away from her. Janice had noticed this happening often lately and was concerned that something was bothering him. She could tell that her friend, who had no children, was hurt; Janice wondered if the problem was serious enough to call her doctor for advice. Joseph looked fine and seemed to show no symptoms of any illness, which made her hesitate.

Later that day, after her friend had left and Joseph was taking his nap, Janice thought about his behavior. If he wasn't sick, what was the problem? She was concerned that he might be so shy that he could have difficulty with relationships and in making friends. She was also bothered that she could be doing something wrong. Joseph was a first child; could she be keeping him too close to her and perhaps subconsciously resisting any contacts he might have with others?

Janice decided to take Joseph shopping the following day and visit the local bookstore. The next morning, Joseph was bright and cheerful and kept himself busy with a toy while Janice browsed through the books in the child-care section of the bookstore. At first she turned to those books that dealt with specific problems, but she didn't know what to look for, so she turned to more general discussions of child rearing. Still not satisfied, she was about ready to give up when she came across a book that described the milestones that occur during infancy.

Sure enough, there it was. "Beginning in the second half of the first year, infants start to show anxiety in the presence of strangers. This behavior is probably the first clue you will notice that tells you attachment is developing." Janice read the words again, relieved and delighted; Joseph was perfectly normal. Now, however, she was determined to learn more about attachment. ■

Beginning at about six months of age, infants show signs of distress when approached by a stranger.

In chapters 1 and 5, we noted that infants are much more actively involved in their own development than we had previously realized. This knowledge has enabled us to examine an infant's psychosocial development from a totally new and exciting perspective. Not only do infants attempt to make sense of their world as they develop cognitively, they also "tune into" the social and emotional atmosphere surrounding them and immediately begin to shape their relationships with others.

There is a growing recognition today that relationships with others are the basis of our social development, commencing in those first days after birth. Joseph's relationship with his mother (the bond with her that's usually called attachment) will become the basis for his relations with others. The attachment that forms between mother and child tells the child what to expect from others. (Usually the bond is between the mother and child, although occasionally it may be between the child and someone else—father, grandparent.)

Think for a moment of what you have read about infants, especially their rapid brain and cognitive development. Infants take in information, and some of this information concerns how they are treated by others, especially their mothers. While they may not grasp the significance of what's going on around them, they can see and hear and understand how they are being treated.

It is what Erikson has called the time of trust, which means that children acquire confidence in themselves as well as others. The degree of trust acquired does not depend as much on nutrition and displays of love as it does on the quality of the mother-child relationship. Parents can encourage a sense of trust by responding sensitively to their infant's needs. Physical contact and comfort are crucial for trust to develop. Gradually, as children master motor control, they learn to trust their bodies, thus increasing their psychological sense of security. With the aid of their parents, and their own growing competence, they begin to think of the world as a safe and orderly place.

An infant's relationships begin with its mother and then extend to father, siblings, friends, and so on. We know now that the entire range of a child's relationships contributes to social development.

Growing knowledge about the emergence of relationships, how they change, and how they affect development has become a key component of developmental research. New findings will create ways to help children master the developmental tasks they face.

Recent research (which we shall discuss) has strikingly demonstrated that everyone in your family is a unique individual, a quite different personality. While growing up, one child might have been exceptionally active while another was shy and quiet. These temperamental differences caused your parents to treat each of you differently.

To help us untangle this important network, in this chapter we'll first present several basic ideas about relationships (such as the active role that infants play in their own development). We'll then analyze the meaning of relationships: what are they and what do they consist of? Next, we'll examine the source of those first relationships and how they develop and influence later relationships. Finally, we'll explore the research on attachment that has attracted so much attention these past few years. Let's begin by looking at several basic ideas about relationships.

The topics just mentioned—attachment, reciprocal interactions, and relationships—"fit" our biopsychosocial model as shown in Table 6.1.

After reading this chapter, you should be able to:

- Describe how theory and research have contributed to our understanding of relationships

Table 6.1 Characteristics of the Period

Bio	Psycho	Social
Personal appearance	Reciprocal interactions	Interactions
Temperament	Sensitive responsiveness	Characteristics of a relationship
	Goodness of fit	First relationships
	Motives	Expanding relationships
	Attachment	
	Intergenerational continuity	

- Assess the role of reciprocal interactions in any relationship
- Distinguish the characteristics of a relationship
- Identify the givens in a relationship
- Trace the origins of relationships
- Evaluate the significance of the mother-infant relationship
- Analyze the impact of attachment on psychosocial development

Relationships and Development

Consider for a moment all that a single relationship incorporates:

- Physical aspects of development such as walking, running, and playing with a peer
- Language aspects, which enable youngsters to share one anothers' lives
- Cognitive aspects, which allow them to understand one another
- Emotional aspects, which permit them to make a commitment to another
- Social aspects, which reflect both socialization and individuation

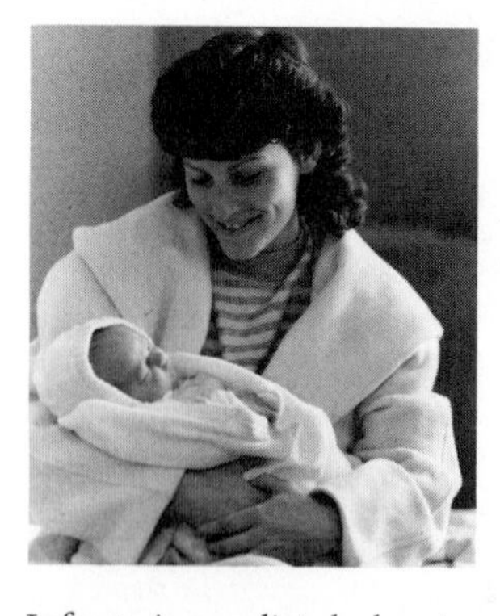

Infants immediately begin to take in information from their environment, and mothers are an important source of this information. From mothers, infants begin to develop a sense of how the world will treat them.

All of these features combine as youngsters interact with a remarkable variety of individuals, from those first crucial days with parents to later interactions with siblings, peers, teachers, and many others. Table 6.2 illustrates the entire range of individuals with whom children interact, from family members to friends.

As you can see from the table, children continue to widen their circle of relationships as they grow. Our concern here, however, is with the origin of relationships in infancy, a concern that has grown out of the changing view of infancy we discussed in chapter 5. How can we link these changes to the current interest in the role of relationships? Let's briefly review several of the developmental concepts we discussed previously.

Background of the Relationship Research

One of the major reasons for the interest in relationships is that developmental psychologists became dissatisfied with studying the relationship between mother and child. Greater precision was demanded: what do we know about the content of the interactions between mother and child? Can we study the quality of these interactions? Is it possible to determine the mother's commitment to the relationship?

Table 6.2 Sources of Pupils' Relationships

Parent-Adult	Sibling-Relatives	School	Peers	Others
Mother-father	Brothers	Principal	Male	Day-care personnel
Mother-stepfather	Sisters	Teacher(s)	Female	Friends (family)
Father-stepmother	Stepbrothers	Classmates	Group	Neighbors
Father	Stepsisters	(older, younger)		Religious
Mother	Extended			Medical
Mother-other	(cousins, etc.)			Psychological
Father-other	Order of birth			Specialists (coaches, counselors, music and art teachers, etc.)
Relatives: aunt, uncle, grandparents				
Foster				
Adopted				
Institutional				

Another important influence has been the acceptance of a **transactional model of development,** which recognizes the child's active role in its development. Also called reciprocal interactions, this model has been described by Brazelton (1984, 1987), Brazelton and Cramer (1990), and Sameroff (1975). Its meaning is quite simple: I do something to the baby and the baby changes; as a result of the infant's change, I change. The process continues indefinitely.

Finally, the recognition that mothers react differently to different children has introduced the concept of **sensitive responsiveness.** For example, we now know that babies are temperamentally different at birth. While most infants like to be held, some dislike physical contact. How will a mother react to an infant who stiffens and pulls away, especially if previous children liked being held? This research has contributed to a greater understanding of the role of relationships in development.

Remember: In any adult-child interactions, infants also exercise some control over the interactions. We, as adults, respond to infants partly because of the way that they have responded to us. An infant's staring, cooing, smiling, and kicking can all be employed to maintain the interactions. Thus these early interactions establish the nature of the relationship between mother and child, giving it a particular tone or style.

We can usually label relationships, using adjectives such as *warm, cold, rejecting* and *hostile.* But we must be cautious. Any relationship may be marked by apparently contradictory interactions. A mother may have a warm relationship with her child as evidenced by hugging and kissing, but she may also scold when scolding is needed for the child's protection. To understand the relationship, we must understand the interactions.

The Importance of Reciprocal Interactions

Thanks to our changed concept of infants (active, not passive), we now recognize the significance of reciprocal interactions in the development of relationships. This new way of looking at infants came about mainly because of the following views, which we have mentioned in chapters 1, 2, and 5.

A Biopsychosocial Model

As we have seen, a new model, a different way of looking at behavior, was needed to understand infant behavior and development. This led to what has become known as the **biopsychosocial model,** which is the model used in this book to

A Multicultural View 

Different Cultures, Different Interactions

In a thoughtful review of the literature analyzing minority infants (African-American, Hispanic-American, Asian-American, Native American, Alaskan native, and Pacific Islander), Coll (1990) notes that during the first three years of life, infants develop those psychomotor, cognitive, and psychosocial skills that enable them to become accepted members of their cultural and social systems. Because of their families' backgrounds, minority infants are exposed to unique experiences that can influence their developmental outcomes in ways not yet understood.

Parents from a particular culture share a common system of beliefs, values, practices, and behaviors that differ from parents in other cultures. Parents from different cultural backgrounds differ in their views of infant competence, how to respond to crying, and what developmental skills are most significant (Coll, 1990). Consequently, infants are subject to different parental behaviors that are shaped by a particular culture.

Many interactions between parents and their infants are universal, such as baby talk, facial expressions, and play. Behaviors such as baby talk are so common that a mother's behavior may seem infantile. Many other maternal behaviors, however, are not that common. Eye contact and face-to-face talk are avoided in some cultures (Field, 1990). Specific examples include Chisholm's study (1983) of the use of cradle boards with Navajo infants. The arousal level and activity level of these infants seems to be lower than infants from other cultures, which affects mother-infant interactions. There are fewer and less intense mother-infant interactions. Chisholm (1983) also found that Navajo infants have fewer contacts with strangers and are less fearful of them in the first year than Caucasian infants, but the pattern reverses in the second year.

Studying mother-infant interactions of 51 low-socioeconomic-status Mexican-American mothers in the Los Angeles area, Zepeda (1986) discovered that tactile stimulation was used more frequently than vocalizations. This study is consistent with other findings that the early interactions of Mexican mothers with their infants are mainly nonverbal. These differences in parental behaviors may have important developmental outcomes, but as yet research has not provided specific answers.

integrate all aspects of development. According to this model, development occurs as the result of an intertwining of biological, psychological, and social forces. For example, an infant who experiences a gastrointestinal problem that produces vomiting has little interest in interacting with others. A biological condition has produced both psychological and social consequences.

Active Processors of Information

A second change came when investigators realized that, from birth, children are active processors of information. They don't only react to stimuli; they see and hear at birth and immediately begin the task of regulating their environment. They fight to control their breathing, and they struggle to balance digestion and elimination.

But brief, calm periods appear after birth when infants take in information from the surrounding world. These fleeting but significant periods are the foundation for the appearance of key developmental milestones during the infancy period. Infants are ready to respond to social stimulation. It is not only a matter of responding passively; infants in their own way can initiate social contacts. Many of their actions (such as turning toward their mothers or gesturing in their direction) are forms of communication. Hinde (1987), too, notes that the interchange between infants and their environments is an active one. Those around infants try to attract their attention, but the babies actively select from these adult actions. In other words, infants begin to structure their own relationships according to their individual temperaments.

Sensitive Responsiveness

The use of a biopsychosocial model and a recognition that children are active processors of information prove helpful if parents respond sensitively to their children. Sensitive responsiveness appears to be essential for parenting. It also seems to be more natural for some adults than others. Still, it is a skill that can be acquired and improved upon with knowledge and experience.

Here are several conclusions that can help parents to understand their child's behavior.

- Children are temperamentally different at birth.
- They instantly tune in to their environment.
- They give clues to their personalities.
- A mother's and father's responses to a child's signals must be appropriate for that child.

Transactions or Reciprocal Interactions

As the preceding three views altered our understanding of the active-passive role of children, a different model of development, called transactional (Sameroff, 1975) or reciprocal interactions (Brazelton & Als, 1979), became widely accepted. The unique temperamental distinctions that infants display at birth cause unique parental reactions. The easy or difficult infant has a decided impact on parents. It doesn't take much effort to visualize how parents respond differently to a crying or cooing infant. (See chap. 1.)

When parents change their behavior according to their child's behavior, they signal pleasure, rejection, or uncertainty about what their child is doing. Children act on their parents and change the parents; these changes are then reflected in how parents treat their children. The tone of the interactions between children and parents assumes a definite structure that will characterize the relationship for the coming years. Figure 6.1 illustrates the back-and-forth reaction of parents and their children.

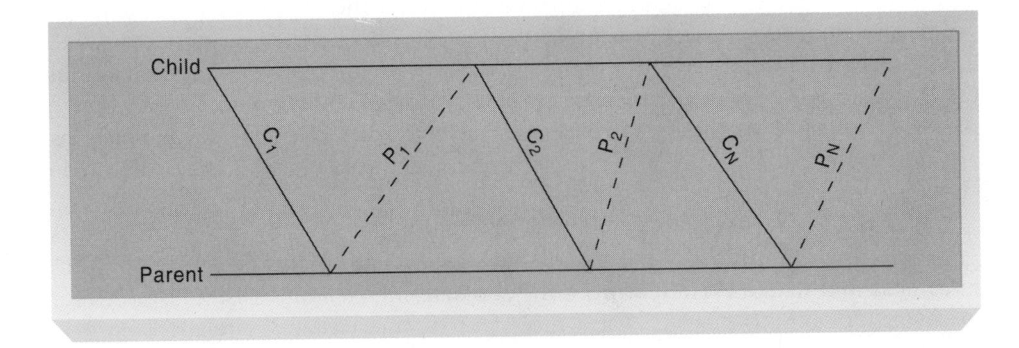

Figure 6.1
Transactional analysis, or reciprocal interactions

Once developmental psychologists accepted these four views, investigations into bonding, attachment, first relationships, peer relationships, and relationships in special situations (such as with irritable children) could be subjected to more precise and meaningful analysis.

Understanding these views, we can now turn to an analysis of the role of relationships in development. First, however, just what do we mean by relationships?

What's Your View?

?

We cannot exaggerate the importance of the initial encounters that infants have with the adults around them. In a particularly significant study that was one of the first to focus on the lingering effects of these early interactions, Osofsky (1976) examined the link between neonatal characteristics and early mother-infant relations in 134 mothers and their 2- to 4-day-old infants.

She observed the infants at a scheduled feeding and in a 15-minute stimulation situation, during which the mothers presented tasks from the Brazelton Neonatal Assessment Scale to their children. The infants were next evaluated (between 2 and 4 days of age) using the full Brazelton scale.

A particularly significant finding showed that the overall pattern of interactions indicates consistent maternal and infant styles that appear soon after birth. Infants who were highly responsive during the Brazelton assessment were also highly responsive during the stimulation periods. She also found strong correlations between the mother's stimulation and the child's responsiveness: More sensitive mothers have more responsive infants.

Again, note the evidence supporting the importance of the mother-infant interactions. Osofsky concluded that both infant and mother contribute to the style of the relationship. With the pattern of interactions formed almost from birth, a clearly defined relationship is set that will undoubtedly shape the course of social development. Do you agree with Osofsky's conclusion that the early mother-infant relationship shapes future relationships? What's your view?

The Meaning of Relationships

Think for a moment about your friends. What type of relationship do you have with them? with your parents? with a husband or wife? with a child? Now consider this classic definition of a **relationship.**

> ***A relationship implies a pattern of intermittent interactions between two people involving interchanges over an extended period of time.*** **(Hinde, 1979, p. 14.)**

If you have a true relationship with someone, that relationship has *continuity.* That is, you can continue to maintain a relationship with a friend you have not seen for years. An extended series of **interactions,** however, does not necessarily constitute a relationship. The cashier you frequently see at the supermarket, the attendant at the gas station, the receptionist at your dentist's office with whom you exchange pleasantries—none of these become partners in a relationship. If the interactions are nothing more than an exchange of money and a thank you, they cannot be classified as relationships.

Although a relationship is identified by the history of these interactions, it is still more than the sum of these interactions. In a relationship, the interactions are integrated in a manner different from the separate interactions. It isn't just a matter of saying, "Good morning. Isn't it a nice day?'" to the cashier. A relationship also involves your perceptions, your mental picture, and your feelings for the other person. What one says or does is significant, but how the partner perceives and judges that behavior is even more important.

These ideas help us to understand the developing relationship between parents and infants. We think today that infants "tune into" their environment from birth. Thus they react to far more than their parents' behavior; the quality of the interactions instantly begins to establish the nature of the relationship.

The Characteristics of a Relationship

Hinde (1979, 1987) argues that eight categories are useful for analyzing relationships: the content of interactions, the diversity of interactions, the qualities of interactions, the relative frequency and patterning of interactions, reciprocity versus complementarity, intimacy, interpersonal perception, and commitment. Table 6.3 describes the meanings of these categories.

Table 6.3 Categories of Interactions

Category	Meaning
1. Content	a. What the partners do together. b. May distinguish different relationships—mother-child, father-child. c. Enables us to label the relationship.
2. Diversity	a. Indicates the types of interactions making up the relationship. b. The number of things mother and child do together contributes to infant's understanding of others.
3. Quality	a. Not only what partners are doing but also how they are doing it. b. A mother may handle an infant gently or roughly.
4. Relative frequency and patterning of interactions	a. Frequency of interactions but in relation to other types of interactions. b. A pattern of warm interaction may demand hugging and scolding.
5. Reciprocity versus complementarity	a. Doing the same things simultaneously or taking turns. Children's play is an example of reciprocity. b. Interactions are complementary when they are different yet blend together, such as a mother changing or feeding an infant.
6. Intimacy	a. Extent to which partners are prepared to reveal themselves to each other—probably never total. b. Meaningless for infants, but changes quickly with development.
7. Interpersonal perception	a. How the partners see each other. b. Since our sense of self is shaped by reactions of others, the importance of first relationships is clearly evident.
8. Commitment	a. Acceptance of relationship—infant and child have little choice. or b. Decision to work toward continuing a particular relationship.

As babies begin to interact with their mothers, a pattern for future relationships is established. The more diverse the interactions that a baby engages in with its mother, the richer the relationship becomes.

The value of Hinde's classification is that we no longer can be satisfied with observing the mother-infant relationship. We can now examine specific features of the relationship (content, quality, etc.), study them, conduct research, and determine how, or if, these eight categories are integrated. Hinde's categories are also valuable for discovering just what might be wrong in a relationship. This work promises to help us achieve much deeper insights into the dynamics of relationships.

An Applied View — Do Infants Prefer Attractive Faces?

In an interesting experiment, Langlois and associates (1991) demonstrated that infants discriminate among adult female faces on the basis of attractiveness. These findings were surprising to many, because infants were not expected to make such subtle discriminations (Langlois & others, 1991, p. 79). In addition, most researchers believed that children only gradually learn what their culture has identified as "attractive."

Working with 60 infants who were 6 months old, the researchers presented them with color slides of 16 adult males and 16 adult females. Half of the slides of each sex were attractive; half were unattractive. (Attractiveness and unattractiveness were rated by 40 undergraduate men and women.) The infants consistently looked longer at the faces judged to be attractive.

In a second phase of their study, Langlois and her colleagues used the faces of black adult women as stimuli. They tested 43 infants who were 6 months old. Both black and white adults rated 16 slides for attractiveness and unattractiveness. The infants looked longer at the faces of attractive black women. In the final phase of their study, the researchers used slides of attractive and unattractive babies as stimuli for 52 six-month-old infants. Once again, the infants preferred the more attractive faces.

The results from these studies showed that infants discriminate attractive from unattractive, and that they prefer the attractive faces of diverse types (Langlois & others, 1991). What is particularly interesting about these studies is that the infants treated attractive faces as distinctive regardless of sex, color, or age. They reacted this way although most of the infants had little experience with some of the types of faces they viewed and little idea of what their culture considered attractive or unattractive. These infants were clearly demonstrating visual preference.

Using the concept of reciprocal interactions to explain how both partners influence a relationship raises the question: What do adults think about the attractiveness of infants? Studying adult responses to infant attractiveness, Ritter and associates (1991) found that adults thought that attractive infants were younger than they actually were. They also assumed unattractive infants were older than they actually were, which led them to expect more advanced behaviors from these infants.

As the researchers noted, less attractive and older-appearing infants are trapped in a vicious cycle of unrealistic expectations and apparently immature behavior. Their failure to behave as expected only increases negative evaluations by adults. You can see how a child's appearance affects parents, teachers, and other adults and influences the relationship either positively or negatively.

The Givens in a Relationship

In the mother-infant relationship (or father-infant), both individuals bring to the relationship physical and biological characteristics ranging from appearance to hereditary endowment. Since we have previously discussed genetic contribution, here we can focus on personal characteristics, such as personal appearance and temperament.

Personal Characteristics

Physical appearance has a powerful effect on a relationship. For good or ill, physical appearance affects the manner in which others react to us. Attractiveness is as important for infants as it is for adults (Langlois & Downs, 1979; Ritter & Langlois, 1988).

You know how appealing the mere sight of a happy baby can be. Their facial expressions and the shape of their features are attractive to most people and help to ensure an infant's survival, given their immaturity and helplessness. The appearance of a sick baby generates concern. We almost instinctively react to their distress. We respond in yet a different manner to an unhappy, crying baby. Tired, frustrated parents may find it difficult to react positively, tending instead to be abrupt and stiff with the baby. Again, an infant's appearance structures our interactions. From what we know of a baby tuning into its environment, we can understand how easily appearance can affect those first relationships.

The first mother-infant interactions quickly establish the style and tone of the relationship (Osofsky, 1976). If the interactions are altered by the infant's physical attractiveness, sensitive responsiveness becomes a matter of prime concern. Given the importance of reciprocal interactions, mothers can initiate a relationship in which an infant quickly realizes that there is something wrong with the quality of the interactions.

Temperament

Defining **temperament** as the individual's behavioral style in interacting with the environment, Carey (1981) states that children's temperaments contribute significantly to their interactions with their environments and immediately begin to structure relationships. Goldsmith and Campos (1990) view temperament as individual differences in tendencies to express the primary emotions, which helps to explain an infant's behavioral patterns. Thus temperament is a critical personality trait, especially in the first days and weeks after birth.

Today's acceptance of the importance of temperament reflects the basic work of Thomas, Chess, and Birch (1970), Thomas (1981), and Chess and Thomas (1987), who believe that in the first half of the twentieth century psychologists overemphasized the role of the environment in development. These authors were struck by the individuality of their own children in the days immediately following birth, differences that could not be attributed to the environment.

To test their hypothesis, Chess and Thomas designed a longitudinal study called the **New York Longitudinal Study.** In 1956 they began collecting data on 141 middle-class children. They observed the behavioral reactions of infants, determined their persistence, and attempted to discover how these behavioral traits interacted with specific elements in the infants' environments. From the resulting data they found nine characteristics that could be reliably scored as high, medium, or low:

- The level and extent of motor activity
- The rhythmicity, or degree of regularity, of functions such as eating, sleeping, and elimination
- The response to a new object or person (approach versus withdrawal)
- Adaptability of behavior to environmental changes
- Sensitivity to stimuli
- Intensity of responses
- General mood or disposition (friendly or unfriendly)
- Degree of distractibility
- Attention span and persistence in an activity

These ratings provided a behavioral profile that was apparent in the children by 2 or 3 months of age. Certain characteristics clustered with sufficient frequency for the authors to identify three general types of temperament:

- **Easy children,** characterized by regularity of bodily functions, low or moderate intensity of reactions, and acceptance of, rather than withdrawal from, new situations (40 percent of the children)

- **Difficult children,** characterized by irregularity in bodily functions, intense reaction, and withdrawal from new stimuli (10 percent of the children)
- **Slow-to-warm-up children,** characterized by a low intensity of reactions and a somewhat negative mood (15 percent of the children)

The authors were able to classify 65 percent of the infants, leaving the others with a mixture of traits that defied categorization. Knowing what kind of temperament their child has can help parents to adjust their style (way of doing things) to their child's. This can be a distinct advantage in forming positive parent-child relationships. The Thomas and Chess work suggests that infants immediately bring definite temperamental characteristics to the mother-infant relationship, characteristics that do much to shape those critical initial interactions.

The Origins of Temperament. Carey (1981) states that temperament appears to be largely constitutional, is observable (at least partially) during the first few days of life, becomes stable at 3 or 4 months, and constantly interacts with the environment. Twin studies illustrate the immediate appearance of temperament. For example, Matheny (1980) studied twin temperament and discovered differences at each age among the non-identical twins. But in his work with identical twins, he found that they were remarkably alike in temperamental characteristics at all ages.

In another significant twin study, Torgersen (1982) studied the nine categories of temperament identified by Thomas and Chess in a sample of 53 same-sex twins at 2 and 9 months. The mothers were also asked about similarities and differences between their twins. All temperamental variables for the identical twins at both ages were more similar than those for the non-identical twins, with more marked similarities at 9 months.

Torgersen (1982) next reported a follow-up study of the same twins at 6 years of age. Again using eight of the categories proposed by Thomas and Chess, Torgersen once more found the identical twins to be more alike than the non-identical. Summarizing the data, Torgersen states (1982) that temperament showed a strong genetic thrust in infancy, and at 6 years of age, a genetic influence was still evident in many of the categories.

Goldsmith (1983), analyzing these and similar studies, points to strong evidence for genetic effects on sociability, emotionality, and activity. Thus evidence continues to indicate a strong genetic role in temperament, with all that implies for the mother-infant relationship (Saudino & Eaton, 1991). (For a different interpretation of the genetic role in temperament, see Riese, 1990.) What are some implications of these temperamental differences, and how do they influence parent-child relationships?

Goodness of Fit. Compatibility between parental and child behavior introduces the concept of **goodness of fit.** As Chess & Thomas (1987) searched for some unifying theme to explain why a child's development was proceeding smoothly or not, they found that a goodness of fit existed when the demands and expectations of parents were compatible with their child's temperament. Goodness of fit signifies the match between the properties of the environment with its expectations and demands, and the child's capacities, characteristics, and style of behaving. Poorness of fit exists when parental demands and expectations are excessive and not compatible with a child's temperament, abilities, and other characteristics. Poorness of fit produces stress and is often marked by developmental problems (Chess & Thomas, 1987, pp. 57–58). A simple way of phrasing this concept is to ask parents and their children how they get along together.

This is the reason Thomas and Chess designed intervention programs to alter the nature of the relationship to one more closely approximating goodness of fit. For intervention to be successful, parents need to be reassured that there is nothing wrong with their behavior (if this is the case). With a child of a different temperament, they might have done quite well.

Parents need to be aware of the link between temperament and an infant's early behavior. For example, Vaughn and associates (1992) attempted to discover any association between security of attachment and an infant's temperament. They studied 555 children ranging in age from 5 to 42 months and found a significant relationship between temperament and attachment at all ages. These researchers conclude that attachment and temperament are not isolated entities, but influence each other.

Although goodness of fit established in infancy produces a good beginning for a child, there is no guarantee that this fit between parental demands and expectations and a child's temperament and capacities will last (Chess & Thomas, 1987). Children change and so do their parents' expectations, and as relationships become more complex (for example, during adolescence) the goodness of fit may also change. The authors give the example of an infant with a low threshold for stimulation—a door closing or a light coming on may disturb her—whose parents have adapted to her temperament. Later in school, however, she may need extra time for homework or lose attention during a long lesson. Unless such environmental demands are made in light of the child's temperament, the goodness of fit could rapidly deteriorate.

Having considered the nature and givens of relationships, we can now ask, what is the source of the interactions that lead to social relationship? How do an infant's first relationships develop?

First Relationships

In an intriguing statement about how early relationships develop, Brazelton and Cramer (1990, p. 3) state:

> ***For all parents-to-be, three babies come together at the moment of birth. The imaginary child of their dreams and fantasies and the invisible but real fetus, whose particular rhythms and personality have been making themselves increasingly evident for several months now, merge with the actual newborn baby who can be seen, heard, and finally held close. The attachment to a newborn is built on prior relationships with an imaginary child, and with the developing fetus which has been part of the parents' world for nine months.***

These interactions help infants to form their first relationships and influence how they relate to others. But how do these first relationships actually begin?

Relationships: Beginnings and Direction

Lisina (1983) has carefully traced the origin and development of the interactions that constitute a relationship by first determining when and if a need for interaction actually exists. She used four criteria:

- *Attention to an adult,* which would signify that the infant recognizes the adult as the object in some activity; that is, the child does something to the adult.
- *Emotion displayed toward the adult,* which would indicate that the infant sees the adult not as a mere object, but as someone who brings pleasure or relieves displeasure.

- *Actions directed at the adult,* which would suggest that the infant is evaluating its role in the relationship—will my overture be accepted?
- *Reactions to an adult's actions,* which would reflect changes in the adult's behavior brought about by the infant's earlier actions—the infant is beginning to evaluate what has happened.

As the number of relationships within the family increases, the interactions among family members grow dramatically. Parents, employing the concept of "sensitive responsiveness," engage in reciprocal interactions with their children, thus influencing the path of psychosocial development.

Infants typically react by general states of excitement and distress, which swiftly focus on the mother as infants recognize their mothers as sources of relief and satisfaction. Mothers rapidly discriminate their infants' cries: for hunger, attention, or fright. Thus infants learn to direct their attention to their mothers. Once the pattern is established, infants begin to evaluate the emerging interactions.

Three motives seem to be at work:

- *Bodily needs*—food, for example—lead to a series of interactions that soon become a need for social interaction. But these basic bodily needs are not the only source of the need for interactions. If you have ever witnessed mother-infant interactions during feeding, you may have been amazed by the infant's intensity in satisfying a basic need. Yet, if you move to a position where the infant can see you, the baby may momentarily stop sucking to attend to the novel sight.
- *Psychological needs* can cause infants to interrupt one of their most important functions, such as feeding. Consequently, in regard to the origin of relationships, the satisfaction of bodily needs is only the beginning. Children, from birth, seem to seek novelty; they require increasingly challenging stimulation. For infants, adults become the source of information as much as the source of bodily need satisfaction.
- *Adult response needs.* The preceding two needs alone are inadequate to explain why, for most of us, relationships with other people are the most important part of our lives. Adults, usually mothers, satisfy needs, provide stimulation, and initiate communication, thus establishing the basis for future social interactions.

The Developmental Sequence

These three influences at work during the infant's early days—bodily need satisfaction, a search for novelty, and adult responses to their overtures—form a basis for social interaction. In Lisina's scheme (1983), the sequence is as follows:

- During the first three weeks of life, infants are not affected much by an adult's appearance. The only exception, as noted, is during feeding periods.
- From about the beginning of the fourth week, infants begin to direct actions at the adults. Emotional reactions also appear at this time, with obvious signs of pleasure at the sight and sound of adults, especially females.
- During the second month, more complex and sensitive reactions emerge, such as smiling and vocalizations directed at the mother, plus animated behavior during interactions.
- By 3 months of age, the infant has formed a need for social interactions. That need continues to grow and be nourished by adults until the end of the second or beginning of the third year, when a need for peer interactions develops.

An Applied View — The Significance of the Mother–Infant Interaction

In a series of sensitive statements describing mother-infant interactions, Stern (1977, 1985) has identified certain characteristics that both mothers and infants bring to the relationship.

The Caregiver's Repertoire

- **Facial expressions.** Facial expressions such as smiling, frowning, showing sympathy, and demonstrating surprise all have social consequences: They are intended to encourage the relationship, or with some mothers and their infants, to avoid interacting.
- **Vocalizations.** Mothers typically exaggerate their speech to their infants and vary its speed. These early speech behaviors seem to have as much a bonding function as an effort to convey information.
- **Gaze.** Mothers and infants gaze at each other for relatively long periods, sometimes introducing vocalizations. These behaviors also further the relationship.
- **Face presentations and head movements.** Peek-a-boo and all its variations seem to have both emotional and cognitive consequences: Infants enjoy them and learn from them.

The Infant's Repertoire

- **Facial expressions.** From the time of Darwin, investigators have been impressed by the variety and revealing nature of facial expressions. Conveying a range of emotions from pleasure to displeasure, infants communicate with those around them, firmly establishing a pattern of human relatedness.
- **Gaze.** With steadily improving sight, infants include both people and objects as stimulation. When this occurs, the nature of interactions also changes, since infants can now exercise more control over their partner.

As these early interactions commence, several characteristics begin to identify the emergence of a successful relationship (Brazelton & Cramer, 1990). The first of these is **synchrony,** which refers to the ability of parents to adjust their behavior to that of an infant. Immediately after birth, infants are mostly occupied by their efforts to regulate such systems as breathing and heart rate, which demands most of their energy and attention. Once parents recognize these efforts, the baby's "language" (Brazelton & Cramer, 1990, p. 121), they can use their own behavior to help their infants adapt to environmental stimuli. This mutual regulation of behavior defines synchrony.

Another characteristic is **symmetry,** which means that an infant's capacity for attention and style of responding influence any interactions. As Brazelton and Cramer (1990, p. 122) note, in a symmetric dialogue parents recognize an infant's thresholds, that is, what and how much stimuli an infant can tolerate.

Other characteristics include **contingency,** which refers to the effects of a parent's behavior on the infant's state. **Entrainment** is a characteristic that identifies the rhythm that is established between a parent's and infant's behavior. For example, when the infant reaches toward the mother, the mother says something like, "Oh, yes, Timmy." The sequence involved in entrainment leads to playing *games* such as the mother making a face at the baby and the infant trying to respond similarly. Once infants realize that they have a share in controlling the interactions (about 6 months of age), they begin to develop a sense of **autonomy.** With these interactions, infants are beginning to form relationships and learn about themselves.

Figure 6.2 illustrates the sequence by which the first interactions, combined with developmental changes, gradually lead to specific relationships.

As the interactions between mother and child increase and become more complex, an attachment develops between the two. With the preceding ideas in mind, we can turn now to the special topic of attachment.

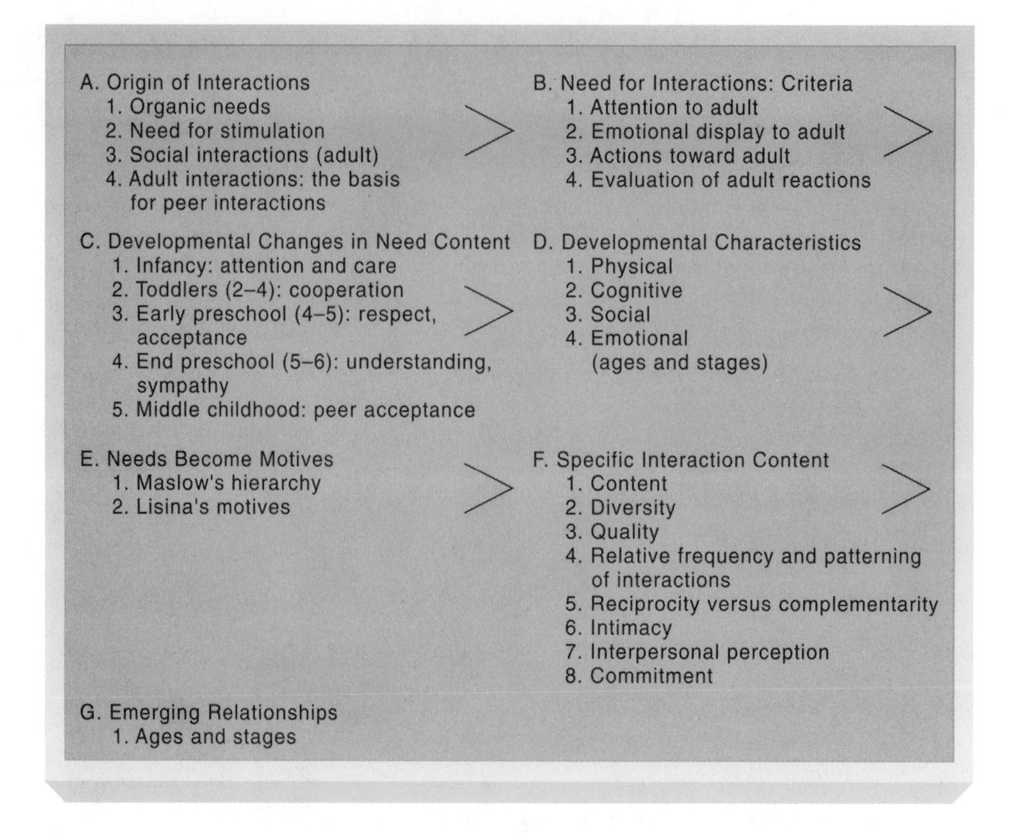

Figure 6.2
The origin and development of relationships

Attachment

Because the roots of future relationships are formed during the first days of life, we may well ask, How significant is the mother-infant relationship in the minutes and hours after birth? We know that infants who develop a secure **attachment** to their mothers have the willingness and confidence to seek out future relationships. One of the first researchers to recognize the significance of relationships in an infant's life was John Bowlby.

Both mother and child bring their own characteristics to the relationship (facial expressions, movements, vocalizations), and as they do, the interactions between the two become more complex and an attachment slowly develops between the two.

Bowlby's Work

The traditional notion of attachment was developed with great insight by John Bowlby (1969). Bowlby's basic premise is simple: A warm, intimate relationship between mother and infant is essential to mental health since a child's need for its mother's presence is as great as its need for food. A mother's continued absence can generate a sense of loss and feelings of anger. (In his 1969 classic, *Attachment,* Bowlby states quite clearly than an infant's principal attachment figure can be someone other than the natural mother.)

Background of the Theory

Beginning in the 1940s, Bowlby and his colleagues initiated a series of studies in which children aged 15–30 months who had good relationships with their mothers experienced separation from them. A predictable sequence of behaviors followed: protest, despair, and detachment.

- *Protest,* the first phase, may begin immediately and persist for about one week. Loud crying, extreme restlessness, and rejection of all adult figures mark an infant's distress.

- *Despair,* the second phase, follows immediately. The infant's behavior suggests a growing hopelessness: monotonous crying, inactivity, and steady withdrawal.
- *Detachment,* the final phase, appears when an infant displays renewed interest in its surroundings, a remote, distant kind of interest. Bowlby describes the behavior of this final phase as apathetic, even if the mother reappears.

From observation of many similar cases, Bowlby defines attachment as follows:

> ***Attachment behavior is any form of behavior that results in a person attaining or maintaining proximity to some other clearly identified individual who is conceived as better able to cope with the world. It is most obvious when the person is frightened, fatigued, or sick, and is assuaged by comforting and care-giving. At other times the behavior is less in evidence.*** **(1982, p. 668.)**

Bowlby also believes that while attachment is most obvious in infancy and early childhood, it can be observed throughout the life cycle. Table 6.4 presents a chronology of attachment behavior.

Table 6.4 Chronology of Attachment Development

Age	Characteristics	Behavior
4 months	Perceptual discrimination; visual tracking of mother	Smiles and vocalizes more with mother than anyone else; begins to manifest distress at separation
9 months	Separation anxiety, stranger anxiety	Cries when mother leaves; clings at appearance of strangers (mother is primary object)
2–3 years	Intensity and frequency of attachment behavior remains constant; increase in perceptual range changes circumstances that elicit attachment	Notices impending departure, signaling a better understanding of surrounding world
3–4 years	Growing confidence; tendency to feel secure in a strange place with subordinate attachment figures (relatives)	Begins to accept mother's temporary absence; plays *with* other children
4–10 years	Less intense attachment behavior but still strong	May hold parent's hand while walking; anything unexpected causes child to turn to parent
Adolescence	Weakening attachment to parents; peers and other adults become important	Becomes attached to groups and group members
Adult	Attachment bond still discernible	In difficulty, adults turn to trusted friends; elderly direct attachment toward younger generation

Source: From John Bowlby, "Attachment and Loss: Retrospect and Prospect", in *American Journal of Orthopsychiatry,"* 52:664–78. Reprinted with permission from the American Journal of Orthopsychiatry.

Although the interactions between a mother and her child and a father and his child may appear quite different, a child will attach to both mother and father.

Other Explanations of Attachment

Bowlby's is not the only explanation of attachment. As you might expect from your reading of the developmental theories in chapter 2, other theorists have commented on the appearance of attachment.

- *Psychoanalytic theorists,* following Freud, hold that infants become attached to those who provide them oral satisfaction. Thus infants attach to their mothers because mothers usually feed them. Erikson, while accepting the traditional psychoanalytic interpretation of the importance of feeding, also believed that a developing sense of trust between mother and infant contributed to attachment between the two.
- *Behaviorists,* following Skinner, emphasize the importance of reinforcement. For behaviorists, feeding is only one form of reinforcement and not terribly important. Physical contact, comforting, and appropriate types of stimulation (visual, vocal) are equally important. The **total range** of reinforcement provided by parent or caregivers explains attachment.
- *Cognitive theorists,* following Piaget and Kohlberg, believe that attachment is more of an intellectual achievement involving cognitive concepts such as object permanence and a developing sense of competence. While Bowlby's work has provided penetrating insights into attachment behavior, these other theorists have also contributed to our understanding.

Attachment Research

Ainsworth (1973, 1979; Ainsworth & Bowlby, 1991), who accepts Bowlby's theoretical interpretation of attachment, devised the *strange situation* technique to study attachment experimentally. Ainsworth defines attachment as follows:

> ***The hallmark of attachment is behavior that promotes proximity to or contact with the specific figure or figures to whom the person is attached. Such proximity-and-contact-promoting behaviors are termed attachment behaviors. Included are signaling behavior (crying, smiling, vocalizing), orienting behavior such as looking, locomotions relative to another person (following, approaching), and active physical contact behavior (climbing up, embracing, clinging).*** **(1973, p. 2.)**

These behaviors indicate attachment only when they are differentially directed to one or a few persons rather than to others. This is especially noticeable when infants first direct their attention to their mothers, which is the infant's way of initiating and maintaining interaction with its mother. Children also attempt to avoid separation from an attachment figure, particularly if faced with a frightening situation.

The Strange Situation Technique

To assess the quality of attachment by the strange situation technique, Ainsworth had a mother and infant taken to an observation room. There, the child was placed on the floor and allowed to play with toys. A stranger (female) then entered the room and began to talk to the mother. Observers watched to see how the infant reacted to the stranger and to what extent the child used the mother as a secure base. The mother then left the child alone in the room with the stranger; observers then noted how distressed the child became. The mother returned and the quality of the child's reaction to the mother's return was assessed. Next, the infant was left completely alone, followed by the stranger's entrance, and then that of the mother. These behaviors were used to classify children as follows:

- *Group A infants (avoidantly attached),* who rarely cried during separation and avoided their mothers at reunion. The mothers of these babies seemed to dislike or were indifferent to physical contact.
- *Group B infants (securely attached),* who were secure and used the mother as a base from which to explore. Separation intensified their attachment behavior; they exhibited considerable distress, ceased their explorations, and at reunion sought contact with their mothers.
- *Group C infants (ambivalently attached),* who manifested anxiety before separation and who were intensely distressed by the separation. Yet on reunion they displayed ambivalent behavior toward their mothers; they sought contact but simultaneously seemed to resist it.

Examining the nature of these mother-infant interactions, Ainsworth (1979) states that feeling, close bodily contact, and face-to-face interactions seem to be equally important in the child's expectations of the mother's behavior. Ainsworth also believes that attachment knows no geographic boundaries. Reporting on her studies of infant-mother attachment in Uganda, Ainsworth (1973) reported that of 28 infants she observed, 23 showed signs of attachment.

She was impressed by the babies' initiative in attempting to establish attachment with their mothers and noted that the babies demonstrated this initiative even when there was no threat of separation or any condition that could cause anxiety. In tracing the developing pattern of attachment behavior, Ainsworth (1973, p. 35) states:

> ***The baby did not first become attached and then show it by proximity-promoting behavior, but rather that these are the patterns of behavior through which attachment grows.***

She reports other studies conducted in Baltimore, Washington, and Scotland indicating that cultural influences may affect the ways in which different attachment behaviors develop. Nevertheless, although these studies used quite different subjects for their studies, all reported attachment behavior developing in a similar manner.

A Special Relationship: The Irritable Infant

In a study that integrates many of the topics we have discussed in this chapter, Crockenberg (1981) examined the relationships between mothers and **irritable infants.** Forty-eight 3-month-old infants were assessed on the Neonatal Behavioral Assessment Scale (Brazelton, 1984) and identified as high or low irritable. Low responsive and high responsive mothers were recognized during home observations. Sources of support were elicited in interviews with the mothers. The youngster's degree of attachment was assessed by the strange situation technique.

The results indicated that security of attachment is associated with the mother's social support (from the father, older children, others). For example, anxious attachment was linked to low social support and has its major impact on highly irritable children. Children low in irritability seemed more immune to the benefits of a support system. Aside from the infants' temperaments, the mothers' responsiveness was linked to the amount of support they received, which then affected security of attachment. Insecure attachment could be predicted at age 3 months for those infants whose mothers were unresponsive to their crying and who themselves had little social support.

Several important implications are apparent. First, this study showed how important reciprocal interactions are in a child's life. Irritable infants, especially those from low-support backgrounds, received less responsive mothering. Second, in a time when more children are being raised by single parents, an adequate support system becomes more significant in any consideration of satisfactory social development.

Supporting Crockenberg's findings, Jacobson and Frye (1991) assigned home visitors to meet with pregnant women once a month until delivery was imminent. The visitors then increased their meetings to once a week. They gradually decreased the visits to once a month until the infants were 1 year old. The researchers found that the mothers in their experimental group (23 mothers and infants) had more securely attached infants than mothers in the control group (23 mothers and infants).

Crockenberg's study, with its far-reaching conclusions for child rearing, also emphasizes that we must accept the partners in a relationship as they are. We are not going to change someone's looks or temperament. All of us—children and adults alike—bring these personal characteristics to any relationship. For example, we have just used irritable infants as an illustration. A mother interacting with an irritable child experiences frustration, perhaps even feelings of hostility, on occasions such as feeding and handling. The mother who feeds, clothes, and talks to her child may well be demonstrating the criteria of adequate mothering. But if the pattern of interactions is examined closely, the mother's resentment about difficult feeding may transfer to the manner in which she treats her child. This may negatively affect the child's perception of the mother as a secure base.

The Klaus and Kennell Studies

Some investigators believe that the initial, intense contacts with the mother have a critical and long-lasting influence on a child's development. Two physicians, Marshall Klaus and John Kennell, both professors of pediatrics at Case Western Reserve University School of Medicine, conducted a series of studies (1976, 1983) concerning the impact of immediate postbirth experiences.

In Klaus and Kennell's initial work (1976), fourteen mothers of newborns were given extended contact with their infants. Heat shields (a panel to provide heat) were placed over the mothers' beds, and they were given their naked infants for one hour of the first two hours after birth and then for five hours on each of the next three days, for a total of sixteen hours. Fourteen other mothers, matched for age, marital status, and socioeconomic status, received the more standard hospital treatment: They were shown the baby immediately after birth, six or seven hours later they briefly held the baby, and they fed their infants about every four hours.

Kennell controlled the conditions as tightly as possible. The women were randomly assigned to each group, given identical explanations for the study, provided with heat shields (both groups), and had no idea that the mother-infant contacts differed during the three days. The question that Klaus and Kennell pursued was: Did the differences in mother-infant contact cause differences in the later mother-child relationship?

The Results

One month after giving birth, all of the mothers returned to the hospital for an interview, a physical examination of the child, and a film of the mother feeding her infant. Klaus and Kennell reported that the extended-contact mothers touched their children more frequently, stood and watched the physical examination more closely, and seemed reluctant to leave their infants with anyone else. Returning at one year after birth for the infant's examination, the extended-contact mothers soothed their youngsters more when they cried. They also seemed to want to remain closer to their infants than the control mothers.

Two years after their child's birth, five mothers from each group were randomly selected and interviewed. The extended-contact mothers employed richer language with their children, using more words and asking more questions than the other mothers. When the children were 5 years old, Klaus and Kennell compared nine of the extended-contact children with ten of the control group children. Children of the extended-contact mothers had significantly higher IQs and higher scores on language tests. (See Table 6.5 for a summary of the results.)

Table 6.5 The Klaus and Kennell Treatment

	Extended Contact	Controls
Number	14	14
At birth	Mothers given child for 1 hour	1. Mother shown the baby 2. Briefly holds baby 6 hours later
Day 1	Given the baby for 5 hours	Feeding every 4 hours
Day 2	Given the baby for 5 hours	Feeding every 4 hours
Day 3	Given the baby for 5 hours	Feeding every 4 hours
Results	1. Closer attachment 2. Greater language involvement 3. Higher IQ 4. Authors believe results confirm existence of sensitive period for attachment immediately after birth	

Source: Marshall Klaus and John Kennell, *Bonding*. The C. V. Mosby Company, Inc., 1983.

The authors of the study believe that the extra sixteen hours of contact during the first three days of life affected maternal behavior for at least one year after birth (possibly longer) and seemed to have important developmental consequences for the infant. They describe this maternal sensitive period as follows:

> ***Immediately after the birth the parents enter a unique period during which events may have lasting effects on the family. This period, which lasts a short time, and during which the parents' attachment to their infant blossoms, we have named the maternal sensitive period....During this enigmatic period, complex interactions between mother and infant help to lock them together.*** **(1976, pp. 50–51.)**

Meaning of the Studies

We should accept the conclusions of Klaus and Kennell with some reservations. Although the authors used careful techniques, one may want to question how different the mothers of the two groups were before the infants' births. Were some mothers more sensitive to their infant's reactions than others? Were the infants temperamentally different at birth, enough to cause different behavioral responses?

The Klaus and Kennell work has received wide public attention. What concerns both investigators and mothers are the inevitable differences that arise in the mother-infant relationship soon, if not immediately, after birth. For example, some neonates require instant specialized treatment (prematures) and must be separated from the mother. Also, what of the untold mothers who never experienced the intimate relationship described in these studies? Have their youngsters suffered unintended consequences? To their credit, Klaus and Kennell have expressed concern about the impact of their conclusions.

> ***We faced a real dilemma in deciding how strongly to emphasize the importance of parent-infant contact in the first hours and extended visiting for the rest of the hospital stay, based on the available evidence. Obviously in spite of a lack of early contact experienced by parents in the past 20 to 30 years, almost all these parents became bonded to their babies. The human being is highly adaptable, and there are many fail-safe routes to attachment. Sadly, some parents who missed the bonding experience have thought that all was lost for their future relationship.***
>
> ***This was, and is, completely incorrect, but it was so upsetting that we have tried to speak more moderately about our convictions concerning the long-term significance of this early bonding experience.*** **(1983, p. 50.)**

The authors' insistence on caution in interpreting their results is laudatory but should not diminish efforts to understand both the nature of relationships and their influence on development.

In an exhaustive review of the literature on bonding, Goldberg (1983) concludes that most of the evidence does not support the existence of a sensitive period for parent-infant bonding, because few careful studies have been done on the actual beginnings of maternal behaviors. A British study conducted by Packer and Rosenblatt (1979) found that the amount of mother-infant interactions in the delivery room was quite low: About 500 of the mothers did not touch their infants in the first 20 minutes; 180 did not even look at their newborn. (The investigators note the conditions were not ideal; the infant was presented to the mother at the convenience of the staff.) This and similar studies cast doubt on the belief that delivery room contacts are characterized by euphoria. As Goldberg (1983) concludes, we do not know much more about initial parent-infant contacts and their effects on later behavior than we knew in the 1970s. But research continues.

Fathers and Attachment

Although we have concentrated on the mother in our discussion of attachment, the father's role in the process has attracted growing interest. As fathers become more involved in child care (in the 1990s, about 90 percent of mothers will work full- or part-time), questions have arisen about an infant's attachment to both mother and father.

Research has focused on demonstrating the differences between mothers' and fathers' behavior (nurturant versus playful), the similarity between parental behaviors (both exhibit considerable sensitivity), and the amount of involvement in the infant's care. Do infants react differently to each parent? The evidence suggests that while the mother usually remains the primary attachment figure, both mother and father have the potential to induce attachment (Fox, Kimmerly, & Schafer, 1991). These authors found that secure attachment to one parent depended on secure attachment to the other.

Continued findings attest to the ability of fathers to bond with their infants. For example, during the birth process fathers are frequently present during labor and delivery, which usually lessens the mother's feeling of distress and may facilitate father-infant bonding. (The evidence is mixed on this conclusion; see Rutter, 1979.) Fathers and mothers display quite similar behaviors when interacting with their newborns, both exhibiting sensitivity to newborn cues. (This finding seems to be consistent across socioeconomic classes.)

At Boston's Children's Hospital, Yogman (1982) discovered that by as early as 6 weeks of age, infants interact differently with their parents than they do with strangers. In the study, adults (mothers and fathers) entered from behind curtains and were told to play with the infant without toys and without taking the infant from the seat. One camera focused on either the mother or father, the other on the infant. Seven minutes of interaction were recorded—two minutes each with the mother, father, and a stranger, separated by 30-second intervals. By 3 months of age, Yogman reported, infants "successfully interacted with both mothers and fathers with a similar, mutually regulated reciprocal pattern as evidenced by transitions between affective levels that occurred simultaneously for infant and parent" (1982, p. 138).

By 7 or 8 months of age, when attachment behavior (as defined by Bowlby and Ainsworth) normally appears, infants are attached to both mothers and fathers and prefer either parent to a stranger. The evidence indicates, then, that fathers possess the potential for establishing a close and meaningful relationship with their infants immediately from birth (Phares, 1992; Cox & others, 1992).

In spite of fathers' increased involvement, most fathers still spend a limited amount of time with their infants and only occasionally are they involved with physical care. As has been well documented, fathers spend more time playing with their children. Fathers tend to be more tactile and physical while mothers are more verbal, which suggests that infants receive qualitatively different stimulation from their fathers (Sroufe, Cooper, & DeHart, 1992).

The interactions between a father and his child tend to be more physical than those between a mother and her child. The qualitatively different types of stimulation a child receives from each parent would seem to suggest implications for the staffing of daycare centers and preschool facilities.

Current Research into Attachment

Does deprivation in one generation lead to problems in the next? Recent research points to the reality of **intergenerational continuity,** that is, the connection between childhood experiences and adult parenting behavior. For example, studies have shown that children raised in unhappy or disrupted homes are more prone to unhappy marriages and divorce. A similar pattern has been found among parents who batter their children: These parents suffered seriously disturbed childhoods marked by neglect, rejection, or violence (Rutter, 1981).

In an attempt to discover if parental behaviors reach across generations, Ricks (1985) has examined the evidence from two perspectives: the impact of separation or disruption, and parental reports of their childhood attachments. Ricks is concerned with the time involved in these studies. Memory, for example, poses an obstacle since our recall of the past may well be affected by our current thinking, our present mood, and by our present status. As Ricks (1985, p. 214) asks: Do childhood memories accurately reflect childhood experiences?

With these cautions, Ricks then examined the separation and disruption studies. Two findings emerged:

- Separation from parents in the family of origin (i.e., the parents' original family) was related to problems in parenting.
- Separation in the family of origin seemed to be associated with marriage problems and with depression in the mother.

A key factor in intergenerational effects was disruption (parents remain together but fight and are basically unhappy). Mothers from disrupted families of origin did not manifest the warm relationships with their children that mothers from nondisrupted families manifested. Preschool children of mothers from disrupted families had poorer language skills than youngsters whose mothers had not come from disrupted families. The early attachment relationship was adversely affected by disruption, divorce, or long-term separation in the mother's family of origin.

In the second phase of her study, Ricks turned to the recollections that mothers had of their own relations with their parents. She consistently found significant relations between a mother's recollection of her childhood attachments and her present ability to serve as a secure base for her child. In the development of social relationships, those first bonds, the initial relationships, have a critical function that may carry across generations.

In more recent research, Fonagy, Steele, and Steele (1991) discovered a link between the way a mother recalls her childhood experiences and the present quality of the relationship with her child. Studying 100 pregnant women, the researchers wanted to assess how the attachment experiences of these women affected attachment with their children at 1 year of age. They found that 75 percent of the women who had been securely attached now had securely attached children. Of the remaining mothers, 23 percent had insecurely attached children. The researchers conclude that the internal representations of childhood attachment seem to carry to the next generation.

Early Emotional Development

As you read about the impact of attachment and early relationships on psychosocial development, you can understand how a child's emotional life is also affected. The study of emotions has had a checkered career in psychology. From peaks of popularity, interest in emotions and their development plunged to depths of neglect. Today we see once again a resurgence of enthusiasm for its study (Leventhal & Tomarken, 1986).

Signs of Emotional Development

One of the first signs of emotion is a baby's smile, which most parents immediately interpret as a sign of happiness. Two-month-old infants are often described as "smilers." While smiles appear earlier, they lack the social significance of the smile that emerges at 6 weeks. Babies smile instinctively at faces—real or drawn—and this probably reflects the human tendency to attend to patterns. Infants gradually learn that familiar faces usually mean pleasure, and smiling at known faces commences as early as the fifth month. Smiling seems to be a key element in securing positive reinforcement from those around the infant.

Smiling has a developmental history and "for no reason" appears soon after birth (Kagan, 1984). These smiles are usually designated as "false" smiles because they lack the emotional warmth of the true smile. By the baby's third week, the human female voice elicits a brief, real smile. By the sixth week the true social smile appears, especially in response to the human face. Babies smile at a conceptual age of 6 weeks, regardless of chronological age.

This 3-month-old infant is responding to its mother's face by smiling. In these interactions we see the roots of a child's psychosocial development.

Why do infants smile? There are several possible explanations:

- Infants smile at human beings around them.
- Infants smile at any high-contrast stimuli, thus eliciting attention from those around them. The infant then links the human face with pleasure.
- Infants smile at discovering a relationship between their behavior and events in the external world.

These behaviors suggest that infants' emotions are much more organized than previously suggested (Tronick, 1989). Several theories of emotional development have been proposed to explain this and other types of emotional development.

Theories of Emotional Development: A Summary

An early theory of emotional development was proposed by Bridges (1930). She believed that neonates demonstrated only one type of emotion: general excitement. Bridges believed that as infants grow, specific positive and negative emotions appear: Distress is shown at 3 weeks, anger grows out of distress at 4 months, disgust follows anger at 5 months, and fear follows disgust at about 6 months. More recently, as researchers have observed specific emotions in the neonate, interest in Bridges' work has faded.

Other theorists, especially Izard (Izard, 1978; Izard & Malatesta, 1987; Izard & others, 1991), believe that infants demonstrate specific emotions. Using an infant's facial expressions as the basis for his work, Izard states that emotions emerge as an infant needs them to adapt, and that emotional development follows a definite pattern:

- The newborn shows startle, interest, disgust, and distress.
- During the next 4 months anger, surprise, and joy appear.
- Fear and shyness emerge during the 6–12 month period.

Izard and his colleagues (1991) believe that emotions are the keystone of adaptation and the motivational component of personality and social relationships. Studying 114 mothers and their infants, these researchers found that the mothers' emotional experiences and emotional behavior predicted the quality of attachment: Those mothers who expressed positive emotions around their children had securely attached infants.

In a thoughtful statement about emotional development, Kagan (1984) states that during the first three or four months infants display reactions that suggest emotions, but most likely these behaviors reflect some kind of internal change we as yet don't understand. For example, widening of the eyes and an increase in heart rate is often referred to as "surprise to novelty." Between four and twelve months, cognitive development (especially memory) helps infants to produce new emotional reactions. For example, unexpected events cause 8-month-old infants to show facial expressions of wariness, a cessation of playing, and perhaps crying (Kagan, 1984, p. 173). As Kagan notes, during these first years, emotions are mainly caused by external events.

First Feelings

As you can tell from this brief excursion into explanations of emotional development, recent interest has yet to be matched by hard evidence. But given the changes in developmental research that we've described in these first five chapters, it can be only a matter of time before the path of emotional development becomes less obscure.

We have previously noted the current view of an infant as an active partner in development and have also described an infant's state with all of its meaning (Tronick, 1989). Today we are aware of the number of abilities an infant brings into the world. Using these insights, Greenspan and Greenspan (1985) have begun to probe into the origins of emotions. They believe that the six emotional milestones they have identified can lead to parental practices that will help infants to establish more satisfying relationships with others:

- *Birth to 3 months.* The Greenspans believe that the major features of this period are self-regulation and interest in the world. For normal infants, each of these tasks supports the other.

- *2 to 7 months.* This is the infant's time of falling in love. The baby begins to focus on its mother and is delighted by her appearance, voice, and actions.
- *3 to 10 months.* During these months, an infant attempts to develop intentional communication. It is a time of reciprocal interactions. The mother and infant are responding to each other's signals. When adults (usually the mother) respond to the baby's signals, the infant learns that its actions can cause a response.
- *9 to 18 months.* This is a time of dramatic observable achievements: standing, walking, talking. An organized sense of self begins to emerge. For example, when the mother returns to her infant, at this stage the baby may no longer look at her. An infant of this age may walk to the mother, touch her, and perhaps say a word or two. The baby has put together several behaviors in an organized manner.
- *18 to 36 months.* Called the time of creating emotional ideas, this is a period of rapid mental growth. Children can form images of the mother in her absence. They now link these cognitive capacities to the emotional world. They remember their mother's reading to them last night—why not tonight? They remember their father wrestling with them last night—why not tonight?
- *30 to 48 months.* The Greenspans believe that the emotional thinking of these months forms the basis for fantasy, reality, and self-esteem. In other words, children can use their ideas to form a cause-and-effect understanding of their own emotions.

If infants are securely attached and have begun to establish positive, emotionally rewarding relationships with those around them, their psychosocial development has had a promising start.

Conclusion

The role of relationships in development has finally achieved a prominent place in our attempts to understand a child's growth. From the initial contacts with the mother to the ever-expanding network of siblings and peers at all ages, relationships exert a powerful and continuing influence on the direction of development.

We are slowly acquiring data about the function of relationships. For example, we have seen how important and persistent are the first interactions with parents. They set a tone for future relationships and set the direction for social and emotional development. Recent research has led to significant findings about the quality of relationships.

In this chapter, you were asked to become familiar with the beginnings of relationships. As you continue your reading and we move into the early childhood period, you'll consider the function of relationships in children's play and their first experiences with school. In middle childhood and adolescence, the effect of peer relationships becomes even more significant. For now, however, we turn our attention to early childhood.

Chapter Highlights

Relationships and Development

Background of the Relationship Research

- Acceptance of the transactional model of development and the idea of sensitive responsiveness has helped us to understand how children form relationships.

The Importance of Reciprocal Interactions

- Infants, as active partners in their development, help to shape their relationships.

The Meaning of Relationships

The Characteristics of a Relationship

- Analyzing relationships by a system of categories such as Hinde's helps to make both theory and research more precise.

The Givens in a Relationship

- Certain characteristics such as appearance and temperament are an intrinsic part of any relationship.
- Appearance affects our initial impression of an individual.
- An infant's temperament immediately affects interactions with adults.
- The work of Chess and Thomas has helped us to understand the concept of goodness of fit—the match between an infant's and parents' temperaments.

First Relationships

The Developmental Sequence

- Developing relationships follow a sequence that incorporates all aspects of development: physical, cognitive, and psychosocial.

Attachment

Bowlby's Work

- Bowlby and his colleagues, studying the separation of children from their parents, identified attachment as an important part of psychosocial development.
- Other explanations of attachment include the psychoanalytic, behavioral, and cognitive.

Attachment Research

- Ainsworth's strange situation technique is designed to assess the security of an infant's attachment.
- Attachment is a cross-cultural phenomenon that knows no geographic boundary.
- Attachment develops early in life and offers clues as to psychosocial development.

- The sensitivity of the infant-mother relationship in the moments following birth has led to considerable controversy.
- A mother's attachment to her infant seems to be influenced by the security of her attachment to her own mother.

Early Emotional Development

Signs of Emotional Development

- Smiling is one of the first clues to emotional development.
- The Greenspans have identified several milestones of early emotional development.

Key Terms

Attachment
Autonomy
Biopsychosocial model
Contingency
Difficult children
Easy children
Entrainment
Goodness of fit
Interactions
Intergenerational continuity
Irritable infants
New York Longitudinal Study
Relationship
Sensitive responsiveness
Slow-to-warm-up children
Synchrony
Symmetry
Temperament
Total range
Transactional model of development

What Do You Think?

1. Probably the basic issue for you to grasp is the extent of an infant's abilities: physical, social, and psychological. Do you think that infants are as we have described them in chapters 5 and 6, or do you think that we have over- or underestimated their competencies?
2. Depending on your answer to question 1, explain how you interpret an infant's participation in developing relationships. That is, given an infant's ability to smile, coo, and make physical responses, how much control do you believe they exercise in their interactions with adults?
3. Think about the role of appearance and temperament in developing relationships. A well-known psychologist once said that some children are so difficult to love that parents may have to fake it. How do you react to this statement? Do you think an infant could detect such parental behavior?
4. Once the Klaus and Kennell studies received wide coverage (which they did in the press and television), many mothers experienced feelings of guilt because their children had not received such treatment immediately after birth. They were afraid that their children would be at a disadvantage. What would you tell them? How would you attempt to relieve their feelings of guilt?

Suggested Readings

Bowlby, J. (1969). Attachment. New York: Basic Books. Here is Bowlby's classic statement about attachment. It is very readable and is a reference you should be familiar with.

Brazelton, T. B., and B. Cramer (1990). *The earliest relationship.* Reading, MA: Addison-Wesley. An excellent, readable account of how relationships develop, presented from both a pediatric and psychoanalytic perspective.

Manchester, W. (1983). *The last lion: Winston Spencer Churchill, 1874–1932.* A particularly revealing look at Churchill's difficult boyhood and his desperate attempt to please indifferent parents.

Maurer, D., and C. Maurer. (1989). *The world of the newborn.* New York: Basic Books. This book contains readable summaries of many of the crucial developmental milestones of infancy that we have discussed.

PART IV

Early Childhood

CHAPTER 7

Physical and Cognitive Development in Early Childhood

Features of the Early Childhood Years 172

Physical and Motor Development 174

Characteristics of Physical Development 174

Continuing Brain Development 176

Physical Development: An Overview 177

Growing Motor Skills 179

The Special Case of Drawing 181

Cognitive Development 182

Piaget's Preoperational Period 182

Intellectual Differences Theory 188

Language Development 192

Language as Rule Learning 193

The Pattern of Language Development 194

Speech Irregularities 194

Conclusion 196

Chapter Highlights 196

Key Terms 197

What Do You Think? 197

Suggested Readings 198

Liz, whom we met in chapter 5, felt she now had a good idea of what perpetual motion meant. Her two children, 2-year-old Maddi and 5-year-old Jackie, engaged in nonstop activities from the time they woke up in the morning until they went to bed—with great reluctance. (Be thankful for small favors, she thought; Maddi still took a long afternoon nap.) At least you could talk to Jackie, who was excited about entering kindergarten in September.

But Maddi—oh, that Maddi! She had discovered the thrill of saying "no" loudly and emphatically, was into everything, and sometimes pretended not to hear Liz' warnings. Liz was worried about what seemed to be a steady stream of threats directed at Maddi. She was concerned that so many negative incidents could affect her relationship with Maddi.

These thoughts came to Liz as she opened a letter from the preschool where Maddi was registered for the fall. Also included was a form for her to fill out, and as she read the items, she became even more concerned.

The first item asked for a description of Maddi. Liz answered honestly that she was healthy, vigorous, active, and seemed to be progressing well. She was toilet trained, helped in dressing herself, and seemed quite independent. Liz thought Maddi was unusually curious, always poking into things, opening drawers, and looking into closets. Liz was often embarrassed by this behavior when she and Maddi visited other homes. On the other hand, Maddi was putting her words together nicely and played fairly well with other children.

In another item, Liz was asked to assess Maddi's relationships with adults. As she thought about her answer, Liz realized that her only negative response applied to her own relationship with Maddi. Once she saw this, Liz grew even more concerned. Was it all her fault? How much did Maddi contribute? Was Liz overreacting? Since she and Maddi were scheduled for an interview at the preschool the next week, Liz decided to mention her worries to the instructor who would be interviewing them. After all, they worked constantly with children of this age.

The following week, while Liz talked with one of the instructors, Maddi played happily with a variety of toys. When the instructor commented that Maddi seemed quite adjusted, Liz had the opportunity to mention how worried she was. The instructor listened and then began to laugh, saying that Maddi seemed to be perfectly normal for her age. She asked Liz if she had ever heard of the "terrible twos." Liz said yes, but she thought it was mainly fiction. Not at all, was the answer. How Maddi was acting was typical for most youngsters of her age. The instructor then went on to describe what Liz could expect for the next year or so. ■

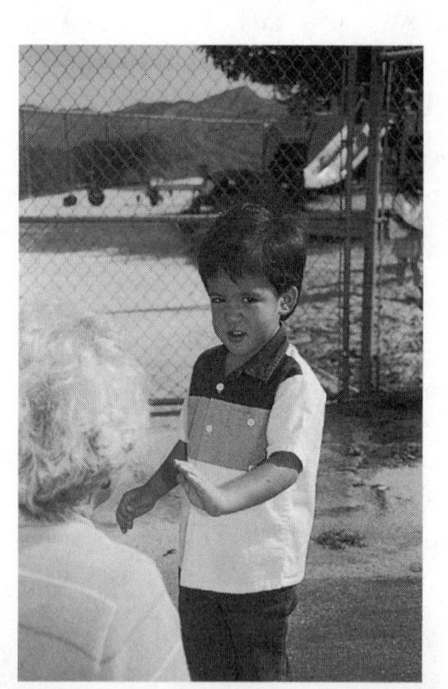

Children pass through periods when they want to test their developing abilities against any restrictions they perceive. The "terrible twos" are a good example.

Liz' concerns and what she was learning about these years is our task for the next two chapters. Bursting with energy, constantly curious, and searching for novelty are all characteristics of children from 2 to 6 years of age. When these traits combine with strong feelings of competence and mastery, parents and other adults working with children of this age face challenges that can try their patience.

The time of infancy is now behind us. By two years, typical children walk, talk, and are eager explorers of their environment. Early childhood youngsters gradually acquire greater mastery over their bodies. The clumsy actions of infancy are replaced by more coordinated, skillful movements. With walking and talking, new directions in development become more obvious. Children's personalities take on definite shadings.

Their interactions with those around them begin to take shape, setting the stage for the kind of relationships they will form in the future. They now begin the process of widening their circle of relationships. Children from 2 to 6 will make new friends and, in our society today, almost inevitably have experience with preschool teachers. Some youngsters of this age are also faced with adjusting to a new sibling (or siblings), which can be a time of great frustration if not handled carefully.

To help you understand the rapid changes that occur during these years, we'll first trace the important physical changes of the period, especially the significance of brain development. We'll then analyze cognitive development, turning to Piaget for an explanation of how these changes occur, and then examine a child's growing representational ability. Finally, during these years children experience the "language explosion" with all that it implies for development, so we'll conclude by discussing this critical phase. These topics fit our biopsychosocial model as seen in Table 7.1.

Table 7.1 Characteristics of the Period

Bio	Psycho	Social
Height changes	Deferred imitation	Play
Weight changes	Symbolic play	
Brain lateralization	Drawing	
Nutrition	Mental imagery	
Health	Representation	
	Language explosion	

After reading this chapter, you should be able to:

- Describe the outstanding physical and cognitive characteristics of the period.
- Analyze the importance of brain lateralization in a child's development.
- Identify the key phases in cognitive development during the preoperational period.
- Contrast Piaget's view of cognitive development with information processing explanations.
- Designate the language milestones of the early childhood years.
- Indicate the speech irregularities that can occur in the early childhood period.

What's Your View?

In regard to the changes that occur during early childhood, you can better understand the potential for conflict between adult rules and children's impulses if you consider how a child's world has dramatically widened. A room or house no longer contains these youngsters. They are mobile and wish (rightly so) to push beyond previous limits. They can walk; soon they are running. They also show expanding intellectual interests, which will culminate in reading and in entering school by the end of the period. It is precisely here that children meet the demands of an adult world. They probably begin to hear "no" for the first time. They want to run out into the streets, and their parents stop them. They reach for that colorful dish on the table, and someone says "don't."

During the early childhood years, peers play an increasing role in social development; parents are no longer the sole source of companionship. As children play together, they express their opinion of each other in a way that provides instant feedback to their personalities. As we know today, a major influence on the development of self is the reaction of others, and these formative years steadily shape the developing self-concept.

Given these youngsters' increased abilities, and their awareness of them, what do you think would constitute one of the major causes of child-adult conflict? What's your view?

Early childhood youngsters find the world a fascinating place. Giving these children the freedom to explore and learn, coupled with sensible restrictions, encourages the development of mastery.

Features of the Early Childhood Years

Early childhood, extending from age 2 to 6, is a time of rapid change: Height increases about 10–12 inches, weight by about 15 pounds; language grows at a phenomenal rate; cognitive changes appear in both thought and language; personality and social development enter a new, distinct phase; feelings of mastery and

Table 7.2 Some Developmental Characteristics of Early Childhood

Age (years)	Height (in.)	Weight (lb)	Language Development	Motor Development
2½	36	30	Identifies object by use; vocabulary of 450 words	Can walk on tiptoes; can jump with both feet off floor
3	38	32	Answers questions, brief sentences; may recite television commercials; vocabulary of 900 words	Can stand on one foot; jumps from bottom of stairs; rides tricycle
3½	39	34	Begins to build sentences; confined to concrete objects; vocabulary of 1,220 words	Continues to improve 3-year skills; begins to play with others
4	41	36	Names and counts several objects; uses conjunctions; understands prepositions; vocabulary of 1,540 words	Walks downstairs, one foot to step; skips on one foot; throws ball overhand
4½	42	38	Mean length of utterance (morphemes) 4.5 words; vocabulary of 1,870 words	Hops on one foot; dramatic play; copies squares
5	43	41	Begins to show language mastery; uses words apart from specific situation; vocabulary of 2,100 words	Skips, alternating feet; walks straight line; stands for longer periods on one foot
5½	45	45	Asks meanings of words; begins to use more complex sentences of 5 or 6 words; vocabulary of 2,300 words	Draws recognizable person; continues to develop throwing skill
6	46	48	Good grasp of sense of sentences; uses more complex sentences; vocabulary of 2,600 words	Jumps easily; throws ball overhand very well; stands on each foot alternately

independence may conflict with parents and teachers. It is Erikson's time of **autonomy** and **initiative** and Piaget's **preoperational period.** Children continue their mastery of toilet training, get ready to read, demonstrate abstract thinking, improve their use of language, and begin to distinguish right from wrong.

Youngsters in early childhood, wanting to do things their way, often clash with the dictates of parents and teachers. Adults often are torn. They want to provide freedom and opportunities, but still must balance the youngster's need to explore and to achieve mastery of self and world against the dangers to a child's safety. The early childhood years require time and adult patience as youngsters explore, test their world, continue to build their sense of trust in adults, and extend trust to others. They constantly absorb new ideas, widening and deepening their cognitive world and expanding their memory. Proper materials and space are critical, since children's movement through space provides the necessary information about self and world that fuels increasing cognitive demands.

We may casually observe these youngsters and comment on their incessant play. But by examining them psychologically, we can see the interplay of movement, activity, and cognitive curiosity that produces constant improvement in competence. Table 7.2 summarizes many characteristics of early childhood youngsters.

Physical and Motor Development

As you can see from Table 7.2, growth in childhood proceeds at a less frantic pace than in infancy. For example, if you double a boy's height at age 2, you will have an approximation of his adult height. But it takes another 14 to 15 years to reach that height. Children during this period grow about another 12 inches and continue to gain weight at the rate of about five pounds a year. Body proportions are also changing, with the legs growing faster than the rest of the body. By about age 6, the legs make up almost 45 percent of body length. At the beginning of this period, children usually have all their baby teeth; and at the end of the period, children begin to lose them. Boys and girls show about the same rate of growth during these years.

Characteristics of Physical Development

Children's rapidly developing motor skills are clearly seen in their drawings from uncontrolled scribbling to controlled "within the lines" attempts to their own creative expressions.

Look at Figure 7.1, the human growth curve. (Note that 10 centimeters equal 4 inches). This curve strikingly illustrates the regularity of physical growth. Most parts of the body (except the brain and the reproductive organs) follow this pattern (Tanner, 1989). With the exception of the two spurts at infancy and adolescence, growth is highly predictable for almost all boys and girls, given satisfactory conditions.

The Sequence of Early Childhood Growth

We are concerned here with physical growth and development. Optimum growth requires proper nutrition, temperature, and rest to stimulate the genetic elements and growth hormones. Tanner (1989) notes that the growth process is self-stabilizing. It is governed by the control system of the genes and fueled by energy absorbed from the environment. If malnutrition or illness deflects children from the normal growth path, but a corrective force (adequate diet or termination of illness) intervenes, the normal course of development will accelerate until the children "catch up"; thereupon, growth slows. These children are often called **developmentally delayed.**

Different cells, tissues, and organs grow at different rates. (Some tissues never lose the ability to grow, such as hair, skin, and nails.) In humans, for example, body length at birth is about four times the length of the face at birth, so the head is relatively large. But the head grows more slowly than the trunk or limbs, so that at age 25 body length is about eight times that of face length.

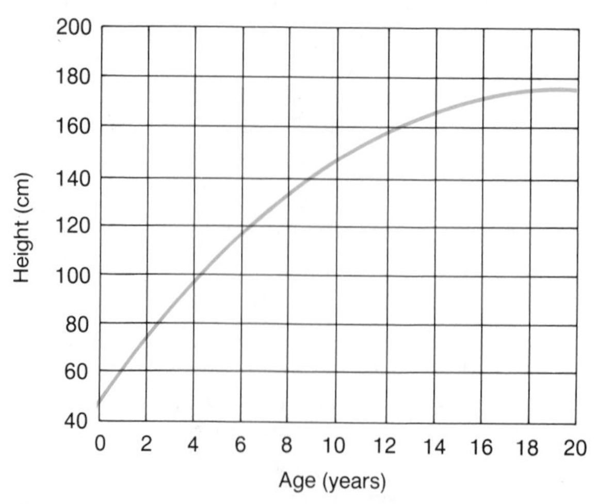

Figure 7.1
The human growth curve.

In their classic interpretation of the first five years of life, Knobloch and Pasamanick (1974) state that healthy development depends on the integration of five kinds of behavior:

- *Adaptive,* including perception, sensorimotor reactions, and eye-hand coordination.
- *Gross motor behavior,* including head balance, sitting, standing, creeping, and walking.
- *Fine motor behavior,* including the use of hands and fingers in grasping and manipulating an object.
- *Language behavior,* including facial expression, gestures, words, phrases, sentences, and comprehension.
- *Personal-social behavior,* including feeding ability, toilet training, self-dependency, and cooperation.

Key Ages of Early Childhood

Using and extending the rich and varied work of Arnold Gesell, Yale's famous child psychologist, Knobloch and Pasamanick believe that the key ages of early childhood are 24, 36, 48, and 60 months. By key ages, the authors mean those stages of development during which children seem to integrate their behavior. These ages can be used to determine if a child's behavior is on a normal developmental schedule. Thus children combine different abilities to produce more organized behavior. Table 7.3 illustrates the developmental accomplishments of these five kinds of behavior.

Table 7.3 Early Childhood Development: Selective Behaviors

Levels of Maturity	Adaptive	Gross Motor	Fine Motor	Language	Personal-Social
24 months	Builds towers, draws circles	Runs well, kicks ball	Can turn pages	Uses phrases, understands directions	Pulls on clothes, plays with toys
36 months	Copies circle, can build bridge of 3 cubes	Stands on one foot, jumps off step	Holds crayon like adult	Uses sentences, answers simple questions	Uses spoon, puts on shoes
48 months	Copies cross, builds with 5 cubes	Skips on one foot, broad jumps	Traces within lines	Uses conjunctions, understands prepositions	Washes and dries face, plays cooperatively
60 months	Counts 10 objects, copies triangle	Skips on alternate feet		Asks "Why?"	Dresses without assistance, asks meaning of words

Source: Data from Hilda Knobloch and Benjamin Pasamanick, *Gesell and Amatruda's Developmental Diagnosis.* Copyright © 1974 Harper and Row Publishers, Inc., New York, NY.

Knobloch and Pasamanick (1974) summarize key early childhood achievements as follows:

- In the second year, walking and running develop, children achieve bowel and bladder control, begin to speak, and acquire some sense of self-identity.
- Between the second and third year, children use language as a vehicle for thought.
- In the fourth year, questioning becomes a way of life, and abstract thinking is apparent. Children should be relatively independent in personal life and home routine.
- In the fifth year, motor control has matured, language is fairly articulate, and social adjustment is apparent.

AN APPLIED VIEW

Height and Weight in the Early Childhood Years

Parents are usually fascinated by the size of their children. Is he small? Is she overweight? How tall will they be when they grow up? While considerable caution is necessary in any interpretation of height and weight data, you may find the following charts interesting. Note that projections begin at the age of 2 years. (These charts are based on data from Tanner, 1989).

You may have heard that if you measure a child's height at age 2 and double it, the result will be the child's adult height. You can see from the chart that this is a good general estimate. There is no such rule of thumb for weight measurements. The meaning of these figures can raise questions. Height measurements seem to be more useful in assessing any developmental disorders or serious illnesses. Weight measurements, however, can fluctuate for many reasons, not all of which suggest a problem.

Continuing Brain Development

During the early childhood years, the brain continues to grow, exercising its powerful control of behavior. Children show a decided preference for using one hand or foot over the other. This preference, called handedness, starts to appear toward the end of infancy and becomes well-established by the age of 5 or 6.

Lateralization

Which hand do you use for writing? If you were to kick a football, would you use the leg on the same side of your body as the hand you use for writing? Pick up a pencil or ruler and assume it is a telescope. Which eye do you use? Are you using the same side of the body that you used for writing and kicking? Your answers to these questions should give you some idea of the meaning of cerebral **lateralization.**

We tend to think of the brain as a single unit, but actually it consists of two halves: the **cerebral hemispheres.** The two halves are connected by a bundle of nerve fibers (the corpus callosum). The left hemisphere controls the right side of the body, while the right hemisphere controls the left side of the body. Although the hemispheres seem to be almost identical, your answers to the preceding questions reveal important differences between the two. These differences are clues to your brain's organization. If you are right-handed, for example, your left cerebral hemisphere is lateralized for handedness and also for control of your speech—you are "left lateralized." Figure 7.2 illustrates the process.

Figure 7.2
Lateralization of handedness.

Much of our knowledge of cerebral lateralization has resulted from studies of brain-damaged patients. Patients with left hemisphere damage, for example, typically have speech difficulties; damage to the right hemisphere frequently causes perceptual and attentional disorders (Rourke & others, 1983). Because as humans we rely so heavily on language, the left hemisphere came to be thought of as the dominant or "major" hemisphere. Today, however, much importance has been placed on the right hemisphere's control of visual and spatial activities.

Physical Development: An Overview

In an excellent overview of physical development, Tanner (1989) discusses how the interaction of heredity and environment produces the rate and kinds of physical growth. Among the chief contributing forces are the following:

- *Genetic elements.* Hereditary elements are of immense importance to the regulation of growth. The genetic growth plan is given at conception and functions throughout the entire growth period.
- *Nutrition.* Malnutrition delays growth and, if persistent, can cause lasting damage. Children in Stuttgart, Germany, were studied each year from 1911 to 1953. From 1920 to 1940 there was a uniform increase in average height and weight, but in the later years of each war (first and second world wars) average height declined as food was curtailed. These children recovered, but there remain questions about the effects of chronic malnutrition. For example, does chronic malnutrition produce permanent brain damage in the fetus and the 1- or 2-year-old child?

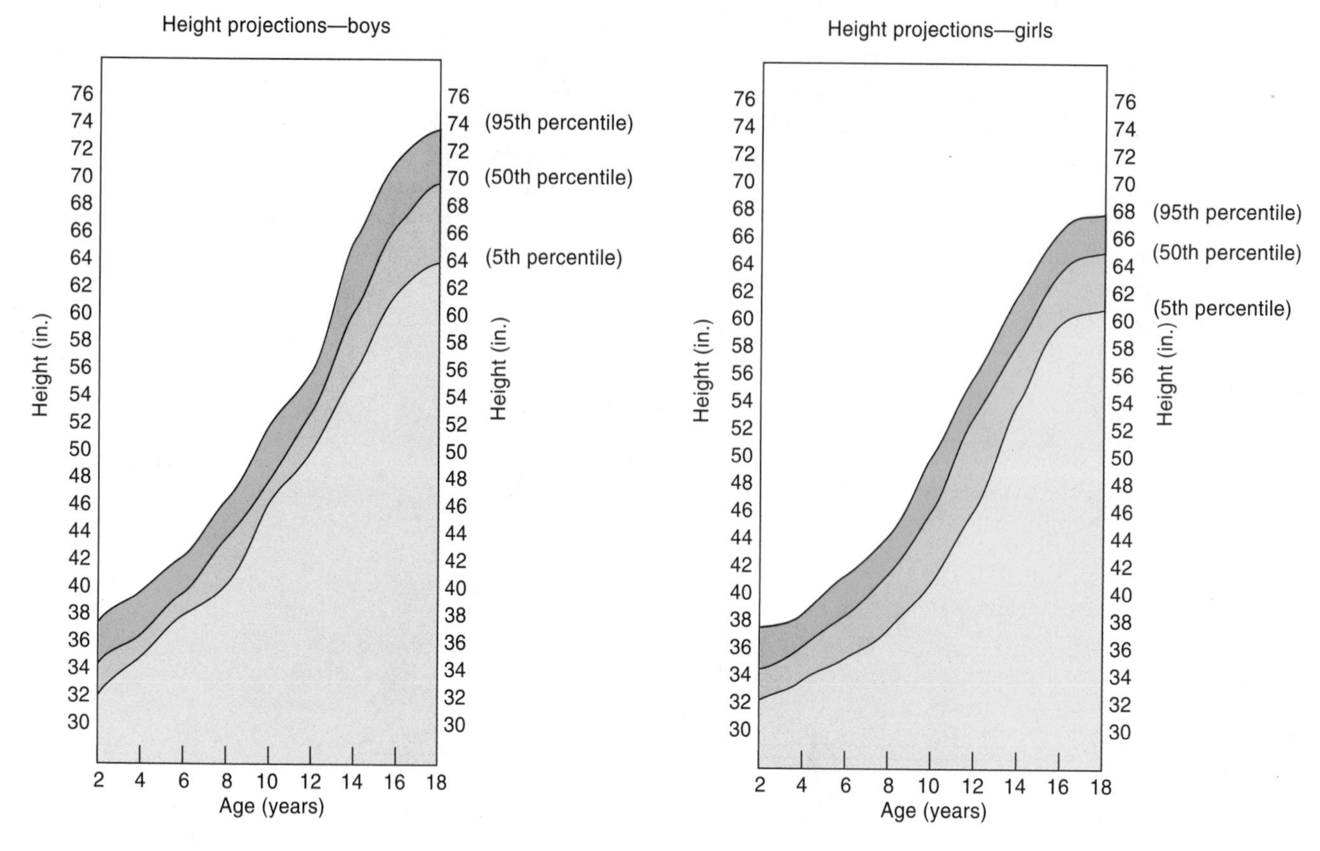

- *Disease.* Short-term illnesses cause no permanent retardation of the growth rate, although they may cause some disturbance if the child's diet is consistently inadequate. Major disease usually causes a slowing of growth, followed by a catch-up period if circumstances become more favorable.
- *Psychological disturbance.* Stress can slow development and occasionally lead to deprivation dwarfism. Small children under uncompromising strain such as divorce seem to "turn off" their growth hormone and become almost dwarfed.
- *Socioeconomic status.* Children from different social classes differ in average body size at all ages. Tanner gives the example of differences in height between British children of the professional class and those of laborers. Children of the professional class are from one inch taller at age 3 to two inches taller at adolescence. There is a consistent pattern in all such studies, indicating that children in more favorable circumstances are larger than those growing up under less favorable economic conditions. The difference seems to stem from nutrition, sleep, exercise, and recreation.
- *Secular trends.* There has been a tendency during the past hundred years for children to become progressively larger at all ages. This is especially true in Europe and America.

This brief overview of physical development again illustrates the importance of the biopsychosocial model. For example, you may be tempted to think that physical growth is essentially biological, mainly determined by heredity. Note, how-

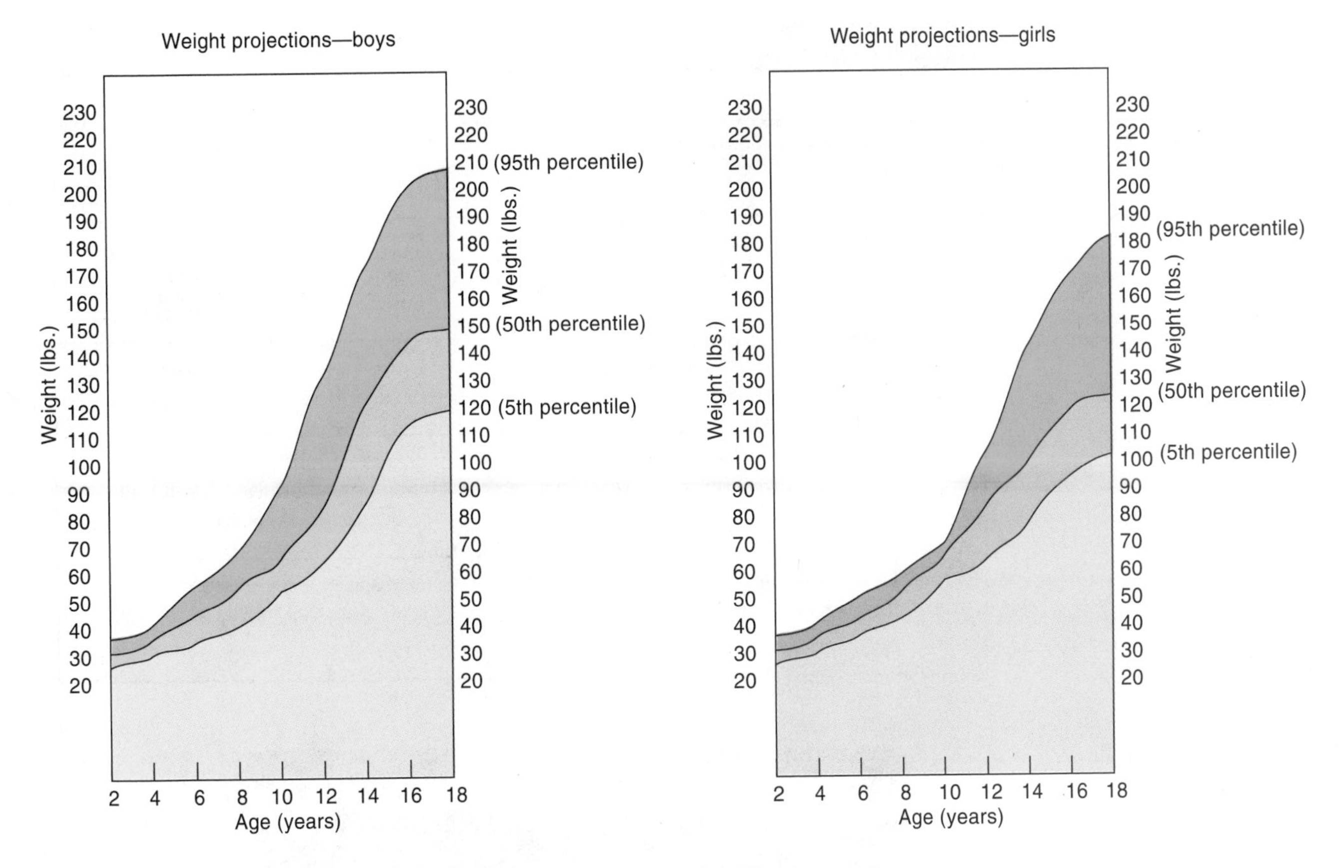

ever, the role played by nutrition and socioeconomic status. The interaction among biological, psychological, and social influences testifies to the power of the biopsychosocial model in explaining development.

Growing Motor Skills

When early childhood children reach the age of 6, no one—neither parents nor teachers—is surprised by what they can do physically. Think back to the infancy period and recall how often we referred to what they *couldn't* do. Stand, walk, run. We tend to take the accomplishments of the 6-year-old for granted, but a great deal of neuromuscular development had to occur before these motor skills became so effortless.

We are concerned here with two types of motor skills: gross (using the large muscles) and fine (using the small muscles of the hands and fingers). The well-known chronicler of children's development, Arnold Gesell (1940), has stated that thanks to perceptual and motor development, 3- and 4-year-old children can hold crayons, copy triangles, button their clothes, and unlace their shoes. Table 7.4 summarizes the development of motor skills.

The energy of the early childhood years is seen in the physical activities of the period: constant motion followed by periods of rest and nutrition.

Cratty (1979) discusses several motor skills as follows:

- *Running*. The 18-month-old child has a hurried walk. The true run appears between ages 2 and 3. By the age of 5 or 6, youngsters run rapidly (about 11.5 feet per second), employing considerable arm action.
- *Jumping*. At about age 18 months, youngsters step off a low object with one foot, hesitating slightly before placing it on the ground. At age 2, youngsters use what Cratty calls "the two-feet takeoff." Soon they begin to jump over low barriers, and by age 5 they are skillful jumpers (they can broad jump three feet and hurdle one-foot objects).

Table 7.4 The Emergence of Motor Skills

Age	Gross Skills	Fine Skills
2	Runs, climbs stairs, jumps from object (both feet)	Throws ball, kicks ball, turns page, begins to scribble
3	Hops, climbs stairs with alternating feet, jumps from bottom step	Copies circle, opposes thumb to finger, scribbling continues to improve
4	Runs well, skillful jumping, begins to skip, pedals tricycle	Holds pencil, copies square, walks balance beam
5	Hops about 50 feet, balances on one foot, can catch large ball, good skipping	Colors within lines, forms letters, dresses and undresses self with help, eats more neatly
6	Carries bundles, begins to ride bicycle, jumps rope, chins self, can catch a tennis ball	Ties shoes, uses scissors, uses knife and fork, washes self with help

This child is using his right hand to dig, signaling that his left cerebral hemisphere is lateralized for handedness and control of speech.

- *Hopping, skipping, galloping.* Hopping may be on one foot in place, using alternate feet, or hopping for distance. Some time after age 3 or 4, youngsters can hop from one to three steps on their preferred foot, and by age 5 they can extend hopping to about ten steps. Girls acquire this skill slightly earlier and more successfully than boys. Skillful skipping and galloping appear between the ages of 6 and 7.
- *Balancing.* Balancing, a measure of nervous system integrity, appears quite early. Three-year-olds can walk a reasonably straight line. Five-year-olds can maintain control while standing on one foot with their arms folded. Girls are slightly superior on this task.

The early childhood years are a time when children show a great love for drawing. Not only are their drawings a sign of motor development, but they also indicate levels of cognitive development and can be emotionally revealing.

The physical picture of the early childhood youngster is one of energy and growing motor skill. Adequate rest and nutrition are critical, and parents should establish a routine to avoid problems. For example, to reconcile a rambunctious child with the necessity of sleep, parents should minimize stimulation through a consistent, easily recognized program: washing, tooth brushing, storytelling, and gentle but firm pressure to sleep. Careful and thoughtful adult care should prevent undue difficulties.

The Special Case of Drawing

Children love to draw. No one has to teach them. In a finely tuned sequence, children move from random scribbles to skillful creations. When something is as natural and fascinating for children as drawing, we can only wonder why the vast majority of youngsters lose this desire and skill.

An early effort to chronicle the skills involved in drawing and writing (and one used widely today) was made by Florence Goodenough (1926). She attempted to match the developmental skills required with the ages at which they appear. The results appear in Table 7.5.

Drawing flourishes during the early childhood years. Brittain (1979) divides early drawings into two levels: **random scribbling** and **controlled scribbling.** Random scribbling lasts until about age 3. Children use dots and lines with simple arm movements. They gradually begin to use their wrists, which permits them to draw curves and loops. They grip the crayon as tightly as possible, usually with the whole hand. Controlled scribbling lasts until age 4. Now children carefully watch what they are doing, whereas before they looked away. They have better control of the crayon and hold it like an adult.

Table 7.5 Ages and Stages in Design Copying

Stage	Age	Description
One	1 year	Scribbling. Motor actions with no visual model and no goal.
Two	2 years	Scribbling. Spontaneous designs; child reacts to them.
Three	2½ years	Scribbling with loops.
Four	3 years	Copying by responding to a visual model.
	4 years	Copying by using angles and curves.
	5 years	Copying and producing a drawing that is organized according to the visual model.

Source: Based on Florence Goodenough, *The Measurement of Intelligence Through Drawing* 1926, Holt, New York.

By age 4 or 5, children begin to paint and hold the brush with thumb and fingers. They hold the paper in place with the free hand. They give names to their drawings and begin to show representation (using one thing for another—see the cognitive section of the chapter).

The well-known psychologist Howard Gardner (1980) has written sensitively about children's drawing and raised several important questions. Noting how drawings develop—from the scribbles of the 2-year-old to the 3-year-old's interest in design to the 4- and 5-year-old's drawing representations—Gardner comments on the liveliness and enthusiasm of their work.

And then suddenly it stops! The end of the early childhood period sees the end of creative expression, except for a select few. Why? We simply don't know. Our ignorance breeds other questions: Do all children possess artistic talent or only the gifted few? Are children's drawings truly creative or are they copies? How much depends on teaching?

Children's drawings are not only good clues to their motor coordination but, as we'll see, provide insights into their cognitive and emotional lives, another example of how a biopsychosocial perspective helps us to understand development.

Cognitive Development

Physical development during early childhood, while observable and exciting, is not the only significant change occurring. Early childhood youngsters expand their mental horizons by their increasing use of ideas and by rapid growth in language. This growing cognitive ability is a fact; explaining it is much more difficult. To help us understand what is happening and how it happens, we turn once more to Piaget.

Piaget's Preoperational Period

Piaget and Inhelder (1969) state that as the sensorimotor period concludes, **representation** (the ability to represent something by a "signifier" such as language) appears. The preoperational period extends roughly from the age of 2 to 7. The child now acquires the ability for inner, symbolic manipulations of reality. Piaget

These children, playing doctor and patients, are furthering all aspects of their development. They are discovering what objects in their environment are supposed to do, they are learning about the give and take of human relationships, and they are channeling their emotional energies into acceptable outlets.

and Inhelder describe the emergence of the symbolic function as the appearance of behavior that indicates representation. They list five of these preoperational behavior patterns:

- **Deferred imitation,** which continues after the disappearance of the model to be imitated
- **Symbolic play,** or the game of pretending
- **Drawing,** which rarely appears before age 2½
- **The mental image,** which is a form of internal imitation
- **Verbal evocation** of an event not occurring at that particular time (A child may point to the door through which the father left and say, "Dada gone.")

Imitation

For Piaget, imitation is mainly accommodation (See chap. 2), as children attempt to change in order to conform to the environment. When the sensorimotor period ends, children are sufficiently sophisticated for the phenomenon of deferred imitation to appear. Piaget gives the example of a child who visited his home one day and, while there, had a temper tantrum. His daughter Jacqueline, about 18 months old, watched, absolutely fascinated. Later, after the child had gone, Jacqueline had her own tantrum. Piaget interprets this to mean that Jacqueline had a mental image of the event.

Play

Piaget argues eloquently for recognizing the importance of play in a youngster's life. Obliged to adapt themselves to social and physical worlds that they only slightly understand and appreciate, children must make intellectual adaptations that leave personality needs unmet. For their mental health, they must have some outlet, some technique that permits them to assimilate reality to self, and not vice versa. Children find this mechanism in play, using the tools characteristic of symbolic play.

Drawing

We have previously discussed drawing as a sign of physical growth. Piaget examines children's drawings for their cognitive significance. Piaget considers drawing as midway between symbolic play and the mental image. In drawing a person, for example, children go through some stages that demonstrate the increasing use of symbols.

1. Initially they make a head with arms and legs, but usually no body. The importance of the drawing seems to be merely the act of doing it.
2. Next is a period of intellectual realism where, even if a profile is drawn, it will nevertheless have two eyes.
3. This is followed by what is called visual realism, in which a profile has only one eye and a pattern appears. Trees, houses, and people are all in correct proportion. Visual realism appears at about age 8 or 9.

Mental Images

Mental images appear late in this period because of their dependence on internalized imitation. Piaget's studies of the development of mental images between the ages of 4 and 5 show that there are two categories of mental images. There are **reproductive images,** which are restricted to those sights previously perceived. There are also **anticipatory images,** which include movements and transformation. At the preoperational level, children are limited to reproductive images.

A good illustration of the difference between the two is Piaget's famous example of matching tokens. Piaget showed 5- and 6-year-old children a row of red tokens and asked them to put down the same number of blue tokens. At this age, children put one blue token opposite each red one. When Piaget changed the arrangement, however, and spread out the row of red tokens, the children were baffled because they thought there were more red tokens than blue. Thus, children of this age can reproduce but not anticipate, which reflects the nature of their cognitive structures and level of cognitive functioning.

Language

Finally, language appears during this period, after children have experienced a phase of spontaneous vocalization, usually between the age of 6 and 10 months; a phase of imitation of sounds at about age 1; and one-word sentences at the end of the sensorimotor period. Piaget and Inhelder (1969) note that from the end of the second year the sequence is that of two-word sentences, short complete sentences, and finally the gradual acquisition of grammatical structures.

In his early writings (1929) Piaget described three fascinating developmental characteristics of preoperational thought:

1. **Realism,** which means that children slowly distinguish and accept a real world. They now have identified both an external and internal world.
2. **Animism,** which means that children consider a large number of objects alive and conscious that adults consider inert. For example, a child sees a necklace wound up and then released; when asked why it is moving, the child replies that the necklace "wants to unwind."

   Children overcome these cognitive limitations as they recognize their own personalities; that is, they refuse to accept personality in things. Piaget also believes that, as with egocentrism, comparison with the thoughts of others—social intercourse—slowly conquers animism. He identifies four stages of animism:

- Almost everything is alive and conscious.
- Only those things that move are alive.
- Only those things that manifest spontaneous movements are alive.
- Consciousness is limited to the animal world.

3. **Artificialism,** which consists of attributing human creation to everything. For example, when asked how the moon began, some of Piaget's subjects replied, "because we began to be alive." As egocentrism decreases, youngsters become more objective, and they steadily assimilate objective reality to their cognitive structures.

   Thus they proceed from a purely human or divine explanation to an explanation that is half natural, half artificial: The moon comes from the clouds, but the clouds come from people's houses. Finally, at about age 9 they realize that human activity has nothing to do with the origin of the moon.

Limitations of Preoperational Thought

Although we see the steady development of thought during this period, there are still limitations to preoperational thought. As the word *preoperational* implies, this period comes before advanced symbolic operations develop. Piaget has stated constantly that knowledge is not just a mental image of an object or event.

To know an object is to act on it, to modify it, to transform it, and to join objects in a class. The action is also reversible. If two is added to two, the result is four; but if two is taken away from four, the original two returns. The preoperational child lacks the ability to perform such operations on concepts and objects.

There are several reasons for the restricted nature of preoperational thought.

Egocentrism In the period of preoperational thought, children cannot assume the role of another person or recognize that other viewpoints exist, a state called **egocentrism.** This differs from sensorimotor egocentrism, which is primarily the inability to distinguish oneself from the world. For example, preoperational children make little effort to ensure that listeners understand them. Thus they neither justify their reasoning nor see any contradictions in their logic.

Centration A striking feature of preoperational thought is the centering of attention on one aspect of an object and the neglecting of any other features—called **centration.** Consequently, reasoning is often distorted. Preoperational youngsters are unable to decenter, to notice features that would give balance to their reasoning.

A good example of this is the process of classification. When youngsters from age 3 to 12 are asked, "What things are alike?" their answers proceed through three stages. First, the youngest children group figurally, that is, by similarities, differences, and by forming a figure in space with the parts. Second, children of about age 5 or 6 group objects nonfigurally. They form the elements into groups with no particular spatial form. At this stage, the classification seems rational, but Piaget and Inhelder (1969) provide a fascinating example of the limitations of classification at this age. If in a group of 12 flowers, there are 6 roses, these youngsters can differentiate between the other flowers and the roses. But when asked if there are more flowers or more roses, they are unable to reply because they cannot distinguish the whole from the part (see Figure 7.3). This understanding does not appear until the third phase of classification, at about the age of 8.

States and Transformations Youngsters concentrate on a particular state or succession of states of an object and ignore the transformation by which one state is changed into another. Their lack of conservation is a good example of the static nature of preoperational thought.

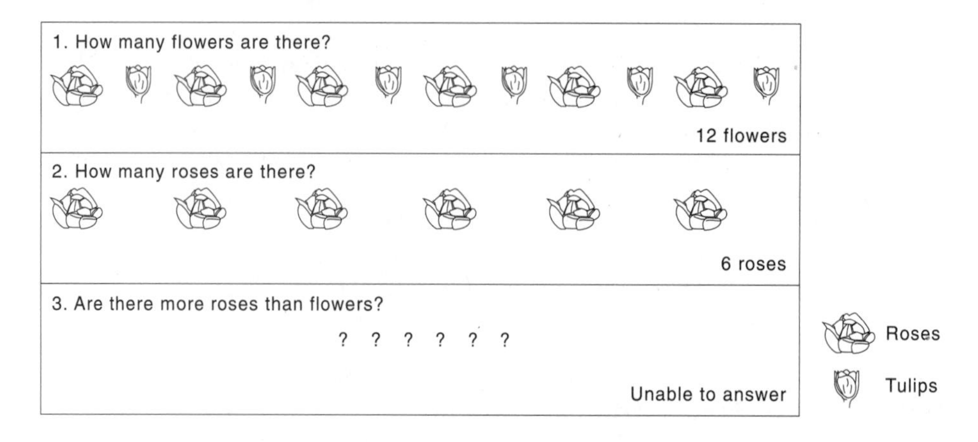

Figure 7.3
Lack of genuine classification.

Conservation means understanding that an object retains certain properties, no matter how its form changes. The most popular illustration is to show a 5-year-old two glasses, each half filled with water. The child agrees that each glass contains an equal amount. But if you then pour the water from one of the glasses into a taller, thinner glass, the youngster now says that there is more liquid in the new glass (see Figure 7.4). Youngsters consider only the appearance of the liquid and ignore what happened. They also do not perceive the reversibility of the transformation. In their minds they do not pour the water back into the first glass.

Irreversibility A truly cognitive act is reversible if it can use stages of reasoning to solve a problem and then proceed in reverse, tracing its steps back to the original question or premise. The preoperational child's thought is **irreversible** and entangles the child in a series of contradictions in logic.

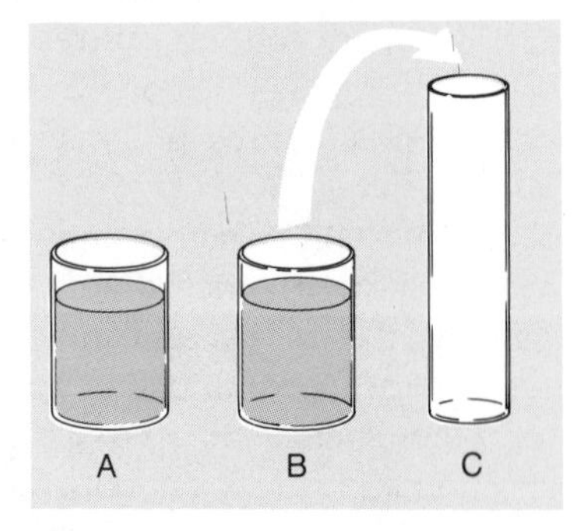

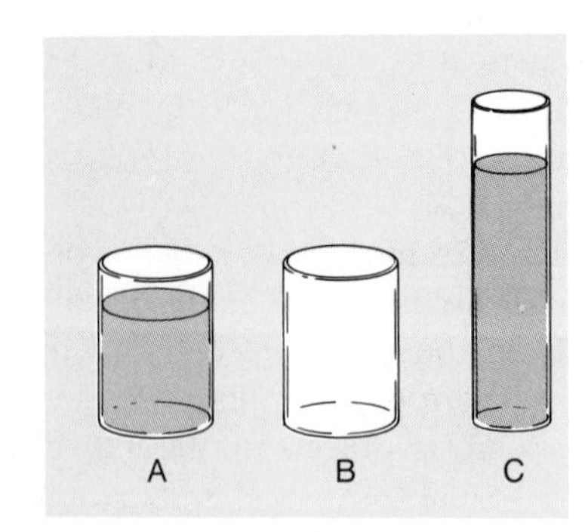

Figure 7.4
Piaget used the beaker task to determine whether children had conservation of liquid. In I, two identical beakers (A and B) are presented to the child; then the experimenter pours the liquid from B into beaker C, which is taller and thinner than A and B. The child is asked if beakers B and C have the same amount of liquid. The preoperational child says no, responding that the taller, thinner beaker (C) has more.

In the water-level problem, for example, the child believes that the taller, thinner glass contains more water. Youngsters cannot mentally reverse the task (imagine pouring the contents back into the original glass). At the conclusion of the preoperational period, children slowly decenter and learn reversibility as a way of mental life.

An Applied View — Children and Their Humor

■ "Why did daddy tiptoe past the medicine cabinet?"
"Because he didn't want to wake the sleeping pills."

Jokes such as these have spurred Paul McGhee (1979, 1988) to analyze children's humor and trace its developmental path. Noting that little is known about how children develop humor, McGhee turned to cognitive development as a possible explanation. While humor also has psychodynamic and social features, McGhee believes that its cognitive properties offer the best basis for unraveling its secrets.

McGhee begins by noting that the basis of most children's humor is incongruity, which is the realization that the relationship between different things just isn't "right." Using this as a basis, he traces four stages of incongruous humor. He treats stages as does Piaget: The sequence of stages remains the same, but the ages at which children pass through them varies.

Stage One: Incongruous Actions Toward Objects. Stage 1 usually occurs sometime during the second year, when children play with objects. They are able to form internal images of the object and thus start to "make-believe." For example, one of Piaget's children picked up a leaf, put it to her ear, and talked to it as if it were a telephone, laughing all the time. One of the main characteristics of stage 1 humor is the child's physical activity directed at the object.

Stage Two: Incongruous Labeling of Objects and Events. Stage 2 humor is more verbal, which seems to be its most important difference from stage 1. McGhee (1979) notes that the absence of action toward objects is quite noticeable. Piaget's 22-month-old daughter put a shell on the table and said, "sitting." She then put another shell on top of the first, looked at them and said, "sitting on pot." She then began to laugh. Children in this stage delight in calling a dog a cat, a foot a hand, and then start laughing.

Stage Three: Conceptual Incongruity. Around age 3, most children begin to play with ideas, which reflects their growing cognitive ability. For example, stage 3 children laugh when they see a drawing of a cat with two heads.

Stage Four: Multiple Meanings. Once children begin to play with the ambiguity of words, their humor approaches the adult level.

■ "Hey, did you take a bath?"
"No. Why, is one missing?"

Children at stage 4 (usually around age 7) understand the different meaning of *take* in both instances. Stage 3 children could not understand the following joke:

■ "Order! Order! Order in the court."
"Ham and cheese on rye, your honor."

Stage 4 youngsters appreciate its ambiguity.

You can see how cognitive development is linked to humor. McGhee (1988) states that the effective use of humor with children can help stimulate new learning and creative thinking, instill an interest in literature, facilitate social development, and enhance emotional development and adjustment.

As we mentioned in our discussion of Piaget's sensorimotor period (See chap. 5), recent research has raised questions about Piaget's assumptions concerning the cognitive abilities of young children. For example, as we explained in the matching tokens problem, children of this age are likely to say that the spread-out row now has more tokens. Piaget believed that the children concentrated on the length of the row because they lacked a concept of number.

When Gelman and Baillargeon (1983) used similar problems only with a smaller number of objects (up to four), however, children of this age successfully answered questions about the number of objects involved. Evidence continues to mount supporting the conclusion that Piaget underestimated the cognitive abilities of young children: They conserve, classify, and overcome egocentrism earlier than Piaget realized (Gardner, 1983). How can we explain these differences? One explanation points to the complexity of Piaget's tasks—they were more difficult than those used by modern psychologists. Also, his reliance on verbal cues may have confused some of the children he tested.

Intellectual Differences

Piaget concentrated on children within the normal range of intelligence. But as you might expect, children reflect the entire range of intellectual ability, from severely retarded to gifted. Both extremes of this intellectual continuum have attracted considerable attention. Most people have an IQ of about 100 on the Binet test. Let's move in opposite directions from this so-called average that claimed Piaget's attention and examine the characteristics of both slow and gifted children.

Slow Learners

Any discussion of slow learners takes us into the world of **mental retardation.** Who are these individuals? A widely accepted definition has been proposed by the American Association on Mental Deficiency:

> ***Mental retardation refers to significantly subaverage general intellectual functioning resulting in or associated with concurrent impairments in adaptive behavior and manifested during the development period.*** **(Grossman, 1983, p. 11.)**

In this definition, *subaverage* refers to an IQ of approximately 70 or below. Estimates are that about 2–3 percent of the population is below average. By *adaptive,* Grossman means the extent to which an individual meets the standards of personal independence and social responsibility expected for age and culture group. Table 7.6 presents the AAMD classification scheme. The meaning of the categories is as follows:

Table 7.6 Classification of Retardation

Mild	52–67 (Binet)	55–69 (WISC—R)
Moderate	36–51	40–54
Severe	20–35	
Profound	Below 20	

Source: The American Association on Mental Deficiency. Copyright American Association on Mental Deficiency, Washington, D.C.

- *Mild:* Development slow. Children capable of being educated ("educable") within limits. Adults, with training, can work in competitive employment. Able to live independent lives.
- *Moderate:* Slow in their development, but able to learn to care for themselves. Children capable of being trained ("trainable"). Adults need to work and live in sheltered environment.
- *Severe:* Motor development, speech, and language are retarded. Not completely dependent. Often, but not always, physically handicapped.
- *Profound:* Need constant care or supervision for survival. Gross impairment in physical coordination and sensory development. Often physically handicapped.

Kessler (1988) reinforces the belief that adaptive behavior changes when she notes that any person (child or adult) may meet the criteria of mental retardation at one time in life and not another. Thus it is possible for both IQ and adaptive functioning to change.

Many children who are exceptional, such as the mildly mentally retarded, with support, can acquire social skills that enable them to function in an appropriate setting.

If you should have a slow learner in the family or occasionally meet them in your neighborhood or in other circumstances, remember that these youths

- Have the same basic needs as the nonretarded
- Demonstrate considerable individual differences
- Are a changing population because of society's evolving perception and treatment

At the other end of the intellectual continuum are gifted children.

The Gifted/Talented

Jimmie is six years and eight months as of this writing and this year he has shown a deep interest in paleontology, meteorology (he has his own weather station), and electronics. He builds his own radio and transmitter sets with little assistance. He has forged ahead with his inventions, too. . . . This year, when he heard about the airplane disasters at Idlewild and LaGuardia airfields, he felt quite dismayed and decided to do something about it. Using a cardboard shoe box, two strips of paper and two magnets, he invented a device which the pilot could use in the cockpit of his plane which would tell him when it is off course on approaching. Needless to say, Thomas Edison is his idol. **(Goertzel & Goertzel, 1962, p. 289.)**

The abilities of this boy, taken from the Goertzels' enduring description of a gifted youngster, nicely define what we mean by gifted. But, though immensely talented, these children frequently grow up in environments (including schools) that don't understand them and often display hostility toward them.

Defining Gifted Even defining the **gifted and talented** has caused considerable controversy. Initially, an IQ of 120 or 140 (or some number) was used to identify the gifted. But these measures were too restrictive. Youngsters who have exceptional talent in painting or music or who seem unusually creative are also gifted. As a result, the Gifted and Talented Children's Education Act of 1978 defined these children as follows:

Thomas Edison's creative genius was not immediately recognized. As with many of the gifted, their talents may go unrecognized or arouse hostility in those around them.

Gifted children may approach tasks quite differently than typical children. Einstein, for example, was thought to be almost incapable of learning because of his delayed language and unique methods of problem solving.

One of history's great figures, Churchill had an inauspicious beginning. School failures and personality clashes with instructors disguised the genius that was to appear later.

The term gifted and talented children means children, and, whenever applicable, youth who are identified at the preschool, elementary, or secondary level as possessing demonstrated or potential abilities that have evidence of high performance capabilities in areas such as intellectual, creative, specific academic or leadership ability, or in the performing and visual arts, and who by reason thereof, require services or activities not ordinarily provided by the school.

Who Are the Gifted and Talented? Estimates are that about 5 percent of school-age children fit this definition. If these children are so talented, why do they require special attention? For every Einstein who is identified and flourishes, there are probably dozens of others whose gifts are obscured. Thomas Edison's mother withdrew him from first grade because he was having so much trouble; Gregor Mendel, the founder of scientific genetics, failed his teacher's test four times; Isaac Newton was considered a poor student in grammar school; Winston Churchill had a terrible academic record; Charles Darwin left medical school.

One major difficulty with the way our society deals with gifted children is that interest in them comes and goes. Initially, interest in gifted children was aroused when the results of IQ tests showed that some children were outstanding. Later, when the Soviets launched their Sputnik satellite, there was concern in the United States for providing opportunities for the gifted that could offer us a chance to catch up (Bloom, 1985; Fox & Washington, 1985).

Today, with our national concern for unusual students of all types, interest in the gifted has increased dramatically. Government reports insist that the gifted are a minority who need special attention. They are indeed a minority, characterized by their special ability, who come from all levels of society, all races, and all national origins, and who include both sexes equally. They are the Einsteins, the Edisons, the Lands.

In spite of a concentrated effort by educators to ensure equality in our schools, minority students remain woefully underrepresented in programs for the gifted (Frasier, 1989). Some writers have speculated that problems in identifying minority gifted children, for example, are the major cause of this underrepresentation (Genshaft, 1991). Other explanations have also been offered for this phenomenon: low IQ test scores, or lack of a stimulating socioeconomic background.

Yet we are now seeing a reconsideration of the role of a pupil's home life in intellectual development. Research has shown that the quality of those homes that foster intellectual achievement is quite similar, regardless of income level (Frasier, 1989; Sue & Okazaki, 1990). Brown (1988), examining model youths, points out the considerable support for intellectual development in the African-American community.

Frasier (1989) suggests that we change the screening procedures, which have inherent limitations built into them for multicultural students. Why not use behavioral traits indicating giftedness, such as the *use* of language rather than a test question *about* language? Parents of multicultural students can help educators reword items on rating scales. Vignettes of successful multicultural students can be used for motivational purposes.

Attempts to identify all children for a gifted program could include

- Seeking nominations from knowledgeable professionals and nonprofessionals
- Using behavioral indicators to identify pupils who show giftedness in their cultural tradition
- Collecting data from multiple sources
- Delaying decision making until all pertinent data can be collected in a case study

Remember: The gifted come from diverse backgrounds, not from any one group, and their talents should not be lost.

The Gifted—A Neglected Group The gifted have been neglected for several reasons. Among them are the following:

- *Failure to be identified.* In a recent survey, the U.S. Office of Education found that 60 percent of schools reported no gifted students. Teachers and administrators simply fail to recognize them.
- *Hostility of school personnel.* Hostility has traditionally been a problem for the gifted. Resentment that they are smarter than teachers, dislike for an intellectual elite, and antagonism toward their obvious boredom or even disruptive behavior have all produced a hostile atmosphere for the gifted.
- *Lack of attention.* The inconsistent interest in these children that we described causes a lack of attention to their needs. Estimates are that only 3–4 percent of the nation's gifted have access to special programs.
- *Lack of trained teachers.* There are remarkably few university programs to train teachers of these children.

Working from different assumptions about the gifted and talented, Benjamin Bloom and his colleagues (1985) reached several interesting conclusions about this group. Defining talent as an unusually high level of demonstrated ability, achievement, or skill in some special field of study or interest, they investigated the development of talent in several fields: psychomotor (including athletic), aesthetic (including musical and artistic), and cognitive.

Selecting Olympic swimmers and world-class tennis players, concert pianists and sculptors, and finally research mathematicians and research neurologists, Bloom and his team subjected the participants to intensive interviewing. They also interviewed the parents and teachers of these individuals, with the subjects' permission. From these interviews, the following general conclusions were made:

- Young children usually view latent ability as play and recreation, followed by a long period of learning and hard work, eventually focusing on one particular learning activity (math, science).
- The home environment structured the work ethic and encouraged a youngster's determination to do the best at all times.
- Parents strongly encouraged children in a specialized endeavor in which they showed talent.
- No one made it alone; families and teachers or coaches were crucial at different times in the development of a youngster's talent.
- Clear evidence of achievement and progress was necessary for a youngster to continue learning more difficult skills.

Bloom concludes that for talent to develop, several qualities must be present, such as a strong interest and emotional commitment to a particular talent field, a desire to reach a high level of attainment, and finally a willingness to expend great amounts of time and effort to reach high levels of achievement in the field of talent.

Remember, however, that these children are similar to other boys and girls in their interests and feelings. Every effort should be made to have them socialize with other youngsters. Terman, for example, found that the gifted have the same interest in games and sports. Also, remember that the intellectually gifted student is not necessarily physically gifted; nor is the artistically talented necessarily mathematically superior.

As you can tell from our discussion thus far, children of these years have marched firmly into a symbolic world, a major part of which is language.

Language Development

Youngsters soon acquire their native language, a task of such scope and intricacy that its secrets have eluded investigators for centuries. During the early childhood period, language figuratively "explodes." Children proceed from hesitant beginnings to almost complete mastery by the end of the period (about seven years). Remember: Children don't learn to speak by imitating adults, nor do most parents reward their children for good grammar (deCuevas, 1990).

All children, however, learn their native language. At about the same age they manifest similar patterns of speech development, whether they live in a ghetto or in some wealthy suburb. Moskowitz (1979) states that within a short span of time and with almost no direct instruction, children completely analyze their language. Although refinements are made between ages 5 and 10, most children have completed the greater part of the process of language acquisition by the age of 4 or 5. Recent findings have also shown that when children acquire the various parts of their language, they do so in the same order. For example, in English, children learn *in* and *on* before other prepositions, and they learn to use *ing* as a verb ending before other endings such as *ed* (Gleason, 1985).

It is a tremendous accomplishment; if you remember how you may have tried to learn a foreign language as an adult, you'll recall how difficult it was to acquire vocabulary and to master rules of grammar and the subtleties of usage. Yet

The attention that adults (especially parents) give children encourages positive interactions and leads to satisfactory and fulfilling relationships. Adult attention will also further language development and enhance a child's self-concept.

preschool children do just this with no formal training. By the time they are ready to enter kindergarten, most children have a vocabulary of about 8,000 words, use questions, negative statements, dependent clauses, and have learned to use language in a variety of social situations (Gleason, 1985).

Language as Rule Learning

By the age of 4 or 5, children will also have discovered several basic rules:

- There are rules for recombining sounds into words.
- Individual words have specific meaning.
- There are rules for recombining words into meaningful sentences.
- There are rules for participating in a dialogue.

What children have acquired naturally, language specialists translate into more technical statements, such as the following:

- The rules of **phonology** describe how to put sounds together to form words.
- The rules of **syntax** describe how to put words together to form sentences.
- The rules of **semantics** describe how to interpret the meaning of words.
- The rules of **pragmatics** describe how to take part in a conversation.

As we trace the path of language development in the early childhood years, you should remember a basic distinction that children quite clearly demonstrate. At about one year (the infancy period), children show an ability for receptive language ("show me your nose"—they receive and understand these words). Now, in early childhood, they produce language themselves, **expressive language.** How do children acquire these language milestones?

The Pattern of Language Development

All children learn their native tongue, and at similar ages they manifest similar patterns of language development. The basic sequence of language organization is as follows (Travers, 1982):

- At about 3 months, children use intonations similar to those of adults.
- At about 1 year they begin to use recognizable words.
- At about 2½ years, most children produce complete sentences, begin to ask questions, and use negative statements.
- At about 4 years they have acquired the complicated structure of their native tongue.
- In about two or three more years they speak and understand sentences that they have never previously used or heard.

Around age 10 months, children begin to use actual words and can usually follow simple directions. At about 1 year, or perhaps slightly older, they seem to grasp that words "mean" things, like people or objects. Sometime between the ages of 12 and 18 months, children begin to use single words as sentences (*holophrases*); by age 2 they utter two-word sentences; and usually by age 4 they produce sentences of several words.

As children grow, it is difficult to specify how extensive their vocabulary is. Do we mean spoken words only? Or do we include words that children may not use but clearly understand? Whatever criteria we use, girls seem to surpass boys by about one year of development until the age of 8.

It is an amazing accomplishment. For example, the vocabulary of every language is categorized; that is, some words are nouns, some are verbs, still others are adjectives, prepositions, or conjunctions. If English had only 1,000 nouns and 1,000 verbs, we could form one million sentences (1,000 × 1,000). But nouns can be used as objects as well as subjects. Therefore the number of possible three-word sentences increases to one billion (1,000 × 1,000 × 1,000).

One billion sentences is the result of a starkly impoverished vocabulary. The number of sentences that could be generated from English, with its thousands of nouns and verbs, plus adjectives, adverbs, prepositions, and conjunctions, staggers the imagination. Estimates are that it would take trillions of years to say all possible English sentences of 20 words. In this context, the ability of children to acquire their language is an astounding achievement. Although most youngsters experience problems with some tasks during this period—difficulty with reading or mathematics—they acquire their language easily and in just a few years. The specific sequence of this accomplishment is seen in Table 7.7. These ages and accomplishments are a good guide to normal language development. Remember, don't confuse youngsters who demonstrate a serious language problem (such as lack of comprehension) with those who experience temporary setbacks.

Speech Irregularities

When speech emerges, certain irregularities appear that are quite normal and to be expected. For example, **overextensions** mark children's beginning words. Assume that a child has learned the name of the house pet—"doggy." Think what that label means: an animal with a head, tail, body, and four legs. Now consider what other animals "fit" this label: cats, horses, donkeys, and cows. Consequently, children may briefly apply "doggy" to all four-legged creatures, but they quickly eliminate overextensions as they learn about their world.

Table 7.7 The Language Sequence

Language	Age
Crying	From birth
Cooing	5–6 weeks
Babbling	4–5 months
Single words	12 months
Two words	18 months
Phrases	2 years
Short sentences and questions	3 years
Conjunctions and prepositions	4 years

Overregularities are a similar fleeting phenomenon. As youngsters begin to use two- and three-word sentences, they struggle to convey more precise meanings by mastering the grammatical rules of their language. For example, many English verbs add *ed* to indicate past tense.

I want to play ball.
I wanted to play ball.

Other verbs change their form much more radically.

Did Daddy come home?
Daddy came home.

Most children, even after they have mastered the correct form of such verbs as *come, see, run,* still add *ed* to the original form. That is, youngsters who know that the past tense of *come* is *came* will still say:

Daddy comed home.

Again, this phenomenon persists only briefly and is another example of the close link between language and thought. We know that from birth, children respond to patterns. They look longer at the human face than they will at diagrams because the human face is more complex. (Remember the Fantz study?) Once they have learned a pattern such as adding *ed* to signify past tense, they have considerable difficulty in changing the pattern.

The path of language development is similar for all children. A particular culture has little to do with language emergence, but it has everything to do with the shape that any language assumes. Children will not speak before a certain time—this is a biological given and nothing will change it. But once language appears, it is difficult to retard its progress. Usually only some traumatic event such as brain damage (which we have discussed) or dramatically deprived environmental conditions can hinder development.

Finally, Anselmo (1987) notes that as children come to the end of the early childhood period, several language milestones have been achieved:

- They become skillful in building words; adding suffixes such as *er, man,* and *ist* to form nouns (the person who performs experiments is an experimenter).
- They begin to be comfortable with passive sentences (the glass was broken by the wind).

An Applied View — The Case of Genie

The startling case of Genie illustrates the durability of language but also demonstrates its vulnerability. Discovered in the upstairs room of a suburban Los Angeles home, Genie was 13½ years old and weighed only 60 pounds when found. She could not stand or chew solid food, nor was she toilet trained (Curtiss, 1977).

From the age of 20 months she had been confined to a small, shaded room where she had been kept in a crib or tied to a chair. Any noises that she made were greeted with a beating. Few, if any, words were exchanged with her parents, nor was there any radio or television set in the home. She was usually fed baby food. Genie also had a congenital hip problem that caused her father to think that she was severely retarded. She had emotional difficulties but had normal hearing and vision. Her language comprehension and usage were almost nonexistent.

After treatment in the Children's Hospital of Los Angeles, Genie was placed in a foster home where she acquired language, more from exposure than any formal training. Estimates are that she has acquired as much language in eight years as the normal child acquires in three. She continues to have articulation problems and difficulty with word order.

While Genie has made remarkable language progress, difficulties persist. For example, she does not appear to have mastered the rules of language (her grammar is unpredictable), she continues to use the stereotypic speech of the language-disabled child, and she seems to understand more language than she can produce. Thus the case of Genie suggests that while language is difficult to retard, sufficiently severe conditions can affect progress in language. Here we see the meaning of a sensitive period when applied to language development.

- By the end of the period, children can pronounce almost all English speech sounds accurately.
- As we have noted, this is the time of the "language explosion" and vocabulary has grown rapidly.
- Children of this age are aware of grammatical correctness.

Conclusion

Thus far in our discussion of the early childhood years, we have seen that while that rate of growth slows somewhat, it still continues at a steady pace. Physical and motor skills become more refined. Cognitive development during these years leads to a world of representation in which children are expected to acquire and manipulate symbols. Language gradually becomes a powerful tool in adapting to the environment.

There are, however, other aspects of development that affect the direction a child's development takes. Youngsters in early childhood seem to "come into their own." Their emerging personalities take on definite dimensions in these years. They must learn to adjust to family members—perhaps a new brother or sister. Many children of this age experience the shock of parental separation and divorce, with its developmental overtones. For many, the idea of family takes on new meaning. These are among the important topics to which we now turn.

Chapter Highlights

Features of the Early Childhood Years

- Children of these years, aware of their growing competence, often clash with the wishes of their parents and teachers.

Physical and Motor Development

- Growth continues at a steady, less rapid rate during these years.
- Brain lateralization seems to be well-established by the age of 5 or 6.
- Height is a good indicator of normal development when heredity and environment are considered in evaluating health.
- Increasing competence and mastery are seen in a child's acquisition of motor skills.

Cognitive Development

- These years are the time of Piaget's preoperational period and the continued appearance of abstract abilities.
- Children's growing cognitive proficiency is seen in their use of humor.
- Although Piaget concentrated on children of normal intelligence, other children are at either end of the intelligence spectrum—slow or gifted.

Language Development

- Children acquire the basics of their language during these years with little, if any, instruction.
- All children seem to follow the same pattern in acquiring their language.

Key Terms

Animism
Anticipatory image
Artificialism
Autonomy
Centration
Cerebral hemispheres
Conservation
Controlled scribbling
Deferred imitation
Developmentally delayed
Drawing
Egocentrism
Expressive language
Gifted and talented
Initiative
Irreversibility
Lateralization
Mental imagery
Mental retardation
Overextensions
Overregularities
Phonology
Pragmatics
Preoperational period
Random scribbling
Realism
Representation
Reproductive image
Semantics
Symbolic play
Syntax
Verbal evocation

What Do You Think?

1. As you can tell from the data presented in this chapter, early childhood youngsters continue their rapid growth, although at a less frantic rate than during infancy. Consider yourself a parent of a child of this age (boy or girl) for a moment. How much would you encourage them to participate in organized, directed physical activities (swimming, dancing, soccer, etc.)? Be sure to give specific reasons for your answer.
2. As you read Paul McGhee's account of the development of children's humor, could you explain his stages by comparing them to Piaget's work? Do you think this is a logical way to proceed? Why?
3. When you consider the tragic case of Genie compared to the enormous language growth of most children, what comes to your mind? Does it change your opinion about how language develops? In what way? What do you think this case implies for the existence of a sensitive period for language development?

Suggested Readings

Fraiberg, S. (1959). *The Magic Years*. New York: Scribners. This little classic provides valuable insights into the early childhood years. It's available in paperback in almost all bookstores.

Moskowitz, B. (1979). The acquisition of language. *Scientific American,* pp. 92–108. If you're like most readers encountering the basics of language development for the first time, you could use a good summary of the process. This article, which you can get in your library and copy, is clear, thorough, and well written. You will find it helpful.

Trelease, J. (1985). *The read-aloud handbook*. New York: Penguin. An excellent source of ideas for capturing children's attention through books. A fine age-related, annotated bibliography is also offered.

CHAPTER 8

Psychosocial Development in Early Childhood

The Family in Development 201

The Changing Family 202

Parenting Behavior 203

Homeless Children 204

Children of Divorce 206

Day Care 209

Early Childhood Education 213

Early Childhood Education: A Positive Outlook 213

Miseducation 214

The Self Emerges 215

The Development of Self 215

The Special Case of Self-Esteem 215

Gender Identity 217

The Importance of Play 222

The Meaning of Play 222

The Development of Play 224

Conclusion 225

Chapter Highlights 226

Key Terms 226

What Do You Think? 227

Suggested Readings 227

Barbara is a former nurse who now takes care of four neighborhood children, ranging in age from 9 months to 4 years. A neat, careful person who loves children (she has two of her own away at school), she is sensitive to their needs and aware of the developmental changes constantly occurring in her four children. Barbara is in great demand by neighborhood families because of these traits, but she refuses to take more than four children at a time.

It's eight o'clock in the morning and Gina, a 4-year-old dynamo, is at the door with her mother, Janice. Janice takes courses at the local college and leaves Gina with Barbara three days each week. Barbara smiles at the thought of Gina. Pretty, bright, and active, with springs in her legs, Gina usually leads the other children (and Barbara) on a merry chase. When you talk to her, Gina constantly bounces up and down.

Janice and Gina walk into the front room, which is spotless and filled with toys arranged attractively along the walls. It's apparent that this home is designed for children: bright, cheerful, and safe. Janice and Barbara talk about Gina for a few minutes; she had a cold the previous day and had a restless night, but Janice says she seems fine now. Upon seeing her 3-year-old friend, Amy, Gina darts over and begins to hug her.

Barbara and Janice watch as Gina asks Amy if she wants to see her new dance. (Gina takes a dance class and will be in a recital soon.) Without waiting for an answer, she goes through her dance and takes a deep bow. Amy laughs happily as Janice and Barbara applaud. Gina then takes Amy's hand and pulls her toward the back yard with its swings and sand box.

It's so easy to observe how Gina feels by observing her play, Barbara thinks. Today she seems to be feeling better, has her energy back, and wants to play physically. At other times, Barbara has seen Gina working out a conflict with Janice by playing with dolls and assigning them mother-daughter roles. In her pretend play world, there were "mommy " dolls, "good" dolls, "bad" dolls, and "napping" dolls. As Barbara watches, she thinks to herself, as she does so often, what a wonderful time these years are. ■

The physical activities of the early childhood years are not only a time for play and games, but also for learning and adjustment.

During the early childhood years, youngsters blossom into distinct personalities. Although temperamental differences are apparent at birth, they now flourish until even the casual observer notices a child's "personality." Early childhood youngsters begin a period in which they must reconcile their individuality with the restrictions of the world around them.

This period is often described as one of socialization versus individuation. **Socialization** means the need to establish and maintain relations with others and to regulate behavior according to society's demands. **Individuation** refers to the fullest development of one's self. These two functions seem to pull in opposite directions. Society has certain regulations that its members must follow if chaos is to be avoided. Yet these rules must not be so rigid that the individual members who constitute the society cannot develop their potential to the fullest.

Any society, ours included, demands resolution of the tension between the two. Each needs the other. In chapter 7, individuation was stressed, the emergence of individual physical and cognitive abilities. In this chapter, we'll explore the expression of those talents within a societal context and trace the possible sources of tension.

But as we examine the great socializing agents of society—family, school, peers, television—a clear warning sign flashes. Is society driving early childhood youngsters too harshly? Are expectations too great? Do we expect our children to grow up too quickly? The popular child psychologist David Elkind thinks so:

> ***What is happening in the United States today is truly astonishing. In a society that prides itself on its preference for facts over hearsay, on its openness to research, and on its respect for "expert" opinion, parents, educators, administrators, and legislators are ignoring the facts, the research, and the expert opinion about how young children learn and how best to teach them.*** **(1987, p. 3)**

What concerns Elkind is the academic pressure that parents and educators are placing on their children. Calling this the "**miseducation** of youth," Elkind believes that these parents are in search of superkids. He believes that such parental behavior reflects parental ego more than parental need and has little to do with a child's needs.

In a thoughtful essay on children in our society today, Neil Postman (1982) argues that childhood has actually disappeared. Citing highly paid 12-year-old models, toddlers in designer jeans, and the growing absence of children's games, Postman laments the disappearance of childhood. In a highly technological society with instant and open communication, nothing is withheld from youngsters. Calling television the "total disclosure medium," Postman believes that the lines separating adulthood from childhood are being erased. This loss of a distinct childhood results in a loss of shame. Everything is revealed to children; there are no secrets.

To come to grips with these issues, in this chapter we'll examine the modern family, tracing the stages through which it proceeds, the impact of parenting on children, the effect of divorce on development, and the growing importance of day care in our society. We'll next explore how these early experiences affect a child's learning by examining the various types of learning. The discussion of learning will lead us to a review of early childhood education—its pros and cons. Throughout these years, a child's sense of personal identity is gradually emerging. Its inherent temperamental characteristics are tempered by the reactions of family members, teachers, and peers. Finally, we'll comment on the significance of play to children of this age. These topics fit the biopsychosocial model as seen in Table 8.1.

Television, one of the major socializing agents in a child's life, can have both prosocial and antisocial consequences. Consequently, parents should carefully monitor what their children watch.

Table 8.1 Characteristics of the Period

Bio	Psycho	Social
Play	Play	Play
Gender	Self	Gender
Self	Gender	Day care
	Learning	
	Divorce	
	Day care	

After reading this chapter, you should be able to:

- Assess the impact of the family on a child's development.
- Evaluate the influence of parenting behavior on a child's development.
- Compare the various types of day-care centers.
- Describe the role of preschool education during the early childhood years.
- Determine the importance of self-esteem in a child's life.
- Appraise the function of play in a child's development.

Our initial task, though, is to examine that great socializing agent—the family.

The Family in Development

This section could have been written for any chapter in a lifespan book, but it is particularly pertinent here because the family is still recognized as *the* great socializing agent. But before we begin our analysis, it would be well to remember that any family is dynamic, not static. Families change, and as they do, they exercise different effects on a child's development—some significant, others not so (Scarr, 1992).

For example, children who remain in an intact family, or who experience the death of a parent, or who go through a parental divorce all undergo unique experiences that must affect development. We need not be so dramatic, however. All families sustain normal change in the course of the lives of their members.

The Changing Family

In our changing society, it becomes increasingly difficult to define "family." As Garbarino (1992) notes, family takes many different forms, both across cultures and within societies. Traditionally, a man and woman marry and raise children. Today, however, we see many variations of this basic theme. Most of us are born into a family—the family of origin—and later in life usually start a family of our own—the family of procreation. You can see, then, that most people spend much of their lives in family units of one type or another.

The nature of that unit has changed dramatically. For example, in 1970, 70 percent of all households in the United States had married couples; in 1990, that figure had dropped to 56 percent. In 1970, 40 percent of married couples had children; in 1990, that figure had dropped to 26 percent. In 1970, 18 percent of families had three or more children; in 1990, that figure had dropped to 7 percent (U.S. Bureau of Census, 1991). About half of all children will spend some time in a single-parent household before they are 18.

Noting that these changes make defining a family a difficult task, Garbarino and Abramowitz (1992, p. 73) use the following criteria: A family is any two people related by blood, marriage, or adoption. Whatever the particular type of family, The National Commission on Children (1991) has identified several characteristics of a strong family:

As social conditions change—working mothers, single-parent families—fathers have become more involved in childcare. Children thus see their parents in roles different from the more stereotypical views of the past.

- Open and frequent communication among family members
- A feeling of belonging to a warm, supportive social unit
- Respect for individual members
- An ability to cope with stressful events
- Well-defined roles and responsibilities

Remember: Any analysis of family life contains two major themes. First is the nature of the family itself. In the late twentieth century, the identification of various forms of family living has sharply altered our view of the traditional family. Yet as we have seen throughout our discussion, different family styles have always existed. Population projections and estimates of family styles present a fairly consistent pattern: Change will continue, but it will represent a modification of current styles. For example, the number of single mothers will increase, but as governmental support also increases, there will undoubtedly be more and better day-care services, increased after-school programs for the older children, and perhaps greater flexibility in the work schedules of working parents. The second major theme is the developmental consequences of these changes. Here we find much less certainty, since it is simply too soon to make definite statements.

As society changes, the manner in which families respond also changes. Children cannot escape the results—positive or negative—of these twists and turns of family living. We also know that one of the most important characteristics of any family is the way that parents treat their children. For example, Dekovic and Janssens (1992) found that popular and rejected children had different experiences. Parents of popular children used an authoritative/democratic style when interacting with their children, relying on verbal persuasion, positive reinforcement, and indirect methods of control. Parents of rejected children tended to be authoritarian/restrictive; that is, they were more critical and controlling.

Parenting Behavior

In a careful analysis of the relationship between parental behavior and children's competence, Baumrind (1967, 1971, 1986) discovered three kinds of parental behavior: authoritarian, authoritative, and permissive. Here is what she means by each of the types.

Authoritarian Parents

Authoritarian parents are demanding, and for them instant obedience is the most desirable child trait. When there is any conflict between these parents and their children (Why can't I go to the party?), no consideration is given to the child's view, no attempt is made to explain why the youngster can't go to the party, and often the child is punished for even asking.

Authoritative Parents

Authoritative parents respond to their children's needs and wishes. Believing in parental control, they attempt to explain the reasons for it to their children. Authoritative parents expect mature behavior and will enforce rules, but they encourage their children's independence and search for potential (see Table 8.2). They try to have their youngsters understand that both parents and children have rights. In the resolution of socialization versus individuation, authoritative parents try to maintain a happy balance between the two.

Table 8.2 Patterns of Child-Rearing Behavior

	Authoritarian		Authoritative		Permissive	
	High	Low	High	Low	High	Low
Control	•		•			•
Clarity of communication		•	•		•	
Maturity demands	•		•			•
Nurturance		•	•		•	

Permissive Parents

Baumrind believes that **permissive parents** take a tolerant, accepting view of their children's behavior, including both aggressive and sexual urges. They rarely use punishment or make demands of their children. Children make almost all of their own decisions. Distinguishing indulgence from indifference in these parents can be difficult.

Beginning with the belief that parents' behaviors are concerned with control, clarity of communication, maturity demands (you can do better than that), and nurturance (love, warmth), Baumrind attempted to link parental characteristics (such as authoritarian, etc.) and qualities in their children (such as outgoing, hostile, etc.). Table 8.2 summarizes these interactions.

In examining Table 8.2, note that authoritative parents are high on control (they have definite standards for their children), high on clarity of communication (the children clearly understand what is expected of them), high in maturity demands (they want their children to behave in a way appropriate for their age), and high in nurturance (there is a warm, loving relationship between parents and children). Thus, according to Baumrind's work, authoritative parents are most desirable.

Today, most developmental psychologists believe that there are two kinds of permissive parents: permissive-indulgent and permissive-indifferent (Maccoby & Martin, 1983; Santrock & Yussen, 1992). *Permissive-indifferent* refers to those parents who are not interested in their children's lives. *Permissive-indulgent* applies to those parents who are quite involved with their children but who are tolerant and accepting and who avoid using their authority and rarely punish their children. What do these behaviors mean for children's characteristics? While no hard answers can as yet be given, Table 8.3 illustrates several characteristics associated with each of the parental styles. Since these characteristics may be quite general and may change as research focuses on this topic, we can be quite certain of one conclusion: regardless of style, parental behavior leaves its mark, often indelibly.

Table 8.3 Parental Behaviors and Children's Characteristics

	Parental Behaviors		
	Authoritarian	Authoritative	Permissive
Children's characteristics	Withdrawn	Self-assertive	Impulsive
	Lack of enthusiasm	Independent	Low self-reliance
	Shy (girls)	Friendly	Low self-control
	Hostile (boys)	Cooperative	Low maturity
	Low need achievement	High need achievement	Aggressive
	Low competence	High competence	Lack of responsibility

Finally, we come to a growing phenomenon in our society, one that affects an increasing number of children—homelessness.

Homeless Children

Traditional definitions of homelessness focused on the lack of a permanent place to live. Researchers now, however, tend to be more specific and concentrate on those who rely on shelters for their residence, or who live on the streets or in parks. The homeless today represent a different population from the days when they were seen as alcoholic men clustered together on skid rows. The so-called new homeless population is younger and much more mixed: more single women, more families, and more minorities. Families with young children may be the fastest-growing segment of today's homeless. On any given night, 100,000 children in this country will be homeless (Walsh, 1992). They are also characterized by few social contacts, poor health, and a high level of contact with the criminal justice system.

Paths to Homelessness

Families seem to move into homelessness in one of three ways. First is the family that quickly shifts from stable housing to homelessness. These families usually have been evicted or have experienced rent difficulties; their rent exceeded a public allowance they were receiving. Second is the family caught in a "slow shift" from stable housing to homelessness. These families usually had lost permanent housing about a year earlier and had doubled up with another family, but could contribute nothing to the rent and had to move to a shelter. Finally, there are those families, usually headed by young mothers, that had never maintained permanent housing (Weitzman, Knickmann, & Shinn, 1990).

The rates of alcohol, drug, and mental disorders are much higher among the homeless than in the general population: Alcohol disorders are found in about two-thirds of the homeless, drug and mental disorders in about half of homeless adults (Fischer & Breakey, 1991). These problems are often found together; alcoholics are frequently drug abusers. When you consider these figures and then realize that 30 percent of the homeless population in the cities are families, you begin to realize the impact that homelessness can have upon a child's development.

The Impact on Development

It is no exaggeration to say that the well-being of homeless children is seriously threatened. Defining the homeless as those in emergency shelter facilities with their families, Rafferty and Shinn (1991) state that these children are particularly vulnerable to health problems, hunger and poor nutrition, developmental delays, psychological problems, and educational underachievement. Before we analyze these potential problems, it is well to remember that homelessness is a composite of several conditions and events: poverty, changes in residence, schools, and services, loss of possessions, disrupted social lives, and exposure to extreme hardship (Molnar & Rubin, 1991). Any one of these conditions, or any combination, may produce different effects on children.

- *Health problems.* Homeless children have much higher rates of acute and chronic health problems, which may have their roots in the prenatal period. For example, homeless women have significantly more low-birth-weight babies and experience greater levels of infant mortality. Their children are more susceptible to asthma, ear infections, diarrhea, and anemia. As you might expect, these children are also subject to immunization delays.

- *Hunger and poor nutrition.* Rafferty and Shinn (1991) summarize recent research on this topic when they describe a homeless family's struggle to maintain an adequate and nutritionally balanced diet while living in a hotel: no refrigerator, no stove, poor food, and lack of food. Homeless children and their families often depend on emergency food assistance. But many times those facilities are themselves suffering from a lack of resources, with the result that the children, and their families, go hungry.

- *Developmental delays.* Homeless children experience, to a significantly higher degree than typical children, coordination difficulties, language delays, cognitive delays, social inadequacies, and a lack of personal skills (did not know how to eat at a table). The instability of their lives, the disruptions in child care, an erratic pattern of schooling, and the manner in which parents adapt to these conditions also impede development (Molnar & Rubin, 1991).

- *Psychological problems.* Homeless children seem to suffer more than typical children from depression, anxiety, and behavioral problems. Again, remember the composites of homelessness discussed earlier that may contribute to these psychological problems. Data are simply lacking that enable us to identify the particular aspect of homelessness that causes a child's anxiety or depression. We must also consider that parental depression affects children, and that children's problems may reflect the parents' feeling of helplessness.
- *Educational underachievement.* Little research has been done on this issue other than to show that homeless children do poorly on reading and mathematics tests. This finding should come as no surprise, given that these children have difficulty in finding and maintaining free public education for substantial periods. Missing educational opportunities is bad enough, but with the frequent moves their families make, these children also miss the remedial work they so urgently need. As Rafferty and Shinn (1991) note, school is especially critical for homeless children because it can produce a sense of stability that is otherwise lacking.

In her interviews with homeless children, Walsh (1992) notes that children, because of their status and lack of power, cannot directly solve the problem of homelessness. Instead they concentrate on coping with the emotions that arise from becoming homeless, perhaps by restructuring the circumstances surrounding their homelessness. Younger children, for example, may unrealistically attribute their problems to some external event that is unrelated to them or their parents. Older children may try to explain away the cause, especially if it pertains to a parent. For example, rather than blame a parent's alcoholism or drug use, a youngster may say that the parent is sick or has problems.

> ***Homelessness and poverty have robbed these children of a good part of their childhood. They have been forced to worry about the things that most children take for granted—food, safety, and a roof over their heads. Some become "little adults" in their efforts to help themselves and their families to survive. And yet, as their stories remind us, they are children. They cherish their toys, they play at the hint of any opportunity, they rush to get lost in the world of fantasy. They think in the magical and concrete ways of children, constructing their world with the logic of childhood. And they make clear in their stories that they would like to be treated as children—to be less burdened by worry and more able to depend on adults for the basics of survival.*** **(Walsh, 1992, p. 178)**

Children of Divorce

Divorce, as an increasingly important aspect of family life, affects children in many ways. Not only is their physical way of life changed (perhaps a new home, or reduced standard of living) but their psychological lives are touched also (Hetherington, Cox, & Cox, 1985). Adults have a tendency to underestimate the circumstances that a child experiences, especially if the divorce is followed by a remarriage, which is true for 75 percent of divorced mothers and 80 percent of divorced fathers.

Another distressing fact children must face is that the divorce rate following a remarriage is higher than that in first marriages. So children can experience the effects of divorce, usually several years in a single-parent home, and then the changed circumstances of a remarriage (Hetherington, Stanley-Hagan, & Anderson, 1989). Before examining these effects, let's look first at the divorce phenomenon.

Facts about Divorce

Nearly half of today's marriages will end in divorce. An estimated 50 percent of our children will live with a single parent before the age of 18. Table 8.4 summarizes many of these changes.

Table 8.4 Children Living with One Parent

Status of Parent	Number
Mothers	
Divorced	5,010,000
Absent spouse	3,371,000
Widowed	838,000
Never married	4,302,000
Fathers	
Divorced	861,000
Absent spouse	443,000
Widowed	132,000
Never married	371,000

Source: Data from the Bureau of the Census, 1988.

Effects on Children

As mentioned earlier, any discussion of the effects of divorce (or any traumatic event) must begin with the child's level of cognitive development. Remember that early childhood youngsters are at Piaget's preoperational level. Their ability to engage in abstract thinking is still limited, and they lack that vital aspect of cognition—the ability to reverse their thinking. This colors their reaction to their parents' divorce.

In about two or three years following the divorce, most children adjust to living in a single-parent home. This adjustment, however, can once again be shaken by what a parent's remarriage means: losing one parent in the divorce, adapting to life with the remaining parent, the addition of at least one family member in a remarriage (Hetherington, Stanley-Hagan, & Anderson, 1989).

Studying the long-term effects of divorce on children, Hetherington and her colleagues (1985, 1989) conclude that divorce has more adverse, long-term effects on boys. Remarriage of a mother who has custody of the children, however, is associated with an *increase* in girls' behavior problems and a slight *decrease* in boys' problems (Hetherington, 1991).

The transition period in the first year following the divorce is stressful economically, socially, and emotionally. Conditions then seem to improve, and children in a stable, smoothly functioning home are better adjusted than children in a nuclear family riddled with conflict. Nevertheless, school achievement may suffer and impulsivity increases. In an interesting comment about the relationship between divorced mothers and their children, Hetherington and Parke (1986, p. 524) state:

Divorced mothers may have given their children a hard time, but divorced mothers got rough treatment from their children, particularly their sons. In comparison with divorced fathers and parents in nuclear families, the divorced mother found that in the first year following divorce her children didn't obey, affiliate, or attend to her. They nagged and whined, made more dependency demands, and were more likely to ignore her. The aggression of boys with divorced mothers peaked at one year following divorce, then dropped significantly, but at six years after divorce was still higher than that of boys in nuclear families.

Other conditions also affect children's reactions to their parents' divorce (Wallerstein & Blakeslee, 1989):

- The bitterness of conflict before the divorce
- The child's reaction to the loss of the parent who leaves
- Any change in the relationship between the child and the departed parent
- The effect the divorce has on the parent who retains custody of the children
- Any change in behavior toward the children on the part of the parent who retains custody

Commonly reported reactions of early childhood youngsters to divorce are shock, depression, and loyalty conflict (Buchanan, Maccoby, & Dornbusch, 1991). They fear that their parents no longer love them and are actually abandoning them. These reactions fit the pattern of preoperational thinking. Their cognitive egocentrism prevents them from seeing the problem from their parents' perspective. They cannot realize that their mother or father loves them just as much even though they are leaving. Thus the divorce centers on them.

Second Chances

Wallerstein and Blakeslee (1989) report findings similar to those of Hetherington and others. Beginning in 1971, Wallerstein studied 60 families that had experienced divorce. She worked with 131 children who were between the ages of 2 1/2 and 18 at the time their parents separated. Her analysis of the 34 youngsters who were in preschool at the time of parental breakup reveals important information about early childhood youngsters.

When the parents separated, their children were seriously upset. Acute separation anxiety appeared. Eighteen months later, about half of the group was still quite troubled, with boys showing the most severe problems. Five years later, the children's adjustment seemed to be tied to the quality of life in the postdivorce or remarried family. About one-third of the group showed signs of depression. Ten years later, the children claimed little memory of the circumstances surrounding the divorce. Wallerstein, however, in analyzing their replies to questions, found a high degree of repression in their conversations. Wallerstein also discovered a phenomenon she called **reconciliation fantasies.** Although the divorce had occurred ten years earlier, many of the children stated poignantly that they wished their parents could get together again. Wallerstein concluded that early childhood youngsters who live through the divorce of their parents are less burdened in the future than those who are older at the time of the divorce.

Does Divorce Cause Childhood Problems?

We have one final comment about the difficulties children have after divorce. In a carefully designed study, Block, Block, and Gjerde (1986) have raised questions about problems that children have after parental divorce. Beginning with data indicating that some youngsters do and others do not have problems, the authors state that one fault of the divorce studies is that they are almost all retrospective (that is, they "look back"). Consequently, they are at the mercy of faulty memories and perhaps an adolescent's or adult's unwillingness to speak of painful experiences.

The authors were able to design a prospective study (looking ahead) of children's personalities before the divorce. They found that 3-year-old boys whose parents eventually divorced already were showing problems, especially in self-control. When they were 7 years old, they were still showing the same behavior. Three-year-old girls whose parents would divorce seemed competent, skillful, and competitive. But by age 4 their behavior had changed: They didn't get along with others, were overly emotional, and tended to withdraw. By age 7, these behaviors were still evident, but the girls also showed continuing cognitive competence.

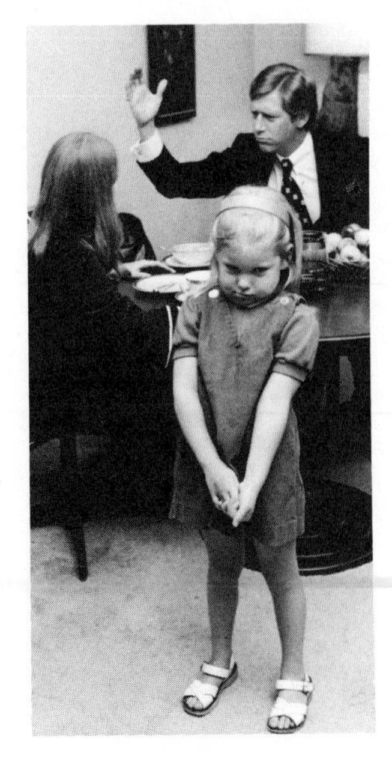

Children develop within a complex network of family relationships. What they are exposed to during these years can have longlasting consequences.

The author's findings point to several important conclusions:

- Researchers cannot focus on the divorce itself as the cause of children's later problems.
- Some youngsters show these same problems years before the divorce.
- Events *preceding* the divorce, especially the stress and conflict involved, may be responsible for problems that appear after the divorce.
- Girls may not be as immune to marital turmoil as we thought. They may simply display their feelings in a manner "appropriate" for girls: anxious, withdrawn, extremely well-behaved.

Hetherington's excellent earlier study (1972) of father absence supports this last conclusion. Girls whose fathers died or who left because of divorce showed little disruption of appropriate sex-role behavior but had later problems in adolescence in relating to males. Daughters of divorced women sought more attention from boys, actively seeking contact with them. Daughters of widows avoided boys and preferred female friends and activities.

Day Care

Almost two-thirds of women with children under age 14, and more than one-half of mothers with children under age 1, are in the labor force. In fact, the single largest category of working mothers is those with children under age 3 (Zigler & Lang, 1991).

What happens to children while mothers are at work? Obviously someone must be taking care of these youngsters, and it is precisely here that questions are raised about **day care.** How competent are the individuals who offer these services? Is the day-care center healthy and stimulating? Is it safe? What are the long-term developmental consequences of day-care placement? In light of recent exposures of the sexual abuse of children in some centers, America's parents are demanding answers to these questions.

Facts about Day Care

Reliable facts about day care are hard to come by, chiefly because of the lack of any national policy that would provide hard data. There are about 35,000–40,000 "places" providing day-care services (principal income is from offering child care). "Places" is perhaps the best way to describe these facilities because of the wide variety of circumstances that exist: Zigler and Lang (1991) refer to them as a "patchwork of arrangements." For example, one mother may charge another mother several dollars to take care of her child. A relative may care for several family children. Churches, businesses, and charities may run large operations. Some may be sponsored by local or state government as an aid to the less affluent. Others are run on a pay-as-you-go basis. Almost everyone agrees that the best centers are staffed by teachers who specialize in day-care services (abut 25 percent of day-care personnel).

The following figures will give you an idea of the types of child-care arrangements and the number of children under 5 years of age in day care with employed mothers. (Based on data from the U.S. Bureau of Labor Statistics, 1988).

Mother cares for child at work	6.7%
Kindergarten	1.2%
Nursery/preschool	6.4%
Group care centers	14.7%
Care in another home	41.3%
• By grandparent	11.5%
• By other relative	6.0%
• By nonrelative	23.8%
Care in child's home	29.7%
• By father	14.2%
• By grandparent	6.7%
• By other relative	2.7%
• By nonrelative	6.0%

Note that the majority of these children are cared for in private homes. This may help to explain why we are uncertain about the influence of day care on development, since most research is done with children from day-care centers.

If we now sort out what is known about types of day-care centers, we can group them as follows (Clarke-Stewart, 1982). (Remember: Most states now have minimum standards for day-care operation.)

- *Private day-care centers run for profit* have no eligibility requirements and will accept almost anyone who can pay the fee. Typically staffed by two or three people (usually not professionally trained), they may operate in a converted store or shop. They probably are minimally equipped with toys and educational activities and usually offer no social or health services.

- *Commercial centers* are also private and run for profit. They may be part of a national or regional chain with uniform offerings and facilities. They usually are well equipped with good food and activities. The centers are a business, much as a McDonald's or Burger King. KinderCare, for example, runs about 900 centers.

- *Community church centers* are often (but not always) run for children of the poor. The quality of personal care (attention, affection) is usually good, but they often have minimal facilities and activities.

Day care has become an important phenomenon in our society as more and more mothers join the work force. Research indicates that developmental outcomes are closely linked to the quality of day care.

Among the variety of day care settings, home day care centers are quite numerous. Often run by a family member or neighbor, they are smaller and more informal than large centers.

- *Company centers* are often offered as fringe benefits for employees. They are usually run in good facilities with a well-trained staff and a wide range of services.
- *Public service centers* are government sponsored, well run, and have high quality throughout. Unfortunately, few of these are available and they are designed to serve children of low-income families.
- *Research centers* are usually affiliated with some university and represent what the latest research says a day-care center should be. Most studies of day care have been conducted in these centers. But since the centers are of the highest quality, the results of such studies do not give a true picture of day care throughout the country.
- *Other centers* include the family-run centers mentioned earlier and cooperative centers where parents rotate responsibility for child care under professional guidance. In another fairly popular form of care, a neighborhood mother, usually a former nurse or teacher, takes care of two or three children, thus avoiding state requirements concerning number of children, facilities, or insurance.

A National Concern

As you can see, the country is moving into a different era with regard to day care. Working mothers are now in the majority, and their numbers promise to increase. Given these events, regulation of day care has taken on new urgency. Legislative battles over standards for day-care centers have been constant, but the best the federal government has been able to do is to propose general standards. Among the most important of these is the suggested ratio of staff to children, which is illustrated in Table 8.5.

Despite the minimal salaries of day-care workers, costs are high because day care is labor intensive. Staff-child ratios usually average from 1:8 to 1:10, much too high considering children's needs and safety. But a new attitude is developing, one with political consequences. National surveys have consistently shown that more than one-half of the voters think there should be a national policy regarding day care.

Table 8.5 Suggested Ratio of Staff to Children

	Age	Maximum Group Size	Staff-Child Ratio
For centers	0–2	6	1:3
	2–3	12	1:4
	3–6	16	1:8
For homes	0–2	10	1:5
	2–6	12	1:6

An Applied View — Desirable Qualities of Day-Care Centers

For parents who want guidelines to identify important features in a day-care setting, here are several specific suggestions by the *National Association for the Education of Young Children* (1986).

1. **Adult Caregivers**
   - Adults should understand how infants and young children grow.
   - There should be an adequate number of adults to meet the individual needs of the children (see Table 8.5).
   - Good records should be kept for each child.
2. **The Program Itself**
   - The setting should facilitate the growth and development of young children working and playing together.
   - Equipment and play materials should be adequate and readily available.
   - Instructors should be sufficiently skilled to aid youngsters in their language and social development.
3. **Relations with Clients and Community**
   - Parents should be actively involved with the center.
   - The community should be aware of activities at the center, and the center should be aware of community resources (recreational, learning centers).
4. **The Ability to Meet the Demands of all Involved**
   - The health of all members—children, staff, and parents—is both protected and promoted.
   - Safety should be a primary concern.
   - There should be adequate space to serve all activities (35 square feet of usable playroom floor space indoors per child and 75 square feet of play space outdoors per child).

Developmental Outcomes

Regardless of what you may have heard or read, there are no definite conclusions concerning the long-term developmental consequences of day care. One reason is that careful follow-up of children from day care is not yet available (Clarke-Stewart, 1989).

Such studies are only beginning to appear. For example, Belsky and Rovine (1988) reported on the results of two longitudinal studies of infants and their families. They found that when infants were 12–13 months old and had twenty or more hours of nonmaternal care per week, some of the infants were insecurely attached. These same infants were more likely to be aggressive and tended to withdraw. Yet this report also contains puzzling data. Over one-half of the infants who had the twenty hours of nonmaternal care were securely attached. These data reveal the uncertainty of our present knowledge and reinforce our conclusion that we must be careful about either positive or negative statements about day care (Lamb & Sternberg, 1990).

Another obstacle has been the use of university-sponsored research centers to study children. These centers are usually lavishly equipped and overstaffed. They simply do not reflect the national norm. Despite this, some conclusions have been drawn.

- The type of care is significant.
- Effects depend on the child's total environment.
- A child's personal characteristics affect the day-care experience (Zigler & Hall, 1989).

These are shaky findings and may very well be reversed by additional research.

Given the changing conditions in which many children grow up, it is not surprising that their experiences affect what and how they learn. Consequently, we shall next turn our attention to education in early childhood.

Early Childhood Education

The number of children in early childhood education programs increased steadily in the past four decades as concerns about preschool education shifted dramatically. As one illustration, while arguments about the value of preschool education have lessened, interest in the "fit" between program and child have increased. One constant has been worry that preschool programs may expose children of these years to too much pressure (Piccigallo, 1988).

The activities in early childhood education programs, if biologically and psychologically appropriate, can encourage all aspects of development.

Early Childhood Education: A Positive Outlook

Yet the help that these programs can bring to many children has been amply demonstrated. Project Head Start, for example, was started in 1965 to aid low-income children. While research has shown mixed results for improving children's achievement, other benefits have been claimed. Lazar and Darlington (1982) studied the long-term results of twelve Head Start programs and reported significant effects on school competence, families, and attitudes about self and school. Among the techniques employed by the various programs were constant communications with parents, training of mothers in the use of educational activities in the home, and periodic home visits. Pooling the data from these studies, Lazar and Darlington reported these findings:

- Children who attended these programs were less likely to be retained in grade and more likely to meet their schools' requirements.
- The children attained higher IQ scores than their controls.
- The children demonstrated higher self-concepts and were proud of their school accomplishment.
- Mothers of program graduates were more satisfied with their child's school performance than control mothers; they also had higher occupational aspirations for their children than control mothers.

These results were obtained several years after the children had left the program and suggest the benefits of positive parental involvement in school affairs.

There are preschool programs for the middle class and advantaged as well as for low-income children. Some of these schools offer only play as the chance to learn how to get along with others. Others, such as the Montessori schools, are much more structured.

Regardless of the kind of school, several features distinguish preschools: low teacher-child ratio, specially trained teachers, availability of resources, and recognition of children's individual differences. These programs are all child centered; that is, they are designed to emphasize individual children and to provide children with enriching, enjoyable experiences suitable for their years.

These program characteristics are especially desirable given the social needs of early childhood youngsters. Most studies of peer relationships during the early childhood period (2–6 years) have been conducted on day-care centers and nursery schools. Note the age progression in developing relationships: Social contacts occur more frequently among 5-year-olds than 3-year-olds. Aggressive interactions are quite common in the period, although aggression decreases in proportion to friendly interactions (especially true among middle-class boys).

Particularly interesting are the changes in quarreling during these years. As in their other social exchanges, older youngsters (4, 5, and 6 years) engage in fewer but longer quarrels with members of their own sex (Hartup, 1983). Boys quarrel more frequently among themselves than girls, usually over objects, gradually changing from physical to verbal aggression toward the end of the period. Although solitary activity persists, older children of this period are more obviously bidding for the attention of their peers.

Miseducation

A different perspective is offered by Elkind (1987). Believing that educating preschoolers too often reflects the parents' needs, Elkind is concerned that many programs are excessively formal and involve inappropriate expectations. Individual differences are sacrificed to some form of attainment, ranging from swimming to reading. Thus pressure can become excessive and the joy and pleasure of these years are lost.

Elkind believes that placement in pressure programs comes from the needs of several classes of parents.

- *Gourmet parents.* They have everything and they want their children to have the same experiences.
- *College-degree parents.* They believe education is the answer to everything, so they want their children to be intellectual "superkids."
- *Gold-medal parents.* These parents want their children to be Olympic class "in something."
- *Prodigy parents.* They made it on their own and think their children can do likewise if they get a fast start in life.

These are only a few of Elkind's classes of parents, but they illustrate his concern that today's preschoolers are placed in educational pressure cookers because of their parents' needs.

Within these changing conditions of society and family, the early childhood youngster must learn to adjust and survive. It is the give and take between pressure and freedom that helps to shape a definite personality.

The Self Emerges

"You are who you are." This oft-repeated statement summarizes what personality means. Psychologists define personality as the "dynamic organization of those psychophysical systems that define a person's behavior and thought." Our concern is how children acquire this sense of self.

In chapter 6 we mentioned the growing belief that humans come into this world with inborn temperamental differences. These differences, then, are the foundation upon which children build their unique personalities. Children form their personalities by using their *genetic endowment* that defines their potential, both physical and intellectual. (Whether the environment helps or hinders the fulfillment of potential is a story we'll consider in chapter 9). Children also learn about themselves from those around them. The reactions of others contribute greatly to what children think of themselves (Damon, 1983).

The Development of Self

To understand development, you must always remember it is an integrated process (Harter, 1983). All aspects—physical, cognitive, social, and emotional—contribute to smooth growth. Nowhere is this more evident than in the analysis of self-development.

- If a youngster has normal physical growth, a sense of mastery develops, increasing self-confidence.
- If cognitive development progresses satisfactorily, children acquire vital knowledge that helps them to adjust to their world, increasing self-confidence.
- If relations with others are mostly pleasant, children learn about others and themselves, increasing self-confidence.

Children distinguish themselves from others at about age 18 months, perhaps even a little earlier. In their famous experiment, Lewis and Brooks-Gunn (1979) put rouge on the noses of children aged 15 to 18 months. When placed before a mirror, the children touched the rouge on their noses, indicating that they knew who they were. They recognized themselves.

Think of what you have learned about cognitive development: Children are becoming less egocentric; they are distinguishing themselves from the world. Emotionally, they have begun the process of separating from their mothers, recognizing that mothers are also individuals. Language is flourishing and others are speaking to them. The convergence of developmental paths helps to further the sense of self. Their growing cognitive ability helps children to represent things, to separate from mothers and the world in general. This leads to acceptance—through cognition, language, and developing relations—of an independent world "out there."

The Special Case of Self-Esteem

Developmental psychologists, educators, parents, and almost everyone who works with children have come to accept the vital role that a child's self-esteem plays in development. What do we mean by self-esteem? A technical definition would be that self-esteem is the evaluative and affective dimension of self-concept (Santrock and Yussen, 1992). A simpler way of defining it would be to say that self-esteem is how children feel about themselves, how they value themselves. These authors

A Multicultural View 

Self and Others

As we have seen in this chapter, one of the accomplishments of the early childhood years is a sense of self. This understanding of self comes partially through interacting with others. When children interact with those from other cultures, their self-identity broadens to accept those with different customs, languages, and ideas.

In commenting upon the growing understanding of others, Ramsey states:

> ***While knowledge of unfamiliar people and lifestyles may reduce children's fears and avoidance of differences, their motivation for reaching beyond cultural, racial, and class barriers largely rests on their self-confidence, their ability to empathize with others' experiences and feelings, and their anticipation of pleasure and satisfaction to be derived from expanding their social relationship.*** **(1987, p. 112.)**

Helping youngsters to achieve this objective are the social goals of multicultural education, which include (Ramsey, 1987):

- Positive cultural, racial, and class identities
- High self-esteem
- Awareness of other emotional, cognitive, and physical states
- The ability and willingness to interact with diverse others
- A sense of social responsibility
- An active concern for the welfare of others

Those working with preschool youngsters have a unique opportunity to further a positive multicultural perspective during these years of enthusiasm and rapid learning. As an example, Yao (1988) offers several suggestions for working with Asian immigrant parents. She begins by urging adults to take the time to familiarize themselves with the social, cultural, and personality traits that make these children and their parents unique. Adults should ask themselves questions such as:

- Do I have any prejudices toward this group?
- What stereotypes do I associate with Asian-Americans? (e.g., they're all superior in math and science)
- What do I know about their culture?
- Will there be any conflict between my values and theirs?

These and similar questions can act as guidelines to help us adapt to those of different cultures and also help children adapt to different ways. We can do no less.

give a good example of self-esteem when they describe the girl who realizes that she is not just a student but a good student, the boy who realizes that he is not just a basketball player but a good basketball player.

There seems to be a close relationship between the way parents treat a child and that child's self-esteem. Coopersmith (1967), in an enduring study, found that high self-esteem in boys was associated with expressions of affection, concern with a child's problems, a happy home life, clearly established rules, reasonable amounts of freedom, and a structured environment that provided help when needed. Remember: These characteristics don't cause high self-esteem; they are related to it.

During the preschool years, children seem to distinguish between two types of self-esteem—social acceptance and competence—but as yet don't identify competence in particular activities. This ability appears at about 7 or 8 years of age and suggests that there are developmental changes in self-esteem. A child coming from a supportive home typically has an inflated sense of self-esteem before beginning school. It isn't until the second grade that pupils' estimation of their self-esteem matches the opinions of those around them; that is, children's estimations correlate with teacher ratings, test scores, and direct observations (Berk, 1992).

Self-esteem seems to be composed of several elements that contribute to a child's sense of worth.

- *A sense of physical safety.* Children who feel physically secure aren't afraid of being harmed, which helps to develop feelings of confidence.
- *A sense of emotional security.* Children who aren't humiliated or subjected to sarcasm feel safe emotionally, which translates into a willingness to trust others.
- *A sense of identity.* Children who know "who they are" have achieved a degree of self-knowledge that enables them to take responsibility for their actions and relate well with others.
- *A sense of belonging.* Children who are accepted by others are comfortable in seeking out new relationships and begin to develop feelings of independence and interdependence.
- *A sense of competence.* Children who are confident in their ability to do certain things are willing to try to learn to do new things and persevere until they achieve mastery (Youngs, 1991).

The infant shown here touching his nose and mouth against the mirror reveals the development of a sense of self, which most infants accomplish by about 18 months of age.

Gender Identity

We realize today that gender identity results from a complicated mix of culture and biology. To clarify the meaning of gender identity, we should make a sharp distinction among gender identity, gender stereotyping, and gender role.

- **Gender identity** is a conviction that one belongs to the sex of birth.
- **Gender stereotypes** reflect beliefs about the characteristics associated with male or female.
- **Gender role** refers to culturally acceptable sexual behavior.

Gender Identity and Biology

Whatever your ideas on gender identity, we start with an unavoidable premise: parts of the agenda are biologically programmed. As Elkin and Handel state:

> ***If any phenomenon in the human world can be considered an incontrovertible fact, it would be this: a human being can either beget a child or bear a child. No human being can do both. This difference in body functioning between males and females is universal; it occurs in all human societies, and there is not a single known exception.*** **(1989, p. 211)**

John Money (1980), working in a Johns Hopkins clinic devoted to the study of congenital abnormalities of the sex organs, believes that sexual differentiation occurs through a series of stages. He has proposed a series of four stages that account for biological sex. (Note: This says nothing about gender role.)

Chromosomal Sex The biological sexual program is initially carried by either the X or Y sex chromosome. This is called **chromosomal sex**. If you recall from chapter 3, the male can carry either the X or Y to join with the mother's egg, which carries only the X sex chromosome. When a sperm with an X chromosome fertilizes the egg, a female is conceived (XX). If the sperm carries a Y chromosome, a male is conceived (XY).

Gonadal Sex Chromosomal sex alone does not determine sex. The XX or XY combination passes on the sexual program to the undifferentiated gonads (Money & Ehrhardt, 1972). This is called **gonadal sex.** If the program is XY, the gonads will then differentiate into testes. This, as you recall, occurs during the sixth or seventh week after fertilization. If the program is XX, the gonads differentiate into the ovaries, starting at about the twelfth week. So during the first six weeks, the dividing, fertilized egg is essentially sexless.

Hormonal Sex Once the testes or ovaries are differentiated, they begin to produce chemical agents called sex hormones. This is called **hormonal sex.** Both testes and ovaries secrete the same sex hormones but in different quantities. Males produce more of the sex hormones called androgens, while females produce more estrogens (the female sex hormone). If no sex hormones were produced, the developing fetus would always be female.

You may have read that aggression seems to be the one personality trait that distinguishes males and females. The male sex hormones, the androgens (especially testosterone) seem to produce a rougher, more aggressive type of behavior. Money (1980) also believes that these prenatal sex hormones may affect brain organization. Though evidence is yet limited, the androgens and estrogens may influence the rapidly developing brain.

Note what has happened so far. The sexual program, originally determined by the chromosomes, has produced a sexual program that dictates body differences between the male and female and also probably a nervous system differentiation.

Genital Sex A baby's sex is determined not only by chromosomes and hormones, but by its external sex organs. This is called **genital,** or morphological, **sex.** Even if males can't produce sperm that are able to fertilize an egg, they are still called male because of their sex organs. The same holds true for women who can't bear children (Elkin & Handel, 1989). Figure 8.1 illustrates these stages.

As you can well imagine, it is genital or morphological sex that determines how society will treat the newly born baby. We shall turn now to gender and socialization.

Figure 8.1
The path of gender identity

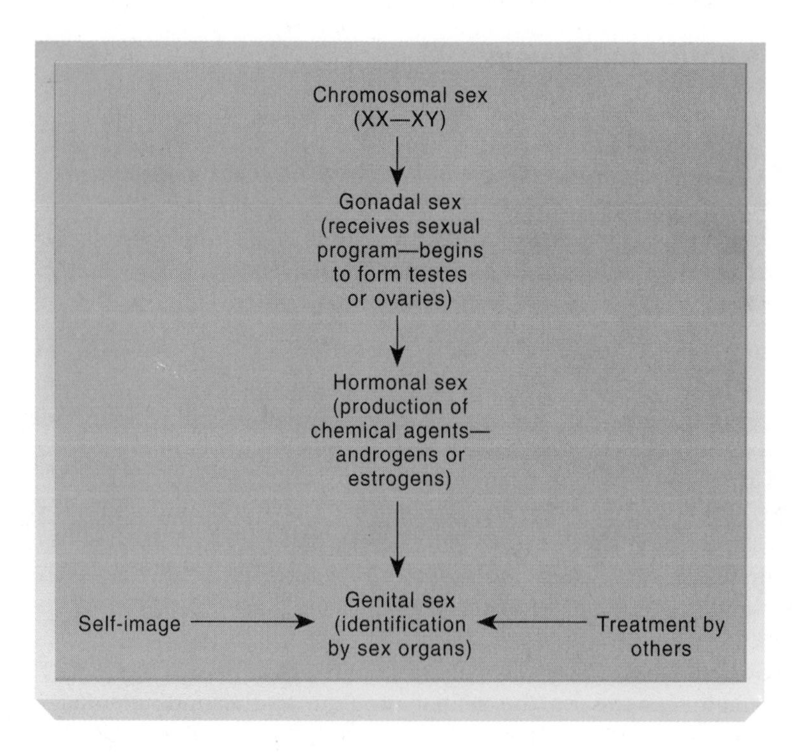

How Children Learn about Gender

It doesn't take the early childhood youngster long to discover which behavior "fits" men and which "fits" women (Fagot, Leinbach, & O'Boyle, 1992). If you think about the competencies that we have traced thus far (especially cognitive) and combine these with the acquisition of a gender identity by age 2 or 3, you can understand children's rapid grasp of appropriate sexual behavior. Lott (1989) reports that by preschool age, most children are well aware of their own gender, which parent they are most like, and the gender of family members and peers.

In acquiring gender understanding children seem to pass through three stages: learning to identify one's sex, understanding that sex is stable over time, and realizing that gender is a fixed and unchanging characteristic (Martin & Little, 1990).

For an idea of how soon children begin to make decisions based on sex, consider Lott's (1989) report on the result of a study that required children ages 2 to 7 to assign various occupations to either a male or female doll. As early as 2 years of age, they assigned male occupations to male dolls. For example, 67 percent of 2- and 3-year-olds chose the male doll for doctors.

Occupational stereotyping seems to be similar for boys and girls and increases with age. These sex stereotypes are not confined to 2- and 6-year-olds. When today's college students were asked to list words that describe male and female, here are some of their choices:

Male	**Female**
Aggressive	Kind
Forceful	Gentle
Confident	Understanding
Strong	People-oriented
Career-oriented	Thoughtful
Independent	Emotional
Dynamic	Nurturing

In an attempt to explain these findings, Martin, Wood, and Little (1990), in a series of studies, found a developmental sequence to the appearance of gender stereotypes. In the first stage, children learn what kinds of things are associated with each sex (boys play with cars; girls play with dolls). From the ages of 4 to 6 years children move to the second stage, where they begin to learn the more complex associations for their own sex (different kinds of activities associated with a toy). By the time of the third stage, roughly 6 to 8 years, children make the same types of associations for the opposite sex.

Where do these ideas come from? As we have seen, boys and girls are exposed to the socialization practices of those around them from birth. Among the most influential of these agents are parents, siblings, peers, and media. (The stereotypic treatment of males and females on television has been well documented and will be discussed at length in chapter 10.)

Parents

Evidence clearly suggests that parents treat boy and girl babies differently from birth. Adults tend to engage in rougher play with boys, give them stereotypical toys (cars and trucks), and speak differently to them. By the end of the second year, parents respond favorably to what they consider appropriate sexual beavior (i.e., stereotypical) and negatively to cross-sex play (boys engaging in typical girl's play and vice versa). For example, Fagot (1985) and Fagot and Hagan (1991) observed toddlers and their parents at home. Both mothers and fathers differentially reinforced their children's

Children often find opposite-sex toys extremely attractive. Many parents encourage the use of such toys to help their children avoid the development of stereotypical sexual attitudes.

behavior. That is, they reinforced girls for playing with dolls and boys for playing with blocks, girls for helping their mothers around the house and boys for running and jumping.

Lips (1993) believes that parents are unaware of the extent to which they engage in this type of **reinforcement.** In a famous study (Will, Self, & Datan, 1976), eleven mothers were observed interacting with a 6-month-old infant. Five of the mothers played with the infant when it was dressed in blue pants and called "Adam." Six mothers later played with the same infant when it wore a pink dress and was called "Beth." The mothers offered a doll to Beth and a toy train to Adam. They also smiled more at Beth and held her more closely. The baby was actually a boy. Interviewed later, all the mothers said that boys and girls were alike at this age and *should be treated identically*.

Perhaps no one has summarized the importance of this differential treatment of sons and daughters better than Block (1983). Noting the reality of this parental behavior, Block states that males and females grow up in quite different learning environments, which has important psychological implications for development.

Siblings

What can we say about the influence of brothers and sisters on gender development? Of one thing we can be certain: A youngster with brothers and sisters has unique experiences in the home. Brothers and sisters differ markedly in personality, intelligence, and psychopathology in spite of shared genetic roots. So circumstances within the family dictate that a variety of experiences will help to shape a child's idea of what is appropriate sex-typed behavior.

About 80 percent of children have siblings, and considering the amount of time they spend with each other, these relationships exercise considerable influence on developmental outcomes (Dunn, 1983). An older brother showing a younger brother how to hold a bat; a younger sister watching her older sister play with dolls; quarrelling among siblings—each of these examples illustrates the impact that sibling relationships have on gender development.

Siblings play a major role in development. For those children with brothers and sisters, older siblings can act as models, help younger brothers and sisters in times of difficulty, and help smooth relations with adults.

Banks and Kahn (1982) state that the sibling relationship helps to shape a child's identity and also provides reassuring feelings of constancy. But it isn't quite that simple. In a classic study of the influence of siblings on sex typing, Brim (1958) found that children with same-sex siblings had more "sex-appropriate" behavior than those with opposite-sex siblings. In another study, Stoneman, Brody, and MacKinnon (1986) discovered that girls with older brothers shared more masculine activities and boys with older sisters were more interested in feminine activities.

Yet other studies (e.g., Tauber, 1979) showed just the opposite. Children with same-sex siblings enjoyed playing with opposite-sex toys. What should you conclude from these conflicting results?

- The quality of relationships among the siblings (positive or negative) may be responsible. If an older brother tends to bully a younger brother, the younger may turn elsewhere.
- Parents may react differently to brothers and sisters. (We have spoken at length about this possibility in chapter. 6.)
- Much more research is needed on this topic.

Peers

When early childhood youngsters start to make friends and play, these activities foster and maintain sex-typed play. Studies show that by the age of 3, children reinforce each other for sex-typed play (Langlois & Downs, 1980). When they engage in "sex-inappropriate" play (boys with dolls, girls with a football), their peers immediately criticize them and tend to isolate them.

Here, again, we see the influence of imitation and reinforcement. During these years, youngsters of the same sex tend to play together, a custom called **sex cleavage** and one that is encouraged by parents and teachers. If you think back on your own experiences, you'll probably remember your friends at this age being either all male or female. You can understand, then, how imitation, reinforcement, and cognitive development come together to intensify what a boy thinks is masculine and a girl thinks is feminine (Dunn, 1983).

Gender Constancy

By now you have probably, and correctly, concluded that gender identification results from a complex mixture of biology and learning. This interplay produces what is called gender constancy. Three steps seem to be involved:

- **Gender identity,** which as we have noted appears anywhere from age 1 1/2 to 3.
- **Gender stability,** which appears from age 3 to 4 and means that youngsters realize that gender normally stays the same throughout life.
- **Gender constancy,** which means that children understand the meaning of gender; even if a woman dresses in a man's suit, shoes, and hat, she remains a woman. Most children acquire gender constancy by age 7 (Frey & Ruble, 1992).

At this point, you may be wondering what developmental processes explain gender development. As usual, there are several explanations.

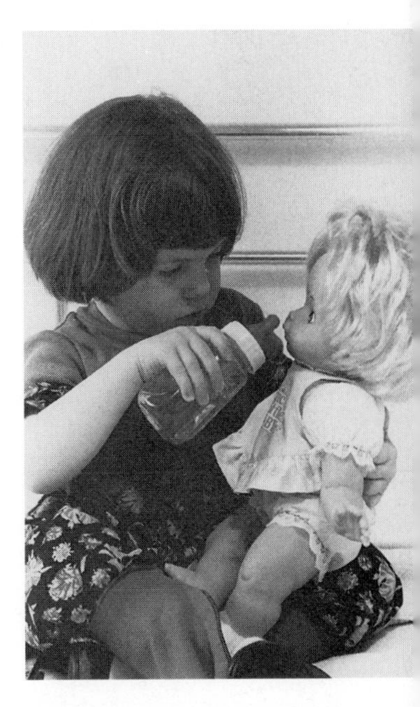

Children quickly learn what objects, activities, and friends are sexually appropriate.

Theories of Gender Development

As Huston notes (1983), two major theories have generated most of the research in gender development during the past ten years: **social learning** theory and **cognitive-development** theory. Since these were treated at length in chapter 2, we will concentrate on their value as explanations of gender development.

Social learning theorists believe that parents, as the distributors of reinforcement, reinforce appropriate sex-role behaviors. By their choice of toys, by urging "boy" or "girl" behavior, and by reinforcing this behavior, parents encourage their children to engage in sex-appropriate behavior. If the parents have a good relationship with their children, they also become models and their children tend to imitate them, acquiring additional sex-typed behavior. Thus children are reinforced or punished for different kinds of behavior. They also learn sex-typed behavior from male and female models who display different kinds of behavior.

A second explanation, quite popular today, is cognitive-development theory, which derives from Kohlberg's speculations about gender development (1966). We know from Piaget's work that children engage in symbolic thinking by about the age of 2. Using this ability, children acquire their gender identity. Then, Kohlberg believes, they begin the process of acquiring sex-appropriate behavior.

A newer, and different, cognitive explanation is called gender schema theory. A schema is a mental blueprint for organizing information. You develop a schema for gender. This schema helps you to combine your gender identity with sex-typed behavior and an appropriate gender role. You develop an integrated schema or picture of what gender is and should be. Note how our discussion of gender illustrates the value of the biopsychosocial model: We can best understand gender development from a combined biological, psychological, and social perspective.

The Importance of Play

"Play can mean many things: a child experimenting with a new toy, the attempts to master a game, acting out fantasies, or imagining new and different characters and situations. As Gardner (1982) notes, play does not include satisfying biological needs (eating, drinking) or problem-solving behavior (trying to complete a task imposed by someone else). Play, then, is a pleasurable activity that children engage in for its own sake (Santrock & Yussen, 1992).

The Meaning of Play

There are two major categories of play:

- *Exploratory play,* in which children use something as a starting point for novel behavior. For example, children with crayons and paper may begin to draw circles (which they have previously done) and then change the pattern, perhaps forming an ellipse. The behavior is not directed to a goal.
- *Rule-governed play,* in which children draw circles, then squares, then diamonds; that is, children follow some type of cultural rules.

These two categories help to separate children's tasks into either play or problem solving. When children try out their own ideas rather than focus on a problem or substitute cultural rules for an implied task, then children are playing (Gardner, 1982).

What's Your View?

What can we say about differences between boys and girls? Do these differences exist? Yes, but our thinking about the reasons for these differences has changed. We no longer say with certainty that biology alone is the cause.

Mathematics is a good example. The myth that "boys are just better at math and science" is slowly crumbling. As ideas about gender roles have changed, so also have parents' expectations for their daughters. Numerous studies have recently shown that the gap between males and females on math scores has steadily declined (Berk, 1992). Nevertheless, there are documented differences, including the following. (Note: We are not speculating about the cause of these differences.)

- *Physical size and strength*. Although almost all girls mature more rapidly than boys, by adolescence boys have surpassed girls in size and strength.
- *Language*. Girls do better on verbal tasks almost immediately, a superiority that is retained. Boys also exhibit more language problems than girls.
- *Spatial skill*. Boys display superiority on spatial tasks, a superiority that continues throughout schooling.
- *Mathematical ability*. We have previously commented on boys' better performance.
- *Achievement motivation*. Differences here seem to be linked to task and situation. Boys do better in stereotypical "masculine" tasks (math, science), girls on "feminine" tasks (art, music). In direct competition between males and females, beginning around adolescence, girls' need for achievement seems to drop.
- *Aggression*. Boys appear to be innately more aggressive than girls, a difference that appears early and is remarkably consistent.

Of all the differences mentioned, only aggression seems to have biological roots, and even that will be shaped by the child's environment: family, friends, school. Do you agree? What's your view?

Play cuts across all aspects of development. But how can we be sure that what we see is actually play? In their exhaustive survey, Rubin, Fein, and Vandenburg (1983) have identified several dispositional features of play:

- Play is not forced on children; they do it because they like it.
- They're not concerned with outcome; they enjoy the act of playing.
- There is an "as if" quality to play; reality is suspended and children get great enjoyment from pretending.
- There are no rules to be followed, which distinguishes play from games.
- When children play they are active; they are not daydreaming.

Children play constantly during these early childhood years, and they play for various reasons. The primary reasons are the following.

Cognitive Development

Play aids cognitive development. Through play, children learn about the objects in their world, what these objects do (balls roll and bounce), what they are made of (toy cars have wheels), and how they work. To use Piaget's terms, children "operate" on these objects through play, and also learn behavioral skills that will be of future use to them.

Social Development

Play helps social development during this period because the involvement of others demands a give-and-take that teaches early childhood youngsters the basics of forming relationships. Social skills demand the same building processes as cognitive skills. Why are some 5- and 6-year-olds more popular with their classmates than others? Watching closely, you can discover the reasons: decreasing egocentrism, recognition of the rights of others, and a willingness to share. These social skills do not simply appear; they are learned, and much of the learning comes through play (Anselmo, 1987).

Emotional Release

Play provides an emotional release for youngsters (Fischer, Shaver, & Carnochan, 1990). There are not the right or wrong, life-and-death feelings that accompany interactions with adults. Children can be creative without worrying about failure. They can also work out emotional tensions through play.

Caplan and Caplan (1984) state that play is a powerful developmental instrument for several reasons:

- Play aids learning because children have the freedom to explore and enjoy.
- Play is investigative because children, lacking knowledge, must search for and discover what works and what doesn't.
- Play is voluntary, with no fixed directions.
- Play can provide an imaginary, escape world, which children sometimes need.
- Play helps to build interpersonal relationships.

Aside from these more formal characteristics, Elkind's ideas seem particularly pertinent.

> ***But children need to be given an opportunity for pure play as well as work. At all levels of development, whether at home or at school, children need the opportunity to play for play's sake. Whether play is the symbolic play of young children, the games with rules and collections of the school-age child, or the more complicated intellectual games of adolescence, children should be given the time and encouragement to engage in them.*** **(1981, p. 97)**

The Development of Play

Does play change over time? Are there age-related features of play that we can identify? It is difficult, if not impossible, to link a specific kind of play to a specific age. But it is possible to link kinds of play with the characteristics of a particular level (early childhood, middle childhood).

Until the age of 18 months to 2 years, children's play is essentially sensorimotor. That is, there is a great deal of repetition involving the body—for example, doing the same things with a toy. Gardner (1983, p. 250) states that at age 2 a great divide is passed since children can engage their world symbolically: letting one thing represent another, adopt different types of roles, and indulge in fantasy and pretend activities. Children may pretend to drink from play cups, or feed a doll with a spoon, or use a clothespin as a doll (Sutton-Smith, 1988).

Play also becomes more social; interactions with other children become more important in play. Gardner (1982) estimates that as much as 25 percent of a 6-year-old's play is **pretend play,** while interactive pretend play increases to about 20 percent. With the beginning of the school years, play becomes more social and rule dominated. Games with rules become important, which reflects children's ability to

Pretend play peaks during the early childhood years. It is a way for children to explore the varieties of familiar events and to speculate about the less familiar.

use abstract thinking in playing games. A strike in baseball, for example, is a ball thrown over the plate, one that a batter swings at and misses, or a "foul ball." School-age children, with their increased symbolic ability, understand these rules and apply them in their games.

Pretend play, however, seems most characteristic of the early childhood youngster. Children of Piaget's preoperational period show an increasing ability to represent. They are better able to engage in abstract thinking, to let one thing represent another. They can pretend. Rubin, Fein, and Vandenberg (1983) believe that pretend play becomes more social with age and seems to entail a three-stage sequence:

1. Pretend play becomes increasingly dramatic until the early elementary school years.
2. Pretend play becomes more social with age.
3. Pretend play gradually declines and is replaced by games—with rules—during middle childhood.

Pretend play begins with simple actions such as pretending to be eating or asleep. But as symbolic ability increases during the early childhood years, the nature of pretending changes. Youngsters will use toy telephones to talk. Later, they will pick up a banana and pretend to talk on the telephone. Pretend play steadily becomes more elaborate. Youngsters will serve tea to a group of dolls or feed soldiers; they also begin to enact the role of others. You can see, then, how play affects all aspects of development.

Conclusion

At the beginning of the early childhood years, most children meet other youngsters. By the end of the period, almost all children enter formal schooling. Their symbolic ability enriches all of their activities, although there are still limitations. Given their boundless energy and enthusiasm, they require consistent and reasonable discipline. Yet they should be permitted to do as many things for themselves as possible to help them gain mastery over themselves and their surroundings.

By the end of the early childhood period, children have learned much about their world and are prepared to enter the more complex, competitive, yet exciting world of middle childhood.

Chapter Highlights

The Family in Development

- The meaning of "family" in our society has changed radically.
- How parents treat their children has a decisive influence on developmental outcomes.
- Baumrind's types of parenting behavior help to clarify the role of parents in children's development.
- Research has demonstrated how divorce can affect children of different ages.
- Divorce plus remarriage produces a series of transitions to which children must adjust.
- Many children attend some form of day care, and the developmental outcomes of these experiences are still in question.

Early Childhood Education

- The number of children attending some form of preschool has increased steadily.
- Care must be taken to avoid exposing early childhood youngsters to undue stress.

The Self Emerges

- The emergence of the self follows a clearly defined path.
- Self-esteem plays a crucial role in a child's development.
- Youngsters acquire their gender identity during the early childhood years.
- Children initially seem to acquire an understanding of gender before they manifest sex-typed behavior.

The Importance of Play

- Play affects all aspects of development: physical, cognitive, social, and emotional.
- The nature of a child's play changes over the years, gradually becoming more symbolic.

Key Terms

Authoritarian parents
Authoritative parents
Chromosomal sex
Cognitive-development
Day care
Gender constancy
Gender identity
Gender role
Gender stability
Gender stereotypes
Genital sex
Gonadal sex
Hormonal sex
Individuation
Miseducation
Permissive parents
Pretend play
Reconciliation fantasies
Reinforcement
Sex cleavage
Socialization
Social learning

What Do You Think?

1. With today's accepted changes in the gender roles of males and females, do you think that a boy or girl growing up in these times could become confused about gender identity? Does your answer also apply to gender roles? Why?
2. Think back on your days as a child. Can you put your parents' behavior in any of Baumrind's categories? Do you think it affected your behavior? Explain your answer by linking your parents' behavior to some of your personal characteristics.
3. In this chapter you read about the "sleeper" effect of divorce (effects show up quite a bit later). Do you agree with these findings? Can you explain them by the child's age at the time of the divorce? (Consider all aspects of a child's development at that age.)
4. You probably have read about child abuse in some day-care centers. Do you think there should be stricter supervision? Why? By whom?

Suggested Readings

Ambrose, S. (1983). *Eisenhower* (Vol. I). New York: Simon & Schuster. Includes the boyhood days of the man who would become soldier, general, and president. A revealing account of the family and community forces that helped to shape his destiny.

Brazelton, T. B. (1981). *On becoming a family: The growth of attachment.* New York: Delacorte. An engrossing account of the development of family relationships by one of America's most renowned pediatricians.

Elkind, D. (1987). *Miseducation: Preschoolers at risk.* New York: Knopf. Typical Elkind—current, perceptive, and a pleasure to read. Using physical and psychological evidence, Elkind builds a strong case against the tendency of some parents to "push" their children. (Available in paperback.)

Johnson, J., Christie, J. and Yawkey, T. *Play and early childhood development.* (1987). Glenview, IL: Scott, Foresman. An excellent overview of the research and theory of the role of play in a child's development.

Wallerstein, J. and Blakeslee S. (1989). *Second chances.* New York: Simon & Schuster. This book is must reading for anyone interested in the effect of divorce on children. It summarizes Wallerstein's ten-year follow-up of the children of divorced parents, and furnishes insights into the entire spectrum of divorce in our society today.

PART V

Middle Childhood

CHAPTER 9

Physical and Cognitive Development in Middle Childhood

Physical Development 232

Physical Changes in Middle Childhood 232

Motor Skills 233

Cognitive Development 235

Piaget and Concrete Operations 235

The Psychometric View of Intelligence 237

New Ways of Looking at Intelligence 239

Thinking and Problem Solving 243

Thinking Skills 243

Decision Making and Reasoning 245

Problem-Solving Skills 247

Moral Development 248

Piaget's Explanation 248

Kohlberg's Theory 250

In a Different Voice 252

Language Development 253

Using Language 254

Bilingualism 255

The Importance of Reading 256

Conclusion 259

Chapter Highlights 259

Key Terms 260

What Do You Think? 261

Suggested Readings 261

It was the week after school opened and 5-year-old Kenny Wilson and his 11-year-old sister Alice had just transferred to the Brackett School District. Mrs. Allan, the principal, took both pupils to their new classrooms and introduced them to their teachers.

"Good morning, Kenny. I'm delighted that you're going to be in this classroom," said Mrs. Groves, the first grade teacher. "Some of the boys and girls in this room live near you. By the way, do you like your new house?"

Kenny, at first nervous, began to respond to Mrs. Groves' warm manner. "Oh, yes. I have my own bedroom. It has big windows."

Mrs. Groves laughed. "Can you see the moon from your window?" she asked.

"You bet. It follows me around the room," said Kenny.

At the same time, Alice was talking with Mr. Gallo, the sixth grade teacher. "Welcome to the Brackett sixth grade, Alice. I think you'll enjoy being here," said Mr. Gallo. "Can you give me an idea of the kind of work you were doing?"

"Well, the social studies teacher was helping us with research skills," said Alice. "We were doing a project on Egypt and we had to pick one specific topic like housing. Then we outlined it and made a report to the class."

"That sounds as if you were doing good work, Alice. You'll like it here; we're doing a lot of the same things."

Both teachers were listening to and observing their pupils as they spoke, searching for clues that would help them work with these new pupils in their classrooms. And the clues were there, as you can tell from both conversations. While each pupil spoke well, there are significant differences in both the thinking and speech of the two students. For example, Kenny's statement that the moon follows him around furnishes clues to his level of cognitive development, which we described in chapter 7 when we discussed cognitive development during the early childhood years.

Alice's skillful explanation of her use of research methods also revealed a level of cognitive development that one would expect of a typical sixth grader. Understanding the cognitive achievements of children Alice's age—in middle childhood—will be our task in the next two chapters. ■

Cognitive theory has made us aware that children's language reveals much about their level of mental functioning.

This chapter, which analyzes the physical, cognitive, and moral milestones of middle childhood, will often take on the appearance of a mystery. Much is happening to 6- to 12-year-old youngsters that eludes initial detection. Cognitive complexities and subtle moral reasoning characterize these years between childhood and adolescence.

To help you discover clues to important developmental achievements, in this chapter we begin by tracing physical and motor development during the middle childhood years. We'll then turn again to Piaget and analyze his concrete operational period, that time of cognitive development when youngsters begin to engage in truly abstract thinking. Here we'll pause and examine the world of intelligence testing, which has caused so much controversy. But today's children are expected to do much more with their intelligence than just sit and listen and memorize: They're expected to develop thinking skills and to solve problems. We'll explore both of these worlds (cognitive and problem solving) and then trace the moral and language development of middle childhood youngsters.

As observers of this phase of lifespan development, we cannot allow ourselves to be deceived by an apparent lull in development. Too much is going on. Deep-seated developmental currents are changing the very process by which children reason and make their moral decisions.

In following the developmental path of children during these years, we can find significant signposts in fiction. Can anyone describe the cognitive ability of a middle childhood youngster better than Mark Twain? Do you recall the memorable scene where Tom Sawyer was desperately trying to avoid whitewashing Aunt Polly's fence? One of his friends, Ben Rogers, passed by, imitating the "Big Missouri" riverboat. When Ben sympathized with him, Tom looked at him and said, "What do you call work?"

His friend asked, "Why ain't *that* work?" Tom neatly dodged the question and asked, "Does a boy get a chance to whitewash a fence every day?" By that time Ben was jumping up and down in his eagerness to paint. When that happened,

> ***Tom gave up the brush with reluctance in face but alacrity in his heart. And while that late steamer Big Missouri worked and sweated in the sun, the retired artist sat on a barrel in the shade close by, dangled his legs, munched his apple, and planned the slaughter of more innocents.***

What else can we do but acknowledge a sophisticated and subtle mind at work?

Do you remember Huck Finn meeting the Duke and the Dauphin? Not believing their stories about lost titles, Huck makes a decision:

> ***It didn't take me long to make up my mind that these liars warn't no kings nor dukes at all, but just lowdown humbugs and frauds. But I never said nothing, never let on; kept it to myself; it's the best way; then you don't have no quarrels and don't get into no trouble.***

As we'll see a little later, such thinking reflects a certain level of moral reasoning. Children of these years use their previous experiences with rewards and punishments and their growing cognitive ability to reach moral decisions.

Think about what you read in this chapter; probe into what lies behind a child's behavior. It will help you to understand the developmental achievements that prepare a youngster to move from these last years of childhood to the different demands of adolescence.

The topics that we just mentioned fit our biopsychosocial model as seen in Table 9.1.

Table 9.1 Characteristics of the Period

Bio	Psycho	Social
Height changes	Conservation	Moral reasoning
Weight changes	Seriation	Language and reading
Sports	Classification	
	Numeration	
	Intelligence tests	
	Theories of intelligence	
	Thinking skills	
	Problem solving	
	Moral reasoning	
	Language and reading	

After reading this chapter, you should be able to:

- Describe the physical development and growing motor skills of middle childhood youngsters.
- Indicate the cognitive accomplishments of the concrete operational period.
- Assess the different theories of intelligence and defend the one that you think is most appropriate.
- Evaluate the various thinking skills programs.
- Discriminate the skills involved in problem solving.
- Contrast theories of moral development.
- Assess language achievement in the middle childhood years.

Physical Development

In contrast to the rapid increase in height and weight during the first years of life, physical development proceeds at a slowed pace during middle childhood. As you can see from Table 9.2, most children gain about two inches in height per year. The same pattern applies to weight gains. By 6 years of age, most children are about seven times their birth weight.

Middle childhood youngsters show steady growth, usually good health, and an increasing sense of competence. Physical growth is relatively slow until the end of the period, when girls' development may spurt. As we have cautioned constantly, variables such as genetic influence, health, and nutrition can cause wide fluctuations in the growth of these children. Two youngsters may show considerable physical variation and yet both be perfectly normal. Table 9.2 illustrates several of the physical features of middle childhood development.

Physical Changes in Middle Childhood

Changes in height and weight are not the only noticeable physical differences. Body proportion changes also. Head size comes more in line with body size. An adult's head size is estimated to be about one-seventh of total body size; the preschooler's is about one-fourth. This difference gradually decreases during the

Table 9.2 Physical Motor Development in Middle Childhood

Age (years)	Height (in.)		Weight (lb.)		Motor Development
	Girl	Boy	Girl	Boy	
7	48	49	52	53	Child has good balance; can hop and jump accurately
8	51	51	60	62	Boys and girls show equal grip strength; great interest in physical games
9	52	53	70	70	Psychomotor skills such as throwing, jumping, running show marked improvement
10	54	55	74	79	Boys become accurate in throwing and catching a small ball; running continues to improve
11	57	57	85	85	Boys can throw a ball about 95 feet; girls can run about 17.5 feet per second
12	60	59	95	95	Boys can run about 18.5 feet per second; dodge ball popular with girls

middle childhood years. Also, the loss of baby teeth and the emergence of permanent teeth change the shape of the lower jaw. By the end of the period, the middle childhood youngster's body is more in proportion and more like an adult's (Tanner, 1989).

Changes in arms, legs, and trunk size also occur. The trunk becomes thinner and longer, and the chest becomes broader and flatter. Arms and legs begin to stretch but as yet show little sign of muscle development. Hands and feet grow more slowly than arms and legs, which helps to explain some of the awkwardness that we see during these years. These children are tremendously active physically and gradually display a steady improvement in motor coordination (Tanner, 1989).

Motor Skills

Healthy, active children of this age—both boys and girls—demonstrate considerable motor skill. Lansdown and Walker (1991) summarize motor development during these years:

- Skill increases with maturity until, by the end of the period, some youngsters are highly skilled and much in demand for various sports.
- Boys are stronger than girls.
- Girls may be more graceful and accurate.
- Sex differences—especially strength—become more obvious toward the end of the period.
- Balance matures by the end of the period.
- Fine motor skills (writing and drawing) improve noticeably.

What's Your View?

?

Desiring social acceptance, middle childhood youngsters strike out in many directions in their endeavor to achieve status. There are many paths: the school paper, dancing, band, and sports. During these years many youngsters, both boys and girls, coordinate their physical abilities and show definite signs of athletic skill. Little league baseball, youth hockey, soccer, Pop Warner football—all become showcases for their athletic talent (Bloom, 1985).

If parents, coaches, and youngsters maintain their perspective, sports can be a healthy vehicle for helping children to mature. But adults who encourage a win-at-any-cost attitude do little to help a 10-year-old maintain perspective (Elkind, 1987). Parents, realizing that they have a skilled athlete in the family and facing formidable college tuition costs, frequently push their children too hard and too fast. And now that the availability of federal monies is linked to equitable athletic opportunities for girls, girls are beginning to feel the same pressures from parents that boys have long experienced.

Sports are an *excellent outlet* for youngsters—but in their proper proportion. What happens to personal values if a youngster becomes totally immersed in sports at a time when schooling becomes even more abstract? The need to read well soars; science becomes an integral part of the curriculum; and mathematics becomes more formidable. The stories of athletes who drop out of college in their senior year, or high school graduates who cannot read and are barely able to write are a grim reminder of a national literacy problem.

Competition in sports is keen, as it is in the classroom, which raises certain issues.

- Competition, as well as cooperation, is a fact of life—in home, in school, in work. Races will be run and scores will be kept.
- Excessive motivation produces negative results. This applies equally as well to competition, which should be a sharp stimulant to motivation. Consider the athlete who, under enormous pressure from some unthinking coach, hides a potentially crippling injury to help the team to win.
- Intense competition with others turns youngsters against one another, and, if prolonged, leads to an undesirable concentration on self—I must win, regardless. Too often under these conditions the end justifies the means, and such behavior becomes routine.
- Competition probably is inescapable, and competition with self should be used as a way of learning to cope with life. What's your view?

The middle childhood years are a time when children demonstrate considerable competence; their continuing mastery of their bodies and their environment lead to emerging skills that are readily observable.

Cognitive Development

During the middle childhood years, children's cognitive abilities become remarkably complicated and sophisticated, reflecting in more observable form the complex mixture of heredity and environment (Cardon & others, 1992). For example, in a series of studies, Rose and associates (1991) found that measures of object permanence made at 1 year of age were significantly related to cognitive performance at 5 years of age, suggesting that some of the roots of later cognition can be found in infancy.

Middle childhood youngsters now enter formal education, and their cognitive abilities should enable them to meet the more demanding tasks set by the school. We shall continue our practice of examining the cognitive abilities of children by beginning with Piaget and his explanation of these years—the **concrete operational period.**

Piaget and Concrete Operations

Recall the early childhood years and the preoperational period. Piaget believed that preschoolers' thinking is still restricted. *Egocentrism* and *irreversibility* still characterize preoperational children (Gelman & Baillargeon, 1983).

During the period of concrete operations, children overcome these limitations. They gradually employ logical thought processes with concrete materials—that is, with objects, people, or events that they can see and touch. They also concentrate on more than one aspect of a situation, which is called **decentering.** They acquire conservation and can now reverse their thinking. Think of the water jar problem (See chap. 2). Now they can mentally pour the water back.

> ***About the age of seven, a fundamental turning point is noted in a child's development. He becomes capable of a certain logic; he becomes capable of coordinating operations in the sense of reversibility, in the sense of the total system of which I will soon give one or two examples. This period coincides with the beginning of elementary schooling. Here again I believe that the psychological factor is the decisive one. If this level of the concrete operations came earlier, elementary schooling would begin earlier.*** **(Piaget, 1973, p. 20)**

Several notable accomplishments mark the period of concrete operations:

- **Conservation** appears. In Piaget's famous water jar problem, children observe two identical jars filled to the same height. While they watch, the contents of one container are poured into another, taller and thinner jar so that the liquid reaches a higher level. Seven-year-old children typically state that the contents are still equal. They conserve the idea of equal amounts of water by decentering, focusing on not only one part of the problem but both. They can now reverse their thinking; they can mentally pour the water back into the original container.
- **Seriation** means that concrete operational children can arrange objects by increasing or decreasing size.
- **Classification** enables children to group objects with some similarities within a larger category. Brown wooden beads and white wooden beads are all beads.

Middle childhood youngsters are at Piaget's stage of concrete operations. Now they demonstrate increasing mental competence, such as classifying.

Features of Concrete Operational Thinking

Children at the level of concrete operations can solve the water-level problem, but the problem or the situation must involve concrete objects. That's the reason for calling the period *concrete operational.* In the water-level problem, for example, children no longer concentrate solely on the height of the water in the glass; they also consider the width of the glass. But as Piaget notes, concrete operational children nevertheless demonstrate a true logic, since they now can reverse operations. Concrete operational children gradually master conservation—that is, they understand that something may remain the same even if surface features change.

Figure 9.1 illustrates the different types of conservation that appear at different ages. For example, if preoperational children are given sticks of different lengths, they cannot arrange them from smallest to largest. Or if presented with three sticks, A, B, and C, they can tell you that A is longer than B and C, and that B is longer than C. But if we now remove A, they can tell you that B is longer than C, but not that A is longer than C. They must see A and C together. The child at the level of concrete operations has no difficulty with this problem.

Figure 9.1
Different kinds of conservation appear at different ages.

Conservation of	Example	Approximate age
1. Number	Which has more?	6-7 years
2. Liquids	Which has more?	7-8 years
3. Length	Are they the same length?	7-8 years
4. Substance	Are they the same?	7-8 years
5. Area	Which has more room?	7-8 years
6. Weight	Will they weigh the same?	9-10 years
7. Volume	Will they displace the same amount of water?	11-12 years

But limitations remain. Piaget (1973) gives the example of three young girls with different colored hair. The question is: Who has the darkest hair of the three? Edith's hair is lighter than Suzanne's but darker than Lili's. Who has the darkest hair? Piaget believes that propositional reasoning is required to realize that it is Suzanne and not Lili. Youngsters do not achieve such reasoning until about 12 years of age.

Concrete operational children can also classify—that is, they can group different things that have something in common. For example, wooden objects may include both a table and a chair. In a classic experiment illustrating mastery of classification, Piaget showed a girl about twenty brown wooden beads and two or three white wooden beads. He then asked her to separate the brown from the white beads. Children at both the preoperational and concrete levels can do this. But then he asked her, "Are there more brown beads or wooden beads?" The preoperational child answered brown, while an older child answered correctly.

Another interesting feature of this period is the child's acquisition of the number concept, or **numeration.** This is not the same as the ability to count. If five red tokens are more spread out than five blue ones, preoperational children, although able to count, still think there are more red than blue tokens. Piaget then constructed an ingenious device that enables a child to trace the blue to the red. The preoperational child can actually move the blue token to the corresponding red, but still thinks there are more red tokens! It is during the concrete operational period that children understand oneness—that one boy, one girl, one apple, and one orange are all one of something (see Figure 9.2).

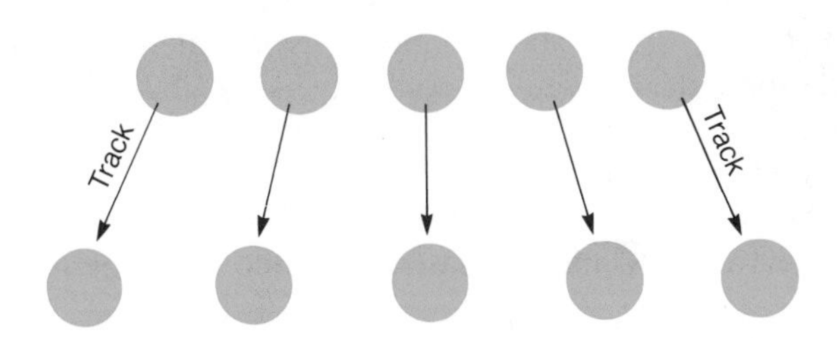

Figure 9.2
Encouraging acquisition of number concept

Piaget's Legacy

As we end our work on Piaget's analysis of cognitive development, what can we conclude? Two major contributions of Piaget come immediately to mind: Thanks to Piaget we have a deeper understanding of children's cognitive development, and he has made us more alert to the need for greater comprehension of how children think (the processes that they use) and not just what they think (the products of their thinking).

There are, as we have seen, serious questions about Piaget's beliefs. By changing the nature of the task (for example, reducing the number of objects children must manipulate—see Gelman & Baillargeon, 1983), by allowing children to practice (for example, teaching children conservation tasks—see Field, 1987), and by using materials familiar to children, researchers have found that children can accomplish specific tasks at earlier ages than Piaget believed (Halford, 1989).

Although Piaget has left an enduring legacy, his ideas are not the only way of analyzing cognitive development. Another technique attempts to measure the different types of tasks that children of different ages can master. This is called the psychometric view of intelligence.

The Psychometric View of Intelligence

Psychologists have followed many routes in their quest to discover the secrets of the human mind. One of these paths led to a belief that intelligence can be measured. Intelligence, that fascinating yet enigmatic something, promised to discriminate the able from the less able.

At the beginning of this century, there was an influx of immigrants to American shores. Some device was needed to classify individuals for education, for work, and ultimately for military service. This helps to explain why the mental testing movement was readily, almost eagerly, accepted. Here at last was a tool, so went the claim, that cut to the heart of a person's innate capacity. The circumstances that led to the development and acceptance of intelligence tests illustrate the methods that investigators adopted to assess native ability.

Binet and Mental Tests

The story of Alfred Binet and his search for the meaning and measurement of intelligence is famous in the history of psychology. Much of his early work on the intellectual and emotional lives of children resulted from studies of his daughters. But in 1904, Binet was asked by the Parisian Minister of Public Instruction to formulate a technique for identifying the children most likely to fail in school. The problem was difficult, since it meant finding some means of separating the normal from the truly retarded, of determining the lazy but bright who were simply poor achievers, and of eliminating the halo effect (assigning an unwarranted high rating to youngsters because they are neat or attractive).

Students should take tests under conditions that enable them to perform as well as possible. Care should be taken that students with different cultural experiences understand exactly what is expected.

The items in Binet's early test reflected these concerns. When an item seemed to differentiate between normal and subnormal, he retained it; if no discrimination appeared, he rejected it. (Binet defined normality as the ability to do the things that others of the same age usually do.) Fortunately for the children of Paris and for all of us, Binet was devoted to his task. The fruits of his and his coworkers' endeavors was the publication in 1905 of the Metrical Scale of Intelligence, which evolved into the several versions of the famous **Stanford-Binet Intelligence Test** (Binet & Simon, 1905).

The Wechsler Intelligence Scales

David Wechsler, a clinical psychologist at New York's Bellevue Hospital, was dissatisfied with previous attempts to measure adult intelligence. In his hospital work, Wechsler needed some reliable means to identify level of intelligence in his examination of criminals, neurotics, and psychotics.

There are three forms of the Wechsler test, designed to measure the intelligence of human beings through the lifespan, beginning at the age of 4:

1. *WAIS, the Wechsler Adult Intelligence Scale,* which is a revised form of his first test, originally published in 1939.
2. *WISC-R, the Wechsler Intelligence Scale for Children—Revised,* which first appeared in 1949 and assesses the intelligence of children from age 5 to 15.
3. *WPPSI-R, the Wechsler Preschool and Primary Scale of Intelligence—Revised,* which is designed to measure the intelligence of children from 4 to 6 1/2 years of age.

Wechsler states (1958) that other important aspects of intelligence include motivation and persistence at a task. For example, individuals with precisely the same score on the same intelligence test may adapt quite differently. One youngster with an IQ of 75 may require institutional care, while another child with an IQ of 75 may function adequately at home. (For an excellent discussion of the nature of intelligence and its many implications, see Sternberg, 1982).

A Multicultural View 

Immigrant Children and Tests

Children's ability to take tests often powerfully influences the results of the test, especially intelligence tests. Tests can scare students or cause them anxiety. Whatever the reason for test anxiety—parental pressure, their own concerns, or the testing atmosphere—merely taking a test can affect performance. This is especially true for pupils with different cultural experiences. Language, reading, expectations, and behavior all may be different and influence test performance (Stigler, Shweder, & Herdt, 1990). One publication, *New Voices: Immigrant Students in U.S. Public Schools* (1988), describes many of these pupils as having experienced wars, political oppression, economic deprivation, and long, difficult journeys to come to the United States.

These children want to succeed, as reflected in interviews with many of them: Almost 50 percent were doing one to two hours of homework every night. (Twenty-five percent of the Southeastern Asian students reported more than three hours each night.) We have spoken throughout our work of the need for sensitive responsiveness. Here is an instance in which being sensitive to the needs of multicultural children can only aid their adjustment and achievement in school.

When children feel comfortable, they do better; this is particularly true for test taking. Multicultural children need information about why the test is being given, when it is being given, what material will be tested, and what kinds of items will be used. These are just a few topics to consider. Language should not be a barrier to performance. For example, they need to understand the terms in the directions of the test—what "analyze" means; what they should do when asked to "compare"; what "discuss" means.

Helping multicultural students in this way means extra time and effort for teachers. But it is teaching, just as teaching English or history is teaching. As more and more multicultural students become users of classroom tests, they shouldn't do poorly because they don't understand the mechanics of the test.

Other psychologists, not entirely happy with either Piaget's views or the psychometric approach, have devised new ways of explaining children's cognitive abilities.

New Ways of Looking at Intelligence

IQ tests, as we have seen, failed to resolve the main issue: how much of intelligence is inherited and how much is learned? For example, advocates of intelligence testing claimed that these instruments measured a person's innate potential for learning. This claim implied that IQ test scores were unaffected by either education or social class. Results, however, have consistently shown that more education and higher social class are the inevitable partners of higher intelligence test scores.

To avoid this pitfall, several recent theories have been proposed. Two of these seem particularly significant.

Sternberg's Triarchic Model of Intelligence

Robert Sternberg (1986) has designed a **triarchic model** of intelligence to answer three questions:

- What is the relationship of intelligence to our internal world? What are the inner processes and strategies that we use?
- What is the relationship of intelligence to our external lives? How does the environment affect intelligence?
- What is the relationship of intelligence to our experiences? How does what we do help to shape our intelligence?

To answer these questions, Sternberg (1986) devised the following threefold model of intelligence:

1. *The components of intelligence.* Sternberg has identified three types of information processing components: **metacomponents,** which help us to plan, monitor, and evaluate our problem-solving strategies; **performance components,** which help us to execute the instructions of the metacomponents; and **knowledge-acquisition components,** which help us to learn how to solve problems in the first place (Sternberg, 1988). These three components are highly interactive and generally act together.

   For example, consider writing a term paper. The metacomponents help you to decide on the topic, plan the paper, monitor the actual writing, and evaluate the final product. The performance components help you in the actual writing of the paper. You use the knowledge-acquisition components to do your research. Figure 9.3 illustrates the manner in which the components work together.

Figure 9.3
The relationship among the components of intelligence

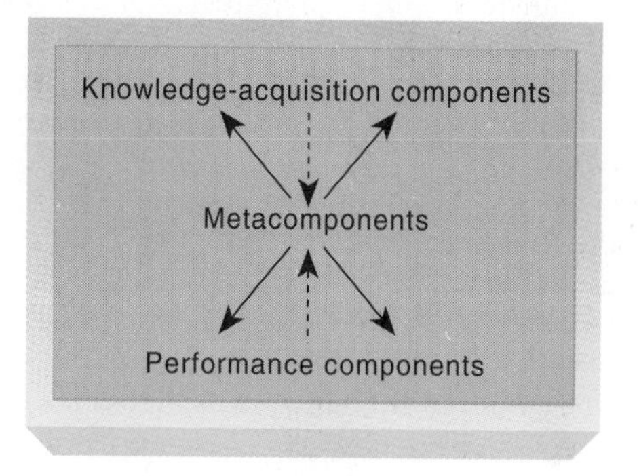

2. *Experience and intelligence.* Sternberg's second aspect of intelligence includes our experiences, which improve our ability to deal with novel tasks and to use pertinent information to solve problems.

3. *The context of intelligence.* The third aspect of intelligence in Sternberg's model refers to our ability to adapt to our culture. It wouldn't make much sense to move a nomadic people to New York City and then test their intelligence on the Wechsler. The major thrust of contextual intelligence is **adaptation.** Adaptation, for Sternberg, has three meanings:

   - Adaptation to existing environments, so that you adjust to current circumstances
   - Shaping existing environments, which implies changing the present environment to meet your needs
   - Selecting new environments

If you apply these descriptions to yourself and others you know, you probably can identify which aspect of intelligence predominates in each person. For example, if you excel in analytical thinking, you probably do quite well on traditional IQ tests.

While Sternberg does not apply his theory directly to children, you can see how his work relates to the developing cognitive competence of the middle childhood youngster. To take one example, with their growing symbolic activity, children can think about their thinking—that is, they are employing metacomponents. Sternberg (1986) has also compiled a list of reasons why we too often fail, which has important implications for children of this age.

Noting that all of us, children and adults, let self-imposed obstacles frustrate us, Sternberg states that what is important is not the level of our intelligence but what we achieve with it. Sternberg has identified the most common obstacles:

- *Lack of motivation.* If children aren't motivated, it really doesn't matter how much talent they have. In a typical classroom situation, for example, the range of intelligence may be fairly narrow and differences in motivation spell success or failure.
- *Using the wrong abilities.* Although children acquire greater cognitive ability, they frequently don't use it or else fail to recognize exactly what is needed. For example, faced with a math test, they may fail to review solutions to word problems (knowledge-acquisition components) and instead depend on their previous experiences.
- *Inability to complete tasks.* Regardless of how skilled children may be in their use of Sternberg's components, if they are unable to sustain their efforts—for whatever reason—they are in danger of failure. The worry here is that this tendency may become a way of life, ensuring difficulty, frustration, and failure.
- *Fear of failure.* Some youngsters may develop a fear of failure early in life. This can prevent them from ever fulfilling their intellectual potential.

These are among the obstacles that adults should be aware of, so they can help middle childhood youngsters develop to the fullest.

Gardner and Multiple Intelligences

Howard Gardner (1982, 1985, 1991) has forged a tight link between thinking and intelligence with his theory of *multiple intelligences.* An especially intriguing aspect is that an individual who is capable of penetrating mathematical insights may be quite baffled by the most obvious musical clues. Gardner attempts to explain this apparent inconsistency by identifying seven equal intelligences.

1. *Linguistic intelligence.* The first of Gardner's intelligences is langauge—**linguistic intelligence.** For example, we can trace the effects of damage to the language area of the brain; we can identify the core operations of any language (phonology, syntax, semantics, and pragmatics); and langauge development in humans has been well documented and supported by empirical investigations. Gardner considers language a preeminent example of human intelligence.

   As we will soon see, during these years children change their use of language to a more flexible, figurative form. Gardner comments on this by noting that middle childhood youngsters love to expand on their accomplishments. They use appealing figures of speech: "I think I'll get lost," meaning that they're about ready to leave.
2. *Musical intelligence.* One has only to consider the talent and career of Yehudi Menuhin to realize that there is something special about musical ability. At 3 years of age, Menuhin became fascinated by music, and by 10 he was performing on the international stage. The early appearance of musical ability suggests some kind of biological preparedness—a **musical intelligence.** The right hemisphere of the brain seems particularly

Relationships with peers, abstract activities, and developing physical skills all occupy middle-childhood youngsters. The success that children achieve in these activities contributes significantly to an emerging sense of competence.

important for music, and musical notation clearly indicates a basic symbol system. Although not considered intelligence in most theories, musical skill satisfies Gardner's criteria and so demands inclusion.

In regard to the middle childhood years, Gardner notes that most children (except for those who are musically talented) cease musical development after school begins. For the talented, up to the age of 8 or 9, talent alone suffices for continued progress. Around 9 years of age, serious skill-building commences, with sustained practice until adolescence, when these children must decide how much of their lives they want to commit to music. In general, society accepts musical illiteracy.

3. *Logical-mathematical intelligence.* Unlike linguistic and musical intelligences, **logical-mathematical intelligence,** Gardner believes, evolves from our contact with the world of objects. In using objects, taking them apart, and putting them together again, children gain their fundamental knowledge about "how the world works." By this process, logical-mathematical intelligence quickly divorces itself from the world of concrete objects. Children begin to think abstractly.

   Gardner then uses Piaget's ideas to trace the evolution of thinking. The development of logical-mathematical thinking, as explored by Piaget, is used as an example of the scientific thinking so characteristic of logical-mathematical intelligence. You may want to review Piaget's ideas concerning the unfolding of intelligence to better understand Gardner's views.

4. *Spatial intelligence.* Brain research has clearly linked spatial ability to the right side of the brain. Here Gardner relies heavily on Piaget, noting that an important change in children's thinking occurs during these years, especially with the appearance of conservation and reversibility. Middle childhood youngsters can now visualize how objects seem to someone

else. During these years, children can manipulate objects using their **spatial intelligence,** but this ability is still restricted to concrete situations and events.

5. *Bodily-kinesthetic intelligence.* As Gardner notes, describing the body as a form of intelligence may at first puzzle you, given that we normally divide the mind and reasoning from our physical nature. This divorce between mind and body is often accompanied by the belief that our bodily activities are somehow less special.

   Gardner urges us to think of mental ability as a means of carrying out bodily actions. Thus thinking becomes a way of refining motor behavior. Our brain's control of our bodily functions has been well documented. Also, the developmental unfolding of bodily movements has been thoroughly recorded. Gardner states that our control of bodily motions and the ability to handle objects skillfully are defining features of an "intelligence"—**bodily-kinesthetic intelligence.**

   We have commented that middle childhood is the beginning of the emergence of bodily skills, to the point where they can dominate some children's lives. Gardner poses an age-old question to which we still have no definite answer: Does increasing mental ability affect the performance of a bodily skill?

6 and 7. *Interpersonal and intrapersonal intelligence.* Gardner refers to **interpersonal** and **intrapersonal intelligences** as the *personal intelligences.* Interpersonal intelligence builds on an ability to recognize what is distinctive in others, while intrapersonal intelligence enables us to understand our own feelings. Autistic children are good examples of a deficit in this intelligence. Often competent in a certain skill, they may be utterly incapable of ever referring to themselves.

   The middle childhood years are a time of greater social sensitivity, a keener awareness of others' motivation, and a sense of one's own competencies. Children begin to develop friendships and devote considerable time and energy to securing a definite place in a circle of friends. This effort can only increase their sensitivity to interpersonal relations. At the same time, they become more aware of themselves, furthering the development of their intrapersonal intelligences. If children of this age do not succeed in establishing harmonious relationships, they may develop feelings of inadequacy and isolation. This fuels a fear of failure that can produce diminished expectations.

Thinking and Problem Solving

Advanced societies demand citizens who can do more with their intelligence than just survive. Rapid change requires the ability to cope. Yet 25 percent of the 14- to 18-year-old group are no longer in school. We can only question how many young people are adapting. Concern about these and similar statistics has prompted renewed interest in two topics that have been with us for years: critical thinking and problem solving.

Thinking Skills

Young people—actually all of us—need the skills and strategies that enable us to be productive members of our society. The famous American educator John Dewey was worried about this years ago. In his classic work, *How We Think* (1933),

WHAT'S YOUR VIEW?

Take a moment and read through Gardner's types of intelligences again. Now rank them by the importance you place on each.

___ Linguistic intelligence

___ Musical intelligence

___ Logical-mathematical intelligence

___ Spatial intelligence

___ Bodily-kinesthetic intelligence

___Interpersonal intelligence

___ Intrapersonal intelligence

- Now look at your rankings. Do you find some more relevant than others? What's your view?
- Check your classmates' lists. How similar are they to yours? How do you explain the differences? Ask members of the group why they ranked the intelligences as they did. Is there anything in their experiences that explains their choices?
- Are there differences within groups? (Age, sex, types of schools attended?)
- Do your answers to these questions tell you anything about your opinions of cognitive development? What's your view?

Dewey remarked that reflective thinking begins with doubt or uncertainty that causes us to search for the materials that will remove the perplexity. Dewey believed that we go through states of thinking, such as the following:

1. We make suggestions as we look for possible solutions.
2. We search for the cause of the uncertainty and attempt to be precise about what is blocking us.
3. We formulate hypotheses, which we use to guide our search.
4. We mentally test these hypotheses.
5. We physically test the hypothesis that seems best suited to resolve our difficulty.

Dewey, of course, was referring to critical thinking. Critical thinking depends on skills that children can learn, such as improving their memory and attending to the important part of any problem. Knowledge is also vital, since the more information you have about something the easier it is to understand and solve the problem.

Several attempts have been made to identify the particular skills young people need. Benjamin Bloom's enduring *Taxonomy of Educational Objectives, Handbook 1: Cognitive Domain* (1956) is one example. Bloom had several goals in mind. First, he believed that the taxonomy would help in identifying specific thinking skills. Second, it would help to specify observable behaviors (recall, define, compare). Third, it would help in the preparation of learning experiences and evaluation devices. Fourth, it could serve as a tool to analyze educational processes.

Bloom's taxonomy, embedded in the history of educational evaluation and precise in its objectives and terminology, has enjoyed widespread acceptance and today forms the core of many thinking skills programs. The taxonomy consists of three major sections: the cognitive, the affective, and the psychomotor domains. Our concern here is with the cognitive taxonomy, which is divided into the six major classes listed below:

1.00 Knowledge—recall of specific facts
2.00 Comprehension—understand what is communicated
3.00 Application—can generalize and use abstract information in concrete situations

4.00 Analysis—breaking a problem into subparts and detecting relationships among the parts
5.00 Synthesis—putting together parts to form a whole
6.00 Evaluation—using criteria to make judgments

A taxonomy of thinking skills provides a useful organization of knowledge about thinking and facilitates answering questions such as:

- What do we know about children—their developmental paths, needs, and interests?
- What kinds of subject matter and materials can help to improve their thinking?
- What does developmental psychology tell us about the appropriate placement of objectives in a thinking skills program? Do the ways that pupils learn suggest a series of steps that must be mastered?

Robert Ennis (1987), noting that critical thinking skills reflect logical thinking, argues that young people should acquire a thinking disposition. A thinking disposition entails answering four questions:

- Are you clear about what is going on? Focus on the question; define what is needed; get more information.
- Do you have a good basis for any judgment you make? Be sure your response is credible; use pertinent data for your decision.
- Are your inferences logical? Again, get the information. Proceed systematically.
- Have you discussed the problem with anyone else? Talk to others. Use new ideas.

Parents and teachers can encourage a middle childhood youngster's cognitive potential to think critically. By appealing to a child's enthusiasms and abilities, adults can encourage these youngsters to begin using "a thinking disposition."

Decision Making and Reasoning

Do you consider yourself logical? How good are you at making decisions? Do you pride yourself on your accurate thinking? While we all want to think that we are models of logical thinking, facts seem to tell us otherwise.

Addressing the question of whether decision-making skill improves with age, Krouse (1986) studied 90 children, 30 each from grades one, three, and six (48 boys, 42 girls). Employing the Concept Assessment Kit (which measures conservation) and a variety of Tversky-Kahneman (1981) tasks, Krouse sought to determine if any relationship existed between children's decision-making behavior (as determined by Tversky-Kahneman tasks) and certain variables: educational level, level of cognitive development, sex of child.

Krouse found that educational level made a difference. Third and sixth graders demonstrated the same variations in decision making as adults, a particularly interesting finding in the light of cognitive theory. Recalling our discussion of attention, perception, representation, classification, and memory, we could conclude, not surprisingly, that children's processing capacities are fairly restricted during the early years. To what do young children attend? Are they capable of discriminating loss of money from loss of object, or is it just something lost?

An Applied View — How Would You Answer These Questions?

1. Imagine that you were lucky enough to obtain two tickets to the great Broadway musical *Phantom of the Opera* for $100. As you walk down the street to the theater, you discover that you have lost the tickets. You can't remember the seat numbers. Would you go to the ticket window and buy another pair of tickets for $100?

2. Imagine that you are on the way to the theater to buy tickets for the Broadway play, *Phantom of the Opera*. They will cost you $100. As you approach the ticket window, you discover that you have lost $100. Would you still pay $100 for the tickets?

These questions, originally posed by Tversky and Kahneman (1981) elicit some interesting answers. How did you answer them? Among their subjects, 46 percent answered yes to question 1, while 88 percent answered yes to question 2. Note that many more people said they would buy new tickets if they had lost the money rather than the tickets. Yet the two situations are almost identical—in each instance you would have lost $100.

How can we explain the difference in the responses? Tversky and Kahneman believe that the way a problem is framed helps to explain our response. As they state:

> ***The frame that a decision-maker adopts is controlled partly by the formulation of the problem and partly by the norms, habits, and personal characteristics of the decision-maker.*** **(1981, p. 453)**

The point here is that our decisions are influenced by the way questions and problems are structured and by our personal characteristics, such as motivation, persistence, and the like. We can help children reach better decisions by teaching them the skills and strategies discussed in this chapter.

As children's decision-making abilities improve with age, they can be further aided by helping them to improve their thinking and problem solving skills.

Younger pupils (preschoolers to grade two) could not judge their mental capabilities as accurately as the older pupils, an ability that gradually and consistently improves during the elementary school years. Whatever the reason, a developmental shift seems to occur around the third grade: Children's decision-making skills come to resemble an adult's.

AN APPLIED VIEW Try This Problem

Do you think you're a good problem solver? Before you answer, try to solve the following problem.

Two motorcyclists are 100 miles apart. At exactly the same moment, they begin to drive toward each other for a meeting. Just as they leave, a bird flies past the first cyclist in the direction of the second cyclist. When it reaches the second cyclist, it turns around and flies back to the first. The bird continues flying in this manner until the cyclists meet. The cyclists both traveled at the rate of 50 miles per hour while the bird maintained a constant speed of 75 miles per hour. How many miles will the bird have flown when the cyclists meet?

Many readers, after examining this problem, immediately begin to calculate distance, miles per hour, and constancy of speed. Actually this is not a mathematical problem; it is a word problem. Carefully look at it again. Both riders will travel for one hour before they meet; the bird flies at 75 miles per hour; therefore the bird will have flown 75 miles. No formulas, no calculations, just a close examination of what is given.

Problem-Solving Skills

Middle childhood youngsters use their newly developed cognitive accomplishments to solve the problems they face in their daily lives. To give you an idea of what we're talking about, see how good you are at solving the problem in the box just above.

Obviously, some people are better at problem solving than others due to intelligence, experience, or education. But anyone's ability to solve problems can be improved—even children's. Some children and adults don't do well with problems because they're afraid of them. "I'm just not smart enough"; "I never could do these." Here is a good example. Group the following numbers in such a way that when you add them, the total is 1,000.

88888888

Unintimidated elementary school children get the answer almost immediately. Some of you won't even bother trying; others will make a halfhearted effort; still others will attack it enthusiastically. What is important is how you think about a problem. Step back and decide what you have to do; decide on the simplest way to get the answer.

In the eights problem, think of the only number of groups that would give you 0 in the units column when you add them—five. Try working with five groups and you will eventually discover that 888 + 88 + 8 + 8 + 8 gives you 1,000.

People who are good problem solvers have several distinct characteristics that are seen in middle childhood youngsters (Hayes, 1989):

- *Positive attitude.* Good problem solvers face problems with confidence, sure that their problems can be solved by careful, persistent analysis.
- *Concern for accuracy.* Good problem solvers seize on the data that are present in the problem. Simply by encouraging children to read more carefully—to reread—and to look for the details that are given, teachers and parents can help children grasp the problem more accurately, thus improving their problem-solving skills.

- *Break the problem into parts.* Good problem solvers divide a problem into parts, study each part, and then put the parts back together again.
- *Avoid guessing.* Good problem solvers refuse to guess; they consistently search for the facts.

Teaching children the basics of problem solving will improve their abilities to recognize and solve problems, both in and out of the classroom.

What can be done to help children improve their problem-solving skills? Here are several helpful suggestions (Hayes, 1989; Bransford & Stein, 1984; Lewis and Greene, 1982):

- Determine just what the problem is.
- Understand the nature of the problem.
- Plan your solution carefully.
- Evaluate both your plan and the solution.

Many of the daily problems children face are vague and ill defined. If children lack problem-solving strategies, their task is next to impossible. This is one of the major reasons there is growing pressure for schools to teach problem-solving skills, either as a separate course or as a part of another course's content.

These new ways of examining intelligence and the renewed interest in critical thinking and problem solving seem particularly pertinent to middle childhood youngsters. With their developing cognitive sophistication, they are equal to the challenge posed by new ideas and deserve the opportunity to sharpen their mental skills.

As middle childhood youngsters think better, reason more maturely, and evaluate their actions, moral development becomes a matter of increasing importance.

Moral Development

The beginnings of moral development emerge as a consequence of learning: Children are rewarded for what their parents believe is right and punished for wrongdoing. Recent research indicates that parents may be more influential than originally thought in a child's moral development. Parents who encourage children to express their opinions about a real dilemma and who themselves present a higher level of moral reasoning help to advance their children's moral development (Walker & Taylor, 1991). With cognitive growth, the time of moral reasoning appears and control of behavior gradually begins its shift from external sources to more internal self-control.

Piaget's Explanation

Youngsters realize that the opinions and feelings of others matter—what I do might hurt someone else. By the end of the period, children clearly include intention in their thinking. For the 6-year-old, stealing is bad because I might get punished; for the 11- or 12-year-old, stealing may be bad because it takes away from someone else. During these years, children move from judging acts solely by the amount of punishment to judging acts based on intention and motivation.

As might be expected, Piaget has examined the moral development of children and attempted to explain it from his cognitive perspective. Piaget formulated his ideas on moral development from observing children playing a game of marbles. Watching the children, talking to them, and applying his cognitive theory to their actions, he identified how children actually conform to rules.

An Applied View — The Role of Memory

One of the most powerful strategies in solving problems is the efficient use of memory. All memory strategies, however, are not equally effective. The appropriateness of the strategy depends on the level of material involved. Try this memory problem.

The following list contains 25 words. Take 90 seconds to study these words. When time runs out, write as many of the words as you can without looking at the list.

paper	fruit	street	wheel
white	step	juice	time
spoke	shoe	car	note
ball	word	judge	run
banana	touch	hammer	table
dark	page	bush	official
walk			

How did you do? Or more importantly from our perspective, how did you do it? Were these among the strategies you used?

1. *Rehearse each word* until you have memorized it: tire, tire, tire.
2. *Rehearse several words*: ball, apple, referee; ball, apple, referee.
3. *Organize the words by category.* Note that several words are related to cars; others could be grouped as fruit; still others could be categorized as relating to books.
4. *Construct a story* to relate as many of the words as possible.
5. *Form images of words* or groups of words.

You may have tried one or a combination of these strategies, but note that you were not told to memorize them in any particular manner.

If you had received specific instructions, each of these strategies would not have been equally effective. When we are asked to remember a particular telephone number, the tendency is to rehearse it for as long as we need to recall it. But if you had been directed to memorize them in a certain order (the way that they were presented), grouping them by categories would not have been efficient. Thinking of a story to link them in the correct order would have been much more efficient.

- In the first stage, children simply played with the marbles, making no attempt to conform to rules. Piaget (1932) referred to this as the *stage of motor rules.*
- In the second stage, at about ages 3 to 6, children seem to imitate the rule behavior of adults, but they still play by themselves and for themselves. Piaget calls this the *egocentric stage.*
- Between ages 7 and 8, children attempt to play by the rules, even though rules are only vaguely understood. This is Piaget's *stage of incipient cooperation.*
- Finally, between the ages of 11 and 12, which Piaget calls the *stage of codification of rules,* children play strictly by the rules.

After youngsters reach the fourth stage, they realize that rules emerge from the shared agreement of those who play the game, and that rules can be changed by mutual agreement. They gradually understand that intent becomes an important part of right and wrong, and their decreasing egocentrism permits them to see how others view their behavior. Peers help here because in the mutual give-and-take of peer relations, children are not forced to accept an adult view.

While observing the children playing marbles, Piaget also asked them their ideas about fairness and justice, what is a serious breach of rules, and how punishment should be administered. From this information, he devised a theory of moral development:

- Up to about 4 years, children are not concerned with morality. Rules are meaningless, so they are unaware of any rule violations.
- At about 4 years, they begin to believe that rules are fixed and unchangeable. Rules come from authority (parents, God) and are to be obeyed without question. This phase of moral development is often called *heteronomous morality* (or moral realism). Children of this age make judgments about right or wrong based on the consequences of behavior, for example, it is more serious to break five dishes than one. They also believe in *immanent justice;* that is, anyone who breaks a rule will be punished immediately—by someone, somewhere, somehow!
- From 7 to about 11 years of age, children begin to realize that social rules are formulated by individuals and can be changed. This phase is referred to as *autonomous morality* (or the morality of reciprocity). At this age, children think punishment for any violation of rules should be linked to the intent of the violator. The person who broke five cups didn't mean to, so should not be punished any more than the person who broke one.

Piaget's ideas about moral development led to the advancement of a more complex theory, devised by Lawrence Kohlberg.

Kohlberg's Theory

Among the more notable efforts to explain a child's moral development has been that of Lawrence Kohlberg (1975, 1981). Using Piaget's ideas about cognitive development as a basis, Kohlberg's moral stages emerge from a child's active thinking about moral issues and decisions. Kohlberg has formulated a sophisticated scheme of moral development extending from about 4 years of age through adulthood.

To discover the structures of moral reasoning and the stages of moral development, Kohlberg (1975) has employed a modified clinical technique called the **moral dilemma,** in which a conflict leads subjects to justify the morality of their choices. In one of the best known, a husband needs a miracle drug to save his dying wife. The druggist is selling the remedy at an outrageous price, which the woman's husband cannot afford. He collects about half the money and asks the druggist to sell it to him more cheaply or allow him to pay the rest later. The druggist refuses. What should the man do: steal the drug or permit his wife to die rather than break the law? By posing these conflicts, Kohlberg forces us to project our own views.

Middle childhood youngsters are at Kohlberg's **preconventional level of morality** for most of the period. Only as they approach ages 10 to 12 do they begin to edge into the **conventional level of morality,** where acts are right because that's the way it's supposed to be (determined by adult authority). The **postconventional level of morality** comes at age 13 and over.

Kohlberg's scheme traces moral development through six stages by successive transformations of cognitive structures (see Table 9.3).

According to Kohlberg, moral judgment requires us to weigh the claims of others against self-interest. Thus youngsters must overcome their egocentrism before they can legitimately make moral judgments. Also, anyone's level of moral development may not be the same as their moral behavior. To put it simply, people may know what is right but do things they know are wrong.

Table 9.3 Kohlberg's Stages of Moral Development

Level I. Preconventional (about 4 to 10 years) During these years children respond mainly to cultural control to avoid punishment and attain satisfaction. There are two stages: *Stage 1.* Punishment and obedience. Children obey rules and orders to avoid punishment; there is no concern about moral rectitude. *Stage 2.* Naive instrumental behaviorism. Children obey rules but only for pure self-interest; they are vaguely aware of fairness to others but only for their own satisfaction. Kohlberg introduces the notion of reciprocity here: "You scratch my back, I'll scratch yours."
Level II. Conventional (about 10 to 13 years) During these years children desire approval, both from individuals and society. They not only conform, but actively support society's standards. There are two stages: *Stage 3.* Children seek the approval of others, the "good boy-good girl" mentality. They begin to judge behavior by intention: "She meant to do well." *Stage 4.* Law-and-order mentality. Children are concerned with authority and maintaining the social order. Correct behavior is "doing one's duty."
Level III. Postconventional (13 years and over) If true morality (an internal moral code) is to develop, it appears during these years. The individual does not appeal to other people for moral decisions; they are made by an "enlightened conscience." There are two stages: *Stage 5.* An individual makes moral decisions legalistically or contractually; that is, the best values are those supported by law because they have been accepted by the whole society. If there is conflict between human need and the law, individuals should work to change the law. *Stage 6.* An informed conscience defines what is right. People act, not from fear, approval, or law, but from their own internalized standards of right or wrong.

Source: Based on L. Kohlberg, "A Cognitive-Developmental Analysis of Children's Sex-Role Concepts and Attitudes," in E. Maccoby (ed.), *The Development of Sex Differences*. Copyright ©1966 Stanford University Press, Stanford, CA.

Lickona (1983) has suggested a variation of the moral dilemma for assessing a child's level of moral reasoning. It also forces youngsters to test the limits of their reasoning.

Assume that you are working with an 8-year-old child. You pose this dilemma to the child: "What would you do if you found a wallet with $10 in it and you needed the money for a ticket to a skating party? But the wallet also has the owner's name and address in it." Question the child about returning the wallet: "You found it, so would that be stealing?" "Why should you return it?" "What if you don't get a reward?" Don't accept the answer of stealing. Ask, "Why is it stealing?" If the child says, "I might get into trouble," you have identified a stage 1 child.

As Lickona notes, if you vary your questions, you will discover the range of the child's moral reasoning. Knowing the upper limits gives you a yardstick for measuring the child's behavior against moral level and helps you to motivate the child to act at the appropriate level.

Not all students of moral development agree with Kohlberg. Strenuous objections have been made to Kohlberg's male interpretation of moral development, especially by Carol Gilligan.

An Applied View — Schools and Character Development

Because there's no way to really know what another person is thinking, we can identify good character only by watching and listening to children (Wynne, 1988). Therefore it makes sense for schools to encourage and reward the good conduct of their students. For example, the *For Character* program developed in the Chicago area publicly acknowledges the good conduct and academic efforts of its students, including the following character-building activities:

- Tutoring peers or students in other grades
- Serving as a crossing guard
- Acting as a student aide
- Acting as a class monitor
- Engaging in school activities
- Raising funds for the school or community
- Joining school or community projects

The program also provides opportunities for motivating students: public recognition through awards and ribbons presented at school assemblies and through mention over the school's public address system, and in school bulletins. Such recognition is given to individual pupils, groups of students, and entire classes. The relationship between academic learning and character development is mutually supportive because educators get back what they put into their work (Wynne, 1988, p. 426).

With teachers and students working together in this way to attain moral objectives, students should begin to develop the following characteristics:

1. *A sense of self-respect* that emerges from positive behavior toward others
2. *Skill in social perspective-taking* that asks how others think and feel
3. *Moral reasoning* about the right things to do
4. *Moral values* such as kindness, courtesy, trustworthiness, and responsibility
5. *The social skills and habits* of cooperation
6. *An openness* to the suggestions of adults

A major objective for those working with children should be to make them sensitive to their relationships with others, to recognize those times when a "helping hand" is needed.

In a Different Voice

Gilligan (1977, 1982) has questioned how accurate Kohlberg's theory is in relation to women. Gilligan believes the qualities associated with the mature adult (autonomous thinking, clear decision making, and responsible action) are qualities that have traditionally been associated with masculinity rather than femininity. (Kohlberg, for example, places most women at the third stage of his hierarchy because of their desire for approval, their need to be thought nice.) The characteristics that traditionally define the good woman (gentleness, tact, concern for the feelings of others, display of feelings) all contribute to their lower scores for moral development.

Noting that most women's moral decisions are based on an ethics of caring rather than a morality of justice, Gilligan argues for a different sequence for the moral development of women. For boys and men, separation from mothers is essential for the development of masculinity; for girls, femininity is defined by attachment to mothers. Consequently, women define themselves through a context of human relationships and judge themselves by their ability to care (Gilligan, 1982).

She notes that a woman's development is masked by a special interpretation of human relationships. A shift in imagery occurs. For example, while cognitive and moral theory clearly trace an 11-year-old boy's thinking, that same theory "casts scant light on that of the girl" (Gilligan, 1982, p. 25). When the 11-year-old boy, Jake, is confronted with the dilemma of the overpriced drug, he has no hesitation—"steal it." The response of the girl, Amy, is quite different. The husband should not steal the drug; rather he should seek other, legal ways of obtaining it. Consequently, Gilligan believes that a new interpretation of a woman's moral development is needed, one that is based on the imagery of a female's thinking.

As a result of her studies, Gilligan has formulated a developmental sequence based on the ethic of care. According to Gilligan, when the outline of women's morality is sketched, the results differ from Kohlberg's speculations about men. Gilligan has described a morality of responsibility based on a concept of harmony and nonviolence and a recognition of the need for compassion and care for self and others (Brabeck, 1983). Gilligan does not argue for the superiority of either the male or female sequence, but urges that we recognize the difference between the two. By recognizing two different modes, we can accept a more complex account of human experience.

Continuing to refine her ideas about the sequence of moral development for women, Gilligan, Lyons, and Hanmer (1990) note that adolescent girls face problems of connections: What is the relationship among self, relationships, and morality? Must women exclude themselves and be thought of as a "good woman," or exclude others and be considered selfish? The answer seems to reside in the nature of the connections that women make with others.

Whether boy or girl, middle childhood youngsters, through their rapid cognitive development, are aware of right and wrong. How do we know this? They write answers to questions; they talk to us. Language development is another clue to their growing maturity.

Language Development

In middle childhood we find children immersed in a verbal world. Language growth is so rapid that it is no longer possible to match age with language achievements. Table 9.4 presents several language accomplishments of middle childhood with other middle childhood characteristics.

Table 9.4 Some Typical Language Accomplishments

Age (years)	Language Accomplishment
6	Vocabulary of about 3,000 words Understands use and meaning of complex sentences Uses language as a tool Possesses some reading ability
7	Motor control improves; able to use pencil Can usually print several sentences Begins to tell time Losing tendency to reverse letters (*b, d*)
8	Motor control improving; movements more graceful Able to write as well as print Understands that words may have more than one meaning (e.g., *ball*) Uses total sentence to determine meaning
9	Can describe objects in detail Little difficulty in telling time Writes well Uses sentence content to determine word meaning
10	Describes situations by cause and effect Can write fairly lengthy essays Likes mystery and science stories Masters dictionary skills Good sense of grammar

Note the steady progression in motor skills: from the initial grasping of a pencil and printing to the writing of lengthy essays by the age of 10 or 11. Greater visual discrimination is apparent in the accurate description of events and the elimination of letter reversals. The growth in cognitive ability is seen in the detection of cause and effect, and the appeal of science and mystery stories.

Using Language

By the age of 7, almost all children have learned a great deal about their language. They appear to be quite sophisticated in their knowledge, but as Menyuk (1982) notes, considerable development is still to come. During the middle childhood years, children improve their use of language and expand their structural knowledge. By the end of the period, they are similar to adults in their language usage.

There are three types of change in language usage during these years (Menyuk, 1982):

- *Children begin to use language for their own purposes,* to help them remember and plan their actions. They move from talking aloud when doing something to inner speech. From about age 7 on, children use language to help them recall things. This applies not only to individual items (such as lists) but also to the relations between objects or actions (psychologists call this **encoding**).
- *Language during these years becomes less literal.* We saw the beginnings of this change in chapter 7, when we discussed children's humor. Now they use language figuratively. On going to bed, an 11-year-old may say, "Time to hit the sack." Children display this type of language by a process called **metalinguistics awareness,** which means their capacity to think about and talk about their language. You can see how this is impossible until children acquire the cognitive abilities that we just discussed.
- *Children are able to communicate with others more effectively.* They understand relationships; they can also express them accurately, using appropriate language. In a sense, more effective communication is the product of the interaction of many developmental forces: physical growth as seen in the brain's development; cognitive development as seen in the ability to use symbols and to store them; and language development as seen in vocabulary development and usage. Language has now become an effective tool in adapting to the environment.

In regard to changes in structural knowledge, most children, especially by the end of the period, begin to use more complexly derived words. Remember when we spoke of the acquisition of morphological rules? Adding *er* to *old* changes its meaning. Children begin to change word stems, which can produce syntactic changes as well.

I really like history.
I really like historical books.

They now understand the rules that allow them to form and use such changes.

The same process applies to compound words. What does your instructor write on? Usually it's a blackboard. Middle childhood youngsters realize that a blackboard in this meaning is not a black piece of wood. The awareness of relations helps them to learn that the same relationship can be expressed in different ways.

Liz slapped Janie.
Janie was slapped by Liz.

A MULTICULTURAL VIEW 

Language, Thought, and Immigrants

As we conclude our discussion of intelligence and thinking, we should pause and consider some of the characteristics of newly arrived immigrants, characteristics that could lead to labeling. We must remember that children entering the United States come into a culture in which almost all citizens speak only English. Consequently, language is an immediate barrier.

Educators have also come to realize that most immigrant students have difficulty with standardized tests for the following reasons:

- Language
- Difficulty in learning a new language if teachers can't use a student's native language for instruction
- English instruction may be too brief
- Any cultural bias in the tests (e.g., asking a math question using an American sport such as baseball)

As First (1988) has stated, if you're going to test immigrant students, at least teach them the things about which you're testing them.

> *Students who are learning a second language, adjusting to a new culture, or recovering from emotional trauma may need more than nine months to complete the learning associated with a given grade level.* (p. 208)

These are common concerns, but from the perspective of our work in this chapter, you can see the potential danger of labeling. Yet immigrants don't always have great difficulty. We are well aware that many Asian-American pupils have a high level of achievement. Nevertheless, immigrant children sound different; they may have difficulty in school; their marks may be low; they may do poorly on standardized tests. Conclusion: They're dumb.

From your reading in this chapter, you now realize that many of these youngsters face a formidable task, and that any judgment about their abilities should be suspended. Ask yourself and others a simple question: If you went to another country where you did not know the language and were unfamiliar with the customs, how would you do? Good luck!

As Menyuk (1982) states, much of children's language development during these years results from their awareness of language categories and the relationships among them.

During middle childhood the relationship of language development (in the sense of mastering a native tongue) to reading becomes crucial. Between the ages of 6 and 10, children must interpret written words wherever they turn: from signs on buses and streets as they go to school, to schooling that is massively verbal, which introduces the topic of bilingualism.

Bilingualism

Many children in the United States do not speak or write English as their primary language, and the number is growing. This situation is becoming more and more common in our schools. All evidence points to the conclusion that our country today is more diverse ethnically and linguistically than ever before. For example, while the country's population increased by 11.6 percent between the years 1970–1980, the Asian-American community increased by 233 percent, Native Americans by 71 percent, Hispanics by 61 percent, and African-Americans by 17.8 percent. Moreover, estimates are that at least 3.4 million pupils are limited in English language skills in a school system primarily designed for those who speak English (Lindholm, 1990).

The rights of language-minority students have come increasingly under analysis with specific implications for the schools and professionals who work in them. Some of the rights of these students are as follows (Garcia, 1990):

- There is a legally acceptable procedure for identifying all students who have problems speaking, understanding, reading, or writing English.
- Once these pupils are identified, there are minimum standards for the educational program that is provided for them. For example, some courts have noted that the teacher must have special training in working with these students, and that the students receive adequate time to acquire English skills.
- A school district is advised, but not compelled, to offer instruction in the student's native language as well as English.
- A school system may not deny services to a student because there are only one or a few students in the district who speak the specific language.

Many misconceptions have surrounded the educational progress of those pupils with an ability to speak two or more languages. For example:

- It is a misconception to assume that introducing a second language hurts the development of the pupil's primary language.
- It is a misconception to assume that bilingual children will become semilingual (going back and forth between the two languages) or confused in their language development.
- It is a misconception to assume that pupils will acquire languages easily. Older students learn syntactic and semantic aspects of a new language faster than younger pupils under appropriate conditions (McLaughlin, 1990).

In the past, a common practice has been to emphasize English and minimize the pupil's primary language, which has caused McLaughlin (1990, p. 74) to note:

> ***Educational programs that do not attempt to maintain the child's first language deprive many children of economic opportunities they would otherwise have as bilinguals. This is especially true of children who speak world languages used for international communication such as Spanish, Japanese, Chinese, and the like. If these children's first languages are not maintained, one of this country's most valuable resources will be wasted.***

When you consider a middle childhood youngster's competence—especially cognitive and linguistic—you conclude that they should be able to read effectively. The topics that engage reading researchers reflect the cognitive abilities of these children.

The Importance of Reading

In a widely read report, *Becoming a Nation Of Readers* (1985), reading was compared to the performance of a symphony orchestra. This analogy illustrates three points. First, like the performance of a symphony, reading is a holistic act. In other words, while reading can be analyzed into subskills such as discriminating letters and identifying words, performing the subskills one at a time does not constitute reading. Reading can be said to take place only when the parts are put together in a smooth, integrated performance. Second, success in reading comes from practice over long periods of time. Third, readers interpret text differently, depending on their background, the purpose for reading, and the context in which reading occurs.

In summarizing the reading research of the last decade, we can make several generalizations:

- *Reading is a constructive process,* which clearly implies that readers construct meaning from what they read (Mason & Au, 1990). The meaning that children glean from their reading also depends upon their previous experiences, which may be rich or deficient. A related problem here is that even if children possess relevant knowledge about what they are reading, they may not use it.

- *Reading must be fluent,* which means that children must be able to decode quickly and accurately. We'll discuss this shortly.

- *Reading must be strategic,* which means that good readers, for example, adapt their reading techniques to the difficulty of the text and the purpose of their reading (Mason & Au, 1990).

- *Reading requires motivation,* which demands that reading materials be interesting and teaching be innovative and challenging.

- *Reading is a continuously developing skill,* which suggests that instruction and materials must match the changing abilities and skills of children.

The relationship between reading ability and academic success is so well documented that you probably need few reminders of this truism. To give you an idea of the current reading status of our children, consider several specific findings of the *National Assessment of Educational Progress in Reading (1990).* Nine-year-old students were better readers in 1988 than students of the same age in 1971 (nevertheless, 7 percent of 9-year-olds in 1988 still could not do basic reading exercises, signalling future school failure). Thirteen-year-olds have shown little change in their reading abilities since 1971, and 17-year-olds show slight gains in reading proficiency from 1971 to 1988. African-American and Hispanic students made substantial improvements, but although African-American and Hispanic students have shown gains, more improvement is necessary because their scores are still substantially below that of same-age white students.

These concerns are mirrored in the kinds of studies occupying today's reading researchers. Among the topics are the following.

Decoding, a controversial and elusive topic, refers to the technique by which we recognize words. There are reading theorists who argue that children should focus on the whole word, and there are those who believe that individual letters must be taught—the phonics method. Research today indicates that early phonics instruction produces the most satisfactory results. Do middle childhood youngsters possess this ability? Absolutely.

Vocabulary, or word meaning, refers to teaching the meaning of a word, not how to pronounce it. Pause for a moment and consider the ramifications of this statement. Knowledge of vocabulary is highly correlated with intelligence, which is highly correlated with reading performance and school success. Is it any wonder, then, that the acquisition of vocabulary is high on any list of reading priorities?

Every effort should be made to instill a love of reading in children, both for academic success and the lifelong pleasure reading affords.

Reading comprehension, which is the ultimate objective in any type of reading instruction, means that a reader not only recognizes words but understands the concepts that words represent. Children of this age should have the capacity to understand the meaning of appropriate words. Wolf and Dickinson furnish a good example.

> ***Imagine if you will, a small group of children ranging in age from three to ten, huddled around a piece of paper with* die Katze *written in bold letters upon it. The youngest child squeals, "A picture! "A slightly older four-year-old child shouts, "Book!" A first grader quickly tries to pronounce it, then says slowly with a puzzled expression," It isn't a picture or a book. It's funny letters. I think it's funny words." The oldest child smiles knowingly and with the slightest disdain says, "It's a word all right, but it isn't in English. I'm pretty sure it means cat, but in Italian, maybe French, too."*** **(1985, p. 227)**

The 10-year-old knows this word is different. To reach this understanding, Chall (1983) believes, children pass through a series of stages:

- In the prereading stage, children up to 6 years of age learn letter and number discrimination and basics of reading.
- In stage 1 (grades 1.5–2.5), the major emphasis is on decoding as it applies to single words and simple stories.
- In stage 2 (grades 2.5–4), reading becomes much more fluent and understanding what is read becomes increasingly important.
- In stage 3 (grades 4–8), reading should be automatic and effort should be put into the comprehension of more complex material.
- Chall's stages 4 and 5 apply to adolescents and adults.

The goal of reading instruction during these years should be to enable children to control their own reading by directing attention where it is needed.

You can understand why reading is so important to children of this age. As they move through the school curriculum, their work becomes increasingly verbal: They read about the history of their country, the story of science, the symbols of mathematics, or any subject you can mention. A youngster who has a reading problem is a youngster in trouble. To the extent that reading problems increase, we all suffer.

Conclusion

From our discussion so far, we know several things about middle childhood youngsters. But are they capable of meeting the problems they face? Yes and no. They can assimilate and accommodate the material they encounter, but only at their level. For example, elementary school youngsters up to the age of 10 or 11 are still limited by the quality of their thinking. They are capable of representational thought but only with the concrete, the tangible. They find it difficult to comprehend fully any abstract subtleties in reading, social studies, or any subject.

According to Erikson, the middle childhood period should provide a sense of industry; otherwise, youngsters develop feelings of inferiority. With all of the developmental accomplishments of the previous six or seven years, youngsters want to use their abilities, which means that they inevitably experience failure as well as success, especially in their school work. The balance between these two outcomes decisively affects a child's self-esteem.

Physically active, cognitively capable, and socially receptive, much is expected of these children, especially in school. Their widening social horizons, with increasingly influential peer input, introduce joy and excitement but also stress and anxiety. We turn to these concerns in the next chapter.

Chapter Highlights

Physical Development

Motor Skills

- These are good years physically as middle childhood youngsters consolidate their height and weight gains.

Cognitive Development

Piaget and Concrete Operations

- Among the major cognitive achievements of this period are conservation, seriation, classification, and numeration.
- Children of this age show clear signs of increasing abstract ability.

The Psychometric View of Intelligence

- Binet and Wechsler gave great impetus to the intelligence testing movement.
- Children from other cultures can be unfairly penalized in taking tests because of cultural and language differences.
- Different views of intelligence have been proposed by Sternberg and Gardner.

Thinking and Problem Solving

Thinking Skills

- Bloom and Ennis have proposed helpful thinking skills programs.
- Children today need critical thinking skills to adapt to sophisticated, technological societies.

Decision Making and Reasoning

- During these years children's thinking skills come to approximate an adult's.

Problem-Solving Skills

- Good problem solvers have several observable characteristics.

Moral Development

Piaget's Explanation

- Piaget formulated a four-stage theory of moral development that is tightly linked to his explanation of cognitive development.

Kohlberg's Theory

- Kohlberg proposed six levels of moral development that follow a child's progress from about 4 years of age to adulthood.

In Another Voice

- Gilligan has challenged the male-oriented basis of Kohlberg's work.

Language Development

Using Language

- Children's language growth during these years shows increasing abstraction and facility in conversing with others.

Bilingualism

- Bilingualism is seen as a valued skill that should be analyzed and encouraged in our schools.

The Importance of Reading

- Children's reading ability has become a source of national concern.

Key Terms

Adaptation
Bodily-kinesthetic intelligence
Classification
Competition
Concrete operational period
Conservation
Conventional level of morality
Decentering
Decoding
Encoding
Interpersonal intelligence
Intrapersonal intelligence
Knowledge-acquisition components
Linguistic intelligence
Logical-mathematical intelligence
Metacomponents
Metalinguistics awareness
Moral dilemma
Musical intelligence
Numeration
Performance components
Postconventional level of morality
Preconventional level of morality
Reading comprehension
Seriation
Spatial intelligence
Stanford-Binet Intelligence Test
Triarchic model
Vocabulary
WAIS
WISC-R
WPPSI-R

What Do You Think?

1. Imagine an 11-year-old boy—let's call him Tom—who lives in the suburbs of a large northeastern city. He shows signs of becoming a great baseball player (a pitcher). His father realizes that his son could eventually win a college scholarship and go on to become a professional if he continues to develop. He decides that Tom should not play pickup games with his friends, because he might hurt himself. He also decides that the family should move to the South so Tom can play ball all year. Neither his wife nor Tom's sister wants to move. As a family friend, you have been asked for advice. What would you suggest?

2. You are a fourth-grade teacher and you turn to Janice and say, "Janice, Barbara is taller than Janie, who is taller than Liz. Who's the tallest of all?" Janice just looks at you. You sigh and think, "Where's Piaget when I need him?" How would you explain Janice's behavior?

3. Billy (9 years old) cuts through the parking lot of a supermarket on the way home from school. In the bike rack by the wall he sees a beautiful racing bike that he really wants. It seems to be unlocked and no one is around. With Kohlberg's work as a guide, what do you think is going through Billy's mind?

Suggested Readings

Gardner, H. (1983). *Frames of mind.* New York: Basic Books. Gardner's book, which is available in paperback, is an excellent example of the new perspective on studying intelligence. Recognizing that we all seem to demonstrate different abilities, Gardner makes an appealing case for the existence of several intelligences.

Lickona, T. (1983). *Raising good children.* New York: Bantam. This little paperback, based on Kohlberg's work, is one of the clearest and most practical guides to the development of children's moral reasoning.

Young, J. (1988). *Steve Jobs: The journey is the reward.* New York: Lynx. A penetrating account of the restless, inquiring mind of the middle childhood years that eventually led to the founding of Apple Computer.

CHAPTER 10

Psychosocial Development in Middle Childhood

Siblings and Development 264

The Developing Sibling Bond 264

How Siblings Help Each Other 265

How Siblings Affect Development 267

The Influence of Peers 268

Children's Friendships 269

Peers in Middle Childhood 270

Schools and Middle Childhood 272

Schools and Social Development 273

Middle Childhood and Educational Change 274

Our Multicultural Schools 276

Bilingual Education 277

Television and Development 278

Television and Cognitive Development 278

Television and Violence 280

Television and Prosocial Behavior 281

Stress in Childhood 282

Abused Children 283

The Nature of the Problem 284

Types of Stress 285

Developmental Effects of Stress 287

Children and Violence 287

The Invulnerable Child 290

Conclusion 293

Chapter Highlights 295

Key Terms 296

What Do You Think? 296

Suggested Readings 296

Tim Owens, who was in the sixth grade at the Brackett School, kept staring at the ground in the school's parking lot. Although all the students in the school were milling around, talking and trying to discover the exact location and cause of the fire, Tim stayed by himself. With a troubled expression, he watched the firemen quickly extinguish the small blaze. He knew who had set the fire.

Earlier in the day, Tony Larson, another sixth grader, had said to Tim, "Let's get something started around here. It's been pretty dull lately. B-o-r-i-n-g. Watch what happens in science today."

"What's up, Tony? You're not going to try anything in Evans' class, are you? (Mildred Evans was the sixth grade science teacher.) She's a pretty good person."

"What difference does that make? Let's liven things up. This place is really dullsville."

When the officials from the fire department gave the all clear and the students returned to the building, Tim walked by the science lab and saw the damage caused by the small fire. He looked around, worried; he was pretty sure he knew what had happened. He liked Mrs. Evans; she had always been fair to him.

Mildred Evans saw him standing at the doorway and said, "This is a tough one, Tim. The fire inspector said it was set. I can't believe anyone would do this. Think of the damage and what could have happened to some of the students. Who could possibly do something like this?"

Tim looked at the teacher. He knew that Tony Larson wasn't really bad; Tony just didn't think that others really liked him, so he was always looking for attention. "Why, he even told me something would happen, just to make sure somebody would know," thought Tim. "If I tell Mrs. Evans what I know, maybe she could straighten him out. But he could be in a lot of trouble, maybe even expelled."

Tim stood there, not sure what to do next.

Torn between loyalty to a friend and yet knowing that his friend's actions are wrong, Tim is faced with a dilemma that has strong implications for psychosocial development. If he decides to confide in the teacher, whom he likes, what will the rest of his friends think? Peer influence at this age is becoming a powerful motivator of behavior. Yet, as you realize from reading Piaget, a pupil of this age knows when something is wrong. Can Tim reconcile his behavior with what he thinks, and at the same time be trusted by his friends? Or will he compromise what he believes to be right to stay "in" with his friends? There are also moral implications, as we have just seen in chapter 9. ■

Loyalty to friends rapidly becomes a powerful influence during the middle childhood years, often causing emotional conflicts and moral dilemmas for children.

Consider the developmental accomplishments that we described in chapter 9. Children's skills—physical, motor, cognitive, linguistic—are beginning to flourish. Now they want an opportunity to demonstrate their prowess and to win recognition from others. But if their efforts meet only failure, the result can be feelings of inferiority. Their experiences with their families, friends, and in school should provide the opportunities for the further development of competence.

With these personal changes come expanding social contacts—perhaps the addition of brothers and sisters, new friends, new adult figures, and school with its challenges and achievements. All of these new contacts—siblings, friends, and teachers—become important influences on a child's development.

One type of positive social contact that a youngster of this age can make is in forming relationships with the elderly. Children 10 or 11 years old visit nursing homes where they find "writing partners," elderly citizens with varied cultural backgrounds and rich personal experiences. The partners write stories and poems that the children use in their language classes. These interpersonal relationships give the children insights into a group they rarely encounter (with the possible exception of grandparents). Such experiences help children to develop those prosocial behaviors so needed in our society: caring, sharing, a sense of responsibility, and a sensitivity to the needs of others. The children also help the elderly. With the enthusiasm and energy of these years, they bring interest and excitement to many whose lives are limited. Here we have an excellent example of what middle childhood youngsters can accomplish and give to others as they begin to move into a wider social context.

In tracing the impact that these different individuals have on development, it is perhaps wise to begin with those closest to home—siblings. We'll then turn our attention to peer influence. Following that, we'll examine those two great socializing agents: school and television. But we know that children are growing up in difficult times, so we should also consider how stress affects development.

These topics fit the biopsychosocial model as shown in Table 10.1. Note the appearance of several topics in each of the columns. Here is an excellent example of the value of the biopsychosocial model: It would be impossible to understand a topic such as stress, which has strong hormonal involvement, which also has psychological consequences, and which can affect relationships, by focusing on only one aspect of the topic, such as the biological. The biopsychological model encourages us to search for a variety of agents that can help to explain how any topic "works."

Table 10.1 Characteristics of the Period

Bio	Psycho	Social
Stress	Sibling influence	Sibling influence
Violence	Peer influence	Peer influence
Invulnerability	Friendship	Friendship
	School	School
	Television	Television
	Stress	Stress
	Violence	Violence
	Invulnerability	Invulnerability

After reading this chapter, you should be able to:

- Assess the influence of siblings on development.
- Analyze the role of peers in middle childhood development.
- Compare the different kinds of influence that schools exercise during the middle childhood years.
- Describe television's effect on the different aspects of development in middle childhood.
- Define the different kinds of stress that can affect children during this period.
- Distinguish the causes of violence in middle childhood youngsters.

Siblings and Development

***Children grow up within a network of relationships—with parents, grandparents, friends, and for 80% of children in the United States and Britain, siblings. How does this experience of being a sibling affect their development?* (Dunn, 1985, p. 787)**

Older siblings can help their younger brothers and sisters and ease many of the normal upsets in development. By being models, offering advice, and interceding with adults, siblings form bonds that survive distance and time.

With these words, one of the leading students of **sibling** relationships, Judy Dunn, emphasizes the importance of brothers and sisters in a child's development. Of one thing we can be very sure: Growing up with brothers and sisters is quite different than growing up without them. Brothers and sisters, because of their behavior toward one another, create a different family environment. As Dunn states (1988), the sibling relationship provides a context in which children demonstrate their varied abilities with frequency and intensity. During their early years, for example, siblings probably spend more time with each other than they do with their parents. An older sibling may spend considerable time taking care of younger brothers and sisters. These relationships last a lifetime, longer in most cases than those between husband and wife or parent and child (Dunn, 1988).

When you examine sibling relationships from this perspective, you can see how they affect development. In the early years, young brothers and sisters can provide security for babies who may feel frightened by strangers or anything different. (The attachment literature we discussed in chap. 5 shows that siblings definitely can attach to each other.) Older siblings are also models for younger children to imitate. They can become sounding boards for their younger brothers and sisters—that is, the younger siblings can try out something before approaching a parent. Older siblings often ease the way for the younger by trying to explain to parents that what happened wasn't all that bad. In this way, bonds are formed that usually last a lifetime (Hinde, 1987).

The Developing Sibling Bond

In their attempt to explain the sibling bond, Banks and Kahn (1982) begin with parents' conflicting guidelines for their children's relationships:

- Be close, but remain distinct individuals.
- Be cooperative, but retain your independence.
- Be admiring, but don't let your brothers and sisters take advantage of you.
- Be competitive, but don't dominate.
- Be aggressive, but not ruthless.
- Be tolerant, but hold on to your own point of view.

These authors believe that several cultural changes have combined to strengthen sibling relationships. Family size, for example, continues to shrink. In the common two-child family, the relationship tends to become intense, with each sibling exercising a strong influence on the other.

Longer lifespans mean that brothers and sisters spend a longer period of their lives together than ever before, as long as seventy or eighty years! When parents have died, their own children have left, and spouses die, siblings tend to tighten the bond and offer each other needed support (Lansdown & Walker, 1991).

Geographic mobility means unavoidable separation. Friendships are broken; schools and teachers change; adjustments to new situations must be made. For many children, the one anchor to be found is a brother or sister.

Divorce and remarriage typically bring unhappiness and hurt. Relationships change, which causes most children to experience emotions ranging from relief to fear. We have previously commented on the developmental effects of divorce, but pause for a moment and consider how relationships can multiply when divorced parents remarry (Banks & Kahn, 1982; Schibuk, 1989).

In what has become a classic analysis, J. P. Scott (1968) has illustrated how family relationships become more complex with each birth. With the addition of one child, the number of relationships increases to three. With the addition of a second child, the number of relationships increases to six, double the original figure. The first child adds two relationships; the second child adds three; the third child adds four. Figure 10.1 shows the increase in relationships with a second child. If a third child adds four new relationships to the family, the total increases to ten.

Third child	4
Second child	3
First child	2
Mother-father	1
Relationships:	10

For younger children, then, the social environment becomes increasingly complex.

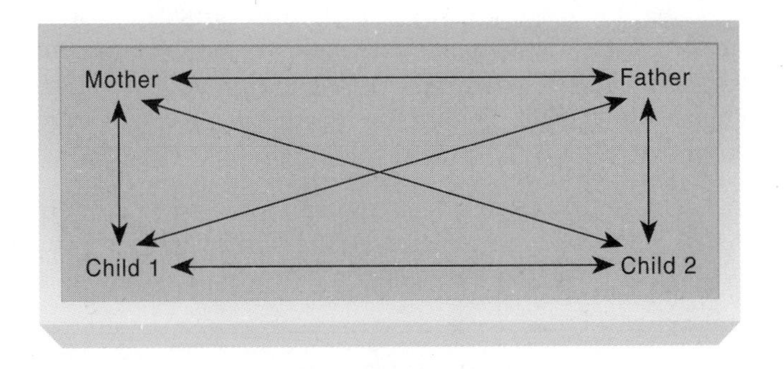

Figure 10.1
Increasing relationships with a second child

How Siblings Help Each Other

As you might conclude from what we've said about siblings, conditions may exist that link siblings in a lifelong relationship. After all, why not? Think of the functions that siblings perform for one another, functions that contribute to the cementing of the bond. Among them are the following (Banks & Kahn, 1982):

1. *Identification and differentiation.* Calling identification the glue of the sibling relationship, Banks and Kahn state that the process by which one youngster learns from a sibling's experiences is a powerful phenomenon.

Siblings perform many functions for each other. Not only do they directly help each other, but by observing older sibling's behaviors, younger siblings decide what to accept and what to reject, thus contributing to their own sense of identity.

Observing, imitating, and tentative trials, on the younger sibling's own terms (i.e., without parental pressure), can become an effective means of acquiring competence. Accepting some of an older sibling's behaviors and rejecting others leads to differentiation, an important and necessary step in developing a healthy self-concept.

2. *Mutual regulation.* Siblings can act as nonthreatening sounding boards for each other. New behaviors, new roles, and new ideas can be tested on siblings, and the reactions, whether positive or negative, lack the doomsday quality of many parental judgments. The emotional atmosphere is less charged. These simple experiments can give children confidence or prevent them from embarrassing incidents (Sroufe, Cooper, & DeHart, 1992).

3. *Direct services.* Cooperative siblings can ease many a burden. From the exchange of clothes as teenagers, to the borrowing of money before a paycheck, to support in life's crises, siblings provide valuable services for each other.

4. *Dealing with parents.* Sibling subsystems are the basis for the formation of powerful coalitions, often called the *sibling underworld.* Older siblings can warn their younger brothers and sisters about parental moods and prohibitions, thus averting problems. Older siblings frequently provide an educational service to parents by informing them of events outside the home (Eisenberg, 1992).

We also know, however, that sibling relationships can be negative. *Rivalry,* whatever the cause, may characterize any bond. An older sibling can contribute to those feelings of inferiority that Erikson so elegantly describes as contributing to the crises of this period. Imagine the difficulty of a first-born sibling forced to share parental attention, especially if the spacing between the children is close (less than two years). If the first-borns must also care for younger children, they can become increasingly frustrated.

The causes for rivalry are not one-sided. Younger children may only see the apparent privileges that are extended to the oldest: a sharing in parental power, authority, and more trivial matters such as a later bedtime or greater use of the television set. As Eisenberg notes (1992), siblings who help each other, share things, and frequently cooperate still have conflicts. Yet the bond remains.

In fact, sibling discord is one of the problems most frequently reported by parents. Researchers now realize that there are multiple causes of sibling conflict (Brody & others, 1992). For example, marital quality and conflict and the family's emotional quality seem to be related to sibling conflict. Studying 152 white children, Brody and his colleagues (1992) attempted to determine whether the way parents treated trouble between their children had any long-term consequences on sibling conflict. They found that sibling conflicts were reduced by family harmony during discussions about sibling problems and by the father's impartiality.

How Siblings Affect Development

In an early and careful investigation, Koch (1960) found that school-age children have unique attitudes toward their siblings. For example:

- Some said they played frequently with their brothers and sisters; other rarely did.
- About one-third of the children said they fought constantly; another one-third said they seldom quarreled.
- Some said they liked playing with a sibling; others much preferred a friend.

When Koch specifically asked the youngsters if they would be happier without their siblings, she received definite answers. One-third replied they would be happier without the sibling. (Could you make her disappear? She's too bossy.) The majority, however, said they preferred keeping the sibling (although they phrased their answer in less than glowing terms: "I'll keep him. He's bad but not that bad.").

Their replies reflected many of the functions provided by siblings—help, money, support. Those who preferred life without a sibling commented on conflicts, bossiness, and abuse. ("He makes me cry." "She's so mean.") The emotional quality of the relationship was apparent in both the positive and negative responses, much more so than when referring to anyone else. Commenting on these findings, Dunn states:

> ***The children talked about affection, comforting, and helping, but also about antagonism and quarreling. And it is striking that these different qualities of the relationship were not closely linked. Children who described their relationship with a brother or sister as very warm, close and affectionate, for instance, were not necessarily those children who experienced little conflict with the sibling or who expressed little rivalry with each other. And the children who fought a great deal with their siblings were not necessarily the children who reported much jealousy about the parents.*** **(1985, p. 51)**

Analyzing how children perceive their relations with their brothers and sisters, Furman and Buhrmester (1985) conducted two studies that attempted to discriminate the quality of these relationships. In the first study of 49 fifth and sixth grade children (using individual interviews), they developed a list of the primary qualities of sibling relationships. They discovered that the most common positive qualities mentioned were companionship (93 percent of the children identified this quality), admiration (81 percent), prosocial behavior (77 percent), and affection (65 percent). The most negative qualities mentioned were antagonism (91 percent) and quarreling (79 percent).

In their second study, they developed a self-report questionnaire to assess children's perceptions of their sibling relationships. Their subjects were 198 fifth and sixth grade children. They found that the questionnaire assessed the nature of the children's interactions in many different social contexts. Four dimensions were particularly significant:

- *Warmth/closeness.* Children felt closer to same-sex siblings, although this finding became less significant as the ages between siblings increased.
- *Relative status/power.* Age had a strong effect on perceptions of status and power. The older siblings reported greater nurturance of and dominance over younger members.
- *Conflict.* Age was a significant factor here. There was more conflict between narrow-spaced siblings than wide-spaced pairings. Children also perceived older siblings of the same sex to be more dominant than older siblings of the opposite sex.
- *Rivalry.* Children reported greater rivalry and parental partiality when siblings were younger. The authors believe that the attention and treatment by others outside of the sibling pair may be more important than in other types of relationships.

What specific developmental implications can we draw from our discussion?

- As with our previous analysis (See chap. 5), the early *affective* quality of the relationship persists through the years. The content of the interactions between siblings will obviously change throughout the years, but its affective quality remains consistent.
- Younger siblings tend to *imitate* older brothers or sisters. Which children imitated gives insight into the relationship. Second-borns imitate most frequently, especially if the first-born had been affectionate. Same-sex pairs imitate each other more frequently than mixed-sex pairs (Dunn, 1985).

Middle childhood youngsters will bring to their interactions with those outside of the family the characteristics that they formed within the family circle. With this in mind, let's turn now to the impact of peers on development.

The Influence of Peers

We typically use the word **peers** to refer to youngsters who are similar in age, usually within 12 months of each other. But as Hartup (1983) notes, equal in age does not mean equal in everything—intelligence, physical ability, or social skills. Also, research shows that many of a child's interactions are with those who are more than 12 months older, although we know little about the nature of these relationships, (Hartup, 1989).

With these cautions, we turn now to the influence of peers during middle childhood. (Here you may want to return briefly to chapter 5 and the analysis of relationships. We'll assume that you understand the basics of a relationship at this point.)

When we turn to same-sex—and mixed-sex—interactions, we can summarize the obvious findings quickly. Children of all ages associate more frequently with members of their own sex. Why? Adults encourage such relationships. Children of the same sex also share more mutual interests, and gender-role stereotypes operate powerfully to reinforce same-sex relationships.

Children's Friendships

In his engaging book, *Children's Friendships,* Rubin (1980) states that the word *friend* reflects the common functions of peer relationships for people of all ages. It refers to nonfamilial relationships that are likely to foster a feeling of belonging and a sense of security. How youngsters think about friends, of course, changes with age.

During middle childhood, peer relationships become increasingly important for social development. Children are attracted to those who share their interests, who play well with them, and who help them to learn about themselves.

Middle childhood youngsters, with the abilities that we have traced, can reach logical conclusions about their friends. Children of this age search for friends who are psychologically compatible with them. For example, does Jimmy share my interests? Does he want to do the same things I do? Children begin to realize, especially toward the end of the period, that friends must adapt to each other's needs.

Children who feel rejected by their peers frequently are plagued by problems. For example, in studying 452 five- to seven-year-old children, Cassidy and Asher (1992) found that children who believed that they had few friends were troubled by feelings of loneliness. Even the younger children in the study recognized that they had peer relationship problems and experienced unhappiness with their rejection. This study, with its focus on loneliness, testifies to the significance of peer relations to normal social development.

These findings were confirmed by Morison and Masten's (1991) longitudinal study of peer reputation as a predictor of later adaptation. Believing that the ability to get along with peers is a critical developmental task of middle childhood, these authors studied 207 third to sixth graders, and seven years later restudied 88 percent of the group. They found that peer reputation assessed in middle childhood was a significant predictor of adolescent adjustment. Disruptive, aggressive, sensitive, and isolated behavior during the early childhood years were associated with antisocial, incompetent behavior in adolescence.

Rubin (1980) believes that friends provide certain resources that adults can't provide. For example:

- *Friends offer opportunities for learning skills.* They teach children how to communicate with one another, which gradually leads to the ability to recognize the other person's needs and interests. (We'll discuss this shortly with Selman's work.) Friendship also means that a child has to learn how to cooperate and how to deal with conflict.
- *Friends give children the chance to compare themselves to others.* "I can run faster than you" is not just a competitive statement. It's also a means of evaluating oneself by comparison with others.
- *Friends give youngsters the chance to feel that they belong to a group.* By age 10 or 11, groups have become important, and, as we saw earlier in our discussion of gender, sex cleavage is the rule. Here children find a social organization that includes not just individual friendships but roles, collective participation, and group support for activities. You can see, then, how being included in a group can further development, and how feelings of isolation and self-doubt can come from being excluded. Friendships are one means of traveling the normal path of social development.

Peers in Middle Childhood

Think of the world these children are now encountering. Physical and cognitive abilities enable them to move steadily, although slowly, toward others such as neighborhood friends. Upon entering school, they increase their contacts and begin to realize that other children have ideas that may differ from their own. The manner in which they react to these different opinions may reflect their own home conditions. Cassidy and her colleagues (1992) discovered that children whose parents are warm, responsive, and consistent disciplinarians have children who are more competent with peers than children whose parents are harsh and rejecting or overly permissive. Cognitive development helps middle childhood youngsters to accept these differences. For Piaget, one of the major obstacles to more mature thinking is egocentrism, that tendency to relate everything to me. With its decline during middle childhood, children gradually see that other points of view exist. This in itself is an important developmental phenomenon, one that has been carefully explored by Robert Selman (1980).

In his efforts to clarify emerging interpersonal relationships, Selman has developed a theory of **social perspective-taking** levels that springs from a social cognitive developmental framework. Selman (1980) states that you can't separate children's views on how to relate to others from their personal theories about the traits of others. Thus, children construct their own version of what it means to be a self or other.

As a result of careful investigations of children's interactions with others, and guided by such theorists as Piaget, Flavell, Mead, and Kohlberg, Selman has identified several levels of social perspective-taking. Table 10.2 presents a summary of thinking at each level and how it affects a child's perception of friends. Note that in the middle childhood years a youngster gradually realizes that those out there are different and have ideas of their own. By the end of middle childhood, a youngster's views of a relationship include self, someone else, and the kind of relationship between them.

Table 10.2 Selman's Theory of Interpersonal Understanding

Level 0: 3–6 years Undifferentiated and Egocentric *Concept of persons:* undifferentiated—does not separate physical and psychological characteristics of persons *Concept of relations:* egocentric—no accurate notion of relations	**Friendship** Friendship depends on physical closeness and functional similarity; admires strength and quickness.
Level I: 5–9 years Differentiated and Subjective *Concept of persons:* differentiates physical and psychological characteristics—intentional acts recognized *Concept of relations:* seen as one-way	**Friendship** Someone does what child wants or child does what other wants—implies recognition of an inner self.
Level II: 7–12 years 2d Person and Reciprocal *Concept of persons:* can look at self objectively and realize that others can too *Concept of relations:* reciprocal in that children realize that others do what they do (i.e., I know that she knows that I know)—sees people this way but not relationships (i.e., not mutual)	**Friendship** Interactions become desirable in themselves—a "meeting of the minds"—but only for specific interests. Still sees interactions as helping self.
Level III: 10–15 years 3d Person and Mutuality *Concept of persons:* 3d person—self and others as subjects and objects; can have mixed thoughts and feelings about something (love and hate) *Concept of relations:* 3d person view of self, others, and system; looks on interpersonal interactions as including self, others, and the relationship	**Friendship** Goal is mutual interest and sharing.
Level IV: 12+ years In-depth and Societal-Symbolic *Concept of persons:* individual seen as complicated, many things going on inside *Concept of relations:* interactions and relationships become complicated because they may reflect deeper levels of communication	**Friendship** Realizes that complex needs can be met by different relationships. Relationships are seen as open and flexible—helps in own self-identity.

We have mentioned cognitive development, diminished egocentrism, and a striving for competence as factors in getting along with others. Youngsters of this age are also better able to communicate with one another and to use reinforcements from their peers to shape their own behavior (Boivin & Begin, 1989). Because of their cognitive, language, and perspective-taking skills, they cooperate better with one another than younger children can, and aggression decreases somewhat. The desire to conform becomes important, especially at the end of the period.

Given the increase in friends during the school years, we can question the role of schooling itself. How does it influence the development of middle childhood youngsters?

Schools and Middle Childhood

Do schools affect development? Perhaps the best way to answer this question is to examine the goals presented in the **America 2000** statement, which includes the national goals adopted by former President Bush and the governors in 1990. The America 2000 strategy is defined as follows:

School is an important milestone for all aspects of development. Children must learn to respond appropriately to authority outside of the family and to get along with peers. It is an important part of psychosocial as well as cognitive development.

> ***An action plan to move America toward the six national goals through a populist crusade, by assuring accountability in today's schools, unleashing America's genius to jump-start a new generation of American schools, transforming a "Nation at Risk" into a "Nation of Students," and nurturing the family and community values essential to personal responsibility, strong schools, and sound education for all children.*** **(p. 25)**

The proposal is a nine-year plan to move toward the following educational goals:

1. ***All children in America will start school ready to learn.***
2. ***The high school graduation rate will increase to at least 90 percent.***
3. ***American students will leave grades four, eight, and twelve having demonstrated competency in challenging subject matter including English, mathematics, science, history, and geography; and every school in America will ensure that all students learn to use their minds well, so they may be prepared for responsible citizenship, further learning, and productive employment in our modern economy.***
4. ***U.S. students will be first in the world in science and mathematics achievement.***
5. ***Every adult American will be literate and will possess the knowledge and skills necessary to compete economically and exercise the rights and responsibilities of citizenship.***
6. ***Every school in America will be free of drugs and violence and will offer a disciplined environment conducive to learning.*** **(p. 9)**

These goals have powerful implications for a child's total—not only educational—development. For example, if all American children are to start school ready to learn, then they must be ready physically—well-nourished, free of illness, unencumbered by the trauma of abuse, conflict, and insecurity. Learning to use their minds well means that they have received, are now receiving, and will continue to receive the vital stimulation needed for optimal cognitive development. In many ways, these national educational objectives are also developmental goals.

Schools and Social Development

To achieve the America 2000 goals, we must recognize that schools are different social contexts at preschool, elementary, and secondary levels. They are organized differently, children perceive them differently, and different aspects of social behavior appear to meet pupils' changing needs (Minuchin & Shapiro, 1983). Preschool experiences are more protective and caring than educational. The children interact with one or two teachers, perhaps an equal number of aides, and several peers.

During the middle childhood years, the elementary school classroom becomes more of a true social unit, with more intense interactions between teacher and pupil and among peers (Travers, 1982). Teachers, as authority figures, establish the climate of the classroom and the kinds of relationships permitted. Peer group relationships stress friendship, belonginess, and status (Adalbjarnardottir & Selman, 1989).

Two different types of interactions can be identified, both of which shape the direction of a middle childhood youngster's growth. First is the relationship with the teacher: usually intense, goal-directed, and subject to evaluation. Second is the relationship with peers, which opens up a new world. When you examine these relationships objectively, the school's role in development for children of these years seems critical.

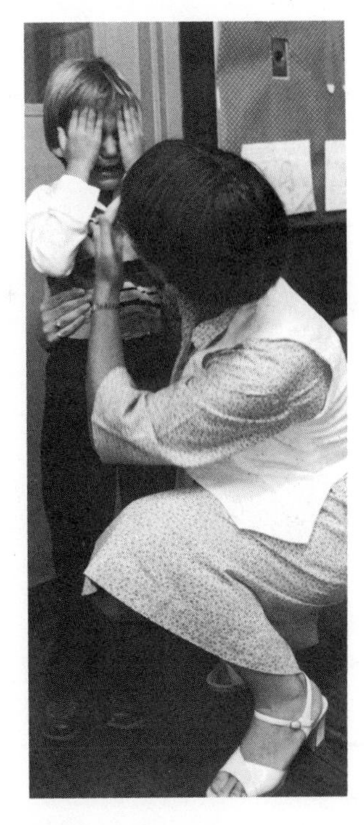

The first day of school is stressful for youngsters. For most children, however, this upset to their psychological equilibrium passes and they adapt quickly.

Think now of the many aspects of development that we have mentioned: the inborn temperamental disposition of children, the pattern of child-rearing behavior parents adopt, the relationships within the family. All of these determine how a child reacts to what happens in a classroom. Does it match the child's needs?

- During the kindergarten and elementary school years, pupils are being socialized. They are learning to respond to teachers and get along with their peers—and being taught the basic skills. Discipline is typically not a major concern, since youngsters of this age usually react well to authority and seek teacher praise and rewards. Adjustment to the school as a major socializing agent and mastery of the fundamentals are the two chief tasks of these years (Hetherington & Parke, 1986).

- Pupils in the middle elementary school grades know a school's routine and have worked out their relationships with their peers. They must concentrate on curricular tasks in a clearly defined classroom atmosphere (Minuchin & Shapiro, 1983).
- The upper elementary years are a time when peer pressure mounts and most youngsters are concerned with pleasing friends. Teachers are authority figures and more challenging to students. Classroom control becomes more of an issue. Children should have mastered the basics and to an extent can function independently. Classroom procedures and rules should be distinct, understandable, and fair (Wittrock, 1986).

Middle Childhood and Educational Change

As children pass through the elementary grades, they will encounter constant change as educators strive to devise methods that will meet the goals of America 2000. For example, a major change in the teaching of reading and writing is a concept known as **whole language.** Rather than teach phonics isolated from meaning, students learn to read by obtaining the meaning of words from context, with phonics introduced as needed. If, while reading a story, a pupil has difficulty with the word *dish,* the teacher stops and sounds it out. Students don't use basal readers; they read appropriate-level literature about themes that interest them and then write about these ideas. Teachers who have begun to use this new technique believe that it motivates their pupils better than the older methods. Not everyone agrees with this approach, however.

International comparisons of students on mathematics achievement tests have repeatedly found Americans at or near the bottom. As a result, the country will soon experience another wave of publicity about a revision of the mathematics curriculum. You may have heard about the "new math" of the 1960s, followed a few years later by the "back-to-basics" movement. One of the reasons that the new math was not an unqualified success was that public school teachers had little to say about it. It was simply imposed on them.

Today's emphasis is less on skills for their own sake and more on thinking about and understanding the meaning of numbers. For example, the mathematician J. Paulos (1988) quotes a couple as saying they're not going to Europe because of all the terrorists. In 1985, 17 of the 28 million Americans who traveled abroad that year were killed by terrorists. That same year, 45,000 people were killed on American highways (that is, the couple had 1 chance in 5,300 of being killed in a car crash). Understanding the numbers involved helps you to evaluate which situation contains the greatest potential danger. The National Council of Teachers of Mathematics recommends that students use calculators at all times, and urges teachers to emphasize problem-solving skills and the practical side of mathematics. The goal is to make mathematics seem less threatening and more useful.

When we turn to science, estimates are that although most of us shake our heads at the "scientific illiteracy" of our youth, less than 10 percent of high school graduates still have the skills necessary to perform satisfactorily in college-level courses. Attempting to combat this trend, many science educators are today turning to a more "hands-on" approach to their teaching. Instead of having their students memorize lengthy formulas, they have them do experiments, starting in the early grades. For example, instead of reading about the principle of buoyancy, students make lumps of clay into various shapes, put them in plastic bags, and discover which shapes float and which sink.

You may argue that there's nothing new in this technique; good teachers have been doing it for years. There are differences, however. Where this approach has been successful, teachers have acted as facilitators, not directors. Teachers are not forced to teach a specific amount of material; in a sense; teaching less can result in teaching more. That is, by teaching generalizable problem-solving strategies along with the concepts of basic subject matter, and by emphasizing that learners should know themselves, teachers can prepare students for a lifetime of learning. Also, these school systems have been totally committed to scientific discovery from the elementary grades through high school.

Consider this possible future scenario. You are meeting with the teacher of one of your children, and you are told that your child should be retained in third grade. How would you react? Making pupils repeat a year's work has come under heavy attack recently, with opponents claiming that it usually doesn't work. After reviewing studies comparing the education of students who were retained with students of comparable achievement and maturity who were promoted, Holmes (1990) concluded that retained students were no better off than those who went on to the next grade. **Grade retention,** either because of immaturity or lack of achievement, is a common practice in our schools that raises many questions (Medway & Rose, 1986):

- Does grade retention produce academic achievement superior to that found in comparable students who are promoted?
- Do students who have been retained drop out of school more frequently than comparable students who were promoted?
- Does a policy of retention discriminate against particular groups of students?
- What evidence does a school use in its decision to retain?
- Does eliminating grade retention mean a return to a policy of social promotion?
- What are the legal ramifications of grade retention?

Although evidence is accumulating that retention has not been a uniformly successful policy, the issue today is widely debated.

Our final change to discuss is **homework,** which is once again enjoying renewed acceptance. (Homework usually refers to school-assigned academic work that is to be completed outside of school, usually in the home.) At the turn of the twentieth century, homework was considered vital; its popularity declined in the 1940s, reemerged in the 1950s (after Sputnik), fell into disfavor in the 1960s because it was seen as a form of useless pressure, and now, with reports of the poor achievement of American students, is once more viewed as essential. Research shows that for high school students two or more hours of homework increases achievement, junior high school students benefit from one to two hours of homework, and there seems to be a slight relationship between homework for elementary school pupils and improved achievement (Cooper, 1989). Homework at the elementary school level, however, brings home and school closer together and encourages pupils to realize that they can learn on their own. Homework should not be a burden for students and their parents, but should be assigned to meet demonstrated needs.

Thus, with growing maturity, middle childhood youngsters face greater demands and higher expectations. From our brief discussion, you can see how the schools should contribute to the youngster's sense of competence, which Erikson identifies as the psychosocial strength of these years. Competence is not confined to academics, but extends into the physical and social worlds, helping middle childhood youngsters develop a needed self-confidence.

If the challenges to these youngsters are great, so are the opportunities. One opportunity for social growth comes with the charge to American schools to educate all the children of all the people (Rutter & Garmezy, 1983).

Our Multicultural Schools

In our pluralistic society, children meet pupils from many different cultures. We know that cultural differences in the classroom can affect both achievement and adjustment. Consequently, an important goal for our schools is to prepare children to enter the larger society and develop positive relationships with those from different cultures. The focus of multicultural education is on helping youngsters develop positive gender, racial, cultural, and class identities and to accept others from different cultures (Ramsey, 1987).

In attempting to avoid misunderstandings that could affect relationships among children, our schools have emphasized several means of integrating youngsters from different backgrounds (Ramsey, 1987).

As cultural differences occur in any society, these changes are reflected in the classroom and schools should act to encourage positive relationships among children of all cultures.

Student Learning Style

Good examples of student learning style can be found in the literature on various cultural groups. For example, studies of Hispanic students have shown that these students tend to be influenced more than other students by personal relationships and praise or disapproval from authority figures. Pueblo Indian children from the American Southwest show higher achievement when instruction utilizes their primary learning patterns (those that occur outside the classroom). These pupils respond well when instruction incorporates the concerns and needs of the community in a global manner. That is, they learn more effectively when a subject, such as mathematics, is taught in a more applied manner—for example, as it is related to specific community needs (Dunn, 1988).

Recognition of Dialect Differences

If dialects hinder academic progress, how should the schools attempt to remedy the problem? Should they use standard (proper) English with these students and risk alienating them or causing them to reject their cultural heritage? Or should English be taught as a foreign language?

Unfortunately, research offers no clear guidance. Students do not seem to gain academically if initially taught in their own dialect. Also, many parents believe that their children should be taught in standard English so that they can master the language skills for future success. Perhaps the best advice here is to adapt to the community's wishes as much as possible. The key to a school's success in teaching these pupils is to establish and maintain close contact with parents.

Bilingual Education

In a landmark decision in 1974 (Lau v. Nichols), the U.S. Supreme Court ruled that LEP (Limited English Proficiency) students in San Francisco were being discriminated against since they were not receiving the same education as their English-speaking classmates. The school district was ordered to provide means for LEP students to participate in all instructional programs. The manner of implementing the decision was left to the school district under the guidance of the lower courts. This decision provided the impetus for the implementation of bilingual education programs in the United States.

Two different techniques for aiding LEP pupils emerged from this decision. The *English as a second language program* (ESL) usually has students removed from class and given special English instruction. The intent is to have these pupils acquire enough English to allow them to learn in their regular classes that are taught in English. With the *bilingual* technique, students are taught partly in English and partly in their native language. The objective here is to help pupils to learn English by subject matter instruction in their own language and in English. Thus, they acquire subject matter knowledge simultaneously with English.

In today's schools, bilingual education has become the program of choice and it's important to remember that programs for LEP students have two main objectives:

- provide these students with the same education that all children in our society have.
- help them to learn English, the language of the school and society.

Bilingual education programs can be divided into two categories. First are those programs (sometimes called "transitional" programs) in which the rapid development of English is to occur so that students may switch as soon as possible to an all-English program. Second are those programs (sometimes referred to as "maintenance" programs) that permit LEP students to remain in them even after they have become proficient in English. The rationale for such programs is that students can use both languages to develop mature skills and to become fully bilingual.

The difference between these two programs lies in their objectives. Transitional programs are basically compensatory-use the students' native language only to compensate for their lack of English. Maintenance programs, however, are intended to bring students to the fullest use of both languages. As you can well imagine, transitional programs are the most widely used in the schools.

Bilingual programs allow students to retain their cultural identities while simultaneously progressing in their school subjects. It also offers the opportunity for pupils to become truly bilingual, especially if programs begin early.

Television and Development

Do you have a television set in your home? Did you laugh at what appears to be a ridiculous question? You probably have at least one, more likely two or three. When you consider that American homes have at least one set and realize the colorful appeal of TV, you can better understand why television has become the school's great competition (Pinon, Huston, & Wright, 1989).

In the Surgeon General's report on television viewing, *Television and Behavior: Ten Years of Scientific Progress and Implications for the Eighties* (1982), one of the questions asked was, who watches television? The answer was simple: almost everyone. Elementary school children watch at least four hours each day. Today television viewing may be the most frequently shared activity among family members (St. Peters & others, 1991).

The beginning of school attendance slows the time spent watching, but at about age 8 the rate increases dramatically. What is particularly interesting is the match between children's ages and program content. For example, Comstock and Paik (1991) report that the popularity of Sesame Street declined between the ages of 3 to 5 (from 30 percent to 13 percent). The Flintstones rose in popularity in these same years, from 11 percent to 36 percent. Are there developmental effects from all this viewing?

The 1982 Surgeon General's report reflects the pattern of development discussed thus far. Babies are briefly attracted by the color and sound; 2- and 3-year-olds watch longer and with some understanding; elementary school children watch for long periods; and teenagers spend less time watching television. These viewing habits are fairly well established. Specifically, the report presented several findings related to development.

Television and Cognitive Development

The moment we concede that children learn from watching television, certain questions arise:

- How active are children in the process?
- To what do they attend?
- How much do they understand?
- How much do they remember?

Answering these questions gives us insight into how TV watching and cognitive development are associated. For example, when you studied Piaget's work, especially his views on operations, you saw how Piaget insisted that children were active participants in their cognitive development. They actively construct their cognitive world.

The same is true of their TV watching. They bring a unique set of cognitive structures to the TV set, structures that reflect their level of cognitive development. Remember what we said about middle childhood youngsters: attention span lengthens, memory improves, and comprehension increases. Children don't drop these

AN APPLIED VIEW — The Extent of Television Watching

For an idea of the role that television plays in our society, try to answer these questions, which have been drawn from several national surveys:

1. What percentage of American homes have a TV set?
2. How long (on the average) is the set on per day?
3. By age 85, how many years of television has the average viewer seen?
4. How many hours of TV does the average viewer watch per week?
5. By age 18, a student has watched how many hours of TV?
6. By age 15, how many killings has a child seen on television?
7. Can television influence a child's behavior?

If you had any doubts about the extent of television viewing, these figures should dispel them.

1. 99%
2. 6 hours
3. 9 years
4. 28 hours
5. 15,000
6. 13,000 killings
7. yes

Televsion has become one of the great means of socialization in a child's life. As such and recognizing the relationship between program content and a child's age, adults should be particularly careful in what children watch.

abilities when they watch TV; they apply them to what they are seeing (Liebert, Sprafkin, & Davidson, 1988). Specifically, we know that:

- *Children remember what is said,* even when they are not looking at the screen. Auditory attention is also at work (voices, sound changes, laughing, and applause).
- *The amount of time spent looking at the set is directly related to age.* By age 4, children attend to TV about 55 percent of the time, even when there are many other distractions in the room.

- *Specific features of programs attract children:* women, movement, and camera angles; they look away during stills and animal shots.
- *Children quickly learn to relate sound to sight:* chase music means a chase scene. (Note the ideal combination of auditory and visual effects that produces powerful attractions.)
- *Comprehension* depends on age and experience (Comstock & Paik, 1991).

This last finding requires additional comment. To understand television, children need three accomplishments:

- They must know something about story form—how stories are constructed and understood.
- They must have world knowledge or general knowledge about situations and events in order to grasp television's content.
- They must have knowledge of television's forms and conventions to help them understand what is happening on the screen. Music, visual techniques, and camera angles all convey information (Liebert & others, 1988).

Consideration of these three requirements helps to put children's viewing in perspective: They simply lack the maturity and experience to grasp fully much of what they are watching. For example, they have the perceptual skills to see and recognize a car moving away. But then the camera may cut to another scene (the sky, a police officer, or the corner of a house). The significance of the cut introduces another theme, embedded in the story, that completely eludes them.

We can conclude, then, that much of what children see on television is not just content. They also learn TV's codes: sound effects, camera techniques, and program organization. Some researchers believe that changes in children's behavior following the viewing of televised violence come from their responding to fast action, loud music, and camera tricks. Children understand television programs according to their level of development, and their level of development is affected by their television viewing.

Television and Violence

If we were to summarize the Surgeon General's report (1982), we would say:

BE CAREFUL OF WHAT CHILDREN WATCH; TELEVISION MAY BE HARMFUL TO THEIR HEALTH.

The key word here is *may.* While research and theory point to television's role in aggressive behavior, there is also almost total agreement that it is *a* cause, not *the* cause of aggressive behavior (Comstock & Paik, 1991). After decades of research, we can safely conclude that televised violence causes aggressive behavior in the children who observe it (Parke & Slaby, 1983). Are children exposed to much television violence? Think of what we've said so far:

- Almost every American home has at least one television set.
- Children watch for many hours each day.
- Much of television's programming contains violence as a common feature. Over 70 percent of all prime-time dramatic shows contain violent scenes.
- Most children watch television with few, if any, parental restrictions. (Our concern here is with middle childhood youngsters who tend to watch adult shows.)

What's Your View?

We have repeatedly noted how different theorists (especially the major theorists discussed in chap. 2) can interpret the same data differently. Television violence is a good example. Two theories have been widely used in attempts to understand television's impact on children: cognitive and behaviorist.

Cognitive theorists, reflecting Piaget's views, believe that children understand what they see according to their level of cognitive development. The cognitive structures that children form can be altered by what they see on television. Children attend to what they see; they learn from what they see; and they remember what they see. They can also apply these new ideas in new settings. Since we have discussed cognitive theory in detail, let's turn to the behaviorists.

Behaviorists, especially in the social learning theory of Bandura (1986), offer a different interpretation. Bandura believes that considerable evidence exists to show that learning can occur by observing others, *even when the observer doesn't reproduce the model's responses.* Referring to this as **observational learning,** Bandura states that the information we obtain from observing other things, events, and people influences the way we act.

On what does Bandura base his conclusions? With his colleagues (Bandura, Ross, & Ross, 1963), Bandura conducted a now-famous experiment. Preschool children observed a model displaying aggression toward an inflated doll under three conditions: In one situation the children saw a film of a human model being aggressive toward the doll. In the next, children witnessed filmed cartoon aggression. Finally, live models exhibited the identical aggressive behavior. The results: Later, all children exhibited more aggression than youngsters in a control group!

Which theory do you think offers the best explanation? What's your view?

- Boys prefer to watch TV violence more than girls.
- Younger children (up to about the third or fourth grade) are less likely to associate violence with motivation and consequences.

Grouping these facts has caused Parke and Slaby (1983) to reach two conclusions:

- Televised violence *increases* children's level of aggression.
- Televised violence *increases* children's passive acceptance of the use of aggression by others.

Can the American public force television producers to decrease the level of violence in their programming, a level that slowly but steadily seems to be increasing? For many reasons, ranging from the appeal of violence to issues of free speech, this is a difficult question to answer. But all responsible adults can take one step with far-reaching consequences: Be alert to what children are watching, because television affects development.

Television and Prosocial Behavior

The Surgeon General's report also refers to television's potential for encouraging **prosocial behavior** in children. Prosocial behavior includes such things as helping, sharing, and comforting behaviors (Eisenberg, 1992).

Testing this potential, Sprafkin, Liebert, and Poulos (1975) selected two episodes from the "Lassie" series. One of these episodes had the lead child character risk his life to save a puppy. The other episode had no such dramatic incident. An episode from the "Brady Bunch" show was also used. The researchers created a situation in which children would have to make a choice between alerting adults that an animal needed help or continuing to work on a task that could win them a prize. If they pushed a help button, they lost time on their task (a game). They had to make a choice, then, between sacrifice or self-interest.

What's Your View?

Should television be used as a prosocial agent? Your initial response to this question is probably a firm yes. Would everyone agree with you? We're not so sure.

In 1975, the United Methodist Church provided money for producing 30-second television spots intended to help children become more cooperative. Psychologists and experienced television personnel combined to make as professional a presentation as possible. The format was identical to that of regular commercials.

Children, ages 4 to 10, watched television while sitting on a comfortable sofa in a relaxed, den-like setting with natural distractions such as toys and books. They were later tested for comprehension of the program and were observed playing.

In the best-known spot, "The Swing," a boy and a girl about 10 years old both run to a swing. They argue and struggle over it, each claiming the first ride. They scowl and look menacingly at each other. Suddenly one of them steps back and suggests that the other go first. They are then shown taking turns, happily swinging through the air. The announcer's voice concludes that "this is how you (children) should behave" (Liebert & others, 1988).

Tens of millions of children all over the world saw "The Swing." After seeing the film, children's cooperation increased and all seemed well. Then a reaction set in amid charges of psychological behavior control. Who has a right to impose values on children? Should all children be taught cooperation? Don't some youngsters need to be aggressive to survive? Isn't this brainwashing children? Are we infringing on that fundamental right we all cherish in our society—the right to freedom?

Would you still answer the first question in the same way? What's your view?

The children who had seen the prosocial "Lassie" episode were more willing to help than those who watched either of the other two programs. They pressed the help button for 93 seconds compared with 52 and 38 seconds for those who had watched the other programs.

Clearly, television can be a negative influence on children; but can it also be a positive influence? If children model and learn from violent television shows and movies, they would be expected to learn positive behaviors and values from shows depicting helpfulness, generosity, cooperation, self-sacrifice, and the like. Recent studies suggest that the media can be used to foster development, although the effects of viewing prosocial television programming appear to be somewhat weaker than the effects of viewing violent programming. **(Eisenberg, 1992)**

A new brother or sister, budding friendships, school challenges, televised violence, inevitable upset at home—put them all together and they spell *stress.*

Stress in Childhood

As youngsters of this age spend more of their time away from home, new contacts and new tasks can upset them. There is no escape; we all have faced them. Were we scarred for life by these encounters? Probably not. Yet for some children, an inability to cope with stress has tragic consequences and today we are more alert to the signs of childhood suicide. For example, among children (5 to 13 years of age) admitted to a child psychiatric unit, 31 percent had recently attempted suicide (Milling & others, 1991); of children who had recently experienced periods of depression, 72 percent expressed thoughts of suicide (Myers & others, 1991); in a study of "normal" children 12 years of age—no prior history of psychiatric care—19 percent reported suicidal thoughts or threats (Pfeffer & others, 1988).

In her excellent and practical analysis of childhood stress, Brenner (1984) describes a spectrum of stressors:

Ordinary	**Moderate**	**Severe**
Jealousy of sibling	One-parent home	Separation from parent (divorce, death, illness)
Typical school anxiety	Multiple parents (biological mother, stepmother, biological father, stepfather)	Abuse Parental alcoholism

But first, let's link what is known about childhood stress to development. We begin by admitting that there is no definition of **stress** that everyone agrees with. Let's use this definition:

Stress is anything that upsets our equilibrium—both psychological and physiological.

Abused Children

Most people consider the United States to be a nation of child lovers. It always comes as a shock, then, to discuss child abuse. Before we begin to discuss this topic, we should agree on what is meant by the term. The 1988 *Study of National Incidence and Prevalence of Child Abuse and Neglect* identified six major types of abuse: physical abuse, sexual abuse, emotional abuse, physical neglect, emotional neglect, and educational neglect.

Child abuse has always been with us, but it has only been a matter of public awareness for the past quarter-century. It still remains an elusive subject that defies precise definition because of the many forms of abuse. While physical and sexual abuses that leave evidence are easy to detect and describe (if they are reported), other forms of abuse that emotionally wound youngsters are perhaps never detected. Professionals believe that abusive behavior involves direct harm (physical, sexual, deliberate malnutrition), intent to harm (which is difficult if not impossible to detect), and intent to harm even if injury does not result.

Another troublesome issue is that of incidence. Figures show tremendous variability. The actual data are only for reported cases and undoubtedly represent only the tip of the iceberg. The true extent of the problem may be staggering. For example:

- In 1990, there were more than 2.5 million reports of child abuse, an increase of 100% since 1980.
- Estimates of national child abuse and neglect range from 35% to 53% of the child population.
- A 1990 state survey of child abuse indicated that 27% of reported abuse cases were due to physical abuse, 45% to neglect, 15% to sexual abuse, and 13% to emotional abuse.
- In 1990, an estimated 1,211 children from 39 states died from abuse or neglect. 90% of these children were under age five, and 50% were infants under age one (Hearings on the Child Abuse Prevention, Adoption, and Family Services Act, 1992).

The Nature of the Problem

Although child abuse is an age-old problem, not until recently did it become widely publicized. In the 1920s, Dr. John Caffey, studying bone fractures and other physical injuries, suggested that parents might have caused the injuries. The skepticism that greeted his conclusions prevented him from officially reporting his findings until the late 1940s. In 1961, C. Henry Kempe and his associates startled the annual meeting of the American Academy of Pediatrics by their dramatic description of the battered child syndrome.

You may well ask, What kind of person could ever hurt a child? While the parental characteristics leading to child abuse are not rigidly defined, several appear with surprising frequency. Here are some of the most frequently found characteristics.

- The parents themselves were abused as children.
- They are often loners.
- They refuse to recognize the seriousness of the child's conditions.
- They resist diagnostic studies.
- They believe in harsh punishment.
- They have unreasonable expectations for the child. (Children should never cry or drop things.)
- They lack control and are often immature and dependent.
- They feel personally incompetent.

A cycle of abuse becomes clear. The most consistent feature of the histories of abusive families is the repetition, from one generation to the next, of a pattern of abuse, neglect, and parental loss or deprivation. In each generation we find, in one form or another, a distortion of the relationship between parents and children that deprives children of the consistent love and care that would enable them to develop fully.

To summarize, there are four categories of parental abuse.

1. The parents have a background of abuse and neglect.
2. Parents perceive the child as disappointing or unlovable.
3. Stress and crisis are usually associated with abusive behavior.
4. No lifeline exists; that is, there is no communication with helpful sources in times of crisis.

But what of the children, themselves? Some observers believe that certain types of children are more prone to be abused than others. Remember: children shape their parents as much as parents shape their children. If a child's actions or appearance irritate a parent predisposed to violence, then the results of the parent-child interaction may be preordained (which doesn't mean that the child is at fault).

Children growing up in a hostile environment feel that to survive they must totally submit to their parents' wishes. They often exhibit continual staring; and a passive acceptance of whatever happens. It is only later in a permissive setting that the pent up fury explodes. They slowly develop complete distrust of others, which often translates into school problems.

When you consider the factors that may trigger abuse—parents, children themselves, poor family relations, socioeconomic conditions, lack of support—it becomes clear that understanding the problem requires considerable and careful research.

The Special Case of Sexual Abuse

Sexual abuse refers to any sexual activity between a child and an adult whether by consent or force. It includes fondling, penetration, oral-genital contact, intercourse, and the use of children in pornography (Kelly, 1986). Estimates are that between 50,000 and 500,000 children are sexually abused each year. Most of the victims are female, but the number of male victims is on the rise.

Abused children feel that they have lost control and are helpless when sexually abused by an adult. All of their lives they have been taught to obey adults so they feel forced to comply. This is particularly sensitive since most abusers are known to the family: a relative, friend, or some authority known to the children.

What are the developmental effects of this violation and betrayal of a child by an adult? Browne and Finkelhor (1986), summarizing several major studies, report the results of both short-term and long-term effects of sexual abuse. Different ages seem to suffer different types of effects. For example, the highest rate of problems was found in the 7-13 year group. Forty percent of the abused children of this age showed serious disturbances; 17 percent of the 4 to 6 year group manifested some disturbance. About 50 percent of the 7 to 13 year group showed greatly elevate levels of anger and hostility compared to 15 percent of the 4 to 6 year-olds. Increase of anxiety, fear and distress were common to all age groups.

Are there long term effects that influence adults who were sexually abused as children? Among the effects are the following.

- Depression (probably the most common finding)
- Above normal levels of tension
- A negative self-concept
- Sexual problems

Sexual abuse is a problem that every reader will find repugnant. Yet we can offer some positive conclusions. We are now better able to identify these children and provide help. Treatment techniques offer hope for the future. As the problem becomes more widely publicized, parents, teachers, and concerned adults are becoming more sensitive to the possibility of its occurrence.

Types of Stress

Different kinds of stress can cause similar reactions. Think of a time when your parents were really angry with you, or when you were faced with a severe challenge (perhaps speaking before a large group for the first time; or just after you had been given some bad news). Some children react in the same way to all of these events, either with high anxiety, fear, avoidance, weakness, or vomiting.

Different kinds of stress also can cause different reactions, of varying intensity. If we group these ideas, we can begin by stating that (Honig, 1986):

- Stress can come from internal sources, usually illness.
- Stress can come from external sources, such as family, school, or peers.
- Stress can be chronic, such as the child who is trapped for years with an alcoholic or abusive parent or with an insensitive teacher.

Table 10.3, based on Brenner (1984), summarizes specific childhood stressors.

When each of us is faced with stress, we react differently. To begin with, not all of us would agree on what stress is. For example, some are probably terrified of flying, while others see it as a pleasant, relaxing adventure. While there are many reasons for these different responses, we can isolate several important individual differences.

Table 10.3 Specific Childhood Stressors

Type	Example
Two-parent families	Changes associated with normal growth: new siblings, sibling disputes, moving, school, working parents
One-parent families	Multiple adults, lack of sex-role model, mother vs. father, financial difficulties
Multi-parent families	New relationships, living in two households
Death, adoption	Parental death, sibling death, possible institutional placement, relationships with different adults
Temporary separation	Hospitalization, health care, military service
Divorce	Troubled days before the divorce, separation, the divorce itself
Abuse	Parental, sibling, institutional; sexual, physical, emotional
Neglect	Physical (food, clothes), emotional (no response to children's needs for attention and affection)
Alcoholism	Secrecy, responsibility for alcoholic parent, suppress own feelings

Sex

As we have repeatedly seen, boys are more vulnerable than girls. This includes their reaction to stress. Select any event that is likely to induce stress, and boys are more susceptible: death, divorce, new sibling, or hospital admission.

Age

Children of different ages respond differently to stress. There is good evidence that infants are relatively immune to the stress of hospital admission; the age of greatest risk is from age 6 months to 4 years. Children above the age of 4 can rationalize that hospitalization, for example, does not mean abandonment by parents. Middle childhood youngsters are less vulnerable to stress caused by a new sibling, while younger children show a great deal of clinging.

The grief reactions of young children are shorter and milder than those of older youngsters. Cognitive level probably explains the difference; younger children can't understand the concept of death itself (Rutter, 1988). Long-term effects may be greater if family breakup causes change in socioeconomic status, and if moving is involved.

Temperament

Recall that we are all born with unique temperaments, differences that affect the way we interact with the environment. Consequently, children's temperament influences the intensity of their reactions after a stressful event—for example, parental separation.

These are but a few of the many factors that help to explain different reactions to stress. But what can we say about their developmental effects?

Developmental Effects of Stress

If we attempt to link development and vulnerability to stress, we reach certain conclusions. For example, Maccoby (1988) notes that we cannot be upset by events whose power to harm us we do not understand. We can't be humiliated by failure to handle problems whose solutions are someone else's responsibility. Maccoby compares why some events are stressful and others are not and formulates several hypotheses:

- *Age changes alone don't explain vulnerability.* While the events that cause stress change with age, we all experience periods of stabilization and destabilization. In other words, we may be more vulnerable to change at certain times (e.g., we may react emotionally to bad news if we have been quite sick).
- *Environmental structure can lessen vulnerability.* Youngsters can handle stress better if all other parts of their lives are stable. For example, entering school is stressful for almost all children. If home conditions are warm and supportive, it helps to ease what can be a difficult transition. On the other hand, if parents have separated during these days, children can find school entrance quite painful.
- *Although middle childhood youngsters face more stressful situations than younger children, they have learned better ways to cope.* Also, as they move away from sole dependence on parents as attachment figures, peers begin to form a strong supportive network, especially toward the end of middle childhood.
- *Recognizing adults as authority figures* gives children a sense of security, which acts as a buffer against stress.
- *With their growing cognitive ability,* children of this age begin to develop coping skills that help them to combat stress. Think of our discussion of this growing cognitive capacity: the ability to think abstractly, solve problems, reach decisions, and to plan ahead. All of these abilities help to combat stress.

Children and Violence

On 3 June 1986, the state of Florida charged 9-year-old Jeffrey Bailey with murder. Making sure that no one else was around, he had pushed a 3-year-old boy who couldn't swim into a pool. He pulled up a chair to watch the boy drown. Later, when the police had pieced together the circumstances of the murder, they arrested Jeffrey, who was described as calm, nonchalant, and enjoying the attention (Magid & McKelvey, 1987).

This startling story is true, and others like it are appearing with increasing frequency in our nation's newspapers. Murder is the most dramatic example of a life gone wrong, but the increasing rate of assaults, robberies, and arson raises a basic question: What has happened to these children to turn them into killers, thieves, and arsonists?

Theories of Early Criminal Behavior

In their massive study of crime and human nature, Wilson and Herrnstein (1985) note that the majority of all young males have broken the law at an early age. Unfortunately, there is clear evidence of the positive association between past and future criminal behavior, which gives rise to this maxim: The best predictor of crime is past criminal behavior. Today's criminologists believe that a tendency to commit crime is established early in life, perhaps as soon as the preschool years (Nagin & Farrington, 1992).

Although ample evidence exists concerning the early appearance of crime, the evidence is merely descriptive, telling us little about the roots of violence. Most theorists have identified a cluster of possible causes that should come as no surprise: constitutional (more males than females commit crimes and more younger males than older males), developmental (broken families), and community (gangs, drugs). No one element can be considered *the* cause, so it is best to think of the causes of crime as complex, multiple, and interactive. One conclusion, however, is possible: No matter the circumstances, an individual, when faced with a choice, will choose the preferred course of action (Wilson & Herrnstein, 1985).

The Causes of Crime

Attempts to identify the causes of crime always begin with the individual child. If two boys are from the same family, with the same opportunities, why does one turn to a life of crime while the other leads a law-abiding life? Was there some subtle genetic transmission? Were the family relations similar for both boys? Were their friends radically different? Examining each case helps us to identify the causes of crime for that child. With sufficient numbers, it becomes possible to categorize the causes of crime as follows:

- Those biological and psychological characteristics that contribute to crime (physical appearance, psychopathology)
- Behavioral characteristics (impulsivity, attitudes, values)
- Home and family conditions (broken homes, family tensions)
- Peers (gangs)
- Media (especially television)

This abbreviated list illustrates the meaning of the statement that the causes of crime are multiple, complex, and interactive. Family conditions may be the cause for one child but not another; that is, one child may experience the environment quite differently than another.

With these ideas in mind, we may well ask whether youthful criminals are different from nondelinquents. What are some characteristics of children who become delinquent? Among the major characteristics found by Wilson and Herrnstein (1985) are the following:

- Constitutional factors seem to be at work; for example, both members of identical twin pairs are more likely to be involved in delinquency than nonidentical twins.
- Males are more prone to criminal behavior than females, and younger males are more likely to commit crimes at a higher rate than older males; age is a major factor in understanding criminal behavior (Regoli & Hewitt, 1991).

An Applied View — A Life of Crime

How does a child become entrapped in a life of crime? In an insider's account of organized crime in the United States, Peter Maas has written a book called *The Valachi Papers*. In a fascinating tale of violence and crime, Valachi (a member of organized crime) portrays the manner in which he was drawn into the criminal network.

He recalls that his "was the poorest family on earth—three rooms, no hot water, only a toilet out in the hall." This was for a family of eight. Valachi was constantly truant from school, and when he was 11 he hit the teacher in the eye with a rock. He was then sent to a disciplinary school, returned to the public schools, and then left school for work after completing only the sixth grade.

It was then that he started to steal, because his father took his money. By the time he was 18, Valachi's petty thefts had earned him membership in a burglary gang. At 19, he was arrested and sent to the notorious Sing Sing prison. Released, rearrested, and returned to Sing Sing, Valachi made his first contact with members of the crime syndicate.

The road Valachi traveled is clear—poor, needing money, truant, petty theft, making the contacts that lead to organized crime, acceptance into the syndicate, and, finally, murder. Most cases are not so clear and dramatic, but the pattern is similar for those for whom the environment is the chief cause.

- Attitudinally, delinquents are hostile, defiant, resentful, suspicious, and resistant to authority.
- Psychologically, delinquents are more interested in the concrete than the abstract and are generally poor problem solvers.
- Socioculturally, delinquents are frequently reared in homes that offer little understanding, affection, stability, or moral clarity. For example, some parents frequently reinforce aggressive behavior in an effort to defuse an explosive situation. The mother may tell a child to do something; the child refuses and becomes aggressive toward the mother. The mother may then "back off" and the child's aggressive behavior has been positively reinforced (Morton, 1987).

We can interpret these conclusions to mean that while any one of these elements can cause an individual's delinquency, usually a high probability of delinquency depends on the interplay of all of them. For example, in the stimulating but culturally inconsistent milieu of underprivileged areas, those with delinquent tendencies express their impulses and desires with little thought of self-control, which can become a type of psychopathology (Crowell, 1987).

Some theorists today are probing into the infancy years, searching for the roots of violence. For example, Magid and McKelvey (1987) state that those who do not form bonds (see the attachment and bonding sections in chap. 6) constitute one of the largest deviant populations in the country. These are the children who cannot form relationships; they describe their reactions to others as making "no connections." The words of convicted serial killer Ted Bundy are revealing:

> ***I didn't know what makes things tick. I didn't know what made people want to be friends. I didn't know what made people attractive to one another. I didn't know what underlay social interactions.***

As children become more and more involved in crime and violence, the need to identify the causes of such behavior and to formulate the means of preventing it become more urgent. The task is difficult, expensive, and lengthy, but well worth it to prevent violence from becoming an accepted part of children's lives.

A warning is necessary here, however. Note that Magid and McKelvey referred to children *who did not form a bond with anyone,* which is not the same as saying that a child has an insecure relationship with parents. In an excellent and timely caution, Fagot and Kavanagh (1990) urge us not to predict adolescent and adult problems with children who have an insecure relationship with parents. Studying 89 children who were clearly classified as insecure/avoidant or securely attached, they found that the parents of both groups reported the same number of problems (defiant, aggressive). Nor do the children in both groups show any differences in problem behaviors at home or in play. The authors conclude that while there may be continuities between attachment and later interactional styles, the data do not warrant predictions or any type of intervention.

There are no guarantees of successful coping, for obvious reasons: the intensity of the stress, the immaturity of the children, and the amount of support they receive. Yet we also know that there is a small number of children who seem oblivious to stress, at least for a time.

The Invulnerable Child

The mother of three children was beset by mental problems. She refused to eat at home because she was sure someone was poisoning her. Her 12-year-old daughter developed the same fear. Her 10-year-old daughter would eat at home only if the father ate with her.

Her 7-year-old son thought they were all crazy and always ate at home. The son went on to perform brilliantly in school and later in college and has now taken the first steps in what looks like a successful career. The older daughter is now diagnosed schizophrenic, while the younger girl seems to have adjusted after a troubled youth. Why? How?

Answering these questions takes us into uncharted territory. We simply don't know much about invulnerability. But what is known points to the ability to recover from either physiological or psychological trauma and return to a normal developmental path.

Identifying Invulnerable Children

Who are these **invulnerable children** who grow up in the most chaotic and adverse conditions, yet manage to thrive? They seem to possess some inner quality that protects them from their environment and enables them to reach out to an adult who can offer critical support (Garmezy, 1987).

The ratio seems to be about 1 in 10; that is, for every ten children who succumb to adverse conditions, one develops normally. In each case studied, an adult was there to offer the emotional support needed: a teacher, aunt, uncle, grandparent (Garmezy, 1987). Invulnerable children seem to possess a winning personality. (Remember the discussion of inborn temperamental differences in chap. 6.) They also seem to have a special interest or talent. For example, some of these youngsters were excellent swimmers, dancers, and artists; others had a special knack for working with animals; some showed talent with numbers quite early. Whatever their interest, it served to absorb them and helped to shelter them from their environment.

Such characteristics—a genuinely warm, fairly easy-going personality, an absorbing interest, and the ability to seek out a sympathetic adult—helped these children to distance themselves emotionally from a drugged, alcoholic, or abusive

Children need emotional support from those in their environment. Even under the most difficult circumstances (divorce, death of a parent, hospitalization), the presence of a "significant other" can help a child to cope, to deal with stress in an appropriate manner for the child's developmental level.

parent (sometimes a parent with all of these problems). Recent studies have shown that children who are emotionally close to a disturbed parent frequently develop problems.

Michael Rutter (1987), a leading researcher in the study of invulnerable children, has attempted to sort out some of the characteristics that help to explain either invulnerability or vulnerability:

- *Sex.* As with any other problem we discuss, boys are more vulnerable than girls. Whether or not psychological vulnerability is innate is difficult to determine. Boys tend to be exposed to psychological stress more than girls, and typically display disruptive behavior.
- *Temperament.* Children with negative personality features are more frequently the target of a disturbed parent. Parents with a problem do not take it out equally on all children.
- *Parent-child relationship.* While not much is known about this characteristic, a good relationship with one parent acts as a buffer for the child.
- *Positive school experience.* The experience could be either academic or nonacademic. The pleasure associated with success seems to help raise a youngster's sense of self-esteem.
- *Early parental loss.* The results of losing a parent, especially a mother, do not become evident until later in life. Even then its effects are not apparent unless combined with another threatening event.

For those working with youngsters who may be exposed to severe adverse conditions, Rutter suggests:

- Try to reduce the risk impact by teaching children methods of stress reduction and by helping them to gain success and pleasure in other activities.
- Try to reduce any negative chain reaction by providing emotional support.
- Help these children to acquire a greater sense of self-esteem by offering a secure relationship and an opportunity for successful achievements.

How Children Cope

Before we begin to discuss coping in childhood, we should remember that for most children psychosocial stress is the villain. That is, a child's hospital admission or a mother's temporary absence is really not the issue here. In most cases of psychosocial stress, the tension is persistent and unrelenting. The child with an alcoholic father or abusive mother has little chance of escaping. Table 10.4 illustrates children's ways of coping with stress.

Table 10.4 Children's Patterns of Coping

Avoiding Stress	Facing Stress
Denial: Children act as though the stress does not exist; may use fantasy (imaginary friends to talk to).	*Altruism*: By helping others, children ease their own pain.
Regression: Children act younger than their years, show greater dependency.	*Humor*: Children joke about their problems.
Withdrawal: Children remove themselves, either physically or mentally; they may run away, or attempt to fade into the background.	*Suppression*: Children push their troubles from their minds; they may play unconcernedly for a while.
Impulsive acting out: Children speak and act impulsively to avoid thinking about reality; by making others angry they attract attention, thus temporarily easing pain.	*Anticipation*: These children plan how to meet stress; they tend to protect themselves and accept what can't be avoided.

Yet some children seem invulnerable, as we noted earlier. Remember, however, that invulnerability does not imply that these are superchildren who can resist all stress.

> ***There is no single set of qualities or circumstances that characterizes all such resilient children. But psychologists are finding that they stand apart from their more vulnerable siblings almost from birth. They seem to be endowed with innate characteristics that insulate them from the turmoil and pain of their families and allow them to reach out to some adult—a grandparent, teacher, or family friend—who can lend crucial emotional support.*** **(Goleman, 1987, p. 82)**

Continuing deprivation or problems will eventually scar them, but these children are better able to function in settings that disable other children. The characteristics of these children are summarized in Table 10.5 (based on work by Farber & Egeland, 1987).

Table 10.5 Characteristics of Invulnerable Children

Age	Characteristics
At birth	Alert, attentive
1 year	Securely attached infant
2 years	Independent, slow to anger, tolerates frustration well
3–4 years	Cheerful, enthusiastic, works well with others
Childhood	Seems to be able to remove self from trouble, recovers rapidly from disturbance, confident, seems to have a good relationship with at least one adult
Adolescence	Assumes responsibilities, does well in school, may have part-time job, socially popular, is not impulsive

We can best summarize all that we know about coping skills and child development in this list of guidelines for adults:

- To encourage coping skills in children, demonstrate them yourself, especially self-control.
- Encourage children to develop self-esteem.
- Be sympathetic to their feelings and learn to recognize when a child is under stress.
- Urge children to adopt a positive attitude, which then helps them to search for solutions.
- Talk to them; get them to examine their problems openly so that they can obtain any available support.

Conclusion

In this chapter, we followed middle childhood youngsters as they moved away from a sheltered home environment and into a world of new friends, new challenges, and new problems. Whether the task is adjusting to a new sibling, relating to peers and teachers, or coping with difficulties, youngsters of this age enter a different world.

But the timing of their entrance into this novel environment is intended to match their ability to adapt successfully, to master those skills that will prepare them for the next great developmental epoch, adolescence. From Tom Sawyer's subtlety to children learning to cope with stress, middle childhood youngsters require those skills that will enable them to deal with their widening social world.

Inevitably, though, they face times of turmoil, which, as we have seen, can come from internal or external sources. For some youngsters, these periods of stress are brief interludes; for others, there is no relief for years. Children cope uniquely using temperamental qualities and coping skills as best they can.

As we complete our work on middle childhood and before beginning to discuss the adolescent years, it is helpful to examine Table 10.6 and review the developmental highlights of the childhood years.

Table 10.6 Milestones in Development and Learning

Age (years)	Developmental Domains				
	Physical	Cognitive	Social	Emotional	Language
Infancy (0–2)	1. Can hear and see at birth 2. Rapid growth in height and weight 3. Rapid neurological development 4. Motor development proceeds steadily (crawling, standing, walking)	1. Seeks stimulation 2. An egocentric view of the world begins to decrease 3. Demonstrates considerable memory ability 4. Begins to process information	1. Need for interaction 2. Smiling appears 3. Reciprocal interactions begin immediately 4. Attachment develops	1. Beginnings of emotions discernible in first months 2. Infant passes through emotional milestones	1. Proceeds from cooing and babbling to words and sentences 2. Word order and inflection appear 3. Vocabulary begins to increase rapidly
Preschool (2–6)	1. Extremely active 2. Mastery of gross motor behavior 3. Refinement of fine motor behavior	1. Perceptual discrimination becomes sharper 2. Attention more focused 3. Noticeable improvement in memory 4. Easily motivated	1. Attachment 2. Beginning of interpersonal relationships a. parents b. siblings c. peers d. teachers 3. Play highly significant	1. Still becomes angry at frustration 2. Prone to emotional outbursts 3. Emotional control slowly appearing 4. Aware of gender 5. Fantasies conform more to reality 6. May begin to suppress emotionally unpleasant memories	1. Goes from first speech sounds (cooing, babbling) to use of sentences with conjunctions and prepositions 2. Acquires basic framework of native language
Middle childhood (7–11)	1. Mastery of motor skills 2. Considerable physical and motor skills	1. Attention becomes selective 2. Begins to devise memory strategies 3. Begins to evaluate behavior 4. Problem-solving behavior shows marked improvement	1. Organized activities more frequent 2. Member of same-sex group 3. Peer influence growing 4. Usually has "best" friend	1. Pride in competence 2. Confident 3. Growing sensitivity 4. Volatile 5. Striving, competitive 6. Growing sexual awareness	1. Rapid growth of vocabulary 2. Uses and understands complex sentences 3. Can use sentence content to determine word meaning 4. Good sense of grammar 5. Can write fairly lengthy essays

Chapter Highlights

Siblings and Development

- Sibling relationships have an enduring and significant impact on development.
- Most children respond positively when questioned about the quality of their sibling relationships.

The Influence of Peers

- During the middle childhood years, children begin to form close friendships.
- Children who have difficulty with their peers are often bothered by personal problems.
- Middle childhood youngsters learn to recognize the views of others.

Schools and Middle Childhood

- Children form and test social relations during these years.
- Children must learn to adjust to a wide variety of classmates, many of whom may be children of different cultures.
- School-age children are encountering considerable change in both curriculum and instructional methods.

Television and Development

- Television is the school's great competitor for children's time and attention.
- Some children spend more time watching television than they do in school-related activities.
- Controversy surrounds the issue of the effects of television violence.
- Television is also credited with the potential for encouraging prosocial behavior.

Stress in Childhood

- Children react differently to stress according to age, gender, and temperament.
- Some children, called invulnerable children, overcome the adverse effects of early stressors.
- Several theories have been proposed to explain how children become violent.
- Researchers are currently examining infant experiences, especially the quality of attachment, for an explanation of early violence.
- Children acquire coping skills that enable them to adjust to stress in their lives.

Key Terms

America 2000
Bilingual education
Grade retention
Homework
Invulnerable children
Observational learning
Peer
Prosocial behavior
Sibling
Social perspective-taking
Stress
Whole language

What Do You Think?

1. Recall your relationship with your brothers and sisters. How would you evaluate the experience, positive or negative? Why? Does your answer reflect some of the topics mentioned in this chapter? If you are an only child, do you think you missed out on something? Why?
2. It is generally accepted that friendships and groups become more important during the middle childhood years. With your knowledge of the developmental features of these years, do you think children of this age are ready for group membership?
3. For individuals to experience stress, they must understand the forces that are pressing on them. Do you think middle childhood youngsters are capable of such an interpretation of the events surrounding them?
4. There is great concern today about the increasing rate of violence among children. Do you think the problem is as serious as the media indicates? From your knowledge of this topic, do you think the predictors of early criminal behavior are useful?

Suggested Readings

Liebert, R.; J. Sprafkin; and E. Davidson (1988). *The early window*. 3d ed. New York: Pergamon. This paperback is one of the best single sources on the impact of television on our society today. Once you read this, you'll understand why TV is considered a major socializing force in a child's life.

Magid, K. and C. McKelvey (1987). *High risk: Children without a conscience*. New York: Bantam. An excellent, readable account of the development of violent children. Valuable case studies.

Parmet, H. (1980). *Jack: The struggles of John F. Kennedy*. New York: Dial. An insightful glimpse of those childhood years when the give-and-take of family interactions was colored by a constant sense of "great expectations."

Ward, G. (1989). *A first-class temperament: The emergence of Franklin Roosevelt*. New York: Harper & Row. To the end of her days, Eleanor Roosevelt felt that no one loved her for herself. In this intriguing account of the young Roosevelts, Ward traces the effect of an alcoholic father and an indifferent mother on the emotional life of a young girl.

PART VI

Adolescence

CHAPTER 11

Adolescence: Background and Context

How Should We Define Adolescence? 299

When Does It Start? 299

Ancient Times 301

The Middle Ages 301

The "Age of Enlightenment" 302

The Twentieth Century 302

G.S. Hall and the Theory of Recapitulation 303

Theories of Adolescence: Anna Freud 305

Theories of Adolescence: Robert Havighurst 306

Theories of Adolescence: Erik Erikson 307

The Search for Identity 308

The Moratorium of Youth 308

Negative Identity 309

Identity Status 310

Changing American Families and Their Roles in Adolescent Life 312

The Loss of Functions 312

The Increase in Age-Related Activities among Older Adolescents 313

The Effects of Divorce 313

The Effects of Gender 315

The Nurturing Parent 315

Interactions with Peers 316

Adult Anxieties about the Adolescent Subculture 316

The Origins of Subcultures 317

Conclusion 319

Chapter Highlights 319

Key Terms 321

What Do You Think? 321

Suggested Readings 321

Who are you?" said the caterpillar. Alice replied, rather shyly, "I—I hardly know, Sir, just at present—at least I know who I was when I got up this morning, but I must have changed several times since then."

Lewis Carroll, *Alice in Wonderland,* 1865

Most American adolescents come through the critical years from ages ten to twenty relatively unscathed. With good schools, supportive families, and caring community institutions, they grow to adulthood meeting the requirements of the workplace, the commitments to family and friends, and the responsibilities of citizenship. Even under less-than-optimal conditions for healthy growth, many youngsters manage to become contributing members of society. ■

Carnegie Corporation, *Adolescence,* 1990, p. 1

Some writers have suggested that like Alice, adolescents experience life as a constant swirl of adjustments. Is adolescence a time of topsy-turvy change, marked by abrupt emotional crises, or is this only a **stereotype,** as the quote above from the Carnegie report indicates? This question has caused a great deal of debate among scientists who study this fascinating period of life.

Another major concern focuses on when (at what age or with what event) this stage of life begins, and when it ends. A third debate addresses the question, "What are the best ways of answering questions about adolescence?" Resolving each of these controversies would help us to agree on a definition of adolescence, and would go a long way toward helping us to understand it. In this chapter we consider each of these questions, starting with the first one. In the last two sections of the chapter, we examine the changing roles of adolescents' families, as well as their interactions with their peers.

As a result of having read this chapter, you should be able to:

- Identify ways in which your own adolescence was different from or similar to that of today's teens.
- Describe how adolescence was viewed in ancient times, during the middle ages, during the "Age of Enlightenment," and in the present century.
- Explain the value of studying adolescence in two stages, early and late, rather than as one stage.
- State what you believe to be the best definitions of the beginning and end of adolescence.
- List G. S. Hall's four stages of development.
- Discuss Hall's notion of "storm and stress."
- State the contributions of Anna Freud and Robert Havighurst to adolescent theory.
- Detail the special importance Erik Erikson's psychosocial theory has for adolescence.
- Name the five main functions that have been lost by the family, as well as the role that remains.
- Explain the increase in age-related activities.
- Describe the effects of divorce on adolescents.

AN APPLIED VIEW — An Average Day in the Life of Some North American Teens

Today (and every other day this year):

- 7,742 teenagers have become sexually active.
- 623 teenagers have gotten syphilis or gonorrhea.
- 2,740 teenagers have gotten pregnant
- 1,293 teenagers have given birth to a child.
- 1,105 teenagers have had an abortion.
- 369 teenagers have miscarried.
- 1,375 teenagers have dropped out of school before graduation.
- 3,288 have run away from home.
- 1,629 teenagers are locked up in adult jails.
- 6 teenagers have died of suicide.
- ? teenagers are being beaten or psychologically or sexually abused.
- ? teenagers have parents who are or soon will be divorced.

On the other hand, today (and every other day this year), teenagers have engaged in many kinds of activities that enrich their own lives and those of the people around them. It is impossible to know exactly how many are involved in each of these activities. (It is interesting that we know much more about teenagers' negative actions, isn't it?) Here are some examples:

- Have joined a service-oriented club (e.g., Scouts, 4-H, Future Farmers of America)
- Became members of Junior Achievement
- Competed in an athletic event
- Became a candy-striper (volunteer nurse's aide)
- Joined Students Against Drunk Driving
- Taught another teen in a peer tutor program
- Served food in a shelter for the homeless
- Volunteered at a day-care center or a nursing home for the elderly
- Answered phones on a drug abuse or suicide hotline
- Delivered newspapers, stocked supermarket shelves, or in some other way earned money at a part-time job

Obviously, we would wish that all adolescents were more interested in the ideals represented by these activities. Can you think of ways that you could help make this happen?

- Identify ways in which friends become increasingly important as the maturing adolescent begins to move beyond the immediate family.
- Explain why parents remain an important source of influence and support to adolescents, even though many parents worry about their declining influence.
- Explain the psychogenic, culture transmission, and behavioristic models of the origins of subcultures.
- Discuss these issues from an applied, a multicultural, and your own point of view.

Is interest in the opposite sex the best sign that a young person has reached puberty? What other indicators could you name?

How Should We Define Adolescence?

In this section we consider the definition of adolescence by reviewing ideas about when it starts, and by examining the history of the concept.

When Does It Start?

At what point did your adolescence begin? Many answers have been offered:

- When you began to menstruate, or when you had your first ejaculation.
- When the level of adult hormones rose sharply in your bloodstream.
- When you first thought about dating.

- When your pubic hair began to grow.
- When you became 11 years old (if a girl); when you became 12 years old (if a boy).
- When you developed an interest in the opposite sex.
- When you (if a girl) developed breasts.
- When you passed the initiation rites set up by society: for example, confirmation in the Catholic Church; bar mitzvah and bas mitzvah in the Jewish faith.
- When you became unexpectedly moody.
- When you became 13.
- When you formed exclusive social cliques.
- When you thought about being independent of your parents.
- When you worried about the way your body looked.
- When you entered seventh grade.
- When you could determine the rightness of an action, independent of your own selfish needs.
- When your friends' opinions influenced you more than your parents' opinions.
- When you began to wonder who you really are.

Although there is at least a grain of truth in each of these statements, they don't help us much in defining adolescence. For example, although most would agree that menstruation is an important event in the lives of women, it really isn't a good criterion for determining the start of adolescence. The first menstruation (called menarche) can occur at any time from 8 to 16 years of age. We would not say that the menstruating 8-year-old is an adolescent, but we would certainly say the nonmenstruating 16-year-old is one.

Probably the most reliable indication is a sharp increase in the production of the four hormones that most affect sexuality: progesterone and estrogen in females, testosterone and androgen in males. But determining this change would require taking blood samples on a regular basis, starting when youths are 9 years old. Not a very practical approach, is it?

Clearly, identifying the age or event at which adolescence begins is not a simple matter. We will need to look at it much more closely, from the standpoints of biology, psychology, sociology (the biopsychosocial model), and several other sciences, and we will do so in other chapters in this book.

These are the best years of your life! You'd better enjoy them now, because before you know it, you'll be weighed down with adult responsibilities!

Can you remember your parents saying this, or something like it? At some time during their teen years, most people are advised not to waste their youth. It used to be a common belief that adolescence is a carefree period, a stage of life when people "sow their wild oats" before settling down to the more rigorous demands of adult maturity.

This view is not so common any more. In fact, some observers believe that it has become the *worst* time of life. Is adolescence an unusually difficult period of life? Are the changes that accompany it more abrupt and disruptive than those of earlier and later stages?

Obviously, adolescence is a *very* bad time for *some* people. But does that mean that adolescents in general are becoming more of an affliction to themselves and to society? Do *average* adolescents have a harder time of it than their predecessors? For an answer, let us use the perspective of history. What follows is a brief summary of the ways adolescence was viewed in earlier times by Western civilizations.

Ancient Times

It appears that teenagers were no more popular with early writers than they are with many people today. Take, for example, this rather cranky statement written in the eighth century B.C. by the Greek poet Hesiod:

> ***I see no hope for the future of our people if they are dependent on the frivolous youth of today, for certainly all youth are reckless beyond words. When I was a boy, we were taught to be discreet and respectful of elders, but the present youth are exceedingly wise and impatient of restraint.***

The famous Greek philosopher and renowned teacher of the young, Socrates, was no great fan either. He wrote this in the fifth century B.C.:

> ***Our youth now love luxury. They have bad manners, contempt for authority; they show disrespect for their elders and love chatter in place of exercise. They no longer rise when others enter the room. They contradict their parents, chatter before company, gobble up their food and tyrannize their teachers.***

Prior to the twentieth century, it appears that children moved directly from childhood into adulthood with no period of adolescence in between.

Socrates' notable student, Plato, had a more positive outlook. In his view of the lifespan, childhood is the time of life when the spirit (meaning life values) develops, and so children should study sports and music. In the teen years, the reason starts to mature, and so youth should switch to the study of science and mathematics. For Plato's student, Aristotle, the teens are the years in which we develop our ability to choose, to become self-determining. This passage is not an easy one, however, and he felt it caused youth to be impatient and unstable.

The Middle Ages

During the Middle Ages, the concept of human development became unrelentingly negative. Children came to be seen as "miniature adults." Children rarely appear in paintings from those times, but when they do, they are always dressed in cut-down versions of their parents' clothes. It was generally agreed that the way to help them become mature adults was through strict, harsh discipline, so that they could overcome the natural evils of the childish personality.

The word *teen,* meaning a person from 11 to 19 years old, is an inflected form of *ten,* used as a suffix (e.g., *four-teen*). In Middle English, however, there was a word *teen* that meant "injury; misery, affliction; grief." The obsolete *teen* and the suffix *-teen* are not related. But coincidentally, *teen* is an accurate description of how youth was looked at in the period from the Romans to the Renaissance in Western culture.

The "Age of Enlightenment"

The beginning of the Age of Enlightenment (from the 1600s to the early 1900s) brought no major change from the previous period in the view of adolescence. For example, Hesiod's and Socrates' observations were echoed by an old shepherd in Shakespeare's *The Winter's Tale* (1609):

> ***I would there were no age between ten and three-and-twenty, or that youth would sleep out the rest; for there is nothing in the between but getting wenches with child, wronging the ancientry [elderly], stealing, fighting.***

This position held sway until the 1700s, when Jean-Jacques Rousseau argued forcibly through his book *Emile* (1762) that children and youth need to be free of adult rules so they can experience the world naturally. He compared childhood to the lives of the American Indians, whom he referred to as "noble savages." He believed that both groups are basically good, and that Indians grow into kind and insightful adults because they are not corrupted by civilization.

In early America, this view did not gain much support. Life for the colonists was not easy, and everyone was expected to work hard, including children. Most youths worked on farms, but as the population grew, more and more went into apprenticeships in the cities. By the nineteenth century, however, a dual pattern began to emerge. By the 1840s, the country was clearly splitting into a large lower and middle class.

The children of the poor continued in the old apprenticeship mold, but middle-class youth began to stay in school longer and longer. The technical demands of the Industrial Revolution called for more extensive education. The reform movement of the "muckrakers" at the turn of the twentieth century, which brought about stricter labor and compulsory education laws, created a more equitable situation between the two social classes. Only in the early 1900s did adolescence, as we know it today, begin.

The Twentieth Century

Now began the age of *empiricism*. Early in our century, those who were interested in understanding youth ceased speculating about the nature of adolescence and began to make careful observations of them. This is what empiricism means.

It should be noted that psychology itself began only in the late nineteenth century. It took as its first task learning how the brain perceives the environment around us, but soon turned to understanding human development. The new science quickly accepted the challenge of explaining the transition from childhood to adulthood. In this task, it was greatly influenced by the writings of G. Stanley Hall and a number of other social scientists who followed him.

A Multicultural View

Is Adolescence a Cultural Phenomenon?

In western cultures, extended schooling keeps children out of full-time productive work so they do not start observing and participating in the adult economic world as they do in, for example, Guatemala. Schooling has become a substitute for adult roles. For instance, a college student spends years studying nursing (chemistry, psychology), but he is not a nurse (chemist, psychologist). Extended schooling, then, artificially stretches the period from childhood to adulthood. This delay or waiting period is unique in human history. Combined with the decreasing age of reaching menarche in middle-class western girls, adolescence can be prolonged more than ten years! Compare this to the Efe, hunters and gatherers in Zaire, who marry and assume adult roles soon after puberty. Is adolescence just a theoretical construct (See chap. 2)? Is it peculiar to cultures with extended schooling? What do you think?

G. S. Hall and the Theory of Recapitulation

G. Stanley Hall (1844–1924) is known as the father of adolescent psychology. Building upon Charles Darwin's theory of evolution, Hall constructed a psychological theory of teenage development, published in two volumes and entitled *Adolescence* (1904).

Hall posited four periods of development of equal duration, which he felt correspond to the four lengthy stages of development of our species: infancy/animal, childhood/anthropoid (humanlike apes), youth/half-barbarian, and adolescence/civilized.

- *Infancy: birth–4 years.* In this stage children recapitulate the animal stage, in which mental development is quite primitive. Sensory development is the most important aspect of this period, together with the development of sensorimotor skills.
- *Childhood: 4–8 years.* Hunting and fishing, using toy weapons, and playing with dolls are common activities of childhood. Language and social interaction begin to develop rapidly, as they did during the nomadic period of the human race.
- *Youth: 8–12 years.* This period corresponds to the more settled life of the agricultural world of several thousand years ago. This is the time when children are willing to practice and to discipline themselves; this is when routine training and drills are the most appropriate—especially for language and mathematics.
- *Adolescence: 12–25 years.* **Storm and stress** typify human history for the past 2,000 years, as well as this developmental period. Adolescence is a new birth, for now the higher and more completely human traits are born.

G.S. Hall was the first American to publish research on the teen years, with his book *Adolescence* (1904).
Archives of the History of American Psychology, University of Akron, Akron, Ohio

Hall believed that each person's development passes through the same four stages as the human species. He thought that all development is determined by physiological (i.e., genetic) factors. Development occurs in an unchangeable, universal pattern, and the effects of the environment are minimal. For example, Hall argued that some socially unacceptable behavior in children, such as fighting and stealing, is inevitable. He urged parents to be lenient and permissive, assuring them that children must have this catharsis, and that when they reached the later developmental stages, these behaviors would simply drop out of existence.

Hall felt that the development of most human beings stopped short of this fourth stage, in which appreciation of music and art are achieved. Most people seemed fixated at the third stage, in the dull routine of work. A social reformer, Hall believed that adolescence is the only period in which we have any hope of improving our species. He felt that placing teenagers in enriched environments would improve their genes, which their children would then inherit. Hence we could become a race of superanthropoids.

Psychologists today argue that **recapitulation theory** is an interesting but quite inaccurate picture of human social development. They believe Hall tried to force reality to fit an outmoded conception of evolutionary development. His theory is considered wrong for several reasons. It most particularly does not present a true picture of adolescence. Although the majority of youth in his time may not have had much appreciation for civilized culture, this was clearly due not to genetic imperfections, but to such factors as having been forced to leave school to work on the farm. In addition, Hall's belief in the genetic transmission of acquired (improved) characteristics is scientifically false. Since Hall looked only at American culture, and since most individuals in the culture did develop similarly, he mistakenly thought that genes were responsible for this similarity. Later studies of other cultures have shown wide differences in developmental patterns. We can conclude that from the standpoint of the biopsychosocial model, Hall overemphasized the biological aspect.

Although Hall is to be admired for his efforts to bring objectivity to adolescent psychology through the use of empiricism, it has been suggested that he had several personal agendas. He was a strong preacher against what he viewed to be teenage immorality, and was especially concerned that educators try to stamp out the "plague of masturbation," which he considered to be running rampant among male youth. Here is a little speech that he recommended high school teachers and clergy give to their youthful charges:

> ***If a boy in an unguarded moment tries to entice you to masturbatic experiments, he insults you. Strike him at once and beat him as long as you can stand, etc. Forgive him in your mind, but never speak to him again. If he is the best fighter and beats you, take it as in a good cause. If a man scoundrel suggests indecent things, slug him with a stick or a stone or anything else at hand. Give him a scar that all may see; and if you are arrested, tell the judge all, and he will approve your act, even if it is not lawful. If a villain shows you a filthy book or picture, snatch it; and give it to the first policeman you meet, and help him to find the wretch. If a vile woman invites you, and perhaps tells a plausible story of her downfall, you cannot strike her; but think of a glittering, poisonous snake. She is a degenerate and probably diseased, and even a touch may poison you and your children.*** **(1904, p. 136)**

WHAT'S YOUR VIEW?

Hall is hardly the only adolescent psychologist who can be accused of bias in his thinking. In a fascinating study, Enright and colleagues (1987) looked at eighty-nine articles published during two economic depressions and two world wars to see if these events had an influence on research. The results were striking:

> In times of economic depression theories of adolescence emerge that portray teenagers as immature, psychologically unstable, and in need of prolonged participation in the educational system. During wartime, the psychological competence of youth is emphasized and the duration of education is recommended to be more retracted than in depression. (p. 541)

Is it likely that youth were viewed as immature during depressions in order to keep them from competing with adults for scarce jobs, and that their maturity is seen as greater during wartime because they are needed to perform such adult tasks as soldiering and factory work? If so, is this bias conscious or unconscious? What do you think?

As you read the other theories in this chapter, see if you can spot what you believe to be biases in them.

To conclude this brief excursion into the historical point of view, let us say that distrust of adolescence has not died in the second half of this century. As noted adolescent sociologist Edgar Freidenberg remarked in 1959:

> ***A great many young people are in very serious trouble throughout the technically developed world, and especially the Western world. Their trouble, moreover, follows certain familiar common patterns; they get into much the same kind of difficulty in very different societies.*** **(p. 6)**

AN APPLIED VIEW — What Were You Like?

Following are some questions about your personality that you might enjoy answering. Pretend you are in the eighth grade. Let your mind drift back to that time, and imagine yourself sitting in your favorite classroom. Look around the room and see who is sitting there. Try to answer these questions as you would have then.

A. Are the following statements true or false?

1. Most of the other kids in the class are stronger than I am. T F
2. I am about as intelligent as anybody in this classroom. T F
3. I am certainly not one of the teacher's favorite students. T F
4. Most people would say I am above average in athletic ability.T F
5. I am probably one of the more attractive students in this class. T F
6. I would say that I am more mature than most of my friends. T F
7. I am more popular than most of my classmates. T F

B. Fill in the blanks in the following sentences:

1. The thing I would like to change most about my life is ______________________________

____
2. My best friend is (give a brief description) ______

____
3. My deepest secret is ______________________________

____
4. My fondest memory is ______________________________

____

There are several things that can be done with this brief questionnaire to enhance your understanding and empathy for adolescents. You may wish to analyze the answers for clues about the definition we are seeking. On the other hand, you might share your answers with a small group of your most trusted friends, or even with your parents and siblings.

We cannot accept these historical views with confidence, however, because each of them suffers from the same critical flaw: They are primarily reflections of subjective opinion, not of scientifically objective measurement. Many more factors must be considered before we can honestly say we have an acceptable definition of adolescence. For an understanding of the most important of these factors, we turn to several experts on adolescence.

THEORIES OF ADOLESCENCE: ANNA FREUD

A trained psychoanalyst like her father, Anna Freud (1895–1983) believed that his definition of adolescence was too sketchy. She suggested (1968) that her father had been too involved with his discovery that sexuality begins not at puberty but in early infancy. As a result, he overemphasized the importance of that earlier stage in the total developmental picture. Anna Freud spent the major part of her professional life trying to extend and modify psychoanalytic theory as applied to adolescence.

Anna Freud saw the major problem of adolescence as being the restoration of the delicate balance between the ego and the id, which is established during latency and disrupted by puberty. Latency, she felt, is the time when children adopt the moral values and principles of the people with whom they identify. Childhood fears are replaced with internalized feelings of guilt that are learned during this period. The id is controlled during latency by the strength of the superego. At puberty, however, the force of the id becomes much greater and the delicate balance is destroyed.

The problems brought about by this internal conflict cause the adolescent to regress to earlier stages of development. A renewed Oedipal conflict (See chap. 2) brings about fears that are entirely unconscious and often produce intense anxiety. Therefore, the unconscious defenses of the ego tend to multiply rapidly, especially the typical ones of repression, denial, and compensation. The problem, of course, is that the use of these defense mechanisms causes new stresses within the individual and tends to further increase the level of anxiety.

Anna Freud described two additional adolescent defense mechanisms:

- *Asceticism,* in which, as a defense against the sexual, "sinful" drives of youth, the teenager frequently becomes extremely religious and devoted to God
- *Intellectualization,* in which the adolescent defends against emotionality of all kinds by becoming extremely intellectual and logical about life

Anna Freud may be seen as emphasizing the psychological aspect of the biopsychosocial model.

Sociologist Robert Havighurst suggested a series of developmental tasks for each stage of life.

Theories of Adolescence: Robert Havighurst

By the 1950s and 1960s, several new theories developed as a reaction to the earlier viewpoints. Robert Havighurst (b. 1900), a sociologist at the University of Chicago, became a major spokesperson of the new view (also see chap. 1). He suggested that there are specific **developmental tasks** at each stage of life, which lie midway between the needs of the individual and the ends of society. He defined these tasks as skills, knowledge, functions, and attitudes that are needed by an individual in order to succeed in life. As with Freudian theory, the inability to negotiate successfully any particular stage interferes with success at all succeeding stages.

For the adolescent period, Havighurst (1951) describes nine developmental tasks:

- Accepting one's physique and accepting a masculine or feminine role
- Forming new relations with age-mates of both sexes
- Achieving emotional independence of parents and other adults
- Achieving assurance of economic independence
- Selecting and preparing for an occupation
- Developing intellectual skills and concepts necessary for civic competence
- Desiring and achieving socially responsible behavior
- Preparing for marriage and family life
- Building conscious values in harmony with an adequate scientific world picture

Although written forty years ago, Havighurst's list holds up rather well today. Research has lent considerable support to Havighurst's theory, and educators and therapists have found his ideas useful. He clearly emphasized the social component of the biopsychosocial model.

An Applied View — The Erikson Psychosocial Stage Inventory

The Erikson Psychosocial Stage Inventory (EPSI) was developed as a research tool to examine adolescents' resolutions of conflicts associated with Erikson's first six stages in psychological development. We remind you that these stages are concerned with basic trust, autonomy, initiative, industry, identity, and intimacy.

Because Erikson regarded adolescence as central to his theory of human development, an investigation of how the adolescent forms an identity is of value. EPSI was tested in a study of 622 adolescents and has 12 items for each of Erikson's stages (Rosenthal, Gurney, & Moore, 1981). On the basis of their extensive research, the authors concluded that the EPSI is a useful measure for studying early adolescence and for "mapping changes as a function of life events" (p. 525). This means that the test can be used to find out the relationships between a person's stage of development and his or her age, IQ, personality traits, and many other characteristics. Here are some sample items, which the respondent is asked to check true or false:

Item Number	Subscale
	Trust
36.	Things and people usually turn out well for me.
	Autonomy
13.	I know when to please myself and when to please others.
	Initiative
34.	I'm an energetic person who does lots of things.
	Industry
60.	I stick with things until they are finished.
	Identity
10.	I've got a clear idea of what I want to be.
	Intimacy
59.	I have a close physical and emotional relationship with another person.

Theories of Adolescence: Erik Erikson

According to Erik Erikson (b. 1902), the main task of the adolescent is to achieve a **state of identity.** Erikson (1958, 1959, 1963, 1968, 1969), who originated the term **identity crisis,** uses it in a special way. In addition to thinking of identity as the general picture one has of oneself, Erikson refers to it as *a state toward which one strives*. If you were in a state of identity, the various aspects of your self-images would be in agreement with each other; they would be identical.

Repudiation of choices is another essential aspect of reaching personal identity. In any choice of identity, the selection we make means that we have repudiated (given up) all the other possibilities, at least for the present. All of us know people who seem unable to do this. They cannot keep a job, they have no loyalty to their friends, they are unable to be faithful to a spouse. For them, "the grass is always greener on the other side of the fence." Thus they must keep all their options open and must not repudiate any choices, lest one of them should turn out to have been "the right one."

Erikson suggests that identity confusion is far more likely in a democratic society because there are so many choices. In a totalitarian society, youths are usually given an identity, which they are forced to accept. The Hitler Youth Corps of Nazi Germany in the 1930s is an example of a national effort backed by intense propaganda to get all the adolescents in the country to identify with the same set of values and attitudes. In democratic societies, where more emphasis is placed on individual decision making, choices abound; some children may feel threatened by this overabundance of options. Nevertheless, a variety of choices is essential to the formation of a well-integrated identity.

Further, it is normal for identity confusion to cause an increase in self-doubt during early adolescence (Seginer & Flum, 1987; Shirk, 1987). Shirk states that such doubts should decrease during the middle teen years, "as social norms for self-evaluation are acquired through role-taking development" (p. 59). He studied self-doubt in 10-, 13-, and 16-year-olds, and found significant decreases with advancing age.

You may recall that in chapter 2 we described Erikson's complete theory as "an amazingly perceptive and at times poetically beautiful description of human life." He probably has done more research and writing on this fifth stage of identity formation than on all the others combined. We believe that most psychologists would call him the foremost theorist on adolescence today. This does not mean, however, that all agree with his view that adolescence is a time of identity crisis.

The Search for Identity

Goethals and Klos (1976) argue that if an identity crisis exists at all, it comes only at the end of adolescence:

> ***It is our opinion that college students do not typically have a firm sense of identity and typically have not undergone an identity crisis. College students seem to be in the process of identity seeking, and experience identity crisis toward the end of senior year and in their early post-college experience. A male or female's disillusionment with their job experience or graduate study, a female's disappointment at being at home with small children, is often the jolt that makes them ask what their education was for, and why they are not as delighted with their lives as they had been led to believe they would be.*** **(p. 129)**

Members of the Hitler Youth Corps were victims of "premature foreclosure," in which their identity was designed for them without their having any choice. They were taught exactly what to wear, how to act, what to think. Some actually turned their parents in to the secret police for what they believed to be violations of Hitler's creed.

Erikson, who himself had an extensive and rather difficult identity crisis in his youth, supposed that "My friends will insist that I needed to name this crisis in everybody else in order to really come to terms with it" (1975, p. 26). Born Erik Homberger, he seems to have rejected his past. It was difficult. His Danish mother remarried a German Jew, and he found himself rejected both by Jewish and Christian children. His identity crisis was resolved by the creation of a brand-new person with a new name, religion, and occupation. Some biographers (e.g., Berman, 1975; Roazen, 1976) have suggested that the surname he chose, *Erikson,* means he is the "son of himself." His theory of human development is no doubt colored by these experiences. At the same time, the intensity and degree of his identity crisis have made him extremely sensitive to the problems that all adolescents go through.

Perhaps the best conclusion we can reach, based on the available evidence, is that while the teen years are definitely a time of *concern* over one's identity, major decisions about it may be postponed by many until they reach early adulthood. This is probably truer today than ever, because of the phenomenon that Erikson calls the **moratorium of youth,** which seems to be lasting longer and longer.

The Moratorium of Youth

Erikson sees adolescence as a period of moratorium—a "time out" period during which the adolescent experiments with a variety of identities, without having to assume the responsibility for the consequences of any particular one. We allow adolescents this moratorium so that they can try out a number of ways of being, the better to come to their own particular identity. The moratorium period does not exist in preindustrial societies. Some have suggested that only Western industrial societies can afford the luxury of a moratorium. Others say that only because the values in Western industrial societies are so conflicted do adolescents *need* a moratorium.

Erikson stated that indecision is an essential part of the moratorium. Tolerance of it leads to a positive identity. Some youth, however, cannot stand the ambiguity of indecision. This leads to **premature foreclosure.** The adolescent who makes his choices too early usually comes to regret them. He or she is especially vulnerable to identity confusion in later life.

Erikson suggested that religious initiation ceremonies such as Catholic confirmation and Jewish bar and bas mitzvah can limit the young, forcing them into a narrow, negative identity. This can happen if the ceremony dogmatically spells out the specific behaviors expected by adults. On the other hand, such ceremonies can suggest to youths that the adult community now has more confidence in their ability to make decisions. The effect depends on the explanation of the goals of the ceremony.

Although some youths tend to be overly idealistic, Erikson believes that idealism is essential for a strong identity. In young people's search for a person or an idea to be true to, they are building a commitment to an ideology that will help them unify their personal values. They need ideals in order to avoid the disintegration of personality that is the basis of most forms of mental illness.

Negative Identity

While most adolescents do not go through changes as great as Erikson did in his youth, many do take on what he calls a **negative identity.** People with negative identities adopt one pattern of behavior because they are rebelling against demands that they do the opposite. An example is the boy who joins a gang of shoplifters, not because he wants to steal, but because he doubts his masculinity and seeks to prove that he is not a coward through the dangerous act of theft. Another example is the sexually permissive girl who is punishing her mother for trying to keep unreasonably strict control over her. Sex is not her goal; proving that she is no longer her mother's baby is.

Erikson suggests that the young Martin Luther was an excellent example of his concept of negative identity. Because of what happened in his youth, he spent his adulthood rebelling against what he had been taught. Here Frau Cotta, the woman who cared for him when he was 11 (in 1494) introduces the shy boy to her family.

In his psychohistorical biography of German religious leader Martin Luther (1483–1546), called *Young Man Luther* (1958), Erikson paints a somber picture of negative identity. Luther's greatness as a leader, says Erikson, was partly built on the enormous anger and unresolved conflict he experienced in his late teens. Luther's decision to become a monk and enter the monastery was the assumption of a negative identity. The choice expressed his rejection of fifteenth-century society rather than his devotion to Catholicism. Luther indulged in further contrariness by trying to be a better monk than anyone else. Luther's strong internal conflict is illustrated by the story of his falling into a faint while performing in the choir. As he fell to the ground, he is said to have cried out, "It isn't me!" Many other incidents also indicate that he couldn't accept being who he was.

Erikson believes that Luther had an extended identity crisis. His monkhood was used as the time and place for working out a positive identity. As his identity evolved, Luther devoted himself without reluctance to God and turned all his fury against the Pope, fomenting the Protestant religious upheaval. Like Erikson, Luther's identity crisis was not resolved until he reached 30.

Erikson recognizes the role of biology in his theory, through his description of eight stages that invariably follow the same sequence for everyone. However, he puts more emphasis on the psychological and social aspects of the biopsychosocial model. Would you agree that his theory is, therefore, better balanced than the others?

Identity Status

Erikson's ideas on adolescence have generated considerable research on identity formation. The leader in this field is James Marcia, who has made a major contribution to our understanding through his research on **identity status.** He and his colleagues have published numerous studies on this topic (Marcia, 1966, 1967, 1968, 1980, 1983; Cote & Levine, 1988; Craig-Bray, Adams, & Dobson, 1988; Dellas & Jernigan, 1987; Kroger & Haslett, 1988; Raphael, Feinberg, & Bachor, 1987; Rogow, Marcia, & Slugowski, 1983; Rowe & Marcia, 1980; Schiedel & Marcia, 1985; Slugowski, Marcia, & Koopman, 1984).

Marcia believes that there are two essential factors in the attainment of a mature identity. First, the person must undergo several *crises* in choosing among life's alternatives, such as the crisis of deciding whether to hold or to give up one's religious beliefs. Second, the person must come to a *commitment,* an investment of self, in his or her choices. Since a person may or may not have gone through the crisis of choice and may or may not have made a commitment to choices, there are four possible combinations, or statuses, for that person to be in:

Status 1. **Identity confusion:** No crisis has been experienced and no commitments have been made.
Status 2. **Identity foreclosure:** No crisis has been experienced, but commitments have been made, usually forced on the person by the parent.
Status 3. **Identity moratorium:** A number of crises have been experienced, but no commitments are yet made.
Status 4. **Identity achievement:** Numerous crises have been experienced and resolved, and relatively permanent commitments have been made.

Table 11.1 summarizes these definitions.

Table 11.1 Summary of Marcia's Four Identity Statuses

	Identity Status			
	Confusion	Foreclosure	Moratorium	Achievement
Crisis	Absent	Absent	Present	Present
Commitment	Absent	Present	Absent	Present
Period of adolescence in which status often occurs	Early	Middle	Middle	Late

Erikson's eight stages (in addition to the six mentioned earlier, there are the stages of generativity and integrity) follow each other in a more or less unchangeable sequence. Research indicates that Marcia's identity statuses have a tendency toward an orderly progression, but not so clearly as Erikson's stages. For example, Meilman (1979) studied males at the ages of 12, 15, 18, 21, and 24. They were rated on attitudes toward occupation, religion, politics, and, for the older subjects, sexual matters. For each of these areas, the older the group, the fewer the individuals in the confusion status and the more in the achieved status.

In fact, most of Meilman's results fit Marcia's theory well. For instance, the number of teens in the achievement category increased progressively through age 24. The largest percentages of those in the identity confusion category were age 12,

AN APPLIED VIEW — Identity Rating

Try placing people you know into one of Marcia's four statuses. Choose ten friends and write their names in the spaces below. Put the number of the identity status you choose for each person after his or her name. Do the rating quickly, without thinking about it too much—this tends to make the rating more accurate.

	Name	Status
1.	______________	______________
2.	______________	______________
3.	______________	______________
4.	______________	______________
5.	______________	______________
6.	______________	______________
7.	______________	______________
8.	______________	______________
9.	______________	______________
10.	______________	______________

Notice how many of your ratings fall into each category. Were most of them in the fourth category, identity achievement? If so, was this because most of the friends you chose for this activity have an achieved identity, or perhaps because you unconsciously chose them on that basis? Have most adults achieved identity? Have you? Once achieved, is identity a permanent state?

68 percent and age 15, 64 percent; at age 18, 48 percent were still seen to be in this category. The foreclosure category was also greater in the younger age brackets: age 12, 32 percent; age 15, 32 percent; and age 18, 24 percent. None of the 12-year-olds were found to be in the achievement category, and only 4 percent of the 15-year-olds were. The moratorium category also increased in each of the age brackets.

Grotevant, Thorbeck, and Meyer (1982) have expanded Marcia's research into the interpersonal realm, including friendships, dating, and sex roles. They feel that before forming intimate relationships, adolescents explore and commit themselves to interpersonal relationships as part of their identity formation.

Carol Gilligan (1982) and others have focused on possible gender differences in identity formation. They have concluded that women are less concerned than men with achieving an independent identity status. Women are more likely to define themselves by their relationships and responsibilities to others. Society gives women the predominant role in transmitting social values from one generation to the next. This role requires a stable identity, and therefore a stable identity appears to be more important to women than it is to men.

In summary, it may be said that the adolescent's personality is undergoing many changes, but they are probably no more traumatic than at any other stage of life. The major concern is to begin to form an adult identity, which means choosing certain values and repudiating others. There is the danger of staying in the moratorium period too long, and of forming a negative identity. It is considered necessary to work one's way from identity confusion through the moratorium to an achieved identity, while avoiding foreclosure. In the next half of this chapter, we will be dealing with the two foundations of the identity process: relationships with family and peers.

Changing American Families and Their Roles in Adolescent Life

It is, after all the simplest things we remember: A neighborhood softball game, walking in the woods at twilight with Dad, rocking on the porch swing with Grandma.

Now, no one has time to organize a ballgame. Our woods have turned to malls. Grandma lives three states away. How will our children have the same kind of warm memories we do?

B. F. Meltz,* Saving the Magic Moments, *1988

Of all the changes in American society in recent years, those affecting families have probably been the most extensive (also see chap. 7). Curiously, researchers have spent little of their time studying adolescent-parent relations—until recently. Steinberg (1987) offers an explanation for the change:

***The reasons for the rekindled interest in adolescents' relations with mothers and fathers are many, but among them surely is the increased public attention that family issues in general (e.g., divorce, stepfamilies, maternal employment, family violence) have received during the past five years.* (p. 192)**

Let us begin with a look at the changing roles of families in modern society.

The Loss of Functions

In 1840 the American family fulfilled six major functions (Sebald, 1977). Table 11.2 lists those functions and suggests which elements of society now perform them.

Table 11.2 The Changing Roles of Families

Former Family Roles	Societal Elements That Perform Them Now
Economic-productive	Factory, office, and store
Educational	Schools
Religious	Church or synagogue
Recreational	Commercial institutions
Medical	Doctor's office and hospital
Affectional	Family

Today the first five functions—economic-productive, educational, religious, recreational, medical—have been taken over by professionals. It appears that the family has been left to provide but one single function—affection for its members. In the nineteenth century, parents and children needed each other more than now, for the following three major reasons (Coleman, 1961):

- *Vocational instruction.* For both males and females, the parent of the same sex taught them their adult jobs. Most men were farmers and most women housewives. Parents knew all the secrets of work, secrets passed on from generation to generation. Today, nearly 100 percent of men work at jobs different from their fathers, and an increasing percentage of women are not primarily housewives, as their mothers were.

- *Economic value.* Adolescents were a vital economic asset on the farm; without children, the farm couple had to hire others to help them. Work was a source of pride to the children. It was immediately and abundantly clear that they were important to the family. Today, instead of being an economic asset, most children are an economic burden on the family's resources.
- *Social stability.* When families almost never moved from their hometowns, parents were a crucial source of information about how to live in the town, knowing all the intricacies of small-town social relationships. One depended on one's parents to know what to do. Today, when the average American moves every five years, the adults are as much strangers in a new place as the children. In fact, with Dad, and now frequently Mom, driving out of the neighborhood to work, the children may well know the neighborhood better than their parents do.

Bronfenbrenner suggests that teenagers today depend more on their friends and less on their families for their values than in earlier times. Possibly because of this, they seem to have less positive feelings toward their parents.

The Increase in Age-Related Activities among Older Adolescents

The change from family to peer group influence during middle adolescence is accelerating these days. This is mainly caused by the specialization of the entertainment industry and the media. Both participatory entertainment, such as sports, and spectator entertainment, such as television, are more and more aimed at specific age groups. Therefore, everyone, teenagers included, tends to watch or participate in recreational activities only with members of their own age group.

Television has been especially powerful in this changeover. When teenagers reach the age of 18, they typically will have watched twice as many hours of television as they will have spent in the classroom. These activities isolate teenagers more and more from adults (who spend less time watching television) and force them to rely on friends for security and values orientation.

Bronfenbrenner (1977) found that as teenagers depend more on their friends, they are more likely to view their parents as lacking in affection and not very firm in discipline. They also show greater pessimism about the future, rank lower in responsibility and leadership, and are more likely to engage in antisocial behavior. Even when such detrimental tendencies are not present, life changes that the child and parents usually are undergoing can make life especially difficult.

The Effects of Divorce

A smoothly functioning family can provide support and nurturance to an adolescent during times of stress. But when the family is itself in a state of disarray, such as during a divorce, not only is the support weakened, but the family often becomes a source of stress.

Divorce has become commonplace in American society. Even with slight decreases in the divorce rate in recent years, there are still over one million divorces every year, which is roughly half the number of marriages performed during the same time (U.S. Bureau of Census, 1986). Divorce tends to occur most in families with a newborn, and second most in families with an adolescent present. It is estimated that as many as one-third to one-half of the adolescent population is affected by divorce (Jurich, Schumm, & Bollman, 1987).

What, then, are the effects of divorce on the development of the adolescent? Unfortunately, conclusions are often based as much on speculation as on research findings, due to problems in the research. Divorcing parents often refuse to let themselves or their children participate in such studies, which makes random samples difficult to obtain (Santrock, 1987).

Nevertheless, it is clear that the divorcing family does contribute additional stress to a developing adolescent. One obvious effect is economic. The increased living expenses that result from the need to pay for two domiciles most often leads to a significant decrease in the standard of living for the children. Most adolescents, particularly young adolescents, are extremely status conscious, and status is often obtained with the things money can buy (clothes, stereos, cars, etc.). Young adolescents may well resent being unable to keep up with their peers in this regard. Older adolescents are better equipped to cope with this type of additional stress, both psychologically and financially, since they can enter the workforce themselves.

Some single parents find life without their spouse quite difficult; others find they like it. Regardless of the effect on the parents, the children in a divorce virtually always suffer.

Another obvious effect of a divorce is the absence of one parent. Often, custodial rights are given to one parent (usually the mother), and so the children are likely to lose an important source of support (usually that of the father). What support the noncustodial parent provides is sometimes jeopardized by the degree of acrimony between the divorced parents. One or both of the parents may attempt to "turn" the adolescent against the other parent. This sometimes results in disturbing, negative tales about a mother or father, forcing adolescents to cope with adult realities while they are still young.

Such distractions also can disrupt the disciplinary process during adolescence. Under any circumstances, administering consistent and effective discipline during this time often requires the wisdom of King Solomon. A difficult job for two parents becomes the primary responsibility of one. Preoccupied parents, perhaps feeling guilty over subjecting the child to a divorce, find it difficult to provide the consistent discipline that the child was used to previously.

Because it is the father who typically leaves a family during divorce, there are often more negative effects for males than for females (Hetherington, 1973). During the teenage years, the father often assumes primary responsibility for disciplining the male adolescents in the family. An abrupt change in disciplinary patterns can lead some adolescents to exhibit more antisocial and delinquent behavior.

Divorce may force adolescents into growing up faster and disengaging from their families. Early disengagement from a family can actually be a good solution for a teen, if she devotes herself to school activities and rewarding relationships with friends or teachers (Hetherington & others, 1989).

Early adolescents often have difficulty accepting remarriage. Older adolescents, because of their greater maturity and self-confidence, seem to have an easier time accepting remarriage, but are likely to confront or question aspects of the new family arrangements. In addition, their acute awareness of sexuality may foster resentment of the new marital closeness of their parents (Hetherington & others, 1989).

Despite the negative aspects we've outlined, you should keep in mind that not all aspects of a divorce have a negative impact on adolescents. Divorce is often a better alternative than keeping a stressful, unhappy family intact. In fact, the few studies that have compared adolescents from the two groups have shown that teenagers from divorced families do better in general than adolescents from intact but feuding families (Hetherington & others, 1989). Obviously the ability of the two parents to resolve their divorce as amicably as possible is an important factor in lessening the burden on the children. And although too many new or inappropriate responsibilities may inhibit ego formation, some added responsibilities may increase self-esteem and independence in the long run. It is also the case that many of the negative consequences of the single-parent family are relieved by a remarriage, though this is by no means always the case (Hetherington & Camara, 1984).

The Effects of Gender

A spate of new studies have investigated the part that gender plays in family life. Three generalizations may be drawn from this research (Steinberg, 1987):

- Boys and girls do not differ markedly in the way they relate to their families in general (Hauser & others, 1987; Hill & Holmbeck, 1987; Montemayor & Brownlee, 1987). An important exception is that for healthy development, many females need to become more emotionally independent from the family, whereas males do better when they maintain close ties (Cooper & Grotevant, 1987; Hakim-Larson & Hobart, 1987; Hill & Holmbeck, 1987; Silverberg & Steinberg, 1987).
- Mothers and fathers relate to their families quite differently. Fathers are more likely to be helpful in family discussion than mothers (Hauser & others, 1987); fathers spend most of their time with their adolescent children playing, while mothers spend about half of their time with them in household matters (Montemayor & Brownlee, 1987); and mothers are more likely to be involved in conflicts (Silverberg & Steinberg, 1987).
- Mother-daughter relationships are the most intense, positively and negatively, and father-daughter relationships are the most bland (Silverberg & Steinberg, 1987; Youniss & Smollar, 1985).

The Nurturing Parent

As we discussed in chapter 8, most family researchers have agreed that there are three styles of parenting (Baumrind, 1986): the **authoritarian, permissive,** and **authoritative parenting styles.** In an extensive study of fifty-six families in which at least one of the adolescents was highly creative (Dacey & Packer, 1992), a picture of a fourth style clearly emerged. The parents in these families were found to be devotedly interested in their children's behavior, but they seldom make rules to govern it. Instead, by modeling and family discussions, they espouse a well-defined set of values, and expect their children to make personal decisions based on these values.

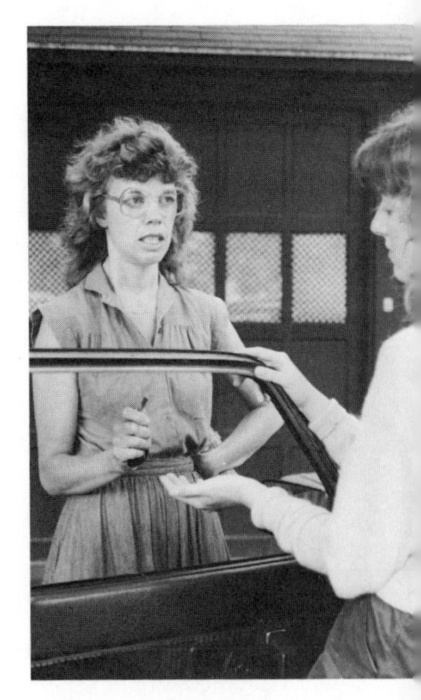

Authoritarian parents insist on strict adherence to the rules they set. The authoritarian parent is unlikely to cultivate her child's creativity.

After the children make decisions and take actions, the parents let the children know how they feel about what was done. Even when they disapprove, they rarely punish. Most of the teens in the study said that their parents' disappointment in them was motivation enough to change their behavior. All of the parents agreed that if their child were about to do something really wrong, they would stop her or him, but that this virtually never happens. We call this **nurturing parenting** (Dacey & Packer, 1992). Only some of the parents in the study are themselves creative, but all appear committed to this approach.

In addition to fostering values formation, nurturing parents also cultivate certain personality and intellectual traits that help children to make sound, insightful decisions. Among the traits are tolerance of ambiguity, risk-taking, delay of gratification, androgyny, problem-solving skills, and balanced use of brain hemispheres.

The success of nurturing parents is based on a well-established principle: People get better at what they practice. These parents provide their children with ample opportunities to practice decision-making skills, self-control, and, most vital of all, creative thinking. They serve as caring coaches as their children learn how to live. This research demonstrates the profoundly positive effects a healthy family can have on the person.

It should be obvious that your interactions with your family make a big difference in how your personality develops. Now let's turn to a consideration of the second aspect of social interaction, the peer group.

Interactions with Peers

> ***[Adolescents spend] only 4.8 percent of their time with their parents and 2 percent with adults who are not their parents—for a total of only 7 percent of waking hours spent with adults!*—Nightingale & Wolverton, 1988, p. 8**

The word **subculture** refers to any group that has its own customs (ways of dressing, for example), but is also part of a larger cultural group. College students might be considered a subculture within the American culture. Some theorists have suggested it is useful to consider adolescents as members of a subculture. In fact, it could be said that there are several such subcultures, for example, gangs, althletic teams, etc.

Adult Anxieties about the Adolescent Subculture

Thoughts of an adolescent subculture bring forth feelings of anxiety among many adults. Parents often worry that they will have less authority over the lives of their adolescent children as the subculture gains increasing influence. One of the worst fears is that under the influence of the youth subculture, teens will be pressured to be delinquent or join a gang (See chap. 13).

Stereotypes of the adolescent subculture are easily found in the media, especially movies and television. For example, adolescents are portrayed as having styles of dress, language, and music all their own. They are viewed as spending most of their time in groups, whether at school, the mall, in concerts, or on street corners. When they are not with their friends, adolescents are pictured as talking on the telephone to each other or listening to a Walkman or "boom box." Adults sometimes have difficulty telling the difference between negative peer influences, such as delinquent behavior and premature sexual activity, and the more superficial signs of the adolescent subculture, such as taste in dress and music.

Gilbert (1986), a social historian, believes that during the 1950s and 1960s almost all adolescent problems were blamed on the peer group. He believes that there were a number of reasons for this. First, it was not until the 1950s that almost all American teenagers attended high school. As a result, adolescents began to mix with other teens from more varied social and economic backgrounds. Some parents feared the influence this might have on their children. In addition, teenagers began to listen to the same music and to dress alike, which made them different from their parents and more similar to one another, despite social class background. Gilbert believes that parents of the 1950s and 1960s were much more anxious about the negative effects of the peer group than was necessary. In fact, only a small percentage of adolescents dropped out of society, becoming beatniks, hippies, or Hell's Angels.

During the 1970s and 1980s, many adolescents were more concerned with career advancement and obtaining material possessions than with causes. Some writers feared that youth were losing their idealism and "selling out" to traditional values (Levine, 1987). In the 1990s, however, an increased commitment to volunteerism and social causes has been observed among many adolescents. While the values of adolescents and their parents are often quite similar on important issues, adolescents in the 1990s continue to differ from their parents in their choices of music, movies, and styles of dress.

While it is clear that the adolescent subculture is alive and well in the 1990s, some social scientists believe it is viewed too simply. Brown (1990), for example, has documented the existence of *many* peer cultures in the United States. Not all adolescents share the same values and patterns of behavior, or even listen to the same music. The subculture adolescents belong to is influenced by many factors, including the adolescents' gender, race or ethnicity, the neighborhood in which they live, their parents' education and income levels, and historical and social events. We have already noted how values of the adolescent subculture have changed from the 1950s to the present. Brown believes that the commonly held view of adolescents united in a hedonistic culture that opposes adult society is clearly false. He argues that social scientists need to study more closely the many peer cultures that exist and change as history and society also change.

The Origins of Subcultures

How do subcultures get started? Why are there so many of them? Three major theories have attempted to explain the origin of subcultures: the psychogenic, the culture transmission, and the behavioristic theories (Sebald, 1977).

The Psychogenic Model

A subculture, any subculture, arises when a large number of people have a similar problem of adjustment, which causes them to get together to deal with the problem and help each other resolve it. Modern teenagers receive a much less practical and more abstract introduction to life than formerly. They see the world as complex and ambiguous. They often feel it is unclear how they fit in and what they ought to be doing. Many try to escape from ambiguity into a world that they create with other teenagers. In the past, this way of creating an identity was used almost solely by delinquent youngsters who were unable to find a respected place in society. Today, escape or avoidance of reality is becoming a much more common reaction to personal difficulty. This is known as the **psychogenic model** because it assumes that today's teens are psychologically disturbed by the world they are living in. Not mentally ill, but disturbed.

The Culture Transmission Model

According to the **culture transmission model,** a new subculture arises as an imitation of the subculture of the previous generation. This takes place through a learning process by which younger teenagers model themselves after those in their twenties. Magazines, movies, and television programs aimed at teenagers have been effective mechanisms for perpetuating the so-called teenage subculture. Thus, though new forms of behavior may *seem* to evolve, in actuality most are only newer versions of the solutions older people found for their problems when they were teenagers. Not surprisingly, this model argues that teenagers today are really not all that different from those of previous decades.

The Behavioristic Model

The **behavioristic model** sees subcultures starting out as a result of a series of trial-and-error behaviors, which are reinforced if they work. It is like the psychogenic model in that a new group is formed by people with similar problems. It differs in that the psychogenic model views teen behavior as innovative, whereas the behavioristic model sees peer group members behaving the way they do because they have no other choice.

According to behaviorism (See chap. 2), teenagers experience adults as "aversive stimuli"; that is, it is painful to interact with adults because in clashes over values, adults almost always win. In an attempt to escape from aversive stimuli, adolescents try out different behaviors with each other. They receive both positive reinforcement (their interactions with their peers make them feel better about themselves) and negative reinforcement (the pain they experience in interacting with adults stops when they stop interacting with the adult world).

A Multicultural View 

Racial Influences on Peer Groups

In a recent study, Steinberg (1990) examined whether parental or peer influence on academic values is stronger for adolescents of different races. Steinberg and his colleagues studied 15,000 high school students from nine different high schools in Wisconsin and California. They found that parental influence on academics was stronger only for the white students. For African-American, Latino, and Asian-American students, the peer group had a greater influence on school attitudes and behavior, including how much time students spent on their homework, whether they enjoyed school, and how they behaved in class. Fortunately for the Asian students, their peers generally valued academic achievement and positively influenced academic achievement. For African-American and Latino adolescents, it was more difficult to find and join a peer group that rewarded academic success. Consequently, these youths often experienced conflict between the positive values of their parents for academic achievement and the negative values held by their peers and did less well in school.

Similarly, Fordham and Ogbu (1986) found that African-American students felt that to be popular, they could not do well in school. When African-American students of high ability attended school with only high-achieving students, they were no longer anxious about losing peer support and were more successful.

Another factor in the perpetuation of youth subculture is **inconsistent conditioning.** For example, teenagers are expected to act responsibly in their spending, but on the other hand, they have to get parental permission for all but the smallest purchases, because they are not legally responsible for their debts. An example of inconsistent conditioning, one that no longer exists in the United States, is when teenagers are asked to fight and possibly die for their country, but are not allowed to vote and help influence their country's policies.

Which of these models best explains the origins of the youth subculture? What would you say? Can you think of a fourth explanation?

Regardless of the orientation of the group, the youth culture tends to force its members into a deeper involvement with each other.

The youth subculture tends to force its members into a deeper involvement with one another. Because teenagers spend so little time with, and derive so little influence from those older and younger than themselves, they have only one another to look to as models. Today, those under 17 make up a much larger percentage of the population than formerly, and they continue their education for a much longer period. As a result, they spend much more time with other youths and much less time with adults.

A proposal by Newman and Newman (1976) recognizes the growing importance of this tendency. They suggest that we divide Erikson's identity stage into two stages: early adolescence (ages 13 through 17), called the *group identity versus alienation stage,* and later adolescence (ages 18 through 22), called the *individual identity versus role diffusion stage.* This division recognizes that it is also necessary to identify with a group or groups in order to achieve a well-resolved personal identity.

Although we agree with their two-stage concept, we disagree with the ages they suggest. On the basis of our view of the research, early adolescence starts at about 11 for females and about a year later for males (this difference will be explained in the next chapter). Middle adolescence starts at about 14, and late adolescence occurs at about 17, leading into early adulthood at 19. We have chosen 19 because this is the age at which youths have usually been out of high school and into jobs or college for one year. We think this year represents a major turning point in the development of most individuals in the Western hemisphere.

Conclusion

Being an adolescent is no easy task these days. It probably never was, but now there are so many choices, and there is a great temptation to try and take them all! It is unclear whether adolescent peer group relationships are all that different from earlier times, but the family in the United States appears to be undergoing major alterations. What is clear is that with every passing year, the peer group plays a greater and greater role in the adolescent's life, while the roles of families and other societal systems (the media, the workplace, etc.) continue to change.

As Nightingale and Wolverton put it in their report to the Carnegie Council on Adolescent Development (1988),

> ***Adolescents have no prepared place in society that is appreciated or approved; nonetheless they must tackle two major tasks, usually on their own: identity formation, and development of self-worth and self-efficacy. The social environment of adolescents today makes both tasks very hard. . . . We must change the view that many people hold of all youth as troubled and harmful to the rest of society.*** **(pp. 1, 16)**

In this chapter, we have given an overview of the personality and social development of adolescents. This is only half of the picture, though. Of at least equal importance are the critical factors of physical and mental development, which we report on in the next chapter.

Chapter Highlights

How Should We Define Adolescence?

- There are fewer adolescents in the 1990s than there used to be, and they make up a smaller proportion of the total population.
- Thinking back on your own adolescence can help you to have a deeper understanding of today's teenagers.
- In ancient times, some philosophers believed that youths were frivolous and irresponsible, while others emphasized their growing intellectual skills and self-sufficiency.
- From the Middle Ages until the start of the twentieth century, strict discipline was believed necessary to force young people to take on adult responsibilities as early as possible.
- Two early twentieth century concepts changed our view of adolescence: compulsory education and juvenile justice.
- Most experts state that the majority of adolescents are happy and productive members of their families and communities (see the Carnegie Report, 1990).
- According to G. Stanley Hall, all human beings pass through four periods of development: birth to 4 years, 4 to 8 years, 8 to 12 years, and 12 to 25 years.
- Hall's interpretation of adolescent development was greatly influenced by his observation that it is a period of *"sturm und drang"* (storm and stress).

Theories of Adolescence: Anna Freud

- Anna Freud believed that the delicate balance between the superego and the id, being disrupted by puberty, causes the adolescent to regress to earlier stages of development.

Theories of Adolescence: Robert Havighurst

- According to Robert Havighurst, each stage in development had specific developmental tasks—skills, knowledge, functions and attitudes—that are needed by a person to succeed in life.

Theories of Adolescence: Erik Erikson

- Human life progresses through eight "psychosocial" stages, each of which is marked by a crisis and its resolution. The fifth stage applies mainly to adolescence.
- While the ages at which one goes through each stage vary, the sequence of stages is fixed. Stages may overlap, however.
- A human being must experience each crisis before proceeding to the next stage. Inadequate resolution of the crisis at any stage hinders development.

Changing American Families and Their Roles in Adolescent Life

- American families have lost five of their six main functions; the only remaining one is providing affection for family members.
- A significant rise in age-related activities has tended to reduce effective communication within families.
- A number of effects of divorce pertain only to adolescents.

Interactions with Peers

- Adults may fear that the subculture of adolescence will negatively influence their teens, and may interpret superficial factors of taste in dress and music as warnings of developing delinquency or premature sexual activity.
- In the 1960s, adolescents were depicted as highly rebellious, but this view has changed. Many teenagers of the 1990s are committed to volunteering in community and social causes.
- There is some evidence that the adolescent subculture in America is not unitary, but consists of many different peer cultures.
- There are three major theories of the origins of subcultures: the psychogenic model, the culture transmission model, and the behavioristic model.

Key Terms

Authoritarian parenting style
Authoritative parenting style
Behavioristic model
Culture transmission model
Developmental tasks
Identity achievement
Identity confusion
Identity crisis
Identity foreclosure
Identity moratorium
Identity status
Inconsistent conditioning
Moratorium of youth
Negative identity
Nurturing parenting style
Permissive parenting style
Premature foreclosure
Psychogenic model
Recapitulation theory
Repudiation
State of identity
Stereotype
Storm and stress
Subculture

What Do You Think?

1. Do you believe you have had an "identity crisis"? If so, what makes you think so?
2. Which explanation in this chapter best explains the origins of the youth subculture? What is your own explanation?
3. What would you say are the characteristics of an adolescent? Try to include examples of actual behavior in your response.
4. In what ways do you think adolescence may have changed over the centuries?
5. Do you know anyone whom you feel has a negative identity? What is this person like?
6. In what ways is your nuclear family different from your mother's or your father's?

Suggested Readings

Auel, J. (1981). *Clan of the cave bear.* New York: Bantam. Auel's wonderful imagination and excellent knowledge of anthropology make this book on the beginnings of the human family a winner. In fast-paced fiction, she describes the relationships between two types of primitive peoples—those who communicate by voice and those who do so with their hands!

Erikson, E. (1958). *Young man Luther.* New York: Norton. Erikson picked Martin Luther as a subject because, in Erikson's view, he was a famous case of negative identity. This book also closely examines the Protestant Reformation and so may appeal to you if you are interested in the beginnings of the Protestant religions.

Goldman, W. (1962). *Lord of the flies.* New York: Coward-McGann. This tale of a group of teenage boys whose plane crashes on a Pacific island, killing the adults, is must reading. You watch the subgroups develop and proceed to the shocking ending.

McCullers, C. [1946] (1985). *Member of the wedding.* New York: Bantam. Twelve-year-old Frankie yearns desperately to join her brother and his bride on their honeymoon. She learns a great deal about the transition from childhood to maturity from the devoted housekeeper.

Moravia, A. (1958). *Two women.* New York: Farrar, Straus & Giroux. This moving, compassionate tale describes the relationship between a peasant mother and her daughter in war-torn Italy. It involves the struggles of the mother to deal with her adolescent daughter's needs under these extremely trying circumstances.

Wright, R. (1945). *Black boy.* New York: Harper & Row. This is a moving autobiography of the novelist's adolescence in the deep South. Its insights are entirely relevant to today's world.

CHAPTER 12

Adolescence: Rapidly Changing Bodies and Minds

Puberty 322
Early Studies 322
Your Reproductive System 323
When Does Puberty Start? 326
The Effects of Timing on Puberty 327
Cognitive Development 332
Variables in Cognitive Development: Piaget 332
Variables in Cognitive Development: Flavell 333
Adolescent Egocentrism 336
Critical Thinking 337
Creative Thinking 338
The Use of Metaphor 339
Creativity, Giftedness, and the IQ 340
Conclusion 342
Chapter Highlights 343
Key Terms 344
What Do You Think? 344
Suggested Readings 345

Gretchen, my friend, got her period. I'm so jealous, God. I hate myself for being so jealous, but I am. I wish you'd help me just a little. Nancy's sure she's going to get it soon, too. And if I'm last, I don't know what I'll do. Oh, please, God. I just want to be normal.

Judy Blume, Are You There, God? It's Me, Margaret, *1970*

If you want to understand adolescence, you will surely need to know quite a bit about puberty. In this chapter, we'll explain its biological basis, the sequence of events that make it up, the contrast of changes for males and females, and the effects of timing. We'll also be examining stages of cognitive development, as well as research on egocentrism and critical and creative thinking.

When you have finished studying this chapter, you will be able to:

- Identify the important parts of the male and female reproductive systems, and explain their functions.
- List the normal sequence of events in puberty for males and females.
- Contrast male and female development in puberty.
- Describe the influence of timing on individual adolescents' emotional reactions to the physical changes of puberty.
- List and identify Piaget's four main stages of cognitive development.
- Explain the cognitive development that takes place in early and late phases of the formal operations stage.
- List Flavell's seven aspects of the transition from childhood thinking to adolescent and adult thinking.
- Describe the major elements of egocentric thinking: the imaginary audience and the personal fable.
- Identity the differences between critical thinking and creative thinking, and discuss the importance of each.
- Show why the use of metaphor is important, especially for creative thinking. ■

PUBERTY

To better understand puberty, we'll answer the following questions in this section: What parts of our bodies are involved? When does puberty start? What are the effects of timing?

Early Studies

***The girls are clearly beginning to look like young ladies, while the boys with whom they have thus far played on scarcely equal terms now seem hopelessly stranded in childhood. This year or more of manifest physical superiority of the girl, with its attendant development of womanly attitudes and interests, accounts in part for the tendency of many boys in the early teens to be averse to the society of girls. They accuse them of being soft and foolish, and they suspect the girls' whispering and titterings of being laden with unfavorable comments regarding themselves.* (King, 1914, p. 13)**

This quaint and somewhat condescending description of the differences between males and females is typical of many of the adolescent theorists of the early twentieth century (e.g., Boas, 1911; Bourne, 1913; Burnham, 1911; King, 1914). Understandably, these writers had far less data available than we do today, and their opinions were largely subjective. For example, King (1914) suggested that the major cause of **delayed puberty** was "excessive social interests, parties, clubs, etc., with their attendant interference with regular habits of rest and sleep" (p. 25). He further stated that the *second* major cause of delayed puberty was "an excess of physical work." It is unlikely that the work itself caused the delay, but rather that poor children, who were most likely to work excessively, were also likely to be poorly fed. It is well known today that malnutrition, whether from poverty or from an eating disorder, will delay puberty.

Your Reproductive System

Much more is known today about many aspects of puberty. We know more about the organs of our reproductive system and how these organs function together. And we are learning how to present this knowledge to adolescents effectively.

How well do you know your own reproductive system? First, take the test in the "Applied View", and then read the following sections on the female and the male sexual systems.

An Applied View — How Well Do You Know Your Reproductive System?

Most of us seem to think we understand the workings of sex and reproduction well enough, yet when asked to define the various parts of our sexual system, we don't do very well. How high would you rate your knowledge?

If you would like to learn how much you really know (and this knowledge is important, if only because the adolescents you deal with may ask you questions about it), take this test. Put an M after each male sex organ, an F after each female sex organ, or M/F if it is both. You can find the correct answers by examining Figures 12.1 and 12.2.

	M, F, M/F		M, F, M/F
Bartholin's glands	_____	Mons pubis (mons veneris)	_____
Cervix	_____	Ova	_____
Clitoris	_____	Ovary	_____
Cowper's glands	_____	Pituitary gland	_____
Epididymis	_____	Prostate	_____
Fallopian tubes	_____	Scrotum	_____
Fimbriae	_____	Testes	_____
Foreskin	_____	Ureter	_____
Glans penis	_____	Urethra	_____
Hymen	_____	Uterus	_____
Labia majora	_____	Vas deferens	_____
Labia minora	_____	Vulva	_____

This test has been given to groups of college sophomores and graduate students. The highest possible score is 24. The sophomores, averaged 19 on the test. The graduate students averaged 13! Why do you suppose the older students got lower scores on the average?

The Female Sexual System

The parts of the female sexual system are defined here and are illustrated in Figure 12.1.

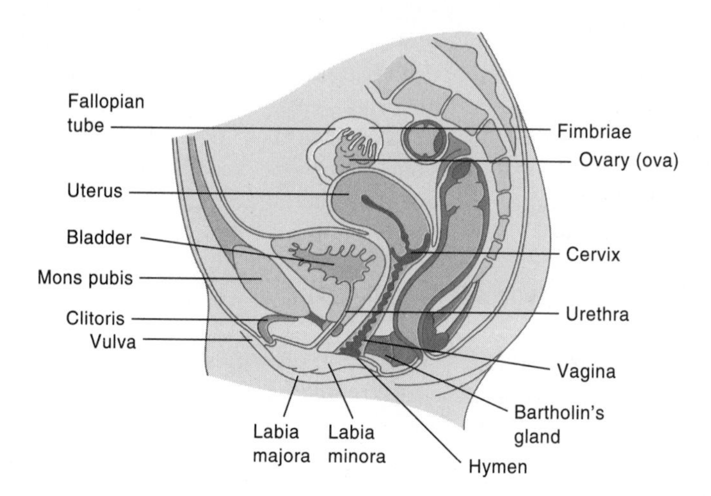

Figure 12.1
The female reproductive system

- *Bartholin's glands.* A pair of glands located on either side of the vagina. These glands provide some of the fluid that acts as a lubricant during intercourse.
- *Cervix.* The opening to the uterus located at the inner end of the vagina.
- *Clitoris.* Comparable to the male penis. Both organs are similar in the first few months after conception, becoming differentiated only as sexual determination takes place. The clitoris is the source of maximum sexual stimulation and becomes erect through sexual excitement. It is above the vaginal opening, between the labia minora.
- *Fallopian tubes.* Conduct the ova (egg) from the ovary to the uterus. A fertilized egg that becomes lodged in the fallopian tubes, called a fallopian pregnancy, cannot develop normally and if not surgically removed will cause the tube to rupture.
- *Fimbriae.* Hairlike structures located at the opening of the oviduct that help move the ovum down the fallopian tube to the uterus.
- *Hymen.* A flap of tissue that usually covers most of the vaginal canal in virgins.
- *Labia majora.* The two larger outer lips of the vaginal opening.
- *Labia minora.* The two smaller inner lips of the vaginal opening.
- *Mons pubis or mons veneris.* The outer area just above the vagina, which becomes larger during adolescence and on which the first pubic hair appears.
- *Ova.* The female reproductive cells stored in the ovaries. These eggs are fertilized by the male sperm. Girls are born with more than a million follicles, each of which holds an ovum. At puberty, only 10,000 remain, but they are more than sufficient for a woman's reproductive life. Since usually only one egg ripens each month from the midteens to the late forties, a woman releases fewer than 500 ova during her lifetime.

- *Ovaries.* Glands that release one ovum each month. They also produce the hormones estrogen and progesterone, which play an important part in the menstrual cycle and pregnancy.
- *Pituitary gland.* The "master" gland located in the lower part of the brain. It controls sexual maturation and excitement, and monthly menstruation.
- *Ureter.* A canal connecting the kidneys with the bladder.
- *Urethra.* A canal leading from the bladder to the external opening through which urine is excreted.
- *Uterus.* The hollow organ (also called the womb) in which the fertilized egg must implant itself for a viable pregnancy to occur. The egg attaches itself to the lining of the uterus from which the unborn baby draws nourishment as it matures during the nine months before birth.
- *Vulva.* The external genital organs of the female.

The Male Sexual System

The parts of the male sexual system are defined here and are illustrated in Figure 12.2.

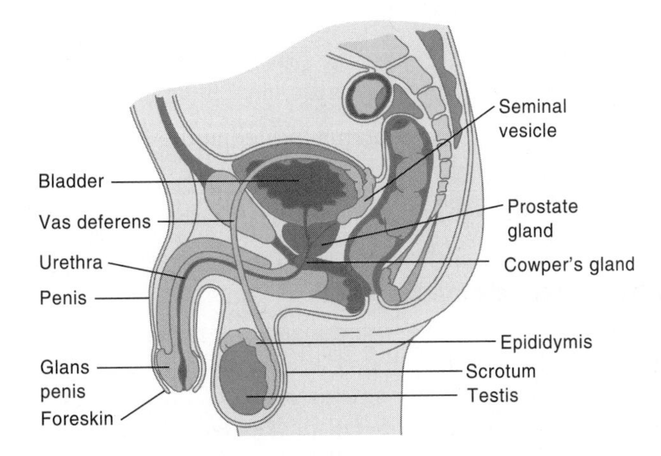

Figure 12.2
The male reproductive system

- *Cowper's glands.* Located next to the prostate glands. Their job is to secrete a fluid that changes the chemical balance in the urethra from an acidic to an alkaline base. This fluid proceeds up through the urethra in the penis, where it is ejaculated during sexual excitement just before the sperm-laden semen. About a quarter of the time, sperm also may be found in this solution, sometimes called *preseminal fluid.* Therefore, even if the male withdraws his penis before he ejaculates, it is possible for him to deposit some sperm in the vagina, which may cause pregnancy.
- *Epididymis.* A small organ attached to each testis. It is a storage place for newly produced sperm.
- *Foreskin.* A flap of loose skin that surrounds the glans penis at birth, often removed by a surgery called *circumcision.*
- *Glans penis.* The tip or head of the penis.

- *Pituitary gland*. The "master" gland controlling sexual characteristics. In the male it controls the production of sperm, the release of testosterone (and thus the appearance of secondary sexual characteristics such as the growth of hair and voice change), and sexual excitement and maturation.
- *Prostate glands*. Produce a milky alkaline substance known as semen. In the prostate the sperm are mixed with the semen to give them greater mobility.
- *Scrotum*. The sac of skin located just below the penis, in which the testes and epididymis are located.
- *Testes*. The two oval sex glands, suspended in the scrotum, that produce sperm. Sperm are the gene cells that fertilize the ova. They are equipped with a tail-like structure that enables them to move about by a swimming motion. After being ejaculated from the penis into the vagina, they attempt to swim through the cervix into the uterus and into the fallopian tubes, where fertilization takes place. If one penetrates an egg, conception occurs. Although the testes regularly produce millions of sperm, the odds against any particular sperm penetrating an egg are enormous. The testes also produce testosterone, the male hormone that affects other aspects of sexual development.
- *Ureter*. A canal connecting each of the kidneys with the bladder.
- *Urethra*. A canal that connects the bladder with the opening of the penis. It is also the path taken by the preseminal fluid and sperm during ejaculation.
- *Vas deferens*. A pair of tubes that lead from the epididymis up to the prostate. They carry the sperm when the male is sexually aroused and about to ejaculate.

When Does Puberty Start?

Is there any one physiological event that marks the beginning of adolescence? The sequence of bodily changes in puberty is surprisingly constant. This holds true whether puberty starts early or late and regardless of the culture in which the child is reared. Table 12.1 lists the sequences of physiological change.

Which of these physical events in the life of the adolescent might we choose as the actual beginning of puberty? Change in hormonal balance is first, but its beginning is difficult to pinpoint. Skeletal growth, genital growth, pubic hair, breast development, voice change, growth spurt—all are inconvenient to measure. Menarche has been suggested as the major turning point for girls, but many women do not recall menarche as a particularly significant event. Sometimes the first ejaculation is suggested as the beginning of adolescent puberty for males, but this too is often a little-remembered (and possibly repressed) event.

Despite the fact that puberty is primarily thought of as a physical change in an adolescent, the psychological impact can be significant. This is especially true with menstruation. Even in these "enlightened" times, too many girls experience menarche without being properly prepared. As a result, an event in a young girl's life that should be remembered as the exciting, positive beginning of the transition to adulthood is instead viewed as a negative, sometimes frightening, experience. Research suggests that better preparation for first menstruation results in more positive attitudes about it (Koff & others, 1980). The same is probably also the case for a male's first ejaculation.

Table 12.1 The Sequence of Physiological Change in Males and Females

Females

- Change in **hormonal balance**
- The beginning of rapid **skeletal growth**
- The beginning of breast development
- The appearance of straight, pigmented pubic hair
- The appearance of kinky, pigmented pubic hair
- **Menarche** (first menstruation)
- **Maximum growth spurt** (when growth is at its fastest rate)
- The appearance of hair on the forearms and underarms

Males

- Change in hormonal balance
- The beginning of skeletal growth
- The enlargement of the genitals
- The appearance of straight pigmented pubic hair
- Early voice changes (voice "cracks")
- First ejaculations (wet dreams, nocturnal emissions)
- The appearance of kinky, pigmented pubic hair
- Maximum growth spurt
- The appearance of downy facial hair
- The appearance of hair on the chest and underarms
- Late voice change (the voice deepens)
- Coarse, pigmented facial hair

Note: For more on these physiological changes, see Muuss (1982).

Given our understanding of the physiology of adolescents and differences in individual psychology and culture, we would have to conclude that *there is no single event marking the onset of puberty but rather a complex set of events,* a process whose effects may be sudden or gradual.

The Effects of Timing on Puberty

In a general sense, the onset of puberty affects all adolescents in the same way. However, the *age* at which these changes begin has some very specific effects on the adolescent's life. (The photo in Figure 12.3, taken in 1912, illustrates how 14-year-old adolescents can differ greatly in their stage of physiological development—and did so even many years ago!)

In this section, eight adolescents (four females and four males) are compared in order to illustrate the differences that often occur among children even though they are all in the **normal range of development.** Each adolescent is 14 years old. The first female and male are early maturers, the second are average maturers, and the third are late maturers. They all fall within the typical range of all adolescents. The fourth female and male represent average adolescents of one hundred years ago.

Figure 12.3
Comparison of male and female growth (King, 1914)

The Early–Maturing Female: Ann

At 5 feet, 5 inches and 130 pounds, Ann is considerably bigger than her age-mates. Her growth accelerated when she was 8 years old, and by the time she was 10 1/2, her maximum growth spurt crested. She is still growing taller, but at a slower rate. Her *motor development* (coordination and strength) had its greatest rate of increase two years ago. She is stronger than her age-mates, but her strength and coordination have reached their maximum.

She started menstruating three years ago, at age 11, and her breasts are already in the secondary (adult) stage. Her pubic and underarm hair are also at an adult stage.

Ann is confused about the way her body looks. She feels conspicuous and vulnerable because she stands out in a crowd of her friends. Her greater interest in boys, and their response, often causes conflicts with other girls. They envy the interest the boys show in her more mature figure. She often has negative feelings about herself because she is "different."

Other girls tend to avoid her now because her early **maturation** makes her older than they are. In later adolescence, she may experience some difficulties; she may find herself in situations (such as with drugs, sex, or drinking) she is not yet ready for. We can say that Ann is experiencing difficulties with her early maturity, but she will begin to feel better about herself as her contemporaries catch up with her developmentally.

Girls who reach puberty late are usually unhappy about it, but not as unhappy as boys who are late. Boys who reach physical maturity early are usually quite happy about it—more happy than girls who are early.

The Average-Maturing Female: Beth

Although Beth is also 14, she is different in almost every way from Ann. She represents the typical adolescent today in the sense of being average in her measurements and physical change. It is clear that there is no "average" adolescent from the standpoint of personality and behavior.

Beth is 5 feet, 3 inches tall and weighs 120 pounds. She reached her maximum growth spurt two years ago and is also starting to slow down. She is presently at the peak of her motor development.

Her breasts are at the primary breast stage; she is beginning to need a bra, or thinks she does. She started menstruating two years ago. She has adult pubic hair, and her underarm hair is beginning to appear.

She feels reasonably happy about her body, and most but not all of her relationships with her peers are reasonably satisfying. Although she does have some occasional emotional problems, they are not related to her physical development as much as are Ann's.

The Late-Maturing Female: Cathy

Cathy is at the lower end of the normal range of physical development for a 14-year-old girl. She is only 4 feet, 8 inches tall, weighs 100 pounds, and is just beginning her growth spurt. She is not too happy about this; she feels that the other girls have advantages in relationships with boys.

Cathy's breasts are at the bud stage; her nipples and encircling areolae are beginning to protrude, but she is otherwise flat-chested. She has just begun menstruation. Pubic hair growth has started, but as yet no hair has appeared under her arms.

Other girls tend to feel sorry for her, but they also look down on her. She is more dependent and childlike than the others. She feels a growing dislike for her body, and she is becoming more and more introverted and self-rejecting because of it. At this stage her immaturity is not a great disability; at least she is more mature than some of the boys her age. As she reaches later adolescence, her underdeveloped figure may be a more serious source of unhappiness for her if she accepts conventional standards of sexual desirability.

The Average Adolescent Female of One Hundred Years Ago: Dorothy

Although records of adolescent physical development of one hundred years ago are less than adequate, we can be fairly certain about some of the data. Dorothy, who was typical for her time, was physically much like Cathy is now. At 4 feet, 7 inches, she was one inch shorter, and at 85 pounds, she weighed 15 pounds less than Cathy.

At age 14, Dorothy would still have had four years to go before her peak of motor development, and she would not have started to menstruate for another year. In all the other physical ways, she looked a great deal like Cathy. The major difference between the two girls is that while Cathy is unhappy about her body's appearance, Dorothy, who was typical, felt reasonably good about hers.

The Early-Maturing Male: Al

Al finds that at 5 feet, 8 inches tall, he towers over his 14-year-old friends. He reached his maximum growth spurt approximately two years ago and weighs 150 pounds. He is now about two years before the peak of his motor development. His coordination and strength are rapidly increasing, but contrary to the popular myth he is not growing clumsier.

As adolescents reach their peak of motor development, they usually handle their bodies better, although adults expect them to have numerous accidents. It is true that when one's arms grow an inch longer in less than a year, one's hand-eye coordination suffers somewhat. However, the idea of the gangling, inept adolescent is more myth than fact.

Al's sexual development is also well ahead of that of his age-mates. He already has adult pubic hair, and hair has started to grow on his chest and underarms. He began having nocturnal emissions almost two years ago, and since then the size of his genitals has increased almost 100 percent.

Because our society tends to judge male maturity on the basis of physique and stature, Al's larger size has advantages for him. He is pleased with his looks, although once in a while it bothers him when someone treats him as though he were 17 or 18 years old. Nevertheless, he uses the advantages of his early maturity whenever possible.

His friends tend to look up to him and to consider him a leader. Because size and coordination often lead to athletic superiority, and because success in school sports has long meant popularity, he has the most positive self-concept of all the adolescents described here, including the females. He has a good psychological adjustment, although he is sometimes vain, and is the most confident and responsible of this group. He engages in more social activities than the others, which also occasionally gets him into trouble, because he is not psychologically ready for some of the social activities in which he is permitted to participate.

The Average-Maturing Male: Bob

Interestingly, Bob is exactly the same height as his "average" female counterpart, Beth, at 5 feet, 3 inches tall. At 130 pounds, he outweighs her by 10 pounds. He is currently in the midst of his maximum growth spurt and is four years away from reaching the peak of his coordination and strength.

An Applied View — Dealing with Early or Late Development

Peterson & others (1988) found an increased risk of sexual abuse for the early-maturing female. Significantly early or late development has been linked to depression and eating disorders in both boys and girls (Rierdan & others, 1988), so it is important for practitioners working with adolescents to be aware of negative reactions to their physical change (or lack of it). If you know a teen who is *significantly* early or late in body development, be on the lookout for psychological problems. If you find evidence of such problems, make arrangements for the youth to have access to appropriate professional attention.

On the other hand, you should understand that "normal" covers a wide band of developmental time. We should help teens who "hate" their bodies because they are not perfectly average to be more accepting. Finally, if you do feel there is a problem, say nothing to the teen until you have an experienced person's advice on the best course of action.

His sexual development began about a year ago with the start of nocturnal emissions, and he is just now starting to grow pubic hair. As yet he has no hair on his chest or under his arms. However, his genitals have reached 80 percent of their adult size.

Bob gets along well with his age-mates. He is reasonably happy with the way his body has developed so far, although there are some activities that he wishes he could excel in. Most of the attributes that he aspires to are already possessed by Al, whom he envies. This causes few problems, since Bob still has every reason to hope his body will develop into his ideal physical image.

The Late-Maturing Male: Chuck

Chuck is also similar in stature to his counterpart, Cathy. They are both 4 feet, 8 inches tall, although at 90 pounds, Chuck is 10 pounds lighter than Cathy. He is a year and a half away from his maximum growth spurt and must wait six years before his motor development will peak.

Chuck's sexual development is also lagging behind those of the other two boys. His genitals are 50 percent larger than they were two years ago, but as yet he has no pubic, chest, or underarm hair. He has not yet experienced nocturnal emissions, although these are about to begin.

He is the least happy with his body of the adolescents described here. His voice has not yet changed, he is not as strong and coordinated as the other boys, and he is much smaller than they are. They tend to treat him as a scapegoat and often ridicule him. He chooses to interact with boys who are younger than himself and is attracted to activities in which mental rather than physical prowess is important, such as chess and band. He avoids girls, almost all of whom are more physically mature than he. This lack of heterosexual experience may later affect his self-concept.

Chuck lacks confidence in himself and tends to be dependent on others. He was of almost average size in grammar school and now feels he has lost prestige. He frequently does things to gain the attention of others, but these actions seldom bring him the acclaim he craves. Probably as a result, he is more irritable and restless than the others and engages in more types of compensating behaviors.

The Average Adolescent Male of One Hundred Years Ago: Dan

At 4 feet, 7 inches, Dan was shorter than Chuck by one inch, and weighed the same, 90 pounds. He trailed Chuck in sexual development by two years, and, at age 14, his genitals had increased only 20 percent in size.

However, the major difference between Chuck and Dan lies in self-satisfaction. While Chuck is extremely unhappy about the way he is developing, Dan was as happy as Bob is now, because he was quite average for that time. Although we cannot know what his relationship with peers was like or how he viewed himself, we can guess that these were quite similar to Bob's.

The preceding descriptions illustrate the great variability in adolescent growth and in adolescent responses to growth. Keep in mind, however, that self-image and peer relationships are not entirely determined by physique. "Average-sized" adolescents do not always lead a charmed life, and many late and early maturers are quite comfortable with themselves. Adolescents who have clarified their values and set their own standards are not likely to be overly affected by pubertal changes or the peer approval or disapproval brought about by them.

The Secular Trend

The small size of Dorothy and Dan, who were average teenagers one hundred years ago, is part of the phenomenon called the **secular trend.** The secular trend refers to the decreasing age of the onset of puberty, including a significant *drop* in the average age at which females in a particular country reach menarche. In Western countries, the average age of menarche has declined about three months per decade over the past hundred years. In the United States, 17 was the average age in the late eighteenth and early nineteenth centuries (Vaughan & Litt, 1990).

Today the average age of onset is 12.5 years. Most researchers feel that improved nutrition, sanitation, and health care are responsible for the trend, and that we are now at a period of leveling off (Brooks-Gunn & others, 1985). We think nutrition is involved because girls must typically achieve a certain proportion of body fat before they can menstruate (Frisch, 1988). Studies of female athletes and dancers have shown that a lack of fat can delay menarche or can stop menstruation after it has begun (Brooks-Gunn, 1987).

Will the secular trend continue in future years? No, it is likely that the trend is a special case that pertains only to the period from the 1700s to today, and is the result of improvements in nutrition, medicine, and health care in general.

Figure 12.4 details the age ranges considered normal for development. Adolescents who experience these changes earlier or later may have no medical problem, but it is probably a good idea to consult a doctor. If a glandular imbalance exists, the doctor can usually remedy the problem with little difficulty.

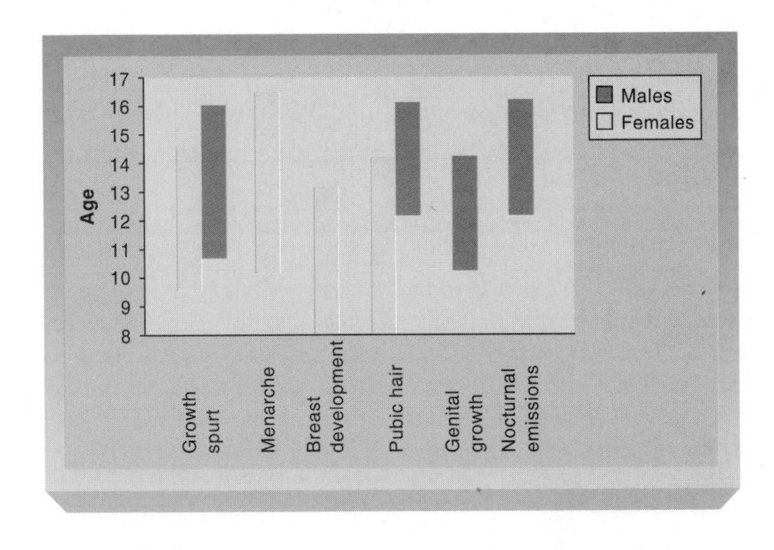

Figure 12.4
Normal age ranges of puberty

In summary, we may say that the vast majority of human bodies proceed toward maturity in the same way, but that in the last few centuries the timing of the process in females has changed radically. Although timing is affected mainly by biology, psychological and social forces clearly influence it, too.

Cognitive Development

I was about twelve when I discovered that you could create a whole new world just in your head! I don't know why I hadn't thought about it before, but the idea excited me terrifically. I started lying in bed on Saturday mornings till 11 or 12 o'clock, making up "my secret world." I went to fabulous places. I met friends who really liked me and treated me great. And of course I fell in love with this guy like you wouldn't believe!

Susan Klein, an eighth-grade student

Adolescence is a complex process of growth and change. Because biological and social changes are the focus of attention, changes in the young adolescent's ability to think often go unnoticed. Yet it is during early and middle adolescence that thinking ability reaches Piaget's fourth and last level—the level of abstract thought (See chap. 2). To understand how abstract thought develops, we have to know more about cognition itself.

Variables in Cognitive Development: Piaget

According to Jean Piaget's theory (as we hope you remember from chap. 2), the ability to think develops in four stages:

The ability to understand that the rules of games can be fairly changed is an aspect of the formal operation stage. Even though these young men are unlikely to change the rules of their chess game, they are now at the intellectual level at which they recognize this is a possibility.

- *The sensorimotor stage (birth to 2 years),* in which the child learns from its interactions with the world.
- *The preoperational stage (2 to 7 years),* in which behaviors such as picking up a can are gradually internalized so that they can be manipulated in the mind.
- *The concrete operational stage (7 to 11 years),* in which actions can be manipulated mentally, but only with things. For example, a child of 8 is able to anticipate what is going to happen if a can is thrown across the room without actually having to do so.
- *The formal operational stage (11 years +),* in which groups of concrete operations are combined to become formal operations. For example, the adolescent comes to understand democracy by combining concepts such as putting a ballot in a box and hearing that the Senate voted to give money for the homeless. This is the stage of abstract thought development.

We are like other animals, especially the primates, in many ways. They, too, can make plans, can cooperate in groups, and may well have simple language systems. But it is the ability to perform formal operations that truly separates us. Piaget's conception of the formal operation is a remarkable contribution to psychology.

It was also Piaget who first noted the strong tendency of early adolescents toward democratic values because of this new thinking capacity. This is the age at which youths first become committed to the idea that participants in a group may change the rules of a game, but once agreed on, all must follow the new rules. This tendency, he believes, is universal; all teenagers throughout the world develop this value.

Having considered Piaget's ideas about adolescent cognition in some detail earlier in this book, let us turn now to a review of the findings of those researchers who have been diligently following him.

AN APPLIED VIEW Defining Democracy

For adolescents to improve their ability to do formal operations, they need to be aware of what abstract thinking is. Furthermore, they need to have plenty of practice in doing it. Suppose you are the teacher of an eighth grade social studies class. You can help your students with both of these tasks by asking them to explain the meaning of such abstract terms as *citizenship, truth, beauty, honor, creativity,* and *adolescence.*

Suppose we use the term *democracy.* Ask your students to suggest examples of democracy; try to get fifteen or twenty examples. Then ask them to group these ideas in three piles: those that are clearly democratic; those that are mostly democratic; and those that are less democratic. Then ask the students to list the criteria that distinguish between the first and last pile. In this way, through induction, they will come to a better understanding of democracy.

As you will see, the definitions of each of the items that the students generate ultimately depend on concrete actions. By helping your students to realize what they are doing when they define these terms, and how this kind of thinking differs from concrete operations, you are giving them opportunities to grow in their formal thinking skills.

Variables in Cognitive Development: Flavell

Psychologist John Flavell (1985), on the basis of his own studies and his review of the literature, has suggested that there are seven aspects of the transition from childhood to adolescent and adult thinking. These seven aspects are considered in the following sections, from the standpoint of an imaginary adolescent situation.

We would like you to imagine the following scenario. Your local television station has decided to sponsor a new program called "Young Women Today." It has already hired a 29-year-old woman to host the show, a person who has had considerable experience in the talk-show format.

In addition, they want to hire a 17-year-old high school girl as an assistant hostess. This has been announced at the local high school. Because the employers are sure that many female students will apply for this job, they have designed a simple application form. With it, they hope to identify the better candidates, whom they will then interview. On this application form, the applicants are to describe two characteristics that would make them an especially good choice for the job. Here are Flavell's seven aspects of the childhood to adolescent transition, analyzed from the standpoint of this scenario.

The Real versus the Possible

For Piaget (1966), probably the most important cognitive change from childhood to adolescence has to do with the growing ability of the adolescent to imagine possible and even impossible situations—the aspect Flavell calls **the real versus the possible.** Elementary school children tend to approach problems by examining the data firsthand and attempting to make guesses about the solution to the problem on the basis of the first piece of information they look at. As Flavell (1985) suggests, the child has "an earthbound, concrete, practical-minded sort of problem approach, one that persistently fixates on the perceptible reality there in front of him" (p. 103). The preliminary approach is a series of guesses as to what information can mean.

This is no longer the case for 17-year-old Ellen, who would very much like to get the television show job. As it happens, she is the daughter of an unemployed actor. She wonders whether she should mention this. She thinks that it may help her, both because her father has been in show business and because her family could use the money. On the other hand, she suspects that the television people might think that if her father has been unsuccessful in show business, she may be, too. Whereas a 9-year-old child might put this fact down on the application, hoping that it will make some difference, Ellen tries to think of all the possible ramifications of this piece of information before deciding whether to use it.

Ellen is obviously a formal operational thinker. For her, as Flavell would describe it, "reality is seen as that particular portion of the much wider world of possibility which happens to exist and hold true in a given problem situation" (p. 103).

Empirico-Inductive versus Hypothetico-Deductive Methods

A further major difference between concrete and formal operational thinkers is in their *method* of problem solving. Younger children use an **empirico-inductive** approach; that is, they are likely to look at available facts and try to induce some generalization from them. Adolescents, on the other hand, are likely to use **hypothetico-deductive** reasoning to hypothesize about the situation and then deduce from it what the facts *should be* if the hypothesis is true. Adolescents tend to look at what might be, in two senses: They attempt to discover several possibilities in a situation before starting to investigate it empirically. Then they try to imagine the possible outcomes of each possibility.

Whereas the younger child might be satisfied with simply putting down two possible characteristics on the application, Ellen is likely to think of what the ramifications are for each of a number of possible characteristics. For example, she might hypothesize that if the station has a reputation for being concerned with poor people, they will be sympathetic with the fact that her father is out of work and needs a job. On the other hand, if the station has a reputation for fierce competition, it might look negatively on this information, thinking that perhaps her whole family are "losers."

Intra-propositional versus Inter-propositional Thinking

Elementary school-age children, especially older ones, may well be able to think of a number of possible outcomes that would result from a single choice. It is only in later adolescence, however, that they are able to think of the ramifications of *combinations* of propositions (called **inter-propositional thinking**).

In Ellen's case, for example, she may decide that her father's unemployment could hurt her, but she may balance this by stating that she herself has maintained a paper route successfully for the past four years. Inter-propositional thinking can become infinitely more complex than **intra-propositional thinking** (thinking about the possible outcomes of a single choice). These complexities usually become part of the repertoire of thinkers as they develop into adulthood.

An important aspect of this complexity is the ability to think logically about statements that may have no relationship to reality whatsoever. As Flavell states, "formal operational thinkers understand that logical arguments have a disembodied, passionless life of their own" (p. 106). Most persons find thinking about abstract concepts—for example, the laws of trigonometry—more difficult than thinking about the construction of a chair, but the laws of logic in each case do not differ. During adolescence and adulthood, the person becomes aware of this.

Combinations and Permutations

Because concrete operational thinkers look at propositions only one way at a time, they are often unable to imagine all of the possible **combinations** or **permutations** of a set of data. During adolescence, the thinker becomes able to realize that systematically generating combinations (A with B, A with C, B with C, etc.) can aid in thoroughly examining the possible solutions of a problem.

It is entirely likely that Ellen would sit down and draw up a list of all her good characteristics before even starting to decide which of these would be best to put on her application. To improve her creativity, she might write all of these characteristics on separate sheets of paper, put them in a hat, and draw them out in pairs to see if she comes across a pair that she thinks will be unusually appealing. Without such a technique, she might never have thought of that particular pair.

Inversion and Compensation

Suppose you have before you two containers with an equal amount of water in each, hanging evenly from each side of a balance scale. Obviously, if a cup of water is added to the container on the right, it will sink to a level lower than the container on the left. If you were asked to make the containers even again, you would probably recognize that it could be done in one of two ways: withdraw a cup of water from the container on the right (**inversion**), or add a cup of water to the container on the left (**compensation**).

Concrete operational thinkers are likely to think of one or the other of these solutions, but not both. They usually will not recognize that there is more than one possible way to solve the problem. Having solved it one way, they no longer pay attention to the task. Because formal operational thinkers are able to imagine both inversion and compensation as being useful in solving a problem, they have a better chance of coming up with other fruitful solutions.

Ellen would be using compensation if she mentioned her newspaper job to balance her father's unemployment. She might be using inversion if she simply left out the fact that her father is unemployed.

Information-Processing Strategies

Older thinkers are not only more likely to have a large array of problem-solving strategies, but are more likely to attempt to devise a plan to use this array. Such a plan is an **information-processing strategy.**

Flavell points to the game of "Twenty Questions" as an example of this difference. In this game, the problem solver is allowed twenty questions of the yes-or-no type to solve a problem. For example, if the problem is to guess what person the questioner is thinking of, an excellent question would be, "Is the person alive?" This eliminates a tremendous number of possibilities. Other questions such as "Is the person a woman?" or "Is the person an adult?" are examples of this effective problem-solving strategy. Concrete operational problem solvers are more likely to name twenty particular individuals and thus (unless they are lucky) lose the game.

Ellen has the problem of attempting to discover what major criteria the television people have in mind in selecting their talk-show host. She might go to several individuals in the community who know about those criteria. Furthermore, she might use a good strategy in asking questions of those informed persons. At any rate, to the extent that she has reached the formal operations stage, she is much more likely to do so.

Consolidation and Solidification

The changes from childhood to adulthood mentioned so far have been qualitative. There are also two quantitative aspects of this transition. The mental gains being made are slowly consolidated—show **consolidation.** Not only are the improved problem-solving techniques learned, but they are employed in a wider variety of situations and with greater skill. Furthermore, these gains seem to be solidified—show **solidification.** That is, the thinker is more certain and confident in the use of the newly gained mental skills and is more likely to use them in new situations.

If Ellen were 12 years old instead of 17, she may have tried some of the same tactics, but she might very well have given up quickly and just put down any two traits that seemed acceptable to her. It is her greater experience with these thinking styles that gives her the motivation to persevere at the task. Table 12.2 offers an overview of Flavell's variables.

Table 12.2 Overview of Flavell's Variable in Intellectual Development

Concrete Operational Thinker		Formal Operational Thinker
1. The real	vs	1. The possible
2. Empirico-inductive method	vs.	2. Hypothetico-deductive method
3. Intra-propositional thinking	vs.	3. Inter-propositional thinking
4. Unable to see combinations and permutations	vs.	4. Able to see combinations and permutations
5. Can use inversion *or* compensation	vs.	5. Can use inversion *and* compensation
6. Poor information-processing strategies (weak plan)	vs.	6. Good information-processing strategies (clear plan)
7. Mental gains not consolidated and solidified	vs.	7. Mental gains consolidated and solidified

Source: From John Flavell, *Cognitive Development*. Englewood Cliffs, NJ: Prentice-Hall, Inc., 1985.

Culture and gender can also influence cognitive development. Piaget's theory seems to assume that the ideal person at the endpoint of cognitive development resembles a Swiss scientist. Most theorists focus on an ideal endpoint for development that, not too surprisingly, ascribes their own valued qualities to maturity.

Piaget (1973) acknowledged that his description of the endpoint might not apply to all cultures, since evidence had showed cultural variation. If one stresses the influence of context (sociocultural and individual) on development as Vygotsky (1978) and Barbara Rogoff (1990) do, then one sees *multiple* directions for development rather than only one ideal endpoint.

Carol Gilligan (1982) says that most theories of development define the endpoint of development as being male only, and that they overlook alternatives that more closely fit the mature female. She believes that if the definition of maturity changes, so does the entire account of development. Using men as the model of development, independence and separation are seen as the goals of development. But if women are used as the models, relationship with others or interdependence are the goals of development. How do you think Ellen's connectedness to others (in the preceding example) might possibly influence her decision making?

Adolescent Egocentrism

Parents often feel frustrated at the seemingly irrational attitudes and behaviors of their adolescent children. One explanation is the reemergence of a pattern of thought that marked early childhood, egocentrism. **Adolescent egocentrism,** a term coined by Elkind (1978), refers to the tendency to exaggerate the importance, uniqueness, and severity of social and emotional experiences. Their love is greater than anything their parents have experienced. Their suffering is more painful and unjust than anyone else's. Their friendships are most sacred. Their clothes are the worst or the best. Developmentally speaking, adolescent egocentrism seems to peak around the age of 13 (Elkind & Bowen, 1979), followed by a gradual and sometimes painful decline.

A Multicultural View 

Following Vygotsky's Model in a Spanish-Speaking Community

Luis Moll and Rosa Diaz (1984) headed a project in a bilingual (Spanish-speaking) community in California in which they collected samples from the community. Writing is social; it is a means of communication and a product of a culture. Accordingly, they asked high school students to uncover social issues to use as the topics of writing.

They then asked students as a homework assignment to survey adults and children in the community about how they felt about speaking both Spanish and English. Homework became a literacy event in the family. The teachers guided the students as suggested in Vygotsky's zone of proximal development theory (See chap. 2). Step by step, the teachers helped them to prepare a questionnaire, gather information, sift through for the important facts, and write up their information. Thus this project looked at the social and cultural contexts relevant to the students' lives, in order to enhance their writing skills.

Elkind sees two components to this egocentrism. First, teenagers tend to create an *imaginary audience*. They are on center stage, and the rest of the world is constantly scrutinizing their behavior and physical appearance. This accounts for some of the apparently irrational mood swings in adolescence. The mirror may produce an elated, confident teenager ready to make his appearance. Then one pimple on the nose can be cause for staying inside the house for days. In fact, school phobia can become acute during early adolescence because of concerns over appearance.

The second component of egocentrism is the *personal fable*. This refers to adolescents' tendency to think of themselves in heroic or mythical terms. The result is that they exaggerate their own abilities and their invincibility. This type of mythic creation on the part of an adolescent can sometimes lead to increased risk-taking, such as drug use, dangerous driving, and disregard for the possible consequences of sexual behavior. Many teenagers simply can't imagine an unhappy ending to their own special story.

Critical Thinking

Although we discussed critical thinking at some length in chapter 9, we should make a number of points about its role in adolescence. Guilford's distinction between convergent and divergent thinking (1975) is helpful here, because critical thinking is made up of these two abilities.

Convergent thinking is used when we want to solve a problem by *converging* on the correct answer. For example, if we were to ask you to answer the question, "How much is 286 times 469," you probably could not produce it immediately. However, if you used a pencil and paper or a calculator, you would almost certainly arrive at the same answer as most others trying to solve the problem. There is only one correct answer. Critical thinking uses convergent thinking. As Moore and Parker (1986) put it, it is "the correct evaluation of claims and arguments."

Divergent thinking, on the other hand, is just the opposite. This is the type of thinking used when the problem to be solved has many possible answers. For example, what are all the things that would be different if it were to rain up instead of down?

Divergent thinking is an important aspect of critical thinking as well as creative problem solving.

Other divergent questions are "What would happen if we had no thumbs?" and "What should we do to prevent ice buildup from snapping telephone lines?" Divergent thinking can be right or wrong, too, but there is considerably greater leeway for personal opinion than with convergent thinking. Not all divergent thinking is creative, but it is more likely to produce a creative concept. To be a good critical thinker, it is not enough to analyze statements accurately. Often you will need to think divergently to understand the possibilities and the implications of those statements, too.

Because adolescents are entering the formal operations stage of intellectual development, they become vastly more capable of critical thinking than younger children. As they move through the teen years, they grow in their ability to make effective decisions. This involves five types of newly acquired abilities (Moore & others, 1985):

Phase 1. Recognizing and defining the problem
Phase 2. Gathering information
Phase 3. Forming tentative conclusions
Phase 4. Testing tentative conclusions
Phase 5. Evaluation and decision making

In the next section, we make a distinction between critical and creative thinking, but it is important that we not make too great a distinction, especially at this age level. Paul (1987) describes this concern well:

> ***Just as it is misleading to talk of developing a student's capacity to think critically without facing the problem of cultivating the student's rational passions—the necessary driving force behind the rational use of all critical thinking skills—so too it is misleading to talk of developing a student's ability to think critically as something separate from the student's ability to think creatively. . . . The imagination and its creative powers are continually called forth.*** **(p. 143)**

Creative Thinking

> ***This is the story about a very curious cat named Kat. One day Kat was wandering in the woods where he came upon a big house made of fish. Without thinking, he ate much of that house. The next morning when he woke up he had grown considerably larger. Even as he walked down the street he was getting bigger. Finally he got bigger than any building ever made. He walked up to the Empire State Building in New York City and accidentally crushed it. The people had to think of a way to stop him, so they made this great iron box which made the cat curious. He finally got inside it, but it was too heavy to get him out of again. There he lived for the rest of his life. But he was still curious until his death, which was 6,820,000 years later. They buried him in the state of Rhode Island, and I mean the whole state.***
>
> ***Ralph Titus, a seventh-grade student***

The restless imagination, the daring exaggeration, the disdain for triteness that this story demonstrates—all are signs that its young author has great creative potential. With the right kind of encouragement, with the considerable knowledge we now have about how to foster creativity, this boy could develop his talents to his own and society's great benefit.

Creative thinking appears to have many elements—divergent thinking, fluency, flexibility, originality, remote associations. We will look more closely at these elements when we get to adult creativity later in this book, but one element that seems to be of special importance in adolescence is the use of metaphor.

The Use of Metaphor

A metaphor is a word or phrase that by comparison or analogy stands for another word or phrase. Common sense suggests a relationship between efficient metaphor use and creativity. Using a metaphor in speech involves calling attention to a similarity between two seemingly dissimilar things. This suggests a process similar to divergent thinking, and there is a growing body of research support for this relationship (Jaquish & others, 1984; Kogan, 1973, 1983; Wallach & Kogan, 1965).

Kogan (1983) believes that the **use of metaphor** can explain the difference between ordinary divergent thinking and high-quality divergent thinking. A creative person must be able not only to think of many different things from many different categories, but to compare them in unique, qualitatively different ways. Although metaphors are typically first used by older children and adolescents, research has looked at the symbolic play of very young children and how it relates to creativity (see Kogan, 1983, for a good review). The early imaginative play of children is now being viewed as a precursor of later metaphor use and creativity.

Howard Gardner and his associates at Harvard University have studied the role of metaphor. Gardner's seminal *Art, Mind, and Brain: A Cognitive Approach to Creativity* (1982) offers many insights into the process. Gardner has based his research on the theories of three eminent theorists: Jean Piaget, Noam Chomsky, and Claude Levi-Strauss. He states that "These thinkers share a belief that the mind operates according to specifiable rules—often unconscious ones—and that these can be ferreted out and made explicit by the systematic examination of human language, action, and problem solving" (p. 4).

Gardner's main efforts have focused on the relationship between children's art and children's understanding of metaphor, both in normal and brain-damaged children. He describes talking to a group of youngsters at a seder (the meal many Jews eat to commemorate the flight of the Hebrews from Egypt). He told them how, after a plague, Pharaoh's "heart was turned to stone." The children interpreted the metaphor variously, but only the older ones could understand the link between the physical universe (hard rocks) and psychological traits (stubborn lack of feeling). Younger children are more apt to apply magical interpretations (God or a witch did it). Gardner believes that the development of the understanding of metaphoric language is as sequential as the stages proposed by Piaget and Erikson, and is closely related to the types of development treated in those theories.

Studying the stories of the Bible is an excellent way to learn about metaphors.

Examining children's metaphors such as a bald man having a "barefoot head" and an elephant being seen as a "gasmask," Gardner and Winner (1982) found clear changes with age in the level of sophistication. Interestingly, there appear to be two opposing features:

- When you ask children to explain figures of speech, they steadily get better at it as they get older. There is a definite increase in this ability as the child attains the formal operations stage.
- However, very young children seem to be the best at making up their own metaphors. Furthermore, their own metaphors tend to be of two types (Gardner & Winner, 1982):

> ***The different patterns of making metaphors may reflect fundamentally different ways of processing information. Children who make their metaphors on visual resemblances may approach experience largely in terms of the physical qualities of objects. On the other hand, children who base their metaphors on action sequences may view the world in terms of the way events unfold over time. We believe that the difference may continue into adulthood, underlying diverse styles in the creation and appreciation of artistic form.* (p. 164)**

It is exciting to think that this discovery by Gardner and Winner may explain why some people become scientists and other writers. If this is so, it certainly is an important key to fostering such talent. Of course, this is not to say that they have the answers to such questions as why children develop one of the two forms of "metaphorizing" (or neither), but their work appears to be a giant step in the right direction.

These researchers believe that the spontaneous production of metaphors declines somewhat during the school years. This is probably because the child, having mastered a basic vocabulary, has less need to "stretch the resources of language to express new meanings" (1982, p. 165). In addition, there is greater pressure from teachers and parents to get the right answers, so children take fewer risks in their language. Gardner and Winner (1982) point to the *Shakespeare Parallel Text Series,* which offers a translation of the bard's plays into everyday English ("Stand and unfold yourself" becomes "Stand still and tell me who you are"), as a step in the wrong direction. "If, as we have shown, students of this age have the potential to deal with complex metaphors, there is no necessity to rewrite Shakespeare" (p. 167).

Creativity, Giftedness, and the IQ

As Feldman (1979) has pointed out, there have been many studies of "giftedness," but only a few of exceptionally creative, highly productive youth. This is a serious omission because, as you will see later in this book, adolescence is a sensitive period in the growth of creative ability. Feldman believes that this unfortunate situation is mainly the fault of "the foremost figure in the study of the gifted," Lewis M. Terman. Terman (1925) was well known for his research on 1,000 California children whose IQs in the early 1920s were 135 or higher. Terman believed these children to be the "geniuses" of the future, a label he kept for them as he studied their development over the decades. His was a powerful investigation, and has been followed by scholars and popular writers alike.

Precisely because of the notoriety of this research, Feldman argues, we have come to accept a *numerical* definition of genius (an IQ above 135), and a somewhat low one at that. Feldman notes that the *Encyclopedia Britannica* now differentiates two basic definitions of genius: the numerical one fostered by Terman; and the concept first described by Sir Francis Galton (1870, 1879): "creative ability of an exceptionally high order as demonstrated by actual achievement."

Feldman (1979) says that genius, as defined by IQ, really only refers to **precociousness**—doing what others are able to do, but at a younger age. **Prodigiousness** (as in child prodigy), on the other hand, refers to someone who is *qualitatively* higher in ability from the rest of us. This is different from simply being able to do things sooner. Further, prodigiousness calls for a rare matching of high talent and an environment that is ready and open to creativity. If such youthful

An Applied View — Tommy's Case

Tommy was an unkempt child, whose physical health was poor due to frequent bouts with viral infection. His teachers reported that the quality of his schoolwork was generally poor, and he had considerable difficulty with spelling and rote learning. He cared little for reading or writing and had a consistently negative attitude toward school in general. He frequently interrupted classes by "asking foolish questions," being rude to the teachers, and playing practical jokes on others.

He set fire to a portion of his home, and when asked why, his reply was, "because I wanted to see what the flames would do." The boy was given a beating by his father, in full view of the neighborhood. He attempted to hatch chicken eggs by sitting on them, and when this didn't work, he encouraged his playmates to swallow some raw. Neighbors frequently noted explosions coming from the basement of the boy's home, where his mother permitted him to play with chemical substances.

The boy's father had little regard for his son's intelligence, and stated that strong disciplinary intervention and obedience training were the best methods for dealing with Tommy. Nevertheless, his mother, a former school teacher, insisted that he be allowed to stay at home and receive the rest of his education from her. Ultimately, she had her way.

When asked their diagnosis of 11-year-old Tommy, many psychology students label him emotionally disturbed, and recommend he be taken from his parents and institutionalized for his own good. Actually, he eventually turned out pretty well—Tommy's full name is Thomas Alva Edison!

prodigies as Mozart in music or Bobby Fischer in chess had been born two thousand years earlier, they may well have grown up to be much more ordinary. In fact, if Einstein had been born *fifty* years earlier, he might have done nothing special—particularly since he did not even speak well until he was five!

So if prodigies are more than just quicker at learning, what is it that truly distinguishes them? On the basis of his intensive study of three prodigies, Feldman states that

> ***Perhaps the most striking quality in the children in our study as well as other cases is the passion with which excellence is pursued. Commitment and tenacity and joy in achievement are perhaps the best signs that a coincidence has occurred among child, field, and moment in evolutionary time. No event is more likely to predict that a truly remarkable, creative contribution will eventually occur.*** **(1979, p. 351)**

Child prodigies are distinguished by the passion with which they pursue their interests. Here we see the young Wolfgang Amadeus Mozart performing for a group of admiring adults. He was not merely precocious–able to perform at levels typical of older children; he was prodigious–able, at a young age, to write music that is still performed by professional musicians.

In summary, critical and creative thinking are similar in that they both employ convergent and divergent production. The main difference between them is that critical thinking aims at the correct assessment of *existing* ideas, whereas creativity is more aimed at the invention and discovery of *new* ideas. While each requires a certain amount of intelligence, creativity also depends on such traits as metaphorical thinking and an independent personality.

Robert Sternberg (1981) says the intelligent person can recall, analyze, and use knowledge, while the creative person goes beyond existing knowledge and the wise person probes inside knowledge and understands the meaning of what is known (See chap. 18). It should be noted that these findings are true not only for adolescence, but for all periods of the lifespan.

An Applied View: Guidelines for Improving Your Own Creativity

Here are some suggestions (Dacey, 1989a) that should help you become a more creative problem solver yourself, and that you could teach to adolescents you work with:

- Avoid the "filtering out" process that blocks problems from awareness. Become more sensitive to problems by looking for them.
- Never accept the first solution you think of. Generate a number of possible solutions; then select the best from among them.
- Beware of your own defensiveness concerning the problem. When you feel threatened by a problem, you are less likely to think of creative solutions to it.
- Get feedback on your solutions from others who are less personally involved.
- Try to think of what solutions someone else might think of for your problem.
- Mentally test out opposites to your solutions. When a group of engineers tried to think of ways to dispose of smashed auto glass, someone suggested trying to *find uses* for it instead. Fiberglass was the result!
- Give your ideas a chance to incubate. Successful problem solvers report that they frequently put a problem away for a while, and later on the solution comes to them full blown. It is clear that they have been thinking about the problem on a subconscious level, which is often superior to a conscious, logical approach.
- Diagram your thinking. Sometimes ideas seem to fork, like the branches on a tree, with one idea producing two more, each of which produces two more, and so on. Diagramming will let you follow each possible branch to its completion.
- Be self-confident. Many ideas die because the person who conceived them thought they might be silly. Studies show that females have been especially vulnerable here.
- Think about the general aspects of a problem before getting to its specifics.
- Restate the problem several different ways.
- Become an "idea jotter." A notebook of ideas can prove surprisingly useful.
- Divide a problem, then solve its various parts.
- Really good ideas frequently require some personal risk on the part of the problem solver. In this we are like the turtle, which can never move forward until it sticks its neck out.

Conclusion

We wish that all children could complete puberty with a normal, healthy body and body image. We wish they could negotiate adolescence so successfully that they become energetic, self-confident adults.

Unfortunately, we know that this is not always the case. Some youths suffer from a negative self-concept because, although they differ from the norm only slightly in their body development, they perceive this as "catastrophic." Others deviate significantly from the norm because of some physiological problem. Of particular concern for females is our society's obsession with thinness. Taken together the various aspects of puberty can cause the adolescent quite a bit of chagrin.

The only solution is to get some perspective on these problems. The good news is that just when they need it, most adolescents develop improved mental abilities that enable them to get a more realistic view of themselves.

Cognitive development is a complex matter, one about which we understood very little prior to this century. Our best evidence is that the intellect develops in stages. Contrary to earlier beliefs, thinking in childhood, adolescence, and adulthood are qualitatively different from each other. Furthermore, there are a number of other aspects of cognitive development: social cognition, information processing, egocentric thinking, critical thinking, and creative thinking.

Although most adolescents are healthy and have relatively untroubled lives, there are some who have problems with substance abuse, mental illness, or delinquent behavior. In the next chapter, we examine the results of these problems.

Chapter Highlights

Early Studies of Puberty

- Theories of adolescence in the early twentieth century were largely based on personal bias, because little empirical data existed.

Your Reproductive System

- Those who work with adolescents need complete knowledge of the reproductive systems of both sexes.

When Does Puberty Start?

- The order of physical changes in puberty is largely predictable, but the timing and duration of these changes are not.

The Effects of Timing on Puberty

- The normal range in pubertal development is very broad, and includes early, on time, and late maturers.
- The adolescent's own perception of being normal has more influence on self-esteem than objective normality.
- Maturity of appearance affects whether adolescents are treated appropriately for their age.
- Early maturing is a positive experience for boys, but may be negative for girls.
- Late maturing is often difficult for both boys and girls.

Cognitive Development

- Piaget focused on the development of the cognitive structures of the intellect during childhood and adolescence.
- The infant and child pass through Piaget's first three stages: sensorimotor, preoperational, and concrete operational.
- Piaget's highest stage of cognitive development, that of formal operations, begins to develop in early adolescence.
- Flavell suggests that there are seven aspects of transition from childhood to adolescent and adult thinking: the real versus the possible, empirico-inductive versus hypothetico-deductive, intra-propositional versus inter-propositional, combinations and permutations, inversion and compensation, information-processing strategies, and codification and solidification.
- Adolescents focus much attention on themselves, and tend to believe that everybody is looking at them. This phenomenon is called the imaginary audience.
- Many adolescents also hold beliefs about their own uniqueness and invulnerability. This is known as the personal fable.

Critical Thinking

- Critical thinking skills include the ability to make inferences from observations, recognize assumptions, think deductively, make logical interpretations, and evaluate weak and strong positions in an argument.
- Critical thinking combines both convergent thinking, in which there is only one correct answer, and divergent thinking, in which there are many possible answers to a problem.
- Effective decision making, a formal operational process, is a part of critical thinking.

Creative Thinking

- Creative thinking includes divergent thinking, fluency, flexibility, originality, and remote associations.
- In adolescence, the understanding and use of metaphor appears to be an important aspect of creative thinking.
- Conventional schooling often has a dampening effect on students' willingness to risk doing creative, metaphorical thinking.
- Criticism of genius, as defined by IQ, holds that IQ indicates only precociousness but cannot account for prodigiousness.

Key Terms

Adolescent egocentrism
Combinations
Compensation
Consolidation
Delayed puberty
Empirico-inductive method
Hormonal balance
Hypothetico-deductive method
Information-processing strategy
Inter-propositional thinking
Intra-propositional thinking
Inversion
Maturation
Maximum growth spurt
Menarche
Normal range of development
Permutations
Precociousness
Prodigiousness
Puberty
Real versus the possible
Secular trend
Skeletal growth
Solidification
Use of metaphor

What Do You Think?

1. Should children be taught about their body functions in school? Should this teaching include sexuality? If so, at what grade should it start?
2. What was the beginning of puberty for you? Why do you think so?
3. In reading about the progress the mind makes during adolescence, could you remember these changes happening to you? Describe some of your memories of those times.
4. Studies show that most people believe they are below average in creativity, which cannot be true, of course. By definition, half of all people are above average. Where do you fit as compared to all of your acquaintances on this trait? Why?
5. Do you believe you can "disinhibit" (free up) your creative abilities? How should you start? Why don't you?

Suggested Readings

Blume, J. (1970). *Are you there, God? It's me, Margaret.* New York: Bradbury. Although written for teens, this book has a wealth of insights into pubertal change, at least for females. Our women friends tell us that Blume really understands.

Clavell, J. (1981). *The Children's Story.* New York: Delacorte. Illustrates the way children and youths tend to accept things without question, whereas adults are more likely to fear change and be suspicious of any deviation from the norm.

Erikson, E. (1969). *Gandhi's truth.* New York: Norton. This is one of the best examples of "psychohistory," which is the biography of a person as seen from the two disciplines of psychology and history. Mahatma Gandhi's quest to free India from British domination makes for good reading. The stories about the forces that influenced his youthful thinking are particularly instructive.

McCoy, K., & Wibbelsman, C. (1987). *The teenage body book.* Los Angeles, CA: The Body Press. An excellent reference book for teenagers and those who work with them.

Potok, C. (1967). *The chosen.* New York: Fawcett. This is the story of a boy whose father is a rabbi in the strict Hassidic (Jewish) religion. It chronicles the struggle he has over his desire to be a good student and still be "normal." Also *The gift of Asher Lev,* 1990.

CHAPTER 13

The Troubled Adolescent

Substance Abuse 347
Prevalence of Use 347
Ethnic Group and Abuse 349
Crime and Abuse 352
Substance Use and Personal Relationships 352
Combating Substance Abuse 353
Mental Health Issues 355
Types of Mental Disorders 356
Eating Disorders 357
Depression 358
The Crisis of Death 361
Suicide 363
Delinquent Behavior 369
The Nonaggressive Offender 369
The Juvenile Delinquent 371
Learning and Delinquency 372
Gangs 372
Conclusion 375
Chapter Highlights 375
Key Terms 377
What Do You Think? 377
Suggested Readings 377

In my dream I saw the world
in a frame of imitation gold.
I heard fear pounding in my ears
And in the white light I could see only black.
Blinded by the sound of darkness
I saw invisible fingers
And heard nonexistent sounds.
I was a nonexistent person
In a nonexistent world.
God help me
As I stab myself with a
Rubber knife.

By a 16-year-old girl living in an adolescent residential center

Although the transition from childhood to adulthood can be difficult, the great majority of adolescents are able to negotiate it successfully. Add the problems of substance abuse, mental health problems, and crime, however, and unhappiness always results for some. (Although most teens have experienced substance abuse, and quite a few have had brushes with crime or mental disturbance, only a small percentage become truly "troubled youths.") As the poem suggests, some teens suffer from two or even all three of these problems. In fact, sad to say, having one of the problems increases the chance that you will suffer from the other two. Therefore, no book on North American adolescence is complete unless it takes a hard look at these distressing areas.

How widespread are these difficulties? The teenagers themselves can probably best answer this question. In a highly scientific poll of 13- to 17-year-olds (Roper, 1987), they were asked how often certain behaviors occurred using ratings of "a lot" and "sometimes" among their fellow students.

The overall numbers are not a cause for rejoicing. Females and those in middle adolescence see the problems as more widespread than their counterparts. Whites are tied or ahead of African-Americans in all categories but crime and pregnancy. Surprisingly, upper and upper-middle-class youths are tied or ahead of the other two groups in all areas but crime and pregnancy (and this difference may be due to their greater access to lawyers and doctors). No region of the country has a monopoly on the problems, but public schools are ahead of private schools in every category. Of course this latter finding is probably more tied to socioeconomic background than to the type of schooling.

In this chapter, we examine how frequently adolescents use alcohol and illegal drugs, and see to what extent sexuality, crime, personal relations, and ethnic groups are involved. Reading the chapter will also help you to become better aware of your own attitudes toward drugs and alcohol. We then investigate the various forms of mental illness that are likely to occur during adolescence. In the final section, we cover delinquent behavior.

When you finish reading this chapter, you should be able to:

- Define basic terminology and phrases related to substance abuse.
- Describe the prevalence of drug use among different groups of adolescents and between different types of drugs, and state any major differences between groups.
- Explain the difficulty involved in trying to examine the connection between ethnicity and substance abuse.

- Compare some common drug abuse prevention methods as to their relative strengths and weaknesses.
- Identify the symptoms associated with anorexia nervosa and bulimia.
- Describe four factors that contribute to the development of eating disorders.
- Identify the symptoms associated with depression and masked depression.
- List the warning signs of adolescent suicide.
- Explain how individual, family, and social factors contribute to adolescent suicide.
- Distinguish between the three major forms of delinquent behavior.
- Explain the connection between running away and prostitution, and describe the reality of life for a teenage female or male prostitute.
- Describe the connection between school performance (including learning disabilities) and delinquent behavior.
- List and describe at least three reasons why youth join gangs.

Let us turn now to a review of the research in each of the three general problem areas: substance abuse, mental health, and delinquent behavior.

There has been considerable change in the choices drug and alcohol users are making in recent years. For example, crack cocaine has become much more popular than hallucinogens such as LSD. Unfortunately, the newer drugs and drink combinations are much more likely to be lethal.

Substance Abuse

There can be no doubt that substance abuse has become a serious problem throughout the world in the second half of the twentieth century. In this section you will find important information, presented from several standpoints.

Prevalence of Use

It is difficult to say precisely how widespread substance abuse is. Studies differ from year to year, from region to region, and disappointingly, from one another (even when year and region are the same). One of the best studies of high school seniors (Office of Technology Assessment, 1991) found a slight decrease in marijuana, stimulants, hallucinogens, sedatives, tranquilizers, cocaine, and crack cocaine, and a leveling off in alcohol (which at 90.7 percent could not get a lot higher than it is). Table 13.1 presents data on substance use for a number of adolescent age groups at the end of the 1980s.

The newest drug to hit the adolescent subculture is MDMA (also known as Adam or Ecstacy). It produces a smoother, longer euphoria than cocaine and is one of the so-called designer drugs (Buffum, 1988; Buffum & Moser, 1986). Its use is definitely on the rise, in part because a number of psychiatrists have said that it can cure psychological ills. The advent of newer drugs should not cloud the fact that the substances reportedly abused most by all age groups are alcohol, followed by cigarettes.

As you can see from Table 13.1, drug use among teens is a serious problem in this country. But as Figures 13.1 and 13.2 indicate, relative to other age groups, adolescents are less likely to report having used alcohol or cigarettes than adults. The same surveys found adolescents reported less use of illicit drugs than those aged 18 to 34 and about the same as those over 35. About 25 percent of adolescents

Table 13.1 Percentage of Surveyed U.S. Adolescents of Different Ages Reporting Ever Having Used Alcohol or Other Types of Psychoactive Substances

	NIDA Household Survey on Drug Abuse, 1988	National Adolescent Student Health Survey, 1987		Monitoring the Future/High School Seniors Survey, 1989
Percentage of respondents reporting ever having used	12- to 17-year-olds[a]	8th graders	10th graders	High school seniors
Alcohol	50.2%	77.4%	88.8%	90.7%
Cigarettes	42.3	50.8	63.5	65.7
Marijuana/hashish	17.4	14.5	35.1	43.7
Smokeless tobacco/snuff	14.9	NA	NA	NA
Inhalants	8.8	20.6	20.6	17.6
Nonmedical use of stimulants	4.2	9.0[b]	15.7[b]	19.1
Nonmedical use of analgesics	4.2	NA	NA	NA
Hallucinogens (all forms)	3.5	2.6	6.7	9.9
Cocaine (all forms)	3.4	3.6	7.7	10.3
Nonmedical use of sedatives	2.4	NA	NA	7.4
Nonmedical use of tranquilizers	2.0	NA	NA	7.6
PCP	1.2	NA	NA	3.9
Crack cocaine	0.9	1.6	2.7	4.7
Heroin	0.6	NA	NA	1.3
LSD	NA	NA	NA	8.3
Needle use	0.4	0.9[c]	0.5[c]	NA

Key: NA = not available.
[a]Sample size is too small to disaggregate by single year of age.
[b]Includes over-the-counter as well as prescription (e.g., amphetamines) drugs.
[c]Information collected on needle use of injection of cocaine only.
Source: Office of Technology Assessment, 1991.

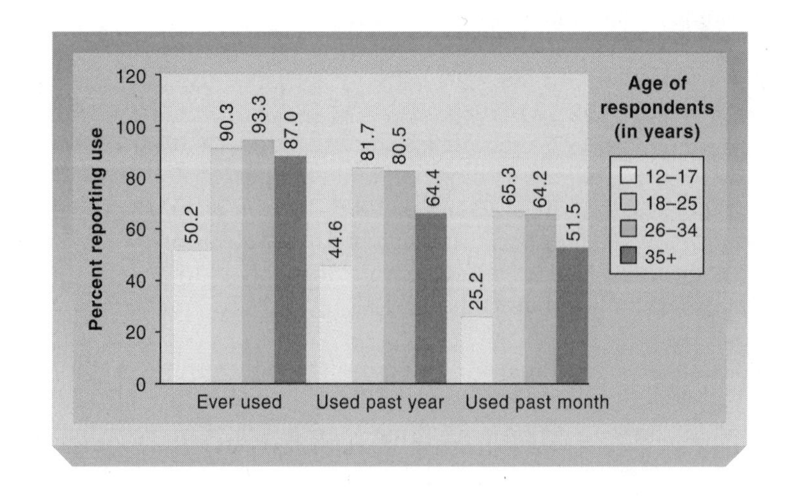

Figure 13.1
Use of alcohol in the United States, by age, 1988

Source: From Office of Technology Assessment 1991, based on data from U.S. Department of Health and Human Services, Public Health Service, Alcohol, Drug Abuse, and Mental Health Administration, National Institute on Drug Abuse, *National Household Survey on Drug Abuse: Main Findings 1988,* DHHS Pub. No. (ADM) 90-1682 (Rockville, MD: 1989).

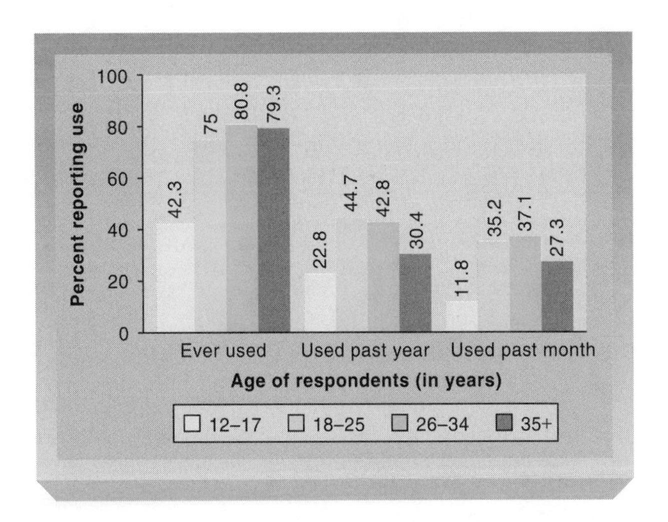

Figure 13.2
Use of cigarettes in the United States, by age, 1988

Source: From Office of Technology Assessment 1991a, based on data from U.S. Department of Health and Human Services, Public Health Service, Alcohol, Drug Abuse, and Mental Health Administration, National Institute on Drug Abuse, *National Household Survey on Drug Abuse: Main Findings:* 1988, DHHS Pub. No. (ADM) 90-1682 (Rockville, MD: 1989).

A Multicultural View

No to Drugs, Yes to Helping Others

In spite of their difficult environment, many inner-city teenagers find ways not only to avoid drug use and the drug culture, but to actively help others. *The Boston Globe* recently described several of these high school students, who were honored at a special dinner for their exceptional contributions as volunteers to neighborhood agencies. Some suburban teens also received honors for giving their time to assist at inner-city agencies. Among the organizations served were planning action committees, an Indian Council, a community comprehensive health center, and a neighborhood community center.

This year, the highest award for volunteer service was given to Julio Martinez, the son of a shoe maker. In the summer of 1991, Julio worked as a volunteer every day at the Tobin Community Center on Boston's Mission Hill. He gave so much time at the center that one of the directors thought he was a salaried employee. Here is an excerpt from his interview with the reporter:

Asked why he volunteers, Julio shrugs.

"My mother encouraged me."

"Lots of mothers encourage sons who don't volunteer."

He shrugs again.

"I just don't like hanging around. I like doing stuff. Some kids do a lot of bad stuff, but I'm not into that."

On troubled streets, he manages to avoid trouble.

"It's not hard. If you put your heart in it, if you say you're not going to do it. I don't do drugs, and I've been offered lots of 'em. But that don't make you a man or nothing. A lot of kids who get into trouble say friends put them up to it, but nobody can make you do anything you don't want to do. That's what I say. . . .

I did do a lot of work, it's true, but there's a lot of other people who did a lot of work, too, and maybe they deserve it more, and so, like, I'm saying, you know, why me?"

reported ever using an illicit drug, as compared to roughly 60 percent of 18- to 34-year-olds and 23 percent of those over 34. Adolescents did, as you might guess, report more drug use within the past month than those 35 and older.

Table 13.2 offers some essential information on the most used controlled substances.

Ethnic Group and Abuse

Although not a great deal is known about the comparative abuse of substances by ethnic groups, one study looked at the number of arrests for drug use per 10,000 members of an ethnic group, with seven groups being considered (Asian-American Drug Abuse Program, 1978). It was found that African-American, Mexican-American, and Native American arrests outnumbered white arrests by three to two.

Table 13.2 Controlled Substances: Uses and Effects

Drugs	Usual Methods of Administration	Possible Effects	Effects of Overdose	Effects of Syndrome
Narcotics				
Opium Morphine Codeine Heroin	Oral, smoked Oral, injected, smoked Oral, injected Injected, sniffed, smoked	Euphoria, drowsiness, respiratory depression, constricted pupils, nausea	Slow and shallow breathing, clammy skin, convulsions, coma, possible death	Watery eyes, runny nose, yawning, loss of appetite, irritability, tremors, panic, chills and sweating, cramps, nausea
Hydromorphone	Oral, injected			
Meperidine (pethidine)	Oral, injected			
Methadone	Oral, injected			
Other narcotics	Oral, injected			
Depressants				
Chloral Hydrate Barbiturates Glutethimide Methaqualone Benzodiazepines Other depressants	Oral Oral, injected Oral, injected Oral, injected Oral, injected Oral, injected	Slurred speech, disorientation, drunken behavior without odor of alcohol	Shallow respiration, cold and clammy skin, dilated pupils, weak and rapid pulse, coma, possible death	Anxiety, insomnia, tremors, delirium, convulsions, possible death
Stimulants				
Cocaine Amphetamines Phenmetrazine Methylphenidate Other stimulants	Sniffed, injected, smoked Oral, injected Oral, injected Oral, injected Oral	Increased alertness, excitation, euphoria, increased pulse rate and blood pressure, insomnia, loss of appetite	Agitation, increase in body temperature, hallucinations, convulsions, possible death	Apathy, long periods of sleep, irritability, depression, disorientation
Hallucinogens				
LSD Mescaline and peyote Amphetamine variants Phencyclidine Phencyclidine analogs Other hallucinogens	Oral Oral, injected Oral, injected Smoked, oral, Injected Oral, injected, smoked, sniffed	Illusions and hallucinations, poor perception of time and distance	Longer, more intense "trip" episodes, psychosis, possible death	Withdrawal syndrome not reported
Cannabis				
Marijuana Tetrahydrocannabinol Hashish Hashish oil	Smoked, oral	Euphoria, relaxed inhibitions, increased appetite, disoriented behavior	Fatigue, paranoia, possible psychosis	Insomnia, hyperactivity, and decreased appetite occasionally reported

Source: Drug Enforcement Administration, *Drug Enforcement Fall 1982*. Washington, D.C.: U.S. Department of Justice.

Arrests for Japanese- and Chinese-Americans were negligible. Of course, these data may reflect biases of the legal system rather than actual use.

One of the most interesting factors in substance abuse has to do with whether a teen's family has arrived in this country in the previous generation or two, as has been the case with so many Asian-Americans. Tessler (1980) suggested a series of interrelationships between these variables, as presented in Table 13.3. This table shows that conflicts between traditional cultural patterns, immigration factors, and the mores of American society often bring about an increase in the use of drugs, particularly alcohol. This is especially so for adolescents, who are in the most vulnerable age group. The table is not meant to represent what always happens, but a high percentage of the time this is what occurs.

Table 13.3 Asian-American Cultural Clashes

Traditional Cultural Patterns	vs. Immigration Factors	vs. American Society
Close-knit family with strong father.	Father has to hold down more than one job to support family and is unable to provide the strong influence tradition demands.	More democratic approach to family structure. Mothers work. Increase of single-parent families.
High parental expectations.	Langauge difficulties; inability to make friends; difference in educational systems.	More tolerance of children who are not high achievers.
Shame associated with having a problem.	Immigration brings many adjustment problems.	More openness regarding problems, and more willingness to seek professional help when necessary.
Pride in one's worth and value to oneself, the family, and the community.	Inability to communicate, obtain a job, achieve a level of academic excellence.	Prejudice, stereotyping, both old and new; Asians are exotic, humble, inscrutable, studious, gang member.
Women are supposed to stay home and raise the children.	Women may have to enter job market to help support family.	Acceptable for women to enter job market and lead a more independent life-style.
Drinking is perceived in many cases as being an acceptable part of family and community life. Alcohol a part of ritual and festivals.	Increased drinking because of pressure may not be considered a problem.	Drinking not perceived as being culturally important

Estrada and associates (1982) reported on the use of alcohol by a group of 107 Latino seventh- and eighth-grade junior high school students (age 13 to 16) in Los Angeles. Subjects responded to a self-administered questionnaire concerning personal issues, social characteristics, and family composition. Also studied were alcohol consumption, school performance, and a number of other behaviors. Findings suggest that the strongest link is between the use of alcohol and marijuana.

Dembo (1981) questioned whether the degree of drug involvement of 1,101 African-American and Puerto Rican seventh graders resulted from their home composition, relationships with parents, attitudes toward school, machismo values, or identification with drug-involved peers. Data were obtained via questionnaires and surveys. Results showed that peer-oriented attitudes and behaviors probably provide the most likely prediction of drug use for these youth.

An Applied View — Do You Have a Drinking Problem?

The following questions may help you to decide whether you have a drinking problem:

- Do you drink to escape from the pressures of life?
- Do you sometimes skip classes or work because of hangovers?
- Do you drink more than your friends?
- Do you drink to escape from reality, boredom, or loneliness?
- Do your friends or loved ones express concern about your drinking?
- Do you drink and get drunk even when intending to stay sober?
- Do you drink when you are alone?
- Do you drink frequently to a state of intoxication?
- Have you had two or more "blackouts" (can't remember some or all of what happened while you were drinking) in the past year?
- Have you gotten into trouble with the police or college officials as a result of your behavior while drinking?

If you must answer *yes* to any of these questions, you may have a problem with alcohol, and perhaps you should seek advice from trained personnel. If you must answer *yes* to more than two of the questions, especially those on the second half of the list, you definitely have reason for concern, and should seek help.

Source: Adapted from Chebator, 1992.

Anti-drug media campaigns have encouraged drug addicts to seek help through drug rehabilitation programs. A number of these programs, such as the Straight Program, have a high success rate in rehabilitating these addicts.

Crime and Abuse

Although it is commonly thought that drug users, especially those who use hard drugs, are regularly involved in criminal activities, reliable data on this aspect of drug use are surprisingly limited. FBI statistics (Federal Bureau of Investigation, 1993) indicate that the drug most associated with crime is alcohol. In 40 percent of assaults and 35 percent of rapes, those convicted had been "under the influence."

Research into the relationship between substance abuse and crime (and other undesirable behaviors) was carried out by Santana (1979). This study looked at nineteen types of undesirable behavior, and compared their occurrence among drug abusers to their occurrence among nonusers. The disturbing finding of this study was that, in every case but one, the users were much more likely to commit undesirable behaviors than nonusers. Even in the single exception, for arson, the behavior was evenly divided. This does not prove that drug use causes crime, but does indicate a strong relationship between the two.

Substance Use and Personal Relationships

There has long been a debate as to whether adolescent drug usage is related more to family circumstances than to interaction with peers (Detting & Beauvais, 1987). Most studies have indicated that the typical drug user has problems with his or her parents. Tudor and associates (1980) found that drug users had conflicts with their parents in the following areas:

- They wanted to be allowed to make their own decisions without having any advice from their parents.
- They did not believe that they should be limited to only those friends their parents endorsed.
- They were less likely to desire affection from their mothers and fathers, nor did they feel close to them.
- They did not desire to imitate their parents.

One other series of studies (Huba & others, 1979, 1980a, 1980b, 1980c) found that although relationships with parents do affect substance abuse in early adolescence, as students get to high school the parental effect drops off:

> ***The finding that by the ninth grade boys and girls have drug use patterns that are related more to perceived patterns in the peer culture than to perceived patterns in the adult culture suggests that adult models become relatively** less important.* **(1980b, p. 464)**

We pointed out in chapter 11 that most adolescents tend to adopt the values of their parents. That drug-abusing teens do not is one way they differ from the norm.

Some adolescents believe that using drugs enhances both their personal relationships and their sexual functioning. The research studies are in remarkable accord: although some substances may briefly disinhibit a person's fears of sexuality, in general they have either no effect or a negative one (Buffum, 1988).

Combating Substance Abuse

The first step toward reducing the effects of drugs and alcohol on our youth is improving our ability to determine, in the early stages, who is at risk. There is good news on this front. A recent study (Christiansen & others, 1989) found that a number of measures administered in the seventh and eighth grade were significant predictors of drinking behavior one year later. These measures included a questionnaire on attitudes toward drinking and such demographic variables as parental ethnic background, religious affiliation, parents' occupations, and parental drinking attitudes.

The effectiveness of treating those identified as having a problem is another story. It is estimated that as many as 15 percent of all American teens need treatment for compulsive drug/alcohol use (Falco, 1988). Unfortunately, there is very little research on treatment programs, and those that do exist, such as the Washington-based Straight Program, are expensive and hard to gain admittance to. Even less is known about educational and prevention programs designed to reduce the *demand* for drugs, which drug abuse experts now believe to be the best approach to this problem. It is even more difficult to reach those youths who have left school and are at even higher risk for substance abuse. In many of our larger cities, nearly half of all teens fall into this category (Falco, 1988).

No prevention program is likely to be effective if the climate in the school is conducive to drug abuse. Tessler (1980) argues that to find out what the climate is in any particular school, it is necessary to administer a survey such as the one in the School Climate Survey (see "Applied View"). The survey is easy to administer to students, teachers, and administrators alike. It takes only a few minutes to complete. Tessler claims that

> ***the results will help everyone get in touch with some of the silent agents responsible for a poor school environment which would make the success of any effective drug education impossible.* (p. 114)**

You may wish to answer the questions in the box in regard to your high school. That would be a good way to remind yourself of the factors that can lead to serious substance abuse. You may also want to send a copy to the high school's principal.

Questions 1 to 4 relate to situations that can cause antisocial or antiauthority behavior. School and police records can also reveal whether this kind of problem

An Applied View — School Climate Survey

Does your school

1. Have racial or ethnic problems? Yes ____ No ____
2. Have a high truancy rate? Yes ____ No ____
3. Have cases of vandalism? Violence? Serious fights? Gangs? Drug problems? Yes ____ No ____
4. Have many cases of student arrests? Yes ____ No ____

   (If you don't know the answer to the above, local law enforcement agencies may be able to give you some information.)
5. Plan events that encourage school unity? Yes ____ No ____
6. Have good recreational and extracurricular activities that are well supported by the student body and staff? Yes ____ No ____
7. Provide good counseling and health services? Yes ____ No ____
8. Give everyone the opportunity to respect their own heritage and those of others? Yes ____ No ____
9. Involve parents in important school decisions and events? Yes ____ No ____
10. Seek a constructive bond with the community: through urban or neighborhood improvement projects, with law enforcement, with the handicapped, with the elderly, and so on? Yes ____ No ____
11. Use community agencies or the expertise of residents to help with school programs? Yes ____ No ____
12. Direct students and parents to community resources that will improve their lives? Yes ____ No ____
13. Provide students with real leadership opportunities (not simply token positions)? Yes ____ No ____
14. Have an "emotional climate" in which students, faculty, and administrators feel free to express their thoughts and feelings? Yes ____ No ____
15. Encourage students to be creative and curious? Yes ____ No ____
16. Help students explore and appreciate their own special talents? Yes ____ No ____
17. Allow students to clarify values? Yes ____ No ____
18. Help students to deal effectively with inner and outer conflicts? Yes ____ No ____
19. Provide problem-solving and decision-making experiences for students? Yes ____ No ____
20. Help students develop goals for the future? Yes ____ No ____

exists in the school. Questions 5 to 9 concern school services and special programs. Response here can indicate a lack of school unity.

Responses to questions 10 to 12 reveal how members of the school community feel the school fits into the life of the larger community it serves. They also indicate whether neighborhood resources are used for the benefit of those attending the school. Questions 13 and 14 deal with whether the three main segments of the school community—students, faculty, and administrators—are able to communicate effectively with each other. Responses to questions 15 to 20 indicate whether there are opportunities for students to develop self-respect and self-awareness. To understand the results, tabulate both the total responses and the three subtotals for the three segments of the school population. If these data indicate a problem in one of the six areas—antisocial responses, services and programs, community relations, communications, affective areas, and general responses—then committees should be established to improve the situation in the problem area. Tessler says that her experience with this approach shows that such changes can go a long way in making the abuse of substances less desirable to students.

It must be admitted that much of the information we have on abuse is not totally reliable, because of the situations in which drug use must be studied. In the next area that we consider, mental health issues, there is better data because the problem is usually studied in the controlled setting of the hospital.

Mental Health Issues

The number of adolescents being admitted for psychiatric care has been skyrocketing in recent years.

A number of psychologists and psychoanalysts (most notably Freud) have suggested that it is normal in adolescence to have distressing, turbulent, unpredictable thoughts that in an adult would be considered pathological. Here is an example of this view:

> ***The fluidity of the adolescent's self-image, his changing aims and aspirations, sex drives, unstable powers of repression, and his struggle to adapt his childhood standards of right and wrong to the needs of maturity, bring into focus every conflict, past and present, that he has failed to solve. Protective covering of the personality is stripped off, and the deeper emotional currents are laid bare.*** **(Ackerman, 1958, pp. 227–28)**

This disruptive state is partly characteristic of the identity stages of confusion and moratorium (See chap. 11). Identity confusion is sometimes typified by withdrawal from reality (Erikson, 1958, 1968). There can be occasional distortions in time perspective. Mental disturbance also often makes intimacy with another person impossible. These characteristics are also seen in the moratorium stage, but they tend to be of much shorter duration.

How common and how serious are these problems? The picture is not clear. Summarizing decades of research, Kimmel and Weiner (1985) conclude that true **psychopathology** (mental illness) is relatively rare during adolescence. It is impossible to determine the frequency of mental illness, however, because of current disagreements over its definition. Weiner (1970) cites numerous studies, which give us considerable reason to believe that "adolescent turmoil," while common, does not really constitute psychopathology.

Studies *do* indicate that when adolescents become seriously disturbed and do not receive appropriate treatment quickly, the chances of their "growing out" of their problems are dim (Walker & Greene, 1987; Wilson, 1987). Weiner (1970) warns that

> ***An indiscriminate application of "adolescent turmoil" and "he'll-grow-out-of-it" notions to symptomatic adolescents runs the grave risk of discouraging the attention that may be necessary to avert serious psychological disturbance.*** **(p. 66)**

It is not clear whether a greater number of adolescents are experiencing mental illness, but the number of them being *admitted* for psychiatric care is skyrocketing. From 1980 to 1984, for example, admissions of teens to psychiatric hospitals increased 480 percent—from 10,764 to 48,375 nationally (Select Committee on Children, Youth, and Families, 1985). Table 13.4 shows the amazing change from 1976 to 1984 in the Minneapolis/St. Paul area. These data may reflect a more positive societal attitude toward receiving help with emotional problems.

Table 13.4 Juvenile Psychiatric Admissions in the Minneapolis/St. Paul Area

Year	Number	Rate per 1,000	Patient days
1976	1,123	91	46,718
1977	1,062	88	53,730
1978	1,268	107	60,660
1979	1,623	142	68,949
1980	1,775	158	74,201
1981	1,745	159	72,381
1982	1,813	165	71,267
1983	2,031	184	76,899
1984	3,047	299	83,015

Types of Mental Disorders

The questions of what kinds of mental illness make up this increase has not received much attention recently. The study by Rosen and colleagues (1965), though dated, examined a very large number of cases. It gives us an idea of the types of mental disorders that adolescents suffer, and there is no reason to assume that these data have changed greatly. Approximately 4 percent of the illnesses of both males and females were accounted for by acute and chronic brain disorders (a malfunction of some part of the brain), 10 percent by mental retardation, and 6.5 percent by schizophrenia (a serious distortion of reality).

In Britain, Rutter (1980) reviewed surveys and found that about 15 percent of 15-year-olds were afflicted by psychiatric disorders, although as many as 6 percent more went undetected. In a more recent study, Horwitz and White (1987) studied differences between male and female adolescents. They found that 11 percent of the males suffered from neurotic disorders, such as anxiety, depression, and obsessive-compulsive reaction, compared to 18 percent of the females.

Although the overall picture is less than clear, we have a great deal of data on specific adolescent mental disturbances. Chief among them are eating disorders and depression and suicide.

Anorexia nervosa has become an increasingly frequent problem among adolescent females.

Eating Disorders

Anorexia Nervosa

Anorexia nervosa is a syndrome of self-starvation that mainly affects adolescent and young adult females, who account for 95 percent of the known cases. Professionals suspect that many males may also be victims (e.g., those who must maintain a low weight for sports), but their illness is covered up. (Larson & Johnson, 1981; Mintz and Betz, 1988). It is characterized by an "intense fear of becoming obese, disturbance of body image, significant weight loss, refusal to maintain a minimal normal body weight, and amenorrhea [suppressed menstruation]. The disturbance cannot be accounted for by a known physical disorder" (American Psychiatric Association, 1985).

Health professionals have seen an alarming rise in the incidence of this disorder among young women in the last fifteen to twenty years (Bruch, 1981; Rosen & others, 1987). Whether there is an actual increase of anorexia nervosa or whether it is now being more readily recognized has yet to be determined.

The specific criteria for anorexia nervosa are

- Onset prior to age 25.
- Weight loss of at least 25 percent of original body weight.
- Distorted, implacable attitudes toward eating, food, or weight that override hunger, admonitions, reassurance, and threats, including
  - Denial of illness, with a failure to recognize nutritional needs
  - Apparent enjoyment in losing weight, with overt manifestations that food refusal is a pleasurable indulgence
  - A desired body image of extreme thinness, with evidence that it is rewarding to the person to achieve and maintain this state
  - Unusual hoarding and handling of food
- No known medical illness that could account for the anorexia and weight loss.
- No other known psychiatric disorder, particularly primary affective disorders, schizophrenia, or obsessive, compulsive or phobic (fearful) neuroses. (Even though it may appear phobic and obsessional, food refusal alone is not sufficient to qualify as an obsessive, compulsive or phobic disorder.)

- At least two of the following manifestations: amenorrhea (loss of menses); lanugo (soft downy hair covering body); bradycardia (heart rate of less than 60); periods of overactivity; vomiting (may be self-induced).

Bulimia Nervosa

Bulimia nervosa is a disorder related to anorexia nervosa and sometimes combined with it. It is characterized by

> ***. . . episodic binge-eating accompanied by an awareness that the eating pattern is abnormal, fear of not being able to stop eating voluntarily, and depressed mood and self-deprecating thoughts following the eating binges. The bulimic episodes are not due to Anorexia Nervosa or any known physical disorder.* (American Psychiatric Association, 1985)**

Bulimia has been observed in women above or below weight, as well as in those who are normal (Lowenkopf, 1982). The specific criteria of bulimia are

- Repeated episodes of binge-eating
- Awareness that one's eating pattern is abnormal
- Fear of not being able to stop eating
- Depressed mood and self-deprecation after binges

Anorectics and bulimics share emotional and behavioral traits, despite their clinical differences. The most characteristic symptoms specific to these disorders are the preoccupation with food and the persistent determination to be slim, rather than the behaviors that result from that choice (Bruch, 1981).

A number of new approaches to treatment and therapy are currently being researched (Scott, 1988). Although success rates are not high, the situation in either disorder is usually so complex and potentially hazardous that only qualified personnel should attempt to treat victims.

Depression[1]

The term *depression* can have many different meanings and manifestations. Originally a word for a pathological symptom, it has found its way into common usage by the general public, and its meaning has been greatly broadened.

Depression may be viewed either as a mood (situationally caused), a syndrome (a complex of behaviors and emotions), or as a clinical disease (Peterson & others, 1993). As you will see, the biopsychosocial model affects each of these approaches. Depression is considered to be a basic affective state that, like anxiety, can be of long or short duration, of low or high intensity, and can occur in a wide variety of conditions at any stage of development. In certain circumstances, such as in reaction to a death in the family, it is a normal and appropriate affective response.

Depression becomes pathological when it occurs in inappropriate circumstances, is of too long duration, or is of such great intensity as to be out of proportion to the cause. Depression is harmful to a person's development when it interferes with the capacity to work, to relate to others, or to maintain the healthy functioning of essential physical needs for sleep and nutrition. Serious depressive conditions can upset a person's functioning in all of these areas and more.

[1] The authors wish to acknowledge the contribution of Dr. David Curran to this section.

Symptoms of Depression

The symptoms of depression may be classified in one of four areas (Beck, 1967):

- *Emotional manifestations*: dejected mood, negative self-attitudes, reduced experience of satisfaction, decreased involvement with people or activities, crying spells, and loss of sense of humor.
- *Cognitive manifestation*: low self-esteem, negative expectations for the future, self-punitive attitudes, indecisiveness, and distorted body image.
- *Motivational manifestations*: loss of motivation to perform tasks, escapist and withdrawal wishes, suicidal thoughts and increased dependency.
- *Physical manifestations*: appetite loss, sleep disturbance, decreased sexual interest, and increased fatigue.

Not all depressed individuals show all of these symptoms, of course, but they are likely to exhibit one or more symptoms from these four categories.

There is some evidence that African-Americans may be more vulnerable to depression than white Americans. Freeman (1982) examined emotional distress, as assessed by the Hopkins Symptom Checklist, among 607 urban African-American high school students 15 to 18 years of age. Subjects reported high distress primarily about feelings of disadvantage, volatile anger, interpersonal sensitivity, and loneliness. Females were significantly more likely than males to suffer emotional distress, although several of the leading distress items were experienced equally by both groups.

Causes of Depression

Adolescents develop depression for a variety of reasons:

- *Capacity for denial.* Young persons appear able to deny the reality of painful conditions or affects with greater effectiveness than adults.
- *Tendency to act out feelings.* The impulsiveness of adolescence makes it more likely that feelings will find expression in actions.
- *The desire to avoid dependence and helplessness.* Most adolescents want to feel independent, strong, and able to control their problems. They do not want to feel dependent on adults or at the mercy of events or feelings.

The causes of depression, both normal and pathological, can usually be generalized under the heading of *loss* (Cantwell & Carlson, 1983; Carlson, 1983; Crumley, 1982; Curran, 1984; Seiden & Freitas, 1980; Shaffer & Fisher, 1981; Tishler & McHenry, 1983). Seriously depressed individuals have usually experienced a series of losses, which may include losses of loved ones through death or relocation.

Depressed adolescents will often recount a history of parental separation, death, or divorce; a series of moves; death or loss of pets; moving away from trusted friends; or express the feeling that childhood was a far better state than adolescence. There ensues a feeling of hopelessness or despair at not being able to regain the lost objects or status. The anger born of this frustration is often turned against the self, with harmful results.

Depression is manifested with increasing frequency during mid and late adolescence. Females are three times more likely to be affected by depression than males. This is probably because girls, when they are unhappy, tend to be intrapunitive, that is, take some action that is harmful to themselves. Boys tend to be extrapunitive and hurt others. Both tendencies clearly seem to result from the gender roles society has assigned to us.

Masked Depression

Toolan (1975) states that "especially in the adolescent we seldom see a clear picture of depression" (p. 407). He refers to this situation as *masked depression.* It is a distinctive quality of adolescent depression that the symptoms listed on page 359 are often not seen or are in some way obscured (Tishler & McHenry, 1983). It is especially true with boys, who "have a need to hide their true feelings, and particularly the softer, tender, weak sentiments" (Tishler & McHenry, 1983, p. 732).

Depressive Equivalents

Adolescents are predisposed, then, toward alternative forms of dealing with depression. These different symptoms for the same disorder are called *depressive equivalents.* They serve the purpose of allowing adolescents to discharge and seek relief for their feelings, and at the same time avoid a recognition of their problems and feelings. Activity of this type distracts teenagers from thinking of their problems and facing the unpleasant image they hold of themselves and their lives.

Examples of depressive equivalents are

- *Concentration difficulty.* Often difficulty in concentrating is the earliest, most frequently cited, and only symptom present and the only one that adolescents are aware of. There is a defensive quality to these problems in concentrating. As the mind seeks to avoid awareness of painfully sad thoughts and feelings, it may skip actively from thought to thought, unable to stay still for fear of it being caught by the waiting depressive alternative. The effect on school performance can be devastating.

Young teens run away from home for a variety of reasons. Sometimes teenagers run away in order to deal with their overwhelming feelings of depression. A recent study found that 160,000 teens are locked out of their homes by their parents, and many wind up living on the streets.

- *Running away.* Depressed teenagers sometimes run away from the family home, foster home, or other residential setting as a means of actively dealing with overwhelming feelings that often originate in family relations. Running away provides a temporary release of tension and gives the feeling that one is in control.

- *Sexual acting out.* The urgent necessity to ward off underlying feelings of being unloved and unwanted may push the adolescent toward promiscuous sexual behavior. Close physical contact with another person provides relief. Females are especially vulnerable.
- *Boredom and restlessness.* Depressed adolescents often manifest their condition by swinging between states of short-lived but unbounded enthusiasm and periods of intolerable boredom, listlessness, and generalized indifference. It is to avoid coming any closer to an awareness of depressive effects that the cycle of excited activity and restlessness is again renewed. "I'm bored" is often an unconscious code phrase for "I'm depressed."
- *Aggressive behavior and delinquency.* Depressed adolescents, especially boys, sometimes carry out angry and destructive behavior, such as vandalism, in place of the depressive feelings. These actions may be designed to counteract the poor self-image and feelings of helplessness by artificially inflating the youth's self-image as a strong, fearless, and clever person.

Masked depression and its depressive equivalents are dangerously unhealthy because they obscure the nature and extent of the individual's distress from the teenager and from significant adults. In some cases these behaviors are very self-destructive and jeopardize normal adolescent development in favor of temporary but ineffectual relief.

In these sections on mental health problems, we have concentrated on psychological and health problems. It should be noted, however, that we are just beginning to learn more about biological factors, too.

The Crisis of Death

Although not a true mental disturbance, the death of a loved one almost always causes teens great disturbance. Teenagers seldom have to deal with their own mortality or the deaths of relatives or friends. How do they become aware of death? The development of children's understanding of death closely parallels other aspects of their cognitive development (Nagy, 1982) as described by Piaget (See chap. 2). Until about 2 years of age (the end of the sensorimotor stage), children lack an understanding of death; they become anxious at the absence of a parent or sibling, but this is a temporary state. As children reach the preoperational stage, from 2 to 7 years, they begin to form a concept of death, but they think of it as quite reversible, as a gradual sleeplike state from which the person will probably soon return. Their sense of loss is similar to the feeling that attaches to the loss of a favorite toy, which can be replaced later by another. Toward the end of this stage, death becomes personified, so that it seems to be an invisible being lurking about. For some children, death becomes frightening at this time.

During the years 7 through 11 (the concrete operational stage), death becomes much more of a reality. Children learn that it is not reversible, and become anxious when thinking of the loss of parents or other loved ones. However, they still cannot think of death in terms of themselves. It is only during adolescence (the formal operations stage) that children realize they are mortal and will die someday. Now they have a sense of their own future and a clearer sense of the time dimension itself. As they begin to plan their lives, the inevitability of death becomes real and they develop a new anxiety about death.

> ***Many adolescents have a clear feeling of the passing of this life stage with the expectation of a specific ending. This sense of the anticipated end of adolescence might be an important parallel to the concept of the finality of death. . . . I would add that ruminations over suicide and forms of actively promoted death encounters would serve to provide the adolescent with a sense of greater temporal definition in his life*** **(Hankoff, 1975, p. 76).**

Adolescents who have a clear sense of their own values, and who have been helped by their parents to form a set of values, are more likely to have a realistic and less fearful attitude toward death. This is not to say that parents can simply give their adolescents an appropriate attitude toward death. Those adolescents in James Marcia's stages of identity diffusion and foreclosure (See chap. 11) are the most likely to have difficulty with the concept of death because of their unwillingness to confront reality.

Kastenbaum (1959) found that youths who are highly religious and who believe in life after death are more likely to show an active interest in the topic and are more likely to conduct their daily lives on the basis of their expectations for the future. They are more apt to have confronted the formal aspects of death.

This is supported by a more recent study by Gray (1987). He administered the *Beck Depression Inventory* to fifty 12- to 19-year-olds who had lost a parent not less than six months and not more than five years before testing. Gray found that those who had achieved a healthy adaptation had had "high levels of informal social support, good relations with the surviving parent prior to loss, a balanced personality style, and presence of religious beliefs" (p. 511).

In her popular book *On Death and Dying,* Elisabeth Kübler-Ross (1969) points out that in "old country" and "primitive" cultures, children have contact with death from an early age. In technological societies, children are shielded from death because it is considered too upsetting. She suggests that when

> ***children are allowed to stay at home where a fatality has stricken, and are included in the talk, discussions, and fears, [it] gives them the feeling that they are not alone in the grief and gives them the comfort of shared responsibility and shared mourning. It prepares them gradually and helps them to view death as part of life, an experience which may help them grow and mature.*** **(p. 6)**

Kübler-Ross has concluded from her studies that most persons who know that their death is imminent pass through a sequence of five stages in their attempts to deal with it: denial and isolation, anger, bargaining, depression, and acceptance. Other scholars have been unable to confirm these findings, however.

Hankoff (1975) has suggested that the absence of initiation rites in technological societies is responsible for the deep fear of death common among adolescents (for more on this, see chap. 15). Initiation rites offer youth a meaningful encounter with this most important aspect of life—death:

> ***In brief, [initiations into] these mysteries involve sexuality, the spirit world, and death. The mystery of death is brought to the initiand through a ritualized performance in which he is symbolically made to die and be born. This initiatory ritual thus is a template for the important human spiritual experience of self-renewal. In traditional cultures the initiatory ritual tangibly and powerfully provides the youth with knowledge of the spiritual world, the mysteries and privileges of adult status, and emotional participation in death and rebirth experiences.*** **(p. 379)**

Suicide

Suicide and attempted suicide among contemporary adolescents is cause for growing concern (see Table 13.5). In this chapter we discuss the psychological causes of its prevalence at this stage.

Table 13.5 Suicide Deaths per 100,000 among U.S. Adolescents Ages 10 to 14 and 15 to 19, 1979–87

	Suicide Deaths/100,000 Population											
	Ages 10 to 14						Ages 15 to 19					
	Total	Male	Female	White	Black	Other	Total	Male	Female	White	Black	Other
1979	0.81	1.09	0.59	0.91	0.26	1.21	8.35	13.32	3.2	8.92	4.38	12.13
1980	0.76	1.21	0.29	0.86	0.33	0	8.51	13.82	3.02	9.29	3.59	11.02
1981	0.89	1.23	0.54	1.01	0.22	0.79	8.63	13.56	3.54	9.45	3.53	11.28
1982	1.09	1.71	0.44	1.13	0.81	1.29	8.7	14.07	3.15	9.58	3.86	8.26
1983	1.09	1.58	0.57	1.1	1.0	1.23	8.7	13.98	3.24	9.4	4.15	12.32
1984	1.28	1.91	0.63	1.38	0.71	1.51	9.01	14.26	3.54	9.92	3.8	10.12
1985	1.61	2.31	0.87	1.76	0.83	1.62	9.97	15.98	3.73	10.81	4.86	12.07
1986	1.5	2.3	0.66	1.58	0.96	2.05	10.18	16.36	3.76	11.28	4.59	8.36
1987	1.5	2.3	0.6	1.6	1.0	0	10.3	16.2	4.2	11.2	5.8	0.7

Source: Office of Technology Assessment, 1991.

The Meaning of Suicide Attempts among Teenagers

Only a small percentage of teenagers who make suicide attempts actually die. This is not true of older age groups. With advancing age, suicide attempts increasingly result in death. The relative infrequency of suicidal deaths in teenagers raises questions about the actual meaning and intent of these apparently self-destructive acts.

Several studies have explored the lethal intent in adolescent suicide attempts. Bancroft (1979) reported that among a general population of those admitted to a hospital emergency room because of self-poisoning, 42 percent stated that they had no intention of dying. Persons in Bancroft's study were considered to have the lowest level of suicidal intent, compared to the 21–35 and the 36-and-over age groups, as learned from the self-reports of subjects. Curran (1984) asked teenagers who attempted suicide if they thought that adolescents who attempt suicide intend to die. Only 16 percent named "wish to die" as the primary motive.

Self-poisoning (usually through drug overdose) is by far the most common mode of attempting suicide among female and younger adolescents in general. Self-poisoning, however, is rarely of high lethality. McIntire (1980) reported that only 12 percent of cases intended to cause death.

It is safe to say that most teenage injuries to self are not attempts to end life. What then is the actual meaning of and reasons for such dramatic acts? What are the hoped-for effects of the suicidal act of low lethality?

Considerable research points to the highly communicative quality of this type of suicidal behavior, particularly in younger and female populations. Further, teenage suicide attempts appear to occur within an interpersonal context (Hawton, 1982a; Topol & Reznikoff, 1984; Wenz, 1979; White, 1974). Often the hope of the suicidal adolescent is to regain a lost love or influence the lover to feel more positively. Bancroft (1979) found that 45 percent of the 16- to 20-year-old suicide attempters studied were "seeking help" by means of their suicide attempt, while 35 percent sought to "influence someone."

Self-poisoning is the most common method of suicide attempts among females and young adolescents.

The finding that teenage suicide attempts are usually of low lethality in no way diminishes the seriousness of the action. The adolescent who attempts suicide is a needy person whose act should be treated with the utmost seriousness. This is also true for those who "only talk about committing suicide." Their remarks should always be referred to qualified personnel.

Adolescent suicide and attempted suicide can derive from a variety of conditions. However, certain common factors have been found. In every case, suicidal behavior occurs as the culmination of multiple, long-standing, significant problems, both within the person and between that person and the environment. The problems involve three major areas: personality problems, family problems, and societal problems.

Personality Problems as Causes of Suicide

Historically speaking, adolescent suicidal behavior has been viewed as behavior of an impulsive nature, often indulged in by relatively normal teenagers (Crumley, 1982; Jacobziner, 1965). It has become increasingly clear, however, that teenage suicide attempters are significantly troubled individuals whose emotional problems are impressive, if not easily labeled. Labeling has proved difficult because of the myriad ways in which teenagers manifest their symptoms and hide or obscure their real feelings to the adult world. Hudgens (1975) stated that "considered by itself, the fact that a teenager has attempted suicide or made a serious suicidal communication tells little about him except that he probably has a psychiatric disorder" (p. 150). We describe the primary personality problems in the following paragraphs.

- **Depression** The feature most often seen in the literature on adolescent suicide is depression. Therapists and other mental health workers have become better at spotting it in its various "masked" forms (Tishler & McKenry, 1983). Further research has suggested that the mental health state of the adolescent suicide attempter and committer is now more disturbed than in previous years. Psychosis, however, is seen in only about 2 percent of the adolescent suicidal population and is especially rare in female adolescents (McIntire, 1980; Schowalter, 1978).

- **Overreliance on Limited Support** Adolescents who attempt suicide tend to overinvest themselves in very few, but very intense, interpersonal relationships. They appear to have a limited capacity to support themselves emotionally or to cope with their lives by means of internal strength alone. Rather, they rely heavily on the support of others, usually peers (Topol & Reznikoff, 1984; Walch, 1976).

Their peer relationships are supercharged but often painfully shortlived. They often overwhelm their friends with their neediness, which leads those friends to distance themselves. Suicidal teenagers often go through a series of failed relationships of this type without the internal strength to support themselves.

- **Communication Skills** Suicidal adolescents tend to express troubled feelings through behavior rather than internal or interpersonal dialogue. They are usually not in counseling at the time of their attempt and usually have never been in counseling. In many instances, the suicide attempt itself is an attempt at communicating to specific individuals or the human environment at large the desperation and loneliness they feel. Adolescents who are aware and tolerant enough of their unhappiness to talk about it are at far less risk of suicide than those who have no other expressive medium available to them.
- **Reality Testing** Adolescents who attempt suicide often lack an adequate capacity to accurately assess their condition in the world. It is difficult for them to put things in perspective. Troubles in one area of their lives are generalized to other areas. This inability to evaluate their world objectively may be a transitory condition in response to a stressful event, or it may be a more chronic condition. This manner of thinking allows them to be overwhelmed by events that may seem trivial to an outsider. It causes them to feel more helpless about their future and at the mercy of events in the present.
- **Hypersensitivity** All adolescents occasionally overreact to situations, but the **hypersensitive youth** has an extreme reaction to situations that would only mildly disturb most people. The disruptions caused by seemingly trivial events may come together in a suicide attempt.
- **Suggestibility** Sabbath (1969) has described the "expendable child," who believes that his parent or parents wish him dead. This parental wish may be conscious or unconscious, spoken or unspoken, true or untrue. But to the extent that children are suggestible, they are likely to comply with this perceived wish. The role of the parents in the suicide attempts of highly suggestible children is clear. Lorand and Schneer (1961) have studied parents whose sadistic behavior conveys to children that they are unwanted. Glaser (1965, 1978) has studied the emotionally detached parent, who is unwilling or unable to demonstrate love. Such parents may care deeply for their children, but suggestible adolescents tend to take their apparent detachment as a wish for their death. Teicher and Jacobs (1966) found that 88 percent of adolescent suicide attempts occur in the home, quite frequently with one or both parents right in the next room, an indication that parental behavior is a factor.

- **Magical Thinking** Many adolescents have an unrealistic view of death's finality and use suicide as a means to radically transform the world and solve problems or to join a loved one who has already died. We call this magical thinking. These feelings are often aided and abetted by the glorification of suicide that sometimes occurs in the media (Garner, 1975).
- **Religious Fanaticism** Some adolescents, whose faith in the omnipotence and omniscience of God is particularly strong, and who are doubtful about whether they are good enough to continue living, decide to attempt suicide and leave the outcome to God. Just as in medieval times when disagreements were settled by jousting matches, extremely religious youths expect God to intervene and save them from death by their own hand if He judges them worthy of life.
- **Lack of Control Over the Environment** Corder, and associates (1974) cite the inability to change one's environment as a frequent cause of attempted suicide. This concept was first studied by Rotter (1971), and a review of the studies of personal control has appeared elsewhere (Dacey, 1976). People tend to fall into one of two categories in terms of their sense of control over their lives. "Internals" see control as self-derived; they have a sense that they can influence what happens to them by their own actions. "Externals" see control as imposed by outside factors; thus, they see life as a matter of chance or luck. Some externals really do not have much control; others only imagine they do not. In either case, external individuals are far more likely to commit suicide than internals.

Family Problems as Causes of Suicide

Considerable research has been devoted in recent years to the constitution, dynamics, and histories of the families of suicidal adolescents (Angle, 1983; Crumley, 1982; Hawton, 1982a, 1982b, 1982c; McKenry and others, 1982). It has been shown that the families of suicidal adolescents experience significantly more dysfunction, disorganization, mobility, and loss than the families of normal teens.

- **Parental Losses** Parental losses tend to occur at an earlier age for the suicidal adolescent than for comparison groups of disturbed, nonsuicidal adolescents. A high incidence of parental deprivation, both physical and emotional, has often been reported (Wade, 1987; White, 1974). Physical abuse in the home has been cited by Green (1978) as a relatively more common element. Jacobs (1971) stressed that suicidal teens often feel they are a burden to their families, and that the family would be better off without them. In many cases this perception was covertly or overtly reinforced by home life.
- **Parent-Child Role Reversals** Parents and children sometimes exchange traditional role behaviors—called **parent-child role reversals.**

> ***In the parent-child interaction, the child adopts some parent-type behavior (for example, care-taking, supporting, nurturing, advising), and the parent acts more as a child would be expected to act (for example, seeking support, acting helpless, or unable to cope). (Kreider & Motto, 1974, p. 365)***

Role reversal occurs frequently in a home where there is one parent or one child. It tends to produce anxiety, pain, frustration, and hostility in adolescents. They are not ready to handle such a burden of adult behavior and may blame

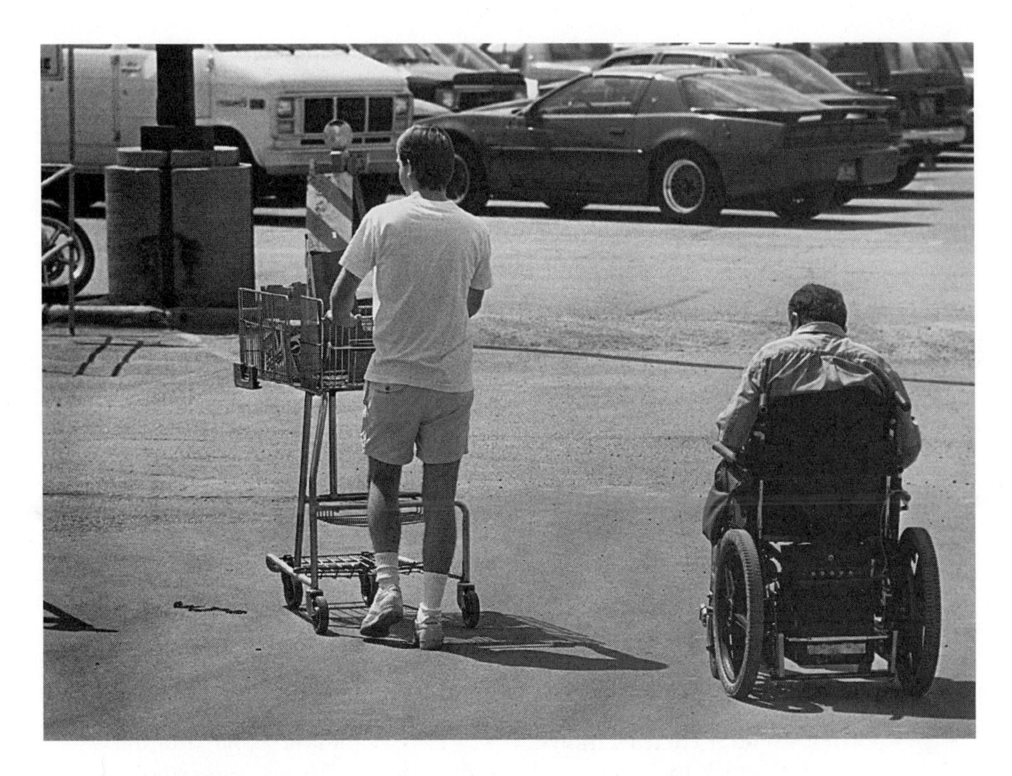

Children who feel they have to take care of their parents have a higher than average incidence of suicide attempts. Here the problem is being needed in ways that are unhealthy.

themselves for their inability. They may seek to alter the family environment, or to escape it, or may collapse under the pressure. They sometimes see suicide as the only way out of the double bind of having to act as both child and adult.

- **The Appearance of Not Being Needed** When we feel that no one needs us, we tend to become lonely and self-centered:

> ***Suicide is not so much the outcome of "pressure," but pressure without social support. Suicide does not automatically mean that a person has not been loved or cared for. It probably does mean that he was not needed by others in an immediate, tangible fashion. 'Needed ' should be understood in the sense we imply when we say we need the first-string member of an athletic team, the paper delivery boy, the only secretary in a small office, or the only wage earner in a family. The person needed must be obviously relied upon by others, and his absence should create a disruptive and foreseeable gap. In this light, it is understandable that one of the highest suicide rates is that of middle-aged bachelors and one of the lowest is that of married women with children. . . . It is nice to know we are loved, but essential to know we are needed.*** **(Wynne, 1978, p. 311)**

- **Communications** Finally, serious impairment of communication between father and daughter is increasingly being noted and treated as a factor in the dynamics of the female adolescent suicide (Angle, 1983; Hawton, 1982a, 1982b, 1982c; McKenry & others, 1982; White, 1974).

Societal Problems as Causes of Suicide

- **Problems with Peers** Peer problems are considered a critical factor in the development of adolescent suicidal behavior (Celotta & others, 1987; Jacobs, 1971; Rohn, 1977; Teicher, 1973; Tishler, 1981; Walch, 1976; Wenz, 1979). This is especially true for disturbed, suicidal adolescents whose family life has often been inadequate. They tend to feel that they have little else in the world to support them, and with the departed friend or lover goes their sense of worth, esteem, and belonging (Hawton, 1982c; White, 1974).

- **Imitation** "Epidemics" of teenage suicides and suicide attempts in single localities recently have caused researchers and laypersons alike to wonder about the contagiousness of adolescent suicidal behavior. The National Center of Disease Control in Atlanta, Georgia, is mounting a major effort to study this. Some excellent research does suggest that well-publicized suicides bring out latent suicidal tendencies in adults and significantly increase the rate of suicide in the geographic area covered by the publicity (Ashton & Donnan, 1981; Bollen & Phillips, 1982; Phillips, 1979). It is reasonable to assume that adolescents are at least as readily influenced as adults.

Curran (1984) has demonstrated that teenagers are quite familiar with suicide as a behavioral alternative to coping with life's problems. He reported that 87 percent of the female high school students questioned knew someone who had attempted or committed suicide. In 55 percent of the cases, it was a person known well by the teenager—a friend, close relative, or family member. Among the males questioned, 57 percent knew someone who had attempted suicide; 29 percent knew the attempter well.

- **Other Societal Factors** Individuals behave within the context of the society in which they live and are influenced by the pressures and changes within that society. It would seem that our society is becoming more suicidogenic for our adolescents. Some of the factors that promote dangerous stresses have been studied by McAnarney (1979). She found that the high suicide rates in the United States appear to be related to the following societal factors:

  Family status. **Family status** refers to the strength of family ties. In societies where family ties are strong, fewer suicides occur. American adolescents who are experiencing the current changing status of the family may suffer in their achievement of identity. Also, when families are not close, parents may not be able to recognize early warning signs and help their teenagers before suicide becomes an alternative.

  Religion. Studies have indicated that in societies where the majority of the people belong to a formal religious organization, suicide attempts are low.

  Transition and mobility. Groups in transition tend to have higher suicide rates than stable ones. For example, many American teenagers experience disorganized urban environments or frequent moves when one or both parents are transferred at work.

  Achievement orientation. Cultures such as the United States, in which achievement is a major priority, have higher suicide rates. Japan and Sweden are other achievement-oriented societies that exhibit high suicide rates.

  Aggression. Sometimes suicide is defined as aggression turned inward. In cultures where the expression of aggressive feelings is discouraged or suppressed, suicides are more prevalent. This may help to explain the higher rate of suicide attempts among females.

Blumenthal and Kupfer (1988) have made an effort to incorporate all these factors into "a new approach to early detection and intervention for the prevention of youth suicide" (p. 1). Their proposal is certainly the most comprehensive to date;

research on its effectiveness will be worth careful scrutiny. A recent review of suicide prevention techniques (Garland & Zigler, 1993) found school-based programs inadequate and possibly harmful. They urge a more integrated approach.

AN APPLIED VIEW **Warning Signs of a Potential Suicide Attempt**

- Abrupt change in school grades
- Withdrawal or moodiness
- Accident proneness
- Change in eating or sleeping habits
- Other significant changes in usual behavior
- Talking about killing oneself
- Talking about "not being" or not having any future
- Giving away prized possessions

DELINQUENT BEHAVIOR

> ***Even in the early 1960s, Leonard Bernstein's West Side Story signaled society's concern that glib sociological excuses would only worsen the handling of behavior disorders. Public opinion and public institutions are turning back to a punishment model. The pendulum has swung away from the idealization of Rousseau's "noble savage," who would be good only if society did not overcontrol and frustrate him, toward today's bleak cynicism of the essential untreatability of the* Lord of the Flies.**
>
> ***John Meeks and Allen Cahill, 1988***

You have probably been a juvenile delinquent. Juvenile delinquency is defined as "any illegal act by a minor." Does this apply to you? Well, it does to your authors, both of whom can remember having committed some petty crime in their youth. Studies indicate that between 90 and 100 percent of high school students admit to having committed some illegal act, such as dime store theft, using a fake ID, or having taken an alcoholic drink while under age. For most teens, delinquency is no problem. Less than 20 percent of all crime in the U.S. is committed by teens (Federal Bureau of Investigation, 1993). Nevertheless, the problem is serious enough for us to review here.

Although juvenile delinquency has usually been differentiated from adult crime only in terms of age (usually 18), in this chapter, adolescent delinquent behavior is distinguished according to three groups: **nonaggressive offenders,** that is, persons who commit crimes that are mainly harmful to themselves, (runaways, prostitutes); the **juvenile delinquent** acting alone, who usually commits theft or destruction to property; and the **aggressive gang,** which engages in a variety of illegal group activities.

Delinquency is a complex issue, reflecting the complexity of problems in 1990s America and including, but not limited to, increased drug traffic, a sagging economy, rising unemployment, and an increasing number of homeless individuals and families. It is our hope to better understand the dynamics of delinquency so that the problems of delinquent youth may be better addressed and, ultimately, prevented.

The Nonaggressive Offender

Persons labeled *nonaggressive offenders* are lawbreakers, technically, but they usually do not do physical harm to another person's body or property. Females constitute the majority of teenagers in this category, although the number of males is growing. About 25 percent of juvenile court cases are of this type (Haskell & Yablonsky, 1982).

The Runaway

According to the U.S. Department of Commerce (1988), the number of children leaving home without permission of their parents has been increasing in recent years. They estimate that one out of ten 12- to 17-year-olds has run away from home at least once. Why do children run away from home? Clashes with family members and the commission of petty crimes are major causes. Almost 70 percent of the runaways interviewed have low achievement and little, if any, involvement with school. Runaways give as reasons the unfairness of laws that make it necessary to get parental permission for such actions as seeing a doctor of their choice or obtaining a job.

The number of female runaways is considered to be slightly higher than male runaways, although this may only reflect the number of runaways who are arrested. Because teenage female runaways are more visible on the streets, they are more likely to be arrested than males and thus counted as runaways (Burgess, 1986). The numbers of adolescent males and females served in runaway shelters are about equal (Rotheram-Borus & others, 1991).

A growing number of runaways are referred to as **system kids** (Rotheram-Borus & others, 1991). These are adolescents who were removed from their parents' homes by social welfare agencies because of parental neglect or abuse. After being placed in foster care or a group home, these adolescents decide to run. According to Robertson (1989), half of the runaway youths come from foster homes, group homes, and delinquent detention facilities.

Although the problem of the runaway has been almost totally disregarded until recently, a number of services to runaways and their parents have been instituted in the past five years. Among these experiments are

- Special services to runaways by youth agencies
- Toll-free phone numbers that are open 24 hours to counsel runaways (See Appendix A at the back of this book.)
- School programs that explain the causes and problems of running away to teenagers
- Conferences, training sessions, and literature made available to the parents of runaways

Prostitution

"It's not so bad, honey," Sherry said. "Flatbackin' ain't the worst thing can happen to ya."

For many years, female runaways have turned to prostitution in order to support themselves. Also of great concern is the finding that the number of teen male prostitutes has been growing rapidly.

Although there are no reliable statistics concerning the number of teenage prostitutes, authorities believe that teenage female prostitution has been growing in recent years, and that male teenage prostitution has been growing at an even faster rate. There is a definite relationship between the number of runaways and the number of teenagers who turn to prostitution to support themselves. It has been difficult to determine the seriousness of this problem because of the variety of legal definitions of prostitution. Statistics may be distorted by the fact that most police officers are reluctant to charge a female teenager with prostitution, but are much less hesitant to do so with a male.

Brown (1979) has studied prostitution among teenage girls. Their common childhood experiences include alienation from family, parental abuse, low level or failure in education, many changes in family and home life, and dismal job prospects. Entrance into prostitution is motivated by desire for financial enterprise and adventure, by delinquent associates, institutionalization, anger and hostility either toward oneself or toward men, sexual promiscuity, and drug abuse. A young girl may also find prostitution a way to gain attention, importance, and the achievement of goals.

Surprisingly, Brown learned that juvenile girls are more seriously punished when arrested for prostitution than are boys (even though boys are more likely to be arrested when caught). As a matter of fact, these girls often receive harsher treatment than do boys who commit more serious crimes. Girls' correctional facilities are generally more rigid than boys' facilities, the fences are higher, and the confinement cells smaller.

Although no extensive survey of the growth in child and adolescent pornography has been conducted by scientists to date, police departments and exposés report a marked increase here, too. Research is badly needed in this area.

Psychologists disagree as to why a teenager enters prostitution. The psychoanalytic school believes that prostitutes have a highly negative self-image, usually because of rejection by the father. Prostitution is a symbolic way to degrade oneself and a way to defend against the need for love. Prostitution also serves as the defense mechanism of compensation (See Chap. 2). It can make the teenager feel free of the internal conflicts and anxieties of being unloved. Erikson's theory explains prostitution as negative identity. It represents a rejection of society's values in general rather than a rejection of self.

Money, drugs, and good times are other attractions of the prostitute's life that seem to be especially appealing to male prostitutes. One male teenage prostitute said that for him, the life offered the appeal of adventure and the risk of danger that is always found in a sexual encounter. He liked the "flashy, fast-paced way you can meet lots of interesting people" (Freudenberger, 1973, p. 11).

The old idea that most prostitutes are nymphomaniacs who simply cannot get enough sex is pretty well discounted by the prostitutes themselves (Kurz, 1977). Kurz interviewed many teenage prostitutes and found that for almost all of them, sex is decidedly unpleasant and something that they do because they see it as the only way to survive on their own.

The Juvenile Delinquent

> ***There ain't nothing more exciting than sneaking into someone's house at night to steal their stuff. God, you can hear with your skin!***
>
> ***Eddie, an ex-juvenile delinquent***

Haskell and Yablonsky (1982) point out that the definition of the "delinquent youth" differs from state to state, town to town, and even among neighborhoods within a town. For example, in most areas the law defines a juvenile offender as someone under 18 years old, but this age can be as low as 16 and as high as 21 (Tolan, 1987). The norms for delinquent behaviors frequently depend on the social class of the youth's family (Cohen & others, 1987).

Do you know the statistics of juvenile crime in your hometown?

When a delinquent act is defined as the violation of legally established codes of conduct, delinquency includes a wide range of illegal behavior from misdemeanors to major crimes against persons and property. Less than 5 percent of American teenagers have committed major crimes, been arrested, or live a consistently delinquent way of life (Federal Bureau of Investigation, 1993).

In every age bracket, the numbers are significantly lower today than in the previous decade (Federal Bureau of Investigation, 1993). This is especially true of three important categories: prostitution, drunkenness, and runaways. It is not easy to explain these findings, but it is possible that the decrease is a result of a movement toward more conservative attitudes, as was found in the 1984 study by the National Association of Secondary School Principals (NASSP, 1984). This study of 1,500 representatively chosen seventh to twelfth graders found marked changes in attitude from a similar study conducted by the NASSP in 1974. The students questioned in 1984 were markedly more conservative in their attitudes toward sex, the

What's Your View?

It has been suggested that because their crimes are victimless, nonaggressive offenders should not be prosecuted. Some argue that because these youths mainly hurt themselves, they are really the only victims of their "crimes." They need help, not punishment by incarceration. In fact, the parents who drive their children from their homes, and the "johns" who solicit the sexual favor of teenagers, are the ones who should be punished by the law.

Parents of runaways or the wives of husbands who catch venereal disease from a prostitute would likely disagree. The parents of the runaway, faced with an "incorrigible child," typically feel helpless. They feel that without police involvement, they would have no way to get their child back. Furthermore, it is argued that prostitutes should be treated criminally because they violate the societal rule that the family must be respected. Anyone who tempts people to disregard their marriage vows must be punished for it. (By the way, why is it almost always males, heterosexuals and homosexuals alike, who seek the services of prostitutes?)

You be the judge. Which side is right, or is there a third possible solution?

family, politics, and a number of other topics, including crime. For example, the great majority agreed that

- The laws concerning the defense of insanity should be much tougher (no doubt reflecting attitudes toward John Hinkley's attempt on former President Reagan's life).
- The death penalty is appropriate.
- Violent crime is the third most important problem facing the nation.

Learning and Delinquency

Numerous studies have found that "chronic underachievement and a poor school record are . . . predictive of rule breaking and antisocial behavior" (Feldman & others, 1984). In fact, the evidence indicates that there is a causal link between problems in achieving academic competence and delinquency (Cullinan & Epstein, 1979; Jerse & Fakouri, 1978; Kauffman, 1981; Perlmutter, 1987; Siegel & Senna, 1981; Whelan, 1982).

Perlmutter (1987) looked specifically at the relationship between learning disabilities and delinquency. He found that learning-disabled adolescents are

> ***more likely to develop severe delinquent behaviors than are their nondisabled peers, but unlikely to exhibit a middle ground between delinquent and nondelinquent behavior. It is hypothesized that this difference is due to the ability of most LD children and adolescents to adapt through developing skills that allow them to compensate for their handicapping conditions.*** **(p. 89)**

To counteract the link between learning problems and delinquency, Rosenberg and Anspach (1973) recommended "educational therapy." This approach includes direct and continuous measurement of student learning activity, individualized instruction, a variety of classroom-wide procedures, and intensive, continuing self-study by administrators and faculty.

Gangs

Gangs often offer youths the fulfillment of basic needs. Some of their functions clearly coincide with those of the larger society. Gangs typically provide protection, recognition of the desire to feel wanted, and rites of passage that mark achievement, status, and acceptance, such as the initiation rite of a potential gang member. Thus we may say that the gang is a kind of subculture, a "subculture of violence" (Hammond & Yung, 1993, p. 145).

According to a study commissioned by the New York City Youth Board (*New York Times,* 1989), urban gangs possess the following characteristics:

- Their behavior is normal for urban youths; they have a high degree of cohesion and organization; roles are clearly defined.
- They possess a consistent set of norms and expectations, understood by all members.
- They have clearly defined leaders.
- They have a coherent organization for gang warfare.

The gang provides many adolescents with a structured life they never had at home. What makes the gang particularly cohesive is its function as a family substitute for adolescents whose strong dependency needs are displaced onto the peer group. The gang becomes a family to its members (Burton, 1978).

Gangs and Social Class

The formation of juvenile gangs typically follows a sudden increase in this country of new ethnic groups due to immigration. The children of new immigrants have a difficult time breaking through cultural barriers such as a new language and racism. Perceiving their prospects of succeeding in the new society as bleak, some of these children form gangs, which provide the structure and security discussed, but also serve as an outlet to attack the society that seemingly will not accept them. In times past, these gangs were composed of Jewish-, Irish-, and Italian- Americans. Today's gangs are frequently formed by Latinos, Asian-Americans, and African-Americans (Burke, 1990; Vigil, 1988).

After a period of decline, urban gangs are again on the rise.

Freidman and associates (1976) studied the victimization of youth by urban street gangs. They found that

> ***rituals of street gang warfare and the practices of victimizing both gang members and nonmembers by having them commit serious crimes and violent offenses may serve to maintain the continuity of the group, to give it structure, and to symbolize the gang's power of life and death over others.*** **(p. 21)**

The gang becomes a vehicle for tearing its members away from the main social structures and authorities, in particular the family and school.

But today's gangs have some disturbing differences from those of years past. They are much more heavily armed and seemingly much more willing to use their weapons. Movies like *West Side Story* (1961) depict gang members carrying knives, chains, and pipes. Today's urban gangs are often armed with AK-47 assault rifles and UZI submachine guns, grenades, and even cluster bombs.

Gang violence has increased dramatically in this decade. Twenty-five percent of all juvenile crimes are committed by urban gangs. Los Angeles, with over 200 gangs with 12,000 members, has perhaps suffered most from this upsurge in violence. That city saw gang-related homicides increase from an already staggering 150 in 1985 to an unfathomable 387 in 1987. Of course, the days of rioting in 1992 eclipse even these statistics.

Law enforcement officials have noted that attacks by gangs on police officers are now quite common and continue to increase (Gates & Jackson, 1990; Sessions, 1990). Many innocent bystanders are also injured or killed by violent gangs, often in **drive-by shootings.** Statistics from the Los Angeles Police Department indicate that 50 percent of the victims of gang violence have no connection at all with any gang activity (Gates & Jackson, 1990). In addition to killing members of rival gangs, gang members also frequently kill one another, even within their own "set" or subdivision. This is especially true of large gangs like the Crips and the Bloods of Los Angeles (Bing, 1991; Ewing, 1990) (although these two gangs have recently sought to establish a peaceful truce between them—Barnicle, 1992).

Gangs are also no longer limited to the large urban areas. Smaller cities and towns in the United States have recently seen an increase in the formation of juvenile gangs (Takata & others, 1987). These gangs are often related to other, well-established gangs from the larger cities. So in effect, a gang such as L.A.'s Crips can set up "franchises" in cities such as Seattle and Shreveport. Suburban gangs have also been on the upswing (Muehlbauer & Dodder, 1983). These suburban gangs usually don't exhibit the same degree of organization or formality. They typically get their "kicks" from the malicious destruction of property.

Characteristics of Gang Joiners

Why do adolescents join gangs? Recent interviews with gang members revealed that companionship, protection, excitement, and peer pressure were the most frequently cited reasons (Hochhaus & Sousa, 1988). Gang members are much more likely to have divorced parents or parents with a criminal history. They are more likely to do poorly in school and score low on IQ tests. Freidman and colleagues (1976) showed that what most differentiates the street gang member from the nonmember is the enjoyment of violence. Female gang members, whose numbers have also increased recently, face many of the same societal barriers that cause males to join gangs, with the additional barrier of sexism (Campbell, 1987).

Gang members also have more unrealistic expectations of success than nonmembers, yet perceive less opportunity to be successful (Burton, 1978). Gangs in effect promise a more equal opportunity than does society for members to succeed at something in life. In general, gang members are found to have more drug-abuse problems, more mental disturbance, and are more angry and violent than the average youth. Many are deeply troubled.

Conclusion

Are adolescents in less or greater trouble than in former years? With the exception of crack, the abuse of drugs and alcohol has not increased significantly but is still at an intolerably high level. Eating disorders, depression, and suicide attempts are very real new threats to adolescents' happiness. There has been some decrease in individual juvenile crimes, but they still account for a considerably higher percentage of overall crimes than would be expected for their segment of the total population.

These facts certainly are causes for alarm. How can they be explained? What can be done to improve the situation? It seems likely that at least part of the answer lies in finding better ways to help them get ready for the transition to adulthood. The stress of this transition no doubt accounts for a lot of the substance abuse, mental illness, and delinquency problem. An important aspect of this transition is adolescent sexual behavior. This is what we will deal with in the next chapter.

Chapter Highlights

Prevalence of Substance Use

- Drug and alcohol abuse is still prevalent, although some important changes have been noted. For example, tobacco use is on the rise among teenagers, while marijuana use is on the decline.

Ethnic Group and Abuse

- Drug arrests of teens of color outnumber white arrests, but may be more related to legal system bias than actual drug use.

Crime and Abuse

- Unsurprisingly, drug use and crime are highly correlated.
- Use of drugs is an effective predictor of violent delinquency.

Substance Use and Personal Relations

- The quality of relationship with parents seems to be a more important factor in patterns of drug use than family structure (divorce, etc.).

Combating Substance Abuse

- Some factors that predict drinking behavior are attitudes toward drinking, parental ethnic background, religious affiliation, parents' occupations, parental drinking attitudes, looking forward to drinking, and expectations of the results of alcohol use.
- One effective method proposed for combating drug use is to identify healthy activities that meet the same psychological needs that teens perceive are being met through drug and alcohol use.

Types of Mental Disorders

- The idea that those who develop mental illness during adolescence will "grow out of it" is not supported by research.
- The rate of emotional disturbance among adolescents is very close to the rate of emotional disturbance among adults.

Eating Disorders

- Two of the most disruptive problems for adolescents are the eating disorders known as anorexia and bulimia nervosa.

- Adolescent girls develop eating disorders more than any other group.
- Developmental, cultural, individual, and familial factors are associated with the development of eating disorders.

Depression

- Among the symptoms of adolescent depression are major change in study, eating, and other behaviors; accident proneness; depressed talk; and giving away prized possessions.
- Teens often mask their depression through depressive equivalents.
- Depression has been explained according to psychoanalytic, cognitive, behavioral, and environmental models.

The Crisis of Death

- The death of a loved one is always hard for everyone, but for adolescents, it often causes serious depression.

Suicide

- Suicide is most common among middle-class whites, but the suicide rate for African-American males has increased in recent years.
- Females are more likely to attempt suicide, but males are more likely to die from it.
- The causes of suicide include personality, family, and societal problems.

The Nonaggressive Offender

- Some common causes for running away are a stressful home environment (verbal or physical abuse), being ordered to leave by parents, getting "lost in the system" of government intervention, and serious personality or emotional problems.
- Runaways typically have poor coping skills that ill prepare them for dealing with the generally higher levels of stress they are exposed to.
- Causes of prostitution include alienation from society, physical or sexual abuse, undereducation, lack of legitimate employment, and family difficulties.

The Juvenile Delinquent

- The number and percentage of crimes committed by juveniles are lower than a decade ago, but are still unacceptably high.
- Chronic academic underachievement, learning disabilities, and school failure are highly correlated with delinquent behavior.
- Drug use, mental health problems, and delinquency are related, but the relationship is complicated.

Gangs

- Gangs typically have a high degree of cohesion and organization, a consistent set of norms, clearly defined leaders, and coherent organization for warfare.
- Gangs have become much more violent in the past decade, probably in part as a response to increased drug trafficking.

Key Terms

Aggressive gang
Anorexia nervosa
Bulimia nervosa
Depression
Drive-by shootings
Family status
Hypersensitive youth
Juvenile delinquent
Magical thinking
Nonaggressive offenders
Parent-child role reversals
Psychopathology
Suggestibility
System kids

What Do You Think?

1. Suppose you had a friend with a substance abuse problem. How could you handle it?
2. How would you know if you were developing a serious addiction? What would you do about it?
3. Why do you suppose people develop eating disorders?
4. Think back to the last time you were "down in the mouth" about something. Now suppose it were five times worse than it actually was. How would you feel? How would you act? How would others treat you?
5. What do you think Hankoff (see "Crisis of Death" section) means when he says "ruminations over suicide and forms of actively promoted death encounters would serve to provide the adolescent with a sense of greater temporal definition in his life"?
6. Try to remember every illegal act you have ever committed (hopefully there won't be too many of them). Can you imagine how any one of those acts could have led you into becoming a juvenile delinquent?

Suggested Readings

Abel, E. L. (1981). *Marijuana: The first twelve thousand years*. New York: Plenum. This book chronicles the use of marijuana throughout history. The author is a psychopharmacist who has done a great deal of research on the subject.

Cohen, S. and Cohen, D. (1986). *A six-pack and a fake I.D.: Teens look at the drinking question*. New York: M. Evans. According to the authors of this book, the decision to drink or not to drink is personal rather than moral. They recognize the tragedy that alcohol can bring into people's lives, but they still "do not see moderate drinking as a problem; indeed, it is often a positive pleasure." They do, however, feel that before coming to conclusions about the use of alcohol, you should have reliable and believable information to help you make the best and most informed decision.

Curtis, R. H. (1986). *Mind and mood: Understanding and controlling your emotions*. New York: Scribner's. According to Curtis, knowing more about emotions and how they affect the body can help in understanding and controlling them. This book has chapters that describe the nervous system and endocrine system, addressing the physiological impact on emotions; a chapter on behavior modification; and a section with personality tests that you can take.

Gibson, M. (1980). *The butterfly ward*. New Orleans: Louisiana State University Press. This set of short stories tells what it is like to be between sanity and insanity. It is a sensitive look at the world of the mentally ill, both in and out of institutions.

Harris, J. (1987). *Drugged athletes: The crisis in American sports*. New York: Four Winds Press. Athletes take drugs to increase speed, strength, and accuracy; to mask pain; to relax muscles; to relieve stress; to improve performance; and to gain pleasure. Harris provides an overview and discusses specific problems of drugs in sports at all levels.

Kennedy, W. (1983). *Ironweed*. New York: Penguin Books. Although this book is about adult addicts, it is such an engrossing look at the problem that we have included it here.

Plath, S. (1971). *The bell jar*. New York: Bantam. This famed book tells the story of Esther Greenwood's painful month in New York City, which leads eventually to her insanity and attempted suicide.

Rebeta-Burditt, J. (1986). *The cracker factory*. New York: Bantam. This novel humorously describes the difficulties of a young woman who takes to drinking because of the pressures in her life and is eventually institutionalized because of an attempted suicide.

Sechehaye, M. (1970). *Autobiography of a schizophrenic girl*. New York: New American Library. Written by a Swiss psychoanalyst, this book describes the method of therapy as it was applied to a case of schizophrenia. Offers many insights into this malady that often starts during adolescence.

CHAPTER 14

Adolescence: Gender Roles and Sexuality

Sexual Identity and Gender Roles 379

Aspects of Gender Role 379

Erik Erikson's Studies 380

Androgyny 381

Sexual Behavior 382

The Sexual Revolution 382

Stages of Sexuality 383

Autosexual Behavior 383

Homosexual Behavior 384

Causes of Homosexuality 385

The Onset of Homosexuality 385

Heterosexual Behavior 386

First Coitus 387

The Many Nonsexual Motives for Teenage Sex 389

The Janus Report 390

Sexual Abuse 391

Sexually Transmitted Diseases 392

AIDS 392

Other Sexually Transmitted Diseases 394

The Teenage Parent 396

Trends in Behavior 397

The Role of Family in Teen Pregnancies 398

Causes of Teenage Pregnancy 400

Conclusion 402

Chapter Highlights 402

Key Terms 404

What Do You Think? 404

Suggested Readings 405

The problem lay buried, unspoken, for many years in the minds of American women. It was a strange stirring, a sense of dissatisfaction, a yearning that women suffered in the middle of the twentieth century in the United States. Each suburban wife struggled with it alone, as she made the beds, shopped for groceries, matched slipcover material, ate peanut butter sandwiches with her children, chauffeured Cub Scouts and Brownies, lay beside her husband at night—she was afraid to ask even of herself the silent question—"Is this all?"

Over and over women heard in voices of tradition and Freudian sophistication that they could desire no greater destiny than to glory in their own femininity. Experts told them how to catch a man and keep him, how to breastfeed children and handle their toilet training, how to cope with sibling rivalry and adolescent rebellion; how to buy a dishwasher, bake bread, cook gourmet snails. . . . They were taught to pity the neurotic, unfeminine, unhappy women who wanted to be poets or physicists or presidents. They learned that truly feminine women do not want careers, higher education, political rights—the independence and the opportunities that the old-fashioned feminists fought for. . . . All they had to do was devote their lives from earliest girlhood to finding a husband and bearing children. ■

Betty Friedan, 1963

These two paragraphs, which opened Betty Friedan's famous book, *The Feminine Mystique* (1963), marked the beginning of a major reexamination of the female gender role. In this chapter, we look at the gender roles of both female and male adolescents. We also turn to three other aspects of adolescence that have undergone many important changes. We look closely at what have been called the three stages of sexuality: autosexuality, homosexuality, and heterosexuality. We also cover two sides of sexuality that no one is happy about: sexually transmitted diseases and teenage pregnancy.

After reading this chapter, you will be able to:

- Distinguish between sexual identity and gender role.
- Define three aspects of gender role, as well as the concept of androgyny.
- Demonstrate awareness of gender role stereotypes and how they influence adolescent behavior.
- Specify the various concerns that adults have about adolescents engaging in sexual intercourse.
- Describe the developmental sequence of human sexuality.
- Discuss three theories of the origin of homosexuality, and their implications for those who work with gay teens.
- Discuss reasons why teenagers engage in premarital sexual activity.
- Explain why some adolescents become runaways, prostitutes, or both.
- List the prevalence, symptoms, consequences, transmission, and treatment of sexually transmitted diseases found in adolescents.
- Discuss factors associated with causes and consequences of teenage parenthood.
- Explain the role of the family in teen pregnancies.

Sexual Identity and Gender Roles

The quotation that introduces this chapter was the opening salvo of the revolution that has made such sweeping changes in gender roles in the latter half of this century. Its early effects were mainly on late adolescent and early adult females, but today there is scarcely a woman in the country (or a man, for that matter) whose life has not been affected by this movement.

The beginning of the "feminist revolution" in the 1960s opened an era of changing views toward gender roles. Do you believe that society's views of gender roles are signifcantly different today from what they were ten years ago?

Contrast it with this statement made over two thousand years ago by the famed Greek philosopher Aristotle: "Woman may be said to be an inferior man." Most of us would disagree with his viewpoint publicly. On the other hand, its underlying attitude is still widespread. People today are far less willing to admit to a belief in female inferiority, but many still act as though it were so. However, the influence of the women's movement, as well as of science and other forms of social change, is profoundly affecting the way we view **sexual identity** and **gender role.**

First, we should make a distinction between the two. Sexual identity results from those *physical characteristics* that are part of our biological inheritance. They are the genetic traits that make us males or females. Genitals and facial hair are examples of sex-linked physical characteristics. Gender role, on the other hand, results largely from the specific traits in fashion at any one time and in any one culture. For example, women appear to be able to express their emotions through crying more easily than men, although there is no known physical cause for this difference.

It is possible for people to accept or reject their sexual identity, their gender role, or both. Jan Morris (1974), the British author, spent most of her life as the successful author James Morris. Although born a male, she deeply resented the fact that she had a male sexual identity and hated having to perform the male gender role. She always felt that inside she was really a woman.

The cause of these feelings may have been psychological—something that happened in her childhood, perhaps. Or the cause may have been genetic—possibly something to do with hormone balance. Such rejection is rare, and no one knows for sure why it happens. Morris decided to have a **transsexual operation** that changed her from male to female. The change caused many problems in her life, but she says she is infinitely happier to have her body match her feelings about her gender role.

Aspects of Gender Role

Some people are perfectly happy with their sexual identity, but don't like their gender role. Gender role itself has three aspects:

- **Gender-role orientation.** Individuals differ in how confident they feel about their sex identity. Males often have a weaker gender-role orientation than females. (This will be discussed later in the chapter.)
- **Gender-role preference.** Some individuals feel unhappy about their gender role, as did Jan Morris, and wish either society or their sex could be changed, so that their gender role would be different. The feminist movement of the last few decades has had a major impact on many of the world's societies in this regard.
- **Gender-role adaptation.** Adaptation is defined by whether other people judge individual behavior as masculine or feminine. If a person is seen as acting "appropriately" according to her or his gender, then adaptation has occurred. People who dislike their gender role, or doubt that they fit it well (e.g., the teenage boy who fights a lot because he secretly doubts he is masculine enough) may be said to be poorly adapted.

(a)

(b)

Some people are so unhappy with their sexual identity that they have it changed surgically. Renee Richards (a) before the surgery, and (b), the female tennis star, used to be a man by the name of Richard Raskind.

Research indicates that while gender *roles* may be modified by differing cultural expectations, sexual *identity* is fixed rather early in a person's development. John Money and his associates at Johns Hopkins University believe there is a critical period for the development of sexual identity, which starts at about age 18 months and ends at about the age of 4. Once sexual identity has been established, it is very difficult to change (Money & Ehrhardt, 1972).

The more traditional view (e.g., Diamond, 1982) holds that sexual identity is the result of interaction between sex chromosome differences at conception and early treatment by the people in the child's environment. However, Baker (1980), in her review of the main body of research on this topic, is unequivocal: She states that sexual identity depends on the way the child is reared, "regardless of the chromosomal, gonadal, or prenatal hormonal situation" (p. 95).

Once culture has had the opportunity to influence the child's sexual identity, it is unlikely to change, even when biological changes occur. Even in such extreme cases of **chromosome failure** as gynecomastia (breast growth in the male) and hirsutism (abnormal female body hair), sexual identity is not affected. In almost all cases, adolescents desperately want medical treatment so they can keep their sexual identities.

Although sexual identity becomes fixed rather early in life, gender roles usually undergo changes as the individual matures. The relationships between the roles of the two sexes also change, and have altered considerably in the past few decades.

Erik Erikson's Studies

Erik Erikson's ideas (see chap. 11) about the biological determinants of male and female gender roles came from his studies of early adolescents (1963). He tossed wooden blocks of various shapes and sizes on a table and asked each child to make something with them. Girls, he found, tend to make low structures like the rooms of a house. Having finished these structures, the girls then use other blocks as furniture, which they move around in the spaces of the rooms. Boys, on the other hand, tend to build towers, which, after completion, they usually destroy.

Erikson likens the roomlike structures of the girls to their possession of a womb, and the towerlike structures of the boys to their possession of a penis. He feels that these differences account for greater aggressive behavior in males. Penis and tower alike are seen as thrusting symbols of power.

What's Your View?

?

Erikson ignores alternative explanations for behaviors that appear to be typical for each gender. For instance, blocks are male gender-typed toys like trucks, so one could assume that boys would have more experience stacking them. Girls, on the other hand, would have trouble stacking dolls!

Erikson's ideas, heavily influenced by Freud, have been severely criticized by representatives of the women's movement as making women victims of their anatomy. Although he has frequently espoused complete economic equality between men and women, Erikson continues to insist that biological differences are the cause of psychological differences, an idea that has not been confirmed empirically. What's your view?

An Applied View — The Young Adolescent's Rigid View of Gender

The term *gender* is used all the time. But what does it mean? Gender refers to our conceptions of what it means to be male or to be female. In a recent article, Ronald Slaby (1990) explored gender as a social category system. By that he means a mental filing system: we use the categories of male and female to organize the information we receive. Slaby argues that gender categories are loaded with meaning. For example, when we hear of a person named Mary, before even meeting her, we hold certain preconceptions of what she will be like. She will be female, so she can't be male. She will (probably) always be female (and there are few traits that remain so stable). If you're female, then a person named Mary shares with you the things you think that all females experience. If you're male, then there are things that you think only males experience that you feel certain a person named Mary does not experience. In this manner, gender acts as a powerful organizing system. As Sandra Bem (1975) describes it, once we have our sexual identity, we begin to view the world through gender "filters." All new stimuli are processed through these filters according to gender roles.

Most adolescents wrestle with what it means for them personally to be male or female. In doing so, they develop their gender categories. Studies of the adolescents' ideal men and women reflect very stereotyped notions of the sexes. According to Slaby, this is due to the inflexibility of their developing gender categories. When children and adolescents are just forming their concepts of gender, they are more likely to use the categories of male and female in a rigid manner. That clarity helps adolescents solidify their understanding of gender. When they become confident and comfortable in their ability to figure out maleness or femaleness, then they become able to use the gender categories more accurately. Then they are able to understand that even though Mary might fit their category of female in most ways, in some ways she may not.

Androgyny

One gender role researcher, Sandra Bem (1975), has argued that typical American roles are actually unhealthy. She says that highly masculine males tend to have better psychological adjustment than other males during adolescence, but as adults they tend to become highly anxious and neurotic, and often experience low self-acceptance. Highly feminine females suffer in similar ways.

Bem believes we would all be much better off if we were to become more androgynous. The word is made up of the Greek words for male, *andro,* and for female, *gyne.* It refers to those persons who have higher than average male *and* female elements in their personalities. More specifically, they are more likely to behave in a way appropriate to a situation, regardless of their sex.

For example, when someone forces his way into a line at the movies, the traditional female role calls for a woman to look disapproving but to say nothing. The androgynous female would tell the offender in no uncertain terms to go to the end of the line. When a baby left unattended starts to cry, the traditional male response is to try to find some woman to take care of its needs. The androgynous male would pick up the infant and attempt to comfort it.

Our gender-role stereotypes define what actions are and are not appropriate for each sex. What would most people say if they saw this young woman chopping wood? What would people have said a hundred years ago?

Androgyny is not merely the midpoint between two poles of masculinity and femininity. Rather, it is a more functional level of gender role identification than either of the more traditional roles. Figure 14.1 illustrates this relationship. In discussing the concept of **androgyny,** we assume that we know what the essential qualities of a male and female are. Critics of Bem have suggested that her concept of androgyny portrays women more like men.

Although some aspects of our gender roles appear to be in a state of flux, such as those having to do with the workplace, there has been even more change when it comes to our sexual behavior. We turn now to examine this aspect of social interaction.

Figure 14.1
Relationships among the three gender roles

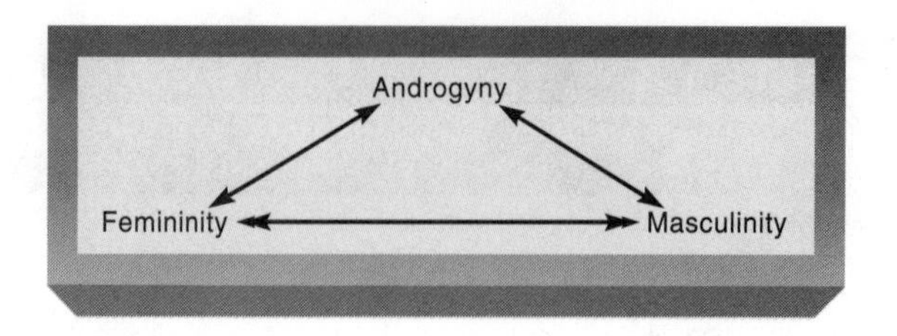

Sexual Behavior

Few aspects of human behavior have changed more in this century than sexual behavior. Until the 1970s, the popular belief about sex was, "They're talking more about it now, but they're not doing anything more about it!" This may have been true earlier in this century, but no longer. The situation has changed so much that it is reasonable to call it a **sexual revolution.**

The androgymous man is not afraid to be seen being kind to an infant.

The Sexual Revolution

Seeing their elders flounder in a sea of confused values, adolescents have begun to consult one another more often on important matters like sex. Edgar Friedenberg, a far-sighted sociologist, saw the beginning of this change as early as the late 1950s. He described these new attitudes in *The Vanishing Adolescent* (1959). The yearning for love and world peace, perennially scorned by some cynical older adults, began to flourish among late teens and young adults in the 1960s. Many middle-age adults came to the disconcerting realization that they were beginning to admire and even emulate the values of their adolescent children. As the spirit of "love among brothers and sisters" grew, so did its consequence, more open sexuality. And a great many adults were no longer sure this was wrong.

After all, that teens should have sexual feelings is entirely normal. As one expert puts it, "sexuality is clearly tied to the developmental tasks of this age period—formation of identity, a growing need for intimacy, and development of social skills" (Mendes, 1992, p. 7). As most adolescents move toward adulthood, interest in sex is biologically, psychologically, and socially inevitable.

Although most teenagers are not ready for mature love, sexual feelings are unavoidable, and for many they are extremely frightening. Now comes one of the most difficult decisions of life: Shall I say yes or no to premarital sex? Parents, clergy, teachers, police, and other adults used to be united in their resistance to it. But now, possibly for the first time in history, adult domination of the values of youth has faltered. As Williams (1989) expressed the change:

> ***Even as adults in America moderate their sexual activity in response to the threat of AIDS and shifting standards of behavior, teen-agers in the last decade have developed a widely held sense that they are entitled to have sex.***

Evidence shows that the forces that traditionally kept the majority of adolescents from engaging in sex are no longer powerful. Available data show that U.S. adolescents are becoming sexually active at increasingly earlier ages. According to a recent analysis of the National Survey of Family Growth (1991), the proportion of 15- to 19-year-old females who report having premarital sexual intercourse has increased steadily since 1970, from 28.6 percent to 36.4 percent in 1975, 42 percent in 1980, 44.1 percent in 1985, and 51.5 percent in 1988. The largest increase was in 15-year-old females (4.6 percent in 1970 to 25.6 percent in 1988). For males, data from 1988 (Sonnenstein, Pleck, & Ku, 1990) found that 64 percent of 15- to 18-year-olds had experienced sexual intercourse, 33 percent by age 15. There may be a decline in early sexual activity recently, however, due to a growing concern about AIDS and other sexually transmitted diseases (Koyle & others, 1989). Results of studies vary, but the most likely percentage of college sophomores who are no longer virgins is about 75 percent for both males and females.

Stages of Sexuality

Many psychologists believe that human sexuality develops in three steps:

1. Love of one's self **(autosexuality)**
2. Love of members of one's own sex **(homosexuality)**
3. Love of members of the opposite sex **(heterosexuality)**

These stages appear to be natural, although some argue that it is as natural to stay in the second stage as to go on to the third.

In the autosexual stage, the child becomes aware of himself or herself as a source of sexual pleasure, and consciously experiments with masturbation. The autosexual stage begins as early as 3 years of age and continues until the child is about 6 or 7, although in some children it lasts for a considerably longer period of time.

When the child enters kindergarten, the homosexual phase comes to the fore (please note that this does not necessarily refer to sexual touching, but rather to the direction of feelings of love). For most children from the age of 7 to about 13, best friends, the ones with whom he or she dares to be intimate, are people of the same sex. Feelings become especially intense between ages 10 and 12, when young people enter puberty and feel a growing need to confide in others. It is only natural that they are more trusting with members of their own sex, who share their experiences. Occasionally these close feelings result in overt sexual behavior (one study found this to be true over one-third of the time). In most cases, however, it appears that such behavior results from curiosity rather than latent homosexuality of the adult variety.

The great majority of teenagers move into the third stage, heterosexuality, at about 13 or 14 years, with girls preceding boys by about a year. We discuss these three phases in the following sections.

Autosexual Behavior

Psychologists have been debating autosexual behavior since the dire warnings of G. S. Hall (See chap. 11). Today, many would disagree with Hall about masturbation, especially females:

> ***From the moment we were born we all began making ourselves feel good by touching and playing with our bodies. Some of these experiences were explicitly sexual. From our parents and later, our schools and churches, many of us learned that we were not to continue this pleasurable touching. Some of us heeded their messages and some of us did not. But by the time we were teenagers, whether we masturbated or not, most of us thought it was bad.*** **(Boston Women's Health Book Collective, 1984, p. 47)**

Masturbation is probably universal to human sexual experience. Although most people still consider it an embarrassing topic, it has always been a recognized aspect of sexuality, legitimate or not. Kinsey, in his 1948 study of male sexuality, found that 97 percent of all males masturbated. As for women, approximately two-thirds have masturbated to orgasm by the time they reach 16 (Gagnon & Simon, 1969). Most 4- to 5-year-olds masturbate, are chastised for it, and stop, then start again at an average age of 14 (Masters & Johnson, 1966). If masturbation is so popular, why has it been considered such a problem?

For one reason, it is believed that the Bible forbids it. Dranoff (1974) points out that the Latin word *masturbari* means "to pollute oneself." For generations, people have taken as a prohibition the passage in Genesis 38:8 in which Onan is slain by the Lord because "he spilled his seed upon the ground." Dranoff argues that Onan was not slain by the Lord for masturbating, but because he refused to follow God's directive to mate with his brother's wife. Instead, he practiced coitus interruptus (withdrawal from the vagina before ejaculation).

In additional to the biblical restrictions, for centuries the medical profession believed that masturbation caused disease. In 1760, Tissot asserted that a common consequence of masturbation is "locomotor ataxia and early insanity." There are many myths about masturbation: It causes one to go mad; it causes hair to grow on one's palms; it causes one to reject sex with anyone else. No research evidence shows that there are any intrinsic bad effects of masturbation. In fact, the American Psychiatric Association has stated that it should not be considered the sole cause of any particular psychiatric problem (American Psychiatric Association, 1985).

Although most psychiatrists feel that there is no intrinsic harm in masturbation and believe it to be a normal, healthy way for adolescents to discharge their sexual drive, some teens (mainly boys) feel such a sense of shame, guilt, and fear that they develop the "excessive masturbation" syndrome. In this case, masturbation is practiced even though the child feels very bad about it. These feelings are reinforced by solitude and fantasy, which leads to depression and a debilitating sense of self-condemnation. Some teens are now being treated for an addiction to making 900 phone line sex calls.

In summary, most psychiatrists argue that masturbation in childhood is not only normal but helpful in forming a positive sexual attitude. It cannot be obsessive at 4, so it should be ignored at that age. However, it can be obsessive at 14, and if the parents suspect this to be the case, they should consult a psychologist.

Homosexual Behavior

Historically speaking, homosexuality has been surrounded by a number of myths:

- Male homosexuals are sissies and will never get involved in a fight.
- Boys with frail physiques and girls with muscular physiques have a strong tendency to become homosexuals.
- Homosexuality results from a mental disorder, usually caused by a hormone imbalance.
- Homosexual men have overprotective mothers and rejecting, inept fathers; in lesbians, the reverse is true.
- Homosexual men frequently attempt to seduce young boys. Since they cannot give birth to children themselves, this is the only way they can replenish their ranks.
- You can always tell the homosexual male because he "swishes" like a woman when he walks; looks at his fingernails with his fingers pointing away rather than toward himself; uses his hands in an effeminate way, with "limp wrists"; usually talks with a lisp; and crosses his legs like a woman.
- You can always tell a female homosexual because she has unusually short-cropped hair; refuses to wear a dress; hates all men; is unusually aggressive; and crosses her legs like a man.

Although one or more of these beliefs are held by many people, there is no evidence, psychological or otherwise, of their validity. Among the most difficult stereotypes confronting homosexuals is the belief that they are "sick." For twenty-three years, until 1973, the American Psychiatric Association (APA) listed homosexuality among its categories of mental illness. In its decision to exclude homosexuality from that category, the APA Board of Trustees argued that because it cannot be said that homosexuality regularly causes emotional distress or is regularly associated with impairment of social functioning, it does not meet the criteria of a mental illness (American Psychiatric Association, 1985). Shortly after this pronouncement, however, one wing of the APA gained acceptance of a category called "Sexual Orientation Disturbance," established for those people, homosexual or otherwise, who suffer anxiety from the sexual choices they have made.

Clearly, some of the stereotypes about homosexuals are untrue and unfair. What generalizations, if any, do you believe can fairly be made about all homosexuals?

Causes of Homosexuality

There have been a number of suggestions about why people become homosexuals. The three most often cited explanations are the **psychoanalytic theory of homosexuality,** the **learning theory of homosexuality,** and the **genetic theory of homosexuality.**

The Psychoanalytic Theory of Homosexuality

Freud's psychoanalytic theory of homosexuality suggested that if the child's first sexual feelings about the parent of the opposite sex are strongly punished, the child may identify with the same-sex parent and develop a permanent homosexual orientation. Because researchers have noted many cases in which the father's suppression of the homosexual's Oedipal feelings was not particularly strong, this theory is not held in much regard today.

The Learning Theory of Homosexuality

The learning theory of homosexuality offers another explanation: Animals that are low on the mammalian scale follow innate sexual practices. Among the higher animals, humans included, learning is more important than inherited factors. According to this theory, most people learn to be heterosexual, but for a variety of little-understood reasons, some people learn to be homosexual.

The Genetic Theory of Homosexuality

There is no *direct* proof that people become homosexual because of genetic reasons. However, in a review of the literature about homosexuality, Buunk and van Driel (1989) note that researchers are looking at what is known about homosexuality in other species for clues. Recently there is interest in the influence of hormones during fetal development (Money, 1987). These theories argue that how the fetus's brain reacts to sex hormones during the second through sixth month of gestation may create a genetic tendency toward homosexuality. They argue that persons born with this tendency (called a predisposition) can be influenced by the environment to either select or avoid homosexuality. In other words, if the biological predisposition is present (a genetic tendency) and certain psychological and social factors (as yet unknown) are in place, then the contention of many homosexuals that their sexual orientation was not a matter of choice would be confirmed.

The Onset of Homosexuality

For a long time, psychologists believed that homosexuality does not manifest itself until adulthood. Recent studies of male homosexuals reviewed in the *Journal of the American Medical Association* (Remafedi, 1988), however, indicate that this belief

What's Your View?

We have the behaviorists and the psychoanalysts giving their explanations of homosexuality. We have many homosexuals who believe that their sexual orientation became clear so early in life that it could only have been caused genetically. Proponents of each of these positions agree that being homosexual is not a matter of choice for the homosexual. Thus it is argued that they should be accepted the same as heterosexuals, or at the very least be given sympathy, for their role in today's society is not an easy one.

There are also those who believe that homosexuality is a matter of free choice, and that those who choose it are behaving in an immoral way. Homosexuals don't have to be that way; they want to. Because they are immoral and because they disrupt the "natural order of things," they deserve society's condemnation. What's your view?

was the result of interviews with teens, most of whom were ashamed or otherwise unwilling to tell about their feelings on the subject. The current studies, using better methods, are in remarkable agreement that at least one-third of all males have had "a homosexual experience that resulted in an orgasm" at least once during their adolescent years. About 10 percent "are exclusively homosexual for at least three years between the ages of 16 and 55" (p. 222).

Most adult homosexuals remember feeling that they were "different" at about 13 years old, the age when most boys are beginning to notice girls. One study followed boys who were seen by medical personnel because of **gender-atypical behavior** (dressing in girls' clothes, playing with dolls, etc.) between the ages of 3 and 6. The majority developed a homosexual identity during adolescence or adulthood.

Remafedi sums up the situation:

> ***Professionals may deny the existence of gay or lesbian teenagers for a number of reasons, some benign and others more malignant. It is both reasonable and judicious to avoid applying potentially stigmatizing labels to children and adolescents. It is also understandable. . . . to adopt a 'wait and see' approach to a teenager's homosexuality, while providing appropriate preventative and acute health care. However, the reluctance of some professionals to acknowledge the existence and the needs of homosexual adolescents is primarily related to the emotionalism surrounding the issue. (p. 224)***

Two other studies reinforce Remafedi's position: those by Harry (1986) and Sullivan and Schneider (1987).

Whatever one believes about homosexuality being a natural stage of sexual development, it is clear that the great majority of people in the United States today do engage in heterosexual behavior sooner or later—and the evidence indicates that they begin much sooner than they used to.

Heterosexual Behavior

At the beginning of this section, we presented some statistics on teen sexuality that may have surprised you. To get a clearer picture of this situation, you will need to look at the data provided by a number of other studies that have researched heterosexual teen behavior.

One societal change that seems to have strongly affected adolescent sexuality is maternal employment. Hansson and colleagues (1981) conducted a study to determine whether maternal employment is associated with teenage sexual attitudes and behaviors, and whether it increases the likelihood of pregnancy. They found that those girls whose mothers are employed outside the home have a greater tendency to begin sexual relations before the age of 19.

A number of studies have looked at the relationship between sexual communications among family members and sexual behavior (Chewning & others, 1986; Darling & Hicks, 1982; Daugherty & Burger, 1984; Fisher, 1986a, 1986b; Moore, Peterson, & Furstenberg, 1986; Wilks, 1986). All have found that parents can have a powerful effect on the children's behavior, including those who are in their late teens, when the parent-child interaction is good, and talk about sexuality is direct.

Wagner (1980) found that sexuality becomes a part of the adolescent's concept of self, regardless of their personal experience or knowledge. She summarizes what we currently know about certain aspects of adolescent sexuality:

- *Knowledge about sexuality.* There is some evidence that teenagers who receive sex information from their parents or someone important to them behave more conservatively and responsibly. Males and females are about equally informed, but neither group knows as much as they need to. For example, 4 out of 10 teens in one study (Zelnick & Kantner, 1977b) thought that the time when you are most likely to get pregnant is during menstruation. Peers and books are the most common sources of information.
- *Attitudes, values, and standards.* Current research reveals a trend toward change in sexual mores among the young. In general, having sex with one person, for whom "love" is felt, is emerging, at least among older teens, as the most popular standard. Adolescent sexuality appears to be affected as much by social change and historical events as by separation and identity formation.
- *Homosexuality.* Most homosexual contact during adolescence is part of a developmental interlude and does not develop into adult homosexual behavior. It happens more among boys than girls, and usually occurs during early adolescence.
- *Male-female differences.* Differences in heterosexual specific practices are more evident in younger than in older adolescents. There is tremendous variability among adolescents in specific sexual practices. More advanced types of sexual behavior (such as petting and intercourse) are occurring at earlier and earlier ages. While there has been a decline in male promiscuity, there has been an increase in female permissiveness.

Wagner concludes that each new sexual experience provides the adolescent with opportunities to test autonomous behavior in a conflict situation. She states that societal changes in attitudes, standards, and behavior have all been reflected in sexuality among adolescents.

First Coitus

Although sexuality develops throughout life, first intercourse is viewed by most as the key moment in sexual development. When do most Americans first experience intercourse? The statistics vary, but all research confirms that this experience occurs at a younger age than it did for previous generations (Scott-Jones & White, 1990; Walsh, 1989; Wyatt, 1989). In 1971, 30 percent of female adolescents aged 15 to 19 had experienced premarital sexual intercourse; it is projected that today that figure is 70 percent (Forste & Heaton, 1988). By the end of adolescence, more than 80

An Applied View: How to Talk to Teens About Sex (or Anything Else, for That Matter)

Adolescents are more likely to talk to adults who know how to listen—about sex, alcohol, and other important issues. But there are certain kinds of responses, such as giving too much advice or pretending to have all the answers, that have been shown to block the lines of communication.

Effective listening is more than just "not talking." It takes concentration and practice. Below are six communication skills that are useful to anyone who wants to reach adolescents. By the way, these skills can also enhance communication with other adults.

Rephrase the person's comments to show you understand. This is sometimes called **reflective listening.** Reflective listening serves four purposes:

- It assures the person you hear what she or he is saying.
- It persuades the person that you correctly understand what is being said (it is sometimes a good idea to ask if your rephrasing is correct).
- It allows you a chance to reword the person's statements in ways that are less self-destructive. For example, if a person says "My mother is a stinking drunk!" you can say "You feel your mother drinks too much." This is better, because the daughter of someone who drinks too much usually can have a better self-image than the daughter of a "stinking drunk."
- It allows the person to "rehear" and reconsider what was said.

Watch the person's face and body language. Often a person will assure you that he or she does not feel sad, but a quivering chin or too-bright eyes will tell you otherwise. A person may deny feeling frightened, but if you put your fingers on her or his wrist, as a caring gesture, you may find that the person has a pounding heart. When words and body language say two different things, always believe the body language.

Give nonverbal support. This may include a smile, a hug, a wink, a pat on the shoulder, nodding your head, making eye contact, or holding the person's hand (or wrist).

Use the right tone of voice for what you are saying. Remember that your voice tone communicates as clearly as your words. Make sure your tone does not come across as sarcastic or all-knowing.

Use encouraging phrases to show your interest and to keep the conversation going. Helpful little phrases, spoken appropriately during pauses in the conversation, can communicate how much you care:

"Oh, really?"
"Tell me more about that."
"Then what happened?"
"That must have made you feel bad."

Remember, if you are judgmental or critical, the person may decide that you just don't understand. You cannot be a good influence on someone who won't talk to you.

percent of the boys and 70 percent of the girls will have been sexually active. In one study of adolescents between the ages of 12 1/2 to 15 1/2, researchers found that approximately one-third of the teens had experienced intercourse (Scott-Jones & White, 1990).

Why are adolescents engaging in sex at earlier ages? Some theorists point to changes in social context (Walsh, 1989). They argue that today's youths learn about sexuality much earlier and from more sources than in the past. Sexually explicit magazines, rock music videos, advertisements displaying sexual situations, and movies depicting sexually graphic material are all part of the everyday culture of teenagers today. In the 1950s such materials weren't commonly available. The women's movement and its focus on double standards about sexuality also contributed to the social context of teens today. Early feminists questioned the **double standard** that engaging in sexual relations was acceptable for males but not for females. Together these factors create a social context that provides the developing adolescent with information about sex beyond what he learns from his peers and family. It is in this new social context that adolescents make decisions about when first to engage in sexual relations.

Factors that increase the likelihood that a teen will engage in premarital sex include coming from a one-parent home, living in poverty, being without religious affiliation, and living in a family where educational discussions about sex never occur. Researchers studying race and premarital sex have found that *by itself,* race or ethnic affiliation is not related to premarital sex (Wyatt, 1990). In many urban areas, it is people of color who live in poverty. Teens from those families are most likely to experience first coitus at an early age (Forste & Heaton, 1988; Wyatt, 1989).

The Many Nonsexual Motives for Teenage Sex

In recent years, researchers have begun to pay more attention to the notion that teens engage in sex for many reasons other than the satisfaction of their prodigious sexual drives. In one of the most enlightening articles on this subject, two adolescent therapists (Hajcak & Garwood, 1989) concluded that for many adolescents, orgasm becomes a "quick fix" for a wide variety of other problems. Among these alternative motives for sex are the desire to

- *Confirm masculinity/femininity.* For some teens, having sex with one or more partners (sometimes called "scoring") is taken as evidence that their sexual identity is intact. This is particularly relevant to those (especially males) who consciously or unconsciously have their doubts about it.
- *Get affection.* Usually some aspects of sexual behavior include physical indications of affection, such as hugging, cuddling, and kissing. To the youth who gets too little of these, sex is not too high a price to pay to get them.
- *Rebel against parents or other societal authority figures.* There are few more effective ways to "get even" with parents than to have them find out that you are having sex at a young age, especially if it leads to pregnancy.
- *Obtain greater self-esteem.* Many adolescents feel that if someone is willing to have sex with them, then they are held in high regard. Needless to say, this is often an erroneous conclusion.
- *Get revenge on or to degrade someone.* Sex can be used to hurt the feelings of someone else, such as a former boyfriend. In more extreme cases, such as "date rape," it can be used to show the person's disdain for the partner.
- *Vent anger.* Because sex provides a release of emotions, it is sometimes used to deal with feelings of anger. Some teens regularly use masturbation for this purpose.
- *Alleviate boredom.* Another frequent motive for masturbation is boredom.
- *Ensure fidelity of girlfriend or boyfriend.* Some engage in sex not because they feel like it, but because they fear their partner will leave them if they don't comply.

Using sex for these reasons often has an insidious result. As Hajcak and Garwood (1989) describe it,

> ***Adolescents have unlimited opportunities to learn to misuse sex, alone or as a couple. This happens because of the powerful physical and emotional arousal that occurs during sexual activity. Adolescents are very likely to ignore or forget anything that transpired just prior to the sex act. Negative emotions or thoughts subside as attention becomes absorbed in sex. . . . The end result is that adolescents condition themselves to become aroused any***

time they experience emotional discomfort or ambiguity . . . sexual needs are only partially satisfied [and] the nonsexual need (for example, affection or to vent anger) is also only partially satisfied, and will remain high. . . . the two needs become paired or fused through conditioning. . . . Indulging in sex inhibits their emotional and sexual development by confusing emotional and sexual needs and, unfortunately, many of these teens will never learn to separate the two. **(pp. 756–58)**

This is not to say that adolescents don't experience genuine sexual arousal. They definitely do, but this does not by itself justify sexual activity. These therapists argue that teens need to be taught to understand their motives, and to find appropriate outlets for them. In fact, this has led some experts to recommend sex education that teaches alternatives to premarital sex.

The Janus Report

The largest, most scientifically designed study of sexuality since the Kinsey Report in the late 1940s was published just as this book went to press. *The Janus Report of Sexual Behavior* (1993) was compiled by Cynthia Janus, MD, and her husband, Samuel Janus, Ph. D. The study covered a wide range of sexual topics, using questionnaire and interview methods. Its sample of nearly 3,000 adults closely resembles the adult population described in the 1990 U.S. Census. Unfortunately, because of the legal problems involved in questioning children and adolescents, the researchers sought answers only from persons 18 or older. A number of questions did involve the subjects' teen years, however. Here are some of the Janus Report's most important findings:

- Nearly 20 percent of men, but only 7.5 percent of women, reported they had had full sexual relations by age 14.
- Younger women responding to the questionnaire reported much younger ages at which they had their first full sexual experience than older women, thus indicating a continuing downward trend.
- Compared to phase 1 of the study (1983 to 1985), in phase 2 (1988 to 1992) 12 percent fewer men and women remained virgins until age 18.
- The South has the earliest ages of sexual initiation and the most reported premarital sex (See Table 14.1).
- Asked whether they had had at least one homosexual experience, 22 percent of men (15 percent lower than Kinsey reported in the late 1940s), 17 percent of women, and twice as many career women as women who were homemakers answered yes.
- An amazing 11 percent of men and 23 percent of women reported having been sexually molested as children.
- Of the women who had had abortions, almost 20 percent had their first before they reached 18 years of age.

The results of this most recent study appear to indicate that several types of sexual experience are occurring even earlier than previous studies show. We will summarize the rest of the major findings of this report in chapters 16 and 20 in the relevant sections.

Table 14.1 Age of First Full Sexual Experience by Section of the United States

	Northeast	South	Midwest	West
N =	558	928	661	573
By age 10	2%	1%	0%	3 %
11–14	12	16	7	11
15–18	44	65	51	51
19–25	39	17	37	34
26+	3	1	5	1
By age 14	14%	17%	7%	14 %
Over age 18	42%	18%	42%	35 %

Source: Janus & Janus (1993) p. 368.

Sexual Abuse

Sexual harassment and abuse during adolescence has been studied by Herold, Mantle, and Zemitis (1979). They surveyed young women in college and found that nearly 85 percent say they have been victims of some sexual offense. The offenses ranged from obscene phone calls (61%) and sexual molestation (44%) to attempted rape (16%) and rape (1%).

Adolescents are typically abused by someone they know and trust. It is often just a continuation of abuse that started during childhood. The most common type of serious sexual abuse is incest between father and daughter (Alexander & Kempe, 1982). This type of relationship may last for several years. The daughter is often manipulated into believing it is all her fault, and that if she says anything to anyone, she will be seen as a bad person, one who may even be arrested and jailed. The outcome is usually another adolescent statistic: a runaway or even a prostitute.

Most sexual offenses were discussed with a friend or with no one. Very few were reported to parents, police, social workers, or other authorities. It has also been found that the effects of abuse may influence a youth's future relationships. Directly following the experience there may be "acting out" behaviors as truancy, running away, sexual promiscuity, and damage to school performance and family relationships.

Gruber, Jones, and Freeman (1982) interviewed a group of female teenagers ranging in age from 13 to 17 who had been sexually abused. These young women were involved in a residential intervention program. Gruber and colleagues found that the victims sustained a diminished self-worth and poorer interpersonal relationships with males. In addition, VanderMay and Neff (1982), in reviewing research and treatment of adult-child incest, concluded that the long-term effects may result in promiscuity, alcoholism, sexual dysfunction, drug abuse, prostitution, depression, and even suicide.

Very few cases of sexual molestation are actually reported to the authorities.

VanderMay and Neff call for improved education to sensitize people to prevent incest, as well as improved reporting systems, legal definitions, and treatment of victims. These may help us better understand and intervene, so that victims can receive professional attention earlier that may reduce the long-term effects of abuse.

It is safe to say that sexuality in the lives of late adolescents and young adults in the last decade of this century is very different from in earlier decades (although perhaps not so different from several centuries ago). What is the relationship between this fact and the problem covered in the next section, sexually transmitted disease? That is a complex question.

Sexually Transmitted Diseases

In this section, we cover research on AIDS and other diseases that are passed on sexually.

AIDS

Not long ago, when people thought about **sexually transmitted diseases (STDs),** gonorrhea came to mind. In the 70s, it was herpes. Today, **AIDS (Acquired Immune Deficiency Syndrome)** causes the most concern (Forestein, 1989).

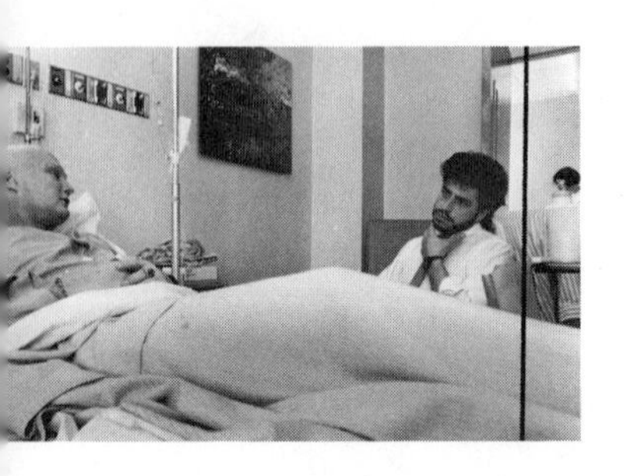

First diagnosed in 1979, *AIDS* has quickly approached epidemic proportions.

AIDS was first diagnosed at Bellevue-New York University Medical Center in 1979 and has quickly approached epidemic proportions. What is known about AIDS is that a virus attacks certain cells of the body's immune system, leaving the person vulnerable to any number of fatal afflictions such as cancer and pneumonia. In addition, the disease can directly infect the brain and spinal cord, causing acute meningitis.

As of July 1989, 100,000 cases of AIDS had been reported since the first diagnosis. Of these 100,000 reported cases, about 60,000 of the patients have since died. AIDS now ranks fifteenth among the leading causes of morbidity and mortality in children and young adults. The first 50,000 cases of AIDS were reported from 1981 to 1987. The second 50,000 were reported in just the following two years (CDC, 1989b). The CDC estimates over 300,000 cases will be reported by 1992 (CDC, 1989c). Keep in mind that these are only *reported* cases of the full-blown AIDS disease. A combination of underdiagnosis and underreporting makes these estimates conservative at best. The CDC estimates that 1–1.5 million people in the United States are currently infected with the initial virus. Studies suggest that about 50 percent of these people will develop the full-blown AIDS disease within 10 years of infection, and that 99 percent will eventually develop the disease (Lifson, Hessol, & Rutherford, 1989).

Trends include increased reporting of AIDS in intravenous drug users, women, children, the elderly, African-Americans, Latino-Americans, heterosexuals, small cities, and rural areas (Catania & others, 1989; CDC, 1989a; Kirkland & Ginther, 1988). The only segment of society in which the incidence of AIDS is decreasing are homosexuals with no history of intravenous drug use, although this group still represents the single largest at-risk group (CDC, 1989b).

The AIDS virus is transmitted through the transfer of substantial amounts of intimate bodily fluids such as blood and semen. The virus is most likely to be transferred through sexual contact, the sharing of hypodermic needles, and, much less likely through blood transfusions (a test for AIDS is now available at blood banks and hospitals). In addition, the virus can be transmitted from an infected mother to an infant during pregnancy or birth. Figure 14.2 shows the concentrations of AIDS in each of these groups, as well as the percentages of cases by race/ethnicity.

In the initial stages of the spread of the disease in this country, the AIDS virus has most often been found in certain segments of the population such as male homosexuals and intravenous drug users, and to a much lesser degree among hemophiliacs. But that could easily change over time. In some Central African countries, where AIDS is thought to originate, the virus is found equally among men and women throughout the population.

Although there is no cure for AIDS, the disease can be effectively controlled through preventive measures. Use of condoms during sexual intercourse and clean, unused needles during intravenous drug use can drastically reduce the risk of contracting the disease. Figure 14.3 reflects the improvement in protection through use of contraceptives. After a slow start, large-scale education efforts by grassroots organizations, as well as by state and federal government agencies, have begun to get these messages out, but the problems remain extremely serious.

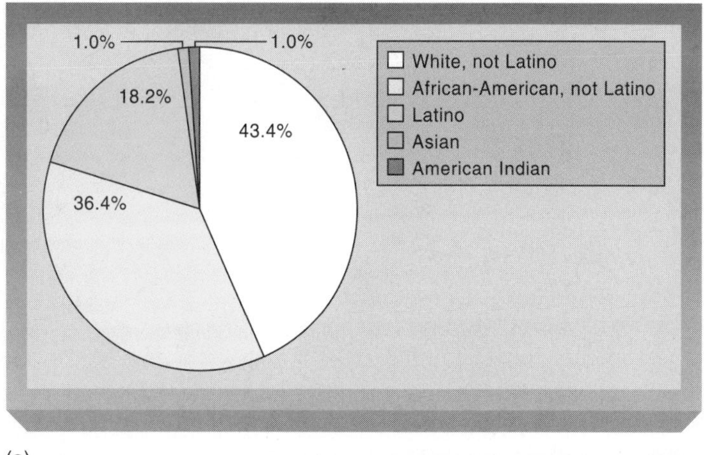
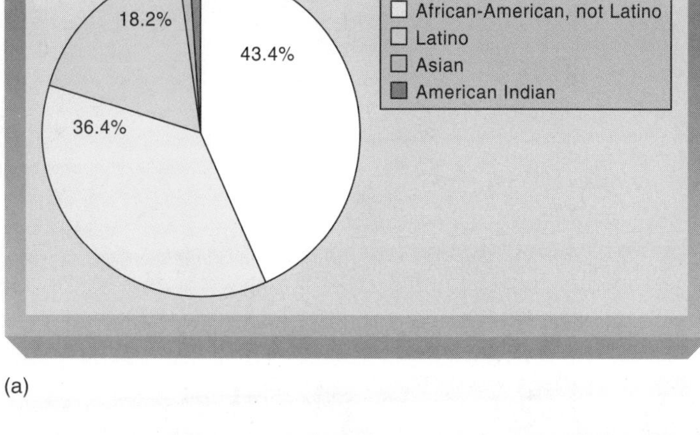

Figure 14.2 (a)
Percentage of 13- to 19-year-old AIDS victims in the United States, by race/ethnicity
Source: From U.S. Department of Health and Human Services, Public Health Service, Centers for Disease Control, Center for Infectious Diseases, Division HIV/AIDS, "HIV/AIDS Surveillance," Atlanta, GA, September 1990.

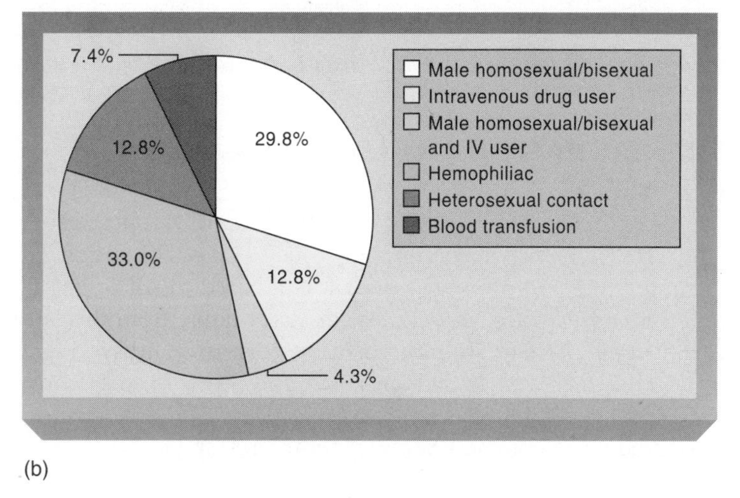

Figure 14.2 (b)
Percentage of 13- to 19-year-old AIDS victims in the United States, by exposure category
Source: From U.S. Department of Health and Human Services, Public Health Service, Centers for Disease Control, Center for Infectious Diseases, Division HIV/AIDS, "HIV/AIDS Surveillance," Atlanta, GA, September 1990.

First, as mentioned, the virus has been identified with a few select groups. If you're not gay or a drug user, you might think you don't have to consider preventive measures. However, a person exposed to the AIDS virus may not show any symptoms for as much as fifteen years. Further, this same person can expose other people to the virus during this incubation phase. Some people have reacted to this by becoming more particular about their sexual partners. Monogamous relationships have been on the rise again during the 1980s, after the "liberated" days of the sexual revolution of the 1960s and 1970s. And the educational message seems to be getting through as condom use increases. But many still ignore the dangers, and the consequences may be years away.

This may be particularly true among adolescents. Adolescents currently constitute only about 1 percent of all diagnosed cases of AIDS in the United States. But given the long incubation period and the research findings that suggest that adolescents are not very well informed about AIDS, many researchers think this may be an underestimation. Adolescents are also more prone than the general public to misconceptions and prejudices generated by the frightening new disease.

For example, there is the misconception that AIDS can be transmitted through casual contact such as kissing or hugging someone with AIDS, or sharing their utensils or bathroom facilities. Such misconceptions unnecessarily increase fear and anxiety in everyone. AIDS prevention efforts aimed at adolescents often have as their main goal the dispelling of such myths (DiClemente, 1987).

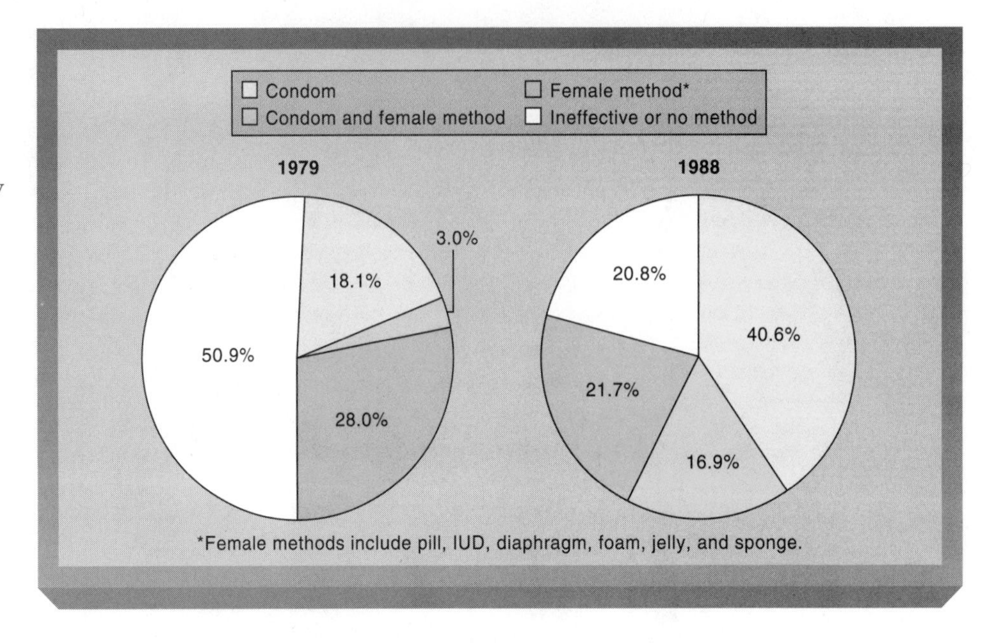

Figure 14.3
Contraceptive use among United States teenage males. The chart shows the percentage of 17- to 19-year-olds who say they used (or didn't use) contraceptives during their last sexual intercourse.

Other Sexually Transmitted Diseases

Often lost in the public focus on the burgeoning AIDS problem is a truly epidemic increase in the prevalence of other STDs. Because of its fatal nature, AIDS gets most of the press and the major funding. But STDs such as gonorrhea, syphilis, chlamydia, and herpes are running rampant compared to AIDS, particularly among adolescents. The effects of such venereal diseases range from the mildly annoying to the life-threatening. There are more than fifty diseases and syndromes other than AIDS that account for over 13 million cases and 7,000 deaths annually (National Institute of Allergy and Infectious Diseases, 1987).

Some of the more common STDs (other than AIDS) are

- *Chlamydial infection.* **Chlamydia** is now the most common STD, with about 5 to 7 million new cases each year (Subcommittee on Health and the Environment, 1987). In one state, black and Hispanic female teens have rates of chlamydia infection over ten times higher than rates reported in white female teens (Massachusetts Department of Public Health, 1991). There often are no symptoms. It is diagnosed only when complications develop. It is particularly harmful for women. It is a major cause of female infertility, accounting for 20 to 40 percent of all cases (Hersch, 1991). Untreated, it can lead to pelvic inflammatory disease (see below). As with all of these diseases, it can be transmitted to another person, whether symptoms are present or not. There is excellent news about this infection. A single-dose antibiotic treatment has been found to be very effective (Martin & others, 1992).

- *Gonorrhea.* The well-known venereal disease **gonorrhea** infects between 1 1/2 and 2 million persons per year. One quarter of the cases reported are adolescents (Klassen, Williams, & Levitt, 1989). Gonorrhea is caused by bacteria and can be treated with antibiotics. When penicillin was introduced in the 1940s, the incidence of gonorrhea declined dramatically. Today, however, the number is rising and has reached the highest level in forty years (Hersch, 1991). The most common symptoms are painful urination and a discharge from the penis or the vagina.

- *Pelvic inflammatory disease (PID).* **Pelvic inflammatory disease** frequently causes prolonged problems, including infertility. It is usually caused by untreated chlamydia or gonorrhea. These infections spread to the fallopian tubes (see diagram in chap. 12), resulting in PID. The scarring caused by the infection often prevents successful impregnation. There are more than 1 million new cases per year in the United States (Washington, Arno & Brooks, 1986). Women who are most likely to get it are those who use an intrauterine device for birth control, have multiple sex partners, are teenagers, or have had PID before. PID is so widespread that it causes $2.6 billion in medical costs per year!
- *Genital herpes.* **Genital herpes** is an incurable disease, with about 500,000 new cases every year. It is spread by a virus during skin-to-skin contact. Its major symptom is an outbreak of genital sores, which can occur as often as once a month. It is estimated that about 30 million people in this country suffer from this infection. Unlike chlamydia, problems associated with herpes are mainly emotional and social rather than medical (Hersch, 1991). People with herpes often experience embarrassment and low self-esteem about their bodies.
- *Syphilis.* Like gonorrhea, **syphilis** is no longer the killer it was before penicillin. However, this sexually transmitted disease still accounts for 70,000 new cases per year. In the state of Massachusetts, 76 percent of these cases were teens of color (Massachusetts Department of Public Health, 1991). It is caused by bacteria. Its first sign is a *chancre* ("shanker"), a painless open sore that usually shows up on the tip of the penis and around or in the vagina. This disease must be treated with antibiotics, or it can be fatal.
- *Hepatitis B.* There were about 200,000 new cases of **Hepatitis B** in the U.S. in 1990, and it is predicted that there will be 300,000 in 1991 (Hersch, 1991). This viral disease is transmitted through sexual contact, and also through the sharing of infected needles. Although a preventive vaccine is available, those who are most at risk for Hepatitis B (intravenous drug users, homosexual men, and inner-city heterosexuals) usually do not have the vaccine readily available to them.

Figure 14.4 depicts the relative percentage of new cases of each type of STD in the United States each year.

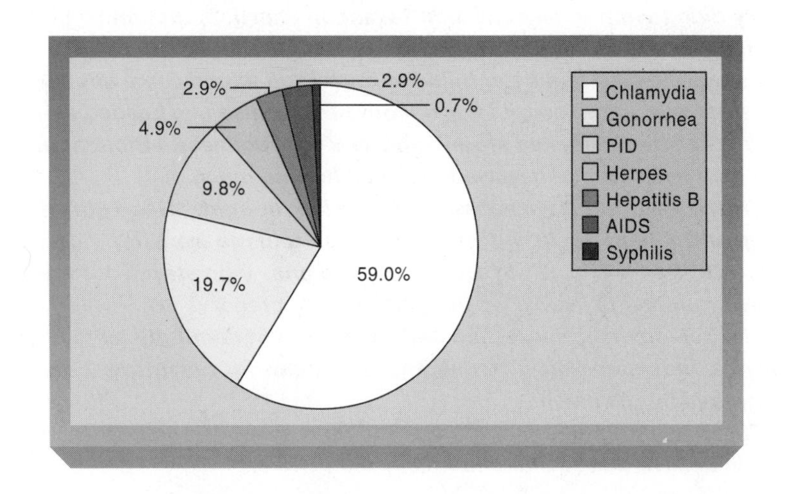

Figure 14.4
The relative percentage of new cases of each type of STD in the United States each year.

Studies have shown that the age group at greatest risk for STDs are individuals between 10 and 19 years old (National Institute of Allergy and Infectious Disease, 1987). This is an age group that is particularly difficult to educate in any area concerning sexuality. The obstacles to education include individuals who refuse to take the information seriously, and parents who won't let the information be taught.

The AIDS crisis and the STD epidemic have several features in common. On the negative side, misconceptions contribute to both problems. Many young people believe that only promiscuous people get STDs, and that only homosexuals get AIDS. Having multiple sexual partners does increase the risk of contracting STDs, but most people do not view their sexual behavior, no matter how active, as being promiscuous. Recent research also suggests *machismo* gets in the way of proper condom use, an effective prevention technique for all STDs. A "real man" doesn't use condoms. And finally, when people do contract a disease, strong social stigmas make accurate reporting difficult.

On the positive side, the preventive and educational measures are basically the same for AIDS and other STDs: dispel the myths, increase general awareness and acknowledgment of the problem, and encourage more discriminating sexual practices. Perhaps some of the educational efforts made on behalf of AIDS prevention and treatment will have a helpful effect on the current STD epidemic. Historically, the health focus on STDs has been on treatment, typically with antibiotics, but recently the Public Health Service has shifted its focus for all STDs to prevention. So perhaps comprehensive efforts of this kind that emphasize all STDs will prove fruitful.

In summary, it seems safe to say that major changes in adolescent sexual practices have occurred in recent decades. Many of them must be viewed with considerable alarm, especially when you consider the tragic increases in STDs and pregnancy.

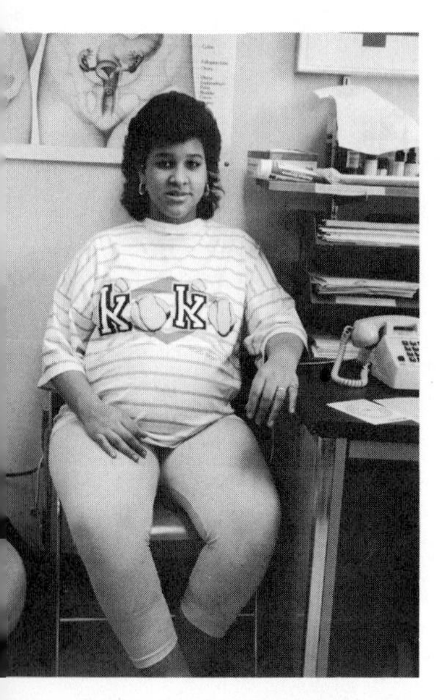

Except for the youngest adolescents, birthrates for adolescents have been dropping in recent years. However, the extent to which young adolescents have been becoming pregnant is certainly a cause for great concern because physically, emotionally, and economically, they are at the greatest risk.

The Teenage Parent

"You're pregnant," the doctor said, "and you have some decisions to make. I suggest you don't wait too long to decide what you'll do. It's already been seven weeks, and time is running out!"

"Look, it just can't be true!" I replied. I was trying to convince myself that the clinic doctor was lying. It wasn't supposed to be like this! I was tired of the bitter quarrel I had been having with the doctor. I resented him with every passion. How could I let myself be seen like this?

I had been fearing this answer. I suppose I knew the truth all along, but I really didn't want to face it. I didn't want an abortion, that much I was sure of. Besides, where would I get the money?

For ages now, I had been thinking my period would come any day. Now the truth was in the open! I walked out of the office and headed aimlessly down the street. I looked around and saw only ugliness. I thought about God and how even He had deserted me. It all hurt so much.

"How could this have happened to me?" I thought. "Good girls don't get pregnant!" All of the things my mother had told me were lies. According to her, only the "fast girls got pregnant." The ones who stayed out late and hung around with boys. I wasn't part of that category!

I looked down at my stuffed belly and thought about my family. Would they be understanding? After all, they had plans for my future. They would be destroyed by the news.

"I'm not a tramp," I said to myself. "Then again, I'm only 16 and who would believe that Arthur and I really are in love?"

The feelings of this unmarried girl are all too typical. Children born of these pregnancies have it even harder (Garn & Petzold, 1983). Harvard biological anthropologist Melvin Kohner (1977) sums it up:

> ***As maternal age drops from age 20, mortality risk for mother and child rise sharply as does the probability of birth defects. Offspring of adolescent mothers, if they survive, are more likely to have impaired intellectual functioning. Poverty, divorce, inept parenting, child neglect, and child abuse are all more frequent in teenage parents.*** **(p. 38)**

Trends in Behavior

As Figures 14.5a and b make clear, out-of-wedlock childbearing is on the increase in all age groups for both white and African-American teenage females. As would be expected, the rate is highest for those under 15. In most states, 10–20 percent of all females have been pregnant at least once by the time they reach their nineteenth birthday. Clearly this is unfortunate, but the rate for 15- to 17-year-olds, which is about one-third as high, might even be described as catastrophic. The largest increases in teen pregnancy rates occurred in those under the age of 15 (Westoff, Calot, & Foster, 1983) and overall, there is more parenthood among unmarried teens, than ever (Furstenberg, 1990).

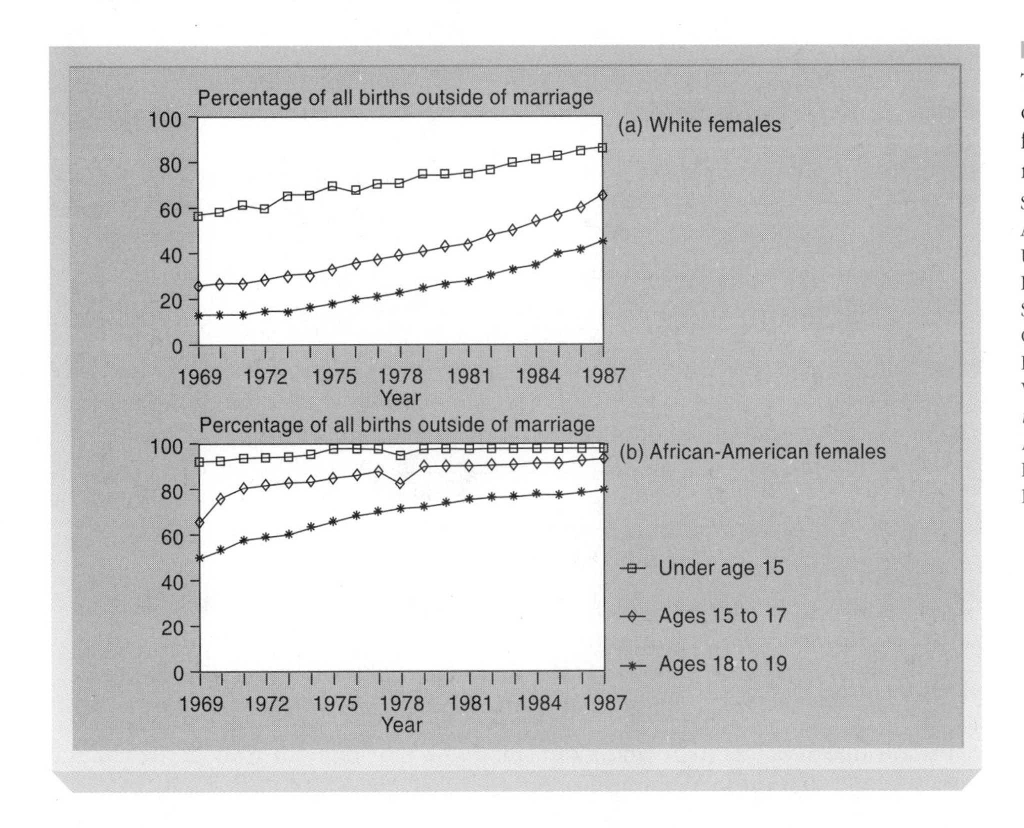

Figure 14.5
Trends in out-of-wedlock childbearing among U.S. females under age 20 by race, 1969–87
Source: Office of Technology Assessment, 1991, based on U.S. Department of Health and Human Services, Public Health Service, Centers for Disease Control, National Center for Health Statistics, Division of Vital Statistics *Vital Statistics of the United States, Volume 1: Natality*. U.S. Government Printing Office, Washington, DC, various years.

Currently over one million teenagers become pregnant in this country every year (Ralph, Lochman, & Thomas, 1984). One in ten girls becomes pregnant before the age of 20 (Foster, 1988). The ramifications of this social problem extend into other social problems. Pregnancy is considered the number one reason for adolescent females dropping out of school (Strobino, 1987). The most likely profile of a pregnant teenager is a girl of color, grew up in a poor, single-parent home, and has low academic and occupational aspirations (Polit, 1985). To further complicate things, the younger the father at the time of the child's birth, the less likely he is to be married (See Table 14.2 for data on this).

Table 14.2 Adolescent Fathers Who Reported Having Fathered a Child Before the Age of 20, by Father's Age at Child's Birth and Marital Status at Conception

Father's age at child's birth	Males who reported having fathered a child before age 20 N=555		Single at conception N=446		Married at conception N=109	
Age 11 to 16	66	(10.1%)	66	(12.7%)	0	(0.0%)
Age 17	85	(15.8)	76	(18.3)	9	(5.8)
Age 18	181	(30.6)	158	(32.8)	23	(22.1)
Age 19	223	(43.5)	146	(36.2)	77	(72.0)
Age 11 to 19	555	(100.0)	446	(100.0)	109	(100.0)

Source: Office of Technology Assessment (1991).

There are many stereotypes about the causes of this epidemic. Some point to earlier menstruation; others talk of the crumbling morals of today's youth. We can say that the images of the fast and easy girl and the sex-obsessed boy are surely false (Herz & Reis, 1987; Kinard & Reinherz, 1987; Klein & Cordell, 1987; Stiffman & others, 1987). Most of these couples have had a substantial relationship prior to the pregnancy, usually for at least six months.

Adolescents in the United States have rates of pregnancy that are among the world's highest; this is especially true of inner-city adolescents (Garcia-Coll, Hoffman, & Oh, 1987; Colletta, 1982; Herz & Reis, 1987; Moore, Jofferth, & Wertheimer, 1979; McKenry, Walters, & Johnson, 1979; Silber, 1980; Stiffman & others, 1987). Six percent of them give birth each year. Of these, one-third give birth out of wedlock, one-third conceive before marriage, and one-third conceive after marriage. In one study, researchers interviewed a thousand 30-year-olds and found that 10 percent of the men and 31 percent of the women had had a child during adolescence (Russ-Eft, Springer, & Beever, 1979).

As Lancaster and Hamburg (1986) point out,

> ***except for the very youngest adolescents, contraception and abortion have lowered the birthrates for adolescents since 1970 to levels that are somewhat lower than those in the 1920s and 1950s. However, the rate of adolescent childbearing outside of marriage has shown steep increases.*** **(p. 5)**

What this means is that more and more children are being born without the cultural approval and support that marriage brings.

The Role of Family in Teen Pregnancies

Rosen (1980) examined the extent to which teenagers involve their parents in decision making in resolving unwanted conceptions. Data were obtained from a questionnaire given to 432 unmarried 12- to 17-year-olds with unwanted conceptions. Although few subjects consulted their parents when they first thought they might be pregnant, more than half did involve their mothers in deciding what to do about the pregnancy. The findings indicate that there may be less of a generation gap between parents and teenagers than is often supposed.

However, there is some doubt that communication about any sexual topics makes much difference. On the basis of their study of 287 African-American and white teens, Furstenberg and colleagues (1986) concluded that

> ***most parents do not want to get directly involved and, certainly, most teenagers are reluctant to encourage involvement. . . .* [*Most parents*] *are relieved to discover that their teenagers are obtaining contraception. Beyond that, it seems that most are either willing or prefer to respect the adolescent's privacy.* (p. 241)**

Held (1981) conducted a study of 62 girls, none more than 17 years old, who were in the third trimester of pregnancy. Twenty-two percent were white, 56 percent African-American, and 16 percent Mexican-American. Each completed the self-esteem inventory, and rated her perception of the reactions of significant people in her life to the pregnancy.

The person on whom these subjects depended most was the mother; the prospective grandmother was the most disapproving. Social support among the three ethnic groups differed: the African-American woman who was keeping her baby had the highest self-esteem, but she and her mother were least likely to rate the pregnancy as "good" or "OK." White women more often rated the pregnancy highly but had lower self-esteem. The ten Mexican-Americans reported the least disapproval.

A Multicultural View 

The Role of Race in Teen Pregnancy

Many studies have reported that teens of color are at great risk of having unwanted pregnancies. To understand this finding, however, it is important to sort out racial values from socioeconomic factors. *Poor teens,* whether African-American, Asian-American, native American, white, or Latino, are three to four times as likely to become unwed teens than economically advantaged teens (Children's Defense Fund, 1989).

The higher rates of teenage parenthood among economically disadvantaged youth are understandable. Teens who are behind in school, who lack basic skills, and who see few opportunities for their future are more likely to become parents as teenagers. For adolescents lacking educational and job opportunities, parenting may be one of the few available ways of achieving adult status. Unfortunately, becoming pregnant as teenagers often makes their lives and the lives of their young children more difficult.

To solve the problem of teenage parenthood, we have to solve problems of education and job opportunities for all teenagers, regardless of race. We must also consider ways to help teenage mothers provide their babies with the needed emotional, intellectual, and physical care, while also enabling the mothers to continue their education. These factors, as you will recall from our discussion of the consequences of teenage pregnancy, often make a difference in the futures of the mothers and their children. Support from family members can be important.

Gabriel and McAnarney (1983) compared the decision about parenthood in two study groups in Rochester, New York: 17 African-American, low-income adolescents (age 15 to 18 years) and 53 white, middle-class adult couples. Their observations showed that the decision to become parents was related to different subcultural values. In contrast to the white adults, the African-American subjects did not see marriage as a prerequisite for motherhood, nor did they view completion of schooling and economic independence as phases of maturation that should precede parenthood.

Instead, these subjects expected that becoming mothers would help them to achieve maturation and acceptance as adults. This may have been because the African-American subjects did not see other adult roles as available, whereas middle-class white couples did. At the same time, adolescent pregnancy and out-of-wedlock motherhood were not viewed negatively among the low-income African-American subjects. Health care programs that encourage birth control to avoid "unwanted pregnancies" may be ineffective because they do not address the needs of African-American clients in terms of the values of their own subculture.

AN APPLIED VIEW **Talking to Teens About Pregnancy**

Nurses, teachers and counselors frequently have opportunities to help teens to clarify their attitudes toward pregnancy, whether they are currently involved with a pregnancy or not. For a number of reasons, these conversations seldom take place, however.

Sometimes those who work with teens feel that discussing pregnancy with them means pushing their own values on them. Others feel that this subject is very delicate, and should not be discussed outside the church or home. Others refrain from discussing it because they feel like they just don't know enough (readers of this book, of course, will not have to be concerned about this problem!). It does seem reasonable to make sure that teens have the *objective facts* about pregnancy and its repercussions for baby, mother, and father. Does this make sense to you?

It should be noted that some of the suggestions for talking to teens about sex in the box on page 388 are useful here, too.

In summary, one note of caution: In an extensive review of the literature on adolescent pregnancy. McKenry, Walters, and Johnson (1979) made an important point about racial difference. They found that while studies of low-income non-white girls tended to focus on *social* factors, studies of white, middle-class girls tended to search for *psychological* explanations for the pregnancies and their outcomes. It is important to remember that while racial and social class differences have been reported in the research, some of these differences may have more to do with the researchers than with the teenagers themselves.

Causes of Teenage Pregnancy

Why do so many adolescents get pregnant? According to McKenry and colleagues (1979), many factors combine to create the situation. They divide the causes into four major categories: physiological, psychological, social, and cognitive abilities. A fifth may be added—moral reasoning abilities.

Physiological Causes

It is presently possible for girls to get pregnant at a younger age because of improvements in the general health among Americans. Menarche occurs at an earlier age than in the past. The American Academy of Pediatrics Committee on Adolescence has learned that the fertility rate among girls under 15 years of age has been rising rapidly. Improved nutrition and health care have also contributed to an increase in the potential for young girls to become pregnant (Waltz & Benjamin, 1980).

Psychological Causes

McKenry and colleagues (1979) have stated that sometimes pregnancy is the result of a teenager's conscious or unconscious desire to get pregnant. These researchers found that the psychoanalytic model is preeminent in psychological explanations of adolescent pregnancy. Ego strength and family relationships are the most commonly cited reasons. Low ego strength or a low sense of personal worth is said to lead to sexual acting out or use of sex as an escape. Highly dependent girls with a great need for affection and those experiencing social or psychological stress are more likely to become pregnant. Also, girls who feel they have little control over their lives are more likely to get pregnant.

Social Causes

Poverty influences the pregnancy rate; insufficient economic and social resources may lead to pregnancy (McKenry & others, 1979). Pressure from peers and the influence of the media are also social precursors to pregnancy during adolescence.

Family situations or problems that have been linked to the incidence of adolescent pregnancy include the following: closeness to father, lack of closeness to mother, generally unstable family relationships, father-absence accompanied by resentment of the mother, and feelings of rootlessness. Although not yet documented, it may also be that there is less shame attached to teen pregnancy than there used to be.

Cognitive Ability Causes

Another major cause involves young girls' lack of the knowledge and maturity required to prevent unwanted conceptions. One must possess the cognitive ability to foresee the consequences of sexual activity. Epstein (1980) indicated that some teenagers believe they are too young to become pregnant or that their sexual encounters are too infrequent. Many teens do not know it is possible to become pregnant during a single act of intercourse. Others are unable to relate risks of pregnancy to their menstrual cycles or are too menstrually irregular to use such information properly.

Moral Reasoning Ability Causes

Differences in moral reasoning may also contribute, especially in the case of male irresponsibility. Herz and Reis (1987), who studied 251 seventh and eighth grade African-American inner-city teens, summarize the situation:

> ***Young men may not be able to perceive a cause-and-effect relationship between their interests in sex, their sexual behavior, and the occurrence of pregnancy, retaining instead an egocentric, childlike belief that they will not be held accountable for what they do.*** **(p. 375)**

They are not able to assume mature reasoning in this case. These authors conclude that there is "a clear need for more factual and practical information on the processes of conception and contraception, as well as decision-making and moral-reasoning skill development" (p. 372).

Finally, in their review of the research, Crockett, Losaff, and Petersen (1986) concluded that we still have much to learn about the causes of teen pregnancy:

> ***Pubertal change and the development of reproductive potential involve several interrelated processes. More importantly, we must attend to the meaning of these changes to girls. Because in our society we have no puberty rites or clear transitional rituals based on maturational status*** **[*see the next chapter*]*****, this change becomes more individually integrated.*** **(p. 170)**

Thus, because we do not teach them society's expectations (either in school or elsewhere), young teens react to puberty and make sexual decisions on a purely personal basis. There is no one "type" who gets premaritally pregnant in adolescence. All suffer from a lack of knowledge that is probably the major culprit.

Conclusion

For each of the aspects of social interaction reviewed in this chapter—sexual identity, gender roles, sexual behavior, sexually transmitted diseases, teenage parenthood—there is one consistent trend with which all adolescents must deal: fast-paced change. Some of this change derives from ground swells in today's society, and some results from the nature of adolescence itself. Thus again we see the biopsychosocial model demonstrated.

The advent of feminism has promoted major innovations in our thinking about the appropriate roles males and females should play. The concept of androgyny has been introduced in the past two decades. Yet perhaps nothing has had a more resounding impact on adolescent life than the recent changes in our attitudes toward sexuality. The five areas of greatest change have been in homosexuality, sexually transmitted disease (including AIDS), the earlier and more widespread participation in sex by teenage females, the increase in child pornography and prostitution, and the increase in pregnancy and childbearing among younger teenagers.

Some observers have suggested that the biggest problem facing adolescents today, and one intertwined with those just listed, is the difficulty in knowing when childhood and youth have ended and adulthood has begun. When the societal lines between these stages of development are blurred, youth cannot be blamed for not knowing how to behave. A clearer induction into adult life has been advocated. This will be the main topic of the next chapter.

Chapter Highlights

Sexual Identity and Gender Roles

- Sexual identity results from those physical characteristics and behaviors that are part of our biological inheritance.
- Gender role, on the other hand, results partly from genetic makeup and partly from the specific traits in fashion at any one time and in any one culture.
- Views of acceptable gender role behaviors have changed considerably during the past twenty years. Androgyny is now considered an acceptable alternative to masculine and feminine gender roles.

Sexuality

- The "sexual revolution" has led to many teenagers becoming sexually active at increasingly younger ages.
- Sexuality develops in three stages, from love of self to love of members of the same gender to love of members of the opposite gender.

Autosexuality

- Masturbation is believed to be a harmless and universal form of human sexual expression.

Homosexuality

- Homosexual behavior has been surrounded by many myths throughout history.
- Several theories suggest different origins of homosexual orientation: psychoanalytic, learning, and genetic.
- Many researchers now believe that homosexual orientation may already be set by adolescence.

Heterosexuality

- Many teens still obtain a great deal of information and misinformation about sex from their peers.
- First intercourse is now occurring earlier than it did in past generations.
- Youths from stable family environments are less likely to engage in premarital sexual relations.
- Effective listening skills are essential for parents who wish to maintain good communication with their adolescents.
- There are many nonsexual reasons why teenagers misuse sex, including a search for affection, rebellion against parents, venting anger, and alleviating boredom.
- Many adolescent runaways and prostitutes are the products of sexual abuse, often by someone they know, a family member or parent.

Sexually Transmitted Diseases

- Today a high prevalence of sexually transmitted diseases (STDs) is found in sexually active adolescents.
- AIDS (acquired immune deficiency syndrome) causes the most concern, because it is currently incurable and often fatal. As yet it is not very common in adolescents, but because it usually lies dormant for ten to fifteen years, it is a cause for great concern.
- Other STDs that affect adolescents are increasing in epidemic proportions, including chlamydia, gonorrhea, genital herpes, syphilis, and Hepatitis B.
- In spite of increased availability and information about contraceptive methods, many teenagers continue to engage in unprotected sex.

Teenage Parenthood

- Teenage pregnancy and parenthood are on the rise among younger teens. With rare exceptions, this situation causes a lot of heartache for the teenage parents, their parents, and their child.

- Teenage mothers who receive emotional support from their families, who continue in school, and who have no additional children while they are still teenagers are likely to have better-adjusted children and avoid the cycle of poverty.
- When teens possess strong self-esteem, feelings of hope concerning the future, and job and academic skills for entry into the job market, they are less likely to become teenage parents.

Key Terms

AIDS (acquired immune deficiency syndrome)
Androgyny
Autosexuality
Chlamydia
Chromosome failure
Double standard
Genetic theory of homosexuality
Genital herpes
Gonorrhea
Hepatitis B
Heterosexuality
Homosexuality
Learning theory of homosexuality
Pelvic inflammatory disease (PID)
Psychoanalytic theory of homosexuality
Reflective listening
Gender role
Gender-role adaptation
Gender-role orientation
Gender-role preference
Gender-role stereotypes
Sexual identity
Sexual revolution
Sexually transmitted diseases
Syphilis
Transsexual operation

What Do You Think?

1. What aspects of traditional male and female gender roles do you think we should try to keep?
2. What is your attitude toward androgyny? Would you describe yourself as androgynous? If not, do you wish you were?
3. Do you agree with the theorists who claim that there are three stages in the development of love and sexuality, and that this development is natural?
4. Why do you think we are seeing such widespread changes in the sexual aspects of our lives?
5. If you were the mayor of a medium-size city, what actions would you take to try to reduce the incidence of sexually transmitted diseases?
6. If you were the mayor of a medium-size city, what actions would you take to try to reduce the incidence of teenage parenthood?

Suggested Readings

Angelou, M. (1970). *I know why the caged bird sings.* New York: Random House. Ms. Angelou recounts her childhood in rural Arkansas. Her strength and resilience model the building of a strong personal and cultural identity.

Auel, J. (1981). *Clan of the cave bear.* New York: Bantam. Auel's wonderful imagination and excellent knowledge of anthropology make this book on the beginnings of the human family a winner. In fast-paced fiction, she describes the relationships, sexual and otherwise, of early humans who, she speculates, were unaware that sex causes pregnancy!

Calderone, M. S., and J. Ramsey (1981). *Talking to your child about sex.* New York: Ballantine. This book offers a creative interpretation of human sexuality in a family setting.

Capote, T. [1948] (1988). *Other voices, other rooms.* New York: Signet. Written when Capote was 23 years old, it is considered by many to be his best work. It is the story of a 13-year-old boy who goes to live with his father in Louisiana and meets many eccentric characters. Through this experience he becomes aware of the adult world and his own homosexuality.

Fromm, E. (1968). *The art of loving.* New York: Harper & Row. Although most of us think of love as a very personal topic, it would be hard to think of anything that has been the subject of more novels, articles, poems, plays, and psychological treatises. Most of these are not particularly helpful, and many are downright corny. Fromm's book is a distinct exception. You will understand what love is and how to give and receive it much better when you have finished reading it.

Jacoby, A. (1987). *My mother's boyfriend and me.* New York: Dial Books. Sixteen-year-old Laurie doesn't know how to handle it when her mother's 27-year-old, handsome, blue-eyed boyfriend starts making advances.

Janus, S., & Janus, C. (1993). *The Janus report on sexual behavior.* New York: Wiley. A wide-ranging survey of attitudes and behavior.

Tannahill, R. (1980). *Sex in history.* New York: Ballantine. This lively book describes the role of sex down through the ages.

Walker, A. (1982). *The color purple.* New York: Harcourt Brace Jovanovich. This disturbing story of African-American teenage pregnancy in the South was hailed by all the reviewers for its insight.

PART VII

Early Adulthood

CHAPTER 15

Passages to Maturity

Initiation Rites 409

Analysis of an Initiation Rite 410

The Passage to Adulthood in Western Countries 412

The Transition to Adulthood in the United States 412

Types of Initiation Activities in the United States 413

The Adolescent Moratorium 413

Implications of the Lack of an Initiation Ceremony 414

Two Proposals 416

Dealing with the Stresses of Adulthood 418

Change as a Source of Stress 419

Stimulus Reduction versus Optimum Drive Level 421

The General Adaptation Syndrome 422

Disease as a Result of Stress 424

Measuring the Relationship between Stress and Physical Illness 426

Risk and Resilience 426

Conclusion 427

Chapter Highlights 428

Key Terms 429

What Do You Think? 429

Suggested Readings 430

All the young men standing in the living room of the old fraternity house had solemn faces. As the fraternity president began intoning the sacred words that would lead to their induction, Dave and Bill looked at each other out of the corners of their eyes. The two "pledges" were in the front row, waiting along with nine other sheepish-looking freshmen. Each knew the other was remembering the same thing—the long ordeal of their pledge period, which had begun at the start of the semester.

For weeks they had had to wear ridiculous beanies on their heads and act as virtual slaves to the fraternity brothers. Despite their best efforts to obey the complex rules, each had incurred numerous violations, which the brothers had noted in their pledge books.

Last night, for the initiation opener, they had endured one blow from a thick magazine, rolled up and taped for the purpose, for each of their rules violations. Bent over and holding their ankles, they had managed to get through the bottom-beatings without crying out, but not without becoming very black and blue. Then they were taken as a pair and dropped off in the middle of a dark woods with instructions to get back to the frat house by 10 A.M. if they wanted to be initiated.

After numerous scares and mishaps, they made it to the main road and hitched into town. On their return to the house, they thought their initiation was finished, but the ordeal was far from over. Now came a series of lesser trials, including

- Being made to lie on their backs while tablespoons of baking soda and then vinegar were poured into their open mouths. They were ordered to close their mouths and keep them that way no matter what. The brothers laughed uproariously when, inevitably, the mixture exploded, shooting a geyser from their tightly-pressed lips high into the air.

- Being blindfolded and made to eat warm "dog manure." Actually they had eaten doughnuts soaked in warm water, but the bag of manure held under their noses made them believe it was the real thing.

- Having a mixture of Liederkranz cheese and rotten eggs smeared on their upper lips, then being made to run around while inhaling the dreadful odor.

Now, as the torture was over, and the final ceremony was underway, both Dave and Bill had the same thought in their minds: I've done it! I've survived! I'm in! ■

This chapter investigates two important factors in adult life. The first section describes the current status of Western passage to maturity, details criticisms of it, and suggests alternatives. The second section, which deals with stress, is placed in this chapter not because young adults are under more stress than other age groups, but because this is the age when people must begin learning how to manage the many sources of stress in life, relying mainly on their own resources.

When you finish this chapter, you will be able to:

- Discuss the purpose of initiation rites and the effects on adolescents of there being no formal rite of passage into adulthood in our culture.
- Describe the components of some initiation rites in preindustrial societies.

- List activities in our own society that may be parallel to these rites.
- Discuss the concept of the adolescent moratorium, its strengths and weaknesses.
- Specify some activities that could serve some of the purposes of an initiation rite.
- Define stress and identify common sources of stress during the teenage years.
- Describe the relationship between stress, physical illness and mental health.
- List the three stages of Selye's general adaptation syndrome.
- Explain the relationship between risk and resiliency.

Initiation Rites

The horrors of the fraternity initiation have been softened by legal restrictions and by more humane attitudes. For example, Baier and Williams (1983) surveyed 440 active members and 420 alumni members of the fraternity system of a large state university. They compared attitudes of these men toward twenty-two **hazing practices** (such as those just described) known to be used by the frats. They found the active members were considerably opposed to more of the practices than the alumni.

Nevertheless, most of us have heard of recent cases of maimings and even deaths of young men who have been put through hazing. Such trials are held by people at all socioeconomic levels and racial and ethnic backgrounds. They are organized by sports teams and criminal gangs, and are not limited to males, either.

Nor is the problem limited to the United States. For example, in his study of French juvenile delinquents, Garapon (1983) saw a "symbolic, sacrificial dimension" to their crimes. He notes that often cars and other stolen goods are either dumped in the canals or burned, which he feels parallels the sacrifices of prized goods that preindustrial tribes carry out with fire and water. He points out other links to tribal initiations: Most crimes are committed at night, and wind up in courtrooms, where there are symbolic costumes such as the judge's robes. In Germany, Zoja (1984) has suggested similar parallels in the path to drug addiction. He states that "Drug addiction can be an active choice, allowing the user to acquire a solid identity and social role, that of the negative hero, as well as access to an esoteric glimpse of an 'other world' " (p. 125).

Why do some people, and the groups they wish to be associated with, seem to enjoy holding **initiation rites** so much? And why are so many adolescents, many of them otherwise highly intelligent and reasonable, willing and eager to endure such pain? Is it simply because they want to join the group, to feel that they belong? There seems to be more to it than that. Throughout the world, adolescents readily engage in such activities because they seem to want to be tested, to prove to themselves that they have achieved the adult virtues of courage, independence, and self-control. And the adults seem to agree that adolescents should prove they have attained these traits before being admitted to the "club of maturity." Compare the activities that Dave and Bill were put through to those of two members of the Kaguru African tribe, Yudia and Mateya (see page 411).

ANALYSIS OF AN INITIATION RITE

Before discussing the implications of American initiation rites (or the lack of them), we'll provide a more detailed description of the purposes and components of such rites.

The first analysis of initiation ceremonies in preindustrial societies such as the Kaguru was completed by anthropologist Arnold Van Gennep in 1909 (Van Gennep, 1960). His explanation is still highly regarded, as can be seen in more recent studies (e.g., Anderson & Noesjirwan, 1980; Brain & others, 1977; Hill, 1987; Kitahara, 1983; Lidz & Lidz, 1984; Morinis, 1985; Ramsey, 1982). Van Gennep argued that the purpose of the initiation rites, as with all rites of passage (marriage, promotion, retirement, etc.) is to cushion the emotional disruption caused by a change from one status to another. As Brain and associates (1977) put it:

After months of grueling training, teenage members of most preindustrial tribes, such as these San Carlos Apaches, are inducted into adulthood.

> ***These rites are seen as a part of general human concern. . . . The particular problem being dealt with is the change from childhood, seen as asexual, to adulthood, seen as sexual. Further, sex is threatening, since it is connected with death and with the unique human knowledge of mortality.*** **(p. 191)**

For males, this transition also involves the end of dependence on their mothers and other older women and the beginning of their inclusion in the world of men.

This ceremony is often scheduled to coincide with the peak in adolescent physiological maturation, and therefore has often been called a **puberty rite.** Van Gennep argues that this is inappropriate, since initiation may be held by one tribe when the children are 8 years old, and in another at 16. Children of 8 have not yet started puberty; those of 16 are halfway through it. Also, the age at which puberty starts differs from individual to individual, and now occurs approximately three years earlier than it did one hundred years ago (See chap. 12). Nevertheless, the initiation rite is usually held at one age within each tribe, regardless of the physical maturity of the individual initiates.

Serving as an introduction to sexuality and separation from mother is one purpose of the initiation rite. Several other purposes have been suggested. In his classic text *Totem and Taboo* (1914/1955), Sigmund Freud offered the psychoanalytic explanation. In his view, such ceremonies are necessitated by the conflict between fathers and sons over who will dominate the women of the tribe. Adolescent males are seen as challenging the father's authority and right to control the women.

To make clear their supremacy in the tribe and to ensure the allegiance of the young males, the adults set a series of trials for the youth at which the adults are clearly superior. The ultimate threat held over the young is castration, the loss of sexual power. Most rites include trials of strength, endurance, prowess, and courage. These usually involve forced ingestion of tobacco and other drugs, fumigations, flagellations, beatings with heavy sticks (running the gauntlet), tattooing, cutting of the ears, lips, and gums, and that most Freudian of inflictions, the circumcision of the foreskin of the penis.

The message is clear: "We, the adult males, are in charge. Join us and be loyal, or else!" Psychologist Bruno Bettleheim (1969) agrees with Freud that there is a fear of castration among the males, but argues that the main role of the initiation rite is to ease the stress of becoming an adult, not to exaggerate it.

In these explanations of initiation rites, biological, psychological, and social factors are emphasized to varying degrees. Can you say what these factors are?

A Multicultural View 

Yudia and Mateya Come of Age

Yudia cannot believe how rapidly her feelings keep changing. One moment she is curious and excited, the next, nervous and afraid. Tonight begins her *igubi,* the rite that celebrates her induction into adulthood. Yudia has longed for this day most of her 11 years, but now she wonders if she really wants the responsibilities of a grown-up.

Though it seems much longer, only a week has passed since the excruciating beginning of her initiation. The memory of it is already dimming: the bright fire, her women relatives pinning her down on the table, her grandmother placing a thin sharp stone against her vulva, the searing pain.

The women had held and consoled her, empathizing fully with her feelings. Each had been through the same agony. For them, too, it occurred shortly after their first menstruation. They had explained to her that this was just the beginning of the suffering she must learn to endure as an adult woman. All during the past week, they had been teaching her—about the pain her husband would sometimes cause her, about the difficulties of pregnancy and childbirth, about the many hardships she must bear stoically. For she is Kaguru, and all Kaguru women accept their lot in life without complaint.

It has been a hard week, but tonight the pleasure of the *igubi* will help her forget her wound. There will be singing, dancing, and strong beer to drink. The ceremony, with its movingly symbolic songs, will go on for two days and nights. Only the women of this Tanzanian village will participate, intoning the time-honored phrases that will remind Yudia all her life of her adult duties.

In a large hut less than a mile from the village, Yudia's male cousin Mateya and seven other 13-year-old Kaguru boys huddle close, even though the temperature in the closely thatched enclosure is a stifling 110 degrees. Rivulets of sweat flow from their bodies and flies dot their arms, backs, and legs. They no longer pay attention to the flies, nor the vivid slashes of white, brown, and black clay adorning all their faces. Their thoughts are dominated by a single fear: Will they cry out when the elder's sharpened stone begins to separate the tender foreskin from their penises? Each dreams of impressing his father, who will be watching, by smiling throughout the horrible ordeal.

Three months of instruction and testing have brought the young men to this point. They have learned many things together: how to spear their own food, how to tend their tiny gardens, how to inseminate their future wives, and most important, how to rely on themselves when in danger.

The last three months have been exhausting. They have been through many trials. In some, they had to prove they could work together; in others, their skill in self-preservation was tested. For most of them, being out of contact with their mothers was the hardest part. They have not seen any of the female members of their families since they started their training. Unlike Yudia's initiation, which is designed to draw her closer to the adult women of the tribe, Mateya's initiation is designed to remove him forever from the influence of the females, and to align him with the adult men.

Now it is evening. Mateya is the third to be led out to the circle of firelight. Wide-eyed, he witnesses an eerie scene. His male relatives are dancing in a circle around him, chanting the unchanging songs. The grim-faced elder holds the carved ceremonial knife. Asked if he wishes to go on, the boy nods yes. Abruptly the ritual begins: The hands of the men hold him tight; the cold knife tip touches his penis; a shockingly sharp pain sears his loins; he is surprised to hear a piercing scream; then, filled with shame, he realizes it comes from him.

Thus far, Yudia's and Mateya's initiations have been different. Mateya's has been longer and harder than Yudia's. She is being brought even closer to the women who have raised them both, but Mateya must now align himself with the men.

The initiations are similar, though, in that both youths have experienced severe physical pain. In both cases, the operations were meant to sensitize them to the vastly greater role sex will now play in their lives. Furthermore, their mutilations made them recognizable to all as adults of the Kaguru tribe.

At this "coming out" ceremony, males and females also receive new names, usually those of close ancestors. This illustrates the continuity of the society. The beliefs of the tribe are preserved in the continuous flow from infant to child to adult to elder to deceased and to newborn baby again. When all is done, Yudia and Mateya can have no doubt that they have passed from childhood to adulthood.

A Multicultural View 

Initiation Rites in the United States

Do we have initiation rites in United States society? Do different ethnic and religious groups have different rites? In the spaces below, write down as many initiation rites as you can think of. Taken together, do these rites you have listed indicate an American definition of maturity? Some activities that may be considered initiation rites are suggested at the bottom of page 413.

1.	6.	11.	16.
2.	7.	12.	17.
3.	8.	13.	18.
4.	9.	14.	19.
5.	10.	15.	20.

The Passage to Adulthood in Western Countries

How are Western youths inducted into adulthood? Are Western initiation rites adequate? These are questions we will now address.

The Transition to Adulthood in the United States

In the industrial past of the United States, it used to be fairly clear when one became an adult. In their late teens, boys and girls usually got married and assumed an adult role. Males were accepted as partners in the family farm or business; females became housewives. This has changed in many ways. What Black (1974) has suggested is still true:

> ***Today, in modern society, initiation of the boy and girl into adult life is far more complicated. Society is fast, heterogeneous, a network of interdependent groups with many different backgrounds, traditions, and outlooks, the products of religious, racial, national, and class differences. In our age of technology, the young have to learn to deal with cars and trains and planes, machines and other electronic equipment, typewriters, television sets, computers, and mass production assembly lines. They face high concentrations of population, high mobility, and relationships on regional, national, and global levels. All this they have to know and understand at a time when customs, laws and institutions are undergoing drastic and rapid change in the midst of a high degree of human differences and human conflict.*** **(p. 25)**

There are no specific rituals comparable to those in preindustrial societies to help Western youth through this difficult period. For example, religious ceremonies like confirmation and bar mitzvah no longer seem to play the role they had in earlier times. Kilpatrick (1975) argues that

> ***at some point we grew too sophisticated, at some point the rituals lost their vitality and became mere ornaments. We may still keep their observance, but they are like old family retainers, kept on in vague remembrance of their past service.*** **(p. 145)**

In his anthology of adolescent literature, Thomas Gregory (1978) points out that many modern writers find the decline of the initiation rite in the United States a noteworthy theme. He describes their thinking as follows:

> ***Today's adolescents are faced with not knowing when they have reached maturity . . . the absence of a formal rite of passage ceremony necessitates a larger and more uncertain transition, with much groping, as adolescents not only try to establish their new adulthood, but also their identity.*** **(p. 336)**

Types of Initiation Activities in the United States

This is not to say that Americans have *no* activities that signal the passage to maturity. We have a number of types of activities, which usually happen at various stages and ages of adolescence. Here is a list of the types and some examples of each:

Religious
Bar mitzvah or bat mitzvah
Confirmation
Participating in a ceremony, such as reading from the Bible
Sexual
Menarche (first menstruation)
Nocturnal emissions (male "wet dreams")
Losing one's virginity
Social
"Sweet Sixteen" or debutante parties
Going to the senior prom
Joining a gang, fraternity, or sorority
Beginning to shave
Being chosen as a member of a sports team
Moving away from one's family and relatives
Joining the armed forces
Getting married
Becoming a parent
Voting for the first time
Educational
Getting a driver's license
Graduating from high school
Going away to college
Economic
Getting a checking or credit card account
Buying one's first car
Getting one's first job

The tuxedo and party dress might be considered costumes in one of America's initiation rites.

The Adolescent Moratorium

In the late twentieth century, the attitude that youths need a "time out" period to explore possibilities and to continue education has become widespread. This phase of life is known as the **adolescent moratorium** (See chap. 11).

Perhaps it is natural that we have discarded rites of passage into adulthood for the more leisurely moratorium. In preindustrial societies, children must take over the responsibilities of adulthood as quickly as possible. The survival of the tribe depends on getting as much help from all individuals as possible. In industrial societies, and even more so in our "information-processing" society, the abundance of goods makes it less necessary that everyone contribute to the society. There is

AN APPLIED VIEW The Components of Maturity

Think of the woman and the man who are the most mature persons you know—people with whom you are personally familiar, or people who are famous. Then ask yourself, "Why do I think these people are so much more mature than others?" In the following spaces, for both the male and the female, create a list of the characteristics that seem to distinguish them in their maturity.

Female	Male
1.	1.
2.	2.
3.	3.
4.	4.
5.	5.
6.	6.
7.	7.
8.	8.
9.	9.
10.	10.

How much do the lists differ? Is male maturity significantly different from female maturity? Which of the two people is older? Which of the two do you admire more? Which of the two are you more likely to want to imitate? Were you able to think of many candidates for this title of "most mature adult," or was it difficult to think of anyone? Are either or both of the people you picked professionals? Are either or both of these persons popular with their own peer group? What is the significance of your answers to you?

also a need for more extensive schooling in preparation for technical types of work. The moratorium, then, comes about because of our advanced economic system. Furthermore, our society values choice. Our children are not expected to follow in their parents' footsteps. It should be noted that poor adolescents, who have fewer choices, also have a shorter moratorium. For these reasons, the initiation ritual has declined considerably since the nineteenth century. Is our society the better or worse for this change?

Implications of the Lack of an Initiation Ceremony

Today we have doubts that this moratorium is turning out to be effective. In fact, it appears that crime is one of the ways that some youth are *initiating themselves* into adulthood. Males especially seem to need to do something dangerous and difficult. Males raised without fathers or father substitutes are especially vulnerable to the attractions of criminality (Dacey, 1986). When they are leaving adolescence, many of them seem to feel they must prove their adulthood by first proving their manhood in risk-taking behavior.

For much of their childhood, males in the United States are also highly dependent upon female attention. This is especially true among those whose fathers are absent. Such a youth sometimes compensates by

> ***joining tribe-like gangs and undergoing harsh initiation rites, all in the service of proving his manhood. Much of the trouble that these youth get into serves the same function as primitive rituals. To compensate for the dominant role of mother in his childhood, the boy needs a dramatic event or a series of them to establish male identity.*** **(Kilpatrick, 1975, p. 155)**

In the 1960s and early 1970s, American youths sought to establish their identities by imitating the very rituals of the preindustrial tribes described earlier in this chapter. Known as "hippies" and "flower children," they attempted to return to a simpler life. Many of them returned to the wilderness, living on farms and communes away from the large cities in which they were brought up. Many totally rejected the cultural values of their parents. The most famous symbol of their counterculture was the Woodstock musical marathon in 1969. With its loud, throbbing music, nudity, and widespread use of drugs, it was similar to many primitive tribal rites. Yet these self-designed initiation rites also seem to be unsuccessful as passages to maturity. Most of the communes and other organizations of the youth movement of the 1960s have since failed. Most American youths have decided "you can't go back again."

Organized sport is another attempt to include initiation rites in American life. The emphasis on athletic ability has much in common with the arduous tasks given to preindustrial youth. In particular, we can see a parallel in the efforts by fathers to get their sons to excel in Little League baseball and Pop Warner football. Fathers (and often mothers) are seen exhorting the players to try harder, to fight bravely, and when hurt, to "act like a man" and not cry.

It has been suggested that sports play the same role in life as the arduous tasks performed by youths in centuries past: learning coordination, cooperation, and the other skills necessary in adult work. Can you think of ways that sports might serve as initiation rites?

Thus, in delinquency, the counterculture, and sports, we see evidence that members of several age groups today yearn for the establishment of some sort of initiation rite. Adolescents and adults alike seem to realize that something more is needed to provide assistance in this difficult transitional period. But what?

Traditional initiation rites are inappropriate for American youth. In preindustrial societies, individual status is ascribed by the tribe to which the person belongs. Social scientists call this an *ascribed identity.* The successes or failures of each tribe determine the prestige of its members. Family background and individual effort usually make little difference. In earlier times in the United States, the family was the prime source of status. Few children of the poor became merchants, doctors, or lawyers. Today, personal effort and early commitment to a career path play a far greater role in the individual's economic and social success. This is called an *achieved identity.* For this reason (and others), preindustrial customs are not compatible with Western youth today.

Roy (1990) suggests that the regular inclusion of family rituals, of which rites of passage are one type, promote communication and healing within the family system. (Examples might be allowing a child to stay home alone, and a parent teaching a child to drive.) Through the marking of change in an individual, the family also realizes that it itself is changing. By traveling the developmental road together, it seems that adolescents and their families are able to adjust to the inherent difficulties of transition more readily.

There are valuable implications for therapists and others through the use of initiation rites. Roy claims that

> ***clergy and therapists can increase their effectiveness by promoting change through rituals. Rites of passage are the most obvious arena for change, individual as well as family. . . . family members*** **[can decode]** ***family disputes around a rite of passage and facilitate the entry or exit of family members from nuclear family units.*** **(p. 63)**

With the apparent conflict that surrounds families with adolescents, it seems a natural option for them to create and employ meaningful rites of passage for the sake of healthy growth and family harmony.

A number of prominent thinkers have suggested that human development could be enhanced if the transition from adolescent to adult were clearer. But how to do it? Here are two suggestions that have been studied.

Two Proposals

The following two ideas on initiation procedure have lately been gaining in popularity. As you read about them, question whether either or both would fulfill the need for some kind of initiation into adulthood.

Outward Bound

The **Outward Bound** program (Outward Bound, 1988) was founded during World War II to help merchant seamen in England survive when their ships were torpedoed. Early in the war, it was learned that many sailors died because they became paralyzed with fear when their ship was hit. Outward Bound was designed to prepare these men to handle their fears in dangerous situations. The program was so successful that after the war it was redesigned for much broader use.

Although they hardly know each other, these boys will find that working together to achieve the goals of this Outward Bound program quickly serves to break down the boundaries society establishes between strangers.

Its basic premise is that when people learn to deal with their fears by participating in a series of increasingly threatening experiences, their sense of self-worth increases and they feel more able to rely on themselves. The program uses such potentially threatening experiences as mountain climbing and rappeling, moving about in high, shaky rope riggings, and living alone on an island for several days. Some of the experiences in the program also involve cooperation of small groups to meet a challenge, such as living in an open rowboat on the ocean for days at a time.

Outward Bound has grown rapidly in recent years, and has installations throughout the country. Each program emphasizes the use of its particular surroundings. For example, the Colorado school emphasizes rock climbing, rappeling, and mountaineering. The Hurricane Island school is Maine uses sailing in open boats on the ocean as its major challenge. The school in Minnesota emphasizes reflection and development of appropriate spiritual needs.

Outward Bound has proven its special worth for teenagers. It originally started with males, but most of its sessions now include equal numbers of males and females. The program operates as a basic rite of passage by offering a chance to prove one's self-worth and to have this feeling validated by others. The philosophy of the program is that participants cannot be told what they are capable of, but must discover it for themselves.

If the Outward Bound experience can effectively reduce **recidivism rates** (the percentage of convicted persons who commit another crime once they are released from prison), it would have widespread implications in treating juvenile delinquents. This program is not punitive as reform schools are, and is considerably less expensive than prison. However, there is reason to believe that the effectiveness of the program decreases significantly after the enrollees have spent some time back in their neighborhoods. Perhaps this just means that they must be brought back for "booster" sessions from time to time.

Although there is a lack of extensive experimental evidence on the effects of Outward Bound, there is no scarcity of testimony from the participants themselves. As one short teenager put it, "Size really doesn't matter up there. What really counts is determination and self-confidence that you can do it!"

Many graduates say that they find life less stressful and feel more confident about their everyday activities as a result of participating in Outward Bound. One of the most positive aspects of the program is its effect on women. Many say that they are surprised to discover how much more self-reliant they have become. Probably the most important result is that most graduates say they feel more responsible and grown-up after having been through the experience.

The Walkabout Approach

In the remote regions of Australia, the aborigines have a rite of passage for all 16-year-old males. It is known as the **walkabout.** In the walkabout, the youth, having received training in survival skills throughout most of his life, must leave the village and live for six months on his own. He is expected not only to stay alive, but to sustain himself with patience, confidence, and courage. During this six-month estrangement from home and family, he learns to strengthen his faith in himself. He returns to the tribe with the pride and certainty that he is now accepted as an adult member.

The Australian aborigines train their teenaged youth, such as this boy learning to spear a ray, to subsist while on a solo "walkabout," an extended initiation rite. The walkabout has been used as a model to induct American youths into adulthood.

According to educator Maurice Gibbons (1974),

> ***the young native faces an extreme but appropriate trial, one in which he must demonstrate the knowledge and skills necessary to be a contributor to the tribe rather than a drain on its meager resources. By contrast, the young North-American is faced with written examinations that test skills very far removed from actual experience he will have in real life.*** **(p. 597)**

As a result of Gibbons' article, a group was set up by Phi Delta Kappa (PDK), the national education fraternity, to see what could be done about promoting walkabouts for boys and girls in this country. A booklet has been produced that makes specific suggestions. In it, the PDK Task Force suggests that

> ***the American walkabout has to focus the activities of secondary school. It does so by demonstrating the relationship between education and action. It infuses the learning process with an intensity that is lacking in contemporary secondary schools. The walkabout provides youth with the opportunity to learn what they can do. It constitutes a profound maturing experience through interaction with both older adults and children. The walkabout enriches the relationship between youth and community.*** **(Task Force, 1976, p. 3)**

There is also a monthly magazine that prints stories about the various types of walkabouts that students in participating schools have devised.

There are three phases in the process: pre-walkabout, walkabout, and post-walkabout. Each of these phases calls for learning specific skills. In the pre-walkabout, adolescents study personal, consumer, citizenship, career, and lifelong learning skills. In the walkabout itself, the skills to be mastered are logical inquiry, creativity, volunteer service, adventure, practical skills, world of work, and cognitive development. The task force suggests a great number of activities that foster learning in each of these skills. Most involve at least six months of supervised study and activity outside the school, such as working in a halfway house for mental patients.

The post-walkabout is a recognition that the student has engaged in a major rite of passage on his or her way to adulthood. It is not enough to recognize this experience in a ceremony where members are confirmed en masse, such as the typical graduation. Instead, an individual ceremony involving the graduate's family and friends is held for each walkabout the student undergoes.

> ***The celebration of transition could take a variety of forms. The ceremonies are varied according to family tastes and imagination, but in each celebration the graduate is the center of the occasion. Parents and guests respond to the graduate's presentation. Teachers drop by to add their comments and congratulations. The graduate talks about his or her achievements, sharing some of the joys and admitting the frustrations.*** **(Task Force, 1976, p. 34)**

What's Your View?

A number of prominent educators believe that schools should concern themselves only with intellectual matters, not with personal and social development. Do you agree? These educators are opposed to ideas like Outward Bound and walkabout. How do you feel about these two ideas? If you like these ideas, can you use them as a springboard to other methods by which we adults might help youth in their "passage to maturity"?

The Outward Bound and walkabout procedures are becoming popular, and they will almost certainly help to alleviate the need for transitional activities, but they are clearly not sufficient in themselves. The complexity of American adulthood requires a variety of such approaches, if we are to develop in our youth the kind of mature women and men we want. For example, there is a growing call to require universal public service of all youth for one year. What do you think about this idea?

Before we can specify initiation activities that would be useful in inducting youth into adulthood, perhaps we need a clearer idea of the successful adult. To put this another way: "What is a mature person?" This question has intrigued thinkers throughout recorded history. It has been variously described as a search for peace, for the knowledge of God, for satori, for nirvana, for self-actualization, or for wisdom. We will be exploring the many aspects of maturation in the chapters that follow, but first we want to consider another aspect of "growing up," dealing with the stresses of adult life in a mature way.

Dealing with the Stresses of Adulthood

In chapter 11, we cited the oft-repeated warning of parents that adolescence is the last chance to have fun, because becoming an adult means taking on the heavy responsibilities of maintaining a job and a family. Well, this is not to say that infancy, childhood, and the teen years are free of stress—far from it! Adolescent psychologist David Elkind (1989) summarizes the current situation:

> ***Young teen-agers today are being forced to make decisions that earlier generations didn't have to make until they were older and more mature—and today's teen-agers are not getting much support and guidance. This pressure for early decision-making is coming from peer groups, parents, advertisers, merchandisers and even the legal system.*** **(p. 24)**

The section that follows applies to these years, too. But we have included this section in the adult part of our book because the stresses of adulthood are different in one important way: More and more, adults are expected to deal with stress *entirely on their own.* True, we adults can and should expect help from others, but there are an increasing number of crises in which independent decisions and actions are called for. For example, many more families today exist in a state of *ongoing, unending crisis.* In these families, there is often only one parent, who is unable to work and therefore is below the poverty level, and who has one or more handicapped or highly disruptive children (Smith, 1990).

Sometimes stress may be due to childhood traumas that begin to manifest themselves only during adulthood. This may be due to psychological defense mechanisms such as denial (See chap. 19). Brown, Bhrolchain, and Harris (1975) studied depression in working-class English women. Among those women who suffered with depression, there was a high incidence of separation from or loss of their mothers by death in early childhood. Many of these women had been born before or during World War II, and, as a result of the Battle of Britain, were sent as children to the countryside or to other countries to protect them from the German bombing of the major British cities. Even today, there are children in the world who must face such trauma. In North America, abuse is a more likely stressor. Knowing how stress works, and how to handle it effectively, is a necessary building block in the process of maturation.

As should be clear by now, there are many sources of stress, regardless of your age. Mainly, stress is due to change. It is the nature of human development to produce inexorable change in every aspect of our existence. This situation is difficult enough when we are young, but at least then we have the support of parents, teachers, and other adults, as well as a more resilient body. There does seem to be more stress as we get older. As we move from early to late adulthood, we must rely more and more on knowledge and insight to avoid having a stressful life. What follows is a detailed description of just what stress is, and how humans try to deal with it. Table 15.1 contains a list of common life events and the ratings they have been given. To evaluate the amount of stress you are under, check the events that you have experienced in the past year. Total your score; it will be explained later in the chapter.

Change as a Source of Stress

> ***If the last 50,000 years of man's existence were divided into lifetimes of approximately 62 years each, there have been about 800 such lifetimes. Of these 800, fully 650 were spent in caves. Only during the last 70 lifetimes has it been possible to communicate effectively from one lifetime to another—as writing made it possible to do. Only during the last six lifetimes did masses of men ever see a printed word. Only during the last four has it been possible to measure time with any precision. Only in the last two has anyone anywhere used an electric motor. And the overwhelming majority of all the material goods we use in daily life have been developed within the present, the 800th, lifetime.*** **(Toffler, 1970, p. 148)**

With the incredible amount of change in this current "lifetime," it is not surprising that the twentieth century has been called the most stressful in which humans have ever lived. So great have been the results of change in terms of stress that sociologist Alvin Toffler (1970, 1984) has labeled it a new disease: **future shock.** Future shock is the illness that results from having to deal with too much change in too short a time. Toffler compares it to culture shock, the feeling we get when arriving in a foreign country for the first time. We become disoriented and anxious, but in the back of our minds, we know that if this discomfort becomes too great, we have only to get on the plane and go back to our own culture where we can feel safe again. Future shock causes the same kind of stressful feeling, except there is no going home to escape from it.

Why is change so stressful? Toffler suggests that the stress results not so much from the direction or even the kind of changes we are faced with in this century, but rather from the incredible *rate* of change in our daily lives. He suggests that there are three major aspects of rate of change, each of which is rapidly increasing:

- **Transience.** There is a lack of permanence of things in our lives. Toffler documents in great detail how much more transient (fast-moving) our lives have become in this century.

Table 15.1 Social Readjustment Rating Scale

Life Events	Life Change Units
1. Death of spouse	100 _____
2. Divorce	73 _____
3. Marital separation	65 _____
4. Jail term	63 _____
5. Death of close family member	63 _____
6. Personal injury or illness	53 _____
7. Marriage	50 _____
8. Fired at work	47 _____
9. Marital reconciliation	45 _____
10. Retirement	45 _____
11. Change in health of family member	44 _____
12. Pregnancy	40 _____
13. Sex difficulties	39 _____
14. Gain of new family member	39 _____
15. Business readjustment	39 _____
16. Change in financial state	38 _____
17. Death of close friend	37 _____
18. Change to different line of work	36 _____
19. Change in number of arguments with spouse	35 _____
20. Mortgage over $10,000	31 _____
21. Foreclosure of mortgage or loan	30 _____
22. Change in responsibilities at work	29 _____
23. Son or daughter leaving home	29 _____
24. Trouble with in-laws	29 _____
25. Outstanding personal achievement	28 _____
26. Wife begins or stops work	26 _____
27. Begin or end school	26 _____
28. Change in living conditions	25 _____
29. Revision of personal habits	24 _____
30. Trouble with boss	23 _____
31. Change in work hours or conditions	20 _____
32. Change in residence	20 _____
33. Change in schools	20 _____
34. Change in recreation	19 _____
35. Change in church activities	19 _____
36. Change in social activities	18 _____
37. Mortgage or loan less than $10,000	17 _____
38. Change in sleeping habits	16 _____
39. Change in number of family get-togethers	15 _____
40. Change in eating habits	15 _____
41. Vacation	13 _____
42. Christmas	12 _____
43. Minor violations of the law	11 _____
TOTAL	_____

Source: From T. H. Holmes and R. H. Rahe, "A Social Adjustment Scale," *Journal of Psychosomatic Research*, 11:213–18. Copyright © 1967 Pergamon Press, Inc., Elmsford, NY.

- **Novelty.** Some changes are more novel than others. New situations in our lives are more dissimilar from old situations than they used to be, and therefore far more stressful.
- **Diversity.** It also matters what percentage of our lives is in a state of change at any one time. People used to maintain stability in most of their lives, allowing only a few aspects to change at any particular point. This stable proportion is now much smaller for most of us.

Although the escalating rate of change in our lives has increased the amount of pressure we are under, research is increasing our understanding of it and so helping us deal with it better.

Stimulus Reduction versus Optimum Drive Level

> ***Many a gifted person has been done in by a promotion.***
>
> **(Cassem, 1975)**

Practically everything we do or attempt to do involves overcoming some type of obstacle. Therefore, to be alive is to be under stress. According to Canadian physiologist Hans Selye (1982), the most successful researcher and theorist in this field, "Crossing a busy intersection, exposure to a draught, or even sheer joy are often enough to activate the body's stress mechanism to some extent" (p. 5).

Freud was the first to espouse the notion that human beings try to avoid stimulation whenever possible. He referred to this tendency as **stimulus reduction.** According to this idea, all our activities are attempts to eliminate stimulation from our lives. Drinking when we are thirsty, sleeping when we are tired, the pursuit of sex—all are efforts to reduce some drive. Thus, Freud believed that the natural state of human beings is **quiescence,** a condition in which we have no needs at all. Some Eastern philosophers have this condition as their main goal in life, a goal they call nirvana.

Most psychologists today believe that Freud's view of stress was wrong. The results of a considerable amount of research (reviewed in Dacey, 1982) indicate that individuals have a level of stimulation that is *optimum* for them. This is referred to as the **optimum drive level** of stimulation. Figure 15.1 compares stimulus reduction theory with optimum drive level theory.

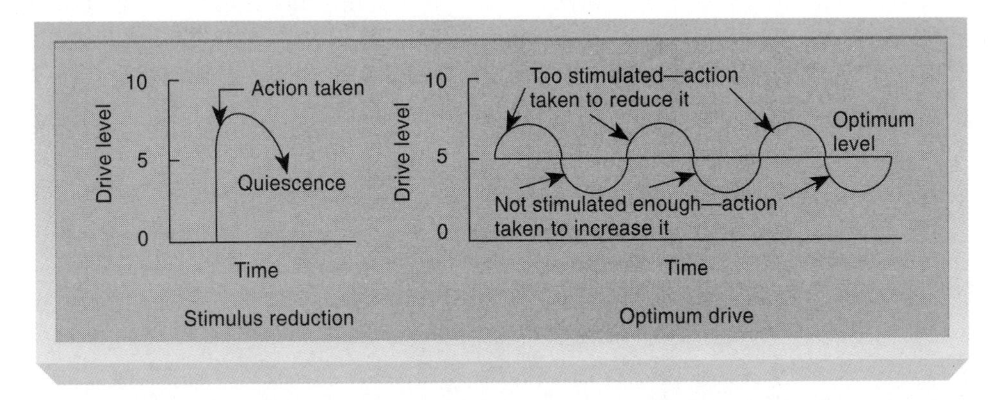

Figure 15.1
Stimulus reduction
Comparison of stimulus reduction and optimum drive theories

According to the concept of optimum drive, when individuals have too much stimulation, they may seek to reduce the stimulation by satisfying the need causing it. This is Freud's position. But at other times, there may be too little stimulation, as in boredom. Under these circumstances, people *seek* new types of stimulation. Thus the goal is not to reduce stimulation to zero, but to maintain it at some optimum level.

What level is optimum differs from individual to individual and from situation to situation. For example, some people enjoy a high level of stimulation in the morning, and become less excitable as the day goes on. Others, who call themselves "night persons," like to be left alone in the morning and only come alive in the evening hours. Some "live wires" demand a great deal of stimulation; others prefer a quieter style of life in general. An excellent book entitled *Flow* (Csikszentmihalyi, 1990; see "Suggested Reading") pursues this theme in a more helpful way.

The General Adaptation Syndrome

In 1936, Hans Selye (the father of stress research) was studying a little-known ovarian hormone. This led to the discovery of the **general adaptation syndrome** (Selye, 1956, 1975). In one of the experiments, hormones from cattle ovaries were injected into rats to see what changes would occur. Selye was surprised to find that the rats had a broad range of reactions:

- The cortex became enlarged and hyperactive.
- A number of glands shrank.
- Deep bleeding ulcers occurred in both the stomach and upper intestines.

Further experiments showed that these reactions occurred in response to all toxic substances, regardless of their source. Later experiments also showed them occurring, although to a lesser degree, in response to a wide range of noxious stimuli, such as infections, hemorrhage, and nervous irritation.

Selye calls the entire syndrome an **alarm reaction.** He refers to it as a generalized "call to arms" of the body's defensive forces. Seeking to gain a fuller understanding of the syndrome, he wondered how the reaction would be affected if stress were present for a longer period of time. He found that a rather amazing thing happens. If the organism survives the initial alarm, it enters a **stage of resistance.** In this second stage, an almost complete reversal of the alarm reaction occurs. Swelling and shrinkages are reversed; the adrenal cortex, which lost its secretions during the alarm stage, becomes unusually rich in these secretions; and a number of other shock-resisting forces are marshalled. During this stage, the organism appears to gain strength and to have adapted successfully to the stressor.

However, if the stressor continues, a gradual depletion of the organism's adaptational energy occurs (Selye, 1982). Eventually this leads to a **stage of exhaustion.** Now the physiological responses revert to their condition during the stage of alarm. The ability to handle the stress decreases, the level of resistance is lost, and the organism dies. Figure 15.2 portrays these three stages. And Table 15.2 lists the physical and psychological manifestations of the two. Table 15.3 details these reactions.

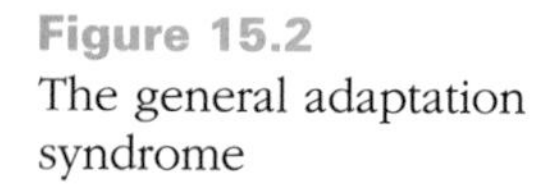

Figure 15.2
The general adaptation syndrome

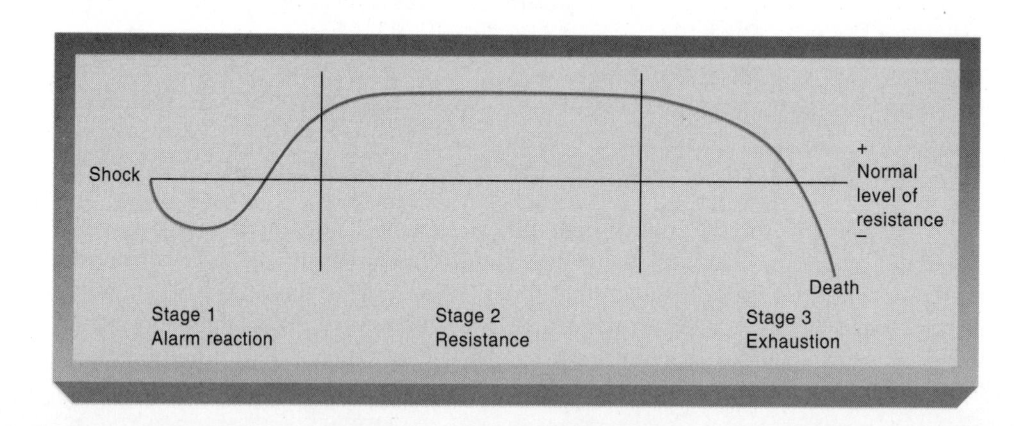

Table 15.2 Selye's Stress Adaptation Syndrome

Stage	Function	Physical Manifestations	Psychological Manifestations
Stage I: Alarm reaction	Mobilization of the body defensive forces.	Marked loss of body weight. Increase in hormone levels. Enlargement of the adrenal cortex and lymph glands.	Person is alerted to stress. Level of anxiety increases. Task-oriented and defense-oriented behavior. Symptoms of maladjustment, such as anxiety and inefficient behavior, may appear.
Stage II: Stage of resistance	Optimal adaptation to stress.	Weight returns to normal. Lymph glands return to normal size. Reduction in size of adrenal cortex. Constant hormonal levels.	Intensified use of coping mechanisms. Person tends to use habitual defenses rather than problem-solving behavior. Psychosomatic symptoms may appear.
Stage III: Stage of exhaustion	Body resources are depleted and organism loses ability to resist stress.	Weight loss. Enlargement and depletion of adrenal glands. Enlargement of lymph glands and dysfunction of lymphatic system. Increase in hormone levels and subsequent hormonal depletion. If excessive stress continues, person may die.	Personality disorganization and a tendency toward exaggerated and inappropriate use of defense mechanisms. Increased disorganization of thoughts and perceptions. Person may lose contact with reality, and delusions and hallucinations may appear. Further exposure to stress may result in complete psychological disintegration (involving violence or stupor).

Source: From C. R. Kneisl and S. W. Ames, *Adult Health Nursing: A Biopsychosocial Approach*. Reading, MA: Addison-Wesley Publishing Company, 1986, p. 20.

As you grow older, your ability to remain in the resistance state decreases. Activity over the years gradually wears out your "machine," and the chances of sustaining life are reduced. As we discuss in chapter 20, no one dies of old age. Rather, they succumb to some stressor because their ability to resist it has become weakened through aging.

Selye (1975) compares his general adaptation stages to the three major stages of life. Childhood, he says, is characteristic of the alarm state: Children respond excessively to any kind of stimulus and have not yet learned the basic ways to resist shock. In early adulthood, a great deal of learning has occurred, and the organism knows better how to handle the difficulties of life. In middle and old age, however, adaptability is gradually lost, and eventually the adaptation syndrome is exhausted, leading ultimately to death.

Selye suggests that all resistance to stress inevitably causes irreversible chemical scars that build up in the system. These scars are signs of aging. Thus, he says, the old adage that you shouldn't "burn the candle at both ends" is supported by the body's biology and chemistry. Selye's work with the general adaptation syndrome has also helped us to discover the relationship between disease and stress.

Table 15.3 Reactions to Stress

Thoughts	Feelings	Bodily Responses	Actions
Initially:			
Attentiveness, increased alertness, focus, discrimination, problem-solving, mental coping devices	Surge of energy, tension, and/or excitement, elation, anxiety, fright, frustration, anger, fulfillment, happiness.	Tachycardia, hypertension, shallow respiration, dry mouth, paleness, perspiring hands, difficulty voiding/ defecating, insomnia, fatigue, tremors, diarrhea, nervousness.	Increased activity level: restlessness, irritability, increased sensitivity and responsiveness, cooperation, alternative plans, compromise.
Later:			
Continued or diminished clarity, focus, discrimination, problem-solving skills, and mental coping devices	Any emotion is possible: ambivalence, loneliness, sadness, helplessness, or hopelessness.	Hypotension, bradycardia, slow respirations, faintness and dizziness, blurred vision, and incontinence.	Problem solving; work, play, exercise; diversification; withdrawal; overuse of drugs, alcohol, and food; excessive sleeping; regression; daydreaming.
With overwhelming stress:			
Impaired perception and cognition; disorganized thinking; minimal focus and discrimination; reality distortions	Panic, detachment.	Fixed, dilated pupils; exhaustion; death.	Disorganization, immobilization.

Source: From S. Jasmin and L. N. Trygstad, *Behavioral Concepts and the Nursing Process*. Copyright © 1979 The C.V. Mosby Company, Inc., St. Louis, MO.

Disease as a Result of Stress

A merry heart doeth good like a medicine.

(Proverbs 17:22)

We are all aware of short-term physical upsets such as fainting, rapid heartbeat, and nausea, caused by the strains of living. Table 15.3 lists the various reactions to stress. You can easily see how many of them are impairing. Only recently have we come to understand more about the relationship between long-term emotional stress and illness.

International studies have shown how widespread this relationship is. For example, it is estimated that in Western industrialized countries, up to 70 percent of all patients being treated by doctors in general practice suffer from conditions whose origins lie in unrelieved stress (Blythe, 1973). The World Health Organization has listed a large number of illnesses, such as coronary disease, diabetes mellitus, and bronchial asthma, as being caused almost entirely by stress. Stress can also cause skin disease, ulcers, nervous tension, dizziness, sore throat, impotence, angina, tachycardia, itchiness, and even accident-proneness.

What makes one person handle difficult life stress better than another person? One answer is practice. Success with similar situations leaves a person with some experience and confidence to draw on in coping with a new stressful situation. The person is less rattled and able to think more clearly and make more realistic responses to the situation.

Social support has also been found to be an important factor in a person's ability to remain composed and to adapt successfully to stressful situations. According to Bowlby (1973), these social supports are created during infancy when a person learns the essential base of security that will carry him or her throughout life. Erikson, you may recall, described this process in his first stage, Trust vs. Mistrust. One study (Nuckolls, Cassell & Kaplan, 1972) looked at the social supports of a group of pregnant women. These social supports were defined as people with whom the women were close, from whom they could obtain affection, and on whom they could rely. The researchers found that women who had many social supports had significantly fewer pregnancy complications than women who had relatively little support. The lack of social support was even more damaging to those women who had high levels of life change. Close family and friends seem to provide a cushioning effect during the stressful times in our lives.

Much of the research looking at the relationship between stress and disease has been completed with adults. Some studies have been completed with adolescents, and the results are similar to the adult findings. Compos and his associates (1987) found that the stressful life events are related to psychological and behavioral problems in adolescents. Not surprisingly, behavioral and psychological problems of adolescence are also related to parents' stress. When parents experience a lot of stress, their children are more likely to have emotional difficulties as well (Compos & Williams, 1990). For African-American adolescent females, a greater number of negative life events was found to be linked to depression, conduct disorder, post-traumatic stress symptoms, and physical illness (Brown & others, 1989).

Other researchers (Daniels & Moos, 1990) found that depressed youths reported more major stressors and daily hassles than healthy youths. Youths with behavioral problems reported more parent and school stressors than healthy youths. At school, highly stressed adolescents are more likely to get into fights, talk back to teachers, play the class clown, and get headaches and stomach aches (Finian & Blanton, 1987).

We now know that being angry a lot of the time (a typical reaction to stress) is highly related to heart attack. There is a growing body of research linking stress with a number of forms of cancer (Cooper, 1984). As a result of the research of Selye and associates, we are beginning to get a much better understanding of this interaction.

A Multicultural View 

Membership in a Minority Group

Dave Mack, an urban youth worker in Chicago, observes that there is a variety of differences among Asian people living in America, and these differences can cause intense stress (cited in Borgman, 1990). Referring to someone as an "Asian" may be misleading. The Chinese are proud of their vast history and the cultures it has spawned. Koreans also possess a rich, though separate, history. Filipinos, Burmese, and Vietnamese, though possibly educated in Chinese schools, nevertheless associate with some Asians while dissociating with others. For instance, among Southeast Asian people, there is a great deal of animosity between Cambodians and the Hmong.

These attitudes of cultural pride provoke endless problems, especially in our large cities. Mack tells of a group of Vietnamese students waiting for a bus in the rain after a basketball game. These students got on the bus when it arrived, just as some Chinese students appeared at the bus stop. Since another bus was not immediately available to transport all of the students, the Vietnamese bus driver ordered his students off of the bus and into the rain so that the Chinese students could get onto the bus first.

Assimilation into urban American culture, itself a smorgasbord of ethnicities all within a few square miles, often proves immensely stressful for all involved. The tension also involves how far one should go in regard to becoming "American." Is there such a thing as an "ethnic American?" Should people who immigrate into America be expected to set aside cultural conflicts for the sake of their new home? At what cost?

Measuring the Relationship between Stress and Physical Illness

Psychologists and psychiatrists have long attempted to measure accurately the relationship between stress and disease. In the early 1900s, Adolph Meyer introduced the concept of *psychobiology,* which emphasized the importance of biographical study in understanding the whole person. He attempted to relate the biology of the person to the likelihood of their getting a variety of diseases. The most successful attempt in this area was by Holmes and Rahe (1967). They developed the *Social Readjustment Rating Scale,* which measures the relationship between events in one's life that require considerable adjustment and the likelihood of getting sick as a result of these crises. The scale is composed of life events that require coping, adaptation, or adjustment. The adult version of this scale was presented at the beginning of this section on stress. Preschool, elementary, and high school versions also exist.

The events in the scale are ranked according to the relative degree of adjustment required by the average individual. A numerical weight, called the **life change unit (LCU),** is assigned to each of the events. The greater the degree of the life change, the higher the unit number.

Holmes and Rahe designed the original adult scale by developing a list of events that could affect psychological well-being. The list was submitted to a sample of adults who rated each item according to the relative amount of adjustment required. The results of this study were then mathematically interpreted, and a numerical value (LCU) assigned to each event. To find out what your score is, look again at Table 15.1. For each life event that you checked as having happened to you within the last year, give yourself the number of points indicated. Then total your score.

Holmes and Rahe's studies show that a score of 200 or more makes some kind of illness likely. Colligan (1975) reports that 86 percent of those who experienced over 300 LCUs in a year developed some serious health problem.

Of course this does not mean that every person who has a high number of LCUs in one year will definitely get sick. It only means that the odds of getting sick are considerably increased. One caution is necessary here: there is the danger of a self-fulfilling prophecy. That is, persons who know they have a higher number of LCUs in a year may believe they are going to become sick, and therefore they do. Nevertheless, it is clear that people under a high level of stress should take especially good care of their bodies and should be ready to check with their doctors quickly if they do develop symptoms of illness.

Risk and Resilience

Individuals who deal well with stress and who have few psychological, behavioral, or learning problems as a result of it are said to have **resilience** (See chap. 10). In recent years, researchers have become interested in studying the characteristics of resilient individuals. The stressors that individuals experience are called **risk factors.** Risk factors include poverty, chronic illness, parental mental illness and drug abuse, exposure to violence through war or some of the tragedies in the inner cities, and the family experiences of divorce and teenage motherhood. Researchers have been interested in identifying **protective factors** (characteristics of resilient individuals that protect them from stress). Three kinds of protective factors have been found so far: family environments, support networks, and personality characteristics (Hauser & Bowlds, 1990).

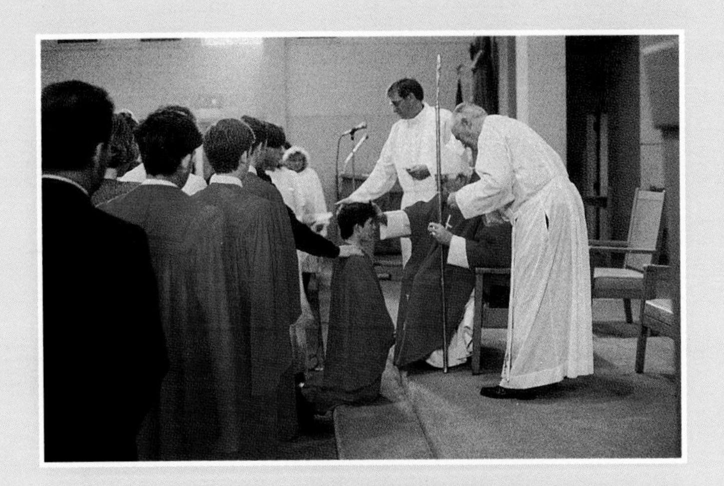

The Passage To Adulthood

In this book we have noted from time to time the variations in human behavior that occur in different cultures. Underlying these behaviors, however, we often find clear similarities in the goals of those cultures, both throughout the world and throughout human history. The need to help our young people move from the status of a child to the status of an adult appears to be a universal aspect of human development.

No societies have ceremonies designed to pass infants into the toddler's world. Movement from early to middle to late adulthood throughout the world is relatively gradual. Why, then, should we feel the need to induct young people, usually at some point in their adolescent years, into adult life?

There have been a number of suggestions offered to explain the "initiation rite." Some believe its goal is to help us move from the world of play

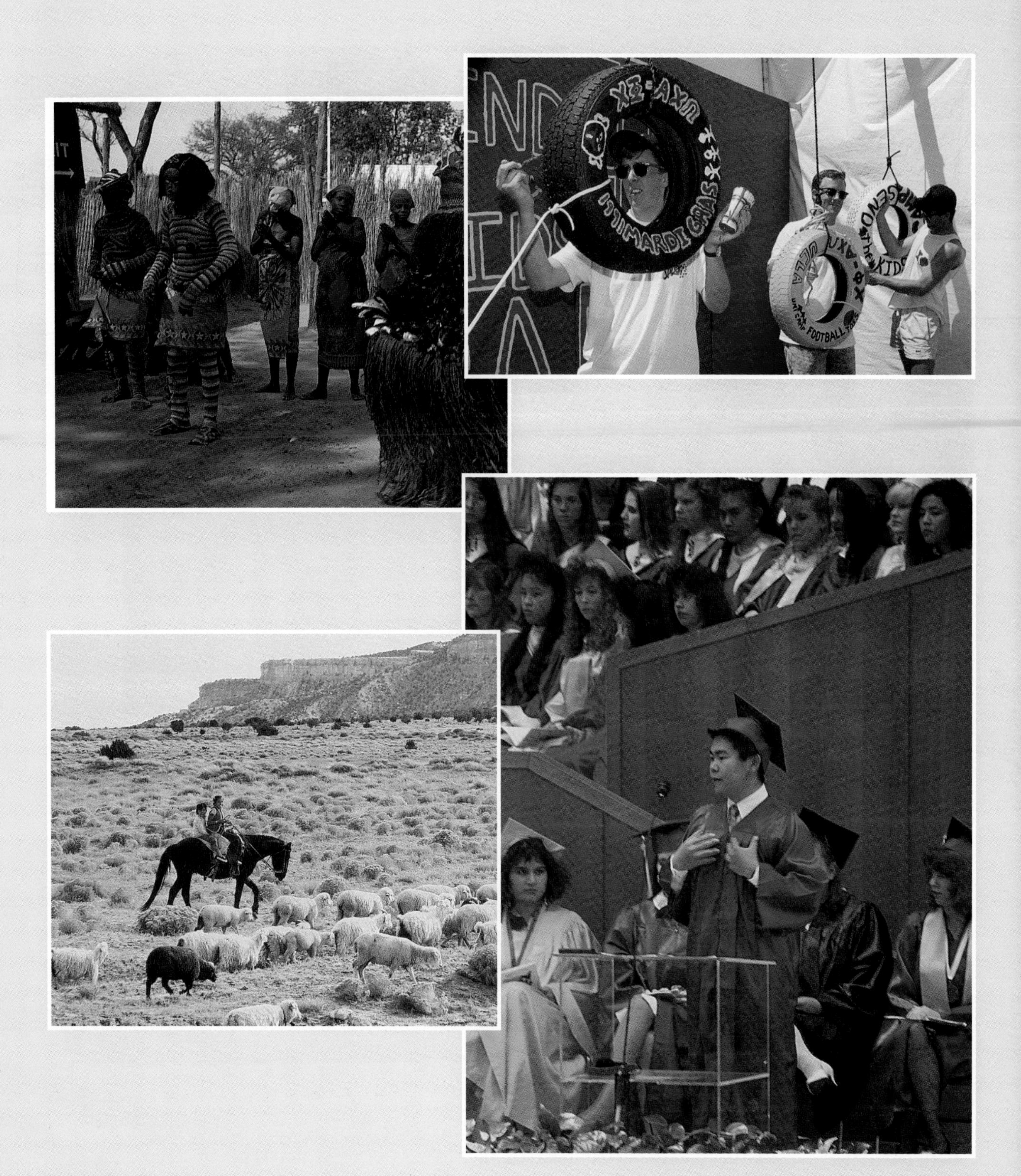

into the world of work. Others argue that the transformation from being cared for as a child to giving care as a parent is so important that it must be formally marked. Reflecting the Freudian position, Brain and associates (1977) state that:

> **The particular problem being dealt with is the change from childhood, seen as asexual, to adulthood, seen as sexual. Further, sex is threatening, since it is connected with death and with the unique human knowledge of mortality (p.191).**

Whatever the reason, there is a great deal of variety that exists among cultures, as they try to assist their young with the passage to maturity (see Chapter 15). A number of these variations are depicted on these pages. As you look at these pictures, try to make up your mind as to why we humans provide these initiation rites. Ask yourself which of them is most effective and which is least effective. Finally, imagine that you are the chief of initiation rites for all the young in our society. You can design any procedure you wish. What would you do?

An Applied View — How Young Cancer Patients Feel When They Learn They Have Cancer

There is a wonderful book of quotes from cancer patients entitled *What it is that I have, don't want, didn't ask for, can't give back and how I feel about it* (Ohio Cancer Information Service, undated). Here are some quotes reflecting how these young people felt when they first learned they had cancer:

- I kept some of my feelings to myself. I knew that my parents were upset, so I tried to be sorry for them. —Kathy
- Mom cried and Dad got mad and I just felt numb inside. —Marsha
- My doctor told me I had cancer, I didn't know what to think. I told him that someone had mixed up the report and had made a mistake. I wish they had. —Dan
- I was very angry. I thought "Why me?" Everybody reassured me it wasn't my fault. —Joyce
- I panicked. I didn't understand. I didn't know who to go to or what to do. Fortunately, my parents were there to help me. —David
- I couldn't believe it! I thought I would wake up in the morning and everything would be just as it was before I got sick. —Kay
- I had my future all planned and it didn't include getting sick. Guess I need to change my plans a little, or postpone them a few years. —Michelle

If you read these quotes carefully, you will get a picture of how people often react to severe stress of all kinds. How could you help a person who is undergoing a period of extreme distress such as learning they have cancer to be resilient? The introduction to this book offers a good suggestion:

> ■ *Your experiences and the feelings you have about your illness are unique to you. No one has ever felt exactly the way you feel. Sometimes it can be hard for you and others to understand what is happening to you. Other patients have gone through similar experiences. Some of them have shared their feelings in this book. Remember, since no two people are alike, their experiences, treatments and feelings might be different than yours.* (p. 1)

Book available from: Ohio Cancer Information Service, 101A Hamilton Hall, 1645 Neil Avenue, Columbus, OH 43210.

Conclusion

We began this chapter by comparing how youths are inducted into adulthood in preindustrial tribes and in the modern United States. We concluded that our situation is much more complex than that of the tribes; and that although there have been several good suggestions for initiation activities in postindustrial societies, the absence of clear initiation rites still causes problems for us.

We then considered the concept of stress, and reviewed Alvin Toffler's concept of "future shock" and Hans Selye's "general adaptation syndrome." It has three stages: alarm reaction, resistance, and exhaustion. We discussed the major warning that comes out of this research: when one is unable to relieve stress over a prolonged period, disease is likely to result. The mature person is one who learns the many skills that help avoid this dire consequence. Finally, we looked at the relationship between risk and resilience.

Is there any way to sum up the concept of maturity? We believe that Erikson's term, *integrity,* probably comes closest. The mature person is one who has achieved an integrated life. But integration of what? That will be the subject of the rest of this book. In the next chapter, we tackle the developmental processes of early adulthood.

CHAPTER HIGHLIGHTS

Initiation Rites

- Initiation rites in other cultures offer a formal ceremony marking the transition from child to adult.
- Our own culture provides no such universal and formal way for taking leave of adolescence and being accepted into the adult community.
- Some researchers have suggested parallels to the initiation process in the crimes juveniles are often required to commit in order to become a member of a gang, and in the sequence of experiences that so often lead to drug addiction.

Analysis of an Initiation Rite

- The purpose of initiation rites in preindustrial societies is to cushion the emotional disruption arising from the transition from one life status to another.
- For the male, these rites also formally end dependence on his mother and other women, and bring him into the group of male adults.
- Initiation rites also serve as an introduction of both genders to the sexual life of an adult.

The Passage to Adulthood in Western Countries

- In our industrial past, the transition to adulthood was fairly clear.
- Entry into adulthood is far more complex for adolescents today, largely due to the increase of sophisticated technologies and the need for many more years of formal education.
- No specific rituals exist in present Western societies to aid youths through this difficult change.
- Four types of activities signal the passage to maturity in America today: religious, social, sexual, educational, or economic.
- The adolescent moratorium, or "time out," allows teens to explore possibilities and extend education. It comes about because of our advanced economic system.
- Some have suggested that the moratorium is ineffective, and this causes teens to try to create their own initiation rites, which may be dangerous.
- Organized sports, the adolescent counterculture, and delinquency all provide some sort of trials that must be passed for acceptance into adulthood.
- The Australian aborigines require all 16-year-old males to spend six months alone surviving in the wilderness. This ritual, called the walkabout, confers adulthood on the youth who completes it. An American form of the walkabout, incorporating varied work in real-world settings, has been suggested.
- The Outward Bound program, with its emphasis on various forms of wilderness survival, serves as a rite of passage for many youth.

Dealing with the Stresses of Adulthood

- Change, especially that caused by future shock, is the major factor in stressful situations.
- Many life events, including major events and daily hassles, contribute to stress in our lives.
- Selye's concept of the general adaptation syndrome includes the stages of alarm reaction, resistance, and exhaustion.
- It is not unusual for disease to result from long-term and varied stressors.
- People who are exposed to many risk factors, but develop few behavioral or psychological problems, are called resilient.

Key Terms

Adolescent moratorium
Alarm reaction
Diversity
Future shock
General adaptation syndrome
Hazing practices
Initiation rites
Life change unit (LCU)
Novelty
Optimum drive level
Outward Bound
Protective factors
Puberty rites
Quiescence
Recidivism rates
Resilience
Risk factors
Stage of exhaustion
Stage of resistance
Stimulus reduction
Transience
Walkabout

What Do You Think?

1. Do you remember any experiences from your own youth that were particularly helpful in your transition to adulthood?
2. The adolescent moratorium seems to be getting longer and longer. Do you think this is a good thing? Why or why not?
3. Can you think of a third plan for initiation along the lines of the walkabout and Outward Bound?
4. How high is your Social Readjustment Rating in total LCUs (life change units)? Do you believe you have anything to worry about? What should you do?
5. Can you name five productive ways of dealing with everyday stress?

Suggested Readings

Benson, H. and W. Proctor (1985). *Beyond the relaxation response.* New York: Berkeley. A stress reduction program that has helped millions of people live healthier lives. Includes his concept of the "faith factor."

Blume, J. (1987). *Letters to Judy: What your kids wish they could tell you.* New York: G.P. Putnam's Sons. Judy Blume offers letters from young adults who confide their concerns over the stresses of friendships, families, abuse, illness, suicide, drugs, sexuality, and other problems. In return, the author shares similar moments from her own life, both as a child and as a parent. She does not hesitate to reveal her own embarrassing situations to help us feel less alone. *A special "Resources" section lists books for additional reading and addresses of special-interest organizations.*

Bly, R. (1990). *Iron John.* Reading, MA: Addison-Wesley. An intriguingly different way to look at male development, told in the form of a Grimm brothers fairy tale.

Csikszentmihalyi, M. (1990). *Flow.* New York: Harper & Row. "This book summarizes, for a general audience, decades of research on the positive aspects of human experience." A fine example of optimal drive.

Erikson, E. (ed.) (1978). *Adulthood.* New York: Norton. A collection of essays on what it means to become an adult. Written by experts from a wide variety of fields.

As I see it, there are four definitions of "adult." First, the *biological* definition: we become adult biologically when we reach the age at which we can reproduce. . . . Second, the *legal* definition: we become adult legally when we reach the age at which the law says we can vote, get a driver's license, marry without consent. . . . Third, the *social* definition: we become adult socially when we start performing adult roles. . . . Finally, the *psychological* definition: we become adult psychologically when we arrive at a self-concept of being responsible for our own lives, of being self-directing. From the viewpoint of learning, it is the psychological definition that is most crucial. But it seems to me that the process of gaining a self-concept of self-directedness starts early in life, . . . and grows cumulatively as we become biologically mature, start performing adult-like roles, . . . and take increasing responsibility for making our own decisions. So we become adult by degree as we move through childhood and adolescence, and the rate of increase by degree is probably accelerated if we live in homes, study in schools, and participate in youth organizations that foster our taking increasing responsibilities. But most of us probably do not have full-fledged self-concepts of self-directedness until we leave school or college, get a full-time job, marry, and start a family. ■

The Adult Learner, Malcolm Knowles, 1989.

Although most adults probably believe that they know what adulthood is, the term is really somewhat difficult to define, as we saw in the preceding chapter. The four definitions mentioned in this quote from Malcolm Knowles are useful, but three others are also essential to an understanding of young adulthood: physical, cognitive, and sexual. In this chapter, we will concentrate on these three.

After reading chapter 16, you should be able to:

- Discuss how early adulthood signifies the peak in physical development.
- Define what is meant by "organ reserve."
- Suggest ways in which life-style choices, such as food, tobacco, and alcohol use and marriage, affect health.
- Discuss Gagnon and Simon's notion of "sexual scripts."
- List Mitchell's six major motives for human sexuality and contrast them with Wilson's sociobiological view.
- Describe young adults' premarital and marital sexual practices.
- List Perry's nine stages of intellectual/ethical development.
- Contrast women's and men's "ways of knowing."
- Identify Sternberg's seven forms of love.
- Define Fromm's notion of "validation" as it relates to love.

CHAPTER 16

Early Adulthood: Physical and Cognitive Development

Physical Development 432
The Peak Is Reached 432
Organ Reserve 433
The Effect of Life-Style on Health 433
Sexuality 439
Freud's Theory 439
Sexual Scripts: Gagnon and Simon 439
Sexual Motivations: Mitchell 440
The Sociobiological View: Wilson 441
Practices of Young Adults 442
Cognitive Development 444
Intellectual/Ethical Development 444
"Women's Ways of Knowing" 446
Physical and Cognitive Aspects of Love 448
The Seven Forms of Love: Sternberg 448
Validation: Fromm 449
Conclusion 450
Chapter Highlights 450
Key Terms 451
What Do You Think? 451
Suggested Readings 452

Physical Development

We address three main questions about physical development in this section: When is peak development reached? What is organ reserve? and What are the effects of life-style?

The Peak is Reached

Early adulthood is the period during which physical changes slow down or stop. Table 16.1 provides some basic examples of physical development in early adulthood.

Table 16.1 Summary of Physical Development in Early Adulthood

Height
Female: maximum height reached at age 18.
Male: maximum height reached at age 20.
Weight (age 20–30)
Female: 14–pound weight gain and increase in body fat.
Male: 15–pound weight gain.[a]
Muscle structure and internal organs
From 19–26: Internal organs attain greatest physical potential. The young adult is in prime condition as far as speed and strength are concerned.
After 26: Body slowing process begins.
Spinal disks settle, causing decrease in height.
Fatty tissue increases, causing increase in weight.
Muscle strength decreases.
Reaction times level off and stabilize.
Cardiac output declines.[b]
Sensory function changes
The process of losing eye lens flexibility begins as early as age 10 and continues until age 30. This loss results in difficulty focusing on close objects.
During early adulthood, women can detect higher-pitched sounds than men.
Nervous system
The brain continues to increase in weight and reaches its maximum potential by the adult years.

Sources: (a) U.S. Bureau of the Census (1986), (b) deVries (1981).

In sports, young adults are in their prime condition as far as speed and strength are concerned. A healthy individual can continue to partake in less strenuous sports for years. As the aging process continues, however, the individual will realize a loss of the energy and strength felt in adolescence.

Early adulthood is also the time when the efficiency of most body functions begins to decline. For example, cardiac output and vital capacity start to decrease (DeVries, 1981). As Table 16.2 clearly shows, these declines are quite steady right through old age. It is important to remember that this chart gives the average

Table 16.2 Approximate Declines in Various Human Functional Capacities with Age

	Percent of Function Remaining		
	30 Years	60 Years	80 Years
Nerve conduction velocity	100	96	88
Basal metabolic rate	100	96	84
Standard cell water	100	94	81
Cardiac index	100	82	70
Glomerular filtration rate	100	96	61
Vital capacity	100	80	58
Renal plasma flow	100	89	51
Maximal breathing capacity	100	80	42

Source: From Alexander P. Spence, *Biology of Human Aging,* © 1989, p. 8. Adapted by permission of Prentice-Hall, Inc., Englewood Cliffs, New Jersey.

changes; there will be considerable differences from individual to individual. (The greatest decline shown is for "male stated frequency of sexual intercourse." Why do you suppose that is?)

Organ Reserve

Although Table 16.2 makes it look all downhill from what is probably your present age, the actual experience of most people is not that bad. This is because of a human capacity called **organ reserve.** Organ reserve refers to that part of the total capacity of our body's organs that we do not normally need to use. Our body is designed to do much more than it is usually called upon to do. Much of our capacity is held on reserve. As we get older, these reserves grow smaller. The peak performance that each of our organs (and muscles, bone, etc.) is capable of declines. A 50–year-old man can fish all day with his 25–year-old son, and can usually take a long walk with him without becoming exhausted; but he has no chance at all of winning a footrace against him.

This is why people are aware of little decline during the early adult years, and often do not experience a sharp decline in most of their everyday activities even into middle age. Our organ reserves are diminishing, but we are unaware of it because we call on them so seldom.

Of course, there are those who regularly try to use the total capacity of their organ reserves. Professional athletes are an example. Here again we see the biopsychosocial model in action. Biology sets the limits, but psychological factors (e.g., personal pride) and social factors (e.g., the cheering crowd) determine whether the person "gives it her all."

The Effect of Life-Style on Health

Young adults are healthier than older adults in just about every way. All of the body's systems reach peak functioning at this age. There is less illness, too. For example, there are fewer hospitalizations and visits to the doctor's office than later, and those that do occur are caused mainly by injuries. Even catching a cold happens more rarely at this stage than at any other stage of life (U.S. Bureau of the Census, 1992).

Good health is clearly related to factors such as genetics, age, and the medical treatment locally available. But these factors are generally beyond the control of the individual. Increasingly, people are beginning to realize that their style of life plays an enormous role in their own health.

The impact of life-style on health is dramatically illustrated by the observations in the book *The Healing Brain* (Ornstein & Sobel, 1987). These researchers determined that the miraculous technological gains in medicine over the last one hundred years have not had as great an impact on health as one might think. It is true that at birth, we can expect many more years of life than people could in the last century. However, a person who has reached the age of 45 today has a life expectancy of only a couple years more than the person who had reached the age of 45 a hundred years ago, in spite of all the money and effort now spent on medicine after that age.

As a counterexample, Ornstein and Sobel offer the people of Nevada and Utah, two states and populations that are similar in geography, education, income, and availability of medical treatment. Yet Nevada's death rate is 40 percent higher than Utah's. Utah is largely composed of Mormons who live a relatively quiet and stable life-style, including very low incidences of smoking and drinking. In Nevada, people drink and smoke much more heavily, and it shows. The rate of cirrhosis of the liver and lung cancer are 100 to 600 percent higher in Nevada than in Utah. A man who reaches the age of 45 in Utah can expect to live 11 years more than the man who reaches age 45 in Nevada. The point is that simple, cost-free choices under the control of the individual are much more effective at improving health than all the expensive, time-consuming medical advances of the last century. Perhaps our priorities are misplaced.

Based on this information, we could conclude that a simple difference in style of life can have more of an effect on health than medical advances. Let's look at the influence on health of some specific life-style choices.

Choices of Foods

Nutrition plays an important role throughout human development, from the neonate to the elderly. By the time we reach middle adulthood, however, increasing evidence demonstrates the influence of nutrition on two major health concerns, heart disease and cancer. Medical science has recently established a link between heart disease and a substance in the blood, **cholesterol.** Cholesterol has been found to leave deposits along the walls of blood vessels, blocking the flow of blood to the heart and resulting in a heart attack. The main culprit in high levels of cholesterol has been found to be diets high in fat. Typically, Americans consume 40 percent of their total calories as fat. The American Heart Association (1984) recommends changes in diet, such as eating fish and poultry instead of red meat, yogurt and cottage cheese instead of cheese, margarine instead of butter, fewer eggs, and drinking skim milk instead of whole milk, as ways of lowering cholesterol in the body. Exercise is also very helpful.

Young adults today are probably more health-conscious than at any other time in history.

A similar link has been found between diet and certain types of cancer, such as cancer of the breast, stomach, intestines, and the esophagus. The American Cancer Society has also come out with a set of recommendations for an improved, healthy diet. They too recommend lowering fat intake, to no more than 30 percent of the daily caloric total, and a diet that is high in fiber, a substance that helps the digestive process. High-fiber foods include leafy vegetables such as cauliflower, broccoli, and brussels sprouts, as well as whole-grain cereals and breads. The proliferation of new products that feature lower levels of fat and higher levels of fiber indicates that the American public is taking this new knowledge to heart (no pun intended).

An Applied View — Dealing with Adult Children of Alcoholics (ACoA)

According to Woititz (1990), there is clear agreement on three aspects of alcoholism: Alcoholism runs in families; children of alcoholics run a higher risk of developing alcoholism than other children; children of alcoholics tend to marry alcoholics. She also finds that adult children of alcoholics usually manifest thirteen characteristics. They

- Are often not clear on what normal behavior is.
- Have difficulty following a project through from beginning to end.
- Lie when it would be just as easy to tell the truth.
- Judge themselves without mercy.
- Have difficulty having fun.
- Take themselves very seriously.
- Have difficulty with intimate relationships.
- Overreact to changes over which they have no control.
- Constantly seek approval and affirmation.
- Usually feel that they are different from other people.
- Are overly responsible or irresponsible.
- Are extremely loyal, even in the face of evidence that the loyalty is undeserved.
- Are impulsive. They tend to lock themselves into a course of action without giving serious consideration to alternative behaviors or possible consequences.

Woititz suggests that although these traits can make life difficult, it is important that the adult children of an alcoholic recognize that they are likely to possess most of them. "They could decide to work on changing aspects of themselves that cause them difficulty, or they could choose not to do so. In either event they have greater self-knowledge, which leads to greater self-understanding, which helps in the development of a sense of self. It is a win-win situation" (p. 98). She urges that: "Being the adult child of an alcoholic is not a disease. It is a fact of your history. Because of the nature of this illness and the family response to it, certain things occur that influence your feelings, attitudes and behaviors in ways that cause you pain and concern. The object of ACoA recovery is to overcome those aspects of your history that cause you difficulty today and learn a better way. The process of recovery for adult children is very disruptive. It means changing the way you have perceived yourself and your world up until now." (p. 101)

Anyone who is 10 percent over the normal weight for their height and build is considered overweight. Anyone who is 20 percent over the normal weight is considered obese. Obesity is now considered a very serious health concern. Over 30 million Americans can be labeled obese (National Institute for Health Statistics, 1986), and the impact on health is significant and widespread. Obesity increases the risk of such diseases as heart disease, diabetes, arthritis, and cancer. In addition, our society places great importance on physical appearance. Overweight persons are likely to suffer from low self-esteem or even depression because of the way they are treated by others.

At the other extreme, there is some disagreement over what is considered to be too thin. Some researchers suggest that it is better to be a little overweight than it is to be underweight. Recent research on animals suggests that a restricted diet that leaves the animal lean and thin might be healthiest (Masoro, 1984; Campbell & Gaddy, 1987). A severely restricted diet leads to some of the complications associated with eating disorders such as anorexia nervosa (See chap. 13). Once again, moderation is the safest course to follow.

White males consume more alcohol than either white females or African-Americans, according to one study. Can you think of any reasons why this might be so?

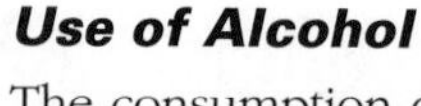

Use of Alcohol

The consumption of alcohol is another great health concern to our society. Alcohol abuse is estimated to cost our economy over $100 billion annually (Holden, 1987). This figure includes medical treatment for cirrhosis of the liver, osteoporosis, ulcers, heart disease, nervous system damage, and certain types of cancer such as breast cancer. It includes the insurance and medical costs incurred by automobile accidents resulting from drinking and driving. It includes the drug treatment necessary to help people control their addiction. And it includes the enormous cost of labor that is lost when heavy drinkers are unable to come in to work.

Ingalls (1983) conducted a survey among college students and reported that 82 percent consumed alcohol and 21 percent were heavy users. Eight percent admitted to having an alcohol problem. It was found that alcohol consumption was directly related to low grade-point average. White males consume more alcohol than women or African-Americans. Most people consume alcohol to attain the relaxed, uninhibited feeling that alcohol tends to produce. The fact is, alcohol dulls the senses. Specifically, it decreases reaction times in the brain and nervous system. Continued drinking may affect the sex life of males, by making it difficult for them to attain and keep an erection.

Alcohol leads to a variety of problems in the family, with the law, and with one's health. Physical problems associated with alcoholism include cirrhosis of the liver, increased changes of cardiomyopathy, a weakening of the heart muscles, stomach and intestinal ailments, as well as cancer of the mouth, liver, and esophagus (Schemeck, 1983).

Ironically, there is some indication that a daily, moderate intake of alcohol may be beneficial (although not for those over the age of 50—see chap. 20). Such small amounts of alcohol seem to produce a protein in the blood that helps lower cholesterol. Unfortunately, the addictive qualities of alcohol make it impossible for many people to drink in only moderate amounts. This is why treatment programs such as the very successful Alcoholics Anonymous ask their clients to abstain totally from alcohol rather than try to control their drinking.

Although the dangers of "passive smoke"—smoke inhaled by those in the vicinity of a smoker—are well established, many people still disregard the no-smoking signs that seem to be springing up everywhere.

Use of Tobacco

Tobacco use is also linked to a variety of health problems. The most common way to use tobacco is to smoke cigarettes. Most people are now aware that heavy cigarette smoking greatly increases the risk of lung cancer. They are perhaps less aware of the links between smoking and cancer of the kidney and stomach, along with the links to other diseases such as heart disease and emphysema (Engstrom, 1986).

The main culprit in tobacco is nicotine. Nicotine is a stimulating drug that the Surgeon General of the United States has concluded is as addictive as heroin or cocaine. There are other ways to be affected by nicotine than by smoking cigarettes. Recent attention has focused on the dramatic increase in the use of smokeless chewing tobacco, primarily in young men. This type of tobacco use leads to a higher risk of cancer of the larynx, mouth, esophagus, and stomach. Awareness of this type of health risk is quickly spreading, as people see news photos of teenage boys who have had to have entire portions of their face removed due to cancer.

Smokers are fourteen times more likely to die from cancer of the lungs, and twice as likely to die of a heart attack as nonsmokers. Other diseases linked to smoking are bronchitis, emphysema, and increased blood pressure. The Federal Office on Smoking and Health reports that 37.9 percent of American men and 29.8 percent of American women smoke. They report that 30.5 percent of men and 15.7 percent of women have quit smoking.

What's Your View?

?

In most articles you may have read about alcohol consumption, it has been associated with a variety of negative biological and psychological effects. Earlier research seemed to indicate that alcohol had no redeeming qualities.

More recently, studies have indicated that there may be some beneficial effects of alcohol, such as a decreased risk of heart attack. According to Criqui (1990), "The consistency of results across studies and the biological plausibility of alcohol's beneficial impact through high-density lipoprotein cholesterol and blood pressure suggest a potential benefit of several drinks per week" (p. 857).

Light social drinking may also encourage cross-cultural interactions. For example, vanWilkinson (1989) attempted to investigate the possible life-style influences on alcohol use among south Texas Mexican-Americans. He surveyed 247 respondents who were classified into six subgroups: working class, urban middle class, farmworkers, farmer/ranchers, migrants, and upper-class Mexican-Americans. All subgroups drank at home and at parties. "Almost all *pachangas* (parties) are held at someone's home and transcend any social class structure. Important for the rich as well as the poor, these events tend to make equals of those who might be unequals."

Do you think the consumption of alcohol can have positive physiological and social effects? What advice would you give to young adults about their drinking habits?

The Department of Health and Human Services has outlined the personality types of smokers versus nonsmokers (Brody, 1984). Adult smokers tend to be risk-takers, impulsive, defiant, and extroverted. Blue-collar workers tend to be the heaviest smokers among men. For females, white-collar workers are the heaviest smokers, and homemakers tend to smoke more than women who work.

Peer pressure from friends is the major reason that young adults smoke. As in adulthood, young people who smoke tend to be extroverted and more disobedient toward authority than nonsmokers. Some young adults see smoking as a way of appearing older and more mature. Young adults who go on to college smoke less than those who do not continue their education. Among young women, smokers are less athletic, more social, study less, get lower grades, and generally dislike school more than nonsmoking females.

You may also involuntarily be exposed to the dangers of tobacco use. A growing body of research is documenting the deleterious effects of passive smoking. Passive smoking is the breathing in of the smoke around you produced by others. For example, a nonsmoker who is married to a heavy smoker has a 30 percent greater risk of lung cancer than someone who is married to a nonsmoker (National Institute for Health Statistics, 1986). The role of smoke in the environment is also covered in chapter 4. There is also evidence that children can be affected when their mothers smoke. This mounting evidence has led to a flurry of legislation prohibiting smoking in certain public areas such as elevators, airplanes, and restaurants, and the designation of smoking areas in workplaces.

The message does seem to be reaching the public. Overall, fewer people smoke now than at any time in the past twenty-five years (with the exception of teenage females). Numerous stop-smoking programs spring up all the time. These programs run the gamut from classic behavioral techniques to hypnosis. But perhaps the most telling evidence that smoking is on the decline in the United States is the reaction of the giant tobacco companies. In recent years these companies have increasingly targeted foreign markets (a growing ethical controversy, since these markets are typically Third-World countries) while diversifying their domestic market with different and healthier products.

Physical Fitness

One of the most popular trends of recent years has been the so-called fitness craze. The signs are all around us: gyms and spas opening everywhere; exercise classes at home on the TV or videocassette, or perhaps even at the workplace; designer clothes and shoes for working out. People in this country are trying to change the old notion of the lazy, fat American.

Health benefits are the obvious reason for this enthusiasm for exercise. Regular, strenuous exercise can increase heart and lung capacity, lower blood pressure, decrease cholesterol in the blood, keep weight at normal levels, enhance cognitive functioning, relieve anxiety and depression, and increase self-esteem (Elsayed, Ismail & Young, 1980; Lee & others, 1981; McCann & Holmes, 1984).

During a physical workout, oxygen travels more deeply into the lungs, and the heart pumps harder to carry more blood into the muscles. Healthy lungs and heart are vital for a long life. Persons who do not exercise and have inactive jobs are at twice the risk for a heart attack compared to persons who do exercise. Individuals who exercise report other health benefits, including better concentration at work and better sleep patterns (VanderZanden, 1989).

Johns Hopkins Hospital in Baltimore reported that a family medical history of heart attacks increases a person's chances of having a heart attack (Findlay, 1983). Among individuals under age 50 from at-risk families, 42 percent have significantly high blood pressure, compared with 20 percent of the general population. Of the same group, 28 percent have dangerously high cholesterol levels, compared to 5–10 percent of the general population. Nearly 25 percent show evidence of "silent" coronary artery disease—double the rate of the general population.

There was a time when only males did calisthenics. Now exercise classes are becoming very popular with female adolescents.

Studies at Harvard show that at-risk individuals do not properly metabolize cholesterol, which collects in the arteries and eventually leads to a heart attack (Bishop, 1983). Carey (1983) explains that young adults who gain a large amount of weight after adolescence are more at risk for heart disease. Scientists feel that with greater public awareness of the dangers of cholesterol, and early detection and treatment, they can decrease heart disease among at-risk individuals.

One study (Ossip-Klein & others, 1989) assigned a number of clinically depressed women to one of three groups: a group that did regular running, a group that lifted weights regularly, or a control group that did nothing special. The women in both exercise groups showed increased self-concept over the women in the control group. The two types of exercise worked equally well.

Many corporations are now providing the time and facilities for employees to build regular exercise into their workday. The reasoning is that work time lost in this way is more than made up for by more productive employees, who end up losing less time due to illness. In one new trend, some health insurance programs are offering free access to exercise facilities, again looking at the long-term gain of having healthier members.

Marital Status

Another life-style factor that appears to affect health is marital status. Despite comedians' jokes, married people seem to be healthier than single, divorced, or widowed people (Verbrugge, 1979). They tend to have less frequent and shorter stays in hospitals. They tend to have fewer chronic conditions and fewer disabilities. Never-married and widowed people are the next healthiest. Divorced and separated people show the most health-related problems. There are a number of possible explanations for this, but clearly one must consider the emotional and economic support that an intact family can provide.

Based on the preceding information, we could conclude that a healthy life-style consists of a diet with reasonable caloric intake that is low in fat and high in fiber, moderate or no consumption of alcohol, no tobacco use, plenty of regular, strenuous exercise, and a supportive spouse. In fact, this accurately describes the Mormon life-style in Utah, which allows the average resident of Utah to live years longer than the average resident of Nevada. This, of course, is no reason to go out and change your religious preference, but it is reason to examine your current life-style and consider some prudent changes.

Health is also an aspect of life in which all three factors of the biopsychosocial model are clearly evident, as you can see. Now let us turn to another complex area, sexuality.

Most people seem to find a strong need to develop a sexual and romantic relationship with another person.

SEXUALITY

Sexuality has important physical and cognitive roles in our lives. Most people who disregard or repress their sexuality suffer for it. Of course, the sex drive, unlike other instinctual drives such as hunger and thirst, can be thwarted without causing death. Some people are able to practice complete chastity without apparent harm to their personalities. The great majority of us, however, become highly irritable when our sexual needs are not met in some way. There is also reason to believe that our personalities do not develop in healthy directions if we are unable to meet sexual needs in adulthood. Why is it so important to us? Although the answer may seem obvious, there are in fact a number of sexual motivations.

Freud's Theory

For Freud, human sexuality is the underlying basis for all behavior (See chap. 2). He held that genitality (the ability to have a successful adult sex life) is the highest stage of development. He believed that those people who do not have adequate adult sex lives fail to do so because they have become fixated at some earlier and less mature level.

Because sexuality is so important and controlled by such deep-seated genetic forces, those who do not fulfill their sexual needs are likely to suffer from mental illness. Women, he believed, are more vulnerable than men to these problems, and he thought his medical practice confirmed his beliefs.

Many researchers (e.g., Chilman, 1974; Masters & Johnson, 1970, 1979) have disputed this view. For example, it has been found that women who have physiological difficulties with their reproductive systems, such as infertility or menstrual disorders, are no more likely than other women to suffer from neurosis. Most probably Freud's theories were influenced by the attitudes and morals in Victorian Vienna, which were quite different from those in the United States today.

Psychologist John Gagnon argues that humans learn to behave sexually according to scripts established by the culture in which they live. For most people, this learning takes place during adolescence and young adulthood.

Sexual Scripts: Gagnon and Simon

Psychologists John Gagnon and William Simon (1987) see sexual behavior as "scripted" behavior. They believe that children begin to learn scripts for sexual attitudes and behavior from the other people in their society. They view sexuality, therefore, as a cultural phenomenon rather than a spontaneously emerging behavior. They believe that

> ***in any given society, [children] acquire and assemble meanings, skills, and values from the people around them. Their critical choices are often made by going along and drifting.*** **(1987, p. 2)**

Learning sexual scripts occurs in a rather haphazard fashion throughout childhood, but this changes abruptly as children enter adolescence. Here much more specific scripts are learned: in classrooms, from parents, from the media, and most specifically, from slightly older adolescents. This process continues throughout early adulthood, through a process of listening to and imitating older adults whom the individual admires. For these researchers, then, the roles sexuality plays in a society come about largely through transmission of the culture by idealized adults.

Sexual Motivations: Mitchell

In a highly respected article, psychiatrist John Mitchell (1972) has suggested the following six major motives for human sexuality:

- *The need for intimacy.* The need for intimacy is one of the deepest that humans experience. It often conflicts with other needs, such as independence and self-protection, but if unmet it can cause intense depression. In the past, sexual interaction usually took place only when two people had achieved a high degree of intimacy.
- *The need of belonging.* The vast majority of adults are able to satisfy their sexual needs with one other person. When a person is unable to do so, it tends to strengthen the unpleasant feelings of aloneness and being different.
- *The desire for power.* Both sexes have a need to feel they are able to exercise some control over another. Sometimes this is expressed as domination, sometimes as manipulation. Its more mature form is the feeling of personal importance each partner can achieve by giving the other partner satisfaction through love-making.
- *The desire for submission.* Submission is the complementary need to the need for power. Just as we sometimes like to be in control, at times we like to be cared for by someone we feel is more powerful than we are.
- *Curiosity.* When their most deep-seated needs are met, most human beings have a desire to explore their environment and their capabilities in dealing with it. It is natural, therefore, for healthy adults to want to find out more about their sexual feelings and to try to increase their competence.
- *The desire for passion and ecstasy.* The Greek word *ekstasis* means "to be outside of oneself." Medical researchers have found that people occasionally need to experience self-transcendence, the feeling of getting outside of and rising above themselves—in other words, a "high." Sebald (1977) suggests that three benefits derive from sexual passion: intense self-awareness, intense awareness of the other person, and a confirmation of the other person as someone who is intensely important to you.
- *The need for socially approved playfulness.* Mitchell does not suggest this seventh motive, but there is good evidence for it (Dacey, 1989a). Although society tends to equate childlike playfulness with being irresponsible and foolish, most of us who are no longer children like to go back to that relaxed and happy period, at least for brief intervals. In sexual play, we can act silly, use baby talk, and just fool around. This kind of activity can leave us better prepared to return to the heavier burdens of adult responsibility.

The Sociobiological View: Wilson

Whatever we see as our reasons for seeking sex, this complicated behavior has obviously evolved through a number of complex stages over which we have had no control. What genetic purposes has the evolution served? E. O. Wilson (1978) and his fellow sociobiologists have a number of unorthodox suggestions:

- *Sex is not designed for reproduction.* Sociobiologists argue that the primary motivation of all human behavior is the reproduction of the genes of each person. Wilson, however, believes that if reproduction were the primary goal of the human species, many other techniques would have been far more effective:

Bacteria simply divide in two (in many species, every twenty minutes), fungi shed immense numbers of spores, and hybrids bud offspring directly from their trunks. Each fragment of a shattered sponge grows into an entire new organism. If multiplication were the only purpose of reproductive behavior, our mammalian ancestors could have evolved without sex. Every human being might be asexual and sprout new offspring from the surface cells of a neutered womb. **(p. 121)**

- *Sex is not designed for giving and receiving pleasure.* Wilson points out that many animal species perform intercourse quite mechanically with virtually no foreplay. Furthermore, lower forms achieve sex without benefit even of a nervous system. Thus, he suggests that pleasure is only one of the means of getting complex creatures to "make the heavy investment of time and energy required for courtship, sexual intercourse and parenting" (p. 122).
- *Sex is not designed for efficiency.* The very complexity of human genital systems makes them subject to a variety of disorders and diseases, such as ectopic pregnancy and venereal disease. The genetic balance brought about by sex is easily disturbed, and if the human being has one sex chromosome too many or too few, abnormalities in behavior and in physiology often result.
- *Sex is not designed to benefit the individual's drives.* If an individual's drive is to reproduce himself or herself, sex is actually an impairment. When sexual reproduction is employed, the organism must accept partnership with an individual whose genes are different. Only with asexual reproduction is multiplication of self totally possible.
- *Sex* does create diversity. Wilson concludes that the only possible reason that evolution brought about the human sexual system is to create a greater diversity of individuals. The purpose of this diversity is to increase the chances of the survival of the species. As conditions have changed throughout history (e.g., during an ice age), some individuals have had a greater chance to survive than others. If there had been only one type of human being with only one set of genes, and that inheritance had not been suited to the changing environment, the species would have become extinct.

Wilson points out that when two different individuals mate, there is the possibility of offspring like individual A, offspring like individual B, and offspring with the characteristics of both. As these individuals mate with others, the possible variations increase exponentially. Diversity, brought about through sex, leads to adaptability of the species, and therefore its greater survival rate.

What's Your View?

?

As you might imagine, the whole theory of sociobiology (which is also covered in other sections of this book) is not without its critics. Obviously, genes do affect our behavior, but not nearly as much as they do in the other species of the animal kingdom. Many experts say that sociobiology has made a valuable contribution to the theory of human sexuality, but that it overemphasizes the role of genes.

We find that many of our students reject sociobiology completely. "Wilson treats people as though we are no better than animals," they say. What do you think? Is there a place for sociobiology in your understanding of human sexuality? If so, what is it?

Why are there not hundreds of sexes instead of just two? Wilson argues that two sexes are enough to create tremendous diversity while keeping the system as simple as possible. Diversity may not only be responsible for variation in the offspring, but contribute to the wide range of values people place on sexuality.

In a recent article, Weinrich (1987) extends the sociobiological theory to explain homosexuality. He suggests that homosexual behavior is the result of a "reproductively altruistic trait." That is, homosexuals give up their right to reproduce their genes. The theory, Weinrich qualifies, is applicable only in societies like ours that do not strictly require marriage and reproductivity.

Practices of Young Adults

Two options are covered in this section: premarital experiences and marital practices.

Premarital Experiences

A number of findings of the new Janus Report (1993; See chap. 14) are relevant to the premarital behavior and attitudes of young adults. Here are some of the most important findings:

- Over 80 percent of respondents were seriously concerned about sexually transmitted diseases, but most respondents reported *increased* rather than decreased sex activity in the past three years.
- Although more caution is being exercised by many singles, more men than women reported they had become more cautious about sex in the past three years.
- Nineteen percent of the single men and 23 percent of the single women surveyed reported using no contraception.
- The majority of singles do not find their life-style gratifying, but only one in three would prefer being married.
- Thirty-eight percent of single men and 45 percent of single women would like to become parents even if they do not marry.
- Middle-income women had had less premarital sexual experience than either low- or high-income women.
- For both men and women, premarital sex experience has increased: from 48 percent to 55 percent for men, and from 37 percent to 47 percent for women.

- Abortion has become more acceptable in the intervening years between phase 1 (1985) and phase 2 (1992)—from 36 percent to 30 percent agreement that abortion is murder, and from 29 percent to 53 percent disagreement.

We have considered evidence that sexual experience has increased among college students. For example, in their study of 793 undergraduates, Earle and Perricone (1986) found "significant increases in rates of premarital intercourse, significant decreases in age at first experience, and significant increases in number of partners" (p. 304).

Kinnaird and Gerrard (1986) examined the premarital sexual activity of unmarried female undergraduates. Young women from divorced and reconstituted families reported significantly more sexual behavior than those from intact families. Family conflict, disruption, and father-abuse were also related to such behavior.

There have been many studies of the premarital experiences of college students, but few studies of persons who have not gone to college. We know from the classic studies that the noncollege educated are almost always more conservative (that is, less experienced) than those who go to college (Kinsey & others, 1948, 1953; Masters & Johnson, 1966, 1970), but that they are subject to the same kinds of societal influence. Therefore we can expect that while noncollege populations have a lower experience rate, theirs, too, is considerably higher than it used to be.

A study of university students by Story (1982) sought a comparison of various sexual outlets (such as masturbation, premarital intercourse with other than future marriage partner, group sex, and sexual experience with person of same sex) over a six-year span. Fifty single males and fifty single females were tested in 1974, and another group of the same size was tested in 1980. Results showed that both males and females in the group tested in 1980 appeared to adhere more to society's conservative or traditional sexual behavior than the group tested in 1974.

Marital Intercourse

Intercourse between a husband and wife is the only type of sexual activity totally approved by American society. Much is expected of it, and when it is unsatisfactory, it usually generates other problems in the marriage (McCary, 1978). Sexual closeness also tends to lessen significantly with the birth of each child, except in cases where the couple has taken specific steps to maintain the quality of their sexual relationship. Two studies (Whitehead & Mathews, 1986) learned that when young couples who are having sexual difficulties regularly attend therapy sessions, they are usually able to resolve their problems. Interestingly, those couples who received placebos (sugar pills) or small doses of testosterone did better than those who did not. This indicates how important the mind is in the area of sexuality.

A number of findings of the new Janus Report (1993; See chap. 14) are relevant to the marital behavior and attitudes of young adults. Among the most important findings are:

- Women who lived with their spouses before marriage are more likely to be divorced than women who didn't; men, less likely. Couples still married had lived together, on the average, for a shorter period of time before marriage than those who are now divorced.
- Among the divorced, men cite sexual problems as the primary reason for the divorce three times more frequently than women. Women cite extramarital affairs twice as frequently as men. But both cite emotional problems as the most frequent cause of divorce.
- Among the divorced respondents, 39 percent of the men and 27 percent of the women reported using no contraception.

Cognitive Development

As with physical development, there has not been a great deal of research on the cognitive development of young adults. Possibly this is because cognitive functioning appears to peak during this period, and so there is less concern over change. A major emphasis of research has been on the relationship between intellectual and ethical growth.

Intellectual/Ethical Development

Perry (1968a, 1968b; 1981) studied the intellectual/ethical development of several hundred Harvard College students, a group of males ages 17 to 22. These students responded to several checklists on their educational views and were interviewed extensively on the basis of their responses. The results of these studies led Perry to suggest a sequence of intellectual and ethical development that typically occurs during the transition from late adolescence to early adulthood. This sequence consists of nine positions, which indicate progress from belief in the absolute authority of experts to the recognition that one must make commitments and be responsible for one's own beliefs.

Perry's nine stages are divided into three broader categories, as follows:

I. **Dualism** ("Things are either absolutely right or absolutely wrong.")

- *Position 1:* The world is viewed in such polar terms as right versus wrong, we versus they, and good versus bad. If an answer is right, it is absolutely right. We get right answers by going to authorities who have absolute knowledge.

- *Position 2:* The person recognizes that uncertainty exists, but ascribes it to poorly qualified authorities. Sometimes individuals can learn the truth for themselves.

- *Position 3:* Diversity and uncertainty are now acceptable, but considered temporary because the authorities do not know what the answers are yet. The person becomes puzzled as to what the standards should be in these cases.

II. **Relativism** ("Anything can be right or wrong depending on the situation; all views are equally right.")

- *Position 4a:* The person realizes that uncertainty and diversity of opinion are often extensive and recognizes that this is a legitimate status. Now he or she believes that "anyone has a right to an opinion." It is now possible for two authorities to disagree with each other without either of them being wrong.

- *Position 4b:* Sometimes the authorities (such as college professors) are not talking about right answers. Rather, they want students to think for themselves, supporting their opinions with data.

- *Position 5:* The person recognizes that all knowledge and values (including even those of an authority) exist in some specific context. It is therefore relative to the context. The person also recognizes that simple right and wrong are relatively rare, and even they exist in a specific context.

- *Position 6:* The person apprehends that because we live in a relativistic world, it is necessary to make some sort of personal commitment to an idea or concept, as opposed to looking for an authority to follow.

III. **Commitment** ("Because of available evidence and my understanding of my own values, I have come to new beliefs.")

- *Position 7:* The person begins to choose the commitments that she or he will make in specific areas.
- *Position 8:* Having begun to make commitments, the person experiences the implications of those commitments and explores the various issues of responsibility involved.
- *Position 9:* The person's identity is affirmed through the various commitments made. There is a recognition of the necessity for balancing commitments and the understanding that one can have responsibilities that are expressed through a daily life-style. Perry (1981) describes this position:

> ***This is how life will be. I will be whole-hearted while tentative, fight for my values yet respect others, believe my deepest values right yet be ready to learn. I see that I shall be retracing this whole journey over and over—but, I hope, more wisely.*** **(p. 276)**

Life is full of situations in which the difference between right and wrong is far from clear.

Some students move through these stages in a smooth and regular fashion; others, however, are delayed or deflected in one of three ways:

- **Temporizing.** Some people remain in one position for a year or more, exploring its implications but hesitating to make any further progress.
- **Escape.** Some people use opportunities for detachment, especially those offered in positions 4 and 5, to refuse responsibility for making any commitments. Since everyone's opinion is "equally right," the person believes that no commitments need be made, and thus escapes from the dilemma.
- **Retreat.** Sometimes, confused by the confrontation and uncertainties of the middle positions, people retreat to earlier positions.

Some have criticized Perry's theory (see Kitchener & King, 1981; Brabeck, 1984). For example, it should be remembered that the subjects of his research were all males. Now let us turn to research on females that was spurred by Perry's work.

"Women's Ways of Knowing"

In a collaborative study, Belenky & associates (1986) set out to answer the question, "Do female ways of knowing develop differently from those of males? If so, how do they come to learn and value what they know?" The study was rooted in Perry's work and the work of Carol Gilligan, whose ground-breaking research on the morality of care and responsibility versus the morality of rights and justice was covered in chapter 9.

Belenky and her associates conducted a series of lengthy and intense interviews with 135 women of diverse socioeconomic background. They found five general categories of ways in which women know and view the world. While some of the women interviewed clearly demonstrated a progression from one perspective to the next, the researchers contend that they are unable to discern a progression of clear-cut stages, as did Perry and Gilligan. The five perspectives are silence, received knowledge, subjective knowledge, procedural knowledge, and constructed knowledge.

1. *Silence.* Females in the **silence** category describe themselves as "deaf and dumb." These women feel passive and dependent. Like players in an authority's game, they feel expected to know rules that don't exist. These women's thinking is characterized by concepts of right and wrong, similar to the men in Perry's first category of dualism. Questions about their growing up revealed family lives filled with violence, abuse, and chaos. The researchers noted that "gaining a voice and developing an awareness of their own minds are the task that these women must accomplish if they are to cease being either a perpetrator or victim of family violence" (p. 38).
2. *Received knowledge.* Women in the **received knowledge** category see words as central to the knowing process. They learn by listening, and assume truths come from authorities. These women are intolerant of ambiguities and paradoxes, always coming back to the notion that there are absolute truths. Received knowers seem similar to the men that Perry described as being in the first stage of dualism, but with a difference. The men interviewed by Perry felt a great affiliation with the knowing authority. The women of this perspective were awed by the authorities, but far less affiliated with them. In contrast to the men of Perry's study, women of received knowledge channel their energies and increased sense of self into the care of others.
3. *Subjective knowledge.* The researchers noted that women in the **subjective knowledge** category often had experienced two phenomena that pushed them toward this perspective: some crisis of male authority that sparked a distrust of outside sources of knowledge, and some experience that confirmed a trust in themselves. Subjectivists value their "gut" or firsthand experience as their best source of knowledge and see themselves as "conduits through which truth emerges" (p. 69). The researchers note that subjectivists are similar to males in Perry's second category of relativism in that they embrace the notion of multiple truths.
4. *Procedural knowledge.* The women in the **procedural knowledge** category have a distrust of both knowledge from authority and their own inner authority or "gut." The perspective of procedural knowledge is

characterized by an interest in form over content (how you say something rather than what you say). They also have a heightened sense of control. This category is similar to Perry's position 4b, where students learn analytic methods that authorities sanction. But it emerges differently in women because they don't affiliate with authorities.

The researchers describe women as having two kinds of procedural knowledge: separate knowing and connected knowing. The terms are reminiscent of Gilligan's work. Separate knowers are analytical and try to separate the self, to reveal the truth. Connected knowers learn through empathy with others.

5. *Constructed knowledge.* Those in the **constructed knowledge** category have integrated the subjective and procedural ways of knowing (types 3 and 4). Women of this perspective note that "all knowledge is constructed and the knower is an intimate part of the known" (p. 137). They feel responsible for examining and questioning systems of constructing knowledge. Their thinking is characterized by a high tolerance of ambiguity and internal contradiction. Indeed, the women whose ways of knowing are of this perspective often balance many commitments and relationships, as well as ideas.

The work of Perry and Belenky and her associates (as well as that of Piaget, Kohlberg, and Gilligan, discussed earlier in this book) has greatly advanced our knowledge of intellectual and ethical development in the late adolescent and early adult years. It has also produced much controversy. Many questions remain to be answered. For example, does socioeconomic level make any difference? What about cultural background? We hope that the research in this area will provide further insights into how we can help youth progress through this period successfully. We could say the same for the next aspect of development we will review, love in young adulthood.

WHAT'S YOUR VIEW?

?

In the following table, you can see that a larger percentage of young adults aged 25 to 29 have a high school diploma and have four or more years of college than do all adults 25 and older. Why do you suppose this is true? Another interesting fact found in this table is that although educational level in general has risen steadily since 1940, there has been a *decrease* in high school graduation among the 25–29–year-olds in recent years. To what do you attribute this surprising finding?

Level of School Completed

	% Ages 25 to 29		% Ages 25 and over	
Year	H.S. or more[1]	4 or more years of college	H.S. or more[1]	4 or more years of college
1940	38.1	5.9	24.5	4.6
1950	52.8	7.7	34.3	6.2
1960	60.7	11.0	41.1	7.7
1970	75.4	16.4	55.2	11.0
1980	85.4	22.5	68.6	17.0
1986	86.1	22.4	74.7	19.4
1987	86.0	22.0	75.6	19.9
1988	85.9	22.7	76.2	20.3
1989	85.5	23.4	76.9	21.1

1. Includes recipients of high school equivalency certificates

Source: National Center for Education Statistics (1992), p. 8.

Physical and Cognitive Aspects of Love

The words written about love over the course of human history are uncountable. In this book, we will limit ourselves to describing the developmental aspects of this emotion: the seven forms of love suggested by psychologist Robert Sternberg, and the definition of the essence of love offered by psychoanalyst Erich Fromm.

The Seven Forms of Love: Sternberg

Sternberg (1986) argues that love is made up of three different components:

- **Passion.** A strong desire for another person, and the expectation that sex with them will prove physiologically rewarding.
- **Intimacy.** The ability to share one's deepest and most secret feelings and thoughts with another.
- **Commitment.** The strongly held conviction that one will stay with another, regardless of the cost.

Each of these components may or may not be involved in a relationship. The extent to which each is involved defines the type of love that is present in the relationship. Sternberg believes that the various combinations actually found in human relations produce seven forms of love (see Table 16.3).

Table 16.3 Sternberg's Seven Forms of Love

Linking	Intimacy, but no passion or commitment
Infatuation	Passion, but no intimacy or commitment
Empty love	Commitment, but no passion or intimacy
Romantic love	Intimacy and passion, but no commitment
Fatuous love	Commitment and passion, but no intimacy
Compationate love	Commitment and intimacy, but no passion
Consummate love	Commitment, intimacy, and passion

Source: From R. J. Sternberg, "The Triangular Theory of Love" in *Psychological Review,* 93:199–235.

This is not to say that the more of each, the better. A healthy marriage will usually include all three, but the balance among them is likely to change over the life of the marriage. For example, early in the marriage, passion is likely to be high relative to intimacy. The physical aspects of the partnership are new, and therefore exciting, while there has probably not been enough time for intimacy to develop fully. This is a dangerous time in the marriage, because when intimacy is moderate, there are many situations in which the couple may misunderstand each other, or may make unpleasant discoveries about each other. Such problems are often much more painful than they are later, when deeper intimacy and commitment have developed.

Passion, Sternberg states, is like an addiction. In the beginning, the smallest gesture can produce intense excitement. As the relationship grows older, however, it takes larger and larger "doses" to evoke the same feelings. Inevitably, passion loses some of its power.

Of course, there are wide differences among couples (Traupmann & others, 1981; Hatfield, Traupmann & Sprecher, 1984). Some never feel much passion, while others maintain at least moderately passionate feelings into old age. Some appear to have strong commitments from the earliest stage of their association (love at first sight?), while others waver for many years.

Sternberg's theory has numerous implications for couples, and for marriage therapists as well. For example, more and more couples are engaging in premarital counseling. In this, they analyze with their counselor the three factors of love, and how each person feels about them. This often helps them to avoid later problems, and to get their relationship off to a good start. For some, it provides information that makes them realize that while their passion is high, their intimacy and commitment may not be, and they wait until these develop, or decide not to get married at all. It is hoped that such counseling will bring about a decrease in our nation's high divorce rate.

Validation: Fromm

In his highly enlightening book on this subject, *The Art of Loving* (1968), Erich Fromm has given us a highly respected understanding of the meaning of love. First, he argues, we must recognize that we are prisoners in our own bodies. Although we assume that we perceive the world around us in much the same way as others, we cannot really be sure. We are the only one who truly knows what our own perceptions are, and we cannot be certain they are the same as other people's. In fact, most of us are aware of times when we have *misperceived* something: We heard a phrase differently from everyone else; had an hallucination when under the influence of a fever, alcohol, or a drug; and so on.

For more and more engaged couples, premarital counseling is becoming a part of their plan to marry.

Thus we must constantly check on the reality our senses give us. We do this thousands, maybe millions, of times every day. Let us give you an example. We assume you are sitting or lying down while you are reading this book. Did you make a conscious check of the surface you are sitting or lying on when you got on it? Probably not. Nevertheless your brain did. You know that if it had been cold, sharp, or wet, you would have noticed. That it is none of these things is something your unconscious mind ascertained without your having to give it a thought.

With some insane people, this is the major problem. Their "reality checker" isn't working right. They cannot tell fact from imagination. They dwell in "castles in the air," out of contact with the real world. We need the feedback from all our senses, doing repeated checks at lightning speeds, in order to keep in contact with reality.

Fromm's point is this: As important as these "reality checks" on our physical environment are, how much more important are the checks on our innermost state—our deepest and most important feelings and thoughts! To check on the reality of these, we must get the honest reactions of someone we can trust. Such a person tells us, "No, you're not crazy. At least I feel the same way, too!" Even more important, they *prove* their insight and honesty by sharing with us their own secret thoughts and feelings. In Fromm's words, they give us **validation.**

Validation is essential to our sanity. We are social animals, and we need to know that others approve of us (or, for that matter, when they don't). When someone regularly makes you feel validated, you come to love them. This is the essence of what Erikson calls intimacy, which we will discuss at length in the next chapter. It fulfills what Maslow calls the need for self-esteem.

It is no accident that the first person outside of our family who validates us is almost always a person of the same gender. It usually happens during adolescence, with your "best girlfriend" if you are female, or "best buddy" if you are male. This first intimate relationship is usually with a person of the same gender because the risk of them misunderstanding you is lower than in a relationship with the opposite gender.

There is, however, great risk in receiving validation. The person who gives it to you is able to do so only because you have let them in on your deepest secrets. This gives them great power, for good or for ill. Because they know you and your insecurities so well, they have the capability to hurt you horrendously. This is why many divorces are so acrimonious. No one knows how to get you better than a spouse with whom you have shared so many intimacies. This is why it is said that "There is no such thing as an amicable divorce."

Nevertheless, we truly need love and the validation that leads to it. As studies of mental illness make clear, those who try to live without love risk their mental health.

Conclusion

Young adulthood is an exciting period in life, during which many changes are taking place. The peak of physical development is reached, and the decline of certain abilities begins. These are almost never apparent, however, because of organ reserve. Our style of life has a powerful effect on this development, including our diet, use of alcohol, drugs and nicotine, and marital status.

There are several quite different explanations of why sexuality develops as it does, and a number of important changes in the current sexual practices of young adults. The major change occurring in the area of cognitive development involves the relationship between intellectual and ethical growth. Love, too, may be variously explained. We looked at the theories of Sternberg and Fromm.

Physical and mental development form the basis of our interactions with the world around us. These interactions, which we refer to as psychosocial development, are covered in the next chapter.

Chapter Highlights

Physical Development

- Early adulthood is the period during which physical changes slow down or stop.
- The human body is designed to do much more than it is usually called upon to do. Much of its total capacity is held in "organ reserve."
- Life-style, food choices, alcohol and tobacco use, physical fitness, and marital status all play an enormous role in the health of a young adult.

Sexuality

- Freud believed that human sexuality is the underlying basis for all behavior.
- Gagnon and Simon see sexual behavior as culturally derived, "scripted" behavior.
- Mitchell suggests six major motives for human sexuality, including the need for intimacy or belonging, the desire for power, submission, and passion, and curiosity. Dacey suggests a seventh motive, socially approved playfulness.

- Wilson concludes that the only possible reason that evolution brought about the human sexual system was to create a greater diversity of individuals in a species.
- Sexual experience has increased among college students.
- Intercourse between a husband and wife is the only type of sexual activity totally approved by American society.

Cognitive Development

- Perry suggested three categories of intellectual and ethical development that typify transition from late adolescence to early adulthood: dualism, relativism, and commitment. By avoiding the three obstacles that might impede their progress (temporizing, escape, or retreat), young adults should be able to achieve commitments that are the hallmark of the mature person.
- Belenky and others suggest there are five perspectives by which women know and view the world: silence, received knowledge, subjective knowledge, procedural knowledge, and constructed knowledge.

Physical and Cognitive Aspects of Love

- Sternberg argues that love is made up of three different components: passion, intimacy, and commitment.
- Fromm states that people need to have their deepest and most important thoughts and feelings "validated" by others who are significant to them.

Key Terms

Cholesterol
Commitment (Perry's term)
Commitment (Sternberg's term)
Constructed knowledge
Dualism
Escape
Intimacy
Organ reserve
Passion
Procedural knowledge
Received knowledge
Relativism
Retreat
Silence
Subjective knowledge
Temporizing
Validation

What Do You Think?

1. What would life be like if we did not reach the peak of our physical abilities in early adulthood?
2. How would you describe your style of life? According to the research you have read about in this chapter, would you say you are more like or unlike the average American?
3. Make a list of ten of your best friends. In which of Perry's three categories would you place each of them?
4. Do you agree that there are major differences in the ways males and females view ethical issues? In what ways?
5. Are you clear on your sexual values? Could you state your principles with precision?
6. Do you agree with Wilsons' sociobiological theory? Why or why not?
7. How can you tell if you are truly in love?

Suggested Readings

Fromm, E. (1968). *The art of loving.* New York: Harper & Row. A highly readable classic in the field. You will never think about love the same way after reading this book.

Geer, J., J. Heiman, and H. Leitenberg (1984). *Human sexuality.* Englewood Cliffs, NJ: Prentice-Hall. This impressively comprehensive book covers virtually all aspects of sex with a depth of understanding.

Lawrence, D. H. [1920] (1976). *Women in love.* New York: Penguin. This novel probes the relationships between two sisters and their lovers. It offers timeless examination of the many aspects of adult interactions. Another of his books, *Sons and Lovers,* is also magnificent on this subject.

May, R. (1975). *The courage to create.* New York: Norton. A remarkably insightful psychoanalyst, May brings to this book his many years of experience helping highly creative people deal with their many stresses.

Peck, Scott M. (1978). *The road less traveled.* New York: Simon & Schuster. "Confronting and solving problems is a painful process, which most of us attempt to avoid. And the very avoidance results in greater pain and the inability to grow both mentally and spiritually." Drawing heavily on his own professional experience, Dr. Peck, a practicing psychiatrist, suggests ways in which confronting and resolving our problems—and suffering through the changes—can enable us to reach a higher level of self-understanding.

Sternberg, R. J. (1986). The triangular theory of love. *Psychological Review 93,* 129–35. This article, too, offers a penetrating view of this most elusive topic.

Woititz, J. (1990). *Adult children of alcoholics.* Lexington, MA: Health Communications, Inc. Janet Woititz describes the characteristics of the adult children of alcoholics, but insists that these are not character defects. "It is my belief that knowledge is freedom and that those who identify can now have new choices. They can decide to work on changing aspects of themselves that cause them difficulty, or they can choose not to do so." This book provides readers with basic tools that will enable them to achieve greater self-knowledge and understanding.

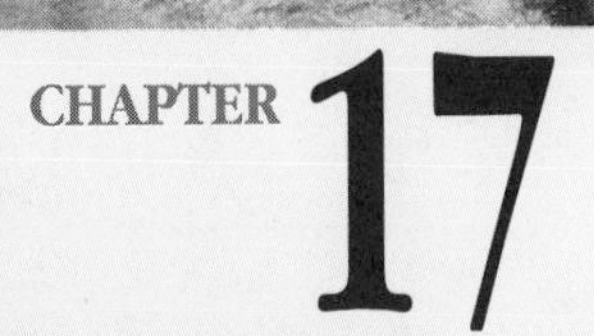

CHAPTER 17

Early Adulthood: Psychosocial Development

Marriage and the Family 454
Changing American Marriages and Families 454
Types of Marriage 457
Patterns of Work 458
Employment Patterns 459
How People Choose Their Careers 459
The Phenomenon of the Dual-Career Family 460
Stages of Personal Development 462
Transformations: Gould 462
The Adult Life Cycle: Levinson 466
Seasons of a Man's Life: Levinson 468
Intimacy versus Isolation: Erikson 470
Male versus Female Identity 471
Conclusion 474
Chapter Highlights 474
Key Terms 475
What Do You Think? 475
Suggested Readings 475

What can today's young adults expect life to be like in the twenty-first century? Research editor Judith Waldrop of *American Demographics* summarizes some of the likely facts and trends:

- As the twentieth century closes, white men will make up less than half of the labor force.
- By 2010, married couples will no longer be a majority of households.
- Asians will outnumber Jews by a margin of two to one, and Latinos will lead African-Americans as the nation's largest minority.
- By the year 2000, more than half of all children will spend part of their lives in single-parent homes.
- By 2010, about one in three married couples with children will have a stepchild or an adopted child.
- Most children will never know a time when their mothers did not work outside the home.
- Households that the Census Bureau now defines as "nonfamilies," including unmarried heterosexual couples, homosexual couples, and friends who live together, eventually will receive legal recognition as families in all states.
- If there is no attempt to stop current trends, half of all children born in New York City this year will be on welfare by 2010. Other big cities will face the same problem.
- As the new century begins, more than 80 percent of women aged 25 to 54 will be in the labor force. Most of the rest will be out of work only temporarily.
- Parental leave and flexible working hours will be the rule for all but the smallest businesses.
- Working won't always mean going to the office, either. With advanced communications equipment, employees on the road or in small satellite offices will be able to work closely with the main office.
- Higher education will be expensive but jobs will be plentiful, so many young people will work before they go to college. By 2000, half of all college students will be aged 25 or older.
- To retire in 2030 on today's equivalent of $1,000 a month, workers will have to save $4,800 a year starting now.
- By the early twenty-first century, the United States will face severe shortages of hospital beds, physicians, and nurses. Staying healthy will become a priority.
- The fastest-growing segment of homemakers will be unmarried men who live alone or head families.
- Before the U.S. population hits 300 million, in 2029, strict national laws will govern recycling, packaging standards, and waste disposal.
- Traffic jams will take to the air. Airlines will serve approximately 800 million passengers in 2000, almost twice as many as in 1990.

Source: Adapted from Waldrop (1990). ■

As you can conclude from this list, life in the next century will certainly involve a great deal of change. No aspects of life will undergo more change than those dealt with in this chapter: marriage and the family, work and leisure, and stages of personal development.

After reading chapter 17, you should be able to:

- Discuss changes that have occurred regarding American marriage and families.
- Define the different types of marriage relationships.
- Appraise Holland's and Super's theories of career selection.
- List the pros and cons of a dual-career family.
- Explain Gould's "transformational" stages of adult development.
- Discuss Levinson's theory of adult development as it relates specifically to men.
- Compare both Levinson's and Gould's theories to Erikson's young adult stage of intimacy and solidarity versus isolation.
- Define what is meant by *individuation* and discuss how it applies differently to women and men.

Marriage and the Family

When two people are under the influence of the most violent, most divisive and most transient of passions, they are required to swear that they will remain in that excited, abnormal, and exhausting condition continuously until death do them part.

George Bernard Shaw

Changing American Marriages and Families

It is not difficult to find critics of marriage (e.g., Kathrin Perutz's book, *Marriage Is Hell*). Not many Americans are paying attention, though. Almost 95 percent of Americans get married at some point in their lives (U.S. National Center for Health Statistics, 1988). To better understand the present situation, let us take a look at the trends in marriage and family relations.

Today, 25 percent of all people who get married for the first time are likely to marry someone who has been married before (Sweet & Bumpass, 1987). In 20 percent of all marriages, both partners were married previously. Trends in the rates of first marriage, divorce, and remarriage since the early twentieth century reflect patterns of change in economic and social conditions in the United States. These changes can be clearly seen in Table 17.1. One of the most interesting trends has been the change in the average age at first marriage. At the turn of the century, the average age for females was almost 22 years, and for males almost 26 years. With the exception of the late Depression and war years, the trend has been toward earlier and earlier marriages.

This trend led Duvall (1971) to predict that marriages in 1990 would come even earlier, at about age 20 for both males and females. Several other investigators (e.g., Neugarten & Moore, 1968) said the same thing in the late 1960s. But these miscalculations only demonstrate the difficulty of predicting the behavior of human beings.

Table 17.1 Median Age at First Marriage, by Sex: 1890 to 1991

Year	Men	Women	Year	Men	Women
1991	26.3	24.1	1955	22.6	20.2
			1950	22.8	20.3
1985	25.5	23.3	1940	24.3	21.5
1980	24.7	22.0	1930	24.3	21.3
1975	23.5	21.1	1920	24.6	21.2
1970	23.2	20.8	1910	25.1	21.6
1965	22.8	20.6	1900	25.9	21.9
1960	22.8	20.3	1890	26.1	22.0

Source: Saluter (1991).

The average age at first marriage in the United States has gradually increased since the record lows during the mid-1950s (Table 17.1). There were small increases in the age at first marriage from 1955 to 1975. However, sharper increases have occurred in the last two decades. One explanation for this trend may be the rising numbers of women who have entered the workforce during this period.

The modern American family is quite different from those of the last century. Those families were larger and much more likely to live near each other in the same city or town. What are the advantages and disadvantages of this tendency?

Another interesting change has been the nearly triple rise in the proportion of young adults who have not married during the past two decades. Eighty percent of men aged 20 to 24 years had never married in 1991, up from 55 percent in 1970. In this same time period, the rate went from 19 percent to 47 percent for men 25 to 29 years old, and from 9 percent to 27 percent for men aged 30 to 34. As expected, since men marry later, men have higher proportions of the never-married in all age groups. For African-Americans, however, the proportions are very similar for both genders (Saluter, 1991). An explanation may be that African-American females may view African-American males as poor marriage prospects because of their lack of employment opportunities (Chapman, 1988).

The rate of divorce and, as a result, the rate of remarriage are higher. In their study of the responses of thousands of women, Norton and Moorman (1987) discovered several trends: "Currently many young adult women (particularly African-Americans) will never marry, remarriage after divorce is becoming less frequent, and data indicate that divorce is leveling [at almost 50 percent]" (p. 3). Interracial marriage represents only one-half of one percent of all marriages (Sweet & Bumpass, 1987). However, this number indicates a rapid increase over earlier periods.

The Bureau of the Census suggests that the recent trend in increased divorced rates is probably caused, at least in part, by four factors:

- Liberalization of divorce laws
- Growing societal acceptance of divorce, and of remaining single
- The reduction in the cost of divorces, largely through no-fault divorce laws
- The broadening educational and work experience of women that has contributed to increased economic and social independence, a possible factor in marital dissolution

Other factors that affect divorce rates are race, the wife's workforce participation, husband's employment status, and residence status (Buehler & others, 1986; Kalter, 1987; Phillips & Alcebo, 1986; South & Spitze, 1986).

What's Your View?

No doubt you have recently heard at least one politician wringing his or her hands over what serious shape the American family is in. In chapter 13, we reported on many of the changes affecting the relationship between parents and children in today's families—for example, the loss of such family functions as job training, and economic dependence. Some even worry that the family itself is on the way out. They prophesy that professionals will have ever-increasing roles in raising our children, as more and more American women seek careers.

Are other countries as concerned about this as we are? There are countries in which the family has always been of tremendous importance—Asian countries such as Japan and China, and Western countries such as Italy and Spain, for example. Is this becoming a big problem for them? What about Russia, where 50 percent of children 1 to 3 years old, and 90 percent of children 4 to 5 years old, are in day care? Do you think they have solved the problem of how to raise children when both parents work full time? What's your view?

A recent Gallup poll asked 1,500 parents "What do you regard as the biggest problem facing you and your family?" Table 17.2 indicates that economic problems are by far the single largest concern of the respondents. Only about one-fifth said they could think of no serious problems.

Table 17.2 Most Significant Problems Facing Today's Family

What Do You Regard as the Biggest Problem Facing You and Your Family?	
Economic problems (total)	56%
Health problems, care	6
Divorce, family well-being	5
Education	3
Dissatisfaction with government	3
Fear of war	3
Old age, Social Security	3
No problems	16
All others	8
No opinion	8
	111%[a]

[a]Total adds to more than 100% due to multiple responses. Copyright © 1988 The Gallup Poll, Princeton, N.J.

Although early and middle-age adults are certainly concerned about every aspect of their family lives, the crucial part of family life for most of them is how they manage their relationship with their spouse. Their major concern is: "What kind of marriage will I have, and how can I make it a happy marriage?"

Although there are confounding effects of cohorts and age with length of marriage, Zietlow and VanLear (1991) found that high marriage satisfaction was associated with less deference in couples who had been married a short time and with more deference and less equality for long-term couples.

More than 40 percent of marriages are remarriages (Wilson & Clarke, 1992). And many of these remarriages involve children: one in five married couples with children had a stepchild in 1985. Stepparenting may involve considerable strain because of feeling excluded from the family or trapped in the role. High expectations, especially for a stepmother who sees herself as a nurturer and caretaker, may contrast sharply with her feelings about her stepchildren (Whisett & Land, 1992). A third of the stepfathers in one study (Marsiglio, 1992) felt that to some degree they are more like a friend than a parent to their stepchildren; 52 percent disagreed somewhat that it is harder to love a stepchild than your own child. Researchers are continuing their investigation of the special characteristics of step families.

AN APPLIED VIEW — Why Marry?

Another way to learn more about types of marriage is to conduct an informal survey of couples from each of three age groups suggested below, asking each person for the major reason he or she decided to marry.

What conclusions can you draw from their answers? Were there important age or gender differences? How many types of marriage did your survey cover?

25–35 years	husband	____________
	wife	____________
	husband	____________
	wife	____________
35–50 years	husband	____________
	wife	____________
	husband	____________
	wife	____________
50 or more years	husband	____________
	wife	____________
	husband	____________
	wife	____________

Types of Marriage

Attitudes toward marriage, and therefore types of marriage, vary greatly throughout the world. Despite many variations in the ways humans begin their married lives, there are basically four kinds of marriage throughout the world: monogamy, polygamy, polyandry, and group marriage.

Monogamy is the standard marriage form in the United States and most other nations, in which there is one husband and one wife. In **polygamy,** there is one husband but two or more wives. In earlier times in this country, this form of marriage was practiced by the Mormons of Utah. There are still some places in the world where it exists, but the number is dwindling. **Polyandry** is a type of marriage in which there is one wife but two or more husbands. The rarest type of marriage, it is practiced only in situations where there are very few females. **Group marriage** includes two or more of both husband and wives, who all exercise common privileges and responsibilities. In the late 1960s, this form of marriage received considerable attention, but it accounts for a miniscule percentage of the world's population and has lost considerable popularity in recent years.

The polygamous family, consisting of one husband and several wives, is almost nonexistent today.

Homosexual marriages, although not officially sanctioned in many parts of the world, are beginning to be accepted in some religions. Such marriages, according to Wyers (1987),

can provide individuals with an intimate, mature relationship. At the same time, though, this type of relationship presents a number of unique challenges to the couple. Aside from prejudice and lack of understanding from society at large, in the United States many social service networks and agencies are unprepared to offer services to gay and lesbian men and women. **(p. 148)**

PATTERNS OF WORK

Waste of time is thus the deadliest of sins. Loss of time through sociability, idle talk, luxury, or more sleep than is necessary for health (six hours) is worthy of absolute moral condemnation. Thus inactive contemplation is also valueless, or even reprehensible if it is at the expense of one's daily work.

Max Weber, *The Protestant Ethic and the Spirit of Capitalism*

Weber, a philosopher and economist, was a leading spokesperson on the role of labor at the turn of the century. How differently we view that role today! Table 17.3 displays a brief summary of the history of work in Western society, and indicates how much our attitudes toward work have changed. Changes in the world of work have been coming more rapidly in recent years than ever before. Working in the United States today is complicated. The rest of this section is devoted to explicating the major trends, their causes, and likely results.

Table 17.3 A Brief History of Work in the Western World

Approximate Years	Primary Role of Work
Early history	Search for food.
8000 B.C.	Cultivation of cereal grains, domestication of animals.
5000 B.C	Greater division of labor, surplus production of goods, trade.
500 B.C.	Work seen as degrading and brutalizing by upper classes; done as much as possible by slaves.
A.D. 500	Serfdom (lord of manor system replaces slavery).
A.D. 1350	Black Death makes workers scarce. Move to towns; guilds of craftspersons formed. Cottage industry, capitalism start.
A.D. 1750	Inventions cause "industrial revolution," demise of small business. Factory system takes advantage of cheap labor.
A.D. 1900	Unions, government regulations, electricity and automation, new management policies greatly improve life of workers.
A.D. 1950	Computers, technology create world of highly skilled, white-collar workers.
A.D. 1970	Age of information processing.

An Applied View — Getting a Job

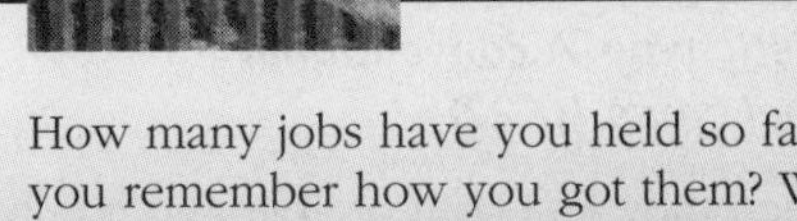

How many jobs have you held so far in your life? Do you remember how you got them? What do the people who hire workers consider when they are hiring? Do blue-collar jobs have significantly different criteria than white-collar jobs?

To find out, you might interview the managers of a bank and a supermarket and ask what they look for in a new employee. Are there differences in behavioral, attitudinal, cognitive, or appearance criteria? Why or why not?

Employment Patterns

Most of the total population 16 years old and older who choose to work are working (U.S. National Center for Health 1992). The figure is decidedly lower for the African-American population—around 89 percent, compared to 93 percent for whites. Not surprisingly, the highest percentage of unemployment is found among persons 16 to 19 years old. For African-Americans in that age group, the percentage is a disconcerting 40 percent, as compared to 15 percent for whites.

Another important aspect of employment patterns has to do with the effects of education. It is obvious that the lower the level of education, the more likely a person is to be unemployed. Unfortunately, this is even more true for African-Americans than for whites and Hispanics. For example, of those with less than four years of high school, 19.9 percent of whites and 15.8 percent of Latinos are unemployed, compared to 35.9 percent of African-Americans.

How People Choose Their Careers

Two theories are described here, those of John Holland and of Donald Super. These two theories have achieved the most acceptance in this field.

Holland's Personality Theory

On the basis of research still considered to be highly reliable, Holland (1973) developed an interesting theory on how people choose their careers. He suggested that in our culture, all people can be categorized as one of six personality types: realistic, investigative, artistic, social, enterprising, or conventional. An individual's personality pattern is estimated by figuring out how much a person's attributes resemble each type.

For example, a person might resemble an artistic type most. This type exhibits "a preference for ambiguous, free, unsystematized activities that entail the manipulation of physical, verbal, or human materials to create art forms or products, and an aversion to explicit, systematic, and ordered activities" (p. 120). This kind of person learns to be competent in artistic endeavors such as language, art, music, drama, and writing.

Our hypothetical person may next most resemble the social type. Such an individual is likely to be cooperative, friendly, generous, helpful, idealistic, insightful, responsible, tactful, and understanding. This person is then rated on the remaining four types in descending order. The six-category composite is the person's personality pattern.

The theory also holds that there are six kinds of environments in which people live. These have the same names as the personality types. According to Holland (1973),

The "artistic" type of person, according to Holland, has a preference for ambiguous, free, unsystematized activities.

> ***each environment is dominated by a given type of personality, and each environment is typified by physical settings posing special problems and stresses. For example, realistic environments are "dominated" by realistic types of people—that is, the largest percentage of the population in the realistic environment resembles the realistic type. A conventional environment is dominated by conventional types.*** **(p. 22)**

People tend to search out environments in which they feel comfortable and competent. Artistic types seek out artistic environments, enterprising types seek out enterprising environments, and so forth. There is a test you can take–the *Strong Vocational Interest Blank*–to see which personality type is most like you.

Super's Developmental Theory

Donald Super has been one of the most influential figures in advancing theories of career choice and development during recent years. Super (1957, 1983, 1990) developed a life stage theory of career development to explain how career identity develops over time and to determine a person's readiness to make a career choice. Super describes five career stages, which he originally associated with different developmental periods. In more recent revisions of his theory (Super & Thompson, 1981), Super suggests that we recycle through each of these stages several times during our lives.

The **exploration stage** is associated primarily with ages 15 to 24. Through school, leisure and part-time work activities, the adolescent or young adult is exposed to a wider variety of experiences. Through these experiences, she further defines her self-concept and has the opportunity to test out her abilities and interests. Early in this stage, initial work-related choices are made by assessing interests, abilities, needs, and values. By the end of this stage, a beginning full-time job is often selected. Super believes that most adolescents are not ready to make definite career choices because they have not yet had the chance to adequately explore available opportunities.

The Phenomenon of the Dual-Career Family

The old family pattern of the husband who goes off to work to provide for his family and the wife who stays home and manages that family is almost extinct. Economic realities and the women's movement have combined to change all that. Now most women do some sort of paid work and contribute 30 percent of the family income. This phenomenon of the **dual-career family** has manifested itself in some unusual ways.

First, women are still considered responsible for the maintenance of the family (Gerson, 1985; Szinovacz, 1984). Therefore, women are often considered unreliable by employers for the more important, higher-paying jobs. Women in general have better access to low-paying jobs with little opportunity for advancement. Women often choose jobs that fit in well with the needs of their family. Women tend to work shorter hours and change the nature of their work more often than men do (Berk, 1985; Moen, 1985).

As Serakan (1989, p. 116) found, women experience increased self-esteem and feelings of self-worth and self-regard when working, critical variables that moderate the relationship among work factors and job satisfaction. Women who spend more time at work, though, experience less job satisfaction because of guilt over not being home than those who work less.

A recent study (Guelzow, Bird, & Koball, 1991) confirmed that multiple roles in dual-career families are not necessarily related to stress, but did find that longer working hours were associated with higher role strain for women; and larger family size and inflexible work schedules were associated with stress for men. Flexibility of work for men was related with role sharing in the household. Hertz (1989) found that when a wife's career became as demanding as her husband's, there was no time for a wife to be a wife; hence, the role of wife was the first to change. Sharing housework became inevitable; as one man put it, "If both of us were relying on the other to fix meals, we'd both starve."

Hertz also found that men have had to adjust psychologically to the increased role of their wives as providers for the family. Men have historically derived much of their self-image and personal satisfaction from their work and their ability to "provide" for their family. Women now contribute much more of the family income than in previous times, and in a small but increasing number of cases, women are the primary breadwinners.

The assumption was that the more conservative blue-collar worker would have more trouble dealing with this new division of labor than the better educated and more "enlightened" upper-class and middle-class husbands. Recent research suggests just the opposite. Upper-middle-class men seem to have the most problem sharing their roles as family providers, unless they make considerably more than their wives (Fendrich, 1984; Hood, 1983). Other research suggests that these men feel cheated because they have no wife at home full time to support them, as their fathers perhaps had.

Working-class men have adapted better to this change, perhaps because the financial reality gives them little choice. The more dependent a family is on the contributions of a wife's income, the more accepting the husband is of the new role of his spouse (Rosen, 1987). All classes of men and women, however, try to continue to portray the husband as the primary provider and the wife as providing secondary support.

According to a recent study by Rosin (1990), "Research indicates, however, that many husbands believe that membership in a dual-career marriage leads to marital satisfaction because of a sense of companionship and partnership with their wives, an increased vitality in the relationship, and a greater feeling of independence" (p. 182).

While women are entering the workforce in increasing numbers, men are in general not participating more in the family work. While some research suggests that men are doing more housework than ever before (Pleck, 1985), others conclude that men do about the same amount of housework as they did in the nineteenth century (Cowan, 1987).

Most wives do about two to three times as much family work as the husbands (Berk, 1985). Women tend to do all the everyday work, including most of the child care, washing of clothes, cooking, and general cleaning. They tend to work alone, during the week and on the weekends, and during all parts of the day.

On the average, the wife in a family does two to three times as much of the family's work as the husband.

Men tend to do the less frequent, irregular work, such as household repairs, taking out the trash, mowing the lawn, and so on. They tend to do family work in the company of others, on the weekend, or perhaps during the evening. In the evening, this work may involve child care while the wife does the after-dinner chores. The tasks that men are most willing to share are the very tasks that women find most enjoyable: cooking and child care. During the inevitable argument over who does what around the house, the husband may point out that he helped out in these areas, not recognizing that he left the more onerous jobs for his wife.

We mentioned that men seem to derive their self-image and identity from their work, and that this is increasingly threatened by the entrance of women into the workplace. One obvious way men could derive more satisfaction is by taking on a greater role in the care of their children. This change, however, does not appear to be occurring at the same pace as women entering the workforce. It is estimated that mothers are actively involved with their children three to five times as much as fathers (Lamb, 1987). Mothers do all the routine chores such as feeding, bathing, and dressing. Fathers primarily play with their children. Fathers tend to spend their time with the children when the mothers are around, while mothers spend much more time alone with the children.

> ***It seems that men in a dual career marriage, no longer cushioned by the traditional wife, need to learn to balance competing demands of work and family. Where these men do not have role models, their changing roles may increase stress or change men's relationship to work, the cornerstone of male identity and self esteem.*** **(Rosin, 1990, p. 175)**

Some male participation occurs regardless of whether the mother is employed. However, as economic necessity forces women to spend more time at a paid job, they are demanding that men help out more equitably. Recent research suggests that men who are forced to participate more in child care may form better relationships with their children. At the same time, this results in significantly more marital distress (Crouter & others, 1987). One of the challenges of the family in the 1990s will be to get husbands to take on voluntarily more of the responsibility of child care (not to mention equitable household management).

> ***The results may lead to increased communication and interaction between husband and wife, enhancing the marriage, as well as an increase in job satisfaction for the working wife and mother who is able to devote more time to work without feeling stress over family obligations.*** **(Serakan, 1989, p. 115)**

Stages of Personal Development

In this section, we discuss the theories of Roger Gould, Daniel Levinson, and Erik Erikson. We also consider gender differences in individuation.

Transformations: Gould

To better understand how adults try to gain maturity, psychoanalyst Roger Gould and his colleagues at the University of California at Los Angeles combined their findings on a large number of outpatients. They discussed the primary concerns of patients at various age levels. Also at that time, Gould administered a questionnaire on this same topic to 524 persons who were not outpatients. Based on the results of these two investigations, Gould generated a theory of adult growth that is reported in his book *Transformations* (1978).

Transformations is not a scientific report of the data, statistics, and research conclusions derived from these studies. Rather, it is a theoretical discussion illustrated by selected case studies, in the manner of Freud (See chap. 2). Gould contends that there are four developmental periods in adult life up through middle age (he does not go into later adulthood). At each of these stages, there is a **major false assumption** remaining from childhood, which must now be reexamined and readjusted by each individual if he or she is to progress in maturity. On the surface, each assumption seems obviously false, but Gould argues that it still controls us subconsciously. A summary of the stages, typical ages, and major false assumptions at each stage is presented in Table 17.4. In this chapter, we will look only at the first three stages, the transition from late adolescence up to the mid-life decade. The fourth stage is covered in chapter 19.

Table 17.4 Gould's Four Stages of Adult Development
***Stage One*—Age 17 to 22—Leaving Our Parents' World**
Major false assumption: "I'll always belong to my parents and believe in their world."
***Stage Two*—Age 22 to 28—I'm Nobody's Baby Now**
Major false assumption: "Doing things my parents' way, with willpower and perseverance, will bring results. But if I become too frustrated, or tired, or am simply unable to cope, they will step in and show me the right way."
***Stage Three*—Age 28 to 34—Opening Up to What's Inside**
Major false assumption: "Life is simple and controllable. There are no significant coexisting contradictory forces within me."
***Stage Four*—Age 35 to 45—Mid-life Decade**
Major false assumption: "There is no evil or death in the world. The sinister has been destroyed."

Source: Data from R. Gould, *Transformations*. Copyright © 1978 Simon & Schuster, New York, NY.

As Gould (1978) defines adult development, "Growing and reformulating our self-definition becomes a dangerous act. It is the act of transformation" (p. 25). Reexamining one's childhood angers and hatred (he refers to them as demons) is often a painful and difficult task. Many adults avoid it or soon give up on the struggle after they have entered it. This prevents them from reaching true maturity.

Gould recommends seven steps, called an **inner dialogue,** which he believes can help in mastering the demons of one's childhood experiences. These steps are:

1. Recognize your tension and confusion.
2. Understand that people are faced with contradictory realities.
3. Give full intensity to the childhood reality; that is, accept the fact that it is real.
4. Realize that contradictory realities still exist (between childhood and adulthood).
5. Test reality. Pick a risk that discriminates one view from another. For example, write a letter to your mother about your childhood concern (that time she was unexplainably mean to you).
6. Fight off the strong urge to retreat when just on the verge of discovery.
7. Reach an integrated, trustworthy view of reality, unencumbered by the demonic past.

At every stage, in addition to major false assumptions, there are several component false assumptions. These steps are aimed at helping us recognize false assumptions and eventually reject them in favor of a more realistic view of the world.

Stage One—Age 17 to 22—"Leaving Our Parents' World"

At stage one, youth begin to recognize the difficulties in accepting their *major false assumption:* "I'll always belong to my parents and believe in their world." As a result of this misunderstanding, there are five other false assumptions, which derive from the first:

- "If I get any more independent, it will be a disaster."
- "I can see the world only through my parents' assumptions."
- "Only my parents can guarantee my safety."
- "My parents must be my only family."
- "I don't own my own body."

Stage Two—Age 22 to 28—"I'm Nobody's Baby Now"

For most young adults, society's sex roles determine their careers and the type of relationships they have.

The major false assumption here: "Doing things my parents' way, with will power and perseverance, will bring results. But if I become too frustrated, confused, or tired, or am simply unable to cope, they will step in and show me the way." Thus even though the direct dependence on parents is, or should have been, mastered at the previous stage, the strong influence of parents must be dealt with now. (As Gilligan points out so well in her book *Making Connections* (1989), interdependence, not dependence, is a more healthy objective.) The following false assumptions need to be recognized and exorcised:

- "Rewards will come automatically if we do what we're supposed to."
- "There is only one right way to do things."
- "My loved ones can do for me what I haven't been able to do for myself."
- "Rationality, commitment, and effort will always prevail over all other forces."

Now one's adult sex role begins to take decisive form. Young adults decide on their careers, their relationships with other persons, and whether to become parents. Their dreams for the rest of their lives now take clear form.

Over nine-tenths of all Americans eventually marry, and the great majority of them marry during this stage. Gould believes that, in part, the motivation for those marriages and the choice of the specific partner result from our inability to deal adequately with our relationship with our parents. As he puts it:

> ***Each and every one of us pick partners that, in subtle ways at least, recreate a parent-child relationship that has not yet been mastered. Our separateness from our parents in our twenties is really just a fiction.* (1978, p. 145)**

Stage Three—Age 28 to 34—"Opening Up to What's Inside"

At stage three, the major erroneous belief of most individuals is that "Life is simple and controllable. There are no significant contradictory forces within me." People at this stage begin to realize that life is really quite relativistic, and that there are very few "eternal verities." The dream of the crack salesman or the fulfilled mother now comes to be questioned. Both men and women seem to need to reevaluate their entire lives as they enter the fourth decade of their existence.

According to Gould, this reevaluation forces an examination of four false assumptions:

- "What I know intellectually, I know emotionally."
- "I am not like my parents in ways that I don't want to be."
- "I can see the reality quite clearly of those close to me."
- "Threats to my security aren't real."

What's Your View?

What follows is a letter written by a 36-year-old woman to her 64-year-old mother. She did this as part of her effort to follow Gould's recommendations for "mastering the demons of her childhood experiences."

Dear Mama,

This is a very difficult letter to write. There are some things you won't want to hear, but I think I should say, because you and I will bear the brunt of them, stated or not.

I am ANGRY. I'm angry with me, but a lot of that anger stems from feelings I have about unfair treatment from my childhood. I know you have always said "Don't blame your parents for your failings as an adult." That philosophy certainly helps an adult to consider changing, but to say your parents have not contributed in some basic ways—good and bad—to the adult you have become, is ignoring what seems to me a very basic truth. Some people have to work harder to become happy, productive adults, and I feel I am one of those people.

For years I have not been at peace with myself. As I write this, I am experiencing one of the most painful times in my life, because I am unable to be nice to myself on any level. If a stranger treated me as badly as I do, I would simply walk away. The sad part is that I must live in constant disapproval of myself. The last seven years have been spent trying to understand my own lack of acceptance. I have recently had several revelations, and I wish you to know about them.

First, I have not always realized the depth and breadth of my self-hatred. Just recently its full intensity has become clear to me. The catalyst was that I thought I was beginning to go through menopause, and even though I am only in my mid-thirties, I immediately thought "Gee, I deserve that!" At that moment, I realized that I have been speaking this way to me for years and didn't realize the extent to which I denigrate myself.

This is where you come in. How did I learn to have so little value for my self? I learned it from the things grownups said to me, and YOU had the strongest influence. For many years I have known that you were not the first in our family to send negative messages to their children on a regular basis. I recall that I did not like Grandma very much after about the age of twelve. It must be just about that time that I figured out she had been unkind to you in many ways. I'm sure her childhood experiences were quite similar, too. However, the fact that upbringings can make generations of unhappy people does not nullify my anger that it also happened to ME.

It is not a supportive and loving parent who praises her child, and then quickly lets her know that there are still faults—so not to think too highly of one's self. There is some level of performance which is *adequate,* and even more than adequate, which is not perfect. (You hope, I realize, that perfect is a personal standard, and therefore no one can be perfect in another's eyes.) This constant reminder that "better" was still to be had, has left me an angry adult, incapable of taking pleasure in my accomplishments.

This cannot be undone, but it can certainly be discontinued, now that its cost is apparent. I intend to set more realistic standards for myself in the future and I wish you would bear this in mind as I continue to share my life with you.

Be sure that you know I am angry about the above. I'm not infallible, and no amount of criticism could ever achieve that. I am angry with YOU because I wanted you to care more about how my behavior affected ME, than what other people thought of me. Other people don't have to live with me and they shouldn't have been more important to you than I was. I am not telling you that I don't love you, only that I am very angry. In fact, if I did not love you, this letter would not be worth the effort—I would just write you out of my life. The exercise has been worthwhile, and I would heartily recommend that you might benefit from the process as well.

I love you always,

Are any of the issues she raises relevant to you? Do you know anyone else for whom these issues are relevant? Do you think she ought to send this letter to her mother, or was having written it enough?

AN APPLIED VIEW — Responsibility for Self

One of the clearest indexes of maturity is the ability to be responsible for your own life and behavior. This exercise asks a number of questions involving responsibility for your own behavior.

1. Name two major purchases you have made in the past year by yourself without a strong influence by anyone else.

   ____

   ____

2. Wherever it is that you live, are you supported by your parents, or do you pay for your own housing?

   ____

   ____

3. Are you completely in charge of what time you come home at night, or do you have to answer to someone else?

   ____

   ____

4. Are you the sole person who decides what clothing you wear?

   ____

   ____

5. To what extent have your parents influenced your career, that is, whether you have chosen to go to college or work, the acceptability of your grades or pay, and so on?

   ____

   ____

6. Are you able to make independent decisions about your sex life? Do you let your parents know what you have decided?

   ____

   ____

You might compare your answers with those of some of your friends to get a relative idea of how responsible for your life you are.

The major cause of conflict at this stage of life is parenthood. In explaining to children what their values ought to be, we are often forced to see how unsure we are of our own values. Divorce increases at this time, too, as spouses attempt to adjust to each other's developing values.

There is a good side to all of this conflict, however. Gould cites Bertrand Russell's statement that this stage was "intellectually the highest point of my life." Gould believes that the age-30 crisis often forces us to come to know ourselves finally for what we really are. "Our confidence in the world increases as we accept the limitations of our powers and the complexity of reality. . . . in short, we come to see that life is not fair" (1978, p. 61).

The Adult Life Cycle: Levinson

Working with his colleagues, Yale psychologist Daniel Levinson has derived a theory of adult development based on intensive interviews with 40 men and 40 women. Rather than depend on questionnaire data from a large number of individuals as Gould did, Levinson decided that intensive interviewing and psychological testing with a small number of representative cases would more likely provide him with the information for a theory of adult development. Because of the number of hours necessary in this study (almost 20 hours were spent on interviews with each subject), Levinson decided to limit the number of cases so that he could get more detailed information.

Key to Levinson's theory of adult development is the notion of **life course** (1986). *Life* refers to all aspects of living—everything that has significance in a life. *Course* refers to the flow or unfolding of an individual's life. Life course, therefore, looks at the complexity of life as it evolves over time.

Equally important to Levinson's theory is the notion of **life cycle.** Building on the findings of his research, Levinson proposes that there is "an underlying order in the human life course; although each individual life is unique, everyone goes through the same basic sequence" (1986, p. 4). The life cycle is a *general* pattern of adult development, while the life course is the unique embodiment of the life cycle by an *individual.*

Through his studies, Levinson further defines parts of the life cycle. He defines the life cycle as a sequence of eras. Each era is **biopsychosocial** in character: it is composed of the interaction of the individual, complete with his or her own biological and psychological makeup, with the social environment. Each era is important in itself and in its contribution to the whole of the life cycle. A new era begins as the previous era approaches its end. That in-between time is characterized as a **transition.**

The intricacies of Levinson's theory of adult life course and life cycle are further elaborated by his concept of the adult **life structure.** Life structure is the underlying pattern or design of a person's life *at a given time.* Levinson notes that "a theory of life structure is a way of conceptualizing answers to a different question: 'What is my life like *now?*' " (1986, p. 6). The primary components of a life structure are the relationships that an individual has with significant others. It is through relationships that we "live out" various aspects of ourselves. Levinson regards relationships as actively and mutually shaped. Life structure may have many components, but generally only one or two components are central in the structure at a given time. The central component(s) is the one that most strongly influences the life structure of the individual.

The evolutionary sequence of the life structure includes an alternating series of **structure-building** and **structure-changing** transitions. During the structure-*building* periods, individuals face the task of building a stable structure around choices they have made. They seek to enhance the life within that structure. This period of relative stability usually lasts five to seven years. During that time, the stability of the life structure affords individuals the freedom to question their choices and to consider modifying their life.

This process of reappraising the existing life structure and exploring new life structures characterizes the structure-*changing* period. This period usually lasts around five years. Its end is marked by the making of critical life choices around which the individual will build a new life structure. Levinson notes that the individual decides at this point, "This I will settle for" (1986, p. 7).

In considering the periods of stability and change in the adult life cycle, Levinson notes, "We remain novices in every era until we have had a chance to try out an entry life structure and then to question and modify it in the mid-era transition" (1986, p. 7). Individuals enter into new stages of adult development as they become focused on certain developmental tasks. You will understand these tasks better when we describe each stage in the following pages.

Levinson gives equal weight to periods of stability and transition. This captures the evolution of the focus of an individual and the flowing quality of adult development. Unlike most theories of child development, in which development takes the form of positive growth, Levinson's study of adult development recognizes a co-existence of growth and decline.

Seasons of a Man's Life: Levinson

Levinson has made two separate studies, one of men (1978) and one of women (in progress—both to be further described in chapter 19). In his first study, which was solely of men, 40 male subjects ranging in age from 35 to 45 were selected, representing four categories (10 each): blue-collar workers paid on an hourly basis, middle-level executives, academic biologists, and novelists.

The hourly workers and the executives were employees of an industrial firearms manufacturer or an electronics plant (about half from each). The biologists were employed at two highly rated universities located between Boston and New York. Of the writers, some were highly gifted novelists whose work had already been accepted by critics; others were less well known, but were regarded as promising and worthy of serious consideration. Of course, this sample cannot be considered to represent the average male in the United States, but the diversity of the people selected in social class origins, racial, ethnic, and religious backgrounds, education, and marital status does make it typical of a great deal of American society today.

The study concentrates on the choices made by each man during his life and how he has dealt with the consequences of his choices, especially as they affect the main components of living: occupation, marriage, and family. After studying these components, Levinson suggested that there are four main seasons of life: (1) childhood and adolescence—birth to 22 years; (2) early adulthood—17 to 45 years; (3) middle adulthood—40 to 65 years; and (4) older adulthood—60 years and older.

Obviously, there is considerable overlap between each of his stages. Between these major stages are substages that help to bring about the transitions necessary for development. Figure 17.1 gives a description of various stages and substages. Levinson himself concentrates on the early and middle adult periods, leaving further consideration of the childhood and late adult periods to others.

He believes that "even the most disparate lives are governed by the same underlying order—a sequence of eras and developmental periods" (1978, p. 64). The purpose of these developmental transitions is to cause greater **individuation.** Individuation refers to our becoming more individual; we develop a separate and special personality, derived less and less from our parents and teachers and more from our own behavior.

Although he hypothesizes more than ten substages in the course of life, Levinson chooses to concentrate on three phases in male development. These are the **novice phase,** the settling-down phase, and the midlife transition (the latter two are discussed in chapter 19).

The Novice Phase

The novice phase of human development extends from age 17 to 33, and includes the early adult transition, entering the adult world, and the age 30 transition (see Figure 17.1). In this phase of life, four major tasks are to be accomplished. The individual in the novice phase should form

- *The dream.* "Most men construct a vision, an imagined possibility that generates and vitality" (p. 91).
- *Mentor relationships.* Each man should find someone who is older, more experienced, and willing to make suggestions at each of the choice points in his early adult life.
- *An occupational decision.* A man should begin to build on his strengths and to choose a vocation that values those strengths.
- *Love relationships.* Each man should make decisions on a marriage partner, the number of children, and the type of relationship that he wants to have with wife and children.

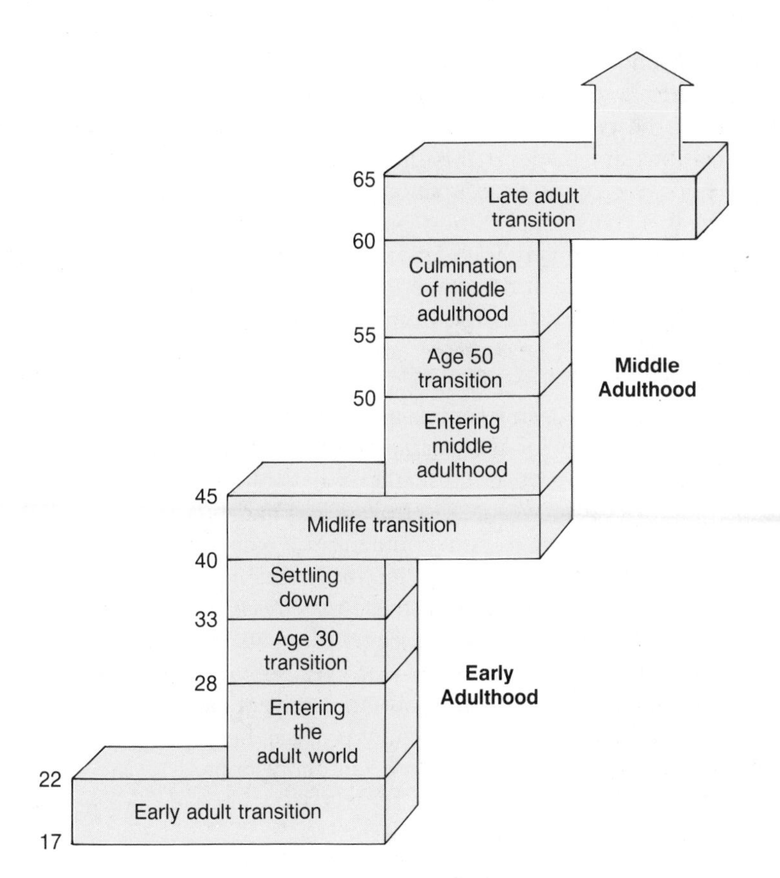

Figure 17.1
Levinson's theory

The Mentor

The concept of **mentoring** has received considerable attention in recent years (e.g., Frey & Noller, 1983; Moyers & Bly, 1990; Noller, 1983; and Rivchun, 1980). Of particular interest are Robert Bly's notions on the subject (which, by the way, he calls "male mothering").

Bly suggests that two events have wreaked havoc with the modern American male's sense of himself. The first, which began in the first half of the nineteenth century, was the Industrial Revolution. As a result of this, fathers were forced to leave home, where they traditionally worked by the side of their sons, and seek employment in factories. The second, which began in the 1960s with "no-fault" divorce, led to the proliferation of single-parent families, about 90 percent of which are headed by females.

With each of these events, much of the teaching and appreciation that boys used to get from their ever-present fathers was lost. Mothers have tried to make up for this loss, but because of deep-seated gender differences, only another male can induct a boy into adulthood successfully (See chap. 16). Bly believes that only those young men who achieve a mentor relationship with some other older man are likely to attain a mature personality. This man may be an uncle, one of the father's friends, or some older man at work. Without such a person, the young man will not be brave enough to confront himself, and he will sink into a defensive, self-deluding life-style. Bly also suggests that because the typical conflicts that exist between sons and fathers are absent in the mentor relationship, the mentor actually can be more helpful to the young man.

A solid relationship with an older mentor is crucial to the career success of a young adult, according to Levinson. Why is it, do you think, that older males are more likely to take a mentoring attitude toward their younger colleagues than are older females toward their younger colleagues?

Levinson found that after each man selects a dream, mentor relationship, occupation, and love relationship, there is a time at around age 30 (plus or minus two years) when he comes to reexamine his feelings about the four major tasks. Important decisions are made at this time, such as an alteration of the dream, a change in mentor, a change in occupation, and sometimes a change in marital status. For some, this transitional period proves to be very smooth. In most cases, however, it challenges the very foundations of life itself. Although he often keeps it to himself, the typical male at this stage undergoes a seriously disturbing period of self-doubt. Fortunately, most emerge from these doubts with a clearer understanding of their strengths and weaknesses, and a clearer view of what they wish to make of themselves.

Thus, for both Gould and Levinson, the transition from late adolescence to early adulthood (and also for the years to come) tends to proceed in stages as orderly as those we have seen in the earlier stages of life. There is more variation as we grow older, because we are controlled less and less by our genetic inheritance and more and more by the environment in which we find ourselves, and by our own individual decisions. This growing independence from our genes and our early experiences is reflected clearly in these two theories. Even if we have had a hard childhood, with alcoholic parents and traumatic accidents, we should be developing the ability to be in charge of our lives. As we grow older, we have the opportunity, and indeed the responsibility, to re-invent ourselves. Of course, how we re-invent ourselves depends on our culture. It has been suggested by some that both Gould's and Levinson's theories are culturally limited to the United States. What do you think? The next personality theorist we will cover, Erikson, would certainly agree.

Intimacy versus Isolation: Erikson

We last talked about Erikson's theory in chapter 11, where we discussed his fifth stage (adolescence), identity and repudiation versus identity confusion. Now we will consider his sixth stage, intimacy and solidarity versus isolation. This stage applies to what he defines as young adulthood, ages 18 to 25.

In his definition of **intimacy,** Erikson states that it should include

1. Mutuality of orgasm
2. with a loved partner
3. of the other sex
4. with whom one is able and willing to share a mutual trust
5. and with whom one is able and willing to regulate the cycles of
    a. work
    b. procreation
    c. recreation
6. so as to secure to the offspring, too, all the stages of a satisfactory development (1963, p. 266).

He points out, however, that it should not be assumed that sexual intercourse is the most important aspect of intimacy between individuals. He is speaking here of far more than sexual intimacy. He is talking about the ability to relate one's deepest hopes and fears to another person and to accept another's need for intimacy in turn.

Those who have achieved the stage of intimacy are able to commit themselves to concrete affiliations and partnerships with others, and have developed the "ethical strength to abide by such commitments, even though they may call for significant sacrifices and compromises" (1963, p. 262). This leads to **solidarity** between partners.

Erikson is fond of quoting Freud's response when asked what he thought a normal person should be able to do well (mentioned as this chapter began): "*Lieben und arbeiten,*"—to love and to work." To Freud, then, sharing responsibility for mutual achievements and the loving feelings that result from them are the essence of adulthood. Erikson fully agrees with this. Thus when Freud uses the term *genitality* to describe this same period, he does not merely mean sexual intercourse; he is referring rather to the ability to share one's deeply held values, needs, and secrets with another through the generosity that is so important in intimacy.

It must be admitted, nevertheless, that Freud was far more concerned with the physical aspects of sex than Erikson, who deserves major credit for moving the school of psychoanalysis away from its fascination with genitalia and toward a greater concern for adult intimacy in general.

The counterpart of intimacy is **distantiation.** This is the readiness all of us have to distance ourselves from others when we feel threatened by their behavior. Distantiation is the cause of most prejudices and discrimination. Propaganda efforts mounted by countries at war are examples of attempts to increase distantiation. It is what leads to **isolation.**

Most young adults vacillate between their desires for intimacy and their need for distantiation. They need social distance because they are not sure of their identities. They are always vulnerable to criticism, and since they can't be sure whether the criticisms are true or not, they protect themselves by a "lone wolf" stance.

Although intimacy may be difficult for some males today, Erikson believes that it used to be even more difficult for females. "All this is a little more complicated with women, because women, at least in yesterday's cultures, had to keep their identities incomplete until they knew their man" (1978, p. 49). Now that there is less emphasis in the female gender role on getting married and pleasing one's husband, and more emphasis on being true to one's own identity, Erikson believes that both sexes have a better chance of achieving real intimacy.

A growing number of theorists, however, many of them feminist psychologists, argue that females *still* have a harder time "re-inventing" themselves, because of the way our society educates them. In the next section, we present their position.

Male versus Female Identity

Identity precedes intimacy for men. . . . For women, intimacy goes with identity, as the female comes to know herself as she is known, through her relationships with others.

Sigmund Freud would have concurred with this statement by Gilligan (1982, p. 12), but for reasons that many modern women disagree with. According to feminist Betty Friedan, he believed "that women need to accept their own nature, and find fulfillment the only way they can, through 'sexual passivity, male domination, and nurturing maternal love' " (Friedan, 1963, p. 43).

The view of the feminists is that because females are trained to believe in the necessity of their maintaining relationships within the family, and because this role often involves self-sacrifice, women find it harder to *individuate*—that is, to develop a healthy adult personality of their own. They argue that for the young adult male, identity formation "involves separating himself from dependency on his family and taking his place in the adult world as an autonomous, independent being, competent, with a clear sense of career and confidence in his ability to succeed" (Wolfson, 1989, p. 11).

For the young adult female, career and self-assurance are not so emphasized. Winning the attention and then the commitment of a man are more important aspects of her identity, and intimacy is the major goal. This is seen as unhealthy. As Wolfson (1989) puts it, "It is difficult to have a sense of yourself that is internally consistent and congruent if your identity is dependent on the identity of someone else. 'I am who you are' is very different from 'I am who I am' " (p. 12).

The 1960s were a time when many grass-roots action groups, such as Students for a Democratic Society, came into being. They were anxious to promote "justice, peace, equality, and personal freedom."

This new point of view arose in the 1960s, primarily with the publication of Betty Friedan's book, *The Feminine Mystique* (1963). The affluent 1950s had "created a generation of teenagers who could forgo work to stay in school. Inhabiting a gilded limbo between childhood and adult responsibility, these kids had money, leisure, and unprecedented opportunity to test taboos" (Matusow, 1984, p. 306). At this time many youths, particularly those of the newly affluent middle class, "wanted to live out the commitments to justice, peace, equality and personal freedom which their parents professed" (Gitlin, 1987, p. 12) but, the young people felt, failed to live up to. Groups such as the Students for a Democratic Society sprang up. There developed programs such as Rennie Davis' Economic Research and Action Project, whose goal was to "change life for the grassroots poor" (Matusow, 1984, p. 315). This movement worked for justice and equality not only for the poor and racial minorities, but also for women.

The movement fostered a new commitment to women's issues, and to studies of women themselves. For example, as a result of her research on the female perspective (See chap. 9), Harvard psychologist Carol Gilligan has come to believe that

> ***for girls and women, issues of femininity or feminine identity do not depend on the achievement of separation from the mother or on the progress of individuation. Since masculinity is defined through separation while femininity is defined through attachment, male gender identity is threatened by intimacy while female gender is threatened by separation.*** **(1982, p. 9)**

Research has also pointed to two other concerns. One has been the tendency of society to "objectify" women, which means seeing them as a particular type of creature or object rather than as individuals. Psychoanalyst Erich Fromm (1955) suggested that "To be considered an object can lead to a deep inner sense that there must be something wrong and bad about oneself " (p. 323).

The second concern involves the ability to admit vulnerability. "Men are taught to avoid, at all costs, showing any signs of vulnerability, weakness, or helplessness, while women are taught to cultivate these qualities" (Wolfson, 1989, p. 44). Psychologist Jean Baker Miller writes that a "necessary part of all experience is a recognition of one's weakness and limitations. The process of growth involves admitting feelings, and experiencing them so one can develop new strengths" (1976, p. 31).

Miller sums up the feminist indictment of Erikson's position. His theory is flawed, she argues, because of his belief that

> ***women's reality is rooted in the encouragement to "form" themselves into the person who will be of benefit to others. . . . This selfhood is supposed to hinge ultimately on the other person's perceptions and evaluations, rather than one's own.*** **(1976, p. 72)**

Obviously, more work is needed to resolve this question of the role of gender in the development of the adult personality. Erikson (1963) himself appears to have endorsed this point of view, by stating that

***there will be many difficulties in a new joint adjustment of the sexes to changing conditions, but they do not justify prejudices which keep half of mankind from participating in planning and decision-making, especially at a time when the other half* [men]*, by its competitive escalation and acceleration of technological progress, has brought us and our children to the gigantic brink on which we live, with all our affluence.* (p. 293)**

In each of the theories described in this section on personality, the role of psychological and social forces is evident. Do you believe that biology also plays a part? For example, might hormones make a difference?

An Applied View — How's Your Individuation Index?

A number of theorists have stressed that maturity involves becoming more and more of an individual as one goes through life's stages. That is, one becomes less dependent on the opinions of one's parents, other relatives, teachers, and friends. The person is better able to determine her or his own personality, using these other influences only as guides. Thus one index of maturity level is the extent to which one is "individuating." Fill in the blanks below to get an idea of your own individuation.

1. Can you think of any occasion within the last month when your friends asked you to do something with them and you refused?

   ____

   ____

2. Can you think of three important decisions you have made within the past year that were definitely not influenced by your parents?

   ____

   ____

   ____

3. Can you name at least two things about yourself that you used to hate, but that you now feel are not all that bad?

   ____

   ____

4. Can you name at least two people who used to have a big influence on your life, but who are no longer able to influence you very much?

   ____

   ____

5. Can you name two things that you now like to do by yourself that you previously didn't like to do?

   ____

   ____

6. Can you name two beliefs or values you hold with which your friends would disagree?

   ____

   ____

7. Can you name three things you do of which your parents would disapprove? Three things of which they would approve?

   ____

   ____

8. Do you think you organize your time differently from most of your friends? Name three ways in which you do things differently from them.

   ____

   ____

9. Are you an "individual"? Suggest five ways in which you are different from everybody else.

   ____

   ____

The answers to these questions do not prove or disprove that you are fully mature. They should help you gain some insight into yourself. Do you like your answers?

Conclusion

By now you can see that the study of human development is mainly the study of change. Few chapters in this book, however, have described a more changing scene than this one.

Americans are getting married later for a larger variety of reasons. They are staying married for shorter periods, having fewer children, and are more reluctant to remarry if they become divorced or widowed. As we discussed in chapter 11, the functions of the family itself have changed tremendously in this century, and some have even predicted the family's demise.

The nation's workers experience a very different environment than their predecessors of fifty years ago, as we leave the industrial age and enter the age of "information processing." There are still many problems to solve—especially the treatment of persons of color and women.

Perhaps the liveliest area in developmental study in recent years has been the field of personality research. More and more we are realizing that, just as in childhood and adolescence, there are predictable stages in adulthood. There appear to be distinct life cycles, the goals of which are individuation and maturity. And differences between male and female development are becoming apparent.

In the next chapter, we return to the topics of physical and mental development, because they are worth understanding for their own sake, and so that you can see the foundations of psychosocial development in the middle adult years.

Chapter Highlights

Marriage and the Family

- Almost 95 percent of Americans get married at some point in their lives.
- There are basically four kinds of marriage throughout the world: monogamy (one husband, one wife), polygamy (two or more wives), polyandry (two or more husbands), and group marriage (two or more of both husbands and wives). Homosexual marriages are also beginning to win acceptance.

Patterns of Work

- Most of the total population 16 years old and older who choose to work are working.
- Holland suggested that, in our culture, all people can be categorized in one of six vocational interest types as to career choice: realistic, investigative, artistic, social, enterprising, or conventional.
- Economic realities and the feminist movement have given rise to dual-career families in which both husband and wife work outside of the home.

Stages of Personal Development

- Gould theorizes four developmental periods, each characterized by a major false assumption, through which adults pass as they mature.
- Levinson suggests that adults develop according to a general pattern known as the life cycle. Each individual's own personal embodiment of the life cycle is known as his or her life course.
- Levinson also describes stages of development unique to men.

- According to Erikson's theory, early adulthood is defined in terms of intimacy and solidarity versus isolation.
- Individuation, the ability to develop a healthy adult personality apart from others, is necessary and occurs differently for men and women.

Key Terms

Biopsychosocial
Distantiation
Dual-career family
Exploration stage
Group marriage
Homosexual marriage
Individuation
Inner dialogue
Intimacy
Isolation
Life course
Life cycle
Life structure
Major false assumptions
Mentoring
Monogamy
Novice phase
Polyandry
Polygamy
Solidarity
Structure building
Structure changing
Transition

What Do You Think?

1. Which of the four types of marriage described in this chapter do your parents have? Your grandparents?
2. Which of the types of marriages described in this chapter are likely to exist 100 years from now?
3. What are some ways of helping the dual-career family meet its responsibilities?
4. How well do Gould's major false assumptions fit the early and middle adult periods?
5. Gould suggests seven steps in what he calls an "inner dialogue." What is your opinion of this approach?
6. Levinson believes that forming a mentor relationship is an essential part of the novice phase. Have you formed such a relationship? What are its characteristics?
7. To be intimate, you must know your own identity. But to achieve an identity, you need the feedback you get from being intimate with at least one other person. How can this catch-22 be resolved?

Suggested Readings

Fowles, J. (1977). *The magus* (rev. ed.). Boston: Little, Brown. This is surely one of the best psychological mystery stories ever written. It concerns a young man who is unable to keep commitments. A secret society decides to try and help him to become more mature.

Friedland, R. and C. Kort. (Eds.) (1981). *The mother's book: Shared experiences*. Boston: Houghton-Mifflin. A moving and realistic collection of essays by mothers about motherhood.

Gordon, M. (1978). *Final payments*. New York: Random House. The heroine of this marvelously revealing novel struggles with numerous obstacles, from within and without, to gain her independence as a responsible adult.

Kilpatrick, W. K. (1975). *Identity and intimacy*. New York: Delacorte. This highly readable examination of Erikson's fifth and sixth stages is brilliant. We guarantee you will see yourself in a new light after reading this book.

Miller, J. Baker. (1976). *Toward a new psychology of women*. Boston: Beacon. A classic in the field of feminist psychology.

Updike, J. (1960). *Rabbit, run*. New York: Knopf. This is the first of four books that chronicles the development of an ordinary man whose nickname is "Rabbit." (He got that name because of his speed as a high school basketball player.) In the book, we follow his efforts to leave behind his exciting life as a sports star and become a responsible family man. It is not an easy trip.

PART VIII

Middle Adulthood

CHAPTER 18

Middle Adulthood: Physical and Cognitive Development

Physical Development 478
Health 479
Muscular Ability 480
Sensory Abilities 480
The Climacteric 482
Sex in Middle Adulthood 484
Cognitive Development 485
Theories about Intelligence 485
New Views of Intelligence 487
The Development of Creativity 488
Information Processing 492
Learning Ability 493
Conclusion 495
Chapter Highlights 495
Key Terms 496
What Do You Think? 497
Suggested Reading 497

The idea that there is a "middle" phase in adult life is a rather modern idea. Most non-Western and preindustrial cultures recognize only a mature stage of adulthood, from about 25 to about 60 years old, followed by a stage of old age decline. Many Western scientists (including the authors of this book) now recognize four stages of adult life: young adulthood (approximate age range from 19 to 34); middle adulthood (approximate age range from 35 to 64); young elderly (approximate age range from 65 to 79), and old elderly (80+).

One non-Western country that does recognize middle adulthood is Japan. The Japanese word for middle age, *sonen,* refers to the "prime of life," the period between early adulthood and senility. Another quite positive word often used for the middle adult years in Japan is *hataraki-zakari,* meaning the "full bloom of one's working ability."

Not all Japanese words for middle adulthood are quite so joyful. The word *kanroku* means "weightiness" or "fullness," both as in bearing a heavy load of authority, and in being overweight.

Why do you suppose industrialized cultures like North America and Japan have several words for middle adulthood, and other cultures have no words at all? ■

After reading chapter 18, you should be able to:

- Identify health concerns of particular importance to middle-aged adults.
- Discuss how muscular and sensory abilities change in middle adulthood.
- Explain the role of hormone treatments in dealing with the climacteric and other aspects of aging.
- Discuss the sexual practices of adults in middle age.
- Contrast the various positions presented regarding the nature of intelligence as people pass through middle age.
- Describe creative individuals and suggest ways in which creativity develops in middle adulthood.
- Explain the information-processing approach to intelligence.
- Decide whether you believe that learning ability declines as people age.

Physical Development

"You're not getting older, you're getting better!"

You may hear middle-aged people saying this to each other. They hope that the changes they are experiencing are minor and not too negative, but let's face it, physical systems do decline with age. As you will see, biological forces greatly influence physical development, but be on the lookout for ways that psychological and social forces are at work as well. Now let us take a closer look at some of these functions.

Health

Weight and Metabolism

As one moves into middle adulthood, weight gain becomes a matter of concern. For example, about one-half of the United States adult population weighs over the upper limit of the "normal" weight range. For some, this is the result of genetic inheritance—about 40 percent of the people with one obese parent become obese, as compared to only 10 percent of those whose parents are not obese (U.S. National Center for Health Statistics, 1986). Others become overweight simply because they do not compensate for their lowering **basal metabolism rate (BMR).**

BMR is the minimum amount of energy an individual tends to use when in a resting state. As you can see from Figure 18.1, this rate varies with age and gender. Males have a slightly higher rate than females. The rate drops most quickly during adolescence, and then more slowly during adulthood. This is caused by a drop in the ratio of lean body mass to fat, "results in a lower BMR, since the metabolic needs of fat tissue are less than those of lean. Even for those who exercise regularly, fewer calories need to be consumed" (Kart, Metress, & Metress, 1988, p. 171).

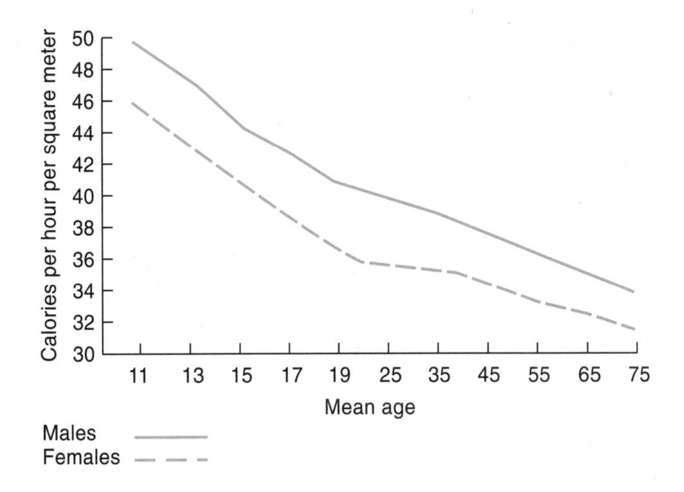

Figure 18.1
The decline of basal metabolism, rate through the life cycle. BMR varies with age and sex. Rates are usually higher for males and decline proportionally with age for both sexes.

Thus if you continue to eat at the same rate throughout your life, you will definitely gain weight. If you add to this a decreased rate of exercise, the weight gain (sometimes called "middle-age spread," referring to wider hips, thicker thighs, a "spare tire" around the waist, and a "beer belly") will be even greater. If you expect this to happen, it probably will (another **self-fulfilling prophecy).**

Many people think that weight gain in middle age is natural, but in fact a number of recent studies (Beneke & Timson, 1987; Clifford, Tan, & Gorsuch, 1991; Troumbley & others, 1990) have clearly demonstrated that increased levels of health problems and risk of death do result from being overweight.

On the other hand, the warnings against "crash diets" should be heeded, especially by middle-aged people. Probably the most popular diet these days includes the use of dietary supplements, and the most popular ingredient in them is an amino acid known as **L-tryptophan.** This acid has been linked to a blood disorder in which white blood cells increase to abnormally high numbers (American Association of Retired Persons, 1990a). This can result in swelling of extremities, severe pain in muscles and joints, fever, and skin rash. To date there have been almost 300 cases in 37 states, and one known death.

Use of Alcohol

Anyone over the age of 50 should "go on the wagon" and stay there, according to Teri Manolio of the National Heart, Lung and Blood Institute (cited in AARP, 1990b, p. 7). Even one or two drinks each day can be dangerous, because they can cause enlargement of the left ventricle of the heart. Such enlargement causes the heart to work harder, and often can cause irregular heartbeats. If a person's heart is already enlarged, the danger is even greater. This data comes from the Framingham (Massachusetts) Heart Study, which has also furnished the following findings:

- The older the person and the larger the number of drinks per day, the greater the risk.
- The risk is smaller for women.
- Men who are obese or have high blood pressure are at the greatest risk.

Muscular Ability

Muscle growth is complete in the average person by age 17, but improvements in speed, strength, and skill can occur throughout the early adult years. In fact, most people reach their peak around age 25—females somewhat earlier than that and males somewhat later. Variations in peak muscular ability depend very much on the type of activity. Consider, for example, the 20-year-old female Olympic gymnast who is thought of as an "old lady" because peak ability in this area tends to come at around age 16.

Between the ages of 30 and 60, there is a gradual loss of strength—about 10 percent on the average (Gimby & Saltin, 1983; Spence, 1989):

- Most loss occurs in the back and legs.
- Muscle tone and flexibility decrease.
- Injuries take longer to heal.
- Muscle is gradually replaced by fat.

A 175-pound man has 70 pounds of muscle at age 30, but typically loses 10 pounds of that muscle to fat by the time he reaches old age (Schulz & Ewen, 1988). However, the current popularity of aerobic activities may be reversing this trend (Buskirk, 1985). Aerobic activity such as swimming and brisk walking, appears to help maintain general health because it demands that the heart pump a great deal of blood to the large muscles of the legs.

Many men in their 30s find it hard to accept that their muscular ability is slowly declining. Unfortunately, this sometimes causes them to overexert themselves, leading to such dire results as serious injury and heart attack.

In the middle period of adulthood (35 to 64) there is a common but unnecessary decline in muscular ability. Unfortunately, the majority of people in this stage of life do not get much exercise. Both males and females may undergo a marked change in life-style. Growing success in one's chosen field often leads to greater leisure and indulgence in the so-called finer things of life.

Sensory Abilities

The five senses—vision, hearing, smell, taste, and touch—are responsible for gathering information about the world around us. Although the stereotyped image of an older person losing hearing or vision is sometimes made fun of, the gradual loss of function of one of the senses is a serious concern for all of us. We often take everyday experiences such as reading a good book, watching television, driving a car to the grocery store, tasting or even smelling a good meal, or just holding a child in our arms for granted. Yet all these experiences depend on one or more of the five senses. What changes in sensory ability can we expect as we pass through middle adulthood? The following trends are only general and vary widely from one individual to another.

Vision

Vision is the sense that we most depend on for information about what's going on around us. The eye begins to change physically at about age 40. The lens becomes less elastic and more yellow. The cornea begins to lose its luster. By age 50 the cornea is increasing in curvature and thickness. At 50, the iris begins to respond less well to light (Eifrig & Simons, 1983; Schulz & Ewen, 1988).

What do these physical changes mean for our vision? In general, our eyes don't adapt to sudden intense light or darkness as effectively as they once did. The ability to focus on nearby objects decreases, leading to a diagnosis of farsightedness and possibly a prescription for bifocals. The ability to detect certain colors can also be hampered by age-related changes in the eye. As the lens yellows, shades of blue and green become more difficult to discern. The ability to detect moving objects may also decrease as we grow older (Leigh, 1982; Warabi, Kase, & Kato, 1984). For most people, these changes in visual ability pose a problem only when lighting is reduced, such as in night driving.

Eye disease is a more serious matter. Increasing age often brings a heightened risk for such diseases. **Glaucoma** results from a buildup of pressure inside the eye due to excessive fluid. The resulting damage can destroy one's vision. Glaucoma is increasingly common after the age of 40, and is the leading cause of blindness by age 70. About 60,000 people in the United States are blind because of it (Johnson & Goldfinger, 1981). Routine eye examinations, however, now usually include a glaucoma test and blindness is often prevented.

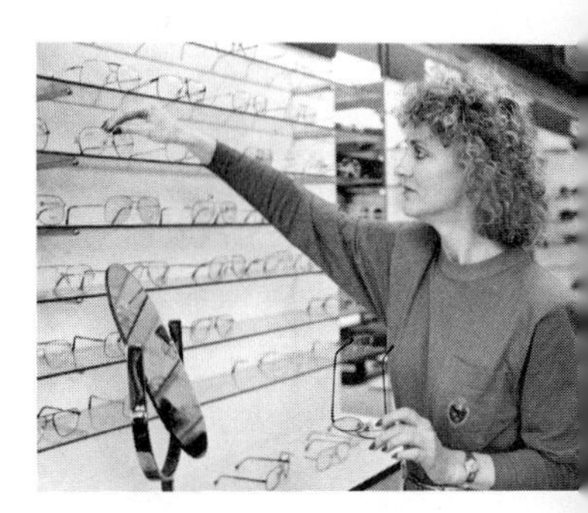

Because of changes in the eye that occur when we reach our 40s, glasses are often necessary.

Less well known but more dangerous is **senile macular degeneration.** This disease of the retina is also a cause of blindness. It begins as blurred vision and a dark spot in the center of the field of vision. Advances in laser surgery have shown promise in treating diseases of the retina such as this.

Hearing

Hearing also seems to be susceptible to decline at about age 40. This is when we begin to lose the ability to detect certain tones. As you progress from middle to old age, certain frequencies, the higher ones in particular, may need to be much louder for you to hear them (Schulz & Ewen, 1988). Our ability to understand human speech also appears to decrease as we grow older. The most likely result is an inability to hear certain consonants, especially *f, g, s, t, z, th,* and *sh* (Marshall, 1981). Cognitive capacity may play a role in the fact that the ability to listen to speech in a crowded environment (e.g., a party) declines faster than the ability to listen to someone alone with no background noise (Bergman & others, 1976; Bergman, 1980).

Our visual system is a more reliable sense than our auditory system. In fact, in contrast to the success that medical science has made in fighting blindness, it appears that deafness in the United States is increasing. In 1940 you could expect to find deafness in 200 out of every 10,000 people in this country. By 1980, you could expect to find it in 300 (Hunt & Lindley, 1989). Bear in mind, however, that these effects are partly a result of choice. Cross-cultural studies have shown that our relatively loud, high-tech culture contributes to our society's general loss of hearing (Bergman, 1980).

One curious aspect of the age-related decline in hearing is that, of the five senses, hearing is easily the most stigmatized loss. People who wouldn't think twice about getting prescription eyeglasses will refuse to get a hearing aid or even admit they are suffering a hearing loss.

Smell

Although not as important a sense as vision or hearing, the **olfactory sense** does a little more than just tell us when dinner is about ready. First, the olfactory sense works closely with our taste buds to produce what we think of as the "taste" of a given food. In fact, it is difficult in studies to separate which sense—taste or smell—is actually contributing to the decline in performance on a certain task. Besides bringing us pleasurable smells, this sense warns us against spoiled foods, smoke or fire, and leaking gas (Stevens & Cain, 1987).

Various studies suggest that our sense of smell decreases as we grow older (Murphy, 1983; Schiffman & Pasternak, 1979; Stevens & Cain, 1987). The decline seems to begin slowly around age 50 and increases rapidly after age 70.

Taste

As we mentioned, our sense of taste is closely tied to our sense of smell, which makes studying age-related effects on taste very difficult. One estimate is that 95 percent of taste derives from the olfactory nerves. Not coincidentally, taste seems to begin to decline around the age of 50 (Cooper, Bilash, & Zubek, 1959; Schiffman, 1977).

Recent studies have suggested that older adults may experience a decline in the ability to detect weak tastes, but retain their ability to discriminate among those foods that have a strong taste, such as a spicy chicken curry (Spence, 1989). The decline in the ability to taste (and smell) may account in part for the decrease in weight that many elderly persons experience.

In general, we can say that while our senses do decline somewhat throughout middle age, we are finding more and more ways to compensate for the losses, so they cause only slight changes in life-style. Another physical concern in our middle years is the climacteric.

The Climacteric

The word **climacteric** refers to a relatively abrupt change in the body, brought about by changes in hormonal balances. In women, this change is called **menopause.** It normally occurs over a four-year period at some time during a woman's 40s or early 50s (Masters & Johnson, 1966). The climacteric also refers to the **male change of life,** whereas the *menopause* refers only to the cessation of menstruation, most often between the ages of 48 and 52. The term **climacterium** refers to the loss of reproductive ability. This occurs at menopause for women, but men tend to be quite old when they are no longer able to produce fertile sperm.

The main physical change in menopause is that the ovaries cease to produce the hormones estrogen and progesterone, although some estrogen continues to be produced by the adrenal glands. Does this decline in hormonal output always cause significant changes in female behavior? The question is difficult to answer. In the first place, women seem to react to menopause in many different ways. As Neugarten (1968) discovered in her extensive study of white mothers, there is not much agreement as to what menopause means or how it feels. Only 4 percent of the women interviewed thought that menopause was the worst thing about middle age, and many found it much less difficult than they had expected. On the other hand, half the women thought that menopause caused a negative change in their appearance, and a third experienced negative changes in their physical and emotional health. The great majority thought that menopause had no effect on their sexual relationships. Many other things are happening in a woman's life at the same time as her menopause, so it is difficult to sort out what is causing what. Undoubtedly a lack of understanding of menopause, coupled with normal fears of growing old, accounts for at least some of the negative feelings about menopause. It should be noted that many women have positive feelings. For example, they no longer have to be concerned about becoming pregnant.

A recent study found no negative mental health consequences from menopause for the majority of middle-aged women sampled (Matthews & others, 1990). Also on the positive side, Sheehy (1992) found that menopause is a time of "coalescence," a time of integration, balance, liberation, confidence, and action. A menopausal woman no longer worries about pregnancy; many feel relief when their children leave the nest.

She moves from the old age of youth to the youth of old age or, in other terms, she moves into a second adulthood. Many women in their fifties may find themselves in the prime of their lives (Fodor & Franks, 1990). They have good health, autonomy, security in their major relationship, freedom, higher income, status, and confidence. Brown (1982) found that many postmenopausal women in a variety of cultures experience greater powers, freedoms, and higher-level responsibilities when their children become adults; but as one study cautions, it may be that only women of higher status feel this increase in power (Todd, Friedman, & Karinki, 1990).

For those women who experience serious problems with menopause, **estrogen replacement therapy (ERT)** can offer considerable relief. When estrogen was first instituted, ERT was found to increase the risk of cancer. Today, however, it is given in quite low levels and is combined with progesterone, which greatly reduces the risk. In addition, a recent study (Myers & Morokoff, 1986) found that postmenopausal women who are receiving ERT demonstrate a higher level of arousal when watching an erotic movie than those who are not receiving ERT.

Estrogen, which eliminates the symptoms of menopause, also stems the deterioration of the cardiovascular, urinary, genital, and nervous systems; slows the aging of the bones and skin; and cuts the death rate from heart disease for women in half. Recent Veterans Administration research, using human growth hormone, initially found that 60 men aged 60 to 80 may have regained the vitality of men 15 to 20 years younger. A study by the National Institute on Aging has begun testing two other hormones, DHEA and testosterone, suspected of retarding age-related symptoms such as loss of strength and vitality and the diminishing size of internal organs in frail, elderly men. Perhaps in the future, doctors may routinely prescribe hormone therapy for both men and women to slow the aging process (Rudman, 1992).

At one time, it was thought that the male hormone balance parallels that of the female. According to a well-designed study conducted by the National Institute on Aging (1979), however, the level of testosterone declines only very gradually with age. Dr. Mitchell Hermann, who conducted the study on men from age 25 to 89, says that his findings contradict earlier results because most of those previous studies were of men in hospitals and in nursing homes who were afflicted with obesity, alcoholism, or chronic illness. All of his subjects were healthy, vigorous men.

Hermann suggests the decrease in sexual potency that men experience in later years is probably not the result of hormone changes, but rather slowing down in the central nervous system, together with a self-fulfilling prophecy (men expect to become impotent, so they do). The effects of hormonal changes on the appearance and emotional state of men are unclear at the present.

It seems likely that as we better understand precisely how hormone balances change and how the different changes interact with each other, the impairments that have been attributed to these changes will decrease (Rowland & others, 1987).

Although middle-aged people are generally healthier than the elderly, chronic health problems are especially prevalent for poor African-American middle-aged women, probably because of the physical labor required for the domestic work so many of them do. African-Americans also sustain 39 percent more work-related injuries and diseases than whites and thus are more likely to drop out of the labor force before retirement (Jackson & Gibson, 1985). These socioeconomic and racial differences are a good place to see how the biopsychosocial model operates. What role would you say each of these three major forces play in causing these differences?

Health also plays an important role in the sexual feelings and behavior of middle-aged adults. In the next section, we examine this and other aspects of sexuality.

Sex in Middle Adulthood

At midlife, minor physiological changes occur in both male and female sexual systems. For the male, there may be lower levels of testosterone, fewer viable sperm, a decrease in sex steroids affecting muscle tone and the cardiovascular system, slight changes in the testes and prostate gland, and a change in the viscosity and volume of ejaculate (Hunter & Sundel, 1989). There is usually a need to spend more time and to give more direct stimulation to the penis in order to attain erection. None of these changes is sufficient to alter significantly the man's interest and pleasure in a sexual and sensual life.

For the female, the reduction in estrogen occurring during and after menopause may cause changes, such as less vaginal lubrication and possible vaginal irritation at penetration, that can eventually affect the ease and comfort of sexual intercourse. Sexual arousal may be somewhat slower after the fifth decade, but orgasmic response is not impaired. As with the males, females may need more time and appropriate stimulation for vaginal lubrication and orgasmic responsivity. Some studies have found a reduction of female interest and desire, while others have found a decrease in frequency of intercourse (Cutler, Schleidt, & Friedmann, 1987). Bretschneider and McCoy (1988) argue that low estrogen levels are associated with decreased sexual interest in women. Hysterectomies can catapult a woman into premature menopause, and can cause her to experience postoperative sexual problems that may require estrogen replacement therapy (Leiblum, 1990).

A study of lesbian women at menopause (Cole & Rothblum, 1990) found very few sexual problems. The 10 percent who did list one or more symptoms qualified their responses to say there were differences, not problems. Masters, Johnson, & Kolodny (1986) found that lesbians make smooth transitions into the middle years, usually in lasting relationships. Peplau (1981) found a similar pattern for gay men.

Many couples find that when they reach middle age, their relationship becomes more romantic.

As you can see, by the time people reach middle adulthood, sexual preferences are clearly a very individual matter. Some couples engage in sexuality a lot right into their old age, and some have a fine marriage without making love very much at all.

The most recent report on frequency of sexual activity among the adults of all ages, the Janus Report (See chap. 14), gives a picture that weakens the stereotype that as we grow older, we become more inactive sexually. As Table 18.1 illustrates, there is a small increase in sexual activity in the middle years, as compared to early adulthood. For men, middle age is the peak sexual time but there is little drop with age, and for women, age makes little difference across the adult lifespan.

The experts pretty well agree on the nature of physical alterations that occur through the middle years of adulthood. There is, however, much less agreement on the course of cognitive development in these years.

Table 18.1 Frequency of Sexual Intercourse by Age

	Ages (Years)									
	18 to 26		27 to 38		39 to 50		51 to 64		65+	
	M	F	M	F	M	F	M	F	M	F
N =	254	268	353	380	282	295	227	230	212	221
a. Daily	15%	13%	16%	8%	15%	10%	12%	4%	14%	1%
b. A few times weekly	38	33	44	41	39	29	51	28	39	40
c. Weekly	19	22	23	27	29	29	18	33	16	33
d. Monthly	15	15	8	12	9	11	11	8	20	4
e. Rarely	13	17	9	12	8	21	8	27	11	22
Active = lines a + b	53%	46%	60%	49%	54%	39%	63%	32%	53%	41%
At Least Weekly = lines a through c	72%	68%	83%	76%	83%	68%	81%	65%	69%	74%

Source: Janus & Janus (1993) p. 25.

Cognitive Development

> ***I am all I ever was and much more, but an enemy has bound me and twisted me, so now I can plan and think as I never could, but no longer achieve all I plan and think.***
>
> ***William Butler Yeats at age 57***

Of course, the "enemy" Yeats refers to is age. Was he right in believing he could think as well as ever or was he just kidding himself? The question of declining intelligence across adulthood has long concerned humans.

Theories about Intelligence

Indeed, no aspect of adult functioning has received more research than intelligence (e.g., Botwinick, 1984; Cooney, Schaie, & Willis, 1988; Nesselroade & others, 1988). Despite this considerable research, investigators still do not agree on whether we lose intellectual ability as we grow old. In fact, there are three basic positions: yes, it does decline; no, it does not decline, and yes, it does in some ways, but no, it doesn't in others. Let's look at the evidence for these positions.

Wechsler's Answer: Yes

Undoubtedly the strongest proponent of the hypothesis that intelligence declines with age is psychologist David Wechsler, who said:

> ***Beginning with the investigation by Galton in 1883 . . . nearly all studies dealing with the age factor in adult performance have shown that most human abilities . . . decline progressively, after reaching a peak somewhere between ages 18 and 25.*** **(1958, p. 135)**

The most widely used test of adult intelligence is the one designed by Wechsler himself (1955).

Terman's Answer: No

The classic Terman study (1925) is an excellent longitudinal study that found an *increase* in intelligence with age. That study was started in 1924 and used the *Stanford-Binet Intelligence Test* for children. The subjects were tested 10 years later,

Lewis Terman began his study of the development of intelligence in the 1920s.

in 1941, and then retested in 1956 with the Wechsler Adult Intelligence Scale (WAIS), which is highly correlated with the Stanford-Binet. At the end of the second 15-year interval, when the subjects were in their twenties, there was an average increase of scores (Bradway, Thompson, & Cravens, 1958).

The subjects were retested in 1969 by Kangas and Bradway (1971). The average subject was then 40 years old. Their scores were found to have increased to an average of nearly 130! Another study supporting the no-decline hypothesis is reported by Owens (1953). In 1919, a group of 363 students entering Ohio State College had their intelligence tested. Thirty years later, Owens retested 127 of them, and all but one of the subjects showed an increase over the 30 years. In 1966, when the individuals were approximately 61 years old, 97 of the subjects were retested and none of the scores had changed significantly.

Canestrari (1963) found that adults in middle age do more poorly on speed tests than do younger subjects. However, when they were given more time, for example, to memorize digits, they did as well as the younger subjects.

Horn's Answer: Yes and No

J. L. Horn believes that intelligence does decline in some ways, but in other ways it does not. The picture is probably more complicated than either of the first two positions reveals. There is evidence that

- One type of intelligence declines, while another does not.
- Some individuals decline while others do not.
- Although decline does eventually occur, it happens only late in life.

Horn has described two dimensions of intelligence: fluid and crystallized. The two can be distinguished as follows: **fluid intelligence** is dependent on the proper functioning of the nervous system. It is measured by tasks that show age-related declines (speeded tasks, tests of reaction time). **Crystallized intelligence** demonstrates the cumulative effect of culture and learning of task performance, and is measured by tests of verbal ability and cultural knowledge (Labouvie-Vief & Lawrence, 1985).

Horn (1975, 1978) hypothesized that whereas crystallized intelligence does not decline and may even increase, fluid intelligence probably does deteriorate, at least to some extent (see Figure 18.2). Horn and his colleagues (1981) have found that this decline in fluid intelligence averages three to seven IQ points per decade for the three decades spanning the period from 30 to 60 years of age.

Figure 18.2
Horn's three types of intelligence

Source: From J.L. Horn, "Remodeling Old Models of Intelligence," in B.B. Wolman (ed.), *Handbook of Intelligence: Theories, Measurements, and Applications*. Copyright © 1985 John Wiley & Sons, Inc., New York, NY.

Some research suggests that some kinds of achievement may rely more on one type of intelligence than on the other (Lehman, 1964). For example, in fields such as mathematics, music, chemistry, and poetry, the best work is usually produced at a relatively young age and therefore may rely most on fluid intelligence. Other major achievements, such as in history, astronomy, philosophy, writing, and psychology, usually occur later in life, which may indicate greater reliance on crystallized intelligence (more on this later).

Earlier work on mental abilities was flawed because the studies were cross-sectional rather than longitudinal. A cross-sectional study looks at different groups of people at different ages and then makes inferences about age-related changes (See chap. 2). The problem is that people of different ages, having grown up at different periods, have necessarily had different life experiences. Therefore the results may be more due to this "cohort" effect than to aging.

A longitudinal study avoids this weakness because it looks at the same group of people over an interval of time. Schaie and Hertzog (1983) conducted a longitudinal study on cognitive abilities and found evidence that they decline as one ages. The evidence for a decline after the age of 60 was strong. They also found evidence that this process starts after the age of 50, although it is probably not observable in everyday life. Making the situation even more complex, Hertzog's latest research suggests that a decline in speed of performance makes declines in intelligence look worse than they are. "It may be the case that a substantial proportion of the age changes actually observed by Schaie in his longitudinal studies is not loss of thinking capacity per se but, rather, slowing in *rate* of intelligent thought" (Hertzog, 1989, p. 650). Perhaps you simply need more time to think and respond when you get older.

New Views of Intelligence

Part of the debate is a matter of definition. What, exactly, is intelligence? How is it measured? It is certainly composed of several different cognitive abilities, such as memory, language, reasoning, and the ability to manipulate numbers. Each of these mental processes can, in turn, be divided into various subprocesses.

Different definitions of intelligence lead to different measures of it. Each measure emphasizes different mental abilities. For example, if you define intelligence as rank in class or school achievement, then cognitive abilities such as verbal comprehension and general information may be much more important than rote memory and perceptual tasks (for a review, see Horn & Donaldson, 1980). In fact, most of the commonly used IQ tests measure only a few of the cognitive abilities that could be measured. Under these circumstances, the question of whether intelligence declines with age becomes unanswerable. Clearly, more specific questions must be asked.

Increasingly, general intelligence is being abandoned as a scientific concept and subject of study. More and more, researchers are proposing several quite distinct cognitive abilities. Horn has now added supporting abilities to the concepts of fluid and crystallized intelligence. These include short-term memory, long-term memory, visual processing, and auditory processing (Horn, 1985; Horn & Donaldson, 1980). Howard Gardner (1983) contends there are seven different cognitive abilities. His list includes linguistic, musical, logical-mathematical, spatial, bodily-kinesthetic, self-understanding, and social understanding abilities. Other lists have been generated by other researchers, (e.g., Sternberg, 1990).

Certainly intelligence cannot be separated from memory. Your mind can process information like lightning, but if you cannot recall the proper information, processing abilities are useless. Does memory decline with age? There is the stereotype of the doddering old person who can't quite seem to remember the names of his grandchildren. But the research does not always confirm the existence of a decline.

We mentioned earlier in this chapter that memory itself can be separated into different types of memory. Horn and his colleagues make a distinction between short-term acquisition and retrieval factors (SAR) and the tertiary storage and retrieval dimension (TSR). Simply put, SAR refers to short-term memory and TSR refers to long-term memory. Short-term memory allows the individual to keep the details of a reasoning problem in awareness so that it can be processed. Long-term memory allows the individual to recall information from the relatively distant past and use it to solve a current problem. Horn and colleagues found that SAR declined with age in much the same way as fluid intelligence, and TSR either did not decline or improve, just as crystallized intelligence did (Horn & Donaldson, 1980).

Baltes and his associates (Baltes, Reese, & Nesselroade, 1977; Baltes & Schaie, 1976) have suggested a resolution to the question of how intelligence develops with age. They have proposed a **dual-process model** of intelligence. There is likely to be a decline in the *mechanics* of intelligence, such as classification skills and logical reasoning, but the *pragmatics* are likely to increase. Pragmatics include social wisdom, which is defined as "good judgment about important but uncertain matters of life" (Baltes, Reese, & Nesselroade, 1977, p. 66).

This seems the most reasonable position. There are just too many famous people whose thinking obviously got better as they got older. To name a few: George Burns, Coco Chanel, Benjamin Franklin, Albert Einstein, Mahatma Gandhi, Helen Hayes, Michelangelo, Grandma Moses, Georgia O'Keefe, Pope John XXIII, Eleanor Roosevelt, Bertrand Russell, George Bernard Shaw, Sophocles, Frank Lloyd Wright, and so on and so on. It is no coincidence that these people also maintained their *creative* abilities well into old age.

The Development of Creativity

As the world changes more and more rapidly, the role of creativity becomes more crucial. In this section, we explore the development of creative ability in the adult years. But first, here is a description of creative individuals.

Traits of the Highly Creative Adult

A number of studies (reviewed in Dacey, 1989a and c, 1992) have compared highly creative and average adults in a number of important traits. In general, highly creative adults

- Like to do their own planning, make their own decisions, and need the least training and experience in self-guidance.
- Do not like to work with others, and prefer their own judgment of their work to the judgment of others. They therefore seldom ask others for opinions.
- Take a hopeful outlook when presented with complex, difficult tasks.
- Have the most ideas when a chance to express individual opinion is presented. These ideas frequently invoke the ridicule of others.
- Are most likely to stand their ground in the face of criticism.
- Are the most resourceful when unusual circumstances arise.
- Can tolerate uncertainty and ambiguity better than others.
- Are not necessarily the "smartest" or "best" in competitions.

In their compositions, creative adults typically

- Show an imaginative use of many different words.
- Are more flexible; for example, in a narrative they use more situations, characters, and settings. Rather than taking one clearly defined train of thought and pursuing it to its logical conclusion, creative adults tend to switch the main focus quickly and easily and often go off on tangents.
- Tend to elaborate on the topic assigned, taking a much broader connotation of it to begin with, and then proceeding to embellish even that.
- Are more original. (This is the most important characteristic. The others need not be evidenced, but this one *must* be.) Their ideas are qualitatively different from the average person's. Employers frequently react to the creative person's work in this way: "I know what most of my people will do in a particular situation, but I never know what to expect from this one!

Now let us turn to the research on the development of creativity.

The "grande dame" of modern dance, Martha Graham's contributions to choreography make her one of the most creative adults in the twentieth century.

Psychohistorical Studies of Creative Achievement

Lehman (1953) examined biographical accounts of the work of several thousand individuals born since 1774. He studied the ages at which these persons made their creative contributions. He compared the contributions of deceased persons with those still living. On the basis of his study, he concluded that

> ***on the whole it seems clear that both past and present generation scientists have produced more than their proportionate share of high-quality research not later than at ages 30 to 39, and it is useless to bemoan this fact or to deny it.*** **(p. 26)**

Figure 18.3 portrays Lehman's general results.

In his report of his own research on this subject, Dennis (1966) criticized Lehman's work, stating that it included many individuals who died before they reached old age. Dennis points out that this biased the study statistically, because we cannot know what proportion of creative contributions these deceased people would have made had they lived longer.

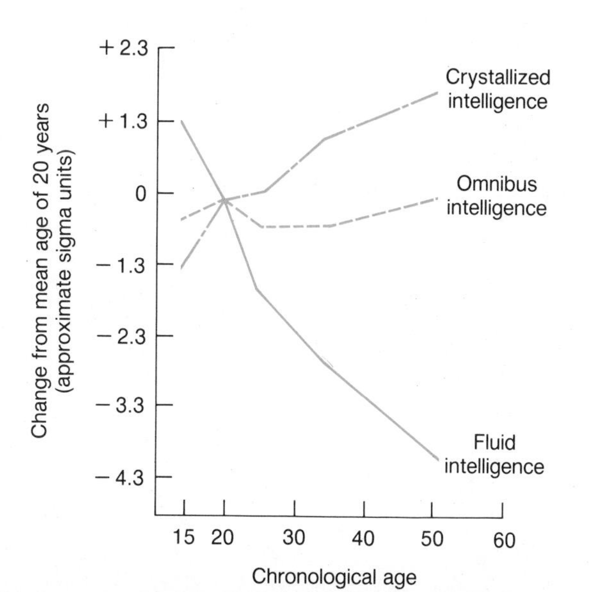

Figure 18.3
Horn's three types of intelligence.

What's Your View?

Think of five people over 35 whom you know and most admire. Write their names here:

______________ ______________
______________ ______________
____

Now, ask yourself, how many of these people are among the most creative people you know? If most are, do you think there is something about growing old that contributes to creativity? If your answer is no, what do you think does contribute? Would you say that, on the average, most people get more creative as they grow older?

Dennis himself studied the biographies of 738 creative persons, all of whom *lived to age 79 or beyond,* and whose contributions were considered valuable enough to have been reported in biographical histories. He did this because he believed "that no valid statements can be made concerning age and productivity except from longitudinal data involving no dropouts due to death" (1966, p. 8).

He looked at the percentage of works done by these persons in each of the decades between the ages of 20 and 80. When creative productivity is evaluated in this way, the results are quite different. He found that scholars and scientists (with the exception of mathematicians and chemists) usually have little creative output in their 20s. For most of them, the peak period is between their 40s and 60s, and most produce almost as much in their 70s as they did in their earlier years. The peak period for artists tended to be their 40s, but they were almost as productive in their 60s and 70s as they were in their 20s. Figure 18.4 depicts these relationships.

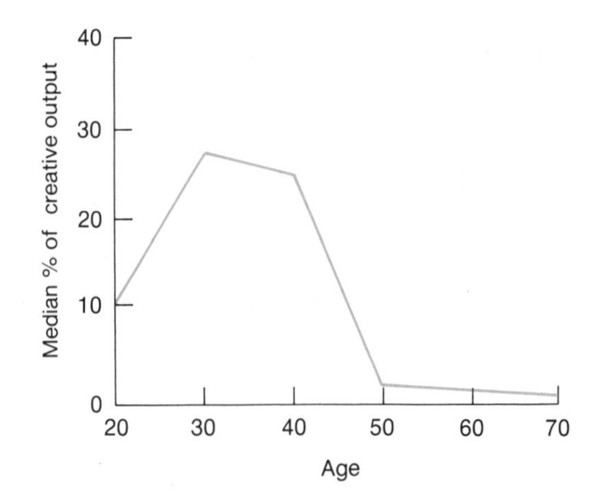

Figure 18.4
Graph of Lehman's findings

Source: Lehman (1953)

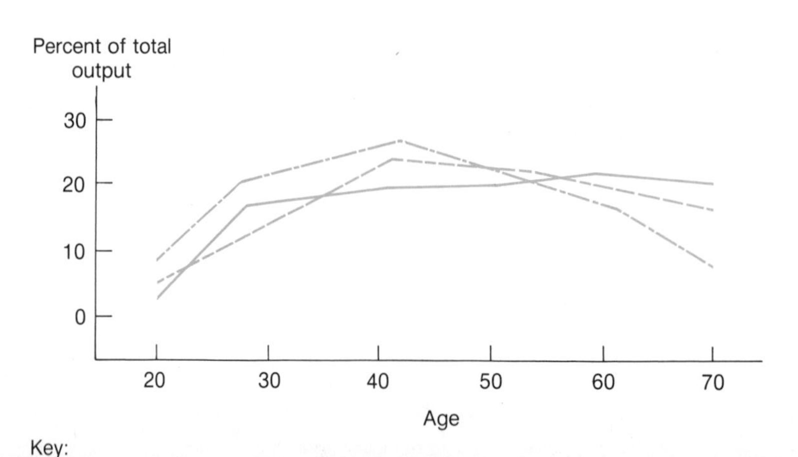

Figure 18.5
Graph of Dennis's findings

Source: Dennis (1966)

An Applied View — Obstacles and Aids to Creativity

We may agree that creativity is a valuable trait and should be fostered, but how? A number of theorists have offered excellent suggestions (e.g., Adams, 1986; Treffinger, Isaksen, & Firestien, 1983), but educator Ralph Hallman's suggestions (1967) on the obstacles and aids to creativity are still classic. According to him, there are several persistent obstacles to creativity:

- *Pressures to conform.* The pressure on the individual to follow standardized routines and inflexible rules is probably the major inhibitor. Authoritarian parents, teachers, and managers who demand order are responsible for destroying a great deal of creative talent.
- *Ridicule of unusual ideas.* This destroys one's feelings of worth and makes one defensive and compulsive.
- *An excessive quest for success and the rewards it brings.* An overconcern with material success is often the result of trying to meet the standards and demands of others in order to obtain the rewards they have to give. In the long run, this distorts one's view of reality and robs one of the strength of character to be creative (Amabile, Hennessey, & Grossman, 1986).
- *Intolerance of a playful attitude.* Innovation calls for playing around with ideas, a willingness to fantasize and make believe, and a healthy disrespect for accepted concepts. Often creative persons are seen as childlike and silly and their activity wasteful, but these are only appearances. As Hallman remarks, "Creativity is profound fun."

In addition to recommending that we avoid these obstacles, Hallman urges that we promote the following aids in ourselves and others:

- *Engage in self-initiated learning.* Most people who are in charge of others (managers, teachers, parents) find it hard to encourage others to initiate and direct their own learning. After all, this is certainly not the way most people were taught. They fear that if their subordinates are given greater freedom to explore reality on their own, they will learn wrong things, or will not learn the right things in the proper sequence. We must put less emphasis on learning "the right facts," and more on learning how to learn. Even if we do temporarily mislearn a few things, in the long run the practice in experimentation and imagination will be greatly to our benefit.
- *Become deeply knowledgeable about your subject.* Only when people make themselves fully familiar with a particular situation can they detach themselves enough to get an original view of it.
- *Defer judgment.* It is important to make wild guesses, to juggle improbable relationships, to take intellectual risks, to take a chance on appearing ridiculous. Refrain from making judgments too early.
- *Be flexible.* Shift your point of view; to dream up new ideas for things, imagine as many possible solutions to a particular problem as possible.
- *Be self-evaluative.* When a person comes up with a creative idea, he or she is always a minority of one. History is replete with examples of ideas that were rejected for years before people began to realize their worth. Therefore, the creative person must be one who knows his or her own mind and is relatively independent of the judgment of others. To become a good judge of their own thinking, people must practice making many judgments.
- *Ask yourself lots of open-ended questions.* One extensive study showed that 90 percent of the time the average teacher asks questions to which there can be only one right answer, which the teacher already knows. Questions that pique curiosity and allow many possible right answers were asked only 10 percent of the time. Realize that you were probably taught that way, and take steps to rectify the tendency.
- *Learn to cope with frustration and failure.* Thomas Edison tried more than 2,000 combinations of metal before he found just the right kind for the electric element in his first light bulb.

Dennis offers an interesting hypothesis to explain the difference in creative productivity between the three groups. The output curve of the arts rises earlier and declines earlier and more severely because productivity in the arts is primarily a matter of *individual* creativity. Scholars and scientists require a greater period of training and a greater accumulation of data than do others. The use of accumulated data and the possibility of receiving assistance from others causes the scholars and scientists to make more contributions in later years than do people in art, music, and literature.

Most of the productive persons in Dennis' study were males. It would be interesting to investigate the patterns of productivity among a comparable all-female group.

More recent studies by Simonton (1975, 1976, 1977a, 1977b) have attempted to resolve the differences between the Lehman and Dennis research. In general, Simonton found evidence that quantity declines with age, which favors Lehman, but that quality does not, which favors Dennis. Unfortunately, because of differences in design and criteria (e.g., differing data sources, differing criteria for inclusion in the studies), it cannot be said that this issue is fully resolved at this time.

We started this section on cognitive development by looking at the measurement of changes in intelligence over the adult years. Next we reviewed what we know about creative development. Now we will consider a third aspect, one that attempts to pull together a variety of intellectual factors.

Information Processing

The information-processing approach to the study of intelligence involves looking at different processing capabilities in areas such as attention, perception, memory, and problem solving (e.g., Craik & Simon, 1980; Kintsch, 1970; Sternberg & Detterman, 1979). This view looks at the flow of information from the moment it is attended to, perceived, placed in memory, organized with previous memory, and then considered and acted upon. We now turn to some of the findings of the information-processing research.

The flow of information through the brain begins at the sensory receptors. At this initial point, the evidence suggests that age-related declines do occur. This part of the process, perhaps more than any other, is affected by the readily observed physical changes brought on by increasing age. For example, the vast majority of sensory information comes through the eyes. It is commonly accepted that as one increases in age, the diameter of the eye decreases and the lens thickens and yellows. These structural changes cannot help limiting the amount of sensory information that can be received at any one time (Fozard & others, 1977).

Once the information is made available to the senses, the question becomes whether one can or will attend to it. First, a distinction has been made between effortful and automatic attention (Hasher & Zacks, 1979). **Effortful attention** refers to processing novel tasks, such as trying to learn a new phone number or a new surgical technique. **Automatic attention** refers to common tasks that require little of a person's limited capacity for attention. Anyone who has worked for many years at a job that initially requires a great deal of attention, such as a pilot or a surgeon, usually finds that they use less and less of their total attention capacity to accomplish their tasks. Effortful attention capacity seems to decline with age, while automatic attention does not (Hoyer & Pludes, 1980). So as one gets older, attending to new tasks may get more difficult but performance on familiar tasks should stay the same.

Once the information is attended to, is it properly stored in memory? Can previous information be recalled in order to help reorganize and process new information? The answer to whether memory declines with age depends on what kind of memory is being considered. Fozard (1980) summarized the findings according to five types of memory: sensory, primary, secondary, working, and tertiary. Table 18.2 provides some examples of these five types of memories.

In general, the findings suggest that secondary and working memory decline with age (Botwinick, 1977; Belbin & Belbin, 1968), while primary and tertiary memory do not (Craik, 1977; Botwinick, 1977).

Table 18.2 Examples of Remembering with Different Types of Memory

Sensory	Shortest time required to identify the letters in the name of someone you have just met.
Primary	Ability to recall the name of someone you have just met immediately after meeting the person.
Secondary	Ability to recall the name of someone you have met after meeting 10 other people.
Working	Ability to recall a rule, such as using only the first names of the 11 people you just met.
Tertiary	Ability to recall information learned long ago, such as vocabulary (what Horn calls *crystallized intelligence*).

Source: From James L. Fozard, "The Time for Remembering" in L. W. Poon (ed.), *Aging in the 1980's: Psychological Issues*. Copyright © 1980 by the American Psychological Association, Arlington, VA.

We also need to consider how the speed of memory processing is affected by age. Fozard's (1980) review found that the time required to recognize a task and prepare a response seems to increase with age with all types of memory. He also found that the time it takes to search memory for a correct response—the time for decision making—increases for most memory types except tertiary.

One implication of this research on memory is that since tertiary or long-term memory does not seem to decline with age, generally the older we get, the larger the knowledge base from which we draw information. An older person, by virtue of a greater number of retained experiences, remembers more things than a younger person.

Of course, all of these processes are interrelated. They can be studied separately but actually they work together. Declines in one process, such as memory, can affect the performance of another, such as problem solving. By the way, the evidence of decline in problem-solving skills is far from clear. Research in this area is relatively recent compared to memory or perception research.

So some cognitive processes seem to decline with age and some do not. This leads us to one final question. When all the pieces are put together, how well can adults learn new things as they proceed through middle adulthood?

Learning Ability

Is there a serious drop in ability to learn from early to late adulthood? Despite some classic studies (e.g., Eisdorfer & Wilkie, 1973; Hulicka, 1967; Knowles, 1984; Taub, 1975), there is still considerable disagreement over this matter.

For example, many studies have shown a marked decline in paired associate learning (the ability to remember associations between two lists of words) (Kimmel, 1974). Obviously, memorizing pairs of words is itself of no great importance, but much essential learning is based on this skill. In most of these studies, however, the measure of how well the associations have been learned is speed of response. Do older persons perform poorly because they are slower to *learn,* or only because they take longer to *show what they have learned?* As Botwinick (1977) reports:

> ***The research strategy has been to vary both the amount of time the stimulus is available for study and the amount of time that is available for response. A general finding is that elderly people need more time for responding than typically is provided; they are at a disadvantage when this***

AN APPLIED VIEW The Lifelong Learning Resource System

As work environments undergo more and faster changes, there is a growing emphasis on learning how to learn, rather than on mastering a specific skill that may become obsolete. Secondary education or even a college education may no longer suffice for an entire career. Workers require "regular booster shots of education and training" (Abeshouse, 1987, p. 23). We need a system for "lifelong learning." Corporations and policymakers are grappling with the issues of retraining technologically displaced workers, especially older workers (Wallace, 1989).

Knowles (1984) makes a marked distinction between teaching aimed at children and youths (**pedagogy**) and the teaching of adults (**androgogy**). Adults differ from younger persons in the following ways:

- *The need to know.* Adults need to know why they need to learn something before undertaking it.
- *The learners' self-concept.* Adults have a self-concept of being responsible for their own decisions, for their own lives.
- *The role of the learners' experience.* Adults come into an educational activity with both a greater volume and a different quality of experience than youths.
- *Readiness to learn.* Adults become ready to learn those things they need to know and be able to do in order to cope effectively with their real-life situations.
- *Orientation to learning.* In contrast to children's and youths' subject-centered orientation to learning (at least in school), adults are life-centered (or task-centered or problem-centered) in their orientation to learning.
- *Motivation.* While adults are responsive to some external motivators (better jobs, promotions, higher salaries, and the like), the most potent motivators are internal pressures (the desire for increased job satisfaction, self-esteem, quality of life, and the like).

time is not available. When sufficient time for response is available, the performance of elderly people is only slightly inferior, or not inferior at all, to that of young people. **(p. 278)**

Does the apparent drop in learning ability from early to late adulthood occur strictly because of ability, or because of factors such as motivation and dexterity?

Another factor in learning ability that has been studied is motivation to learn (Botwinick, 1977). It has been suggested that persons in middle and late adulthood are less motivated to learn than younger people. Further, it appears that often they are aroused and anxious when placed in a laboratory situation for testing their learning ability. To the extent that this is true, their ability to learn is underestimated.

It has also been found in laboratory experiences that older adults are more likely than younger adults to make the "omission error." That is, when they suspect they may be wrong, they are more likely to refrain from responding at all, and therefore are scored as having not learned the task. But when asked what they think the answer is, they are often right.

Also, the meaningfulness of the task affects motivation. It is clear that the motivation for middle-aged and elderly adults is different from that of younger adults, and many experiments have not taken this into consideration in studying learning ability.

Decline in cognitive ability due to aging can often be offset through motivation and new learning experiences. This is shown by the growing number of adults who return to formal education in middle adulthood.

Why are more and more adults going back to school? In the past, children could be educated to deal with a society and workplace that would remain essentially unchanged for their entire lives. Today the rate of change and innovation is too rapid. Fifty-year-old men and women who were born before the computer was being invented, and certainly didn't learn about computers in school as children, are now routinely asked to use them at work. It is now commonplace for people to return to school in order to advance or even maintain their present career. Also, sometimes an adult returns to school in order to learn a new skill or hobby that will be enjoyable during retirement.

Women in particular are going back to school in large numbers. Middle-aged women of today were not encouraged to pursue higher education when they were young. Some women are trying to catch up in order to be competitive with their male counterparts. Other women may return to school, not to study accounting or computer science, but to study literature or history, just to broaden their knowledge and for their own personal enjoyment. Many need occupational training to support themselves and their families after a divorce.

The education system is having to adapt to this change. Most adults cannot afford to quit their current jobs to go back to school. More schools are now offering evening and part-time programs. Many schools offer courses that can be taken at home, in some cases taking advantage of technology such as TV or computers.

Not all new learning by middle-aged adults takes place in a formal school setting. Employees in many different types of work settings are asked to keep abreast of new innovations. Other social experiences, such as practicing a religion, going to a library, watching television, and serving as a volunteer, involve new learning. Learning is now more than ever a lifelong task.

Conclusion

What is true of personality and social development appears to be especially true of physical and mental development: *What you expect is what you get.* If you expect

- Your weight to go up
- Your muscles to grow flabby and weak
- Your senses to dull
- Your climacteric to be disruptive
- Your intelligence to drop
- Your creativity to plummet
- Your sexual interest and ability to decline they probably will.

This is called the self-fulfilling prophecy. People who take a positive outlook, who enthusiastically try to maintain their bodies and minds, have a much better chance at success. They are also better able to deal with the stresses that life naturally imposes on us all.

This is not to say that we can completely overcome the effects of aging. It means that through our attitudes, we can learn to deal with them more effectively. In the next chapter we look at the psychosocial features of middle age, which depend so much, as we have said before, on physical and cognitive development.

Chapter Highlights

Physical Development

- Health concerns in middle adulthood include increasing weight and lower metabolism.
- In the middle period of adulthood there is a common but often unnecessary decline in muscular ability, due in part to a decrease in exercise.

- Sensory abilities—vision, hearing, smell, and taste—begin to show slight declines in middle adulthood.
- The climacterium, the loss of reproductive ability, occurs at menopause for women, but at much older ages for most men.

Sex in Middle Adulthood

- Frequency of sexual intercourse appears to increase over the middle years, as well as marital satisfaction.
- Some minor changes in sexual physiology occur for both sexes, but usually these need not hamper sexual satisfaction.

Cognitive Development

- Horn suggests that while fluid intelligence deteriorates with age, crystallized intelligence does not.
- Baltes has proposed a dual-process model of intelligence, which suggests a decline in the mechanics of intelligence (classification skills, logical reasoning), yet an increase in the pragmatics of intelligence (social wisdom).
- Creativity, important in a rapidly changing world, manifests itself at different peak periods throughout adulthood.
- The information-processing approach to the study of intelligence, employing computer analogies, examines different processing capabilities in attention, perception, memory, and problem solving.
- Learning in middle adulthood can be enhanced through motivation, new learning experiences, and changes in education systems.

Key Terms

Androgyny
Automatic attention
Basal metabolism rate (BMR)
Climacteric
Climacterium
Crystallized intelligence
Dual-process model
Effortful attention
Estrogen replacement therapy (ERT)
Fluid intelligence
Glaucoma
Male change of life
Menopause
Olfactory sense
Pedagogy
Self-fulfilling prophecy
Senile macular degeneration

What Do You Think?

1. When you look at the physical shape your parents and grandparents are in, and their attitudes toward the subject, do you see evidence of the self-fulfilling prophecy?
2. If you are a female and have not yet gone through menopause, what do you anticipate your feelings will be about it?
3. If you are a male, can you imagine what women facing menopause must be feeling?
4. What are some ways that our society might foster the cognitive abilities of its adult citizens?
5. What are some ways that our society might foster the creative abilities of its adult citizens?
6. What are some ways that our society might foster the learning of its adult citizens?
7. What is your attitude toward your parent's sexuality?

Suggested Readings

The Boston Women's Book Collaborative. (1984). *Our bodies, growing older.* Boston: self-published. An excellent update of the popular *Our bodies, our selves.* This book is for middle-aged and elderly women.

Gardner, H. (1983). *Frames of mind: The theory of multiple intelligences.* New York: Basic Books. Offers a comprehensive view of the numerous faces of intelligence.

The Journal of Creative Behavior. Pick up any volume of this fascinating journal at your library and browse through it. A wide variety of interesting topics are covered, and more often than not the articles are written creatively.

Nilsson, L. and J. Lindberg. (1974). *Behold man: A photographic journey of discovery inside the body.* New York: Delacorte. A book of photographs, many of them pictures enlarged thousands of times. This is a magnificent description of the human body.

Wolfe, T. (1987). *The bonfire of the vanities.* New York: Farrar, Straus & Giroux. With his customary verve, Wolfe looks inside the heads of five men and two women, all New Yorkers in early middle age, and shows us how they think.

CHAPTER 19

Middle Adulthood: Psychosocial Development

Marriage and Family Relations 499

Marriage at Middle Age 499

Relationships with Aging Parents 500

Relationships with Siblings 500

The Middle-Aged Divorcée 502

Personality Development: Continuous or Changing? 504

Continuity versus Change 504

Transformations: Gould 505

Seasons of a Man's Life: Levinson 506

Seasons of a Woman's Life: Levinson 511

Adaptations to Life: Vaillant 512

Generativity versus Stagnation: Erikson 514

Continuous Traits Theory 515

Patterns of Work 516

Special Problems of the Working Woman 517

The Mid-Career Crisis 518

Some Suggestions for Dealing with the Mid-Life Crisis 520

Conclusion 520

Chapter Highlights 520

Key Terms 521

What Do You Think? 522

Suggested Readings 522

Recently, a middle-aged woman friend shared some reflections with me:

I remember walking to school one day in the second grade, chatting with my girlfriend's mother as she escorted us. I told her that I had noticed how much more quickly the day seemed to pass than it used to. Seven seemed a very advanced age to me then, so I was sure this phenomenon was related to being finally grown-up. Later that year we moved into a new house in a new community. Moving day was very exciting. The real grown-ups were very busy, so the most entertaining thing I had to do was to sit around and think about my life. Moving seemed to have wrapped up the first part of my life into a discrete little package. And it came to me that there I was, almost *eight* years old, and I didn't have a *feeling* for all that time. I promised myself, as I sat in our old, soft maroon chair, holding some of my accumulated possessions dislocated by their recent journey, that five years later to the day I would sit again in the same spot, in the same position, holding the same objects. Then, I figured, with all the awareness born of old age, I would really know what five years would *feel* like. And five years later I did just that.

I am still trying to comprehend or capture a sense of time passing. Now only the units have changed. Every once in a while I hold very still and try to catch twenty years. Twenty years feel like those five did long ago. Twenty from now, if I'm lucky, I'll be staring my death in the face. It's all so odd. Somewhere inside I *was* all grown-up when I was seven. That "me" hasn't really aged or changed much, and it's still watching as the world wrinkles on the outside. Days are minutes, months are weeks, and years are months. I'll probably be menopausal in the morning!

Quoted by Lila Kalinich, MD, in "The Biological Clock," 1992 ■

After reading chapter 19, you should be able to:

- Explain what is meant by "emotional divorce" and the "empty nest syndrome."
- List some components of a happy marriage.
- Discover how relationships with parents and siblings change in middle adulthood.
- Discuss the positive and negative effects of divorce on middle-aged adults and their families.
- Determine your position on whether personality development is continuous or changing.
- List Levinson's three major developmental tasks for middle-aged men, along with the four polarities men must confront.
- Describe the differences between Levinson's research with men and his current research with women.
- Explain and compare Gould's Stage IV Transformation, Vaillant's concept of "adaptive mechanisms," and Erikson's stage, "generativity versus stagnation."
- Define the NEO model of personality.
- Highlight particular employment concerns among middle-aged adults.
- Describe the five major problem areas for working women.
- Discuss what is meant by a "mid-career crisis," as well as ways of dealing with it.

Marriage and Family Relations

In this section, we consider four important aspects of middle-aged life today: relationships in marriage, relationships with aging parents, relationships with brothers and sisters, and divorce. Although not all would agree, we define this part of life as going from 35 to 64 years of age.

Marriage at Middle Age

Middle age is often a time when husbands and wives reappraise their marriage. The mid-life transition (a period during which people seriously reevaluate their lives up to that time) often causes a person to simultaneously examine current relationships and consider changes for the future. Often, whatever tension that exists in a marriage is suppressed while the children still live at home. As they leave to go off to college or to start families of their own, these tensions are openly expressed.

Sometimes couples learn to "withstand" each other rather than live with each other. The only activities and interests they share are ones that revolved around the children. When the children leave, they are forced to recognize how far apart they have drifted. In effect, they engage in **emotional divorce** (Fitzpatrick, 1984).

But most couples whose marriages have lasted this long have built the type of relationship that can withstand reappraisal, and they continue for the rest of their lives. Census data suggest that the highest rates for separation and divorce occur about five years after the beginning of the marriage (Sweet & Bumpass, 1987).

In fact, some studies suggest that this period after the children leave home is like a second honeymoon (Campbell, 1975; Rhyne, 1981). After the initial period of negative emotions that follow this disruption of the family, often called the **empty nest syndrome,** married couples can evaluate the job they have done with their children. They can pat themselves on the back for a job well done and then relax now that a major life goal has been accomplished. They then realize that they have more freedom and privacy and fewer worries. They usually have more money to spend on themselves. And this period after the children leave home is now much longer than it used to be. Husbands and wives now can look forward to spending twenty or thirty years together as a couple rather than as a large family.

The period in a marriage after which there is no longer a need to care for children is often a time for pursuing dreams that were previously impractical.

The Happy Marriage

What does research have to say about the components of a happy, or at least a lasting, marriage? Gottman and Krokoff (1989) conducted a longitudinal study looking at the types of interactions between husband and wife and the effect on marital satisfaction. Earlier research suggested that there was always more negative interaction in unhappy marriages than in happy marriages.

Gottman and Krokoff decided to look at the effect of different types of negative interactions rather than one global category. They found that certain types of conflict may in fact be *positive* factors in a happy, lasting marriage. They also found that certain types of conflict, particularly defensiveness, stubbornness, and withdrawal on the part of the husband, indicated that a marriage was in trouble.

Gottman and Krokoff assign to the wife the role of manager of marital disagreements, and suggest she get her husband to "confront areas of disagreement and to openly vent disagreement and anger" (p. 50). Most husbands tend to try to avoid relational confrontations (Moyers & Bly, 1990). Therefore, overcoming this reluctance can have extremely beneficial long-term effects on the marriage.

The Unmarried

About one in twenty people in middle age have never been married (Sweet & Bumpass, 1987). In general, a person who has never married by middle age will not get married. Such people tend either to have very low or very high education levels. At the low extreme, of those who have less than five years of school, one person in seven has never married. The factors that kept these people from completing school, such as mental illness or other handicaps, are probably the same ones that make them less likely to get married. At the other extreme, 13 percent of middle-aged women with seventeen or more years of education have never been married.

There are a number of possible explanations for this. These women may choose higher education and a career over marriage. They may believe marriage will hold them back. A cultural factor may be at work, since some men feel threatened by women with more education than they have. Perhaps women who delay marriage to pursue an education end up having a smaller pool of available men to choose from.

Relationships with Aging Parents

Relationships and roles among the generations are always changing.

Middle age is also a time when most people develop improved relationships with their parents. Middle-aged children, most of them parents themselves, gain a new perspective on parenthood and so reevaluate the actions taken by their own parents (Farrell & Rosenberg, 1981). Also, grandchildren can strengthen bonds that may have weakened when people left their parents' home during early adulthood.

In many cases, however, the relationship begins to reverse itself: As elderly parents grow older, they sometimes become as dependent on their middle-aged children as those children once were on them. Most people fail to anticipate the costs and emotional strains that the aging of their parents can precipitate. This can lead ultimately to new sources of tension and rancor in the relationship.

The most frequently cited problem of middle-adult women is not menopause or aging, but caring for their aging parents and parents-in-law (Massachusetts Women's Health Study, 1988; James, 1990). This is not so surprising when we realize that it is virtually always the daughter(s) in the family who is responsible for the care of the elderly parents (Brody & Schoonover, 1986; Kendig, Hashimoto, & Coppard, 1992; Matthews, 1987). Brody and associates (1986) found that the daughter who is the primary caretaker relies on siblings, especially her sisters if she has them, for emotional support. These sisters typically feel guilty about not doing enough. On the average, brothers provide less help and feel less guilty about it. It seems that the demands on women to fulfill the role of family nurturer are deeply rooted and powerful. Of course, as more women enter the workplace, and as more of them become assertive about equitable family responsibilities, this pattern may well change, with males assuming more of the burden.

There are two other features of family life in middle adulthood that we will deal with here: relationships with siblings and the problem of divorce.

Relationships with Siblings

Developmental psychology has, for some time now, recognized the importance of sibling relationships for a child's cognitive and social growth. But do these special relationships stop contributing to a person's development after adolescence? Does the relationship slowly decline in importance as one ages? Are the characteristics of the relationship the same in middle adulthood as they were in childhood? Psychological research has recently begun to focus on some of these questions.

The fact is, sibling relationships have the potential to be the most enduring that a person can have. You don't usually meet your spouse until young adulthood or at least adolescence. Your parents usually pass away before you do. You usually pass away before your children do. But most siblings are born within a few years of each other, and such a relationship can last 60, 80, or even 100 years!

Nevertheless, sibling relationships do change. As Cicirelli (1980), a prominent researcher on adult sibling relationships, notes:

> ***At the beginning it is one of intimate daily contact and sharing of most experiences, including the socializing influence of the same parents. Throughout the school years, siblings may have different teachers and different friends and peer groups, but they still have their home experiences in common. Later, when they leave their parents' home to pursue a career or to marry and establish families of their own, they tend to separate from each other as well. They may live in different cities or even different countries. Contact becomes voluntary except on certain ritual occasions, and most life experiences are no longer shared. Still later, they may share the obligations of caring for their parents during their declining years. With the death of the parents, sibling contact returns to a more voluntary level until the end of life.*** **(p. 455)**

What then are some of the characteristics and effects of such a long and evolving relationship?

Although the research on adulthood sibling relationships is considerably less abundant than that on childhood and adolescent sibling relationships, there are some partial answers to this question. Some researchers have measured change of the relationship with age by looking at how close siblings live to one another and how often they see one another (Rosenberg & Anspach, 1973). They reason that such contact is necessary for a relationship to exist. This research indicates that sibling relationships do decline with age. However, other researchers (Allan, 1977; Cicirelli, 1979) suggest that measures of feelings of closeness and affection are better indicators of a relationship than proximity and frequency of contact. This research supports the notion of strong sibling relationships even in old age. One consistent finding is that the relationship between sisters tends to be stronger than the relationship between brothers.

What is the nature of adult sibling relationships? As in childhood, there is often rivalry, in addition to closeness. Some researchers have found this rivalry to be very common, especially among adult brothers (Adams, 1968). Other researchers suggest that childhood rivalry dissipates in adulthood (Allan, 1977), largely because the siblings have less contact with each other as they grow older.

One would hope that a growing maturity would also lessen rivalry. Certainly adult siblings are faced with more serious and important tasks than are childhood siblings. For example, most middle-aged siblings must make mutual decisions concerning the care of their elderly parents and eventually deal with the aftermath of their death.

Another consideration is the effect of changing family patterns on sibling relationships. Couples now are having fewer or even no children. Children will therefore be less available to parents for companionship and psychological support. On the other hand, parents and their siblings will be living longer, more active lives. The obvious conclusion is that sibling relationships will become more and more important in the future.

The Middle-Aged Divorcée

Because divorce rates have been increasing for those over 40, mid-life divorce will become an important focus in the future (Uhlenberg, Cooney, & Boyd, 1990). The divorce rates are higher for second or subsequent marriages, for African-Americans, and for those who are less educated and who have lower earning ability. Although mid-life is a time when a divorce is less likely to occur (see Figure 19.1), the proportion of divorced persons in mid-life is relatively high, since many who divorced earlier have never remarried. Men and women over 40 experience significantly more turmoil and unhappiness than younger people at the same stage of divorce (Chiriboga, 1989). The reasons are clear: length of time of marriage, complex economic and property linkages, a web of social relationships, and a generally higher standard of living.

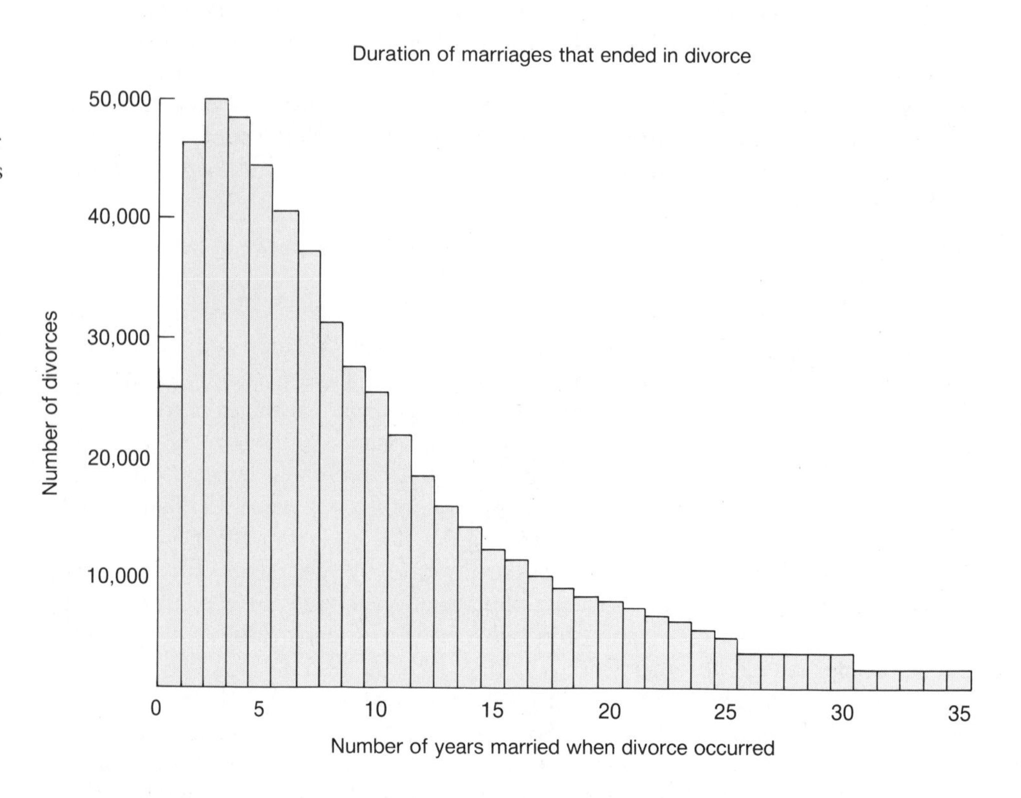

Figure 19.1
The chart of divorces in the United States shows that most divorces occur within the first five years of marriage, with the peak at three years. Interestingly, the same pattern is found in most other societies, ranging from contemporary Sweden to the hunting and gathering groups of southwest Africa.

In respect to the last point, it should be noted that women, especially at mid-life, are often hard hit economically by divorce. The National Longitudinal Survey found they experience significant declines in income, increased rates of poverty, and dramatic life-style changes upon separation, divorce, or widowhood; they do not recover unless they remarry (Arendell, 1987; Hoffman & Duncan, 1988). Furthermore, since most women who divorce at mid-life are already working, there is little room to maneuver for more income after divorce. This is especially true for African-American women, since the only notable increase in workforce participation after divorce is for white women (Morgan, 1991).

More and more children are living in homes without fathers. Over half of single-parent mothers are not receiving full child support, and half of those are not receiving any. While 68 percent of white women were awarded child support, only 35 percent of African-American women and 41 percent of Latino women received such awards. It's no wonder that marriage has been found to be an important predictor of well-being at mid-life, especially for both African-American and white women (Crohan & others, 1989).

A discomforting finding of White (1992) is that children of divorce, even after they have left home and their parents enter middle age, receive significantly less support (child care, advice, transportation, loans, and so forth) than do children whose parents have remained married. Remarriage does not seem to increase or decrease this support deficit.

In general, new divorce laws were considered to have a liberating effect on women. Until the liberalization of divorce laws and attitudes, women usually had no choice but to endure a difficult and sometimes even abusive marriage. Cultural norms allowed men much more moral latitude, the so-called double standard. Women basically just had to put up with it.

Before liberalization, it was necessary to establish "reasonable grounds" for divorce, such as adultery or physical abuse. Women who were granted divorces were almost always awarded larger alimony settlements than they receive today, in order to make up for the loss of income. This was necessary because women lagged far behind men in their earning potential, and often women stayed at home to raise the children and run the household while the husband advanced a career. A woman could not survive a divorce if it left her and her family destitute.

In 1970, the divorce laws were liberalized in California, a change soon followed across the country. This was the introduction of a **no-fault divorce.** Weitzman (1985) contends that the new divorce laws have impoverished divorced women and their children. A major facet of the new laws is supposed to be their gender neutrality. Men and women are treated equally under the divorce law, and this includes the division of property and alimony.

But perhaps these laws were premature. Men and women did not in 1970, and still do not today, live in an economically equitable system. Men still possess a competitive advantage in earning potential, and women are still held responsible by the culture for raising the children and running the household. When a divorce judge divides up the property equally, grants little or no alimony, and provides inadequate child support, it is the woman and children who suffer.

This is particularly true of the middle-aged homemaker. The courts seldom give much recognition to the years spent running the household while the husband invested time in advancing a career. Such a woman cannot reasonably be expected to compete with others who have been gaining experience in the marketplace while she stayed at home.

In the year following a divorce, a woman and her children can expect a 73 percent decline in their standard of living, according to Weitzman's research. The former husband can expect a 42 percent increase. After five years, the woman's standard of living will likely be 30 percent lower than what it was during the marriage, while the husband's will be 14 percent more.

Another effect is that no-fault divorce laws mandate an even division of property. This often means that the family house must be sold. Before these laws, the house was almost always given to the woman and her children. Besides being an economic burden, the loss of the house has psychological consequences for the development of the children. The dislocation often requires that the children leave their school, neighborhood, and friends. And this usually comes just after the divorce, a period of tremendous stress for any child.

There are some indications that reform is on the way. In California, the rules for division of property and spousal support are being revised. More things are now being considered family assets, to be divided equally, such as the major wage earner's salary, pension, medical insurance, and future earning power. In addition, some states have now become aggressive at pursuing husbands who are delinquent on child support payments. Their names are being published and a part of their wages is being withheld and given to the ex-wife. Massachusetts is even considering paying spouses the money lost through delinquent payments, and reimbursing itself through taxes and garnishments of the other spouse.

Divorce will probably never be the "civilized" process that the early proponents of no-fault divorce laws hoped it would become. At best, we can expect the suffering of spouses to become more equal, and the suffering of the children to be reduced.

Marriage and family life are but two of the factors that have a marked effect on personality development, to which we now turn.

Personality Development: Continuous or Changing?

The controversy over whether personality is stable or changing is relevant in middle adulthood, too.

Continuity versus Change

All of us have heard someone say, "Oh, he's been like that ever since he was a baby!" Such a comment doesn't sound like a philosophical statement, but think about what it implies. It implies that a person can remain basically the same throughout his or her lifespan. This is the fundamental question addressed by the issue of *change versus continuity of development.* Do human beings ever really change, or do we all stay pretty much the same? It is the focus of great debate by child-development and lifespan psychologists alike because of its implications.

If we assume that people remain the same regardless of what happens to them as their life continues, then the period of early childhood takes on great meaning (Brim & Kagan, 1980; Kagan, 1984; Rubin, 1981). Several of the developmental theorists we have discussed (e.g., Freud and Piaget—see chap. 2) have focused much energy on the early years of childhood in the belief that what happens to a person during childhood determines much of what will happen to him or her in the future.

Conversely, others (e.g., Erikson and other adult development theorists) believe that because people are constantly changing and developing, all life experiences must be considered important. In that case, early childhood becomes a somewhat less significant period in the whole of development, and adolescence and adulthood come more into focus. It also implies that getting children "off on the right foot" is not enough to assure positive development. These are just some of the implications of the debate about personality continuity versus change.

In the study of adulthood, the issue of continuity versus change gets even more complex. In general, there are two distinct theoretical positions in the study of adult personality. There are those who feel that adults remain basically the same—that the adult personality is stable. This is *continuity* in adult development. There are other theorists who view the adult as constantly in a process of change and evolution. That is what the position of *change* refers to.

The study of continuity versus change in adulthood is complicated by the many ways that the issue is studied. Some researchers look at pieces of the personality (personality traits) as measured by detailed questionnaires. They think that the answers to such questionnaires assess adult personality. These researchers are known as **trait theorists.**

Others note that such questionnaires measure only parts of the personality. They argue that adult personality is much more complicated than any list of personality traits. What is interesting to them is how those traits fit with the whole of the person. Beyond that, they are also interested in how an adult's personality interacts with the world around him or her. They believe that research based on personality traits is too narrow in focus, and that we must also look at the stages of change each person goes through. They are known as **stage theorists.** The old saying, "you can't see the forest for the trees," sums up their position—the parts prevent you from seeing the whole.

The differences in how to go about measuring adult personality complicate the study of continuity and change in adulthood. Researchers use extremely different methods of measuring adult personality and then relate their findings to support either the position of continuity or change. In general, trait theorists have found that the adult personality remains the same: their work supports *continuity* (McCrae & Costa, 1984). McCrae and Costa summed up their research in the title of an article: "Still Stable after All these Years." Researchers looking at the whole of the adult personality through extensive interviews have found support for the notion of *change* in adulthood (Erikson, 1963, 1975; Gould, 1978; Levinson, 1978, 1986, in press; Vaillant, 1977).

The study of continuity versus change in adult psychology has important implications, just as it does in childhood psychology. The findings of personality studies add to our knowledge of what "normal" adult development means. Yet as we have seen, the studies vary in their definition of what adult personality is and how it should be measured. Not surprisingly, they also differ in what their studies tell us about normal adult development. They have different answers to the question: "If nothing very unusual happens (like a catastrophe), how will the adult personality develop?"

Trait theorists like McCrae and Costa might say, "If nothing unusual happens, then the adult personality will stay relatively the same. Normal adult personality development is really the *maintenance* of personality." Gould, Levinson, Vaillant, and Erikson, looking at the whole of the adult, would respond differently (see also chap. 17). They might say, "The adult personality naturally and normally develops through change. Normal adult personality development is a continual process of growth and change."

Who is right? We suggest you read our summary of the studies that each camp provides, and try to make up your own mind. We begin with Roger Gould's theory, as it applies to middle age.

Transformations: Gould

The part of Roger Gould's theory of personality development that deals with middle adulthood is labeled Stage IV. Let us remind you that the basis of his theory is the concept that all of us unconsciously learn assumptions about the nature of life. In his study of adult women and men, he found that healthy personality development depends on rejecting these "major false assumptions."

The mid-life decade (ages 35 to 45) is often a time in life when one's parents, other relatives, and some close friends go through serious illness and death. It is also the time when our self-deceptions and the lies of others can have repercussions far more serious than ever before, because of the greater power that middle-aged persons usually have. Therefore, the major false assumption that tends to be rejected at this time of life is "There is no evil or death in the world. Everything sinister has been destroyed."

Five false assumptions accompany this erroneous belief, all of which should now be abandoned:

- "The illusions of safety can last forever."
- "Death can't happen to me or to my loved ones."

- "It is impossible to live without a protector." (a substitute "mother" or "father" who is usually the person's spouse)
- "There is no life beyond this family."
- "I am innocent."

Even if one's parents are still living and well at this time, most adults experience a role reversal with them. Parents who are in their sixties and seventies often become dependent on their middle-aged children, thus bringing the cycle full circle. There is a growing sense of vulnerability, of the passage of time, and a realization of what is truly important. Men at this time say that they no longer fear their bosses or idealize their mentors. Women begin to realize that male protectors really are not all that necessary. Those who come to this realization, both male and female, have a feeling of freedom never before experienced.

Gould realizes, of course, that there are developments in life after the mid-life transition, but his research ends at this period.

Seasons of a Man's Life: Levinson

This section continues a discussion of Levinson's (1978) theory that was begun in chapter 17. Figure 19.2 reproduces that part of his theory that applies to middle adulthood.

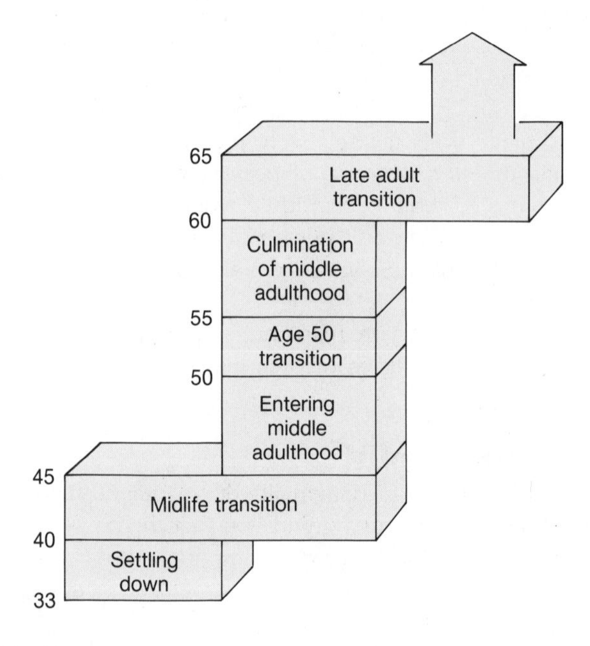

Figure 19.2
Levinson's theory—middle age

Settling Down

The settling-down phase usually extends from age 33 to 40. At this time, most men have pretty well decided what occupation they choose to pursue. During this period, most men attempt to achieve two tasks: (1) establish a niche in society, and (2) advance up the ladder of the occupational group. During this phase, the male attempts to overcome his dependency on his mentor, and slowly is able to "become his own man." This is a step in the direction of greater individuation, in that the man thinks less and less as others want him to, and more and more as his own views dictate. He is now ready to go into the third phase of his adulthood, the mid-life transition.

The Mid-Life Transition

The **mid-life transition,** which usually lasts for five years, generally extends from age 40 to 45. It involves three major developmental tasks:

- The review, reappraisal, and termination of the early adult period.
- Decisions as to how the period of middle adulthood should be conducted.
- Dealing with the polarities that are the source of deep division in the man's life at this stage. These polarities, which represent the continual struggle toward greater individuation, are (1) young/old (2) destruction/creation, (3) masculinity/femininity, and (4) attachment/separation.

The so-called midlife crisis—do you think most middle-aged men go through one?

Young/Old

Young/old is the major polarity to be dealt with during the mid-life transition. Levinson refers to Freud's disciple, Carl Jung. Jung (1966) suggested that the experience of thousands of generations of human beings has gradually produced deep-seated ideas that each of us must learn to deal with. The major one is that although we begin to grow old even at birth, we are also interested in maintaining our youth, if only to avoid the ultimate consequence of our mortality: death.

In tribal symbolism, the word *young* represents fertility, growth, energy, potential, birth, and spring. *Old,* on the other hand, encompasses termination, death, completion, ending, fruition, and winter. Until the age of 40, the man has been able to maintain his youthful self-image. Through producing and raising children, and in some cases through a creative product such as a book, invention, or painting, he has been able to see himself as part of a new and youthful recycling of life. Subconsciously, at least, he has been able to maintain the myth of immortality.

At about the time of his 40th birthday, he is confronted with evidence of his own declining powers. He is no longer able to run, play tennis, or shoot basketballs as effectively as he could in his twenties. He sometimes forgets things, and his eyesight may not be as good. Even more damaging to his hope for immortality is the illness of his friends. There are more frequent heart attacks, strokes, and other serious illnesses among people of his age. In many cases, his parents suffer serious illnesses, or even die. These events lead to one inevitable realization: He is going to die, and perhaps in the not-too-distant future. Even the 32 years left him (on the average) do not seem like much, because over half of life is now already past.

Now there is a sense of wanting to leave a legacy. Most individuals want to feel that their life has made some difference, and they want to leave something behind them that can be remembered. Therefore, it is typical at this time that the individual becomes more creative and often works harder than he has in the past to make a contribution considered worthwhile by those who follow him.

Destruction/Creation The male going through a mid-life transition realizes not only the potential of the world to destroy, but also his own capacity to be destructive. He recognizes the evil within himself and his own power to hurt, damage, and injure himself and others. He knows, if he is honest with himself, that he has not only hurt others inadvertently, but sometimes with clear purpose. He sees himself as both victim and villain in the "continuing tale of man's inhumanity to man" (Levinson, 1978, p. 224).

The more honest he is with himself, the more he realizes how tremendous is his capacity to destroy. There is a bonus to this honesty, however: In recognizing his power to be destructive, he begins to realize how truly powerful he can be in creating new and useful forms of life. As with the young/old polarity, he now attempts to strike a new balance between his destructive and creative sides.

Masculinity/Femininity Levinson again borrows from Jung, using his concept that all persons have a masculine and feminine side, and that they emphasize one over the other because of the demands of society. This emphasis often costs us greatly. A rich adulthood can be achieved only by compensating for that part of us that was denied during our childhood. In most males, the feminine side has typically been undernourished, and now must come to the fore if they are to be all they are capable of being.

According to Levinson, in early adulthood, femininity has a number of undesirable connotations for the male. To the young man, masculinity connotes bodily prowess and toughness, achievement and ambition, power, and intellectuality. To him, the feminine role represents physical ineptness, incompetence and lack of ambition, personal weakness, and emotionality. Now is the time when the polarity between these self-concepts must be seen and reconciled.

The male who is to achieve greater individuation now recognizes that these dichotomies are false, and that he does indeed have a feminine side that must also be nourished. The mature male is able to allow himself to indulge in what he before has disparaged as feminine aspects of his personality. Such a male feels secure enough in his masculinity to enjoy his ability to feel, to nurture, to be dependent. Levinson suggests such men are now freer to assume more independent relationships with their mothers, to develop more intimate love relationships with peer women, and to become mentors to younger men and women alike.

Attachment/Separation By the attachment/separation dichotomy, Levinson means that each of us needs to be attached to our fellow members of society, but also to be separate from them. As the human being develops, there is a fascinating vacillation between each of these needs. In childhood, there is a clear-cut attachment to mother and later to family. Children need support because of their incompetence in dealing with the complexity of the world around them.

During adolescence, this need switches toward an emphasis on separateness from family, as the individual proceeds through the identity moratorium. During this time, most adolescents need to separate themselves not only from their parents, but from the entire society around them in order to try out new ways of being. This need vacillates back toward attachment during early adulthood. The ultimate goal, of course, in *interdependence* (See chap. 6).

Throughout their twenties and thirties, most men are involved in entering the world of work and in family and have a strong attachment to others who can help them be successful in these goals. Now, in the mid-life transition, a new separateness, perhaps a second adolescence, takes place. The man, especially the successful man, begins to look inside and to gain greater awareness of his sensual and aesthetic feelings. He becomes more in touch with himself by being temporarily less in touch with the others around him.

Because the men interviewed so extensively by Levinson and his colleagues were between the ages of 35 and 45, their study of adult development ends at the mid-life transition. Levinson recognizes, however, that there is still a great deal to be learned about development after this stage. He ends his book by encouraging those who are attempting to develop theory and research on the periods following this stage of life.

Farrell and Rosenberg's Findings

Contrary to Levinson's position, some researchers feel that although mid-life crises do occur in some individuals, they are not a universal part of adult development (Costa & McCrae, 1980). Further, they believe that because middle-aged adults have

AN APPLIED VIEW — The Mid-Life Crisis Scale

Unless noted otherwise, all questions are answered with this six-point scale:

1. Strongly Agree 2. Agree 3. Slightly Agree
4. Slightly Disagree 5. Disagree 6. Strongly Disagree

1. Marriage is as rewarding and enjoyable after 15 or 20 years as it is in the earlier years.
2. Many men I know are undergoing what you would call a change of life or middle-age identity crisis.
3. Almost any job or occupation becomes routine and dull if you keep at it for many years.
4. I am still finding new challenges and interest in my work.
5. In some ways, I wish my children were young again.
6. When your child grows up, he is almost bound to disappoint you.
7. Many people claim that middle age is one of the most difficult times of life. Has it been (or do you think it will be) that way for you? (1. Very Much So; 2. Somewhat; 3. Perhaps; 4. I Doubt It; 5. Not at All)
8. I wish I had the opportunity to start afresh and do things over, knowing what I do now.
9. Many of the things you seek when you are young don't bring true happiness.
10. I find myself thinking about what kind of person I am and what I really want out of life.
11. A person must remain loyal to his commitments if they do not turn out the way he expected.
12. How would you characterize your relationship to your wife now? (1. Very Close; 2. Close; 3. Neither Close Nor Distant; 4. Distant; 5. Far Apart)

Physical Health Questions

1. Do you have any particular physical or health trouble?
   If yes, what is it? ____________________
   ____
2. Have you ever had the following diseases: Asthma
   If yes, when was that? ____________________
   ____
3. Hay fever
   If yes, when was that? ____________________
   ____
4. Skin trouble
   If yes, when was that? ____________________
   ____
5. Stomach ulcer
   If yes, when was that? ____________________
   ____
6. Do you feel you are bothered by all sorts of pains and ailments in different parts of your body?
   ____
7. Have you ever felt you were going to have a nervous breakdown? ____________________

Source: From M. P. Farrell and S. D. Rosenberg, *Men at Midlife*. Copyright © 1981 Auburn House, Dover, MA.

such a strong tendency to deny the experience of crises of any kind, it is difficult to confirm or deny the existence of a mid-life crisis simply by taking adults' responses at face value. When interviewed or when completing questionnaires, adults may consciously or unconsciously give the socially desirable responses. To get a true understanding of adult development, it may be necessary to look beyond the answers and narratives given by adults.

The work of Farrell and Rosenberg (1981) is a good example. They interviewed and gave a battery of personality tests to two groups of men: those between the ages of 25 and 30 and those between the ages of 38 and 48. Included in that battery was a mid-life crisis scale developed for the purpose. They asked the participants about tasks central to mid-life development, such as:

- Assuming the role of "patron" of the family
- Becoming a source of financial/emotional stability for younger and older generations
- Learning to live with changing physical abilities and accepting unfulfilled dreams of youth

These tasks probably sound familiar, since they are similar to some in Levinson's theory.

Farrell and Rosenberg analyzed the responses of the men in their study in a manner radically different from Levinson. They listened to the responses and then assessed the responses on two dimensions: ability to confront stress, and amount of life satisfaction. Figure 19.3 is a summary of the four personality types that result from their findings.

Figure 19.3
Typology of responses to middle-age stress

	Denial of Stress	Open Confrontation with Stress
Dissatisfied	**IV Punitive-Disenchanted** 1. Highest in authoritarianism 2. Dissatisfaction associated with environmental factors 3. Conflict with children	**I Anti-Hero** 1. High alienation 2. Active identity struggle 3. Ego-oriented 4. Uninvolved interpersonally 5. Low authoritarianism
Satisfied	**III Pseudo-Developed** 1. Overtly satisfied 2. Attitudinally rigid 3. Denies feelings 4. High authoritarianism 5. High on covert depression and anxiety 6. High in symptom formation	**II Transcendent-Generative** 1. Assesses past and present with conscious sense of satisfaction 2. Few symptoms of distress 3. Open to feelings 4. Accepts out-groups 5. Feels in control of fate

The researchers confirmed the idea that many middle-aged men work hard at denying feelings of weakness or distress. Glaring differences across socioeconomic lines were also found: Those who typically confronted their stressful feelings (types I and II) were much more likely to be affluent than those who had not (types III and IV).

Men who deny stress showed manifestations of coping with it in different ways. Those whose responses were grouped in the category of Pseudo-Developed (III) represented themselves as being similar to the Transcendent-Generative, with "cheery self-confidence," when really they were not. Responses of this type were interpreted to be "masks" of true feelings. "We call the men's reports of their experiences 'masks' to emphasize our suspicion that the subjective experience of the men and their presentations of self are not necessarily congruent with each other" (Farrell & Rosenberg, 1981, p. 31). The researchers hypothesize that these men create a highly structured life, one that leaves little room for self-exploration or expression.

The fourth type of response, termed Punitive Disenchanted (IV), was also one that avoided confronting the stress of mid-life. Responses of this type reflect a faulty interpretation of feelings. Men in this category interpreted personal feelings of dissatisfaction to be feelings of dissatisfaction with others and the world around them. This transformation of feelings is called projection (See chap. 2).

In contrast to the findings of other theorists, Farrell and Rosenberg's findings do not suggest the existence of a universal mid-life crisis, per se. They suggest, instead, that there are universal *stresses* at mid-life, and thus each man must create a buffer from or resolution of those stressors. "For most men, then, the movement toward midlife is a process of self-insulation" (Farrell & Rosenberg, 1981, p. 212).

Either most men experience a distinct crisis during their early forties that alters their self-perceptions and behavior (as Levinson says), or most react to middle age by attempting to "insulate" themselves from reality (as Farrell and Rosenberg argue). It cannot be both ways. What do you think?

Seasons of a Woman's Life: Levinson

In recent years, Levinson (in press) has turned his attention to female progress toward maturity. For his research, he selected three groups of women between the ages of 35 and 45. One-third are homemakers whose lives have followed the traditional family-centered pattern, one-third are teachers at the college level, and one-third are businesswomen. He sees these women as representing a continuum from the domestic orientation to the public orientation, with the college teachers being somewhere in between. Each of the 45 women has been seen eight to ten times by the research staff (half of whom are female) for a total of 15 to 20 hours of interviewing.

Of greatest importance is the finding that females go through a sequence of stages very similar to the stages experienced by the males who were studied. Each gender may be seen as going through an alternating series of structure-building and structure-changing stages. Levinson found, for example, that men in their late thirties want to "become their own man." He also found that at just this time, women desire to "become their own woman"; that is, they want greater affirmation both from the people in their world and from themselves.

Although male and female growth toward maturity may be in similar stages, Levinson and his associates also believe that there are major differences between the genders within these similar stages. There are important sociohistorical differences. For women, the central themes are **gender splitting,** the **traditional marriage enterprise,** and the emerging **gender revolution.**

Gender Splitting

All societies support the idea that there should be a clear difference between what is considered appropriate for males and for females: Gender splitting appears to be universal. Women's lives have traditionally been devoted to the domestic sphere; men's to the public sphere. Human societies have seen a need for females to stay at home to protect the small number of offspring (compared to other species), while the male goes about being the "provisioner" (getting the resources the family needs outside of the home).

The Traditional Marriage Enterprise

In the final analysis, everyone gets married because they believe they can have a better life by doing so. There may be exceptions, but they are probably rare. At any rate, the main goal is to form and maintain a family. Gender splitting is seen as contributing to this goal. Being supportive of the husband's "public" role—that is, getting resources for the family—is seen by the woman as a significant part of her role. When she goes to work, this goal is not largely different. Levinson reports that it is still a source of conflict when a female does get to be the boss. Women pay a heavy price for the security this role affords. Many find it dangerous to develop a strong sense of self.

The Gender Revolution

For women, it appears that there is an important transition at about 30 years of age. The transition involves a period of self-evaluation that usually leads to greater satisfaction with life.

But the meanings of gender are changing, and becoming more similar. This is because there is so much more work to be done by young and middle-aged adults. The increase in life expectancy has created a large group, the elderly, who consume more than they produce. This, together with the decrease in birth rate, has brought many more women out of the home and into the workplace. Two other factors have also been at work: the divorce rate, which has reached 50 percent, and the increase in the educational levels of women.

In his study of women, Levinson found support for the existence of the same stages and a similar mid-life crisis as for men (Levinson, 1986). A major study of women's development (Reinke & others, 1985) used a methodology similar to Levinson's and found important transitions in the lives of women, but not exclusively clustered around the mid-life period (ages 40 to 45). Instead, women described important transitions in their lives at ages 30, 40, and 60.

Interestingly, Reinke found a universal turning point for women to occur between the ages of 27 and 30. Among women in their twenties and thirties, 80 percent of those with preschool children experienced a major transition, and of these, 50 percent were between the ages of 27 and 30. Women who manifested the transition were more likely to have been employed outside their home in their mid-twenties (63 percent) than women who did not manifest the transition. The 27-to 30-year-old transition period was characterized by personal disruption, reassessment, and reorientation. The transition lasted an average of 2.7 years, and generally ended in increased life satisfaction. Many women in their early forties also manifested a transition that included decreases in marital satisfaction and increases in assertiveness. The transition was not as widespread as the transition occurring at age 30 and seemed to be tied to children growing up.

Their data led the researchers to conclude that transitions in a woman's life may be integrally tied to family life cycle. "My research on women suggests that the course of relationships exerts a greater press on women's development than does chronological age" (Reinke, 1985, p. 275).

We are, in Levinson's opinion, at a cultural crossroads. The old division of female homemakers and male providers is breaking down, but no clear new direction has yet appeared. Researchers will be watching this dramatic change closely.

Both Gould and Levinson are psychologists. The next theory is that of a psychiatrist, most of whom are trained in the Freudian tradition. As we shall see, that makes for a rather different view of adult personality development.

Adaptations to Life: Vaillant

The subtitle of George Vaillant's book *Adaptation to Life* (1977) is "How the Best and the Brightest Came of Age." Vaillant's claim that the subjects of his study were among the smartest young men of their time seems to be justified. He has investigated mountains of data collected on a carefully selected group of students from Harvard University's classes of 1939 through 1944. The investigation included 260 white males.

The young men were selected because of the superiority of their bodies, minds, and personalities. A major consideration was that each subject be highly success oriented. Although their intelligence was not greatly higher than that of other students at Harvard, almost two-thirds of them graduated with honors (as compared to one-fourth of their classmates), and three-fourths went on to graduate school.

Almost all had solid, muscular builds and were in excellent health. Their average height was 70 inches, and their average weight was 160 pounds. Interestingly, 98 percent were right-handed. If the current theory is correct (Dacey, 1989a), persons dominated by the left side of the brain (which is indicated by right-handedness) tend to be intelligent but not very imaginative. Left-handed people are thought to be more creative. It would be interesting to know how creatively productive this group of men has been, but because he was interested only in their mental health, Vaillant has not addressed this question.

Almost 20 hours of physical, mental, and psychological tests were administered to the men. Their brain waves were recorded, and anthropologists measured each man to determine his body type, although these last two measurements had little bearing on the results of the study. Finally, the family history of each of the subjects was carefully recorded. Using this voluminous data, Vaillant set out to describe the personal development of these special people.

As in the theories of Sigmund Freud and Erik Erikson, defense mechanisms are important in Vaillant's explanation of the mental health of these subjects. He believes that everyone uses defense mechanisms regularly. Thus defense mechanisms can range all the way from serious psychopathology to perfectly reasonable "adaptations to life." As he puts it:

> ***These intrapsychic styles of adaptation have been given individual names by psychiatrists (projection, repression, and sublimation are some well-known examples.). . . In this book, the so-called mechanisms of psychoanalytic theory will often be referred to as coping or adaptive mechanisms. This is to underscore the fact that defenses are healthy more often than they are psychopathological.*** **(1977, p. 7)**

On the basis of his data, Vaillant concluded that adaptive mechanisms, as he calls them, are as important to the quality of life as any other factor. Nevertheless, he does not challenge Freud's definition: these mechanisms are subconscious defenses of the ego (See chap. 2). Vaillant makes a number of generalizations about these mechanisms on the basis of his observations. He believes that they

- Are not inherited
- Do not run in families
- Are not related to mental illness in the family
- Cannot be taught
- Are discrete from one another
- Are dynamic and reversible

Vaillant believes that four other generalizations also emerge from the data:

- Life is shaped more by good relationships than by traumatic occurrences during childhood.

- The constantly changing nature of human life may qualify a behavior as mentally ill at one time and adaptive at another.
- To understand the healthiness or psychopathology of the individual, it is necessary to understand what part these adaptive mechanisms play in the healing process. Furthermore, there is a natural tendency in most people to progress to higher-level mechanisms as they grow toward maturity.
- Since human development continues into adulthood, it is necessary to have a longitudinal study such as this to understand that development.

Generativity versus Stagnation: Erikson

Generativity, Erikson's term for the major goal of the middle years of adulthood, includes coming to understand those who are different from you, and desiring to make a lasting contribution to their welfare.

Let us turn now to a theory that is considered a classic. In this final section, we will put forth Erikson's explanation of personality development in middle adulthood: generativity versus stagnation.

Generativity means the ability to be useful to ourselves and to society. As in the industry stage (See chap. 2), the goal here is to be productive and creative. However, productivity in the industry stage is a means of obtaining recognition and material reward. In the generativity stage, which takes place during middle adulthood, one's productivity is aimed at being helpful to others. The act of being productive is itself rewarding, regardless of recognition or reward. Erikson adds that generativity is

> ***that middle period of the life cycle when existence permits you and demands you to consider death as peripheral and to balance its certainty with the only happiness that is lasting: to increase, by whatever is yours to give, the good will and the higher order in your sector of the world.*** **(1978, p. 124)**

Although Erikson certainly approves of the procreation of children as an important part of generativity for many people, he does not believe that everyone needs to become a parent in order to be generative. There are, of course, people who from misfortune or because of special and genuine gifts in other directions cannot apply this drive to offspring of their own, and instead apply it to other forms of altruistic concern and creativity (1968).

It is at this stage of adulthood that some people become bored, self-indulgent, and unable to contribute to society's welfare; they fall prey to **stagnation.** Such adults act as though they were their own only child. People who have given birth to children may fail to be generative in their parenthood, and come to resent the neediness of their offspring.

Although generativity can provide great satisfaction to those who reach it, several theorists (e.g., Roazen, 1976) have suggested that the majority of adults never do. Many males appear to become fixed in the industry stage, doing their work merely to obtain the social symbols of success—a big car, a fancy house, a huge television. Many women, they suggest, may become fixed in the identity stage, confused and conflicted about their proper role in life. They rarely achieve intimacy and therefore rarely reach the stage of generativity.

Becoming generative is not easy. It depends on the successful resolution of the six preceding Eriksonian crises we have described in this book. People who are able to achieve generativity have a chance to reach the highest level of personhood in Erikson's hierarchy: integrity. We will examine that stage in chapter 21.

As you can see, all of the theories described thus far hold that as we age, we go through many important changes. Let us turn now to the other side of the coin: the position that the adult personality is made up of traits that remain continuously stable, in most cases, throughout adulthood.

Continuous Traits Theory

In their extensive study of men at mid-life, McCrae and Costa (1984) found no evidence of personality change over the adult years, nor any evidence for the existence of any mid-life crisis. The research measured the stability of several different personality traits over a period of six years (longitudinal study) and also looked at those same personality traits in a cross-age population (cross-sectional analysis).

At the first testing, McCrae and Costa administered several personality inventories to men ranging in age from 17 to 97. When they combined the data they gathered from the inventories, they defined three major personality traits that they feel govern the adult personality: neuroticism, extroversion, and openness to experience. Each of those three traits is supported by six subtraits or 'facets' (see Figure 19.4). The three major traits together are termed the **NEO model of personality.** The researchers found relative stability of those traits throughout male adulthood.

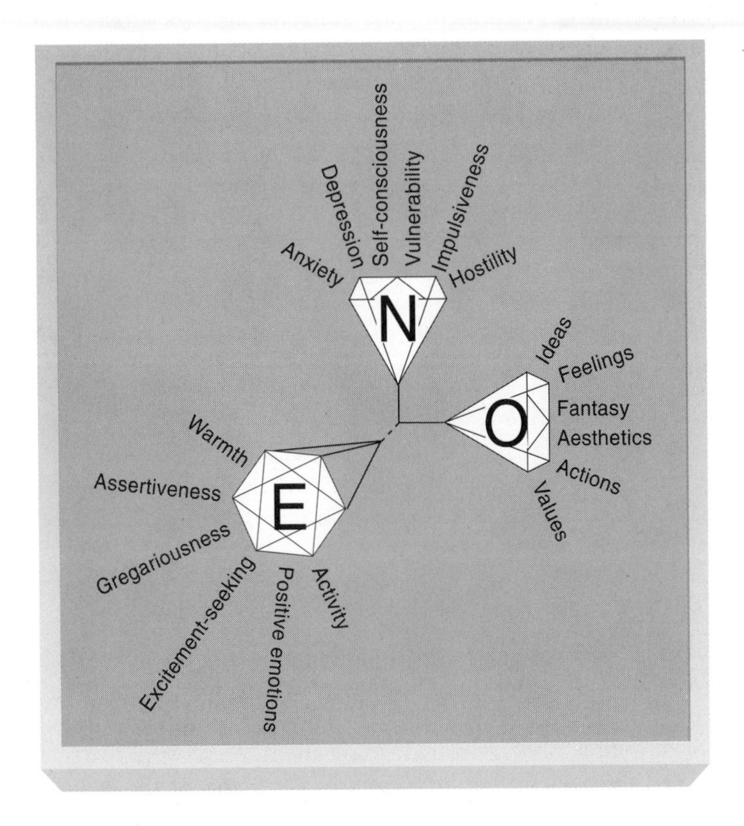

Figure 19.4
Schematic representation of the three-dimensional, 18-facet NEO (neuroticism, extroversion, openness) model

Neuroticism is described as an index of instability or a predisposition for some kind of breakdown under stress. Behaviors associated with this trait include

- A tendency to have more physical and psychiatric problems (without medical problems)
- Greater tendencies to smoke and drink
- More general unhappiness and dissatisfaction with life

Extroversion, the tendency to be outgoing and social, is the second global trait found by McCrae and Costa. Extroversion seems to spill over into almost all aspects of the personality, since it implies an overriding interest in people and social connections.

Openness to experience, the third general trait, is exactly that: an openness to new ideas, fantasies, actions, feelings, and values. Overall, a lack of rigidity in regard to the unfamiliar characterizes openness to experience.

The researchers note that while habits, life events, opinions, and relationships may change over the lifespan, the basic personality of an individual does not (McCrae & Costa, 1984). They feel that people who experience a mid-life crisis probably have the personality makeup that biases them toward that behavior. The Costa and McCrae research, however, lacks consideration of the reciprocal nature of influence of a person with his or her environment. Even so, the rigor of McCrae and Costa's work warrants careful consideration from those interested in personality development across the lifespan.

Another recent study that has looked at stability of traits across adult periods has been conducted by Dorothy Eichorn and her associates at Berkeley (1981; see also Sears & Sears, 1982). These researchers examined stability by looking at the relationship between traits in adolescence and in middle age. They found that middle-age blood pressure, goodness of health, IQ, IQ gain, political ideology, and drinking problems can be predicted fairly well in adolescence. Other traits, however, such as the status of marriage and psychological health, are less stable.

What's Your View?

As you can see, there is a clear conflict between the trait researchers and the positions of the four stage theorists. It appears that they cannot both be correct, although the truth may lie somewhere in between. We have devoted much more space to the stage theories. In part this is because they have received so much attention in the popular press, as well as by scholars. It is also because we are biased toward them. (It is certainly obvious throughout this book that we have a great deal of admiration for the ideas of Erikson.)

Whatever the case may be, the important thing is that, through discussion, further reading, and observation, you try to come to your own opinion. What do you think?

Patterns of Work

As the "baby boom" generation enters middle age, many changes are beginning to occur in the workplace. Fewer and fewer new young workers will be available to enter the workforce. Employers will find themselves with a larger number of older workers. This shrinking labor pool is leading many employers to pay more attention than ever to the welfare of their employees. Some of the new issues of the 1990s are child care, elderly care, home-based work, nontraditional work schedules, and the spiraling cost of health care benefits.

Employers are also attempting more creative approaches to development and training. In the past, employers have tended to spend the majority of their training and development resources on younger employees. Perhaps this is just a matter of employers being unaware of the problems and concerns of their older employees. Or perhaps money and time invested in younger employees were considered better spent than that on older employees.

But in recent years, a trend has emerged among employers to recognize some of the concerns of middle-aged employees. Among them are

- Awareness of advancing age and awareness of death
- Awareness of body changes related to aging
- Knowing how many career goals have been or will be attained

A Multicultural View 

Work in the Lives of African-American and White Women

What does a career contribute to a woman's identity? "The power to earn one's way has a profound effect on one's inner landscape" (Baruch, Barnett, & Rivers, 1983, p. 152). For both African-American and white women, occupational status was positively associated with perceived control. African-American women with higher than average personal earnings reported greater feelings of control, and white women with higher than average earnings reported greater life satisfaction (Crohan & others, 1989). The self-esteem and personal worth of middle-aged African-American women has been found to be determined by their ability to be good providers (Coleman, Antonucci, & Adelman, 1987). This is because there are few African-American women who are solely homemakers. Furthermore, an even higher percentage of African-Americans are single parents.

This does not mean that African-American women are more eager to work than white women (Crohan & others, 1989). It is the degree of control they feel over their jobs that contributes to the well-being of both groups of women (Adelman, 1987). The major concern of the educated mid-life women interviewed by Grossman & Chester (1990) was their economic vulnerability, despite the fact that most were working.

- Search for new life goals
- Marked change in family relationships
- Change in work relationships
- Sense of skills and abilities becoming outdated
- Feeling of decreased job mobility and increased concern for job security

Businesses have responded to these issues with continuing education, seminars, workshops, degree programs, and other forms of retraining. As the labor pool shrinks, the welfare of the older, established workforce becomes more valuable.

Special Problems of the Working Woman

It is probably not news to you that the problems facing women in the workplace are different from those of men. But what actual difference does being a woman make? Some recent research has been unable to establish clear differences (Northcott & Lowe, 1987), but many more reveal important differences. The difficulties working women face have been studied intensively in recent years (Amato, 1987; Galambos & Lerner, 1987; Smart & Ethington, 1987). A review of this literature reveals the following five major problem areas.

Sexual Harassment on the Job

Sexual harassment can take many forms (Garvey, 1986), such as verbal sexual suggestions or jokes, leering, "accidentally" brushing against your body, a "friendly" pat, squeeze, or pinch, or arm around you, catching you alone for a quick kiss, explicit propositions backed by the threat of losing your job, and forced sexual relations (Faier, 1979).

Although harassment of professional women has received considerable media attention in recent years (for example, Justice Thomas' nomination hearings), blue-collar women have even more difficulty with this problem. Those who work in the trades report that their marriages are strained by their jobs; not only do their husbands not like their nontraditional occupation, but they are unwilling to support them against the sex discrimination and harassment they suffer (Dinnerstein, 1992; Grossman & Chester, 1990).

Most blue-collar women have been verbally harassed (88 percent), and many have been pinched, fondled, and otherwise physically assaulted (28 percent). They have little support from their supervisors; in fact, 20 percent of the perpetrators *are* the supervisors. These women also suffer discrimination: Their competence is questioned, the job requirements are stiffer, they are denied advancement, and necessary job information is withheld. They have little support for their complaints (Schroedel, 1990).

Equal Pay

The Equal Pay Act was passed in 1963, and states that men and women in substantially similar jobs in the same company should get the same pay. Under this act, the complainant may remain anonymous while the complaint is being investigated.

In spite of these legal protections, women make less money on the average than men. Why? In addition to male prejudice, women may be absent more from work due to illness of children, women seem to have greater anxiety about using computers, and females take much less math in school than males.

Career and/or Family

The research reported on the dual-career family on p. 460 of chapter 17 applies here as well.

"Fiscal Fitness"

Financial advisor Froma Joselow (1979) has suggested that many women lack **fiscal fitness;** that is, they are not experienced in managing money. The lack of financial knowledge has definitely impeded the progress of many women up the ladder to success in business.

Travel Safety

Working women are exposed to a considerable number of hazards to which the average housewife and mother is not subjected. Crimes against women are no longer limited to the inner city, now occurring in suburban and even rural areas with a high frequency.

These five problems are gradually being recognized in the workplace, and there is hope that they will be alleviated. Another type of problem, which is more likely to affect men than women, is the mid-career crisis.

The Mid-Career Crisis

Considerable attention is now being given to the crisis many people undergo in the middle of their careers. For some it is a problem of increasing anxiety, which is troublesome but no serious problem. For others the difficulty is truly threatening.

A number of changes in the middle years are not caused by work: the awareness of advancing age, the death of parents and other relatives, striking changes in family relationships, and a decrease in physical ability. Other changes are entirely work related.

Coming to Terms with Attainable Career Goals

By the time a person reaches the age of 40 in a professional or managerial career, it is pretty clear whether she or he will make it to the top of the field. If they haven't reached their goals by this time, most people adjust their level of aspirations and, in some cases, start over in a new career. Many, however, are unable to recognize that they have unrealistic aspirations, and suffer from considerable stress.

Even people whose career patterns are stable, such as Catholic priests, often have a mid-career letdown. Nor is this crisis restricted to white-collar workers. Many blue-collar workers, realizing that they have gone about as far as they are going to go on their jobs, suffer from depression.

This is also the time when family expenses, such as college education for teenage children, become great. If family income does not rise, this obviously creates a conflict, especially if the husband is the sole provider for the family. If the wife goes back to work, other types of stress often occur.

The Change in Work Relationships

Relationships with fellow employees obviously change when one has come to the top of one's career. Some middle-aged adults take a mentoring attitude toward younger employees, but others feel resentful toward the young because they still have a chance to progress. When people reach their forties and fifties, they often try to establish new relationships with fellow workers, and this contributes to the sense of conflict.

A Growing Sense of Outdatedness

In many cases, an individual has to work so hard just to stay in a job that it is impossible to keep up-to-date. Sometimes a younger person, fresh from an extensive education, will join the firm and will know more about modern techniques than the middle-aged person does. These circumstances usually cause feelings of anxiety and resentment in the older employee because he or she is afraid of being considered incompetent.

Often employers are unwilling to spend money on the training of older workers.

Inability to Change Jobs

Mayer (1978) finds that age discrimination in employment starts as early as age 35 in some industries, and becomes pronounced by age 45. A federal law against age discrimination in employment was passed in 1968 to ease the burden on the older worker, but to date the law has been poorly enforced on both the federal and state levels. Many employers get around the law simply by telling older applicants that they are "overqualified" for the available position.

The Generativity Crisis

Erikson suggests that people in the middle years ought to be in the generativity stage (discussed earlier in this chapter). This is a time when they should be producing something of lasting value, making a gift to future generations. In fact, this is definitely a concern for middle-aged workers. The realization that the time left to make such a contribution is limited can come with shocking force.

The generativity crisis is similar to the identity crisis of late adolescence in many ways. Both often produce psychosomatic symptoms such as indigestion and extreme tiredness. Middle-aged persons often get chest pains at this time. These symptoms are rarely caused by organic diseases in people younger than 50, and usually are due to a depressed state associated with career problems (Levinson, 1978, in press).

The resolution of this crisis, and of other types of crises that occur during early and middle adulthood, depends on how the person's personality has developed during this period. In the following section, we continue our examination of this topic, which began in chapter 17.

Some Suggestions for Dealing with the Mid-Life Crisis

Levinson and others have found that a fair number of workers experience a stressful mid-life transition. One major way of dealing with it is for the middle-aged worker to help younger employees make significant contributions. Furthermore, companies should take a greater responsibility for fostering continuing education of their employees.

We should continue the type of job transfer programs for middle-aged people that we now have for newer employees. Although sometimes the changes can be threatening, the move to a new type of job and the requisite new learning experiences can bring back a zest for work.

Perhaps the federal government should consider starting mid-career clinics. Such clinics could help workers reexamine their goals, consider job changes, and provide information and guidance. Another solution might be the establishment of "portable pension plans" that would move with workers from one company to another so they would not lose all they have built up when they change jobs.

Can you imagine any other remedies?

Conclusion

Dealing with change is as serious a challenge in mid-life as at any other time. Learning new ways to get along with one's spouse, parents, siblings, and children is necessary at this time of life. There is considerable debate over whether personality changes much during this period. Some think it goes through a series of predictable changes; others view it as continuous with earlier life. Workers often experience a "mid-life crisis" at this time, and women workers have an additional set of burdens to handle.

Yet with any luck, we make it safely through middle age and move on to late adulthood. In the next two chapters, we investigate the pluses and minuses of this development.

Chapter Highlights

Marriage and Family Relations

- Middle age offers a time for marriage reappraisal, which proves positive for most couples.
- In general, a person who has not married by middle age will never marry. Middle age is also a time when most people develop improved relationships with their parents, though in many cases, the relationship begins to reverse itself when those parents become dependent upon their middle-aged children.
- Sibling relationships have the potential to be the most enduring that a person can have, with the relationship between sisters being strongest. It is usually one of these sisters who cares for the elderly parents.
- Liberalization of divorce laws have improved the position of women following divorce; however, Weitzman argues that serious inequity still exists.

Personality Development: Continuous or Changing?

- Research by trait theorists such as Kagan and McCrae & Costa generally supports the notion that human beings remain fairly stable throughout life.
- By contrast, theorists such as Erikson Gould, Levinson, and Vaillant argue that human beings are best described as constantly changing and developing throughout life.
- According to Roger Gould, the major false assumption to be rejected during middle adulthood is that "there is no evil or death in the world. Everything sinister has been destroyed."
- Levinson suggests that most men go through a mid-life transition in which they must deal with the polarities between young and old, masculinity and feminity, destruction and creation, and attachment and separation.
- Levinson also suggests that females go through a similar experience to that of males, with some notably different influences.
- Vaillant, in his study of the "smartest" young men of their time, concluded that adaptive mechanisms are very important to quality of life.
- Erikson's theory places middle adulthood within the stage labeled "generativity versus stagnation". *Generativity* means the ability to be useful to ourselves and to society without concern for material reward.
- McCrae and Costa defined three major personality traits that they feel govern the adult personality: neuroticism, extroversion, and openness to experience.

Patterns of Work

- In recent years, a trend has emerged among employers to recognize some of the concerns of middle-aged employees.
- Five major problem areas are related to working women: sexual harassment on the job, equal pay, career and/or family, "fiscal fitness," and travel safety.
- Considerable attention is now being given to the crisis many people undergo in the middle of their careers.
- One major way of dealing with the mid-life crisis is for the middle-aged worker to help younger employees make significant contributions.

Key Terms

Emotional divorce
Empty nest syndrome
Fiscal fitness
Gender revolution
Gender splitting
Generativity
Mid-life transition
NEO model of personality
No-fault divorce
Stage theorists
Stagnation
Traditional marriage enterprise
Trait theorists

What Do You Think?

1. What is the best way for middle-aged people to take care of their ailing, elderly parents?
2. What changes would you make in our divorce laws?
3. Would you agree that almost everyone has a mid-life crisis, but they just deny it?
4. What are the main differences between male and female personality development?
5. Who's right, the stage theorists or the trait theorists?
6. What laws would you make to improve the workplace for women?

Suggested Readings

Breslin, J. (1986). *Table money.* New York: Ticknor & Fields. An empathetic tale of matrimony, alcoholism, and the struggle to attain maturity, focusing on a poor middle-aged Irish-American couple.

Guest, J. (1976). *Ordinary people.* New York: Ballantine. The evocative story of the relationships among a middle-aged couple, their teenage son, and his psychiatrist.

Hansberry, L. (1959). *A raisin in the sun.* A moving drama portraying the inner lives of an African-American family.

Updike, J. (1981). *Rabbit is rich.* New York: Knopf. The third in Updike's series about an ordinary American male. In this volume, Rabbit reaches middle age.

PART IX

Late Adulthood

CHAPTER 20

Late Adulthood: Physical and Cognitive Development

Must We Age and Die? 526
Physiological Aspects of Aging 526
Genetic Aspects of Aging 529
Effects of the Natural Environment on Aging 529
Other Modifiers of Ability 530
Physical Development 531
Reaction Time 531
Sensory Abilities 534
Other Body Systems 535
Health 536
Appearance 539
Cognitive Development 541
Cognitive Decline in the Elderly: Tests versus Observations 541
Terminal Drop 542
Creativity 543
Conclusion 546
Chapter Highlights 547
Key Terms 548
What Do You Think? 548
Suggested Readings 548

And just as young people are all different, and middle-aged people are all different, so too are old people. We don't all have to be exactly alike. . . . There are all kinds of old people coping with a common condition, just as kids in puberty when their voices start cracking. . . . Every one of them is different. We old folks get arthritis, rheumatism, our teeth fall out, our feet hurt, our hair gets thin, we creak and groan. These are the physical changes, but all of us are different. And I personally feel that you follow pretty much the pattern of your younger years. I think that some people are born young and some people are born old and tired and gray and dull. . . . You're either a nasty little boy turned old man or a mean little old witch turned old or an outgoing, free-loving person turned old.

As for myself, I feel very excited about life and about people and color and books, and there is an excitement to everything that I guess some people never feel. . . . I have a lot of friends who are thirty-year-old clods. They were born that way and they'll die that way. . . . But me, I'm happy, I'm alive, and I want to live with as much enjoyment and dignity and decency as I can, and do it gracefully and my way if possible, as long as possible. ■

Author M. F. K. Fisher, who died recently at 82
Quoted in Berman and Goldman (1992)

Even when he was very old and ill, Winston Churchill's fabled sense of humor never left him. In addition, it was only in his later years that he found time to sharpen his artistic skills, which brought him world renown in that realm. He was an excellent example of how fruitful the last third of life can be.

A special session of the British House of Lords was being held to honor Winston Churchill on his ninetieth birthday. As he descended the stairs of the amphitheater, one member turned to another and said, "They say he's really getting senile." Churchill stopped, and leaning toward them, said in a stage whisper loud enough for many to hear, "They also say he's deaf!"

This vignette portrays old age rather negatively. Must growing old mean decline? Must the teeth and hair fall out, the eyes grow dim, the skin wrinkle and sag? Must intelligence, memory, and creativity falter? Must old age be so awful?

Or is there actually only a relatively slight decline in capacity, a decline greatly exaggerated by our values and presumptions? Many elderly people seem to be having the time of their lives! Could it be that aging is largely a matter of self-fulfilling prophecy (people expect to deteriorate, so they stop trying to be fit, and then they *do* deteriorate)? Could most of us in our later years be as capable as those famous few who seem to have overcome age today?

These questions concern most of the 23 million men and women over 65 who constitute 10 percent of our population today. In 1900, the over-65 population was 3 million, only 3 percent of the total. By the year 2040, it will have reached 20 percent (see Figure 20.1). Such questions should also concern those of us who hope to join their ranks some day. In the next section we look at the answers to these questions, as revealed by considerable new research.

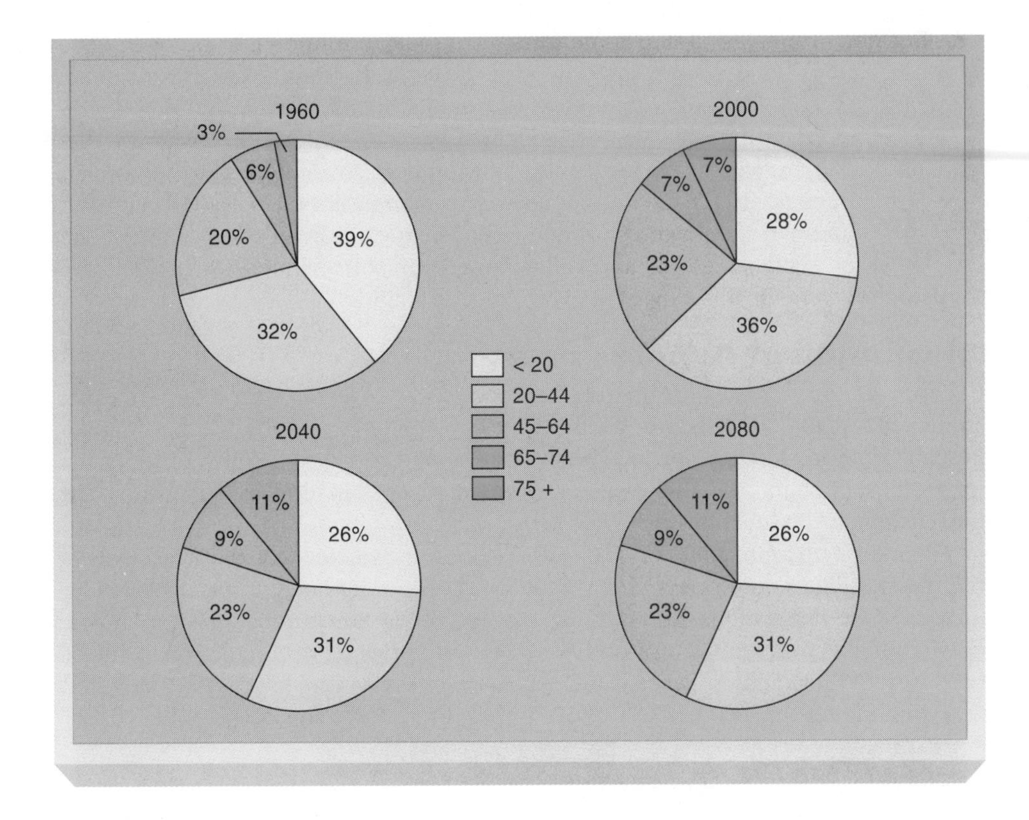

Figure 20.1
Age distribution of United States population, selected years

Source: Social Security Administration

After reading chapter 20, you should be able to:

- Compare the major physiological theories of aging.
- Determine the contribution of genetics and the environment to aging.
- Explain changes that occur in late adulthood related to reaction time and sensory abilities.
- Describe the ways in which the skeletal system, skin, teeth, and hair alter in appearance as people age.
- Evaluate the nature of the relationship of hormone balance to physical ability.
- Define Alzheimer's disease and identify several theories about its cause.
- List Gerson and associates' findings regarding the relationship between mental and physical health.
- Explain other factors in the relationship between cognition and aging.

- Appraise the theory known as terminal drop.
- Discuss Dacey's theory of creativity and list critical periods of life during which creative ability can be cultivated most effectively.

Must We Age and Die?

This olive tree found in Greece over 2,500 years old. Do you think it's possible that it might live forever?

"Nothing is inevitable except death and taxes," the old saying goes. But is death inevitable? True, no one so far has attained immortality; the oldest known person in the United States, as certified by the Social Security Administration, is 114 years old. And, of course, the vast majority of the people who have ever lived are dead.

But not all organisms die. Some trees alive today are known to be over 2,500 years old; they have aged, but show no sign of dying. Bacteria apparently are able to live indefinitely, as long as they have the requirements for their existence. The fact is, we are not sure why we age, and until we are, we cannot be certain that aging and death are absolutely inevitable. If fact, recent cover stories in the popular press (e.g., Darrack, 1992) have trumpeted as yet unpublished research that offer hope that aging can be slowed, stopped, or even reversed! Let's look now at the three types of explanations of aging that have been documented (Spence, 1989): physiological, genetic, and environmental aspects of aging.

Physiological Aspects of Aging

It is apparent that organisms inherit a tendency to live for a certain length of time. An animal's durability depends on the species it belongs to. The average human lifespan of approximately seventy years is the longest of any mammal. Elephants, horses, hippopotamuses, and asses are known to live as long as fifty years, but most mammals die much sooner.

Some doctors are fond of stating that "no one ever died of old age." This is true. People die of some physiological failure that is more likely to occur the older one gets. One likely explanation of aging, then, is that the various life-support systems gradually weaken. Illness and death come about as a cumulative result of these various weaknesses. Seven different physiological factors have been suggested as accounting for this process.

Wear and Tear Theory

The **wear and tear theory** seems the most obvious explanation for aging, but there is actually little evidence for it. To date, no research has clearly linked early deterioration of organs with either hard work or increased stress alone.

A complex interaction, however, may be involved. It is known that lower rates of metabolism are linked to longer life, and that certain conditions, such as absence of rich foods, cause a lower metabolism. Therefore, the lack of the "good life" may prolong life. On the other hand, it is also known that poor people who seldom get to eat rich foods tend to die at an earlier age than middle-class or wealthy people. Therefore, the evidence on this theory is at best conflicting.

Aging by Program

According to this theory, we age because it is programmed into us. It is hard to understand what evolutionary processes govern longevity (if any). For example, the vast majority of animals die at or before the end of their reproductive period, but human females live twenty to thirty years beyond the end of their reproductive cycles. This may be related to the capacities of the human brain. Mead (1972) argued that this extra period beyond the reproductive years has had the evolutionary value of helping to keep the children and grandchildren alive. For example, in times when food is scarce, older people may remember where it was obtained during the last period of scarcity. On the other hand, it may be that we humans have outwitted the evolutionary process, and due to our medical achievements and improvements in lifestyle are able to live on past our reproductive usefulness.

Another enduring hypothesis, proposed by Birren (1960), is known as the "counterpart" theory. According to this concept, factors in human existence that are useful in the earlier years become counterproductive in later years. An example is the nonreplaceability of most cells in the nervous system. The fact that brain cells are not constantly changing enhances memory and learning abilities in the earlier years, but it also allows the nervous system to weaken as dead cells are not replaced.

Spence (1989) suggests that the hypothalamus may well be an "aging chronometer" (p. 17), a "timer" that keeps track of the age of cells and determines how long they should keep reproducing. Although research on **aging by program** is in its infancy, some findings indicate that older cells may act differently from younger cells. Although cells in the nervous system and muscles do not reproduce themselves, all the other cells in the body do reproduce, at least to some extent. But these cells are able to reproduce only a limited number of times, and they are more likely to reproduce imperfectly as they get older. There is also reason to believe that reproduced older cells do not pass on information accurately through the DNA. This weakens the ability of older cells to continue high-level functioning.

Homeostatic Imbalance

Comfort (1964) has suggested that it may be a failure in the systems that regulate the proper interaction of the organs, rather than wear of the organs themselves, that causes aging and ultimately death. He states that aging results from "an increase in homeostatic faults" (p. 178). These homeostatic (feedback) systems are responsible, for example, for the regulation of the sugar and adrenalin levels in the blood. Apparently there is not a great deal of difference in the systems of the young and the old when they are in a quiet state. It is when stress is put on the systems (death of a spouse, loss of a job, a frightening experience) that we see the effects of the elderly **homeostatic imbalance.** The older body simply isn't as effective, qualitatively or quantitatively, in reacting to these stresses. Figure 20.2 shows graphically the relationship between problems with the homeostatic systems and the competence of the organism to react effectively to dangers in the environment.

Cross-Linkage Theory

The proteins that make up a large part of cells are themselves composed of peptides. When cross-links are formed between peptides (a natural process of the body), the proteins are altered, often for the worse. For example, **collagen** is the major connective tissue in the body; it provides, for instance, the elasticity in our skin and blood vessels. When its proteins are altered, skin and vessels are adversely affected. This is known as the **cross-linkage theory.**

Accumulation of Metabolic Waste

Although the connection has not been clearly established as yet, it may be that waste products resulting from metabolism build up in various parts of the body and contribute to the decrease in competence of those parts. Examples of this effect of the **accumulation of metabolic waste** are cataracts on the eye, cholesterol in the arteries, and brittleness of bones.

Autoimmunity

With increasing age, there is increasing **autoimmunity,** the process by which the immune system in the body rejects the body's own tissue. Examples of this are rheumatoid arthritis, diabetes, vascular diseases, and hypertension. It may be that, with age, the body's tissues become more and more self-rejecting.

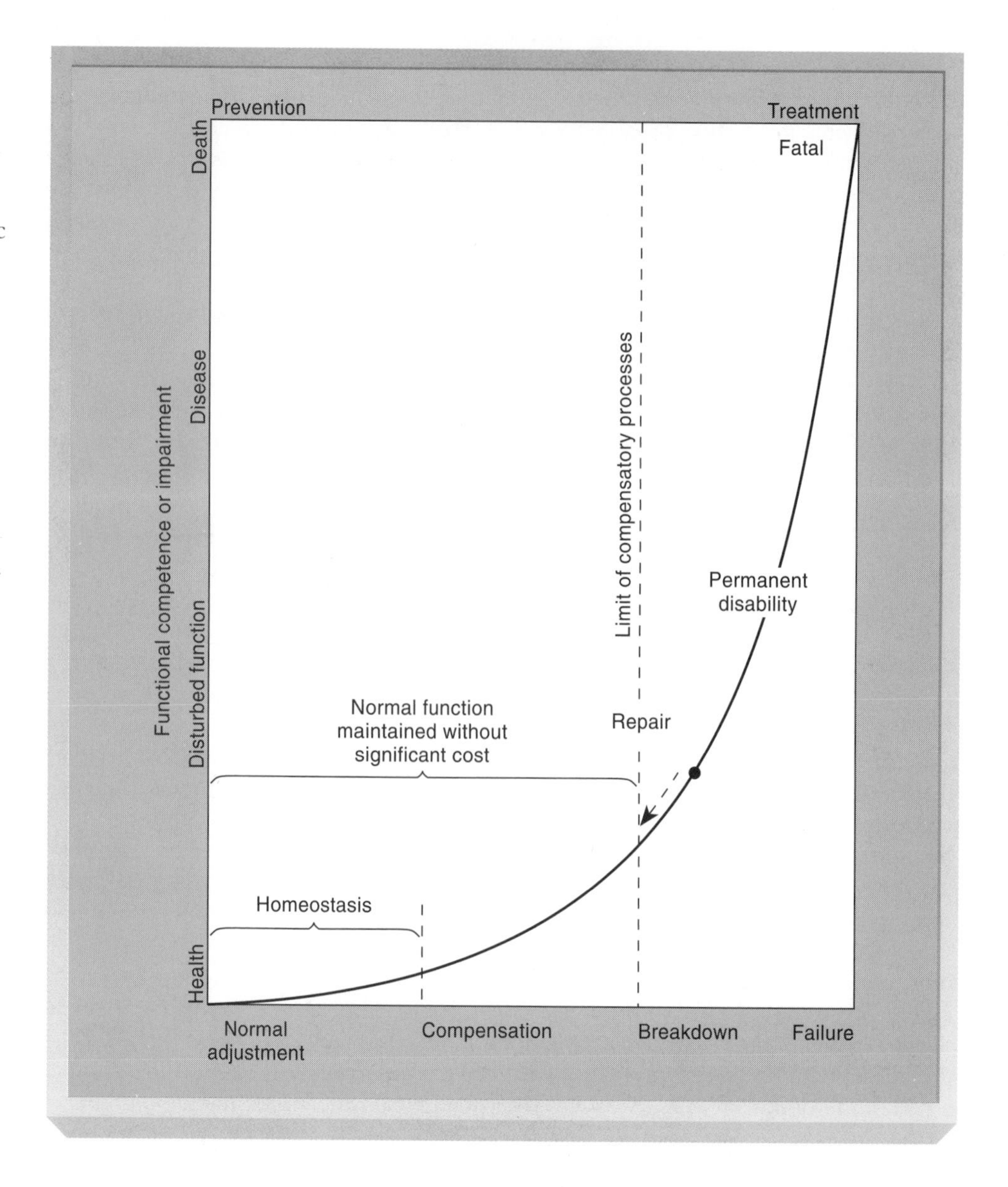

Figure 20.2
Homeostasis and health. Progressive stages of homeostasis from adjustment (health) to failure (death). In the healthy adult, homeostatic processes ensure adequate adjustment in response to stress, and even for a period beyond this stage compensatory processes are capable of maintaining overall function without serious disability. When stress is exerted beyond compensatory capacities of the organism, disability ensues in rapidly increasing increments to severe illness, permanent disability, and death. When this model is viewed in terms of homeostatic responses to stress imposed on the aged and to aging itself, a period when the body can be regarded as at the "limit of compensatory processes," it is evident that even minor stresses are not tolerable, and the individual moves rapidly into stages of breakdown and failure.

This may be the result of the production of new **antigens,** the substances in the blood that fight to kill foreign bodies. These new antigens may come about for one of two reasons:

- Mutations that cause the formation of altered RNA or DNA.
- Some cells may be "hidden" in the body during the early part of life. When these cells appear later, the body does not recognize them as its own, and forms new antigens to kill them. This in turn may cause organ malfunction.

Accumulation of Errors

As cells die, they must synthesize new proteins to make new cells. As this is done, occasionally an error occurs. Over time, these errors mount up. This **accumulation of errors** may finally grow serious enough to cause organ failure.

Genetic Aspects of Aging

Gene theory also suggests that aging is programmed, but says that the program exists in certain harmful genes. As Spence (1989) explains it,

> ***perhaps there are genes which direct many cellular activities during the early years of life that become altered in later years, thus altering their function. In their altered state, the genes may be responsible for the functional decline and structural changes associated with aging.*** **(p. 19)**

Whatever the reason, there is little doubt that genes affect how long we live. Kallman and Jarvik's (1959) research into identical twins still offers strong evidence of this. They found that monozygotic twins (those born from the same egg) have more similar lengths of life than do dizygotic twins (those born from two eggs). This effect is illustrated in Figures 20.3 and 20.4.

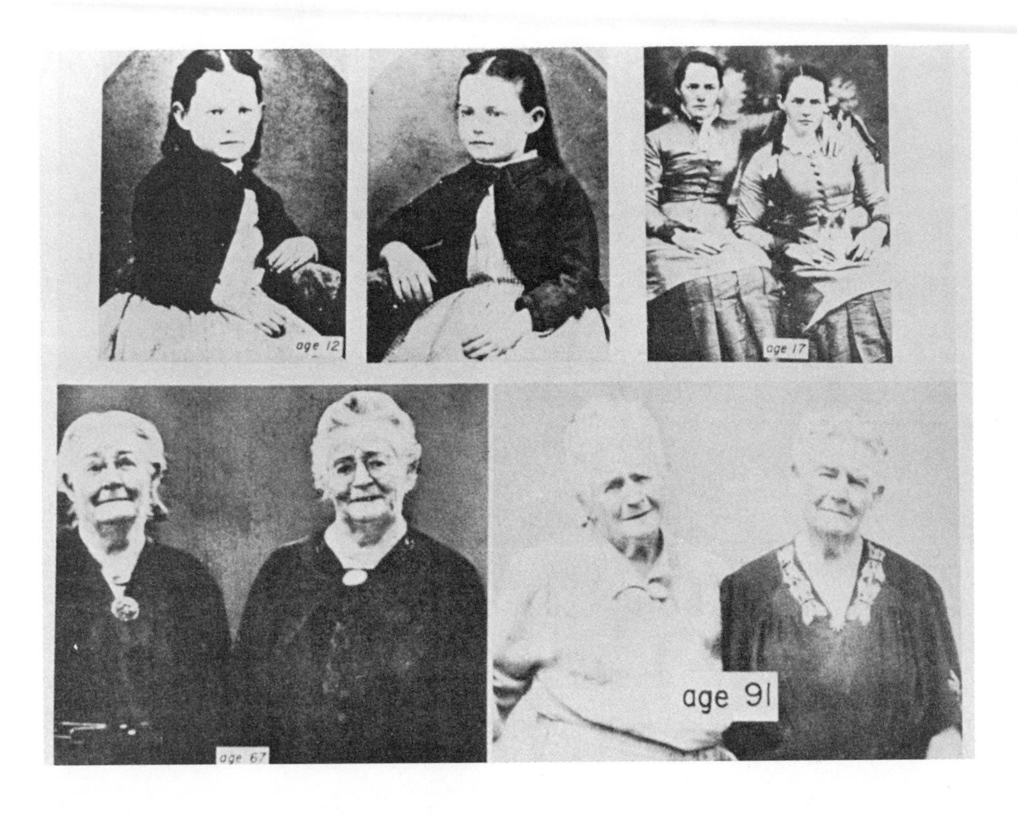

Figure 20.3
One-egg twins at the ages of 12, 17, 67, and 91 years (Note the long separation of these twins between the ages of 18 and 66).

Effects of the Natural Environment on Aging

Our genes and the physiology of our organs greatly affect the length of our lives. This may be seen in the extreme accuracy with which insurance companies are able to predict the average number of deaths at a particular age. A mathematical formula to predict the number of deaths within a population was produced as early as 1825 by Gomertz. The formula is still quite accurate except for the early years of life.

Today there are many fewer deaths in early childhood, since vaccines have eliminated much of the danger of diseases at this age. This shows that the natural environment is also an important factor in the mortality rate. Many of today's elderly would have died in childhood had they been born in the early part of the nineteenth century.

Considering mortality rates for specific cultures rather than world population shows the effects of the environment more specifically. Starvation in Africa and earthquakes in Guatemala obviously had a tragic effect on the mortality rates of

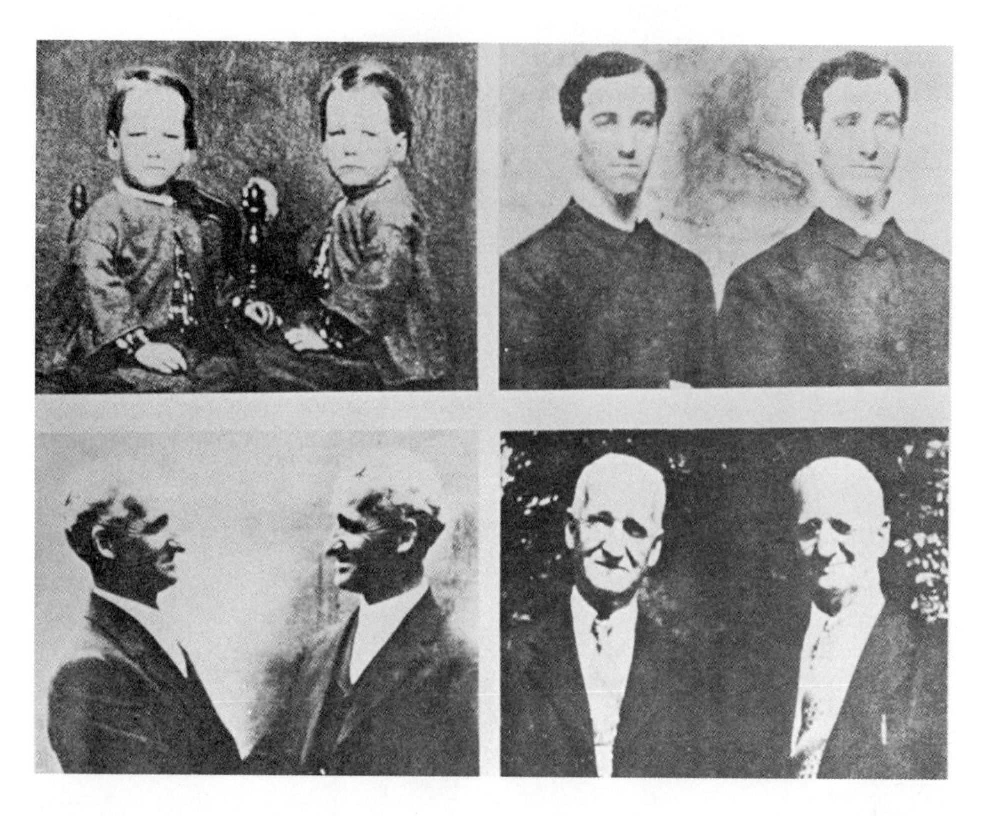

Figure 20.4
One-egg twins at the ages of 5, 20, 55, and 86 years

those two countries. A number of authors have reported on the effects of radiation (Demoise & Conrad, 1972; Spence, 1989). There is reason to believe that the nuclear testing in the Pacific in the 1950s is affecting the aging process of some of the residents there. There is little or no evidence, however, that the radiation we are all exposed to every day is having any impact on aging. It remains to be seen if events such as the nuclear accident at Three-Mile Island will have adverse effects.

Other Modifiers of Ability

In addition to genetic, physiological, and natural environmental factors that indirectly affect the individual's rate of aging, there are factors that can modify a person's level of ability more directly. Many of these modifiers interact with one another in complex ways. Some of the major modifiers are training, practice, motivation, nutrition, organic malfunction, illness, injury, stress level, educational level, occupation, personality type, and socioeconomic status.

For example, although it appears that social networks (the number and extent of friendships a person has) do not directly affect the *survival* of elderly people, they are related to their *quality of life*. Within nursing homes, those who tend to be aggressive and verbally agitated have poor social networks, and generally lack intimacy with their fellow patients. It is less clear whether the behavior causes the poor quality of network, or vice versa (Cohen-Mansfield & Marx, 1992). Socioeconomic factors such as amount of income and quality of housing do promote superior networks, however (Shahtahmasebi, Davies, and Wenger, 1992).

Figure 20.5 summarizes the relationships between the hypothesized factors that affect aging of physical and mental systems. These include two genetic, five physiological, and five environmental factors, as well as twelve other modifiers of human abilities. You could find no clearer example of the biopsychosocial model at work!

Before reading the next section, try your hand at the true-false test on the topic of aging in the following box.

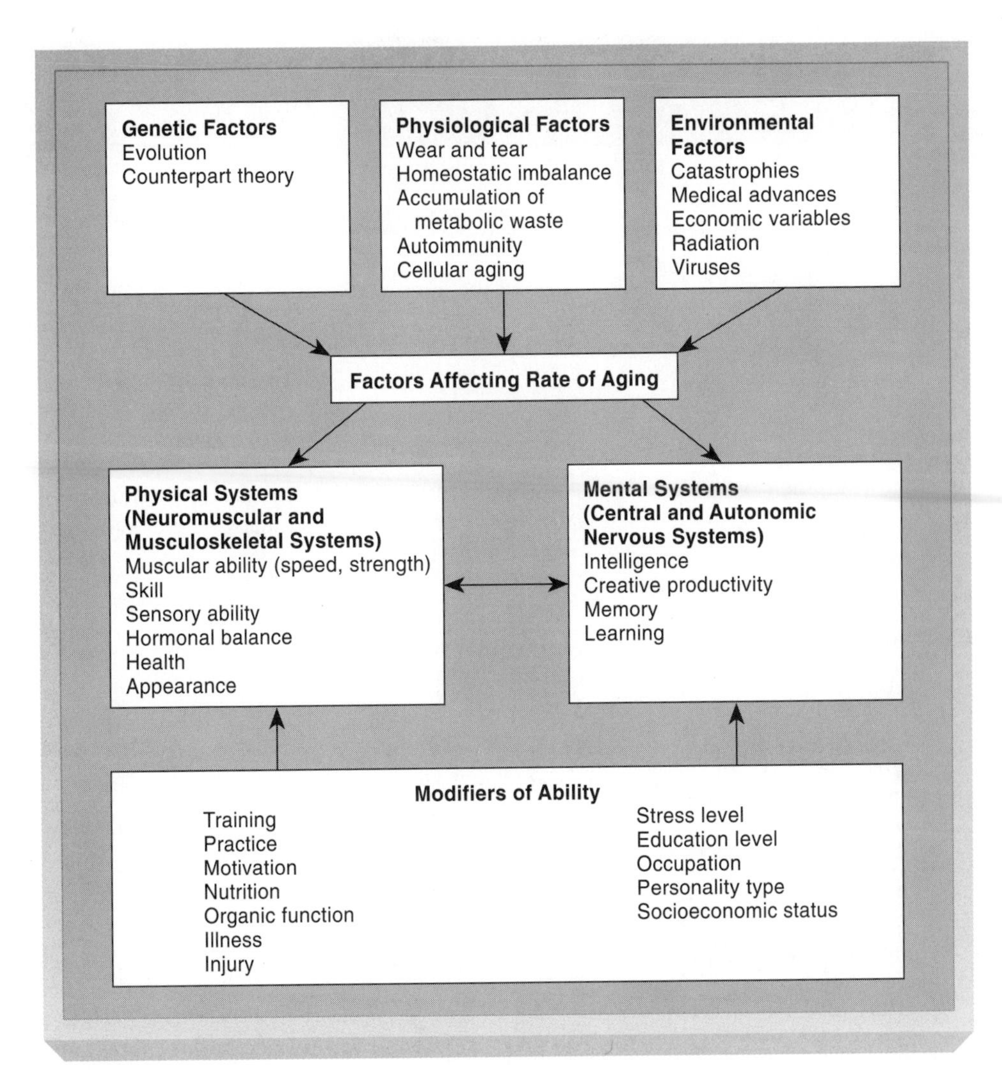

Figure 20.5
Influences on adult mental and physical systems

Physical Development

In this section, we consider development from the standpoints of reaction time, sensory abilities, other body systems, health, and appearance.

Reaction Time

It is obvious that physical skills decline as people grow older. This appears to be especially true of manual dexterity (Ringel & Simon, 1983). As one psychologist (Troll, 1975) puts it, "After about 33, hand and finger movements are progressively more clumsy" (p. 20). Although older people are able to perform short, coordinated manual tasks, a long series of tasks such as playing a stringed instrument becomes increasingly difficult for them. Famed Spanish guitarist Andrés Segovia and pianist Vladimir Horowitz are notable exceptions.

Is this because the human nervous system is deteriorating through aging? To answer this question, psychologists have completed numerous studies of "reaction time," the time between the onset of a stimulus and the actual muscle activity that indicates a reaction to it. Studying reaction time is a scientific way of separating the effects of the central nervous system from the ability of the rest of the body to perform manual tasks.

Seniors usually find that their coordination declines with the years, and so they often quit playing sports. Now there is a strong movement to reverse this trend.

An Applied View — The Facts on Aging Quiz

Mark the items *T* for true or *F* for false.

1. A person's height tends to decline in old age.
2. More older persons (over 65) have chronic illnesses that limit their activity than younger persons.
3. Older persons have more acute (short-term) illnesses than persons under 65.
4. Older persons have more injuries in the home than persons under 65.
5. Older workers have less absenteeism than younger workers.
6. The life expectancy of African-Americans at age 65 is about the same as whites'.
7. The life expectancy of men at age 65 is about the same as women's.
8. Medicare pays over half of the medical expenses for the aged.
9. Social Security benefits automatically increase with inflation.
10. Supplemental Security Income guarantees a minimum income for needy aged.
11. The aged do not get their proportionate share (about 11%) of the nation's income.
12. The aged have higher rates of criminal victimization than persons under 65.
13. The aged are more fearful of crime than are persons under 65.
14. The aged are the most law abiding of all adult groups, according to official statistics.
15. There are two widows for each widower among the aged.
16. More of the aged vote than any other age group.
17. There are proportionately more older persons in public office than in the total population.
18. The proportion of African-Americans among the aged is growing.
19. Participation in voluntary organizations (churches and clubs) tends to decline among the healthy aged.
20. The majority of the aged live alone.
21. About 3% less of the aged have incomes below the official poverty level than the rest of the population.
22. The rate of poverty among aged African-Americans is about three times as high as among aged whites.
23. Older persons who reduce their activity tend to be happier than those who remain active.
24. When the last child leaves home, the majority of parents have serious problems adjusting to their "empty nest."
25. The proportion widowed is decreasing among the aged.

The key to the correct answers is simple: Alternating pairs of items are true or false (i.e., 1 and 2 are true, 3 and 4 are false, 5 and 6 are true, etc.; and 25 is true).

In their excellent review of this research, Elias and associates (1977) present a decidedly positive picture. They report that in simple discrimination tasks, where the subject is asked to respond to a sensory cue, there is essentially no decrease in reaction time with age. Even when tasks are made more complicated, such as when a series of responses are called for or a number of stimuli must be matched, increases in reaction time with age are not great.

An example of this research can be seen in the study by Elias and Elias (1976). As Figure 20.6 indicates, there is a greater increase in reaction time scores between simple and complicated tasks than between the young and the elderly. This finding holds for both males and females. If the ability of older people to respond quickly does not decline greatly, why does it appear that their actual performance is slower and less capable? Elias and associates (1977) point to a number of variables that cause this decline. Among the most important are the following:

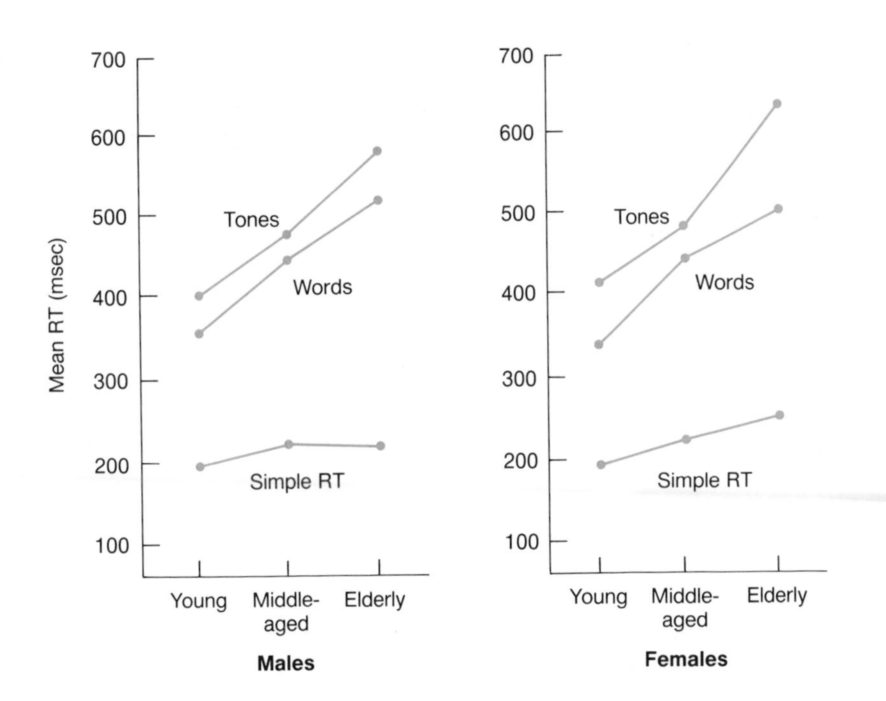

Figure 20.6
Research on reaction time. Mean RT for a simple RT task; subjects were asked to respond to a tone as quickly as possible. In the matching RT task, subjects were asked to identify a pair of stimuli (nonverbal or verbal) as "same" or "different" as quickly as possible.

- *Motivation.* Because they have had many experiences, older adults are often less likely than younger people to believe that doing well on tests in general is important. Therefore they probably do not try as hard to do well.
- *Depression.* The elderly are frequently seen as having lost mental abilities, when in fact they are slow to respond because they care less about everything—a standard effect of being depressed.
- *Anxiety.* Some seniors find being tested an unusual experience, which makes them afraid.
- *Response strategies.* Over the years, we have made progress in teaching our children how to respond to problems. The older adult probably was not taught these strategies.
- *Response style.* The older we get, the more we tend to use a single trusted response to all problems. But the best problem solvers are those who have at their command a number of styles for dealing with different types of problems.
- *Ageism.* It has been shown that young people process and respond to the speech of older people in stereotypical ways such as "baby talk" or pandering (Giles & others, 1992). This type of prejudice is known as **ageism.** Because they are often treated in a condescending way, elderly persons sometimes behave in a less competent manner than they are capable of.

Ageism can be a prejudice that can seriously affect the elderly, lowering the quality of their lives. What are some things you could do to fight this unfair attitude? For example, how might you help change the stereotype that senility is normal in the elderly?

We can conclude that variables other than sheer neural or motor activities account for most change in physical skills over time. Older adults can improve their ability in any of these areas if they desire.

What's Your View?

By 2020, up to one out of every five Americans will be 65 or older, and the vast majority will possess a driver's license. This is of concern because statistics show that accidents caused by drivers over the age of 75 equal or surpass those of teenagers, considered the most dangerous group of drivers. A number of states are moving to monitor older drivers more aggressively. However, in those cases where a license is denied or revoked, the drivers may find themselves stranded without crucial transportation (Harvard Health Letter, 1991).

Does society have the responsibility to provide transportation for people when it takes away their licenses? If not, does government have any responsibility to help such people? What's your view?

Sensory Abilities

Each of the five senses undergoes change in late adulthood. We look at some of those changes in the following paragraphs.

Vision

A number of problems may affect the eye as the adult ages (Eifrig & Simons, 1983). The lenses may become less transparent, thicker, and less elastic. This tends to result in farsightedness (Corso, 1971; McFarland, 1968). The illumination required to perceive a stimulus also increases with age (McFarland, 1968). Timiras (1972) stated that retina changes almost never occur before the age of 60, but Elias and associates (1977) argued that conditions that affect the retina, such as glaucoma, are often not detected, so problems may occur without causing noticeable changes in the eye. At any rate, with the improvements in recent years in ophthalmology and optical surgery (Kornzweig, 1980), the declining ability of the human eye need not greatly affect vision in older age (with the single exception of poor night vision, which hampers driving).

Cataracts are the result of cloudy formations on the lens of the eye. They form very gradually and inhibit the passage of light through the eye. Cataracts are most common after the age of 60. If they become large enough, they can be removed, usually by laser surgery.

Hearing

Decline in hearing ability may be a more serious problem than decline in vision. However, we seem to be much less willing to wear hearing aids than we are to wear glasses, perhaps because we rely on vision more.

Most people hear fairly well until late adulthood, but men seem to lose some of their acuity for higher pitches during their middle years (Marshall, 1981). This difference may occur mostly in men who are exposed to greater amounts of noise in their occupations and in traveling to and from their jobs.

Smell

Some atrophy of olfactory fibers in the nose occurs with age. When artificial amplifiers are added, however, the ability of the older people to recognize foods by smell is greatly improved.

Taste

It is difficult to study the effects of aging on the sense of taste because taste itself is so dependent on smell. About 95 percent of taste derives from the olfactory nerves. It is clear, however, that there is decline in tasting ability among the elderly

(Grzegorczyk, Jones, & Mistretta, 1979). Rogers (1979) reported that whereas young adults have an average of 250 taste buds, 70-year-olds have an average of no more than 100.

Touch

The tactile sense also declines somewhat after the age of 65 (Turner & Helms, 1989), but the evidence for this comes strictly from the reports of the elderly. Until scientific studies are performed, this finding must be accepted with caution.

Other Body Systems

Included in the category of body systems are the skeletal system, skin, teeth, hair, and locomotion (ability to move about).

Skeletal System

Although the skeleton is fully formed by age 24, changes in stature can occur because of the shrinking of the discs between spinal vertebrae (Mazess, 1982). As was mentioned earlier, collagen changes. Frequently this causes bone tissue to shrink (Hall, 1976; Twomey, Taylor, & Furiss, 1983). Thus the bones become more brittle. Because the entire skeletal system becomes tighter and stiffer, there is frequently a small loss of height in the aged. Diseases of the bone system such as arthritis are probably the result of changes in the collagen in the bones with age. Diseases such as osteoporosis and arthritis can be very crippling.

A number of factors can cause the body to become shorter and stiffer with age.

Skin

Collagen is also a major factor in changes in the skin. Because continuous stretching of collagen causes it to lengthen, the skin begins to lose its elasticity with age. Other changes are a greater dryness and the appearance of spots due to changes in pigmentation (Dotz & German, 1983; Robinson, 1983; Walther & Harber, 1984). The skin becomes coarser and darker, and wrinkles begin to appear. Darkness forms under the eyes, more because of a growing paleness of the skin in the rest of the face than because of any change in the under-eye pigmentation. Years of exposure to the sun also contribute to these effects.

Another important cause of the change in skin texture has to do with the typical weight loss among the elderly. There is a tendency to lose fat cells, which decreases the pressure against the skin itself. This tends to cause sagging, folds, and wrinkles in the skin. Older persons not only need less nourishment, but they also lose the social motivation of eating at meetings and parties that younger persons have (Carnevali & Patrick, 1986). As the elderly eat less, their skin becomes less tight and wrinkles appear.

Teeth

The loss of teeth is certainly one physical aspect that makes a person look older (Pizer, 1983). Even when corrective measures are taken, the surgery involved and the adjustment to dentures have a major impact on the person's self-image. In most cases, the loss of teeth is more the result of gum disease than decay of the teeth themselves. It appears likely that education, the use of fluorides, the use of new brushing techniques, and daily flossing will make this aspect of aging far less of a problem in the future.

Hair

Probably the most significant signature of old age is change of the hair. Thinness, baldness, grayness, stiffness, and a growing amount of facial hair in women are all indications of growing old. Hormonal changes are undoubtedly the main culprit here. Improvements in hair coloring techniques and hair implants may make it possible to avoid having "old-looking" hair (for those to whom this is important).

Locomotion

It is generally assumed that the decrease in locomotion by older adults is due to aging. However, a recent study suggests that while older people do indeed walk more slowly, the pattern of coordination between their limbs remains essentially the same as for younger adults. Thus if people maintain good health and an active life-style, there is reason to believe that this locomotive decline can be kept to a minimum (Williams & Bird, 1992).

Health

It is well known that health declines when one reaches the older years. This decline, however, usually does not occur until quite late in life (Kart, 1988). Table 20.1 shows rates of death per 100,000 people. (There are a number of interesting comparisons to be found in this table—can you spot them?)

Table 20.1 Death rates, by Age, Sex, and Race: 1960 to 1989
[Number of deaths per 100,000 population in specified group.]

Sex, Year, and Race	All ages	Under 1 yr. old	1–4 yr old	5–14 yr. old	15–24 yr. old	25–34 yr. old	35–44 yr. old	45–54 yr. old	55–64 yr. old	65–74 yr. old	75–84 yr. old	85 yr. old and over
Male												
1960	1,105	3,059	120	56	152	188	373	992	2,310	4,914	10,178	21,186
1989	922	1,077	47	32	152	203	302	628	1,570	3,415	7,950	17,695
Female												
1960	809	2,321	98	37	61	107	229	527	1,196	2,872	7,633	19,008
1989	817	891	41	21	54	76	142	338	888	1,997	5,083	14,070

Source: U.S. National Center for Health Statistics, *Vital Statistics of the United States* (1992).

Illness is not a major cause of death until one reaches the thirties, and only then among African-American females. This is probably because they receive fewer preventive health services, and because medical treatment is frequently delayed until the later stages of disease. Not until adults reach the age of 40 does ill health, as opposed to accidents, homicide, and suicide, become the major cause of death. Arteriosclerotic heart disease is the major killer after the age of 40 in all age, sex, and race groups. Table 20.2 summarizes the diseases and other conditions for older adults according to type of condition, age, sex, and family income.

Alzheimer's Disease

Probably the single greatest scourge of the elderly, and in many ways the most debilitating, is Alzheimer's disease. Alzheimer's was discovered by the German neurologist Alois Alzheimer in 1906, but it did not gain much attention until the 1970s. Researchers recently announced that they have found a gene that is implicated in the cause. This does not, however; mean they have found a cure for it. Alzheimer's is difficult to diagnose and even more difficult to treat. Relatively few of its symptoms respond to any type of treatment, and then only in the earliest stages. The single exception is that the depression that normally accompanies the disease can be treated effectively (Gierz, Haris, & Lohr, 1989). Alzheimer's disease usually follows a 6- to 20-year course.

The normal brain consists of billions of nerve cells (neurons), which convey messages to one another chemically by way of branchlike structures called dendrites and axons. Neuron groups generate specific chemical transmitters that travel

Table 20.2 Chronic Conditions among Elderly Persons

Age, Sex, and Family Income	Type of Chronic Condition and Year: Prevalence per 1,000 Population								
	Arthritis	Asthma	Chronic Bronchitis	Diabetes	Heart Conditions	Hypertensive Disease	Impairments of Back or Spine (except paralysis)	Hearing Impairments	Vision Impairments
Average	380.3	35.8	41.2	78.5	198.7	199.4	67.1	294.3	204.6
Sex									
Male	287.0	42.3	47.3	60.3	199.3	141.2	54.6	338.2	183.1
Female	450.1	31.1	36.6	91.3	198.3	240.9	76.3	262.1	220.4
Family Income									
Less than $5,000	411.7	41.4	45.4	82.0	219.0	216.1	78.7	232.0	232.0
$5,000–$9,999	353.3	32.6	37.2	76.1	190.0	179.5	57.3	271.6	163.2
$10,000–$14,999	310.9	•	27.4	81.1	158.9	192.6	39.3	247.3	181.3
$15,000 or more	300.8	•	40.7	62.7	174.8	161.4	48.5	259.2	169.2

Source: From C. S. Kart, et al., *Aging, Health and Society.* Copyright © 1988 Jones and Bartlett Publishing Company, Boston, MA.

between cells at the synapses. While Alzheimer's victims look normal externally, their brains are undergoing severe changes. One particular trait of Alzheimer's is the breakdown of the system in the brain that produces the chemical transmitter acetylcholine. A decrease in the enzyme that tells the system to produce acetylcholine is directly related to insanity. Brain autopsies unmask severe damage, abnormalities, and even death of neurons. There is often as little as 10 percent of the normal amount of acetylcholine, and the degree of its loss corresponds closely to the severity of the disease (Gelman, 1989).

There are several theories as to the causes of Alzheimer's. Some feel that a virus causes the disease, while others feel that environmental factors are involved. A popular theory is that the disease is caused by an overabundance of metal accumulation in the neurons, mainly aluminum. Some dispute this theory, since although we all ingest aluminum, our body rejects it by refusing to digest it.

It now appears that Alzheimer's may be transmitted genetically, since we know that a sibling of an Alzheimer's patient has a 50 percent chance of contracting the disorder. Almost all persons with Down syndrome, a genetic disorder, will develop Alzheimer's disease if they live long enough. It may be that a defective gene makes a person likely to get it if environmental conditions are present.

The President of the Alzheimer's Association reports that the cost of caring for an Alzheimer's patient in the home is $18,000–$20,000 per year, while nursing home care may reach as high as $36,000 per year (Cowley, 1989; Kantrowitz, 1989). The financial burden is usually placed on the patient's family, since Medicare does not cover the cost and Medicaid covers only families who are in poverty. Sadly, a family must exhaust all its resources before becoming eligible for Medicaid.

Another alternative to chronic care is the adult-care facility. These programs provide help with daily tasks such as feeding, washing, toileting, and exercising. The average cost of adult-care centers is $27 per day, including meals. One drawback to many of the adultcare facilities is that the staff often does not include medical personnel.

At this time the outlook is bleak for patients and families suffering with Alzheimer's. If you desire further information on what is known about the disease, write to the Alzheimer's Association, 70 East Lake Street, Suite 600, Chicago, Illinois, 60601.

An Applied View — Facts about Alzheimer's

- Approximately 10% of the 65-and-over population may have Alzheimer's. Forty-seven percent over 85 already have the disease.
- The National Institute of Health allocated $5.1 million for Alzheimer's research in 1978, and $123.4 million for research in 1989. Though the dollar amount allocated for research has increased each year, more money is needed.
- Heredity is the cause in 10%–30% of all Alzheimer's cases.
- There are always at least two victims of Alzheimer's: the afflicted person and the primary caregiver. The stress of caring for Alzheimer's patients makes the primary caregiver much more vulnerable to infectious disease, as well as vulnerable to all the problems that go with dealing with high levels of stress.
- Although caring for the Alzheimer's patient may often seem to be a thankless task, there is evidence that sensitive listening and attempts to enter the patient's reality may help prevent antisocial behavior and anxiety-related outbursts (Bohling, 1991).

The Relationship between Physical and Mental Health

In their study of the general health of 1,139 elderly persons living in an urban community, Gerson, Jarjoura, & McCord (1987) made a number of interesting findings that another study (Lohr and others, 1988) has supported:

- There is a strong relationship between physical and mental health.
- Married men have better health than unmarried men (married women are not healthier than unmarried women). This may be because their wives take care of them, or because they feel responsible to their families to maintain their health.
- Social resources are strongly associated with mental health for women, but less so for men. Probably women are more likely than men to *use* their social resources to help them maintain their mental health. Many men hate to admit to their friends that they are having a mental disturbance, because it is not considered "manly."
- Economic resources are clearly related to mental health. As was pointed out, this is especially true for African-American women.

There is a stereotype that the older people get, the less health-conscious they become. The picture is usually more complex than this. For example, compared to those in their sixties and eighties, centenarians (those over 100) ate breakfast more regularly, avoided diets, ate more vegetables, and relied on their doctor more than on the news media for nutrition information (Johnson, Shulman, & Collins, 1991). They also were found to be more dominant and imaginative (Martin & others, 1992). On the other hand, they did not tend to avoid fats or comply with nutritional guidelines designed to reduce the risk of chronic disease (Johnson & others, 1992). Here again we see the complex interaction of biopsychosocial factors as they affect health and the aging process.

A Multicultural View 

Gateball and the Japanese Elderly

Gateball, developed in post World War II Japan, is a team sport combining elements of golf and croquet. It employs strategies that exercise both mind and body. While the object is to score the most points, a great deal of emphasis is placed on the individual's contributions to the team as opposed to individual achievement.

In addition to these benefits, gateball also provides a social outlet, building relationships around a common interest. Gateball has become immensely popular among "silver agers," as older people are called by the Japanese. In a modern, crowded, industrial country like Japan, with early retirement and the highest life expectancy in the world, gateball provides a way for seniors to continue their lifelong pattern of group participation in something worthwhile.

Observations of gateball players in action found people who share information, laughter, and a relaxed sense of belonging. When a player who hadn't been there for some time reappeared, he or she was warmly welcomed back. Exchange of food occurs routinely on the break, and people often encourage others to take some home with them.

In America, this kind of community spirit among seniors is evidenced in some sports (such as golf and bowling), but the stereotype of the elderly person in the rocking chair remains pervasive. Sports provide a means by which older people can remain healthy as well as connected to other people. It is another important lesson we might glean from Japanese culture.

Health and Retirement

Muller and Boaz (1988) state: "Many studies of the retirement decision have found that poor health increases the probability of retirement. Yet doubts have been expressed whether self-reported deterioration of health is a genuine cause of retirement, rather than a socially acceptable excuse" (p. 52). They studied the actual health status of retirees, and found support for the idea that the majority of those who say they are retiring for health reasons are telling the truth.

There is no question that health affects the physical abilities of adults across the age range. Yet it is also clear that social circumstances are the main factor in health in the older years. Thus, the healthiness of adults appears to be more a result of the cultural conditions in which they find themselves than of their age.

Appearance

All of the physical factors treated earlier have an effect on the person's self-concept, but none has a stronger impact than physical appearance. Appearance is of great importance in the United States. A youthful appearance matters a great deal, and to women more than to men. Because women are affected by the "double standard" of aging, they are more likely to use diet, exercise, clothing, and cosmetics to maintain their youthful appearance. Nevertheless, there are significant changes in physical appearance with age that cannot be avoided.

Sales of products that promote youth and beauty are booming.

As you can see from the data presented in this chapter, there is a certain amount of physical decline in later life, but it does not occur until most people are quite advanced in age. And with the new medical technologies and marked trend toward adopting more healthy living styles, you can expect to see a much more capable elderly population. Will this happy news also be true for cognitive development?

An Applied View — A Comparison of Physical Abilities

One of the best ways of comprehending adult development is to do a miniseries of experiments of your own. In this activity, you can compare the physical development of friends and relatives with just a few simple pieces of equipment. Pick at least three people in at least two age groups. The results might surprise you.

You may find that you need to alter the instructions of the following experiments to make it more convenient for you to do the study. This is perfectly all right, as long as you make sure that each test is the same for every person who takes it. The greater the variety of the adults you enlist in the study, in factors such as age, sex, and religion, the more interesting your results will be.

A. *Muscular strength*. Several of the following tests require a heavy table. Simply have each subject grasp a leg of the table with one hand and raise the leg off the ground three inches. Subjects who can hold it up three inches for 30 seconds have succeeded in step 1. Now have them try to hold it up another three inches, and if they succeed they get credit for step 2 (each step must be done for 30 seconds). In step 3 they must hold it at a height of one foot, in step 4 they must hold it at a height of one and a half feet, and in step 5 at a height of two feet. Record one point for each step successfully completed.

B. *Vision*. The subject should be seated at the table with chin resting on forearms and forearms resting on the table. Set a magazine at a distance of 2 feet and ask the subject to read two lines. Next 3 feet, then 4 feet, then 5 feet, and finally at 6 feet. One point is given for each step.

C. *Hearing*. The subject is seated with his or her back to the table. You will need some instrument that makes the same level of noise for the same period of time. For example, a portable radio might be turned on a brief moment and turned back off, each time at the same volume. A clicker might be used. Whatever you use, make the standard noise at 2 feet behind the subject, and have him or her raise a hand when the sound is heard. Then make the sound at 4 feet, 6 feet, 10 feet, and 20 feet. You may need to vary these distances, depending upon the volume of the instrument you are using. Be sure to vary the intervals between the noises that you make, so that you can be sure the subject is really hearing the noise when you make it. Give points for each appropriate response.

D. *Smell,* E. *Taste,* and F. *Touch*. For these three tests, you will need five cups of equal size. The cups should hold milk, cola, soft ice cream, yogurt, and applesauce. Each food should be chilled to approximately the same temperature. Blindfold subjects and ask them to tell you what the five substances are.

G. *Reaction Time—I*. For this test you will need three squares, three circles, three triangles; one of each is one inch high, two inches high, and three inches high. Make them of cardboard, plastic, or wood. Arrange the nine figures in random order in front of the subject. Tell the subject that when you call out the name of one of the figures (e.g., a large square), he or she is to put a finger on it as quickly as possible. Call out the name of one of the figures, and time the response. Repeat this test five times, calling out a different figure each time, and compute the total time required for the five trials.

H. *Reaction Time—II*. Again arrange the nine pieces randomly in front of the subject. You will call out either the shape ("circle") or a size ("middle-size"). The subject puts the three circles or the three middle-size pieces in a pile as quickly as possible. Repeat this experiment five times, using a different designator each time, and compute the total time the task requires.

I. *Health—I*. Ask subjects to carefully count the number of times they have been in the hospital. Give them one point if they have been hospitalized 10 times or more, two points for 5 through 9 times, three points for 3 times, four points for no more than one time, five points if they have never been hospitalized for any reason (pregnancy should not be counted).

J. *Health—II*. Ask subjects to carefully remember how many times they have visited a doctor within the past year. Give one point for four or more times, two points for three times, three points for two times, four points for one time, five points for no visits at all.

Compare the scores for each test for persons within the age groups you have chosen. Do you see any patterns? Are the results of aging evident? What other conclusions can you draw?

Cognitive Development

In chapter 18, we described the changes in intelligence that develop across the adult lifespan, focusing on middle age. In this chapter we take a closer look at old age. The major question is whether there tends to be a significant decline in cognitive competence as we become elderly. A review of the research on this question reveals a serious conflict. Salthouse (1990) states it well:

> ***Results from psychometric tests and experimental tasks designed to assess cognitive ability frequently reveal rather substantial age-related declines in the range from 20 to 70 years of age. . . . On the other hand, adults in their 60s and 70s are seldom perceived to be less cognitively competent than adults in their 20s and 30s, and in fact, many of the most responsible and demanding leadership positions in society are routinely held by late middle-aged or older adults.*** **(p. 310)**

Cognitive Decline in the Elderly: Tests versus Observations

So where lies the truth: in the results of tests, or in our observations of actual performance? As we said, the answer is not simple. What factors could account for this discrepancy? Currently there are four hypotheses, each of which play a part (Salthouse, 1990):

- *Differences in type of cognition.* Intelligence tests tend to measure specific bands of cognitive ability, whereas assessments of real-life activities probably also include noncognitive capacities, such as personality traits. Thus some aspects of IQ may decline without causing lowered performance on the job, for example.
- *Differences in the representativeness of the individuals or observations.* There are many examples of elderly persons who can perform admirably even into their nineties, but do these individuals really represent the average elderly person? Probably not. It is also probable that only the most competent individuals are able to survive in such demanding situations. Those who are less competent will have dropped out of the competition at an earlier age. Therefore, when we examine the abilities of successful older persons, we may be studying only the "cream of the crop." Finally, it seems likely that observed competence represents only one type of cognitive ability (balancing the company's books, reading music), whereas intelligence testing involves several (verbal, math, reasoning, and other abilities).
- *Different standards of evaluation.* Most cognitive tests tend to push individuals to their limits of ability. Assessments of real-life tasks, those with which people are quite familiar (such as reading the newspaper), may require a lower standard of testing.
- *Different amounts of experience.* Doing well on an intelligence test requires one to employ traits that are not used every day, such as assembling blocks so that they resemble certain patterns. The skills assessed in real-life situations are more likely to be those the individual has practiced for years. For example, driving ability may remain high if the person continues to drive regularly as she ages.

Each of these hypotheses seeks to explain the discrepancy between tested ability and actual performance as being the result of faulty research techniques. Many other writers have suggested that perhaps the elderly really are inferior in cognitive competence to younger persons, but that they find ways to *compensate* for their lost cognitive abilities. Thus older persons are able to maintain performance levels even as ability is waning. Unfortunately, no one has suggested *how* such compensation might be achieved, so this hypothesis is doubtful.

Another question about the decline of cognitive ability in the elderly is "Can lowered capacity be recovered through training?" Here there is good news. Numerous methods have been successful in improving elderly cognitive skills. For example, important improvements have been demonstrated in:

- Spatial orientation and inductive reasoning (Willis & Schaie, 1986)
- The flexibility of fluid intelligence and problem-solving strategies (Baltes, Sowarka, & Kliegl, 1989)
- Cognitive plasticity of memory skills (Kliegl & others, 1989, 1990)
- Long-term effects of fluid ability (in the old-old) (Willis & Nessleroade, 1990)

It seems likely that, through the use of new training techniques and possibly through new drugs, the decline in cognitive ability in old age (whatever it may actually be) will submit to reductions and even reversals in the years to come.

Another aspect of cognitive decline has proven of great interest to psychologists. It is referred to as terminal drop.

Terminal Drop

Birren (1964) defines **terminal drop** as the period preceding the person's death, from a few weeks up to two years prior. This research looks at the relationship, not between IQ and age (which is the number of years from birth till the time of testing), but between IQ and survivorship (which is the number of years from the time of testing till death). The hypothesis is that the decline in intelligence in the later years may be caused by the person's perception, consciously or unconsciously, of impending death. This perception causes the person to begin withdrawing from the world. Consequently, performance on an IQ test drops markedly.

According to Birren,

Researchers have come to several different conclusions concerning the effect of age on intelligence level.

> ***the individual himself may or may not be aware in himself of diffuse changes in mood, in mental functioning, or in the way his body responds. Terminal [drop] may be paced by a disease initially remote from the nervous system, such as a cancer of the stomach. Thereafter, over a series of months, a rapid sequence of changes might be observed in overt behavior or in measured psychological characteristics of the individual. It is as though at this point the physiology of the individual had started on a new phase, that the organism is unable to stop.*** **(p. 280)**

There has been considerable support for the terminal drop hypothesis. Siegler (1975) reviewed eight studies of the phenomenon and discovered a strong positive relationship between the length of survivorship and high level of intellectual ability. That is, the higher the IQ, the longer the person is likely to live after the time of testing.

Some researchers have questioned whether terminal drop pervades all cognitive abilities, or is restricted to specific ones. In a study designed to answer this question, White and Cunningham (1988) found terminal drop to apply to a shorter period of time than five years, and to be more restricted than was previously thought. They reached the following conclusions:

> ***Only vocabulary scores for those who died at age 70 or less and within two years of testing were affected by terminal drop. . . . Thus the terminal drop phenomenon may be limited to abilities that typically are relatively unaffected by age, such as vocabulary or other verbal abilities. Furthermore, the effects may be restricted to a time period much closer to death than had been originally proposed.* (p. 141)**

One final warning: Although virtually all studies described here use IQ tests to measure intelligence, the two are not identical. IQ involves school-related abilities, whereas intelligence is made up of these and many other abilities (e.g., spatial ability and "street smarts"). It is not possible at present to discern how much this discrepancy affects the research. Another mental trait that is hard to study, but is even more vital to understand and cultivate, is creativity.

Creativity

In chapter 18, we concluded that quantity of creative production probably drops in old age but that quality of production may not. This is based on studies of actual productivity. But what about *potential* for production? Might it be that the elderly are capable of great creativity but that, as with IQ, factors like motivation and opportunity prevent them from fulfilling this ability? That is the question addressed by the next set of studies we'll discuss.

Cross-Sectional Studies of Creative Productivity

The first large-scale study to look at the creative productivity of *typical* people at various ages *who are still alive* was completed by Alpaugh and associates (1976; also Alpaugh & Birren, 1977). They administered two batteries of creativity tests to 111 schoolteachers aged 20 to 83. Their findings support the idea that creativity does decline with age. One major criticism of their research is that the tests they used are probably not equally valid for all age groups. The younger subjects are more likely to have had practice with these types of materials than the older ones.

Jaquish and Ripple attempted to evaluate the effects of aging while avoiding the problem of using age-related materials (1980). These researchers gathered data on six age groups across the lifespan: 10–12 years (61 people); 13–17 years (71 people); 18–25 years (70 people); 26–40 years (58 people); 40–60 years (51 people); and 60–81 years (39 people). There were a total of 350 subjects in the study.

The definition of creativity in this study was restricted to the concepts of fluency, flexibility, and originality; these are collectively known as divergent thinking abilities. These three traits were measured through the use of an auditory exercise, which was recorded on cassette. Known as the Sounds and Images Test, it elicits responses to the "weird" sounds presented on the tape. Responses are then scored according to the three traits of divergent thinking. The researchers believe that this test is so unusual that no age group is likely to have had more experience with it than any other.

A number of interesting findings have resulted from this study, as you can see in Table 20.3. On all three measures of divergent thinking, the scores generally increased slightly across the first five age groups. Scores for the 40- to 60-year-old group increased significantly, while scores for the 61- to 84-year-old group decreased significantly below the scores of any of the younger age groups. Furthermore, when decline in divergent thinking did occur, it was more pronounced in quantity than in quality. That is, there were greater age differences in fluency (a measure of quantity) than in originality (a measure of quality). Finally, it is not known to what extent the oldest subjects were affected by hearing loss, particularly of high and low tones. Obviously, this could be an alternative explanation of the findings for this oldest group.

Table 20.3 Means for Divergent Thinking and Self-Esteem Scores

Age Group	Fluency	Flexibility	Originality	Self-Esteem
	Mean	Mean	Mean	Mean
18–25	31	19	19	34
26–39	30	18	19	37
40–60	36	21	20	38
61–84	22	15	15	32

Source: From G. Jaquish and R. E. Ripple, "Cognitive Creative Abilities Across the Adult Life Span" in *Human Development*. Copyright © 1980 Educational Center for Human Development, Cambridge, MA.

Probably the most important finding of the study had to do with the relationship between divergent thinking and self-esteem, which was measured by the Coopersmith Self-Esteem Inventory. Table 20.3 indicates that self-esteem follows a pattern quite similar to the other three traits.

Of special interest is the relationship between self-esteem and divergent thinking for the oldest group: The correlations were moderately high in every case. This indicates that self-esteem may have a positive effect on creative abilities over the years, or that creativity may enhance self-esteem, or both. Jaquish and Ripple (1980) conclude that

> ***there is much more plasticity in adult development than has been traditionally assumed. Such an interpretation should find a hospitable audience among those people concerned with educational intervention. If the creative abilities of older adults were to be realized in productivity, it would be difficult to overestimate the importance of the formation of new attitudes of society, teachers in continuing adult education programs, and in the older adults themselves.*** **(p. 152)**

The Stages of Life During Which Creativity May Best Be Cultivated

In this final section we present Dacey's theory (1989b) that there are certain critical periods in life during which creative ability can be cultivated most effectively. Its relevance to the study of late adulthood will be obvious. Table 20.4 presents a list of these periods for males and females.

Table 20.4 Peak Periods of Life During Which Creativity May Most Readily Be Cultivated

For Males	For Females
1. 0–5 years old	0–5 years old
2. 11–14 years old	10–13 years old
3. 18–20 years old	18–20 years old
4. 29–31 years old	29–31 years old
5. 40–45 years old	40 (37?)–45 years old
6. 60–65 years old	60–65 years old

Source: From J. S. Dacey, *Fundamentals of Creative Thinking*. Copyright © 1989 D. C. Heath/Lexington Books, Lexington, MA.

The basic premise of this theory is that a person's inherent creativity can blossom best during a period of crisis and change. The six periods chosen in Table 20.4 are ages at which most people experience stress due to life changes.

The table presents a new theory, and thus it must be considered speculative. Nevertheless there is some direct evidence from research to support it. There is also some excellent research from the fields of personality and cognitive development to indicate that these periods are more volatile than any others.

Included in the theory is the concept that major gains in creative performance are less likely with each succeeding period. That is, what happens to the person in the early years is far more influential than what happens in the later years. The older people become, the less likely they are to have a sudden burst of high creative production.

The Sixth Peak Period

Evidence for the first five peak periods has been presented earlier in this book. The rationale for the sixth period, from 60 to 65, is reviewed here.

Research indicates that the time right after retirement can be a period of creative growth, as the individual turns from the demands of a work schedule to the opportunities offered by an artistic endeavor.

For most men and for a growing number of women, this is the period in which retirement occurs. Even if a woman has not been in the labor force, she has many adjustments to make because of her husband's retirement. Thus most adults are faced with a major adjustment of self-concept at this time in their lives.

Although some do not adjust well and begin withdrawing from society, others take advantage of the change to pursue creative goals that had previously been impossible for them. Obviously a majority of the "young old" (the new term for those who are 60 to 70) do not suddenly become creative, but a substantial number do. Of the several thousand highly productive people he studied, Lehman (1953, 1962) found over 100, or almost 5 percent, whose major productivity began in the years after 60.

In addition to highly visible contributions, many of the elderly become creative in less newsworthy ways. Gerontologist Jack Botwinick, who presents an excellent analysis of this topic in his book *We Are Aging* (1981), suggests that many elderly persons exercise a newfound creativity by mentoring younger people. Though largely unheralded, such guidance and encouragement of younger creators unquestionably has invaluable benefits for society.

This is not to say that maintaining or increasing creative performance in one's later years is easy. There are inherent problems that are not readily overcome. Psychologist B. F. Skinner (1983), who has a 60-year-long career of highly creative achievement, states that productivity is difficult for the elderly because they tend to lose interest in work, find it hard to start working, and work more slowly:

> ***It is easy to attribute this change to them, but we should not overlook a change in their world. For motivation, read reinforcement. In old age, behavior is not so strongly reinforced. Biological aging weakens reinforcing consequences. Behavior is more and more likely to be followed by aches and pains and quick fatigue. Things tend to become "not worth doing" in the sense that the aversive consequences exact too high a price. Positive reinforcers become less common and less powerful. Poor vision closes off the world of art, faulty hearing the enjoyment of highly fidelitous music. Foods do not taste as good, and erogenous tissues grow less sensitive. Social reinforcers are attenuated. Interests and tastes are shared with a smaller number of people.*** **(p. 28)**

It is increasingly clear that creativity may blossom at any age. This theory of "the peak periods of creative growth" is not meant to disparage that fact. There is solid evidence, however, that the best opportunities lie in the six periods identified by the theory. Are there also periods in the development of gender roles in later life that have special importance? of sexuality? of family relations? of work and retirement? We'll examine these questions in the next chapter.

One final point: What is the relationship between intelligence and creativity, and how do they relate to the highest level of human cognition, wisdom? In Table 20.5, Robert Sternberg's (1990) comparison of the three is presented.

Table 20.5 Summary, Simplified Comparison among Wisdom, Intelligence, and Creativity

	Construct		
Aspect	**Wisdom**	**Intelligence**	**Creativity**
Knowledge	Understanding of its presuppositions and meanings as well as its limitations	Recall, analysis, and use	Going beyond what is available
Processes	Understanding of what is automatic and why	Automatization of procedures	Applied to novel tasks
Primary intellectual style	Judicial	Executive	Legislative
Personality	Understanding of ambiguity and obstacles	Eliminating ambiguity and overcoming obstacles within conventional framework	Tolerance of ambiguity and redefinition of obstacles
Motivation	To understand what is known and what it means	To know and to use what is known	To go beyond what is known
Environmental context	Appreciation in environment of depth of understanding	Appreciation in environment of extent and breadth of understanding	Appreciation in environment of going beyond what is currently understood

Source: Sternberg (1990).

Conclusion

At the beginning of this chapter we asked, "Must growing old mean decline?" The answer is that some decline is inevitable, but the picture is much less gloomy than we have been led to believe. The loss of mental and physical abilities is, on the average, relatively slight; some individuals experience only moderate physical loss and no cognitive loss at all. For many older adults, compensatory skills and abilities may replace lost capacities. The same is true for personal and social development.

Here once again we run into the bugaboo of all human development: the self-fulfilling prophecy. Because American society has been changing rapidly for many decades, older adults are frequently viewed as incompetent—their experience appears to have little relevance in "modern times." Yet carefully controlled laboratory measurements of their abilities make clear that their losses may be relatively slight.

Perhaps as we learn to understand the aging process, and as the process is better understood by the public in general, the majority of adults will not assume that their abilities must undergo severe decline. In such a situation, the quality of life of the elderly will surely improve greatly! As you'll see in the next chapter, such improvements are already taking place in the social and personal lives of the elderly.

Chapter Highlights

Must We Age and Die?

- There are a variety of physiological theories regarding aging and death. These include aging by program, homeostatic imbalance, and cross-linkage theories.
- Gene theory also suggests that aging is programmed, but says that the program exists in certain harmful genes.
- The natural environment is also an important factor in mortality (for example, there are now fewer elderly mortalities due to the spread of influenza).
- Major modifiers of ability such as training, nutrition, illness, stress level, and personality type also affect one's rate of aging.

Physical Development

- While reaction time appears to decline with age, Elias and others have pointed to several reasons for this decline.
- We may conclude that variables other than sheer neural or motor activities account for most change in physical skills over time.
- Changes in sensory abilities, the skeletal system, skin, teeth, hair, and locomotion are noticeable in late adulthood.
- Even though there is a slowing down of hormone production during late adulthood, a detriment in one area is often compensated for by some other gland.
- Probably the single greatest scourge of the elderly, and in many ways the most debilitating, is Alzheimer's disease.
- There is a strong relationship between physical and mental health.

Cognitive Development

- A number of factors have been suggested as explaining the difference between tested and observed changes in elderly cognition. These include differences in type of cognition, the representativeness of the individuals or observations, standards of evaluation, and amounts of experience.
- The decline in intelligence in the later years may be caused by the person's perception, consciously or unconsciously, of impending death. This is called terminal drop.
- Creativity is evidenced in late adulthood and in some cases may be strongest during this time. While quantity of creative production probably drops in old age, the quality of creative production and potential ability probably do not.

Key Terms

Accumulation of errors
Accumulation of metabolic waste
Ageism
Aging by program
Antigens
Autoimmunity
Collagen
Cross-linkage theory
Gene theory
Homeostatic imbalance
Terminal drop
Wear and tear theory

What Do You Think?

1. Must we grow old and die?
2. Which factors do you believe most strongly affect aging: physiological, genetic, environmental, or some others?
3. Regarding the decline of physical systems, which has the greatest impact on the person's life: reaction time, sensory abilities, other body systems, hormonal balance, health, or appearance? Why?
4. Regarding the decline of cognitive systems, which has the greatest impact on the person's life: intelligence, creativity, memory, or learning? Why?
5. What is the nature of wisdom?

Suggested Readings

Cunningham, W. and J. Brookbank. (1988). *Gerontology*. New York: Harper & Row. An up-to-date general reference on the psychological, biological, and sociological factors in aging.

Olsen, T. (1961). *Tell me a riddle*. New York: Dell. The superb title story details the difficulties of old age and terminal illness as they occurred to a working-class woman during the Great Depression.

Sarton, M. (1973). *As we are now*. New York: Norton. A novel in the form of the diary of a retired schoolteacher, this is a powerful portrayal of her experiences when she is put in a nursing home by her relatives.

Skinner, B. F. and M. E. Vaughan. (1983). *Enjoy old age: A program of self-management*. New York: Norton. In this book, the grandfather of behaviorism explains how to use behavior modification to better handle the problems of aging. A good read, whether you are elderly or plan to help someone who is.

Spence, A. (1989). *Biology of human aging*. Englewood Cliffs, NJ: Prentice-Hall. A highly detailed look at the systems of the body and how they are affected by aging.

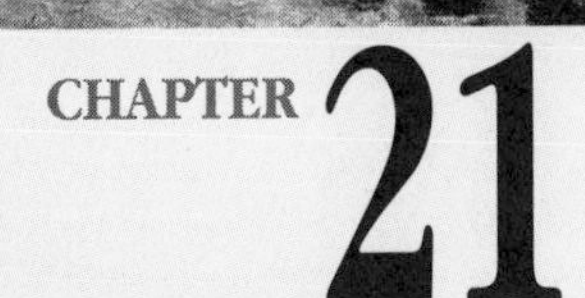

CHAPTER 21

Late Adulthood: Psychosocial Development

Social Development 550
Gender Roles 550
Sexuality 550
The Elderly and Their Families 555
The Older Worker 558
Retirement 559
Personal Development 562
Personal Development: Committee on Human Development 562
Personal Development: Erikson 565
Conclusion 566
Chapter Highlights 566
Key Terms 567
What Do You Think? 567
Suggested Readings 567

Getting married in 1948 I married into a world that had a very definite definition of marriage. I married into a man-oriented society, where the man was the provider, the center, the whatnot, and the woman circled around the man. But I married Ruby Dee; Ruby had some other ideas about marriage. [Laughs.] There were one or two other extra things on her agenda that I didn't know about, but when they came up I saw no reason to challenge her. For example, after our first baby was born and we moved down to Mount Vernon, New York, I remember Ruby standing washing dishes one day and saying to me, quite confidently, "You know, I'm not going to do this the rest of my life." I said, "No? What are you going to do?" She said, "Well, I'm going to be an actor. I'm still going to act." I said, "Yes." She said, "I'm going to go to acting school." And I said, "Well, okay." And at that time we didn't have much money, so we decided that Ruby would go to acting school and I would stay home and sometimes wash the dishes and take care of the baby. And when Ruby came home she would teach me what she learned at the acting class.

Now, when I stepped into the marriage I didn't think of that, but that happened. And over the years I think the central thing that I've learned is how to more easily open my life and let Ruby in. I think women are much more generous in letting people into their lives than men—although I might be wrong. I think that men come with a sense that they are the complete embodiment of what God intended should represent value, virtue, power, and all that sort of thing, and that women, to some degree, are there to service and serve them—that's how women fulfill themselves. Gradually my wife and the circumstances of the time led me to a different understanding, and I'm glad for that, because it made me a broader and a deeper and a much richer person. It has enriched my spirit, my spirituality. It means also—and this is the best part of it all—that I learned a lesson a long time ago, and the lesson is, "The way to make a man rich is to decrease his wants." I learn more and more every day what not to want and how not to want it. To me that is the regimen of the spirit.

Actor and biographer Ossie Davis, age 75
Quoted in Berman and Goldman (1992) ■

After reading chapter 21, you should be able to:

- Define what is meant by "crossover" regarding gender roles in late adulthood.
- Contrast the social, emotional, and physical aspects of sexuality for men and women in late adulthood.
- Analyze White's summary comments regarding sexuality in late adulthood.
- Identify the four basic phases through which most adults pass in relation to their families.
- Describe characteristics of the older worker.
- Assess the relationship between retirement and leisure.
- Evaluate Williamson and associates' seven phases of retirement.
- Summarize Neugarten's findings regarding the effects of aging on personal development.
- Contrast activity theory with disengagement theory.
- Appraise Erikson's final stage of human development: integrity versus despair.

Social Development

This section is devoted to five aspects of social development: gender roles, sexuality, families, the older worker, and retirement.

Gender Roles

> ***When people enter their sixties, they enter a new and final stage in the life cycle. At this point they confront the loss of many highly valued roles, the need to establish a new life structure for the remaining years, and the undeniable fact of life's termination. Widowhood and retirement are the central role transitions likely to occur at this time, but the death of friends and relatives also diminishes one's social network. Although people are aware of the inevitability of these role losses as they enter old age, their often abrupt reality may result in severe role discontinuity.*** **(Sales, 1978, p. 185)**

To sum up, the major concern for gender roles among the elderly is **role discontinuity.** See the accompanying box for several questions you should keep in mind as you read the theories and research summaries that follow.

What's Your View?

Role discontinuity occurs when people experience an abrupt change in their style of life and their role in it. Is this a natural part of growing old? Should we expect our world to shrink and our power to erode? Or is this just a stereotype of old age?

Is role discontinuity a problem only for the poor, who have less control over their lives than the wealthy? Does high intelligence make a difference? How about gender? What's your opinion?

A number of gerontologists have noted that people in late adulthood experience a **crossover** in gender roles. Older men become more like women, and older women become more like men. They do not actually crossover—they just become more like each other. This is what Neugarten (1968) found in her studies of aging men and women. She states that "women, as they age, seem to become more tolerant of their own aggressive, egocentric impulses; whereas men, as they age, [become more tolerant] of their own nurturative and affiliative impulses" (p. 71). In Gutman's terms (1973), men pass from "active to passive mastery," and women do just the opposite.

The "cross-over effect" concerns the tendency of men to do more things that are considered feminine, such as washing the dishes, and for women to do more things that are considered male, such as taking charge of repairs to the home.

The differences between men and women, so many of which seem to be based on sexuality, are no longer as important. With the barriers breaking down, older men and women seem to have more in common with each other, and thus may be more of a solace to each other as they deal with the disruptive changes of growing old. This is not to say that men and women reverse gender roles. Rather, they move toward androgyny (See chap. 14), accepting whatever role, male or female, is appropriate in the situation.

On the basis of data obtained by University of California at Berkeley, Norma Haan (1976, 1981, 1989) concludes that the gender-role changes that result from aging generally lead toward greater candor with others and comfort with one's self. For the most part, she says, "people change, but slowly, while maintaining some continuity" (1985, p. 25).

Sexuality

Until the 1980s, most reports on sex among the elderly agreed that sexual practices drop off sharply in old age. For example, Pearlman (1972) reported that only 20 percent of elderly males have sex two times or more per month. There are serious doubts, however, about the reliability of these reports. Society disapproves of the idea of sex among the elderly, so it may be that many do not report what actually goes on.

A study at Duke University centered primarily on the social and emotional aspects of sexuality in later adulthood, but included the physical component as well (Williamson, Munley, & Evans, 1980). Since 1960, when the study began, researchers at Duke University have interviewed and medically evaluated 270 people over age 60. Their main goal was to define interest in sex and amount of sexual activity as people age. Here are the study's main conclusions:

- At age 68, about 70 percent of males engage in intercourse, and one out of five are sexually active in their eighties.
- For women, activity does not decrease with age as it does for men.
- Older men are more sexually active and more interested in sex than older women.
- Interest decreases for both genders, but not as much as sexual activity does.
- Sexual interest is positively linked to health for males. Healthier individuals are more interested in sex.
- For women, interest is dependent upon the enjoyment and quality of sex in the past.

The Janus Report

The most recent report on frequency of sexual activity among the elderly, the Janus Report (See chap. 14), gives a picture that weakens the stereotype of the sexually inactive elder even more. As Table 18.2 makes clear (See chap. 18), in the last decade of this century, there is little difference in the practices of young adults and elderly adults! In fact, this study found rather minor differences among any of its four age groups. Either the adults in the Janus study were more honest in reporting what they actually do, or there have been some great changes in sexual attitudes among older Americans in recent years.

Widowhood

Gender differences in attitudes and interest in sex are accounted for by a number of factors. For example, women outlive men by approximately seven years. Most married women will become widows because they marry men nearly four years older than themselves. In contrast, most men in society will not become widowers unless they reach age 85. (National Center for Health Statistics, 1992). Due to this imbalance of elderly males and females, it is more difficult for women to find sexual partners in their aging years.

Whether or not a woman is sexually active depends mainly on her marital status. In contrast, single men in the Duke study were just as sexually active as their married counterparts. Men have generally had more opportunity for extramarital sexual relations than women, due to the **differential opportunity structure** (Glenn, 1978). This means that due to social disapproval and more rigid rules enforced by parents, peers, and the legal system, women have not had the same access to sex as men.

Impotency

One of the biggest fears in males of increasing age is **impotency.** Physical changes, nonsupportive partners and peers, and internal fears may be enough to inhibit or terminate sexual activity in males. It may become a self-fulfilling prophecy.

Sleep laboratory experiments have shown that many men in their sixties to eighties who have labeled themselves as impotent regularly experience erections in their sleep (Rubin, 1979). In many cases a man is capable of having intercourse, but a physical condition such as diabetes impedes it. New types of prothetic devices can remedy a variety of psychological and physical problems.

One pervading myth is that surgery of the prostate gland inevitably leads to impotency. Many elderly men experience pain and swelling of this small gland, and a **prostatectomy** (removal of all or a part of this gland) is necessary. Most impotency that results from the removal of this gland is psychological rather than physical.

Most sexual problems experienced by women are due to hormonal changes. The vaginal walls begin to thin, and intercourse may become painful, with itching and burning sensations. Estrogen pills and hormone creams relieve many of these symptoms (Butler & Lewis, 1977). Further, if women *believe* that sexual activity ceases with menopause and aging, it probably will. Although women have fewer concerns about sex, they are often worried about losing their attractiveness, which can also have a negative effect on their sex lives (McCary, 1978).

Starr and Weiner (1981) surveyed 800 adults between the ages of 60 and 91, drawn from all parts of the country and representing all ethnic and racial groups. Half were married, a third were widowed, 11 percent were divorced, and 4 percent had never married. The group responded to fifty open-ended questions about their sexual lives and then mailed back the questionnaire anonymously. Table 21.1 presents their surprising findings.

Table 21.1 Findings of the Starr-Weiner Study of Sexual Activity in Later Life

Frequency of intercourse is 1.4 times per week. (Kinsey had reported a frequency of once every two weeks for 60-year-olds.)

- The ideal or fantasized lover for most, particularly women, is close to their own age.
- Most see their sex lives remaining pretty much the same as they grow even older.
- Most have a strong continuing interest in sex.
- They believe that sex is important for physical and mental well-being.
- The perception of most of the respondents is that sex is as good as when they were younger.
- For a large number, both male and female, sex is *better* in the later years.
- Orgasm is considered an essential part of the sexual experience.
- Most of the women are orgasmic and always have been.
- The orgasm for many is stronger now than when they were younger.
- Masturbation is an acceptable outlet for sexual needs.
- For a majority, living together without marriage is acceptable.
- An overwhelming number of respondents, including widows, widowers, divorcees, and singles, are sexually active.
- Most are satisfied with their sex lives.
- Many vary their sexual practices to achieve satisfaction.
- For a surprising number of older people, oral sex is considered the most exciting sexual experience.
- Respondents typically show little embarrassment or anxiety about sex.
- Most enjoy nudity with their partners.

Source: Data from B. Starr and M. Weiner, *The Starr-Weiner Report on Sex and Sexuality in the Mature Years,* 1981, p. 241.

Sex in Nursing Homes

In their study of the sexual behavior of sixty-three residents in nursing homes, Wasow and Loeb (1979) reported the following findings:

> ***The aged interviewees believed that sexual activity was appropriate for other elderly people in the homes; they personally were not involved, chiefly because of lack of opportunity. Most of them admitted having sexual thoughts and feelings. Medical and behavioral personnel showed great reluctance to discuss the subject. It would seem that, if the quality of life in old age is to be improved, there should be some provision in nursing homes for those who desire appropriate sexual activity.*** **(p. 73)**

The question of whether to allow residents of nursing homes to engage in sex has been a growing problem as elders' attitudes toward sexual mores change with the rest of us.

Sex and the "Old-Old"

In a fascinating study restricted to people 80- to 102-years old by Bretschneider and McCoy (1988), we get a look at the sexual activities of those elderly who used to be thought totally inactive. Table 21.2 gives us some surprises! In their past lives, the men claim to have engaged in intercourse more than the women, but there was no difference in enjoyment. In their present lives, 62 percent of the men and 30 percent of the women say they have intercourse at least sometimes, and 76 percent of the men and 39 percent of the women say they enjoy it at least mildly. Another interesting finding of this study is that the men reported having their first intercourse at an average age of 22, and women at age 25. What a difference from today's figures (See chap. 14)!

Nevertheless, it is quite likely that there is a real decline in sexual activity among the elderly. Men are often concerned about their ability to consummate intercourse. They also worry about their loss of masculinity, in terms of looks and strength. Nevertheless, the literature on sex among the elderly shows a new attitude emerging. Datan, Rodeheaver, & Hughes (1987) describe very well a difference between **generative love** and **existential love:**

> ***We believe that existential love, the capacity to cherish the present moment, is one of the greatest gifts of maturity. Perhaps we first learn this love when we first confront the certainty of our own personal death, most often in middle adulthood. Generative love is most characteristic of parenthood, a time during which sacrifices are gladly made for the sake of the children. However, it is existential love, we feel, that creates the unique patience and tenderness so often seen in grandparents, who know how brief the period of childhood is, since they have seen their own children leave childhood behind them.***
>
> ***We have not yet awakened to the potential for existential love between old women and old men, just as we are not yet prepared to recognize the pleasures of sexuality as natural to the life span, particularly to the postparental period.***
>
> ***Those old people who have had the misfortune of spending their last days in nursing homes may learn that love can be lethal. We have been told of an old woman and an old man who fell in love. The old man's children thought this late flowering was "cute"; however, the old woman's children thought it was disgraceful, and over her protests, they removed her from the nursing home. One month later she registered her final protest: she died.*** **(p. 287)**

Out of all the hundreds and hundreds of studies of monkeys, there is one finding that applies to every type: From the largest gorilla to the tiniest spider monkey, they all spend about four hours a day in "grooming." Grooming refers to their

Table 21.2 Reported Frequency and Enjoyment of Sexual Intercourse by 80- to 102-Year-Old Men and Women in the Past (Younger Years) and in the Present

	Entire Sample		Frequency: Never (1)		Sometimes (2–3)		Often (4–5)		Very Often (6–7)		
	N	%	n	%	n	%	n	%	n	%	x^2
Past											
Men	92	92	3	3	3	3	60	65	26	28	
Women	90	88	2	2	10	11	70	78	8	9	14.2
Present											
Men	80	80	30	38	27	34	21	26	2	3	
Women	80	78	56	70	16	20	8	10	0	0	18.5

	Entire Sample		Enjoyment: None (1)		Mild (2–3)		Moderate (4–5)		Great (6–7)	
	N	%	n	%	n	%	n	%	n	%
Past										
Men	91	91	3	3	5	6	23	25	60	66
Women	92	90	5	5	9	10	34	37	44	48
Present										
Men	79	79	19	24	11	14	21	27	28	35
Women	82	80	50	61	10	12	13	16	9	11

Source: From J. G. Bretschneider and N. L. McCoy, "Sexual Interest and Behavior in Healthy 80- to 102- Year Olds" in *Archives of Sexual Behavior,* 17(2):109–29. Copyright © 1988 Plenum Press, New York, NY.

different ways of touching—stroking, removing bugs, hugging, sex. It is obviously genetic. It seems likely that we humans have something in common with them. We all need to be touched, too. The elderly get less touching than the rest of us, perhaps because they are not seen as being attractive. But they need physical contact just as much as everyone else. It is hoped that a new attitude will spread and make their lives that much happier.

White (1981) has reviewed the literature on sexuality and offers the following summary.

- Males are more active than females, except for very old age groups (those 85 or older) in which males and females do not significantly differ.
- When sexual interest or activity ceases or declines in the aging female, it is usually due to declining interest or illness in her male partner.
- Males do show a gradual decline in sexual activity with advancing age, though this decline may be a cohort difference, because some males actually show an increased interest in sex with age. In the absence of longitudinal data, a cohort explanation cannot be ruled out.

- Physiological changes in the sexual organs in advanced age cause some difficulties for some individuals but do not adequately explain decreased or nonexistent sexual activity in either sex.
- It is difficult to find research on aging and sexuality that does not suffer from sample bias and methodological problems.

The Elderly and Their Families

The familial relationship undergoes changes in membership, organization, and role during the aging process. Due to improved health care, individuals can expect to live longer, which means that married couples will have more years together after their children leave home. Though there are exceptions (those who are divorced, childless, or who never married), most middle-aged couples go through similar stages in the life cycle (Williamson & others, 1980). There are four basic phases:

1. The child-launching phase
2. The childless preretirement period
3. The retirement phase
4. Widowhood

The duration of each stage in the life cycle, and the ages of the family members for each stage, vary from family to family. Child-bearing patterns have a lot to do with life in the late stages of life. Couples who complete their families in their early years will have a different life-style when their last child leaves home than couples with "change of life" babies, who may have a dependent child at home when they are ready to retire. This can pose serious economic problems for those retirees on fixed incomes, trying to meet the staggering costs of education. In addition, with children in the home, it is difficult to save for retirement.

Retirement may bring about changes for both spouses, but it may be particularly stressful for wives who have not properly prepared themselves emotionally and financially for retirement. Retirement generally signifies a decrease in income and a lowering of the standard of living, but it may take a while before some of these problems are noticed. Household duties may change, with the husband generally helping more.

An Applied View — Decisions Most Older Couples Must Make

Most older couples need to make a number of decisions that will be vital to their family lives (Cox, 1988), including whether or not to

- Remain in their current home with its history and memories, or move to a new home or apartment
- Remain in the same community or move to a different one, or perhaps move to a retirement community
- Remain active in current organizations, join new ones, or simply not be bothered with organizational affiliations
- Try to locate near children and close friends or move to a different section of the country
- Seek activities satisfying to both husband and wife, or participate independently

Obviously the decisions they make can have a major impact on their families and themselves. Each of these decisions has the capacity to cause considerable stress for all the family's members. Being aware of them, and confronting them openly, perhaps with a counselor, can greatly reduce the stress.

Widowhood

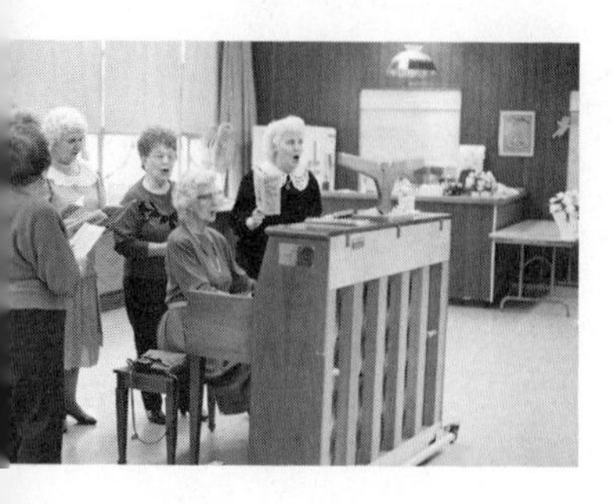

Older women, who are more likely to be widowed than men because the men die earlier, often turn to activities with other widows for enjoyment.

Widowhood is the final phase in the life cycle. With women outliving men by large margins, the wife is most often the survivor. Only half of women over 65 are living with a partner. Lopata's (1973) study of Chicago area widows sought to find what changes occur with the death of a spouse. She found that "widowhood means the loss, reorganization, and acquisition of social roles" (p. 6). A widow forfeits the role of her partner's "nurse, confidant, sex partner and housekeeper."

Many widows have to take on additional duties, including managing household finances and janitorial tasks, and some will have to seek employment. Widowhood affects social relationships with family and friends. Often a widow is the "fifth wheel" in social settings, and former relationships may dissipate. Fortunately, new social activities and friends emerge and replace the old ones. Many of Lopata's widows came to realize some compensations in widowhood, including increased independence and a decline in their work load.

Children may play an integral part in their mother's adjustment to widowhood in three ways: by taking over some of the father's responsibilities; by supplanting the father as the mother's center of attention; and by being supportive and maintaining relationships (Lopata, 1973).

Remarriage is an alternative to the loneliness that most widows and widowers feel after losing their spouse. However, remarriage rates for senior citizens are low, and it is not an option for most. The reasons for not remarrying include

- Many of the elderly view it as improper.
- Children may oppose remarriage.
- Social Security laws penalize widows who remarry.
- There are three single women for every single man over age 65.

Seniors who choose to remarry, however, enjoy much success if the ingredients of love, companionship, financial security, and offspring consent are present.

An important aspect of happiness in the elderly person's family life is whether he or she is living with a spouse, with children, or with another relative. The first has been more common in recent years (Turner & Helms, 1989), probably as a result of better health among the elderly, and better support systems for them. In most cases, living with one's spouse is preferred, so this is probably contributing to an increase in happiness in our senior citizens.

Care of Elderly Parents

Elderly people identify their adult children, when they have them, as the primary helpers in their lives. When these people have both an adult son and an adult daughter, elderly people most often name the son as the primary helper (Stoller, Forster, & Duniho, 1992). This is surprising, because as we pointed out in chapter 19, care of elderly parents is almost always undertaken by *daughters,* if the elderly persons have any.

In the past, elder care was most often done by unmarried daughters, if there were any. They were expected to do this because it was assumed that the work would be easier for them, since they had no responsibilities for husband or children. In fact, married women report that their married status makes caregiving for their elderly parents easier. They have less depression, higher incomes, and other forms of socioeconomic support than unmarried women (Brody, Litvin, & Hoffmann, 1992).

A Multicultural View 

Caregiving for Elderly in Swarthmore, PA, and Botswana, Africa

Draper and Keith (1992) were perplexed by the question, "Why is care for the elderly such a problem for Americans?" Draper had recently returned from studying the life-style of the !Kung people of Botswana, and decided to do a comparison study with Keith of elder care in Swarthmore, a Philadelphia suburb.

Older residents of Swarthmore are extremely worried about their care, especially with regard to loss of health, which generates feelings of fear. The researchers learned that for many of the elders, need for care is a primary reason for moving into Swarthmore. They do so in order to be near a child or relative who could supervise their eventual move into a retirement home, should professional care prove necessary. The problem is that whether they moved into Swarthmore (to be near relatives), or moved out of Swarthmore (to enter a retirement community), the costs to the older person usually involve loss of ties to their communities.

In contrast, elders of the !Kung villages "age in place." They do not retire, relocate, or enter age-graded elder care institutions. Indeed, they have no other place to go. The !Kung were asked, "For an old person, what makes a good life?" One third responded, "If you have a child to take care of you, you have a good life." Care is almost always provided by one's children and community.

For the people of Swarthmore, technologically superb care is available, but its benefits must be weighed against the loss of community ties and personal autonomy. For the !Kung, social needs and physical care are compatible.

What about the situation in which the elders have no children, or at least none who are willing or able to care for them? Research indicates that elderly individuals who have no kin tend to substitute a close friend whom they persuade to take the place of the absent relative. Nearly 40 percent of a group of elders surveyed could actually identify such a person who filled this role (MacRae, 1992).

Clearly, most family and close friends still feel that they ought to take care of the elderly in their own homes if possible. In their study of the outcome of several types of elder care, however, Strawbridge and Wallhagen (1992) conclude that "While important, family care for frail elders is not always appropriate and should be but one option in long-term care" (p. 92).

The Changing Role of the Grandparent

In today's world, with increased life expectancy, grandparenthood has become a unique experience within the family system (Smith, 1989). Grandparenthood is positively linked to the mental health and morale of elderly persons. Kornhaber and Woodward (1981) suggest that there is a "vital connection" between grandparent and grandchild.

In a classic study, Neugarten and Weinstein (1964) examined styles of grandparenting and created categories for five general styles: "formal," "the fun seeker," "the surrogate parent," "the reservoir of family wisdom," and "the distant figure." An interesting finding was that the fun seeker and the distant figure emerged as the most popular styles of grandparenting. Both exclude an emphasis on authority. Many grandparents preferred a grandparent-grandchild relationship in which their role was simply to enjoy being with their grandchildren rather than feeling responsible as co-parents with their adult children.

In the American culture of today, the importance of grandparents is intensified, due to the many roles they feel they must play: providing emotional support and financial assistance to their children and grandchildren in divorce and substance abuse situations (Smith, 1989) and acting as gender-role socialization agents

An Applied View — Do You Know Your Grandparents?

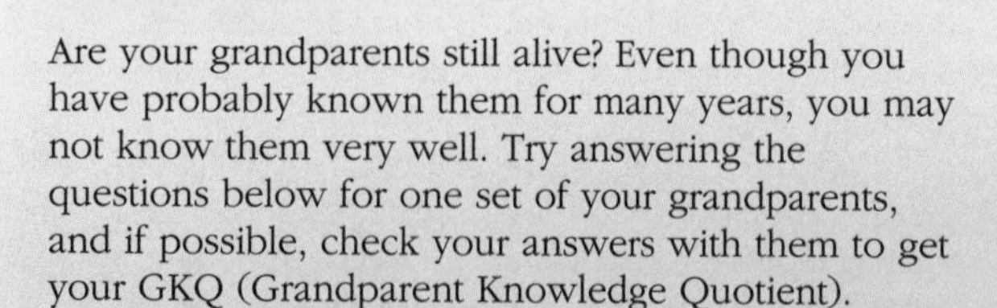

Are your grandparents still alive? Even though you have probably known them for many years, you may not know them very well. Try answering the questions below for one set of your grandparents, and if possible, check your answers with them to get your GKQ (Grandparent Knowledge Quotient).

1. What's your grandmother's favorite activity?
2. In total, how many rings do your grandparents wear?
3. What color are your grandfather's eyes?
4. Where does your grandfather eat lunch?
5. What is your grandmother's favorite TV show?
6. In what year did your grandparents meet?
7. For whom did your grandmother vote for president in 1992?
8. Does your grandfather know how to prepare asparagus?
9. Who is your grandfather's favorite relative?
10. Name some of either of your grandparents' favorite movie stars.

You might try making up a test like this about yourself, and asking your grandparents, parents, siblings, and friends to respond to it. It should be interesting to see which of them knows *you* best.

(Thomas & Datan, 1983). Bengston and Robertson (1985) include some "symbolic functions" of the grandparent role in today's culture: acting as the "family watchdog"; behaving as arbitrators; and merely "being there."

Erikson (1963) explained, in his stage of "generativity," the significance of grandchildren to grandparents. Erikson believed that generativity referred to providing a better life for future generations, and that not having reached the stage of generativity would cause "stagnation and self-absorption" in the individual. (See chap. 19) Thomas and Datan (1983) highlight the reciprocal nature of "generativity" by arguing that the personal development of grandparents is furthered by their close rapport with younger generations and vice-versa.

The Older Worker

Only a small percentage of all older adults are in the labor force—about 11 percent. Much of this is due to forced retirement (Sommerstein, 1986). However, even with the extension of the mandatory age of retirement from 65 to 70 by Congress in 1978, not many people want to continue working past 65. For example, a large steel corporation that has never had mandatory retirement finds that less than one percent of their 40,000 workers stay on past 65. The average age of retirement at this company is below 62.

Interest in Work

Older people clearly care less about working than younger people. For example, Cohn (1979) reports that "Toward the end of the period of labor force participation, the satisfactions men derive from work are transferred from the actual experience of work to its consequences" (p. 264). It seems likely that these percentages are even lower today. That is, no more than half of older workers may be willing to work for its own sake. Of course, many younger workers do it only for the pay, too. The social value of the work itself is probably the most significant differential.

Discrimination

Discrimination against the older worker can be subtle and hard to detect (Findley, 1979). For example, one 61-year-old female designer was told by the company she worked for that they were going out of business and that she would not be able to collect her company pension. Also losing their jobs were two other persons who had the same job as her, ages 32 and 29. After the company folded, the older woman learned that the owners had formed a new company and had reemployed the two younger designers. She complained to the United States government, who took the case to court and won $125,000 in benefits for her.

Performance

As this society ages through greater longevity, the aging of the baby boom generation, and decreased birth rates, some of the stereotypes about aging are coming under closer scrutiny. One stereotype is that work performance necessarily declines with age (Rhodes, 1983). This stereotype is getting more research attention because the number of workers in the last two decades of their careers will *grow* 41 percent while the number of workers 16 to 35 years old will *decline* slightly (Johnston, 1987).

The stereotype is bolstered by research on aging that demonstrates a decline in abilities such as dexterity, speed of response, agility, hearing, and vision. If all these abilities decline, then surely job performance must decline with age. However, McEvoy and Cascio (1989) recently conducted an extensive meta-analysis (a study that compiles the results of many other studies) of ninety-six studies and found no relationship between age and job performance. It made no difference whether the performance measure was ratings or productivity measures, nor whether the type of job was professional or nonprofessional.

What explanation is there for these results? How can a worker deal with declining physical abilities? Experience is one answer. There is said to be no substitute for it, and it is certainly valued by employers. Other reasons cited are that older workers have lower absenteeism, turnover, illness, and accident rates (Kacmar & Ferris, 1989; Martocchio, 1989). They also tend to have higher job satisfaction and more positive work values than younger workers (Rhodes, 1983). These qualifications seem to offset any decreases in physical ability caused by increasing age.

We are making strides in understanding the special problems of the older worker, especially in measuring their individual needs. The key seems to lie in improvements in career counseling. A number of new programs are being set up for this purpose (Brady & Gray, 1988). When older workers are adequately advised, they can remain a useful and satisfied part of the work force (Bove, 1987; Bornstein, 1986; Cahill & Salomone, 1987). More and more, we find that counselors in this field employ the biopsychosocial model in their work. Can you see how?

Retirement

"I just don't want to retire," said Charlie, a 65-year-old shipping clerk. "But you've worked hard, and you should get the fruits of your labor," said his boss. "Fruits of my labor, my backside! I know lots of guys, as soon as they retire, they get sick or something and then they die. I know if I retire I'm gonna die. I'm gonna die!" Despite his protestations, Charlie was retired. Three months later he was dead.

For many people, retirement is a welcome relief from a frustrating and boring job. For others, it is just as difficult as being unemployed. Retirement requires changing the habits of an adult lifetime. This probably explains why over 11 percent of those 65 and over are employed.

Nevertheless, the great majority of the elderly do not choose to work. Money is probably not the major factor in that decision (Hayward, 1986). The decision is a complicated one, but most people now feel they have enough financial security so that they need not work. Health may be the biggest factor. Crowley (1986) found that the well-being of 1,200 retirees was highly dependent on the state of their health at the time of retirement.

Retirement seems to be harder on males than females. Many feel that they have nothing to do, while their wives still have a job. The home still must be taken care of, the meals cooked, the clothes cleaned. Most older wives have already adjusted to a reduction in their roles because their children have left home. For men, the change usually comes all at once.

Women's attitudes toward retirement are clearly different from those of men (Campione, 1987). For example, Barfield and Morgan (1978) found that even in families where the husband is planning to retire early, the wife often is not. Campbell (1979) has suggested that because women make up a larger percentage of the older population, extending the retirement age to 70 will benefit them more. It may also alleviate some of the strain on programs such as Social Security.

Retirement and Leisure

How you view retirement depends on the work and leisure experiences you have had up to the point of retirement. The leisure activities pursued throughout life play a crucial role in your social adjustment later on. The relationships among work, retirement, and leisure can be seen in Figure 21.1.

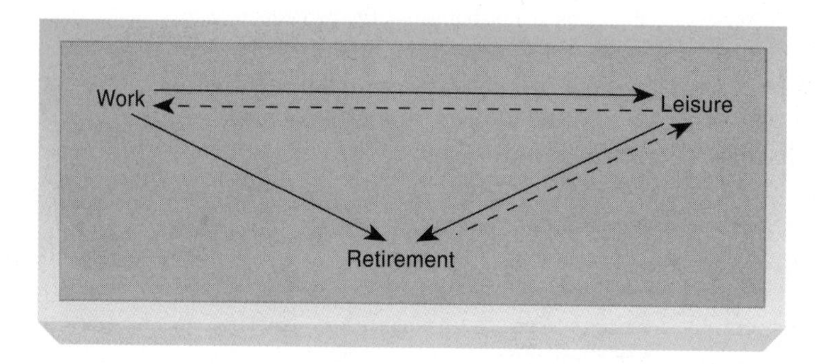

Figure 21.1
The interrelatedness of work, leisure, and retirement. Solid lines denote the direct influence of (1) work on leisure, and (2) work and leisure upon retirement. Broken lines suggest a possible feedback influence whereby preferences for uses of leisure time may affect choice of jobs, and the availability of more time in retirement may affect content of leisure activities.

The type of work you do directly affects how you spend leisure time, in two ways. The scheduling of work affects when leisure time is available. A person who works second or third shift might be unable to take part in activities that are often thought of as evening activities, such as dining and dancing. Secondly, the content of work may influence how much time will be left over for leisure activities. Persons with physically draining positions may be too tired to do anything but nap after work. On the other hand, a person with a desk job may choose physically challenging activities during leisure hours.

The "stuffed briefcase" syndrome refers to people who feel they must work even when they are pursuing such relaxing activities as watching television.

When leisure is no more than an extension of work experiences and attitudes, it performs what leisure theorists call a *spillover function*. For example, some people try to relax by performing some of their easier tasks at home—the "stuffed briefcase" syndrome. Conversely, when leisure time is engaged in to make up for disappointment and stress at work, it is described as a *compensatory function*. An example would be excessive partying at the end of the work day or week. An overabundance of compensatory leisure will leave one as unprepared to face retirement as an overabundance of work-related activities.

The Seven Phases of Retirement

Williamson and associates (1980) have defined seven phases that most retirees pass through. These are illustrated in Figure 21.2 and explained in Table 21.3.

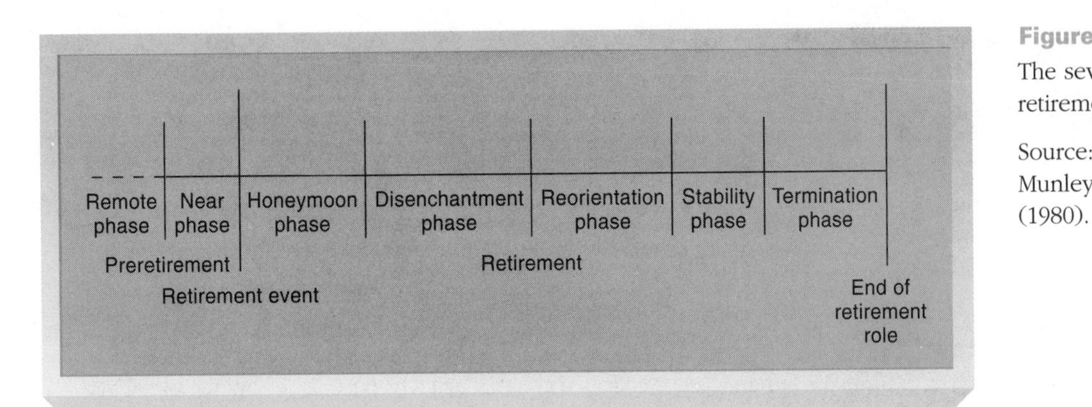

Figure 21.2
The seven phases of retirement

Source: Williamson, Munley, and Evans (1980).

Table 21.3 The Seven Phases of Retirement

1. *Remote phase.* A stage of denial in which the person does little to prepare himself or herself for the transition into retirement.
2. *Near phase.* Some workers attend preretirement programs offered by the employer. Preretirement programs help the worker to understand pension and benefit plans along with the legal, health, and leisure aspects of retirement.
3. *Honeymoon phase.* A time of euphoria in which individuals try to overcompensate for all the things they never had time for. Persons who were involuntarily forced to retire do not go through the honeymoon phase. This euphoria eventually leads to disenchantment.
4. *Disenchantment phase.* A sense of loss or disappointment as the pace of life lessens, due to unrealistic expectations of what retirement should be like.
5. *Reorientation phase.* The retiree tries to rebuild a satisfying, stable, and realistic life-style.
6. *Stability phase.* The individual learns to adjust to change because his or her life becomes balanced and routine.
7. *Termination phase.* The retirement role is cancelled if the retiree accepts a job or becomes disabled.

The seven phases of retirement have no specific timing, because people retire at different ages and retire under different circumstances. Some individuals adapt rather quickly and leap from the honeymoon stage to the stable phase. Obviously, a person's progress through the seven phases depends on his or her attitude toward departure from the work force.

Making Retirement More Enjoyable

There is a growing belief that retired persons are an important resource to the community. Numerous efforts have been made in recent years to tap this powerful resource. A number of national programs now make an effort to involve retired persons in volunteer and paid work of service to society.

Recent surveys seem to indicate that the "golden years" are not so golden for many retired persons. But with improving health conditions (See chap. 18), improved understanding of the nature of life after 65, and a considerable increase in government involvement, the lives of retirees have a far better chance of being fruitful.

AN APPLIED VIEW — Pet Ownership by Elderly People

Most pets are important to their families, and they have proven especially important to older people. Pets play a role in the lives of elderly people who find retirement boring, or are lonely because of the death of family and friends. In a recent study, a sample of older people indicated that their pets were a factor in their choice of housing (Smith & others, 1992). Unfortunately, past practices often discriminated against pet owners, who often were forced either to give up their pet or seek alternative housing. Younger people usually have more housing options, but those options for older people are often limited, due to reduced incomes or health concerns.

Fortunately, things are changing. In 1983, President Reagan signed a law prohibiting discrimination against elderly and handicapped pet owners in federally assisted housing (Public Law 98–181). This law has been enforced unequally because of clumsy state guidelines. Nevertheless, several private housing facilities for semi-independent elderly people in the Chicago area have recently changed their policies to permit pet ownership. This, along with other similar actions across the country, is encouraging.

Since pets are often vital companions for older people, can you think of other things that might be done to help the elderly obtain pets, and to help make ownership easier?

Contrary to the stereotype, getting old need not, and usually does not, mean being lonely. In fact, the elderly, most of whom have a good deal of free time, often use it to develop their social lives. We now move to a consideration of their personal development.

PERSONAL DEVELOPMENT

Two points of view have received most interest in the field of personality development in the elderly: those of the Committee on Human Development and those of Erik Erikson.

Personal Development: Committee on Human Development

What are the effects of aging on personal development? The Committee on Human Development has addressed this question for many years. Gerontologist Bernice Neugarten and her associates at the University of Chicago have been responsible for the highly respected research of this committee (e.g., Havighurst, Neugarten, & Tobin, 1968; Neugarten, 1968).

Adults between the ages of 54 and 94 who were residents of Kansas City were asked to participate in a study of change across the adult lifespan. The sample is somewhat biased in that it represents only white persons who were living on their own (i.e., not institutionalized) at the time of the study. They were somewhat better educated and of a higher socioeconomic group than is typical for this age group. Nevertheless, the sample is reasonably well balanced, and because of the thoroughness with which these persons were studied, this research has become a classic in the field of adult psychology. Neugarten summarizes the findings:

- In the middle years (especially the fifties), there is a change in the perception of time and death and the relationship of the self to them. Introspection, contemplation, reflection, and self-evaluation become important aspects of life. People at this age become more interested in their "inner selves."

- Perception of how well one can control the environment undergoes a marked change:

> ***Forty-year-olds, for example, seem to see the environment as one that rewards boldness and risk-taking, and to see themselves as possessing energy congruent with the opportunities perceived in the outer world. Sixty-year-olds, however, perceive the world as complex and dangerous, no longer to be reformed in line with one's wishes, and the individual as conforming and accommodating to outer-world demands.* (1968, p. 140)**

- **Emotional energy** declines with age. Tests showed that intensity of emotion invested in tasks undergoes a definite decline in the later years.
- **Gender-role reversals** also occur with age (discussed earlier in this chapter).
- **Age-status** becomes more rigid with development. Age-status refers to society's expectations about what is normal at various ages. These expectations change not only with advancing years, but according to the particular society and to the historical context. For example, in 1940 a woman who was not married by age 22 was not considered unusual, but if she had not married by the time she was 27, people began to worry. In the late 1980s, the expected age of marriage is much less rigid. This holds only for the United States; in Samoa, for example, there is concern if a person is not married by age 15.

What is the optimum pattern of aging in terms of our relationships with other people? This question has also been investigated by the Committee on Human Development. For many years, there have been two different positions on this question, known as the **activity theory** and the **disengagement theory.**

According to the activity theory, human beings flourish through interaction with other people and through keeping physically active. They are unhappy when, as they reach the older years, their contacts with others shrink as a result of death, illness, and societal limitations. Those who are able to keep up the social activity of their middle years are considered the most successful.

Disengagement theory contradicts this idea. According to this position, the belief that activity is better than passivity is a bias of the Western world. This was not always so; the Greeks, for example, valued their warriors and athletes but reserved the highest distinction for such contemplative philosophers as Sophocles, Plato, and Aristotle (Bellah, 1978). Many people in the countries of the Eastern hemisphere also hold this view. According to disengagement theory, the *most mature* adults are likely to gradually disengage themselves from their fellow human beings in preparation for death. They become less interested in their interactions with others, and more absorbed in internal concerns. They accept the decreasing attention of a society that views them as losing power.

In a second aspect of the study by the Committee (Havighurst & others, 1968), a distinction is made between social and psychological disengagement. Social disengagement refers to restricting interactions with other human beings; psychological disengagement has to do more with concentrating one's attention within oneself. In a sense, they are opposite sides of the same coin. The results were quite clear: Both social and psychological disengagement increase with age. The investigators found that psychological disengagement is more prevalent in a person's fifties, apparently as a precursor to social disengagement, which becomes more apparent in 60- and 70-year-olds.

An Applied View — National Retirement Programs

- *The Foster Grandparents Program.* People over 60 are eligible for part-time work in infant homes, schools for the retarded, and convalescent homes. They do not replace regular staff, but establish a one-to-one relationship with the people they serve.
- *The Retired Executives' Service Corps.* In this program, retired executives offer consultation on a part-time basis, usually to new or struggling companies.
- *Senior Worker Action Program.* Offered under the Office of Economic Opportunity, this program offers senior citizens work as school aides, babysitters, companions, handymen, homemakers, and seamstresses. It also supplies training to obtain new skills or brush up on old ones such as typing.
- *The National Council of Senior Citizens.* This large organization has twenty-one projects throughout the country that offer various kinds of employment to senior citizens.
- *Mature Temps.* This is the free employment service offered by the American Association of Retired Persons. It has offices in at least thirteen major cities, and helps to place older people in temporary paid jobs.
- *Senior citizen's centers.* There are over 1,200 senior citizen centers in the United States, offering a wide variety of services to retirees.
- *Other employment services.* On a more local basis, retirees can get employment help from Senior Placement Bureaus, Senior Home Craftsmen, Good Neighbor Family Aides, Red Cross Centers, the Grey Panthers, and the home economics departments of many state universities.

Clearly the group on the top is enjoying itself more than the woman on the bottom, but does that mean that the disengagement theory is wrong? Isn't it only natural for people to slowly disengage from society as they approach death?

Does this mean that the tendency toward disengagement is more normal than the tendency toward activity? Not necessarily. Researchers found that as disengagement increases, the individual's feelings of happiness also usually decrease. This seems to indicate that activity is more desired by the elderly, but they simply cannot achieve it. On the other hand, the unhappiness that results from disengagement is really quite moderate, which may support the disengagement theory.

Havighurst and associates (1968) found that, in terms of "successful aging," a person who stays at home and pursues a hobby can be just as happy and adjusted as one who joins many postretirement activities and work. The different personality types have their unique ways of adjusting to life stresses and changing life occurrences.

In summary, neither the activity theory nor the disengagement theory of optimal aging is totally supported by the Committee. The authors conclude that some older persons

> ***accept this drop in activity as an inevitable accompaniment of growing old; and they succeed in maintaining a sense of self-worth and a sense of satisfaction with past and present life as a whole. Other older persons are less successful.*** **(Havighurst & others, 1968, p. 171)**

Neugarten (1968) concludes that most elderly persons are conflicted. They wish to remain active in order to maintain their sense of self-worth, but they also wish to withdraw from social commitments in order to protect themselves from the pain of loss, caused by the deaths of people they care about and by thoughts of their own death. She suggests that this conflict is resolved in different ways by people of different personality types; therefore, neither the activity theory nor the disengagement theory can completely explain the process of aging.

Nevertheless, it is unfortunate that some administrators of retirement programs and nursing homes have used the disengagement theory to justify restricting the activities of some elderly. Cath (1975) believes that because some of these individuals

What's Your View?

?

If you hope to be the kind of person who ends his or her life with a sense of integrity, the time to begin is right now. You don't get to be a happy elderly person unless you plan and work for it. There are many ways to do this, but we have some suggestions to pass along to you.

Whenever you are about to make an important decision, follow these four steps:

1. Come to some tentative conclusion about what you should do.
2. Close your eyes and picture yourself as a 75-year-old woman or man.
3. Imagine yourself explaining your decision to that old person, and try to picture her or him telling you what she or he thinks of what you have decided.
4. If you don't like what you hear, rethink the decision and go through this process again.

Use this technique to get a better perspective on your thinking, and we think you will be surprised by how much wisdom you already have.

have "unconscious gerontophobic [fear of the elderly] attitudes . . . , they interpreted [disengagement] according to their personal motivation and limitation—often colored by financial considerations" (p. 212).

Personal Development: Erikson

For Erikson, resolution of each of the first seven crises in his theory should lead us to achieve a sense of personal **integrity.** Older adults who have a sense of integrity feel their lives have been well spent. The decisions and actions they have taken seem to them to fit together—their lives are *integrated.* They are saddened by the sense that time is running out and that they will not get many more chances to make an impact, but they feel reasonably well satisfied with their achievements. They usually have a sense of having helped to achieve a more dignified life for humankind.

The acceptance of human progress, including one's own, is essential for this final sense of integrity. This is the path to wisdom, which Erikson defines as "the detached and yet active concern with life itself in the face of death itself" (1978, p. 26).

When people look back over their lives and feel that they have made many wrong decisions, or more commonly, that they have frequently not made any decisions at all, they tend to see life as lacking integrity. They feel **despair,** which is the second half of this last stage of crisis. They are angry that there can never be another chance to make their lives make sense. They often hide their fear of death by appearing contemptuous of humanity in general, and those of other religions and races in particular. As Erikson puts it, "Such despair is often hidden behind a show of disgust" (1968, p. 140).

Erikson believes that the character of the old Swedish doctor in Ingmar Bergman's famous film, *Wild Strawberries,* excellently reflects the eight stages he proposes.

Erikson (1978) has provided a panoramic view of his life-cycle theory in an analysis of Swedish film director Ingmar Bergman's famous film *Wild Strawberries.* In this picture, an elderly Swedish doctor goes from his hometown to a large city, where he is to be honored for fifty years of service to the medical profession. On the way, he stops by his childhood home and, resting in an old strawberry patch, begins an imaginary journey through his entire life, starting with his earliest memories. In the ruminations of this old man, Erikson sees clear and specific reflections of the eight stages he proposes.

Erikson demonstrates through Bergman's words that, like his other seven stages, this last stage involves a life crisis. Poignantly, the old doctor struggles to make sense out of the events of his life. He is ultimately successful in achieving a sense of integrity. The film is well worth seeing, and Erikson's analysis makes it an even more meaningful experience.

Conclusion

Many changes happen during later adulthood. There is a "crossover" effect in gender roles, sexual activity changes and the nature of family relationships undergoes several alterations.

As we leave the industrial age and enter the age of "information processing," the nation's older workers experience an incredibly different environment from that of their predecessors of fifty years ago. Even retirement is a much-changed adventure. As in each of the preceding chapters, life appears as never-ending change, but change with discernible patterns. There is, perhaps, more stress in old age than before, but there are also more effective techniques for dealing with it.

In the next, final chapter, we examine the final experience of life, one that is itself often a very significant stressor—dying. We also review research on the principal method of dealing with both stress and death, and a vital part of life for most Americans: spirituality.

Chapter Highlights

Social Development

- A number of gerontologists have noted that people in late adulthood experience a "crossover" in gender roles: Older men become more like women, and older women become more like men.
- White summarizes that in late adulthood males are generally more sexually active than females, but that declining activity in the female is usually attributable to declining interest or illness in the male partner.
- Research suggests four basic phases through which most middle-aged and aged couples pass: child-launching, childless preretirement, retirement, and widowhood.
- While older people show less interest in working than younger people, discrimination and stereotypes about older workers underestimate their desire to help.
- Williamson and associates have proposed seven phases of retirement. Retirement can be enhanced through leisure activities and maintenance of a strong social life.

Personal Development

- Neugarten's work on the effects of aging on personal development reveals that elderly people become more interested in their "inner selves," perceptions of the environment change, psychic energy declines, gender-role reversals occur, and age-status becomes more rigid.

- The Committee on Human Development investigated patterns of aging, including activity theory and disengagement theory.
- For Erikson, resolution of each of the first seven stages in his eight-stage theory should lead people to achieve a sense of integrity in the last stage. If not, they experience despair.

Key Terms

Activity theory
Age-status
Crossover
Despair
Differential opportunity structure
Disengagement theory
Emotional energy
Existential love
Gender-role reversals
Generative love
Impotency
Integrity
Prostatectomy
Role discontinuity

What Do You Think?

1. Which would you rather be, an old man or an old woman? Why?
2. What is your position on sex in old-age institutions? Should there be any restrictions at all?
3. Which aspect of your family life do you most fear change in as you get old?
4. Which do you favor, the activity theory or the disengagement theory?
5. Some have said that whereas Erikson's first seven stages describe crises in which action should be taken, the last stage, integrity versus despair, is merely reactive. All he has the elderly *doing* is sitting in a rocking chair and looking back over their lives. Do you believe that this criticism is valid?

Suggested Readings

Cowley, M. (1980). *The view from 80.* New York: Penguin. Critic Donald Hall describes this book as "Eloquent on the felt disparity between an unchanged self and the costume of altered flesh." Cowley wrote this intensely personal book when he himself was 82.

de Beauvoir, S. (1971) *Coming of age.* This magnificent psychological study gives one an intense look at what it means rather than how it feels to become old.

Publications of the American Association of Retired Persons (AARP). With about 30 million members, the powerful AARP is a rich source of information on the elderly. Write to them in Washington, DC.

Roff, L. and C. Atherton (1989). *Promoting successful aging.* Chicago: Nelson-Hall. This readable book is half about theory and research and half about the specific strategies for dealing with the needs of the elderly. It is a bible for those who want to work with older adults.

CHAPTER 22

Dying and Spirituality

The Role of Death in Life 568

What Is Death? 569

Four Types of Death 569

The Legal Definition of Death 570

Dealing Successfully with the Death of Others 571

Grief Work 571

Pathological Grieving 572

The Role of Grief 573

The Role of the Funeral 574

Dealing Successfully with One's Own Death 576

Kübler-Ross: The Stages of Dying 576

Suicide: The Saddest Death 579

The Overall Picture 579

The Influence of Race and Place of Residence on Suicide 580

The Influence of Gender on Suicide 580

"Successful" Dying 581

The "Death with Dignity" Law 581

The Hospice: "A Better Way of Dying" 582

Spirituality 582

Religious Participation 584

Frankl's Theory of Spirituality 585

Jung's Theory of Spirituality 586

Wilson's Theory of Spirituality 587

Fowler's Theory of Spirituality 588

Conclusion 590

Chapter Highlights 590

Key Terms 591

What Do You Think? 591

Suggested Readings 592

As I said, you got to check out some time; you can't stay here forever. . . . Well, you just hope you go with a little dignity, that you don't have to suffer from Alzheimer's or die slowly in a rest home alone somewhere. I thought it was very interesting when I was in Bora Bora for six months. I was down there making a movie. . . . I did not live with the company, but found a house by myself up a road. There was a family next door—Polynesian natives—and the old lady of the house was there and she was dying. She suddenly said, "I'm going to die," and went to bed for three days and died. I remember at her funeral that everybody was dressed in white; it was like a wedding almost. And I was talking to them about it and they said, "Oh, yes, the fathers and the mothers are the most respected. They just go and die when they're ready." I said, "Geez, does that really go on like this all the time?" They said, "That's the way this society works." There are no old age homes, no old people wandering around, you know, suffering to death. Isn't it a strange thing, I thought to myself. What a respect for age that we don't have. . . . Like I say, life is like a hotel. We all check in and we all check out. Seems to me that the Polynesian approach to checking out is a great deal better than ours.

Actor Jason Robards, age 69
Quoted in Berman and Goldman (1992) ■

The Role of Death in Life

Americans, who have long been accused of abhorring the subject of death, are now giving it considerable attention. From the ways we teach the young about death (Duncan, 1979; Williams, 1979) to the ways in which we bury our dead (Cassell, 1979; Pine, 1979; Schneider, 1984; Swanson & Bennett, 1983), we seem to have switched to an eager confrontation of the problem.

In this section, we review research and theory from the social and physical sciences to examine three major concerns: What is death? How do we deal with the death of others? How do we deal with our own death?

After reading chapter 22, you should be able to:

- Explain the four types of death.
- Give the legal definition of death.
- Explain what is meant by grief work, and its place in Lindemann's stages.
- Explain what is meant by "pathological grieving," and how it prevents successful conclusion of a life crisis.
- Discuss the positive effects of grief and the funeral on dealing with another's death.
- Define Kübler-Ross' stages of dying.
- Discuss suicide among the elderly along with the influence of race, residence, and gender.
- Evaluate information on alternative ways of dying "successfully."
- Present reasons for the growth in religious participation among adults.
- Compare the theories of spirituality presented by Frankl, Jung, and Wilson.
- List and describe Fowler's six stages of faith development.

What Is Death?

The matter of my friend rests heavy upon me.
How can I be salved?
How can I be stilled?
My friend, who I loved, has turned to clay.
Must I, too, lie me down
Not to rise again for ever and ever?

Gilgamesh, c. 2,000 B.C.

In the movies, death is almost invariably portrayed in the same sequence: Dying people make a final statement, close their eyes, fall back on the pillow or into the arms of a loved one, and are pronounced dead. In fact, death rarely occurs that way (Veatch, 1981, 1984). In most cases, the person dies gradually. "Death is a process, not a moment" (Schulz, 1978, p. 90).

Since the first historian, Homer, began recording the lives of the Greeks, the meaning of death has been a central issue.

Ascertaining when people are truly and finally dead has been a medical problem for centuries. For example, in the early 1900s, Franz Hartmann claimed to have collected approximately 700 cases of premature burial or "close calls" (Hardt, 1979). In 1896 the "Society for the Prevention of Premature Burial" was founded. Fear of premature burial was so strong that in 1897 in Berlin, a patent was granted for a life signal device that sent up a warning flag and turned on a light if there was movement inside the coffin.

We no longer have any serious problem in determining whether a person is dead. But determining exactly *when* death occurred has come to be of even greater importance because of organ donations. All the body's systems do not cease simultaneously, so there are disagreements over which system is most significant in judging whether a person is dead.

Four Types of Death

Today, four types of death are recognized: clinical death, brain death, biological or cellular death, and social death.

Clinical Death

In one sense, the individual is dead when his or her respiration and heartbeat have stopped. This is the aspect of dying referred to in the movies when the doctor turns and sadly announces, "I'm sorry, but he's gone." Actually, **clinical death** is the least useful to the medical profession and to society at large because it is unreliable.

Since the advent of **cardiopulmonary resuscitation** (CPR), many individuals whose lungs and heart had ceased to function were saved. In other cases, spontaneous restarting of the heart and lungs has occurred after failure.

Brain Death

Death of the brain occurs when it fails to receive a sufficient supply of oxygen for a short period of time (usually 8 to 10 minutes). The cessation of brain function occurs in three stages: First the cortex stops, then the midbrain fails, and finally the brain stem ceases to function. When the cortex and midbrain stop operating, **brain death** has occurred, and the person enters an irreversible coma. The body can remain alive in this condition for a long time, because the autonomic processes are governed by the brain stem. Consciousness and alertness, however, will never be regained.

Biological Death

The cells and the organs of the body can remain in a functioning condition long after the failure of the heart and lungs. **Biological death** occurs when it is no longer possible to discern an electrical charge in the tissues of the heart and lungs.

Social Death

Sudnow (1967) has suggested the concept of **social death,** which "within the hospital setting, is marked at that point in which a patient is treated essentially as a corpse although perhaps still 'clinically' or 'biologically' alive" (p. 74). He cites cases in which body preparation (e.g., closing the eyes, binding the feet) were started while the patient was still alive, in order to make things easier for the staff. Also, in some cases, autopsy permissions have been signed by the family members while the patient was still alive.

So when is a person really dead? If a person has suffered brain death, but the heart is still beating, should the heart be removed and used for a transplant operation?

One complex case (Goldsmith, 1988) involved the Loma Linda University Medical Center, famous for its work in the area of organ transplants. They attempted to initiate a program that would provide scarce organs for transplants. In this program, healthy hearts and other organs would be taken from babies born with anencephaly, a condition in which part or all of the brain is missing at birth. Ninety-five percent of these babies die within one week. When the anencephalic infants were born, they were flown to Loma Linda and given traditional "comfort care" (warmth, nutrition, and hydration). In addition, they were put on artificial breathing support for a maximum of seven days. The hospital maintained that they were put on respirators until a technical definition of brain death could be ascertained. But critics contended that this was done because the organs needed the time to mature for a successful transplant to occur. A storm of controversy ended the program before any donations could be made. Critics accused the hospital of "organ farming." The hospital argued that it was not only trying to increase the number of organ donors for infants but also giving the families of anencephalic babies an opportunity to "turn their tragedy into something good." With the advent of more advanced technology, the distinction between life and death becomes blurred, and the ethical considerations grow increasingly complex.

It is possible to ensure that all or part of your body be donated for use after your death, usually by a designation on your driver's license. Have you done this? If not, what are your reasons for refraining?

The Legal Definition of Death

In 1968 the "Harvard Ad Hoc Committee to Examine the Criteria of Brain Death" suggested the following criteria for **legal death:** "Unreceptivity and unresponsivity, no movements or breathing, no reflexes, and a flat **electroencephalogram** (EEG) reading that remains flat for 24 hours" (Schulz, 1978, p. 98).

Such criteria preclude the donation of organs in most cases, because the organs would probably suffer irreparable damage in the 24 hours needed to check the EEG. Others have suggested that the time at which the cerebral cortex has been irreparably damaged should be accepted as the time when organs can be removed from the body.

Organ donation is not the only difficulty involved. An increasing number of cases illustrate the ethical problem created by maintaining the life of comatose individuals with the support of technical equipment. A number of medical personnel, philosophers, and theorists have suggested that maintaining life under these conditions is wrong. What do you think?

Although scientists may disagree on the exact nature of death itself, we have been learning a great deal about how people deal, and how they *should* deal, with the death of their loved ones. We turn to that subject now.

Dealing Successfully with the Death of Others

Representative Bryan had the true passion of a runner. It's a shame he's gone, but I can't help feeling that he might not have regretted the manner of his death. Outdoors on a brisk autumn afternoon, in the company of a friend, with a feeling of life in all of the nerve ends that a long life excites. **(McCarthy, 1979, p. 120)**

In modern Western societies, death comes mostly to the old. For example, 55 percent of all males are 65 or over when they die, and almost a third are past 75. The mortality rate has declined in our country from 17 per 1,000 population in 1900 to slightly less than 9 per 1,000 today. For the first time in history, a family may expect to live twenty years (on the average) without one of its members dying.

The causes of this lower mortality are clear: virtual elimination of infant and child mortality, and increasing control over the diseases of youth and middle age. For this reason, the subject of death became more and more taboo in the first half of this century. Probably as a result, social scientists spent little time studying our reactions to it or trying to find better ways to help us deal with it. Fortunately, in recent decades, this has changed.

Grief Work

No one ever told me that grief felt so like fear. I am not afraid, but the sensation is like being afraid. The same fluttering in the stomach, the same restlessness, the yawning. I keep on swallowing. **(Lewis, 1963, p. 7)**

In fact, grief has a great deal in common with fear, and most grieving people really are afraid, at least unconsciously. They are frightened by the strength of their feelings, and they often fear that they are losing their sanity. They feel that they cannot go on, that their loss is so great that their own lives are in danger.

In fact, in some cases they are. Seligman (1975) notes documented cases in which persons have died as a direct cause of grief. Most people who have experienced a loss of a loved one have felt that they never want to love that deeply again. Fortunately, when the grieving is over at last, most people find a renewed desire to love, together with a deepened capacity to do so. As Kübler-Ross (1975) puts it, "If we choose to love, we must also have the courage to grieve" (p. 96).

Grief not only *follows* death; when there is advance warning, it frequently precedes it. Fulton (1977) finds that there are four phases in anticipatory grief: depression, a heightened concern for the ill person, rehearsal of death, and finally an attempt to adjust to the consequences that are likely to occur after the death.

The topic of anticipatory grief has received increasing attention lately. There have been major debates about its exact nature. For example, Hill and associates (1988) found that those widows who had anticipated their husband's death reported a higher level of mental health than those who had not. Most researchers agree that a forewarning of the impending death of a loved one can have therapeutic consequences (Parke & Weiss, 1983):

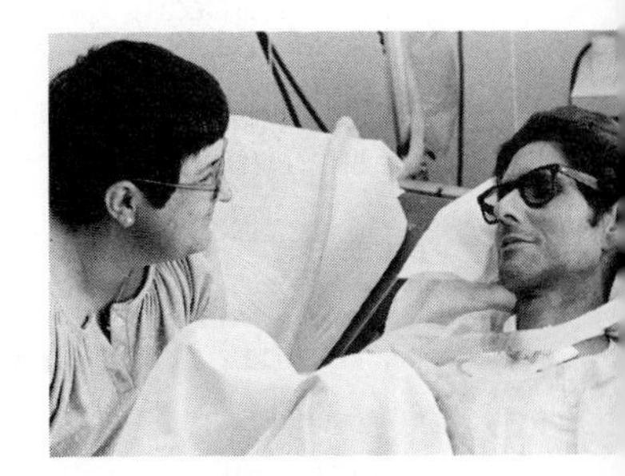

Anticipatory grief, which precedes the sick person's death, is usually very difficult. It often provides a number of benefits to the grieving, however.

- Avoiding the shock and fear that often accompanies sudden death
- Being able to make plans for the future that won't be regarded as betrayals
- Expressing thoughts and feelings to the dying loved ones, thereby avoiding a sense of lost opportunity
- A time to prepare for the changes ahead

But researchers disagree over whether an actual grief process is experienced before the death. Some say that true grief and an actual confrontation of the realities of the death of a loved one can come only after the death has occurred (Silverman, 1974; Parke & Weiss, 1983). Other scientists contend that there is some type of grieving process at work when a person is forewarned of an impending death. One need only look at the anguish of a parent or spouse to recognize it. This process should not be expected to take the same form as post-death grieving and surely won't fully reconcile those who grieve to the realities ahead (Rando, 1986).

Rando and others have begun to devise therapeutic intervention techniques to facilitate the unique aspects of anticipatory grief. As medical technology increasingly makes it possible to stall death, such work will continue to grow in importance.

Pathological Grieving

The stages of grief, painful as they are, are experienced by most individuals. In some cases, however, morbid grief reactions occur that prevent the successful conclusion of the life crisis. There are three types: delayed reaction; distorted reaction; and pathological mourning.

Delayed Reaction

In some cases, the intense reaction of the first stage is postponed for days, months, and in some cases years. In these cases, it is common for some seemingly unrelated incident to bring to the surface an intense grieving, which the individual does not even recognize as grief.

Lindemann (1944) gives the example of a 42-year-old man who underwent therapy to deal with an unaccountable depression. It was soon discovered that when he was 22, his 42-year-old mother committed suicide. Apparently, the occurrence of his own forty-second birthday brought to the surface all the feelings that he had managed to repress since that time.

Distorted Reactions

In most cases, distorted reactions are normal symptoms carried to an extreme degree. They include adopting the behavior traits of the deceased, such as aspects of the deceased's fatal illness and other types of psychosomatic ailments, particularly colitis, arthritis, and asthma.

An example we know of is a young man whose mother died of lymphomic cancer. At her death, she had large boils on her neck. Some weeks after she died, her son discovered lumps on his neck that quickly developed into boils. On examination, it was found that they were benign. In fact, it was determined that they were entirely psychosomatic. That is, the doctors decided that the only explanation of their existence was the great stress in the young man's mind over the loss of his mother.

Pathological Mourning

In pathological mourning the stages are not skipped, but they are prolonged and intensified to an abnormal degree. Frequently, the person suffering from pathological mourning tries to preserve every object of the deceased in perpetual memory.

An example of this illness is the man who worked hard with his wife to renovate an old cottage they had bought in order to live by a lake not far from their home. A few days before they were going to move in, she died of a heart attack. Some months later, friends noticed that he would disappear for several days at a time. One friend followed him to the cottage, and found that he had created a

An Applied View: A Personal Experience of One of the Authors

I am stepping out of my role as an author to relate an experience of mine that is relevant to this discussion of the function of grief. In April 1957, I joined the United States Navy and sailed to the Mediterranean for a six-month tour of duty on an oil supply ship. In early November I returned home to a joyful reunion with my family. After this wonderful weekend at home, I returned to my ship. Two days later I received a telegram informing me of a tragedy: My mother, two younger brothers, and two younger sisters had been killed in a fire that had destroyed our house. My father and three younger brothers and a sister had escaped with serious burns.

On the long train ride home from the naval port, I recall thinking that, as the oldest, I should be especially helpful to my father in the terrible time ahead. I was also aware of a curious absence of dismay in myself.

In our medium-sized upstate New York town, the catastrophe was unprecedented, and expressions of grief and condolences were myriad. People kept saying to me, "Don't try to be so brave. It's good for you to let yourself cry." And I tried to, but tears just wouldn't come.

At the funeral, the caskets were closed, and I can remember thinking that maybe, just maybe, this was all just a horrible dream. I distinctly remember one fantasy about my brother Mike. He was born on my first birthday and in the several years before the fire, I had become especially close to him. I imagined that he had actually hit his head trying to escape and had wandered off to Chicago with a case of amnesia, and that no one was willing to admit that they didn't know where he was. I knew this wasn't true, but yet I secretly clung to the possibility. After a very difficult period of time, our family gradually began a new life. Many people generously helped us, and eventually the memories faded.

Several times in the years that followed, I went to doctors because of a stomach ache, a painful stiff neck, or some other malady that couldn't be diagnosed physically. One doctor suggested that I might be suffering from an unresolved subconscious problem, but I doubted it.

Then one night in 1972, fifteen years after the fire, I was watching "The Walton's Christmas," a television show in which the father of a close family is lost and feared dead. Although dissimilar from my own experience, this tragedy triggered an incredible response in me. Suddenly and finally it occurred to me: "My God, half my family is really gone forever!" I began sobbing and could not stop for over three hours. When I finally stopped, I felt weak and empty, relieved of an awful burden. In the days that followed I believe I went through a clear-cut "delayed grief" reaction.

Therefore, the answer to the question, at least in my experience, is clear: Grief work really is essential, and we avoid it only at the cost of even greater pain. My father died some years ago, and my grief was immediate and intense. I cannot help but feel that my emotional response this time was considerably more appropriate and healthy.

John Dacey

shrine to her memory in it. Her clothes and other possessions were laid out in all the rooms, and her picture was on all the walls. Only after extensive therapy was he able to give up the shrine.

The Role of Grief

Most psychologists who have examined the role of grief have concluded that it is an essential aspect of a healthy encounter with the crisis of death. They believe that open confrontation with the loss of a loved one is essential to accepting the reality of a world in which the deceased is no longer present. Attempts to repress or avoid thoughts about the loss are only going to push them into the subconscious, where they will continue to cause problems until they are dragged out and accepted fully.

And yet, dealing with grief is also costly. For example, Fulton (1977) found that the mortality rate among grieving persons is seven times higher than a matched sample of nongrieving persons. The first five items on Holmes and Rahe's *Social Readjustment Rating Scale* (See chap. 17) all involve separation and loss from loved ones. These five most stressful events, in descending order, are death of a spouse, divorce, marital separation, going to jail, and death of a close family member. These events and the grief attached to them are most likely to cause illness.

On the other hand, anthropologist Norman Klein (1978) has suggested that psychologists may be too insistent that our grief be public and deep:

> ***In our own society, faddish therapies stress the idea that expressing sorrow, anger, or pain is a good thing, and the only means for "dealing with one's feelings" honestly . . . yet it is surely conceivable that some Americans can work through grief internally and privately, without psychological cost. It is even more conceivable that whole cultural subgroups may have different ways of conceding and responding to such experience.*** **(p. 122)**

Klein goes on to cite the Japanese, who are most reticent about public grief and yet seem to suffer no ill effects from this reticence. The Balinese frequently laugh at the time of death, because, they say, they are trying to avoid crying; yet they seem to be psychologically healthy. Some cultures employ "keeners" who wail loudly so that the bereaved will not have to do so themselves.

Is the expression of grief essential? At this time, the studies of social and medical scientists do not offer us a conclusive answer.

The Role of the Funeral

In your opinion, what are the appropriate arangments for a funeral? Are you in favor of formal religious services? Why or why not?

One of the hardest aspects of dealing with the death of a loved one is deciding how the funeral (if there is to be one) is to be conducted. Funerals have always been an important part of American life, whether the elaborate burial rituals practiced by Native Americans, or the simple funerals of our Colonial forbears. Some research indicates that the rituals surrounding funerals have a therapeutic benefit that facilitates the grieving process (Bolton & Camp, 1989; Kraeer, 1981; Rando, 1986).

Once the intimate responsibility of each family, care for the dead in the United States has been transferred to a paid service industry. The need for this new service was brought about by changes in society during the first part of this century. The more mobile, urbanized work force had less family support and less time to devote to the task of caring for the dead. In a relatively short time, funeral homes and funeral directors became the accepted form of care for one's dead relatives (Fulton & Owen, 1988).

This commercialization of care for the dead has had mixed results. During the 1950s and 1960s, funeral homes came under stinging criticism for their expense and their lack of sensitivity to the needs of the surviving family members (Bowman, 1959; Mitford, 1963). The bereaved often felt that the funeral directors were more interested in dramatic and expensive presentations of the body than in what might be best for the family members.

A more current survey has revealed that only about 42 percent of funeral directors have had any formal education in the physical and psychological effects of the death of a loved one, and most of those who had some education felt it was inadequate (Weeks, 1989). Recent trends such as cremations and memorial services without the body (often because some body parts have been donated for transplants or science research) have relieved people of the more unpleasant and expensive aspects of funeral services. Some families are now involving a professional grief counselor in the process.

The Funeral in Other Times and Countries

Looking at the funeral practices of former cultures shows us not only how they buried their dead but also something about their values.

Ancient Egypt

Upon the death of the head of the house in ancient Egypt, women would rush frantically through the streets, beating their breasts from time to time and clutching their hair. The body of the deceased was removed as soon as possible to the embalming chambers, where a priest, a surgeon, and a team of assistants proceeded with the embalming operation. (The Egyptians believed in the life beyond; embalming was intended to protect the body for this journey.) While the body was being embalmed, arrangements for the final entombment began. When the mummified corpse was ready for the funeral procession and installation in its final resting place, it was placed on a sledge drawn by oxen or men and accompanied by wailing servants, professional mourners simulating anguished grief, and relatives. It was believed that when the body was placed in an elaborate tomb (family wealth and prestige exerted an obvious influence on tomb size), its spirits would depart and later return through a series of ritualistic actions.

Ancient Greece

Reverence for the dead permeated burial customs during all phases of ancient Greek civilization. Within a day after death the body was washed, anointed, dressed in white, and laid out in state for one to seven days, depending on the social prestige of the deceased. Family and friends could view the corpse during this time. For the funeral procession, the body was placed on a bier carried by friends and relatives and followed by female mourners, fraternity members, and hired dirge singers. Inside the tomb were artistic ornaments, jewels, vases, and articles of play and war. Like the Egyptians, the ancient Greeks prepared their tombs and arranged for subsequent care while they were still alive. About 1000 B.C. the Greeks began to cremate their dead. While earth burial was never entirely superseded, the belief in the power of the flame to free the soul acted as a strong impetus to the practice of cremation. A choice of burial or cremation was available during all the late Greek periods.

The Roman Empire

Generally speaking, the Romans envisioned some type of afterlife and, like the Greeks, practiced both cremation and earth burial. When a wealthy person died, the body was dressed in a white toga and placed on a funeral couch, feet to the door, to lie in state for several days. For reasons of sanitation, burial within the walls of Rome was prohibited; consequently, great roads outside the city were lined with elaborate tombs erected for the well-to-do. For the poor, there was no such magnificence; for slaves and aliens, there was a common burial pit outside the city walls.

Anglo-Saxon England

In Anglo-Saxon England (approximately the time when invading Low German tribes conquered the country in the fifth century), the body of the deceased was placed on a bier or in a hearse. On the corpse was laid the book of the Gospels as a symbol of faith and the cross as a symbol of hope. For the journey to the grave, a pall of silk or linen was placed over the corpse. The funeral procession included priests bearing lighted candles and chanting psalms, friends who had been summoned, relatives, and strangers who deemed it their duty as a corporal work of mercy to join the party. Mass was then sung for the dead, the body was solemnly laid in the grave (generally without a coffin), the mortuary fee was paid from the estate of the deceased, and liberal alms were given to the poor.

Colonial New England

Burials and funeral practices were models of simplicity and quiet dignity in eighteenth-century New England. Upon death, neighbors (or possibly a nurse if the family was well-to-do) would wash and lay out the body. The local carpenter or cabinetmaker would build the coffin, selecting a quality of wood to fit the social position of the deceased. In special cases, metal decorations imported from England were used on the coffin. In church, funeral services consisted of prayers and sermons said over the pall-covered bier. Funeral sermons often were printed (with skull and crossbones prominently displayed) and circulated among the public. The funeral service at the grave was simple, primarily a brief prayer followed by the ritual commitment of the body to the earth. The filling of the grave, with neighbors frequently supplying the necessary labor because there were no professional gravediggers, marked the formal end of the early-colonial funeral ceremony.

Let's hope it won't happen, but what would you do if you were called upon to organize a funeral tomorrow? Would you know whom to notify? How would you arrange for the preparation and disposition of the body? What kind of ceremony, religious or otherwise, would you ask for? What would you do about a funeral home, a cemetery plot, and the will and death benefits of the deceased?

Perhaps you could discuss this with some of your friends or classmates, to see how they would feel. Notice of how you feel about opinions that differ from yours.

Source: Turner & Helms (1989, pp. 492–93).

Dealing Successfully with One's Own Death

***Having nearly died, I've found death like that sweet feeling that people have that let themselves slide into sleep. I believe that this is the same feeling that people find themselves in whom we see fainting in the agony of death, and I maintain that we pity them without cause. If you know not how to die, never trouble yourself. Nature will in a moment fully and sufficiently instruct you; she will exactly do that business for you; take you no care for it.* (Michel de Montaigne, "Of Physiognomy," 1585–88)**

Why is the acceptance of death so painful to so many of us? Why does it come up in every developmental stage, only to be partially resolved and partially denied?

Many people find dying a much harder experience than Montaigne would have us believe it is. Many dying patients feel seriously depressed before their deaths, and a large number have suicidal feelings. Among the reasons for these depressions are the following:

- Medication-induced mood alterations
- Awareness of how little time is left
- Feelings of isolation from relatives and friends who are withdrawing
- Feelings of grief for the losses that are close at hand
- Feelings of disillusion and resentment over injustice

The increased feeling of depression was the greatest difference between a group of terminal and nonterminal cancer patients (Schulz, 1978). This depression is sometimes described as cognitive withdrawal, because many patients have a decreasing ability and motivation to process stimuli as death nears (see *terminal drop,* chap. 20). Also common is a strong sense of fear and a deep sense of sorrow (Feifel, 1977).

Must dying, then, always be such an unhappy experience? It is really more complicated than that, according to the European psychiatrist Elisabeth Kübler-Ross.

Kübler-Ross: The Stages of Dying

Dr. Kübler-Ross is the most famous student of the process of death. Her three books on the subject have all been best sellers: *On Death and Dying* (1969), *Death—The Final Stage of Growth* (1975), and *To Live Until We Say Goodbye* (1978).

Kübler-Ross discovered that, far from wanting to avoid the topic of death, many dying patients have a strong urge to discuss it. She interviewed hundreds of terminally ill persons. On the basis of these interviews, she developed a five-stage theory, describing the emotions that underlie the process of dying. The stages in her theory are flexible, in that people can move through them quickly, slowly, or not at all. There is some fluctuation between the stages, but by and large people tend to move through them in sequential order. Because the recognition of these stages has had a great impact on professionals and others, we'll describe the stages in some detail here. They are portrayed in Figure 22.1.

1. *Shock and denial.* The defense mechanism of denial is a very basic one. It protects the individual by filtering out information that threatens the person's equilibrium. As with the first stage of grief, denial allows time for the person to gather strength and to marshall other defenses. Dr. Laurens Weisman (quoted in Feifel, 1977) suggests that

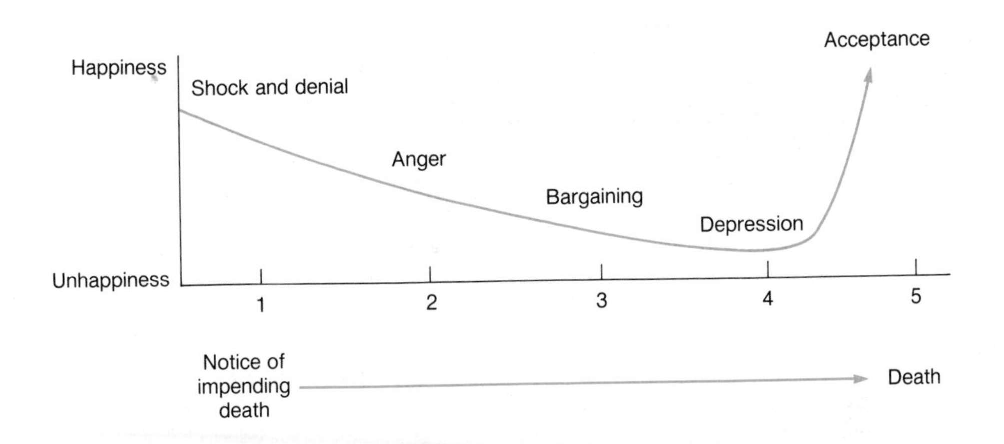

Figure 22.1
Kubler-Ross' stages of dying

Source: E. Kubler-Ross, On Death and Dying. Copyright © 1969 Macmillan Publishing Company, New York, NY.

most of us have come to view denial as a bad thing, and to see those persons who use denial as making some sort of error. A long series of patients, who have taught me so much else, have also convinced me that denial is not necessarily a mistake, particularly in patients who have to face certain realities. **(p. 100)**

Apparently, denial helps prevent many persons from being overwhelmed by initial shock. Typically, such persons go from doctor to doctor hoping to receive a different diagnosis. They hope desperately that a phone call will come, explaining that the test results were mixed up, and that the correct diagnosis will show only a minor illness.

Dying persons also often use the defense mechanism of *compartmentalization*. That is, they simultaneously hold contradictory beliefs. For example, a patient may admit that she is going to die soon, but may still discuss her long-term plans for the future. Another patient may say that he believes that death is imminent, while at the same time he doubts it.

2. *Anger*. The most common feeling at this stage is, "Why me?" Strong feelings of resentment toward others, and especially envy of healthy individuals, prevail. Kübler-Ross suggests that this is usually the most difficult stage for everyone involved, because "this anger is misplaced in almost all directions and projected onto the environment at times almost at random" (1969, p. 50). It is not unusual for patients to suffer full-blown paranoia at this time, suspecting that someone is "trying to get them." Nevertheless, the anger stage is essential and usually passes into stage 3 rather rapidly.

3. *Bargaining*. Now the individual tries to think of ways to postpone death. Most of these bargains are made with God. The person makes some kind of deal in which, if spared, he or she will undertake some unusual sacrifice to help others. This kind of thinking is similar to the so-called magical thinking often seen in young children, when they feel responsible for causing events they could not possibly have control over. When the individual realizes that there is no hope in this direction, a period of depression almost always sets in.

4. *Depression.* As stated earlier in this section, depression is common in dying persons, and serious depression has been found in approximately half the cases studied. Kübler-Ross refers to this as *preparatory depression,* because it usually functions as an attempt to anticipate the future loss of one's life. In this, it is similar to anticipatory grief. It is also a clear instance of disengagement (See chap. 20).

   Family members and others often work at cross purposes to the patient at this time. The patient is trying to get ready to leave, and the family keeps encouraging him or her to hang on a little longer. The patient is often confused by a jumble of feelings. For some, especially those in pain, there is guilt at wanting to die.

   There is considerable evidence that people have at least minimal control of the exact time when they die. For example, some patients, even those in considerable pain, manage to linger on, waiting for a birthday, holiday, or some significant person to arrive before letting go of life.

   Kavanaugh (1972) suggests that there are two prerequisites at this stage for peaceful, dignified dying:

> ***First, the dying person needs to receive permission to pass away from every important person he will leave behind. Only then can he go on to the other problem, the need to let go of every person whose possession he holds dear.*** **(p. 75)**

   A friend of ours who is a surgeon tells us that even those poor souls who have no one close to them at the time of their dying seem to need permission to go. In such cases, he says, he finds he is the only one who can say it is all right, and so he does.

5. *Acceptance.* In the final stage of dying, a peaceful time almost devoid of feeling sets in. The patient withdraws from others and from worldly concerns, and more calmly awaits death. Kübler-Ross describes this acceptance as being neither happy nor resigned but as a quiet time of contemplation. There is a certain degree of quiet expectation.

   Erikson (1963) believes that dying persons can be categorized into one of two groups. He says that some are characterized by a sense of *despair,* which is exemplified by their strong fear of death. They act as though they cannot accept the life they have lived:

> ***The lack of this accrued ego integration is signified by fear of death: the one and only life cycle is not accepted as the ultimate of life. Despair expresses the feeling that time is now too short, too short to start another life and to try out alternate roads to integrity.*** **(pp. 268–69)**

   Others feel that their lives have been well lived. They are suffused with a sense of *integrity,* and they can view their impending death with equanimity.

Kübler-Ross' stage theory was the first to counter the common assumption that it is abnormal to have strong emotions during the dying process. However, there has been no empirical confirmation that this particular sequence of stages is universal. Her theory overlooks the effects of personality, ethnic, or religious factors. It also does not take into account the influence of the specific illness and treatment, nor the availability of social support (Kastenbaum & Kastenbaum, 1989). So, should we accept Kübler-Ross' model or not? The biological, psychological, and social aspects of the process of dying are obviously very complex. We will need to know much more—knowledge gained through careful research—before we can answer this question. We can say that her theory, like all good theories, has at least provided us with five constructs (the five stages) to help guide that research.

No matter how well prepared we are, death is always sad and stressful. Most of the time, though, we know that there is nothing we can do about it, and so we must accept the inevitable. How different it is when the death was not inevitable, but was chosen by the person because life had become no longer worth living.

Suicide: The Saddest Death

As you will see, we are beginning to learn more about how and why suicides happen. This information will surely help us in our efforts to prevent such unfortunate deaths.

The Overall Picture

Suicide and attempted suicide among adolescents are growing national problems (Holinger, Offer, & Ostrov, 1987) and an increasingly common response to stress and depression among young persons (Kienhorst & others, 1987). Suicide now ranks as the second leading cause of death among persons age 15 to 19, and many experts believe that if no suicides were "covered up," it would be the leading cause. Teenagers have become not only more suicidal but apparently more reckless and self-destructive in general. As the suicide rate has risen steadily over the past twenty years, so too has the rate for motor vehicle accidents (the leading cause of death), accidents of other types, and homicides (Bem, 1987; Vital Statistics of the U.S., 1982).

Although suicide rates for teenagers have risen 72 percent since 1988, the rate for most other age groups has decreased. It would be safe to say that while the United States as a whole has become slightly less suicidal, teenagers and young people in general (age 30 and under) have become dramatically more suicidal. The increase has risen most steadily and most consistently among teenagers. It should be noted, however, that teen suicidal deaths rank high because teens have a relatively low rate of death from other causes.

There are two groups that are much more prone to suicide, however: the so-called middle-old and old-old groups. While the suicide rate for 15- to 24-year-olds is high—12.5 per 100,000—the rate for those 75 and older is almost double that, with males accounting for almost all the difference (U.S. National Center for Health Statistics, 1986) (also see Table 22.1). In fact, the older single white male is by far the person most likely to die of suicide. Woodruff-Pak (1988) states that adolescent suicides are paid more attention to because they are usually in good health. For the elderly, poor health is often both a cause of suicide and a way of covering it up. Loneliness is a second major factor.

Woodruff-Pak relates the story of an elderly retiree from the police force who had made an excellent adjustment to leaving the force. When he discovered that he had a large brain tumor, however, he became very depressed at the thought of leaving his wife, children and nine grandchildren, to all of whom he was deeply attached. One day, after asking his wife to go to the store for his favorite candy, he shot himself. His wife discovered him and threw away his pistol. She told the paramedics she saw a man fleeing from her house. The "case" remained unsolved, and she was able to bury her husband in a Catholic cemetery.

Table 22.1 Suicide Rates, by Sex, Race, and Age Group: 1970 to 1988 (Lives lost per 100,000)

Age	Total			Male						Female					
				White			African-American			White			African-American		
	1970	1980	1988	1970	1980	1988	1970	1980	1988	1970	1980	1988	1970	1980	1988
All ages	11.6	11.9	12.4	18.0	19.9	21.7	8.0	10.3	11.5	7.1	5.9	5.5	2.6	2.2	2.4
10–14 years old	0.6	0.8	1.4	1.1	1.4	2.1	0.3	0.5	1.3	0.3	0.3	0.8	0.4	0.1	0.9
15–19 years old	5.9	8.5	11.3	9.4	15.0	19.6	4.7	5.6	9.7	2.9	3.3	4.8	2.9	1.6	2.2
20–24 years old	12.2	16.1	15.0	19.3	27.8	27.0	18.7	20.0	19.8	5.7	5.9	4.4	4.9	3.1	2.9
25–34 years old	14.1	16.0	15.4	19.9	25.6	25.7	19.2	21.8	22.1	9.0	7.5	6.1	5.7	4.1	3.8
35–44 years old	16.9	15.4	14.8	23.3	23.5	24.1	12.6	15.6	16.4	13.0	9.1	7.4	3.7	4.6	3.5
45–54 years old	20.0	15.9	14.6	29.5	24.2	23.2	13.8	12.0	11.7	13.5	10.2	8.6	3.7	2.8	3.8
55–64 years old	21.4	15.9	15.6	35.0	25.8	27.0	10.6	11.7	10.6	12.3	9.1	7.9	2.0	2.3	2.5
65 years and over	20.8	17.8	21.0	41.1	37.5	45.0	8.7	11.4	14.0	8.5	6.5	7.1	2.6	1.4	1.6
65–74 years and over	20.8	16.9	18.4	38.7	32.5	35.4	8.7	11.1	12.9	9.6	7.0	7.3	2.9	1.7	2.0
75–84 years and over	21.2	19.1	25.9	45.5	45.5	61.5	8.9	10.5	17.6	7.2	5.7	7.4	1.7	1.4	1.3
85 years and over	19.0	19.2	20.5	45.8	52.8	65.8	8.7	18.9	10.0	5.8	5.8	5.3	2.8	—	—

Source: Bureau of the Census (1992).

The Influence of Race and Place of Residence on Suicide

Suicidal behavior remains, as it has consistently for decades, a behavior in which whites (Vital Statistics of the U.S., 1982) and the middle class (Jacobziner, 1965; Tishler, 1981; Weisman, 1974) are overrepresented. A comparative chart for white versus African-American suicide rates is presented in Table 22.1.

In regard to place of residence, since the late 1960s persons from rural areas have had a higher rate of suicide than those from urban areas (Wilkinson & Isreal, 1984).

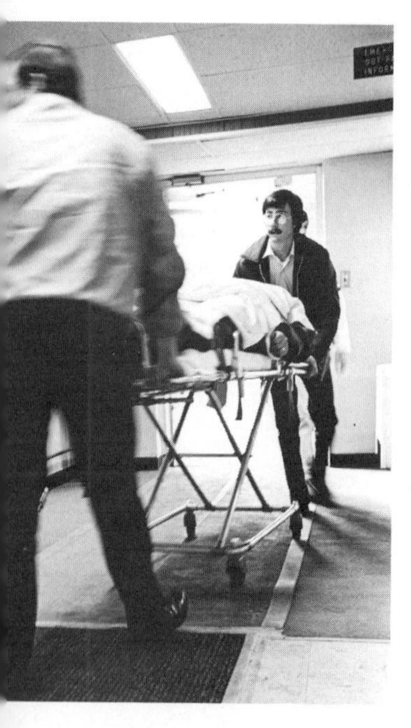

Males are three to four times more likely to die of suicide attempts than are females.

The Influence of Gender on Suicide

Table 22.1 also presents the suicide rates by gender. As you can see by looking at this table, major gender differences exist. The rate for males is much higher because of the type of suicidal behavior engaged in, the methods used, the lethality of the attempt, and the degree of psychiatric disturbance present. Males and females are two very different suicidal types. Universally, males are about four times more likely to die of suicide than females (Vital Statistics of U.S., 1982).

Attempt rates show even more dramatic gender differences, but in the opposite direction. *Failed* attempts at suicide among females are much higher than for males (Woodruff-Pak, 1988). The literature consistently has cited female-to-male ratios of at least 3 to 1. Jacobziner (1965), Weiner (1970), and White (1974) reported ratios of four to one. More recent studies show a far greater number of females among teenage suicide attempters: 5 to 1 (Curran, 1984), 9 to 1 (Hawton 1982b, 1982c; McIntire, 1980), 9.5 to 1 (Birtchnell & Alarcon, 1971), and 10 to 1 (Toolan, 1975).

A major reason for the high survival rate among females is the method used. While males often resort to such violent and effective means as firearms and hanging, females tend to choose less violent and less deadly means, such as pills. Male suicidals are considered significantly more disturbed than female (Hawton, 1982a, 1982b, 1982c; Otto, 1972; Teicher, 1973). Males are usually more committed to dying and therefore succeed far more often.

AN APPLIED VIEW **My Attempts at Suicide (Anonymous)**

My first psychiatrist told my parents that my psychological tests indicated that I was potentially suicidal. I was 14 then. At 22, I had made five suicide attempts and had been in six mental institutions, which add up to 29 months as a mental patient and five years of intensive therapy. My diagnosis was borderline schizophrenia, chronic depression, and sadomasochism. Why? How had I become so obsessed with suicide?

When I flash back on my adolescent days, I remember feeling ugly, socially awkward, stuck away in an all-girls' boarding school reading Camus and Hesse, unpopular, and stupid! In fact, I was not quite as dreadful as all that, but in my mind I was. I felt different. I once wrote, "I'm at the bottom of an upside-down garbage can and it's so ugly." The world was horrible, but I was the worst part of it.

Suicide was my escape. Unsuccessful suicide attempts put me in the care of others who delicately forced me to confront my feelings of sadness and anger. I had to learn to share with others and sometimes that was what I secretly wanted. Two of my attempts, however, were calculated, purposeful acts. Despite what shrinks may say, I wanted to be dead, not taken care of.

What did death mean to me? One of my earliest memories is sitting on moss-covered ground in a grove of pines, reading *The Prayer for the Dead* with my basset hound curled up beside me. Suicide meant escape from hell on earth. No other purgatory could be worse than this one. Even if I were reincarnated, I would end up being some "lowly animal" with the kind of mind that could not plague me with frightening, lonely, depressing thoughts. I clung to my friends and family, but it only increased my anger and self-contempt. I treated those people as my keepers who temporarily saved me from being left alone with my tormenting mind.

The final blow hit in Boston. I gradually withdrew from the few friends I had as well as my family. Death had grown so close that I no longer felt that I had much time. It was impossible to commit myself to anyone or anything. I was reserved, yet few people could sense how obsessed I was with death. Signs of affection terrified me because I knew I could not let anyone count on me. I needed death if life became too unbearable.

It finally did. I had become so passive that I no longer made contact with people. They had to call me. So much time had elapsed since I had felt close to someone that it seemed my "disappearance" would not really upset anyone. In addition to this, I was convinced that I was too stupid to handle academics or even a menial job (even though I had two jobs at the time). On a day when I knew no one would try and reach me, I took three times the lethal dosage of Seconal.

I was found 24 hours later and came out of a coma after 48 more. My arm was paralyzed. This time, I was placed in a long-term hospital. Another try at life began. With the help of an excellent therapist and the patient love of those whom I had thus far rejected, I have started once more. It has been two years since I took the pills. I think I know why people bother to live now.

Successful Dying

Thus far, this chapter has been pretty depressing, we know. Death and dying, by whatever means, are not one of our favorite topics, but they do have an important place in our study of life. In the next section of this chapter, however, we will be talking about two of the ways we humans have to deal with death in a mature and satisfying way.

The "Death with Dignity" Law

On 30 September 1976, Governor Edmund Brown, Jr., of California signed the "Natural Death Act," the first death-with-dignity law in the nation. The statute states the right of an adult to sign a written directive instructing her or his physician to withhold or withdraw life-sustaining procedures in the event of a terminal condition. The law contains specific definitions for "terminal condition," "life-sustaining procedure," and "qualified patient." The directive must be drawn up in the form set forth by the statute. It must be signed and dated by an adult of sound mind and witnessed by two persons not related by blood or entitled to the estate of the declarant.

Since then, all but ten states have established such procedures. In June 1990, the Supreme Court ruled that because the desires of a Missouri woman who has been lying in a coma since 1983 had not been made explicit (through a signed document), she could not be allowed to die despite her parents' wishes. However, the Court has ruled that when procedures established within each state are followed, the "right to die" would be constitutional.

Almost 80 percent of Americans die in hospitals, and 70 percent of those deaths involve some aspect of medical technology such as breathing, feeding, and waste elimination equipment. Hence it is essential that those who do not want to be maintained on life-support systems if they become terminal to put their wishes in writing according to their state's laws.

What's Your View?

Many things spread in popularity even if they are not always good for people. Although the "death with dignity" concept does appear to grant greater control over life to the person whose life it is, it has been argued that many individuals will invoke the law only when they believe they are dying. It won't occur to most people to think about it before then.

When people know they are likely to die, they are frequently in a depressed state. They may feel that now they are becoming worthless, and so should not be a burden on those around them. They may feel that they "just want to get it over with." Opponents of the law say that this is no time for a person to be making judgments about what should happen if death appears imminent. Their judgment is impaired by the depression. As Attorney Thomas Marzen has remarked, "People who are not dying are being denied treatment. The family doesn't object, the doctor doesn't object, and no one seems to care" (Gest, 1989, p. 36). What do you think?

The Hospice: "A Better Way of Dying"

The "death ward" in most hospitals is not a nice place to be. The atmosphere is one of hushed whispers and fake smiles. No children below the age of 12 are allowed. Medications to control the pain are usually given on a schedule rather than as needed. Machines are used to continue life at all costs, though the patient may desire death. Viewing a typical American death ward made British historian Toynbee concluded that "Death is un-American."

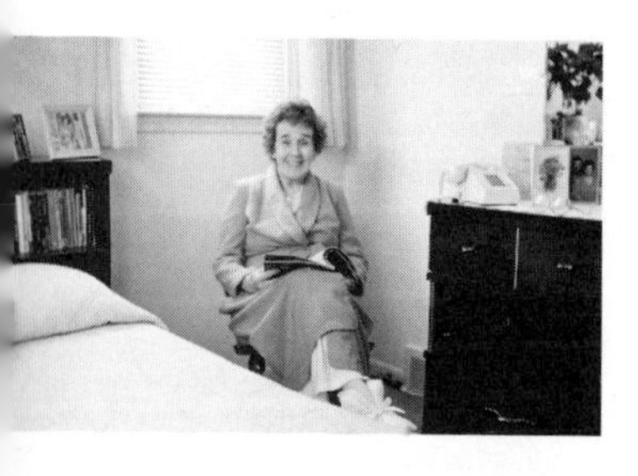

The modern hospice is organized to afford a more "natural death" to the dying. Although this woman knows she does not have long to live she is able to maintain a positive attitude with the help of the hospice staff.

In reaction to this, some have reached back to the Middle Ages, when religious orders set up havens where dying pilgrims could come to spend their last days (Stoddard, 1977). The modern **hospice** was established to provide for a more "natural" death for those who are terminally ill. The first United States hospice opened in New Haven, Connecticut, in 1971. Since then the National Hospice Organization has been formed to help promulgate this movement. A 1981 survey by the Joint Commission on Accreditation of Hospitals identified over 800 hospice programs in various stages of development (Falknor & Kugler, 1981). By 1984, this number had risen to 1,429 (Joint Commission of the Accreditation of Hospitals, 1984).

The hospice is not a new type of facility; it is a new philosophy of patient care in the United States (although not new in Europe). "What people need most when they are dying is relief from the distressing symptoms of their disease, security of a caring environment, sustained expert care, and the assurance that they and their families won't be abandoned" (Craven & Wald, 1975, p. 1816).

Hospices do pioneering work in such neglected areas as the easing of pain and psychological counseling of patients and their families. Even before patients begin to suffer pain, which in diseases such as cancer can be excruciating, they are given a mixture of morphine, cocaine, alcohol, and syrup so that they come to realize that pain can be controlled. Known as "Brompton's Mixture," this concoction is used only when the patient's need is severe; it is very effective in alleviating both pain and the fear accompanying it. A major goal of the hospice is to keep the person's mind as clear as possible at all times.

Whenever advisable, the hospice allows the patient to remain at home, and provides daily visits by staff nurses and volunteers. Jane Murdock, a California school teacher,

> ***recalls how her dying mother at first refused to see her grandchildren after she was brought home from the hospital. But when the visiting hospice team began reducing her pain and reassuring her and her family in other ways, a new tranquility set in. Finally the woman even let the youngsters give her medication, and assist her about the house. Says Murdock: "I felt when she died that it was a victory for all of us. None of us had any guilt."***
> **("A Better Way of Dying," Alban, 1978, p. 66)**

The hospice program movement has now grown to the point that it supports its own journal. Articles in the *Hospice Journal* often provide supplementary information about special issues concerning the terminally ill. For example, one recent article on hospice care for patients with AIDS (Schofferman, 1987) delved into additional issues of concern: irrational fear of contagion, homophobia (fear of homosexuals) by friends and relatives, and special difficulties in caring for substance abusers.

One of the major questions now being considered is whether the hospice should continue as a separate facility run solely for that purpose or become a standard part of all major hospitals. Currently, hospice programs in the United States are primarily home-based care, with the sponsorship of such programs evenly divided between hospitals and community agencies (Torrens, 1985).

Whatever the case, it seems clear that this movement will continue to grow considerably. This is because hospices have already proved that they are better places to die. As Cronin and Wald (1979) put it, they are places "where we can finally come to terms with the self—knowing it, loving it reasonably well, and being ready to give it up" (p. 53); in short, they are places where we have an opportunity for successful dying.

For most of us, though, dying is frightening and, ultimately, very hard to understand. To come to terms with it, we almost always rely on our spiritual rather than our cognitive powers.

SPIRITUALITY

> ***There is but one true philosophical problem, and this is suicide: Judging whether life is or is not worth living amounts to answering the fundamental question of philosophy.***
>
> ***Albert Camus, 1955***

> ***Religion is for those who are afraid to go to Hell. Spirituality is for those who have already been there.***
>
> ***A recovering alcoholic***

An Applied View: If You Had Your Life to Live Over Again, What Would You Do Differently?

This question was recently asked of 122 retired persons by DeGenova (1992). She discovered that her sample chose the pursuit of education more than any other area. This emphasis on education among retirees may be because they feel their lack of education led to missed or limited opportunities. While these people indicated they would have spent more time doing a variety of things, they said that they would have spent less time worrying about work.

This poem describes the feelings of one 81-year-old (Nadine Stair) on the topic.

> If I Had My Life To Live Over
> I'd like to make more mistakes next time.
> I'd relax, I would limber up. I would be sillier than I have been this trip. I would take fewer things seriously. I would take more chances.
> I would climb more mountains and swim more rivers. I would eat more ice cream and less beans. I would perhaps have more actual troubles, but I'd have fewer imaginary ones.
> You see, I'm one of those people who live sensibly and sanely hour after hour, day after day. Oh, I've had my moments, and if I had it to do over again, I'd have more of them. In fact, I'd try to have nothing else. Just moments, one after another, instead of living so many years ahead of each day.
> I've been one of those persons who never goes anywhere without a thermometer, a hot water bottle, a raincoat, and a parachute. If I had to do it again, I would travel lighter than I have.
> If I had my life to live over, I would start barefoot earlier in the spring and stay that way later in the fall. I would go to more dances. I would ride more merry-go-rounds.
> I would pick more daisies!

Religious life is of great importance to many of the elderly.

Spirituality is concerned not only with whether life is worth living but why it is worth living. It may involve the attempt to better understand the reasons for living, through striving to know the intentions of a Supreme Being. An example is reading inspired books such as the Bible. Another is trying to discern the purposes and goals of some universal life force by, for example, examining historical trends in biological changes of species. In any case, spirituality includes all of our efforts to gain insight into the underlying, overriding forces of life. For many, it is the only justification for morality.

How important a role does spirituality play in the lives of typical American adults? One way this question has been investigated is through looking at religious participation and at religious attitudes. These factors are sometimes misleading, but together they offer one fairly reliable answer to the question.

Religious Participation

Americans have always been highly religious. This statement is supported by numerous Gallup polls (Gallup, 1988). A majority of the elderly consider religious practice to be of major importance in their lives. According to surveys, 96 percent of those over 65 believe in God, and 82 percent report that religion plays a significant role in their lives (Koenig, Kvale, & Ferrell, 1988).

Furthermore, the influence of religion appears to be highly related to a sense of well-being in elderly persons. Studies of the relationship, however, may be criticized. It is known that those who attend services regularly are healthier than those who do not, so perhaps it is good health rather than religion that is causing the good feelings (George & Landerman, 1984; Lawton, 1984; Levin & Markides, 1986).

In their study of 836 elderly persons, Koenig and associates (1988) attempted to clear this up. They looked at nonorganizational practices (prayer, Bible reading, etc.) and subjective religious feelings, as well as organized religious practices. They found moderately strong correlations between morale and all three religious measures. The relationships were especially strong for women and for those over 75.

WHAT'S YOUR VIEW?

?

Religious Preference, Church Membership, and Attendance: 1957 to 1989 (in percent)

The following chart lists a number of facts about American religious practices.

Year	Religious Preference					Church/synagogue members	Persons attending church/synagogue	Age and Region	Church/synagogue members, 1989
	Protestant	Catholic	Jewish	Other	None				
1957	66	26	3	1	3	73	47	18–29 years old	61
1967	67	25	3	3	2	73	43	30–49 years old	66
1975	62	27	2	4	6	71	41	50 years and over	76
1980	61	28	2	2	7	69	40	East	69
1985	57	28	2	4	9	71	42	Midwest	72
1988	56	28	2	2	9	65	42	South	74
1989	56	28	2	4	10	69	43	West	55

Source: Bureau of the Census (1992).

These facts suggest a number of interesting questions. For example:

- The percentage of our population who belong to Protestant religions has been dropping, and the number who say they have no religious preference has been growing. Is this a real change, or are people becoming more honest? If the latter, why?
- Church/synagogue membership was dropping but has recently rebounded. Why?
- Older people participate more in religious services than younger people. Why?
- The West has the least religious participation, the South the most. Why?

What are your opinions on these questions?

Spirituality does appear to develop with age. A number of theories have been offered as to how and why this is so. Among them, four views have come to receive highest regard: those of Viennese psychoanalysts Viktor Frankl and Carl Jung, sociobiologist E. O. Wilson, and theologian James Fowler.

Frankl's Theory of Spirituality

Frankl (1967) describes human life as developing in three interdependent stages according to the predominant dimension of each stage.

1. *The somatic (physical) dimension.* According to the **somatic dimension,** all persons are motivated by the struggle to keep themselves alive and to help the species survive. This intention is motivated entirely by instincts. It exists at birth and continues throughout life.

2. *The psychological dimension.* Personality begins to form at birth and develops as a result of instincts, drives, capacities, and interactions with the environment. The **psychological dimension** and the somatic dimension are highly developed by the time the individual reaches early adulthood.

3. *The noetic dimension.* The **noetic dimension** has roots in childhood but primarily develops in late adolescence. It is spiritual not only in the religious sense but also in the totality of the search for the meaningfulness

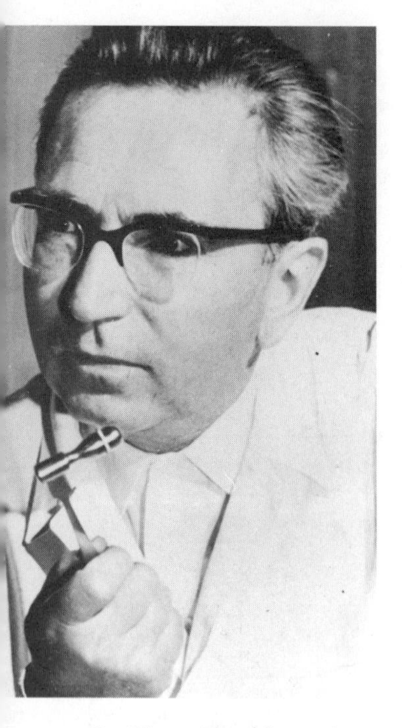

Dr. Victor Frankl, survivor of six years in a Nazi concentration camp, has made many contributions to the field of psychology and spirituality.

of life. This aspect distinguishes humans from all other species. The freedom to make choices is the basis of responsibility. Reason exists in the noetic realm. Conscience, which greatly affects the meaningfulness that we discover in life, resides in the noetic.

Frankl believes that development in the physical and personality dimensions results from the total sum of the influences bearing upon an individual. The noetic, on the other hand, is *greater than the sum of its parts.* This means that we as adults are responsible for inventing (or reinventing) ourselves! Whatever weaknesses our parents may have given us, we can and should try to overcome them. They need not govern our lives.

It is in the noetic dimension that the person is able to transcend training and to aspire to the higher levels of spiritual thought and behavior. For example, Teilhard de Chardin (1959) urges that if a person uses this noetic capacity to perceive the "Omega" or final point toward which human life is moving, frailties and foolishness can be overcome, and a high level of morality becomes natural.

Jung's Theory of Spirituality

Jung (1933, 1971), a student of Freud's, agreed to a large extent with his mentor's description of development in the first half of the human life. But he felt that Freud's ideas were inadequate to describe development during adulthood. Jung saw spiritual development as occurring in two stages.

The First Half of Life

In each of the Jungian functions of thinking, feeling, sensing, and intuiting, the personality develops toward individuation. (See chap. 17). Most people are well individuated by the middle of life, at approximately age 35. That is, we are most different from each other at this age.

The Second Half of Life

The goal of human development in the second half of life is just the opposite. Here a movement toward wholeness or unity of personality is the goal. Somewhere around mid-life, the individual should begin to assess the various systems of his or her personality and come to acknowledge the disorganized state of these systems. There should be a turning inward or self-inspection that marks the beginning of true adult spirituality. The goals of this introspection are

- Discovering a meaning and purpose in life
- Gaining a perspective on others, determining values and activities in which one is willing to invest energy and creativity
- Preparing for the final stage of life—death

Dr. Carl Jung, a student of Freud's, introduced many concepts into psychology, including the anima and animus.

Spirituality in the second half of life develops as a complement to the first half of life. By nourishing one's undeveloped side of life, one comes to a recognition of the spiritual and supernatural aspects of existence. For men, this means developing the **anima,** and for females the **animus.**

In contrast to the self-determination of spirituality seen in Frankl's and Jung's psychological points of view, sociobiology sees spirituality as determined almost entirely by instinct, that is, as a function of the genes.

Wilson's Theory of Spirituality

> ***The predisposition to religious belief is the most complex and powerful force in the human mind and in all probability an ineradicable part of human nature.*** **(Wilson, 1978, p. 169)**

Harvard sociobiologist Edward O. Wilson, the leading spokesperson for the sociobiological point of view (See chaps. 14 and 16), argues that religion and spirituality are inseparable. Together they grant essential benefits to believers. He argues that all societies, from hunter-gatherer bands to socialist republics, have religious practices with roots that go back at least as far as the Neanderthal period.

For example, he argues that even modern Russian society, which is still largely articlergy, pro- socialism, is as much a religious society as any other. Wilson (1978) cites one of Lenin's closest disciples, Grefori Pyatakov, who describes what it is to be "a real Communist"—one who

> ***will readily cast out from his mind fears in which he has believed for years. A true Bolshevik has submerged his personality in the collectivity, the "Party," to such an extent that he can make the necessary effort to break away from his opinions and convictions and can honestly agree with the "Party"—that is the test of a true Bolshevik.*** **(p. 184)**

Wilson says that in this statement we see the essence of religious spirituality. Humans, he argues, have a need to develop simple rules for handling complex problems. We also have a strong need for an unconscious sense of order in our daily lives. We strongly resist attempts to disrupt this order, which religion almost always protects.

Religion is one of the few uniquely human behaviors. Rituals and beliefs that make up religious life are not seen among any other animals. Some scientists, notably Lorenz and Tinbergen (Hinde, 1970), argue that animal displays, dances, and rituals are similar to human religious ceremonies. Wilson believes this comparison is wrong; animal displays are for the purpose of communicating (sexual desire, etc.), but religious ceremonies intend far more than mere communication.

Their goal is to "reaffirm and rejuvenate the moral values of the community" (Wilson, 1978, p. 179). Furthermore, religious learning is almost entirely unconscious. Most religious tenets are taught and deeply internalized early in life. Early teaching is a necessity if children are to learn to subvert their natural self-interests to the interest of society.

The sociobiological explanation of spirituality, then, is that through religious practice, the survival of practitioners is enhanced. Those who practice religion are more likely to stay alive (or at least they were in the past) then those who do not.

Wilson believes that even a person's willingness to be controlled may be genetic. Although all societies need some rebels, they also require that the vast majority of people be controllable, typically through religious and political beliefs. Therefore, Wilson believes that over the long run genes that favor willingness to be controlled have been favored by natural selection.

The potential for self-sacrifice can be strengthened in this manner, because the willingness of individuals to relinquish rewards or even surrender their own lives will favor group survival. The Jonestown mass suicide, for example, appears to have occurred because of the group's hope to remain united in afterlife. This concept of self-sacrifice is also the basis for Weinrich's (1987) assertion that homosexuality is an altruistic behavior (See chap. 16).

Religions usually favor survival of their believers. This is not always true, though, and is true to differing degrees. It has been estimated that there have been over 100,000 different religious faiths since humankind began. Obviously most have failed.

Some are even contrary to the survival needs of their believers. The Shaker religion, which disallows sexual intercourse to any of its participants, is an example. Shakerism lasted in this country for no more than two centuries. It flourished in the nineteenth century, but there are only a very few believers today. With no new recruits, it seems doomed to extinction. This, Wilson argues, is always the case for those religions that do not somehow enhance the vitality and hardiness of the groups that support them. The constant pursuit of a better chance for survival is why new ones are started.

Religions, according to the sociobiological point of view, develop through three steps:

1. *Objectification.* First, a perception of reality is described. **Objectification** occurs, which includes images and definitions that are easy to understand. Examples are good versus evil, heaven versus hell, and the control of the forces of nature by a god or gods.

2. *Commitment.* People devote their lives to these objectified ideas. Out of this **commitment,** they are willing under any circumstances to help those who have done the same.

3. *Mythification.* In **mythification,** stories are developed that tell why the members of the religion have a special place in the world. These stories are rational and enhance the person's understanding of the physical as well as spiritual world. The stories include explanations of how and why the world, as well as the religion, came to be. In earlier, less sophisticated religions, the faith is said to have been founded at the same time as the beginning of the race. These rarely include all-powerful or all-knowing gods. In less than one-third of the known religions is there a highly placed god, and in even fewer is there a notion of a moral god who has created the world. In the later religions, God is always seen as male, and almost always as the shepherd of a flock.

Not surprisingly, Wilson sees science as taking the place of theology today, because science has explained natural forces more effectively than theology. In fact, he asserts that science has explained theology itself. Although he sees theology as being phased out, he argues that the demise of religion is not at all likely. As long as religions make people more likely to survive and propagate themselves, Wilson suggests, they will enjoy worldwide popularity.

Wilson's theory attempts to explain spiritual development within societies. Theologian James Fowler has offered a description of the development of faith throughout the life cycle of individuals, without regard to the culture in which faith forms.

Fowler's Theory of Spirituality

James Fowler (1974, 1975a, 1975b) offers a theoretical framework built on the ideas of Piaget, Erikson, and Kohlberg. He believes strongly that cognitive and emotional needs are inseparable in the development of spirituality. Spirituality cannot develop faster than intellectual ability, and also depends on the development of personality. Thus Fowler's theory of faith development integrates the role of the unconscious, of needs, of personal strivings, and of cognitive growth.

Fowler sees faith developing in six steps. He says that the stages in faith development can be delayed indefinitely, but the person must have reached at least a certain *minimal age* at each stage in order to move on to a succeeding stage. His six stages are as follows:

1. *Intuitive-projective faith.* For **intuitive-projective faith,** minimal age is 4 years. In this stage, the individual focuses on surface qualities, as portrayed by adult models. This stage depends to a great extent on fantasy. Conceptions of God or a supreme being reflect a belief in magic.

2. *Mythical-literal faith.* For **mythical-literal faith,** minimal age is 5 to 6 years. Fantasy ceases to be a primary source of knowledge at this stage, and verification of facts becomes necessary. Verification of truth comes not from actual experience but from such authorities as teachers, parents, books, and tradition. Faith in this stage is mainly concrete, and depends heavily on stories told by highly credible storytellers. For example, the traditional story of Adam and Eve is taken quite literally.

3. *Poetic-conventional faith.* For **poetic-conventional faith,** minimal age is 12 to 13 years. The child is entering Piaget's codification stage. Faith is still conventional and depends on a consensus of opinion of other, more authoritative persons. Now the person moves away from family influence and into new relationships. Faith begins to provide a coherent and meaningful synthesis of these relationships.

   There is an awareness of symbolism and a realization that there is more than one way of knowing truth. Learned facts are still taken as the main source of information, but individuals in stage 3 begin to trust their own judgment and the quality of selected authorities. Nevertheless, they do not yet place full confidence in their own judgment.

4. *Individuating-reflective faith.* For **individuating-reflective faith,** minimal age is 18 to 19 years. Youths in stage 3 are unable to synthesize new areas of experience, because depending on others in the community does not always solve problems. Individuals in stage 4 begin to assume responsibility for their own beliefs, attitudes, commitments, and life-style. The faith learned in earlier stages is now disregarded, and greater attention is paid to one's own experience. In those individuals who still need authority figures, there is a tendency to join and become completely devoted to clubs and cults.

5. *Paradoxical-consolidation faith.* For **paradoxical-consolidation faith,** minimal age is 30. In this stage such elements of faith as symbols, rituals, and beliefs start to become understood and consolidated. The person begins to realize that other approaches to dealing with such complex questions as the supernatural and supreme being can be as valid as her or his own. The individual at this stage considers all people to belong to the same universal community and has a true regard for the kinship of all people.

6. *Universalizing faith.* For **universalizing faith,** minimal age is 40 years. As with Kohlberg's final stage, very few people ever reach this level. Here the individual lives in the real world but is not of it. Such persons do not merely recognize the mutuality of existence; they act on the basis of it. People at this stage appear to be truly genuine and lack the need to "save face" that exists at the lower stages.

Stage 6, as described by Fowler, compares closely to a hypothetical stage 7 of morality proposed by Kohlberg (1973). Although he never found anyone at a stage 7 level of morality, Kohlberg believed that theoretically there should be a stage for those few persons who rise above the purely cognitive and achieve a place where they transcend logic. These individuals, who are rare indeed, come to understand why one should be just and ethical in a world that is unjust. A burning love of universal humankind presses them always to act in truly moral ways.

The development of spirituality and morality appears to be parallel all along the sequence, especially in the Kohlberg and Fowler models. At the early levels, the orientation is basically selfish; ethical thinking and behavior are virtually nonexistent. The child is "good" only in order to please more powerful persons.

At the second two levels, concern for the opinion of the community in general takes over. "What will people think!" is uppermost in religion as well as in moral decisions. Only if and when the highest levels are reached do true spirituality and morality emerge. And for a few individuals at the highest level, the distinction between the moral and the spiritual no longer exists.

Why do people vary so much in level of spirituality? Can the biopsychosocial model help us answer this question? Do you believe that Fowler's ideas would apply to Eastern cultures? What's your opinion?

Conclusion

"The distinction between the moral and the spiritual no longer exists." What a wonderful goal to choose as a means of living the good life. It is probably also the best way to ensure a successful death. We sincerely hope that reading this book will contribute in some small way to your achievement of those two preeminent goals.

Chapter Highlights

The Role of Death in Life

- Today, four types of death are recognized: clinical death, brain death, biological or cellular death, and social death.
- In modern Western societies, death comes mostly to the old.
- Grief both follows and can precede the death of a loved one.
- In some cases, morbid grief reactions occur that prevent the successful conclusion of the life crisis. These are known as delayed reactions, distorted reactions, and pathological reactions.
- Most psychologists who have examined the role of grief have concluded that it is a healthy aspect of the crisis of death.
- Funerals have always been an important part of American life. Research has indicated that the rituals surrounding funerals have therapeutic benefit that facilitates the grieving process.
- Kübler-Ross has offered five stages of dying: shock and denial, anger, bargaining, depression, and acceptance.

Suicide: the Saddest Death

- Suicide rates for those over 75 is double that of adolescents, with the older white male most likely to take his life.

- Males and females are very different regarding suicide, with men being more likely to die than women.

"Successful" Dying

- The hospice movement and "death with dignity" legislation have provided people with more control over their own death, making it a bit easier to accept.

Spirituality

- In recent decades, participation in religious activities has been changing in a number of ways.
- The elderly practice their religions to a greater degree than other adults.
- Theories of spirituality have been presented by Frankl, Jung, Wilson, and Fowler.
- Fowler has incorporated the work of Erikson, Piaget, and Kohlberg in his theory of the development of faith, which proceeds through six stages.

Key Terms

Anima
Animus
Biological death
Brain death
Cardiopulmonary resuscitation (CPR)
Clinical death
Commitment
Electroencephalogram (EEG)
Hospice
Individuating-reflective faith
Intuitive-projective faith
Legal death
Mythical-literal faith
Mythification
Noetic dimension
Objectification
Paradoxical-consolidation faith
Poetic-conventional faith
Psychological dimension
Social death
Somatic dimension
Universalizing faith

What Do You Think?

1. What do you think is the best way to define death?
2. Most of us think of grief as something that happens to us. Do you think it makes sense to describe it as "work"?
3. How do you feel about the idea of "successful death"?
4. Are there some old people—those who have lost their spouse and all their friends, or those who are undeniably terminal—who should be allowed to take their own lives? Should these people be helped to have a "ceremony of death"?
5. In this chapter we have suggested two ways of making death more dignified. Can you think of any others?
6. Are you satisfied with your own level of religious participation? What do you think you should do differently?
7. How would you define your own spirituality?
8. How does Fowler's use of the term "commitment" compare to Sternberg's and Perry's (chap. 16)?

Suggested Readings

Agee, J. [1957] (1971). *A death in the family.* New York: Bantam. This Pulitzer Prize winning novel focuses on the effect of a man's death on his young son.

Becker, E. (1973). *The denial of death.* New York: Free Press. This is a brilliant analysis of the human failure to acknowledge death. Looks into the theories of Freud, Rank, Jung, Fromm, and others. Becker was awarded the Pulitzer Prize for this work.

Caine, L. (1987). *Widow.* New York: Bantam. Describes the feelings of a woman about her husband's death and the ways she finds to deal with her grief.

Freud, S. (1950). *Totem and taboo.* New York: Norton. One of Freud's most famous works, it explains how psychoanalysis looks at death and dying.

Kübler-Ross, E. (1969). *On death and dying.* New York: Macmillan. Briefly described in this chapter, this is a classic in the field.

Tolstoy, L. [1886] (1981). *The death of Ivan Illich.* New York: Bantam. Upon learning that he has terminal cancer, a man starts a lonely journey into understanding of the meaning of life and the ability to accept his own death.

GLOSSARY

A

Accommodation: When we modify our **schemata** to meet the demands of the environment; that is, we make our minds fit reality (Ch. 2).

Accumulation-of-Errors Theory: As cells die, they must synthesize new proteins to make new cells. As this is done, occasionally an error occurs. Over time, these errors mount up and may finally grow serious enough to cause organ failure (Ch. 20).

Accumulation of Metabolic Waste: Waste products resulting from metabolism build up in various parts of the body, contributing greatly to the decreasing competence of those parts. Some examples of this effect are cataracts on the eye, calcification in the arteries, and brittleness of bones (Ch. 20).

Acrosome: Tip of the **sperm** (Ch. 3).

Activity: Human beings flourish through interaction with other people. They are unhappy when, as they reach the older years, their contacts with others shrink as a result of death, illness, and societal limitations. Those who are able to keep up the social activity of their middle years are considered the most successful (Ch. 21).

Adaptation: Piaget states that all human beings tend to adapt to the environment. Adaptation consists of two complementary processes: **assimilation** and **accommodation** (Ch. 2). Also refers to how an individual, child or adult, adjusts to his or her environment (Ch. 9).

Adhesion: The prepared surface of the uterus and the outer surface of the fertilized **egg,** now called the **trophoblast,** touch and actually "stick together" (Ch. 4).

Adolescent Egocentrism: The reversion to the self-centered thinking patterns of childhood that sometimes occurs in the teen years (Ch. 12).

Adolescent Moratorium: A "time-out" period during which the adolescent experiments with a variety of identities, without having to assume the responsibility for the consequences of any particular one (Ch. 15).

Afterbirth: The period following birth in which the **placenta** and other membranes are discharged (Ch. 4).

Age Cohorts: Groups of people born at about the same time (Ch. 1).

Ageism: The prejudice that the elderly are inferior to those who are younger (Ch. 20).

Age-Status: Refers to society's expectations about what is normal at various ages. These expectations change not only with advancing years, but according to the particular society and to the historical context (Ch. 21).

Aggressive Gang: An organized group of **juvenile delinquents** who engage in a variety of illegal activities (Ch. 13).

Aging by Program: The vast majority of animals die at or before the end of their reproductive period, but human females live twenty to thirty years beyond the end of their reproductive cycles. Nevertheless, all animals seem to die when their "program" dictates (Ch. 20).

AID: An acronym for "Artificial Insemination by Donor" (Ch. 3).

AIDS (Acquired Immune Deficiency Syndrome): A condition caused by a virus that invades the body's immune system, making it vulnerable to infections and life-threatening illnesses (Ch. 4, 14).

Alarm Reaction: Selye's term for a generalized "call to arms" of the body's defensive forces (Ch. 15).

Allele: Alternate forms of a specific **gene**; there are genes for both blue and brown eyes (Ch. 3).

America 2000: The national educational goals adopted by former President Bush and the governors in 1990 (Ch. 10).

Amniocentesis: Entails inserting a needle through the mother's abdomen, piercing the **amniotic sac,** and withdrawing a sample of the amniotic fluid (Ch. 4).

Amniotic Sac: The sac that is filled with amniotic fluid and in which the embryo and fetus develop (Ch. 4).

Androgyny: Refers to those persons who have higher than average male *and* female elements in their personalities. Such persons are more likely to behave in a manner appropriate to a situation, regardless of their gender (Ch. 14).

Anima: The female side of the personality. Males tend to repress it until later in life (Ch. 22).

Animism: Children consider a large number of objects as alive and conscious that adults consider inert (Ch. 7).

Animus: The male side of the personality. Females tend to repress it until later in life (Ch. 22).

Anorexia Nervosa: A syndrome of self-starvation that mainly affects adolescent and young adult females. It is characterized by an "intense fear of becoming obese, disturbance of body image, significant weight loss, refusal to maintain a minimal normal body weight, and **amenorrhea.** The disturbance cannot be accounted for by a known physical disorder" (Ch. 13).

Anoxia (lack of oxygen): If something during the birth process should cut the flow of oxygen to the fetus, there is the possibility of brain damage or death (Ch. 4).

Anticipatory Images: Piaget's term for images (which include movements and transformation) that enable the child to anticipate change (Ch. 6).

Antigens: The substances in the blood that fight to kill foreign bodies (Ch. 20).

Apgar: A scale to evaluate a newborn's basic life signs; administered one minute after birth and repeated at three-, five-, and ten-minute intervals; it uses five life signs: heart rate, respiratory effort, muscle tone, reflex irritability, and skin color (Ch. 5).

Apnea: Brief periods when breathing is suspended (Ch. 5).

Apposition: The fertilized **egg,** now called a **blastocyst,** comes to rest against the uterine wall (Ch. 4).

Artificialism: Consists in attributing everything to human creation (Ch. 7).

Assimilation: When we perceive the environment in a way that fits our existing schemata; that is, we make reality fit our minds (Ch. 2).

Attachment: Behavior intended to keep a child (or adult) in close proximity to a significant other (Ch. 6).

Authoritarian Parenting Style: The parents strive for complete control over their children's behavior by establishing complex sets of rules. They enforce the rules through the use of rewards and, more often, strong discipline (Ch. 11).

Authoritarian Parents: Baumrind's term for parents who are demanding and want instant obedience as the most desirable child trait (Ch. 8).

Authoritative Parenting Style: In this, the most common style, parents are sometimes authoritarian and sometimes permissive, depending to some extent on the mood they happen to be in. They believe that both parents and children have rights, but that parental authority must predominate (Ch. 11).

Authoritative Parents: Baumrind's term for parents who respond to their children's needs and wishes; believing in parental control, they attempt to explain the reasons for it to their children (Ch. 8).

Autoimmunity: The process by which the immune system in the body rejects the body's own tissue. Examples are rheumatoid arthritis, diabetes, vascular diseases, and hypertension (Ch. 20).

Automatic Attention: Refers to common tasks that require little of a person's capacity for attention (Ch. 18).

Autonomy: Erikson's term for a child's growing sense of independence (Ch. 6).

Autosexuality: The love of oneself; the stage at which the child becomes aware of himself or herself as a source of sexual pleasure, and consciously experiments with masturbation (Ch. 14).

B

Babbling: The sounds that children make that sound like speech (Ch. 5).

Basal Metabolism Rate (BMR): The minimum amount of energy an individual tends to use when in a resting state (Ch. 18).

Behavioral Theory: The theory of human learning that says that all learning is ultimately determined by forces outside the control of the organism. What we learn and how we learn it are determined completely by genetic inheritance and by the influences of our environment, past and present (Ch. 2).

Behavioristic Model: This model sees subcultures starting out as a result of a series of trial-and-error behaviors, which are reinforced if they work. Peer group members behave the way they do because they have no other choice (Ch. 11).

Bilingual Education Act: Schools must make provisions for students who find instruction incomprehensible because they do not understand English (Ch. 10).

Binocular Coordination: Three-dimensional vision; appears around age 4 months (Ch. 5).

Biological Death: Death occurs when it is no longer possible to discern an electrical charge in the tissues of the heart and lungs (Ch. 22).

Biopsychosocial: A term for the idea that development proceeds by the interaction of biological, psychological, and social forces (Ch. 1, 17).

Blastocyst: Name of the fertilized egg after initial divisions (Ch. 3).

Bodily-Kinesthetic Intelligence: The control of bodily motions and ability to handle objects skillfully; one of Gardner's multiple intelligences (Ch. 9).

Brain Death: Death of the brain occurs when it fails to receive a sufficient supply of oxygen for a short period of time (usually eight to ten minutes). The cessation of brain function occurs in three stages: first the cortex stops, then the midbrain fails, and finally the brain stem ceases to function (Ch. 22).

Breech Birth: About four out of every hundred babies are born feet first, or buttocks first, while one out of a hundred are in a cross-wise position (transverse presentation) (Ch. 4).

Bulimia Nervosa: This disorder is characterized by "episodic binge-eating accompanied by an awareness that the eating pattern is abnormal, fear of not being able to stop eating voluntarily, and depressed mood and self-deprecating thoughts following the eating binges. The bulimic episodes are not due to **anorexia nervosa** or any known physical disorder" (Ch. 13).

C

Capacitation: A process by which the layer surrounding the **sperm** is removed; may be done externally or naturally in the woman's genital tract (Ch. 3).

Cardiopulmonary Resuscitation (CPR): A technique for reviving an individual's lungs and/or heart that have ceased to function (Ch. 22).

Centration: A feature of preoperational thought—the centering of attention on one aspect of an object and the neglecting of any other features (Ch. 7).

Cerebral Hemispheres: The two halves of the brain; the left side of the brain controls the right side of the body, the right side of the brain controls the left side of the body (Ch. 7).

Cesarean Section: Surgery to deliver the baby through the abdomen when a vaginal delivery is impossible (Ch. 4).

Chlamydia: Now the most common **STD,** with about 5 to 7 million new cases each year. There often are no symptoms; it is diagnosed only when complications develop (Ch. 14).

Cholesterol: A substance in the blood that comes to adhere to the walls of the blood vessels, restricting the flow of blood and causing strokes and heart attacks (Ch. 16).

Chorionic Villi Sampling (CVS): A catheter (small tube) is inserted through the vagina to the villi, and a small section of the villi is suctioned into the tube (Ch. 4).

Chromosomal Sex: The biological sexual program initially carried by either the X or Y sex **chromosome** (Ch. 8).

Chromosome Failure: A genetic abnormality such as gynecomastia (breast growth in the male) or hirsutism (abnormal female body hair) (Ch. 14).

Chromosomes: Stringlike bodies that carry the **genes** (Ch. 3).

Classification: Concrete operational children can group objects with some similarities within a larger category (e.g., brown wooden beads and white wooden beads are all beads) (Ch. 9).

Climacteric: Refers to a relatively abrupt change in the body, brought about by changes in hormonal balances (Ch. 18).

Climacterium: Refers to the loss of reproductive ability (Ch. 18).

Clinical Death: The individual is dead when his or her respiration and heartbeat have stopped (Ch. 22).

Closed Adoption: Form of adoption in which biological parents are removed from the lives of their adopted child (Ch. 3).

Cognitive Development: The growth of those mental processes by which we gain knowledge, such as perception, memory, and representation (Ch. 8).

Cognitive Structures Theory: The theory that a child's intelligence develops through the gradual addition of specific quantitative and qualitative mental abilities (Ch. 2).

Collagen: The major connective tissue in the body; it provides the elasticity in our skin and blood vessels (Ch. 20).

Color Blindness: A condition affecting the ability to detect some colors; caused by a defective **gene** on the X **chromosome** (Ch. 3).

Commitment (Fowler's term): For Fowler, the third step in the birth of a religion, in which people devote their lives to objectified ideas. They are willing under any circumstances to help those who have done the same (Ch. 22).

Commitment (Perry's term): The third phase in Perry's theory. The individual realizes that certainty is impossible, but that commitment to a certain position is necessary, even without certainty (Ch. 16).

Commitment (Sternberg's term): One of Sternberg's three aspects of love; the strongly held conviction that one will stay with another, regardless of the cost (Ch. 16).

Compensation: The cognitive ability to recognize an inequality of quantity and then add to the lesser amount to create an equality (Ch. 12).

Competition: Efforts to excel by achieving a particular goal; may come at the expense of others (Ch. 9).

Conservation: Concrete operational children gradually master the idea that something may remain the same even though its surface features change (Ch. 7).

Consolidation: A quantitative cognitive change from childhood to adulthood whereby improved problem-solving techniques are employed in a wider variety of situations and with greater skill (Ch. 12).

Constructed Knowledge: Belenky's fifth phase of women's thinking; characterized by an integration of the **subjective** and **procedural** ways of knowing (types 3 and 4) (Ch. 16).

Contingency: The effects of a parent's behavior on the infant's state (Ch. 6).

Continuity: The lasting quality of experiences; development proceeds steadily and sequentially (Ch. 1).

Controlled Scribbling: Drawing in which children carefully watch what they are doing, whereas before they looked away; they have better control of the crayon and hold it now like an adult (Ch. 7).

Conventional Level of Morality: Kohlberg's stage of moral development during which children desire approval from individuals and society (Ch. 9).

Coordination of Secondary Schemes: Infants combine secondary schemes to obtain a goal (Ch. 5).

Crawling: Locomotion in which the infant's abdomen touches the floor and the weight of the head and shoulders rests on the elbows (Ch. 5).

Creeping: Movement in which a child is on hands and knees and the trunk does not touch the ground; most youngsters creep at about 9 months of age (Ch. 5).

Cross-Linkage Theory: A theory of aging stating that the proteins that make up a large part of cells are themselves composed of peptides. When cross-links are formed between peptides (a natural process of the body), the proteins are altered, often for the worse (Ch. 20).

Crossover: Older men become more like women, and older women become more like men. They do not actually cross over–they just become more like each other (Ch. 21).

Cross-Sectional Studies: Research method comparing groups of individuals of various ages at the same time in order to investigate the effects of aging (Ch. 1).

Crystallized Intelligence: Involves perceiving relationships, educing correlates, reasoning, abstracting, concept of attainment, and problem solving, as measured primarily in unspeeded tasks involving various kinds of content (semantic, figural, symbolic) (Ch. 18).

Culture Transmission Model: A new subculture arises as an imitation of the subculture of the previous generation. This takes place through a learning process by which younger teenagers model themselves after those in their twenties (Ch. 11).

Cystic Fibrosis: A chromosomal disorder that produces a malfunction of the exocrine glands (Ch. 3).

Cytomegalovirus (CMV): A virus that can cause damage such as mental retardation, blindness, deafness, and even death. A major difficulty in combating this disease is that it remains unrecognized in pregnant women (Ch. 4).

D

Day Care: A child spends part of the day outside of his or her own home in the care of others (Ch. 8).

Decentering: Concrete operational children can concentrate on more than one aspect of a situation (Ch. 9).

Decoding: The technique by which we recognize words (Ch. 9).

Defense Mechanisms: Freud's concept of unconscious mental mechanisms that protect our conscious minds from becoming aware of truths we would rather not know (Ch. 2).

Deferred Imitation: Imitation that continues after the disappearance of the model to be imitated (Ch. 7).

Delayed Puberty: The stages of pubertal change do not begin until a significant time after the normal onset (Ch. 12).

Depression: A condition marked by a sorrowful state, fatigue, and a general lack of enthusiasm about life; children are now thought to be subject to this problem. Depression is not a specific syndrome or illness in itself. It is a basic affective state that, like anxiety, can be of long or short duration and of low or high intensity, and can occur in a wide variety of conditions at any stage of development (Ch. 13).

DES (Diethylstilbestrol): In the late 1940s and 1950s, DES (a synthetic hormone) was administered to pregnant women, supposedly to prevent miscarriage. It was later found that the daughters of the women who had received this treatment were more susceptible to vaginal and cervical cancer (Ch. 4).

Descriptive Studies: Information is gathered on subjects without manipulating them in any way (Ch. 1).

Despair: The counterpart to **integrity** in the last stage of Erikson's theory. When people look back over their lives and feel that they have made many wrong decisions, or more commonly, that they have frequently not made any decisions at all, they tend to see life as lacking integrity. They are angry that there can never be another chance to make their lives make sense (Ch. 21).

Developmentally Delayed: A condition in which a child's growth rate slows; it need not cause any permanent retardation if corrected. Its causes may be either physical or psychological (Ch. 7).

Developmental Risk: A term used to describe those children whose well-being is in jeopardy due to a range of biological and environmental conditions (Ch. 4).

Developmental Tasks: Havighurst suggests these specific tasks at each stage of life, which lie midway between the needs of the individual and the ends of society. These tasks, such as skills, knowledge, functions, and attitudes, are needed by an individual in order to succeed in life (Ch. 1, 11).

Differential Opportunity Structure: Due to social disapproval and more rigid rules enforced by parents, peers, and the legal system, women have not had the same access to sex that men have had (Ch. 21).

Difficult Children: Children whose temperament causes conflicts with those around them (Ch. 6).

Discontinuity: Behaviors that are apparently unrelated to earlier aspects of development (Ch. 1).

Disengagement Theory: According to this position, the belief that activity is better than passivity is a bias of the Western world. In disengagement theory, the *most mature* adults are likely to gradually disengage themselves from their fellow human beings in preparation for death. They become less interested in their interactions with others, and more concerned with internal concerns (Ch. 21).

Distantiation: The readiness of all of us to distance ourselves from others when we feel threatened by their behavior. Distantiation is the cause of most prejudices and discrimination (Ch. 17).

Diversity: According to Toffler, stress is increased by what percentage of our lives is in a state of change at any one time (Ch. 15).

Dizygotic Twins: Twins that result from two ripened **eggs** being fertilized by separate **sperm.** Their **genes** are no more alike than those of siblings born of the same parents but at different times (Ch. 3).

DNA: (Deoxyribonucleic acid); often referred to as the structure of life (Ch. 3).

Dominant: The **gene** that tends to be expressed in a trait (Ch. 3).

Double Standard: The belief that the standards for female sexual behavior should be higher than those for males (Ch. 14).

Down Syndrome: Genetic abnormality caused by a deviation on the twenty-first pair of **chromosomes** (Ch. 3).

Drawing: The ability to form lines into objects that reflect the world; a physical activity that often reveals cognitive and emotional development (Ch. 7).

Drive-by Shootings: Shootings committed by someone riding in a car past the victim (Ch. 13).

Dual-Career Family: A family in which the wife does some sort of paid work and also manages the family's functions (Ch. 17).

Dual-Process Model: A model of intelligence that says there may be a decline in the *mechanics* of intelligence, such as **classification** skills and logical reasoning, but that the **pragmatics** are likely to increase (Ch. 18).

Dualism: Perry's initial phase of ethical development, in which "things are either absolutely right or absolutely wrong" (Ch. 16).

E

Easy Children: Children whose temperament enables them to adjust well and to get along with others (Ch. 6).

Ectoderm: The outer layer of the embryo that will give rise to the nervous system, among other developmental features (Ch. 4).

Ectopic Pregnancy: The fertilized **egg** attempts to develop in one of the **fallopian tubes**; sometimes referred to as a *tubal pregnancy* (Ch. 4).

Effortful Attention: Refers to processing novel tasks, such as trying to learn a new phone number or a new surgical technique (Ch. 18).

Egg: The female sex cell which, when fertilized by the male, grows into the fetus (Ch. 3).

Ego: One of the three structures of the psyche, according to Freud; it is in contact with reality, and mediates between the **id** and **superego** (Ch. 2).

Egocentrism: Refers to children's belief that everything centers on them; *sensorimotor egocentrism* refers to the inability to distinguish oneself from the world (Ch. 5, 7).

Electroencephalogram (EEG): A graphic record of the electrical activity of the brain (Ch. 22).

Embryonic Period: Third through the eighth week following **fertilization** (Ch. 4).

Emotional Divorce: Sometimes a couple learns to "withstand" each other, rather than live with each other. The only activities and interests that they shared were ones that revolved around the children. When the children leave, they are forced to recognize how far apart they have drifted; in effect, they are emotionally divorced (Ch. 19).

Empirico-Inductive Method: A method of problem solving used by young children in which they look at available facts and try to induce some generalization from them (Ch. 12).

Empty Nest Syndrome: Refers to the feelings parents may have as a result of their last child leaving home (Ch. 19).

Encoding: Translating objects and events into language (Ch. 9).

Endoderm: The inner layer of the embryo that will give rise to the lungs and liver, among other developmental features (Ch. 4).

Endometriosis: A condition in which tissue normally found in the uterus grows in other areas, such as the **fallopian tubes** (Ch. 3).

Escape: Perry's term for refusing responsibility for making any commitments. Since everyone's opinion is "equally right," the person believes that no commitments need be made, and so escapes from the dilemma (Ch. 16).

Esteem Needs: Refers to the need for positive reactions of others to us as individuals and also to the need for a positive opinion of ourselves. One level of Maslow's need hierarchy (Ch. 1).

Estrogen Replacement Therapy (ERT): A process in which estrogen is given in low levels to a woman experiencing severe problems with **menopause** (Ch. 18).

Existential Love: The capacity to cherish the present moment, perhaps first learned when we confront the certainty of our own personal death (Ch. 21).

Exosystem: An environment in which you are not present, but which nevertheless affects you (e.g., your father's job) (Ch. 1).

Exploration Stage: According to Super's theory, that period in a person's career, usually from ages 15 to 24, during which a variety of work experiences are chosen (Ch. 17).

Expressive Language: Children move from purely receptive speech to expressing their own ideas and needs through language (Ch. 7).

External Fertilization: Fertilization that occurs outside of the woman's body (Ch. 3).

Extinction: According to behavioral theory, refraining from **reinforcing** behavior is the best way to extinguish it; extinction differs from **punishment** in that no action is taken (Ch. 2).

Extragenetic: Sources of information other than **genetic** (Ch. 3).

Extrasomatic: Sources of information outside of the body (e.g., computers, books) (Ch. 3).

F

Failure-to-Thrive (FTT): The weight and height of FTT infants consistently remain far below normal (the bottom 3% of height and weight measures); there are two types of FTT cases: organic and nonorganic (Ch. 5).

Fallopian Tubes: Passageway for the **egg** once it is discharged from the ovary's surface (Ch. 3).

Family Status: An indication of how close and strong family ties are. A strong family status is associated with a lower incidence of adolescent suicide (Ch. 13).

Fertilization: Union of **sperm** and **egg** (Ch. 3).

Fetal Alcohol Syndrome (FAS): Refers to babies affected when their mothers drink alcohol during pregnancy; they manifest four clusters of symptoms: physiological functioning, growth factors, physical features, and structural effects (Ch. 4).

Fetal Hydrocephalus: Surgery to cure a condition in which the brain regions fill with fluid and expand (Ch. 4).

Fetal Period: The period extending from the beginning of the third gestational month to birth (Ch. 4).

Fetoscopy: A tiny instrument called a fetoscope is inserted into the amniotic cavity, making it possible to see the fetus (Ch. 4).

Fiscal Fitness: The idea that many women lack experience in managing money, and need to become "fiscally fit" (Ch. 19).

Fluid Intelligence: Involves perceiving relationships, educing correlates, maintaining span of immediate awareness in reasoning, abstracting, concept formation, and problem solving, as measured in unspeeded as well as speeded tasks involving figural, symbolic, or semantic content (Ch. 18).

Forceps Delivery: Occasionally, for safety, the physician withdraws the baby with forceps during the first phase of birth (Ch. 4).

Fragile X Syndrome: A sex-linked inheritance disorder in which the bottom half of the X **chromosome** looks as if it is ready to fall off; causes mental retardation in 80% of the cases (Ch. 3).

Free Will: The concept that some of our decisions are entirely controlled by us, without the influence of outside forces (Ch. 2).

Future Shock: The illness that results from having to deal with too much change in too short a time (Ch. 15).

G

Gamete Intrafallopian Transfer (GIFT): The **sperm** and the **egg** are placed in the **fallopian tube** with the intent of achieving fertilization in a more natural environment (Ch. 3).

Gender Constancy: Refers to children understanding the

unchanging nature of gender (Ch. 8).

Gender Identity: Refers to children cognitively realizing that they are males or females. (Ch. 8).

Gender Revolution: Levinson's term; the meanings of gender are changing and becoming more similar (Ch. 19).

Gender Role: A pattern of behavior that results partly from genetic makeup and partly from the specific traits that are in fashion at any one time and in any one culture. For example, women are able to express their emotions through crying more easily than men, although there is no known physical cause for this difference (Ch. 14).

Gender-Role Adaptation: Defined by whether the individual's behavior may be seen as in accordance with her or his gender (Ch. 14).

Gender-Role Orientation: Individuals differ in how *confident* they feel about their sexual identity. Those with low confidence have a weak orientation toward their gender role (Ch. 14).

Gender-Role Preference: Some individuals feel unhappy about their gender role, and wish either society or their gender could be changed, so that their gender role would be different (Ch. 14).

Gender-Role Reversals: Older men see themselves and other males as becoming submissive and less authoritative with advancing years. Conversely, older women see themselves and other women as becoming more dominant and self-assured as they grow older (Ch. 21).

Gender-Role Stereotypes: Behavior common in our culture that can be identified as typically either male or female. Some stereotypes are more or less true, some are clearly false, and some we are not sure about (Ch. 14).

Gender Splitting: Levinson's term; all societies support the idea that there should be a clear difference between what is considered appropriate for males and for females; gender splitting appears to be universal (Ch. 19).

Gender Stability: Refers to children understanding that gender typically remains the same throughout life (Ch. 8).

Gene: The part of the **chromosome** that carries the hereditary characteristics (Ch. 3).

General Adaptation Syndrome: Selye's theory about the three stages of reaction to **stress** (Ch. 15).

Generative Love: Most characteristic of parenthood, a time during which sacrifices are gladly made for the sake of the children (Ch. 21).

Generativity: Erikson's term for the ability to be useful to ourselves and to society. As in the industry stage, the goal is to be productive and creative, but in the generativity stage, which takes place during middle adulthood, one's productivity is aimed at being helpful to others (Ch. 19).

Gene Theory: The theory that aging is due to certain harmful genes (Ch. 20).

Genetic: Refers especially to the information **genes** contain (Ch. 3).

Genetic Theory of Homosexuality: The theory that homosexuality is caused by some factor in a person's DNA (Ch. 14).

Genital Herpes: An incurable sexually transmitted disease, with about 500,000 new cases every year. With no cure, about 30 million people in this country now experience the recurring pain of this infection (Ch. 14).

Genital Sex: Sex is determined not only by **chromosomes** and hormones but by external sex organs (Ch. 8).

Genotype: An individual's genetic composition (Ch. 3).

German Measles (Rubella): A typically mild childhood disease caused by a virus; pregnant women who contract this disease may give birth to a baby with a defect: congenital heart disorder, cataracts, deafness, or mental retardation. The risk is especially high if the disease appears early in the pregnancy (Ch. 4).

Germinal Choice: Refers to attempts to match the **sperm** and **egg** of selected individuals (Ch. 3).

Germinal Period: The first two weeks following **fertilization** (Ch. 4).

Glaucoma: Results from a buildup of pressure inside the eye due to excessive fluid. The resulting damage can destroy one's vision (Ch. 18).

Gonadal Sex: The XX or XY chromosome combination passes on the sexual program to the undifferentiated gonads. If the program is XY, the gonads differentiate into testes. If the program is XX, the gonads differentiate into the ovaries, starting at about the twelfth week (Ch. 8).

Gonorrhea: Well-known venereal disease accounting for between one and a half and two million cases per year. One quarter of those were reported among adolescents. The most common symptoms are painful urination and a discharge from the penis or the vagina (Ch. 14).

Goodness of Fit: Compatibility between parental and child behavior; how well parents and their children get along (Ch. 6).

Grade Retention: Retaining students in a grade (Ch. 10).

Group Marriage: A marriage that includes two or more of both husbands and wives, who all exercise common privileges and responsibilities (Ch. 17).

H

Hazing Practices: The often dangerous practices used by some fraternities to initiate new members (Ch. 15).

Hemophilia: A genetic condition causing incorrect blood clotting; called the "bleeder's disease" (Ch. 3).

Hepatitis B: A viral disease transmitted through sex or shared needles (Ch. 14).

Herpes Simplex: An infection that usually occurs during birth; a child can develop the symptoms during the first week following the birth. The eyes and nervous system are most susceptible to this disease (Ch. 4).

Heterosexuality: Love of members of the opposite sex (Ch. 14).

Holophrase: Children's first words; they usually carry multiple meanings (Ch. 5).

Holophrastic Speech: One word to communicate many meanings and ideas (Ch. 5).

Homeostatic Imbalance: The theory that aging is due to a failure in the systems that regulate the proper interaction of the organs (Ch. 20).

Homework: Refers to school-assigned academic work that is to be completed outside of school, usually at home (Ch. 10).

Homosexuality: Love of members of one's own sex (Ch. 14).

Homosexual Marriage: Though not accepted legally, the weddings of homosexuals are now accepted by some religions (Ch. 17).

Hormonal Balance: One of the triggering mechanisms of puberty that may be used to indicate the onset of adolescence (Ch. 12).

Hormonal Sex: Once the testes or ovaries are differentiated, they begin to produce chemical agents called sex hormones (Ch. 8).

Hospice: A facility and/or program dedicated to assisting those who have accepted the fact that they are dying and desire a "death with dignity." Provides pain control and counseling, but does not attempt to cure anyone (Ch. 22).

Human Genome Project: The attempt to identify and map the 50,000 to 100,000 **genes** that constitute the human genetic endowment (Ch. 3).

Hypersensitive Youth: The hypersensitive youth has an extreme reaction to situations that would only mildly disturb most people. The disruptions caused by seemingly trivial events may come together in a suicide attempt (Ch. 13).

Hypothetico-Inductive Method: A method of problem solving used by adolescents in which they hypothesize about the situation and then deduce from it what the facts should be if the hypothesis is true (Ch. 12).

I

Id: One of the three structures of the psyche, according to Freud. Present at birth, it is the source of our instinctive desires (Ch. 2).

Identical Twins: Twins whose **genes** are identical; they share the same **genotype** (Ch. 3).

Identity Achievement: Marcia's final status, in which numerous crises have been experienced and resolved, and relatively permanent commitments have been made (Ch. 11).

Identity Confusion: Marcia's initial status, in which no crisis has been experienced and no commitments have been made (Ch. 11).

Identity Crisis: Erikson's term for the situation, usually in adolescence, that causes us to make major decisions about our identity (Ch. 2, 11).

Identity Foreclosure: One of Marcia's statuses, in which no crisis has been experienced, but commitments have been made, usually forced on the person by the parent (Ch. 11).

Identity Moratorium: One of Marcia's statuses of adolescence, in which considerable crisis is being experienced but no commitments are yet made (Ch. 11).

Identity Status: Refers to Marcia's four types of identity formation (Ch. 11).

Implantation: When a fertilized egg becomes embedded in the uterine wall (Ch. 3, 4).

Impotency: The inability to engage in the sexual act (Ch. 21).

Inconsistent Conditioning: A situation in which children are sometimes expected to behave in a certain way, but at other times are not (Ch. 11).

Individuating-Reflective Faith: The fourth developmental step of Fowler's theory of faith. Individuals in stage four begin to assume responsibility for their own beliefs, attitudes, commitments, and life-style. The faith learned in earlier stages is now disregarded, and greater attention is paid to one's own experience (Ch. 22).

Individuation: Refers to our becoming more individual; we develop a separate and special personality, derived less and less from our parents and teachers and more from our own behavior (Ch. 8 and 17).

Information Processing: The study of how children (and adults) perceive, comprehend, and retain information (Ch. 7).

Information-Processing Strategy: A cognitive problem-solving plan (Ch. 12).

Initiation Rites: A cultural and sometimes ceremonial task that signals the entrance to some new developmental stage. Such rites can indicate the passage from adolescence to adulthood (Ch. 15).

Initiative: Erikson's term for children's ability and willingness to explore the environment and test their world (Ch. 7).

Inner Dialogue: Gould's seven steps, which he believes can help in mastering the demons of one's childhood experiences (Ch. 17).

Integrity: The resolution of each of the first seven crises in Erikson's theory should lead us to achieve a sense of personal integrity. Older adults who have a sense of integrity feel their lives have been well spent. The decisions and actions they have taken seem to them to fit together (Ch. 21).

Interaction: Behaviors involving two or more people (Ch. 6).

Inter-Propositional Thinking: The ability to think of the ramifications of *combinations* of propositions (Ch. 12).

Intergenerational Continuity: The connection between childhood experiences and adult behavior (Ch. 6).

Internal Fertilization: A natural process in which fertilization occurs within the woman (Ch. 3).

Interpersonal and Intrapersonal Intelligences: Gardner's personal intelligences; interpersonal intelligence builds on an ability to recognize what is distinctive in others, while intrapersonal intelligence enables us to understand our own feelings (Ch. 9).

Intimacy: Erikson's stage that represents the ability to relate one's deepest hopes and fears to another person and to accept another's need for intimacy in turn (Ch. 17).

Intra-Propositional Thinking: The ability to think of a number of possible outcomes that would result from a *single* choice (Ch. 12).

Intrauterine Device: Usually a plastic loop inserted into the uterus as a contraceptive device (Ch. 3).

Intuitive-Projective Faith: The first developmental step of Fowler's theory of faith. In this stage, the individual focuses on surface qualities, as portrayed by adult models. This stage depends to a great extent on fantasy. Conceptions of God or a supreme being reflect a belief in magic (Ch. 22).

Invasion: During invasion, the **trophoblast** "digs in" and begins to bury itself in the uterine lining (Ch. 4).

Inversion: The cognitive ability to recognize an inequality of quantity and then subtract from the greater amount to create an equality (Ch. 12).

In Vitro Fertilization: Fertilization that occurs "in the tube" or "in the glass"; an external fertilization technique (Ch. 3).

Invulnerable Children: Children who sustain some type of physical or psychological trauma yet remain on a normal developmental path (Ch. 10).

Irreversibility: A child's inability to reverse thinking; a cognitive act is reversible if it can utilize stages of reasoning to solve a problem and then proceed in reverse, tracing its steps back to the original question or premise (Ch. 7).

Irritable Infants: Infants who are generally more negative in their behavior and more irregular in their biological functioning than typical children (Ch. 6).

Isolation: The readiness all of us have to isolate ourselves from others when we feel threatened by their behavior (Ch. 17).

J

Juvenile Delinquent: A minor who commits illegal acts (Ch. 13).

K

Knowledge-Acquisition Components: Sternberg's term for those components that help us to learn how to solve problems in the first place (Ch. 9).

L

L-tryptophan: This amino acid has been linked to a blood disorder in which white blood cells increase to abnormally high numbers (Ch. 18).

Laparoscope: A thin, tubular lens used to identify mature **egg(s)** (Ch. 3).

Lateralization: There is a preferred side of the brain for a particular activity (if you are right-handed in writing, you are left-lateralized for writing) (Ch. 7).

Learning Disability: A general term that refers to multiple disorders leading to difficulties in listening, speaking, reading, writing, reasoning, or mathematical abilities (Ch. 10).

Learning Theory of Homosexuality: The belief that homosexuality is the result of learned experiences from significant others (Ch. 14).

Legal Death: Condition defined as "unreceptivity and unresponsivity, no movements or breathing, no reflexes, and a flat **electroencephalogram** (EEG) reading that remains flat for 24 hours" (Ch. 22).

Life Change Unit (LCA): Changes in life (e.g., divorce) that are rated in units according to the degree that they tend to cause stress. The higher the life change is rated, the more stress, and possible disease, the change is likely to cause (Ch. 15).

Life Course: Levinson's term. *Life* refers to all aspects of living—everything that has significance in a life; *course* refers to the flow or unfolding of an individual's life (Ch. 17).

Life Cycle: Levinson's term. The life cycle is a *general* pattern of adult development, while the life course is the unique embodiment of the life cycle by an *individual* (Ch. 17).

Lifespan Psychology: A study of development as a life-long process beginning at conception and ending in death (Ch. 1).

Life Structure: Levinson's term. The underlying pattern or design of a person's life *at a given time* (Ch. 17).

Linguistic Intelligence: One of Gardner's multiple intelligences (Ch. 9).

Logical-Mathematical Intelligence: One of Gardner's multiple intelligences (Ch. 9).

Longitudinal Studies: The experimenter makes several observations of the same individuals at two or more times in their lives. Examples are determining the long-term effects of learning on behavior; the stability of habits and intelligence; and the factors involved in memory (Ch. 1).

Love and Belongingness Needs: Refers to the need for family and friends. One level of Maslow's need hierarchy (Ch. 1).

M

Macrosystem: A term indicating society at large (Ch. 1).

Magical Thinking: Many adolescents have an unrealistic view of death's finality and use suicide as a means to radically transform the world and solve their problems or to join a loved one who has already died. These feelings are often aided and abetted by the glorification of suicide that sometimes occurs in the media (Ch. 13).

Main-Effects Model: A model of development that attributes growth mainly to the contribution of one influence, e.g., the environment or the body (Ch. 1).

Major False Assumptions: Remaining from childhood, these beliefs must be reexamined and readjusted by each individual if he or she is to progress in maturity (Ch. 17).

Male Change of Life: Change in hormonal balance and sexual potency (Ch. 18).

Manipulative Experiments: The experimenter attempts to keep all variables (all the factors that can affect a particular outcome) constant except one, which is carefully manipulated (Ch. 1).

Maturation: The process of physical and mental development due to physiology (Ch. 12).

Maximum Growth Spurt: The period of adolescence when physical growth is at its fastest (Ch. 12).

Meiosis: Division of the germ cells, resulting in 23 chromosomes; also called *reduction division* (Ch. 3).

Menarche: The onset of **menstruation** (Ch. 12).

Menopause: The cessation of **menstruation** (Ch. 18).

Menstrual Cycle: The interaction between the pituitary gland and the ovaries that occurs in four-week phases (Ch. 3).

Mental Imagery: Mental processing involving visual depiction of people, events, and objects (Ch. 7).

Mentoring: The act of assisting another, usually younger, person with his work or life tasks (Ch. 17).

Mesoderm: The middle layer of the embryo that gives rise to muscles, the skeleton, and the circulatory and excretory systems (Ch. 4).

Mesosystem: The relationship among **microsystems** (Ch. 1).

Metacomponents: Sternberg's term for those components that help us to plan, monitor, and evaluate our problem-solving strategies (Ch. 9).

Metalinguistics Awareness: A capacity to think about and talk about language (Ch. 9).

Microsystem: The immediate setting that influences development, such as the family (Ch. 1).

Mid-Life Transition: Levinson's term for the phase that usually lasts for five years and generally extends from age 40 to 45. It involves three major developmental tasks (Ch. 19).

Miscarriage: When a pregnancy ends spontaneously before the twentieth week (Ch. 4).

Miseducation: David Elkind's term for excessive academic pressure placed on young children (Ch. 8).

Mitosis: Cell division in which each **chromosome** is duplicated, maintaining the number at 46 (Ch. 3).

Monogamy: The standard marriage form in the United States and most other nations, in which there is one husband and one wife (Ch. 17).

Monozygotic Twins: Identical twins (Ch. 3).

Moral Dilemma: A modified clinical technique used by Kohlberg to discover the structures of moral reasoning and the stages of moral development; a conflict leads subjects to justify the morality of their choices (Ch. 9).

Moratorium of Youth: A "time-out" period during which the adolescent experiments with a variety of identities, without having to assume the responsibility for the consequences of any particular one (Ch. 11).

Musical Intelligences: One of Gardner's multiple intelligences (Ch. 9).

Mutation: A change in the structure of a **gene** (Ch. 3).

Mythical-Literal Faith: The second developmental step of Fowler's theory of faith. Fantasy ceases to be a primary source of knowledge at this stage, and verification of facts becomes necessary. Verification comes not from actual experience, but from such authorities as teachers, parents, books, and traditions (Ch. 22).

Mythification: Stories are developed that tell why members of a religion have a special place in the world. These stories are rational and enhance the person's understanding of the physical as well as the spiritual world (Ch. 22).

N

Naturalistic Experiments: In these experiments, the researcher acts solely as an observer and does as little as possible to disturb the environment. "Nature" performs the experiment, and the researcher acts as a recorder of the results (Ch. 1).

Negative Identity: Persons with a negative identity adopt one pattern of behavior because they are rebelling against demands that they do the opposite (Ch. 11).

Negative Reinforcement: Any event that, when it *ceases to occur* after a response, makes that response more likely to happen in the future (Ch. 2).

NEO (neuroticism, extrovertism, openness) Model of Personality: McCrae and Costa's theory that there are three major personality traits, which they feel govern the adult personality. Each of those three traits is supported by six subtraits or "facets" (Ch. 19).

Neonate: Term for an infant in the days immediately following birth (Ch. 5).

Neurological Assessment: Identifies any neurological problem, suggests means of monitoring the problem, and offers a prognosis about the problem (Ch. 5).

New York Longitudinal Study: Long-term study by Chess and Thomas of the personality characteristics of children (Ch. 6).

Noetic Dimension: Frankl's third stage of human development has roots in childhood, but primarily develops in late adolescence. It is spiritual, not only in the religious sense but in the totality of the search for the meaningfulness of life. This aspect makes humans specifically different from all other species (Ch. 22).

No-Fault Divorce: The law that lets people get divorced without proving some atrocious act by one of the spouses. This new law recognized that people could, in the course of their lives, simply grow apart from each other to the point that they no longer made good marriage partners. In legal language, this is known as an irretrievable breakdown of a marriage (Ch. 19).

Nonaggressive Offenders: Anyone who commits a crime in which there is no aggressive intent. Usually refers to prostitutes and runaways (Ch. 13).

Normal Range of Development: The stages of pubertal change occur at times that are within the normal range of occurrence (Ch. 12).

Novelty: Toffler's term for the dissimilarity of new situations in our lives (Ch. 15).

Novice Phase: Levinson's initial phase of human development extends from ages 17 to 33, and includes the early adult transition, entering the adult world, and the age-30 transition (Ch. 17).

Nucleotides: The small blocks of the **DNA** ladder (Ch. 3).

Numeration: Concrete operational children grasp the meaning of numbers, the oneness of one (Ch. 9).

Nurturing Parenting Style: The style of parenting in which parents use indirect methods such as discussion and modelling rather than punishment to influence their child's behavior. Rules are kept to a minimum (Ch. 11).

Nurturist: One type of **main-effects** model that attributes development to the environment (Ch. 1).

O

Objectification: Fowler's term for the first step in the birth of a religion in which a perception of reality is described. This includes images and definitions that are easy to understand. Examples are good versus evil, heaven versus hell, and the control by a god or gods of the forces of nature (Ch. 22).

Object Permanence: Refers to children gradually realizing that there are permanent objects around them, even when these objects are out of sight (Ch. 5).

Observational Learning: A term associated with Bandura, meaning that we learn from watching others. Also called social learning (Ch. 2).

Olfactory Sense: The sense of smell, which uses the olfactory nerves in the nose and tongue (Ch. 18).

One-Time, One-Group Studies: Studies that are carried out only once on only one group of subjects (Ch. 1).

Open Adoption: Form of adoption in which biological parents may remain in contact with their adopted child (Ch. 3).

Operant Conditioning: The theory that when operants (actions that people or animals take of their own accord) are reinforced, they become conditioned (more likely to be repeated in the future) (Ch. 2).

Operations: Cognitive events that take the place of actual behavior (Ch. 2).

Optimum Drive Level: The level of optimum stimulation for an individual (Ch. 15).

Organization: Our innate tendency to organize causes us to combine our **schemata** more efficiently. The schemata of the infant are continuously reorganized to produce a coordinated system of higher-order structures (Ch. 2).

Organ Reserve: Refers to that part of the total capacity of our body's organs that we do not normally need to use. For example, when you walk up the stairs, you probably use less than half of your total lung capacity (Ch. 16).

Organogenesis: The formation of organs during the embryonic period (Ch. 4).

Outward Bound: A program in which people learn to deal with their fears by participating in a series of increasingly threatening experiences. As a result, their sense of self-worth increases and they feel more able to rely on themselves. The program uses such potentially threatening experiences as mountain climbing and rappeling, moving about in high, shaky rope riggings, and living alone on an island for several days (Ch. 15).

Overregularities: At a certain point in their language development, children inappropriately use the language rules they have learned—for example, "I comed home" (Ch. 7).

Ovulation: That time when the **egg** is released from the ovary's surface (Ch. 3).

Ovum: Another term for the **egg** (Ch. 3).

P

Paradoxical-Consolidation Faith: The fifth developmental step of Fowler's theory of faith. In this stage, such elements of faith as symbols, rituals, and beliefs start to become understood and consolidated. The person begins to realize that others' approaches to dealing with such complex questions as the supernatural and supreme being can be as valid as her or his own (Ch. 22).

Parent-Child Role Reversals: Parents and children sometimes exchange traditional role behaviors. In the parent-child interaction, the child adopts some parent-type behavior (e.g., caretaking, supporting, nurturing, advising), and the parent acts more like a child is expected to act (Ch. 13).

Passion: Sternberg's term for a strong sense of desire for another person, and the expectation that sex with them will prove physiologically rewarding (Ch. 16).

Pedagogy: The science of teaching children (Ch. 18).

Peer: Refers to a youngster who is similar in age to another child, usually within 12 months (Ch. 10).

Pelvic Inflammatory Disease (PID): Disease that often results from **chlamydia** or **gonorrhea,** and frequently causes prolonged problems, including infertility. Symptoms include lower abdominal pain and a fever (Ch. 14).

Performance Components: Sternberg's term for those components that help us to execute the instructions of the **metacomponents** (Ch. 9).

Permissive Parenting Style: The parents have little or no control over their children, and refrain from disciplinary measures (Ch. 11).

Permissive Parents: Baumrind's term for parents who take a tolerant, accepting view of their children's behavior, including both aggressive and sexual urges; they rarely use **punishment** or make demands of their children (Ch. 8).

Permutations: The act of altering a given set of objects in a group in a systematic way. One of the abilities described by Flavel (Ch. 12).

Phenotype: The observable expression of gene action (Ch. 3).

Phenylketonuria: Chromosomal disorder resulting in failure to break down the amino acid phenylalanine (Ch. 3).

Phonology: Describes how to put sounds together to form words (Ch. 7).

Physiological Needs: Refers to those dominant needs (hunger, thirst) that must be satisfied if higher levels of motivation are to become active. One level of Maslow's need hierarchy (Ch. 1).

Placenta: The placenta supplies the embryo with all its needs, carries off all its wastes, and protects it from danger (Ch. 4).

Plasticity: Resiliency; the ability to recover from either physiological or psychological trauma and return to a normal developmental path (Ch. 1, 10).

Poetic-Conventional Faith: The third developmental step of Fowler's theory of faith. Faith is still conventional and depends on a consensus of opinions of other, more authoritative persons. Now the person moves away from family influence and into new relationships. Faith begins to provide a coherent and meaningful synthesis of these relationships (Ch. 22).

Polyandry: A marriage in which there is one wife but two or more husbands (Ch. 17).

Polygamy: A marriage in which there is one husband but two or more wives (Ch. 17).

Positive Reinforcement: Any event that, when it *occurs after* a response, makes that response more likely to happen in the future (Ch. 2).

Postconventional Level of Morality: Kohlberg's stage of moral development, during which individuals make moral decisions according to an enlightened conscience (Ch. 9).

Postnatal Depression: Many women feel "down" a few days after giving birth; this is fairly common and is now thought to be a normal part of pregnancy and birth for some women (Ch. 4).

Power Theory of Intelligence: Perkins' term; the view that intelligence is solely dependent on the neurological efficiency of the brain; a genetic interpretation (Ch. 9).

Pragmatics: The rules of pragmatics describe how to take part in a conversation (Ch. 7).

Precociousness: The ability to do what others are able to do, but at a younger age (Ch. 12).

Preconventional Level of Morality: Kohlberg's stage of moral development, during which children respond mainly to cultural control to avoid **punishment** and attain satisfaction (Ch. 9).

Premature Foreclosure: A situation in which a teenager chooses an identity too early, usually because of external pressure (Ch. 11).

Prematurity: About 7 out of every 100 births are premature, occurring less than 37 weeks after conception; prematurity is defined by low birth weight and immaturity (Ch. 4).

Prepared Childbirth: A combination of relaxation techniques and information about the birth process; sometimes called the *Lamaze method,* after its founder (Ch. 4).

Pretend Play: A characteristic of the early childhood youngster; these children show an increasing ability to let one thing represent another, a feature that carries over to their play (Ch. 8).

Primary Circular Reactions: Infants repeat some act involving their bodies (Ch. 5).

Procedural Knowledge: Belenky's fourth phase of women's thinking; characterized by a distrust of both knowledge from authority and the female thinker's own inner authority or "gut" (Ch. 16).

Prodigiousness: The ability to do *qualitatively* better than the rest of us are able to do; such a person is referred to as a prodigy (Ch. 12).

Prosocial Behavior: Refers to behaviors such as friendliness, self-control, and being helpful (Ch. 10).

Prostatectomy: The removal of all or part of the male prostate gland (Ch. 21).

Protective Factors: Characteristics of **resilient** persons that protect them from the problems related to stress (Ch. 15).

Psychoanalysis: Sigmund Freud's explanation of the psychic development of humans. Also his method of psychological therapy (Ch. 2).

Psychoanalytic Theory: Freud's theory of the development of the personality (Ch. 2).

Psychoanalytic Theory of Homosexuality: Freud's theory suggests that if the child's first sexual feelings about the parent of the opposite sex are strongly punished, the child may identify with the same-sex parent and develop a permanent homosexual orientation (Ch. 14).

Psychogenic Model: Teenagers form a subculture because they are avoiding or escaping from reality (Ch. 11).

Psycholinguistic: Language theory that attempts to identify the psychological mechanisms by which individuals learn their native language (Ch. 5).

Psychological Dimension: The second stage of Frankl's theory of human development, in which personality begins to form at birth and develops as a result of instincts, drives, capacities, and interactions with the environment. This and the **somatic dimension** are highly developed by the time the individual reaches early adulthood (Ch. 22).

Psychopathology: Mental illness, relatively rare during adolescence (Ch. 13).

Psychosocial Theory: Erikson's theory of the development of the personality (Ch. 2).

Puberty: A relatively abrupt and qualitatively different set of physical changes that normally occur at the beginning of the teen years (Ch. 12).

Puberty Rites: An initiation ceremony often scheduled to coincide with the peak in adolescent physiological maturation (Ch. 15).

Punishment: In behavioral theory, any action that makes a behavior less likely to happen. Differs from extinction in that an action is taken (Ch. 2).

Q

Quiescence: A condition in which an individual has no needs at all (Ch. 15).

R

Random Scribbling: Drawing in which children use dots and lines with simple arm movements (Ch. 7).

Reading Comprehension: Understanding the concepts and relationships that words represent (Ch. 9).

Realism: Children distinguish and accept the real world surrounding them, meaning that they now have identified an external as well as internal world (Ch. 7).

Real versus the Possible: The growing ability of the adolescent to imagine possible and even impossible situations (Ch. 12).

Received Knowledge: Belenky's second phase of women's thinking; characterized by being awed by the authorities, but far less affiliated with them (Ch. 16).

Recessive: A **gene** whose trait is not expressed unless paired with another recessive gene; for example, both parents contribute genes for blue eyes (Ch. 3).

Recidivism Rates: The percentage of convicted persons who commit another crime once they are released from prison (Ch. 15).

Reciprocal Interactions: We respond to those around us and they change; their responses to us thus change and we in turn change (Ch. 1); similar to **transactional model**; recognizes the child's active role in its development; "I do something to the child, the child changes; as a result of the changes in the child, I change" (Ch. 5).

Reconciliation Fantasies: When children wish their parents could get together again following divorce (Ch. 8).

Reflective Listening: A method of talking to others; you rephrase the person's comments to show you understand (Ch. 14).

Reflex: When a stimulus repeatedly elicits the same response (Ch. 5).

Reinforcement: Increasing the probability that a response will recur under similar conditions (Ch. 8).

Relationship: A pattern of intermittent interactions between two individuals over a period of time (Ch. 6).

Relativism: The second phase in Perry's theory. An attitude or philosophy that says anything can be right or wrong depending on the situation; all views are equally right (Ch. 16).

Representation: Recording or expressing information in a manner different from the original; the word *auto* represents the actual car (Ch. 7).

Reproductive Image: Mental images that are faithful to the original object or event being represented; Piaget's term for images that are restricted to those sights previously perceived (Ch. 7).

Repudiation: Choosing an identity involves rejecting other alternatives (Ch. 11).

Resiliency: *See* plasticity (Ch. 10).

Respiratory Distress Syndrome (RDS): This problem is most common with prematures, but it may strike full-term infants whose lungs are particularly immature; RDS is caused by the lack of a substance called *surfactant,* which keeps the air sacs in the lungs open (Ch. 5).

Retreat: According to Perry's theory of ethical development, when someone retreats to an earlier ethical position (Ch. 16).

Rh Factor: Term describing Rh incompatibility, an incompatibility between the blood types of mother and child; if the mother is Rh-negative and the child Rh-positive, miscarriage or even infant death can result (Ch. 4).

Risk Factors: Characteristics of individuals who are prone to suffer serious problems when under stress (Ch. 15).

Role Discontinuity: Abrupt and disruptive change caused by conflicts among one's various roles in life (Ch. 21).

S

Safety Needs: Refers to the importance of security, protection, stability, freedom from fear and anxiety, and the need for structure and limits. One level of Maslow's need hierarchy (Ch. 1).

Schemata: Patterns of behavior that we use to interact with the environment (Ch. 2).

Secondary Circular Reactions: Infants direct their activities toward objects and events outside themselves (Ch. 5).

Secular Trend: The phenomenon (in recent centuries) of adolescents entering puberty sooner and growing taller and heavier (Ch. 12).

Self-Actualization Needs: Refers to the tendency to feel restless unless we are doing what we think we are capable of doing. One level of Maslow's need hierarchy (Ch. 1).

Self-fulfilling Prophecy: Making an idea come true simply by believing it will (Ch. 18).

Semantics: The rules of semantics describe how to interpret the meaning of words (Ch. 7).

Senile Macular Degeneration: This disease of the retina is a leading cause of blindness, beginning as blurred vision and a dark spot in the center of the field of vision. Advances in laser surgery have shown promise in treating diseases like this (Ch. 18).

Sensitive Periods: Certain times in the lifespan when a particular experience has a greater and more lasting impact than at another time (Ch. 1).

Sensitive Responsiveness: Refers to the ability to recognize the meaning of a child's behavior (Ch. 6).

Sequential (Longitudinal/Cross-Sectional) Studies: A cross-sectional study done at several times with the same groups of individuals (Ch. 1).

Seriation: Concrete operational children can arrange objects by increasing or decreasing size (Ch. 9).

Sex Cleavage: The custom that youngsters of the same sex tend to play together (Ch. 8).

Sex-Linked Inheritance: Sex-linked inheritance is explained by the fact that the female carries more **genes** on the twenty-third **chromosome** (Ch. 3).

Sexual Identity: Sexual identity results from those *physical characteristics* and *behaviors* that are part of our biological inheritance. They are the traits that make us males or females. Examples of sex-linked physical characteristics are the penis and testes of the male. A corresponding behavior is the erection of the penis when stimulated (Ch. 14).

Sexual Revolution: The extraordinary change in human sexual behavior that occurred in the 1960s and 1970s (Ch. 14).

Sexually Transmitted Diseases (STD): A class of diseases, such as **AIDS, gonorrhea, herpes,** and **chlamydia,** that are transmitted through sexual behavior (Ch. 14).

Sibling: A brother or sister (Ch. 10).

Sickle-Cell Anemia: A chromosomal disorder resulting in abnormal hemoglobin (Ch. 3).

Silence: Belenky's first phase of women's thinking, characterized by concepts of right and wrong; similar to the thinking of men in Perry's first stage (Ch. 16).

Skeletal Growth: The development of the bone structure in the body (Ch. 12).

Sleeping Disorders: Problems with a child's sleeping pattern, such as an inability to sleep or sleepwalking (Ch. 5).

Slow-to-Warm-Up Children: Children whose reactions are initially mildly negative but then show slow adaptation (Ch. 6).

Social Death: That point at which a patient is treated essentially as a corpse, although perhaps still "clinically" or "biologically" alive (Ch. 22).

Social Learning: Another name for **observational** learning; children (and all of us) learn from watching others. The term is associated with Albert Bandura (Ch. 8).

Social Perspective-Taking: Children's views on how to relate to others emerge from their personal theories about the traits of others (Ch. 10).

Socialization: The need to establish and maintain relations with others and to regulate behavior according to society's demands (Ch. 8).

Solidarity: Erikson's term for the personality style of persons who are able to commit themselves in concrete affiliations and partnerships with others, and have developed the "ethical strength to abide by such commitments, even though they may call for significant sacrifices and compromises" (Ch. 17).

Solidification: Cognitive growth in which the thinker is more certain and confident in the use of newly gained mental skills and is more likely to use them in new situations (Ch. 12).

Somatic Dimension: The first stage of Frankl's theory of human development, in which all persons are motivated by the struggle to keep themselves alive and to help the species survive. This intention is motivated entirely by instincts. It exists at birth and continues throughout life (Ch. 22).

Spatial Intelligence: One of Gardner's multiple intelligences (Ch. 9).

Sperm: The germ cell that carries the male's 23 **chromosomes** (Ch. 3).

Spina Bifida: A genetic disorder resulting in the failure of the neural tube to close (Ch. 3).

Stability: A belief that children's early experiences affect them for life (Ch. 1).

Stage of Exhaustion: The third stage in Selye's theory of **stress,** caused by a gradual depletion of the organism's adaptational energy. The physiological responses revert to this condition during the stage of alarm. The ability to handle the stress decreases, the level of resistance is lost, and the organism dies (Ch. 15).

Stage of Resistance: The second stage in Selye's theory of **stress.** If the organism survives the initial alarm, an almost complete reversal of the alarm reaction occurs. Swelling and shrinking are reversed, the adrenal cortex that lost its secretions during the alarm stage becomes unusually rich in these secretions, and a number of other shock-resisting forces are marshalled. During this stage, the organism appears to gain strength and to have adapted successfully to the stressor (Ch. 15).

Stage Theorists: Researchers who believe that research based on personality traits is too narrow in focus, and that we must also look at the stages of change each person goes through (Ch. 19).

Stagnation: According to Erikson, the seventh stage of life (middle-aged adulthood) tends to be marked either by **generativity** or by stagnation—boredom, self-indulgence, and the inability to contribute to society (Ch. 19).

Stanford-Binet Intelligence Test: Usually refers to the intelligence test originally designed by Alfred Binet, then brought to America and revised on subjects near Stanford University (Ch. 9).

State of Identity: According to Erikson, the main goal of adolescence (Ch. 11).

Stereotype: Conventional, oversimplified belief about how a person should behave in a particular situation. For example, women should console a crying child, but men should not (Ch. 11).

Stillbirth: After the twentieth week, the spontaneous end of a pregnancy is called a *stillbirth* if the baby is born dead (Ch. 4).

Stimulus Reduction: Freud's notion that human beings try to avoid stimulation whenever possible. According to this idea, all our activities are attempts to eliminate stimulation from our lives (Ch. 15).

Storm and Stress: Hall's description of the rebirth that takes place during adolescence (Ch. 11).

Stress: Anything that upsets our equilibrium—both psychological and physiological (Ch. 10).

Structure Building: Levinson's term. During structure-building periods, individuals face the task of building a stable structure around choices they have made. They seek to enhance the life within that structure. This period of relative stability usually lasts five to seven years (Ch. 17).

Structure Changing: Levinson's term. A process of reappraising the existing life structure and exploring the possibilities for new life structures characterizes the structure-changing period. This period usually lasts for around five years. Its end is marked by the making of critical life choices around which the individual will build a new life structure (Ch. 17).

Subculture: A group of people who have social, economic, ethnic, or age characteristics distinctive enough to distinguish them from others within the same culture or society (Ch. 11).

Subjective Knowledge: Belenky's third phase of women's thinking; characterized by some crisis of male authority that sparked a distrust of outside sources of knowledge, and some experience that confirmed a trust in women thinkers themselves (Ch. 16).

Sudden Infant Death Syndrome (SIDS): About 10,000 2- to 4-month-old infants die each year from this syndrome; thought to be brain-related (Ch. 5).

Symbolic Play: The game of pretending; one of five preoperational behavior patterns (Ch. 7).

Symmetry: An infant's capacity for attention and style of responding influence any interactions (Ch. 6).

Synchrony: The ability of parents to adjust their behavior to that of an infant (Ch. 6).

Suggestibility: Adolescent behavior resulting from the perceived wishes of others. Some adolescents may attempt suicide out of the perception, true or not, that their parents wish them dead (Ch. 13).

Superego: One of the three structures of the psyche, according to Freud (Ch. 2).

Syntax: The rules of syntax describe how to put words together to form sentences (Ch. 7).

Syphilis: A **sexually transmitted disease,** which presents a great danger in that in its early stage there are no symptoms. Its first sign is a chancre ("shanker"), a painless open sore that usually shows up on the tip of the penis and around or in the vagina. After a while this sore disappears, and if no treatment follows, the disease enters its third stage, which is usually deadly (Ch. 14).

System Kids: Children who have been removed from their families and placed in orphanages or foster homes for so long they are described as belonging to the "system," in other words, the state government. Often these children are runaways (Ch. 13).

T

Tay-Sachs Disease: A fatal disease in which a **gene** fails to produce proper enzyme action (Ch. 3).

Telegraphic Speech: Initial multiple-word utterances, usually two or three words (Ch. 5).

Temperament: A child's basic personality, which is now thought to be discernible soon after birth (Ch. 6).

Temporizing: An aspect of Perry's theory of ethical development, in which some people remain in one position for a year or more, exploring its implications but hesitating to make any further progress (Ch. 16).

Teratogens: Any agents that can cause abnormalities, including drugs, chemicals, infections, pollutants, and the mother's physical state (Ch. 4).

Terminal Drop: The period of from a few weeks up to two years prior to a person's death, during which his or her intelligence is presumed to decline rapidly (Ch. 20).

Tertiary Circular Reaction: Repetition with variation; the infant is exploring the world's possibilities (Ch. 5).

Thalidomide: During the early 1960s, thalidomide was a drug popular in West Germany as a sleeping pill and an anti-nausea measure that produced no adverse reactions in pregnant women. In 1962, physicians noticed a sizeable increase in children born with either partial or no limbs. Feet and hands were directly attached to the body. Other outcomes were deafness, blindness, and occasionally mental retardation. Investigators discovered that the mothers of these children had taken thalidomide early in their pregnancy (Ch. 4).

Total Range: The behaviorist's explanation of attachment that focuses on all the reinforcements provided by the environment (Ch. 6).

Traditional Marriage Enterprise: Levinson's term. The main goal of this type of marriage is to form and maintain a family (Ch. 19).

Trait Theorists: Researchers who look at pieces of the personality (personality traits), as measured by detailed questionnaires (Ch. 19).

Transactional Model of Development: Similar to **reciprocal interactions**; recognizes the child's active role in its own development; "I do something to the child, the child changes; as a result of the changes in the child, I change" (Ch. 1, 6).

Transience: Toffler's term for the lack of permanence of things in our lives that leads to increased **stress** (Ch. 15).

Transition: Levinson's concept that each new era begins as an old era is approaching its end. That "in-between" time is a transition (Ch. 17).

Transsexual Operation: An operation that changes the physical characteristics of an individual to those of the opposite sex (Ch. 14).

Triarchic Model: A three-tier explanation of intelligence proposed by Robert Sternberg (Ch. 9).

Trophoblast: The outer surface of the fertilized egg (Ch. 4).

Type T Personalities: (T for Thrills)—Adolescents who seem to have inherited a proneness to taking risks (Ch. 1).

U

Ultrasound: The use of sound waves to produce an image that enables a physician to detect structural abnormalities (Ch. 4).

Umbilical Cord: A cord that connects the mother and fetus. Contains arteries and veins that supply the fetus with blood (Ch. 4).

Universalizing Faith: The final developmental step of Fowler's theory of faith. Here the individual lives in the real world, but is not of it. Such persons do not merely recognize the mutuality of existence; they act on the basis of it (Ch. 22).

Use of Metaphor: The ability to think of a word or phrase that by comparison or analogy can be used to stand for another word or phrase (Ch. 12).

Validation: Fromm's term for the main ingredient of love (Ch. 16).

Variable Accommodation: Focusing on objects at various distances; appears at about 2 months (Ch. 5).

Verbal Evocation: Using language to indicate somebody or something not present or occurring at that time (Ch. 7).

Vocables: Consistent sound patterns to refer to objects and events (Ch. 5).

Vocabulary: Knowing the meaning of words, not just how to pronounce them (Ch. 9).

W

WAIS: The Wechsler Adult Intelligence Scale, an intelligence test devised by David Wechsler (Ch. 9).

Walkabout: Originally an Aborigine initiation rite, the American version attempts to focus the activities of secondary school by demonstrating to the student the relationship between education and action (Ch. 15).

Wear and Tear Theory: The theory that aging is due to the cumulative effects of hard work and lifelong **stress** (Ch. 20).

Whole Language: Students learn to read by obtaining the meaning of words from context, with phonics being introduced when needed (Ch. 10).

WISC-R: The Wechsler Intelligence Scale for Children (revised), a children's intelligence test devised by David Wechsler (Ch. 9).

WPPSI-R: The Wechsler Preschool and Primary Scale of Intelligence (revised), an intelligence scale for children age 4 to 6 1/2 years, devised by David Wechsler (Ch. 9).

Z

Zone of proximal development: The distance between a child's actual developmental level and a higher level of potential development that can be attained with adult guidance (what children can do independently and what they can do with help) (Ch. 2).

Zygote: The fertilized egg (Ch. 3).

Zygote Intrafallopian Transfer (ZIFT): The fertilized **egg** (the zygote) is transferred to the **fallopian tube** (Ch. 3).

References

Abeshouse, R. P. (1987). *Lifelong learning, Part I: Education for a competitive economy*. Washington, DC: Roosevelt Center for American Policy Studies.

Ackerman, J. W. (1958). *The psychodynamics of family life*. New York: Basic Books.

Adalbjarnardottir, S. & Selman, R. (1989). How children propose to deal with the criticisms of their teachers and classmates: Developmental and stylistic variations. *Child Development, 60,* 539–50.

Adams, B. N. (1968). *Kinship in an urban setting*. Chicago: Markham.

Adams, C. Labouvie-Vief, G., Hobart, C. J., & Dorsey, M. (1990). Adult age group differences in story recall style. *Journal of Gerontology, 45*(1), 17–27.

Adams, J. (1986). *Conceptual blockbusting* (3rd ed.). Reading, MA: Addison-Wesley.

Adelman, P. (1987). Occupational complexity, control, and personal income: Their relation to psychological well-being in men and women. *Journal of Applied Psychology, 72,* 529–37.

Ainsworth, M. D. S. (1973). The development of infant-mother attachment. In B. Caldwell & H. Ricciuti (Eds.), *Review of child development research*. Chicago: University of Chicago Press.

Ainsworth, M. D. S. (1979). Infant-mother attachment. *American Psychologist, 34,* 932–37.

Ainsworth, M. D. S. & Bowlby, J. (1991). An ethological approach to personality. *American Psychologist, 46,* 333–41.

Alban, D.F. (1978, June 5). A better way of dying. *Time,* 66.

Alexander, A. & Kempe, R. S. (1982). The role of the lay therapist in long-term treatment. *Child Abuse and Neglect, 6*(30), 329–34.

Alexander, T. (1987). *Make room for twins*. New York: Bantam Books.

Allan, G. (1977). Sibling solidarity. *Journal of Marriage and the Family, 39,* 177–84.

Alpaugh, P. & Birren, J. (1977). Variables affecting creative contributions across the adult life-span. *Human Development, 20,* 240–48.

Alpaugh, P., Renner, V., & Birren, J. (1976). Age and creativity. *Educational Gerontology. E.G., 1,* 17–40.

Amabile, T. M., Hennessey, B. A., & Grossman, B. S. (1986). Social influences on creativity. *Journal of Personality and Social Psychology, 50,* 14–24.

Amato, P. R. (1987). Maternal employment: Effects on children's family relationships and development. *Australian Journal of Sex, Marriage and Family, 8*(1), 5–16.

American Association of Retired Persons. (1990a, January). FDA warns against dietary supplement. *A. A. R. P., 31*(1), 7.

American Association of Retired Persons. (1990b, January). Study links alcohol to heart damage. *A. A. R. P., 31*(1), 7.

American Heart Association. (1984). *Eating for a healthy heart: Dietary treatment of hyperlipidemia*. Dallas: American Heart Association.

American Psychiatric Association. (1985). *Diagnostic and statistical manual of mental disorders* (3rd ed.). Washington, DC: Author.

Anderson, A. M. & Noesjirwan, J. A. (1980). Agricultural college initiations and the affirmation of rural ideology. *Mankind 12*(4), 341–47.

Angle, C. (1983). Adolescent self-poisoning: A nine-year follow-up. *Developmental and Behavior Pediatrics, 4*(2), 83–87.

Anselmo, S. (1987). *Early childhood development*. Columbus, OH: Merrill.

Apgar, V. (1953). A proposal for a new method of evaluation of the newborn infant. *Anesthesia and Analgesia, 32,* 260–67.

Arendell, T. J. (1987). Women and the economics of divorce in the contemporary United States. *Signs, 13,* 121–35.

Aries, P. (1962). *Centuries of childhood*. New York: Knopf.

Arieti, S. & Bemporad, J. (1978). *Severe and mild depression*. New York: Basic Books.

Ashton, J. & Donnan, S. (1981). Suicide by burning as an epidemic phenomenon: An analysis of 82 deaths and inquests in England and Wales in 1978–9. *Psychological Medicine, 11*(4), 735–39.

Asian-American Drug Abuse Program. (1978). *Pacific Asians and alcohol.* Los Angeles: Author.

Aslin, R. (1987). Visual and auditory development in infancy. In J. Osofsky (Ed.), *Handbook of infant development*. New York: Wiley.

Avery, M. E. & Litwack, G. (1983). *Born early*. New York: Little, Brown.

Baier, J. L. & Williams, P. S. (1983, July). Fraternity hazing revisited: Current alumni and active member attitudes toward hazing. *Journal of College Student Personnel, 24*(4), 300–5.

Baillargeon, R. (1987). Object permanence in 3 1/2 and 4 month old infants. *Developmental Psychology, 23,* 655–64.

Baker, S. & Henry, R. (1987). *Parents' guide to nutrition*. Reading, MA: Addison-Wesley.

Baker, S. W. (1980, Autumn). Biological influence on human sex and gender. *Signs, 17,* 80–96.

Baltes, P. B., Reese, H. W., & Nesselroade, J. R. (1977). *Life-span developmental psychology: Introduction to research methods.* Monterey, CA: Brooks/Cole.

Baltes, P. B. & Schaie, K. W. (1976). On the plasticity of intelligence in adulthood and old age—Where Horn and Donaldson fail. *American Psychologist, 31,* 720–25.

Baltes, P. B., Sowarka, D., & Kliegl, R. (1989). Cognitive training research on fluid intelligence in old age: What can older adults achieve by themselves. *Psychology and Aging, 4*(2), 217–21.

Bancroft, L. (1979). The reasons people give for taking overdoses: A further inquiry. *British Journal of Medical Psychology, 52,* 353–65.

Bandura, A. (1986). *Social foundations of thought and action.* Englewood Cliffs, NJ: Prentice-Hall.

Bandura, A., Ross, D., & Ross, S. (1963). Imitation of film-mediated aggressive models. *Journal of Abnormal and Social Psychology, 66,* 3–11.

Bandura, A. & Walters, R. (1963). *Social learning and personality development*. New York: Holt, Rinehart and Winston.

Banks, S. & Kahn, M. (1982). *The sibling bond*. New York: Basic Books.

Barfield, R. & Morgan, J. (1978). Trends in planned early retirement. *Gerontologist, 18,* 13–18.

Barnicle, M. (August 3, 1992). The spread of gangs. *The Boston Globe,* 34.

Baruch, G., Barnett, R., & Rivers, C. (1983). *Lifeprints: New patterns of love and work of today's women.* New York: McGraw-Hill.

Baumrind, D. (1967). Child-care practices anteceding three patterns of preschool behavior. *Genetic Psychology Monographs, 75,* 43–88.

Baumrind, D. (1971). Current patterns of parental authority. *Developmental Psychology Monographs, 4,* 1–103.

Baumrind, D. (1986). *Familial antecedents of social competence in middle childhood.* Unpublished manuscript.

Beaconsfield, P., Birdwood, G., & Beaconsfield, R. (1983). The placenta. *Scientific American, 34,* 94–102.

Beck, A. T. (1967). *Depression: Clinical, experimental, theoretical aspects*. New York: Hoeber Medical Division, Harper & Row.

Becoming a Nation of Readers: The Report of the Commission on Reading. (1985). Washington: National Academy of Education, National Institute of Education, and Center for the Study of Reading.

Beidelman, T. (1971). *The Kaguru: A matrilineal people of East Africa.* New York: Holt, Rinehart & Winston.

Belbin, E. & Belbin, R. M. (1968). New careers in middle age. In B. L. Neugarten (Ed.), *Middle age and aging*. Chicago: University of Chicago Press.

Belenky, M., Clinchy, B., Goldberger, N., & Tarule, J. (1986). *Women's ways of knowing*. New York: Basic Books.

Bellah, R. (1978). To kill and survive or to die and become. In E. Erikson (Ed.), *Adulthood* (pp. 61–80). New York: W. W. Norton.

Belsky, J. & Braungart, J. (1991). Are insecure-avoidant infants with extensive daycare experience less stressed by and more independent in the strange situation? *Child Development, 62,* 567–71.

Belsky, J. & Rovine, M. (1988). Nonmaternal care in the first year of life and the security of infant-parent attachment. *Child Development, 59,* 157–67.

Bem, A. (1987). Youth suicide. *Adolescence, 22*(86), 271–90.

Bem, S. (1983). Gender schema theory and its implications for child development: Raising gender-aschematic children in a gender-schematic society. *Signs, 8,* 598–616.

Bem, S. L. (1975). Androgyny vs. the light little lives of fluffy women and chesty men. *Psychology Today, 9*(4), 58–59, 61–62.

Beneke, W. & Timson, B. (1987). Some health risk benefits of behavioral weight-loss treatments. *Psychological Reports, 61*(1), 199–206.

Bengston, V. L. & Robertson, J. L. (Eds.). (1985). *Grandparenthood.* Beverly Hills, CA: Sage.

Benson, H. & Proctor, W. (1985). *Beyond the relaxation response.* New York: Berkley.

Berg, W. K. & Berg, K. (1987). Psychophysiological development in infancy: State, startle, and attention. In Joy Osofsky (Ed.), *Handbook of infant development.* New York: Wiley.

Bergman, M. (1980). *Aging and the perception of speech.* Baltimore, MD: University of Baltimore Press.

Bergman, M., Blumenfield, V. G., Cascardo, D., Dash, B., Levitt, H., & Margulios, M. K. (1976). Age-related decrements in hearing for speech: Sampling and longitudinal studies. *Journal of Gerontology, 31,* 533–38.

Berk, L. (1992). *Child development.* Boston: Allyn & Bacon.

Berk, S. F. (1985). *The gender factory: The apportionment of work in American households.* New York: Plenum Press.

Berman, M. (1975, March 30). [Review of *Life history and the historical movement* by E. Erikson.] *New York Times Magazine,* 2.

Berman, P. L. & Goldman, C. (1992). *The ageless spirit.* New York: Ballantine Books.

Berndt, T. (1992). *Child development.* New York: Harcourt, Brace, Jovanovich.

Bettelheim, B. (1969). *The children of the dream.* New York: Macmillan.

Bettelheim, B. (1976). *The uses of enchantment.* New York: Knopf.

Binet, A. & Simon, T. R. (1905). The development of intelligence. *L'Annee Psycholique,* 163–91.

Bing, L. (1991). *Do or die.* New York: Harper/Collins.

Birren, J. (1960). Behavioral theories of aging. In N. Shock (Ed.), *Aging.* Washington, DC: American Association for the Advancement of Science.

Birren, J. (1964). *The psychology of aging.* Englewood Cliffs, NJ: Prentice-Hall.

Birren, J., Cunningham, W., & Yamamoto, K. (1983). Psychology of adult development and aging. In M. Rosenzweig & L. Porter (Eds.), *Annual review of psychology.* Palo Alto, CA: Annual Reviews.

Birren, P., Renner, V., & Birren, J. (1976). Age and creativity. *Educational Gerontology, 1,* 17–40.

Birtchnell, J. & Alarcon, J. (1971). The motivation and emotional state of 91 cases of attempted suicide. *British Journal of Medical Psychology, 44,* 45–52.

Bishop, J. E. (1983, December 23). New gene probes may permit early predictions of disease. *Wall Street Journal,* 11.

Bishop, J. E. & Waldholz, M. (1990). *Genome.* New York: Simon & Schuster.

Black, A. (1974). *Without burnt offerings.* New York: Viking.

Blackman, J. (1984). *Medical aspects of developmental disabilities in children birth to three.* Rockville, MD: Aspen Publications.

Blasband, D. & Peplau, L. H. (1985). Sexual exclusivity versus openness in gay male couples. *Archives of Sexual Behavior, 14*(5), 395–412.

Block, J. (1983). Differential premises arising from differential socialization of the sexes: Some conjectures. *Child Development, 54,* 1335–54.

Bloom, B. (Ed.) (1956). *Taxonomy of educational objectives.* New York: McKay.

Bloom, B. (1964). *Stability and change in human characteristics.* New York: Wiley.

Bloom, B. (1985). *Developing talent in young people.* New York: McGraw-Hill.

Blumenthal, S. J. & Kupfer, D. J. (1988). Overview of early detection and treatment strategies for suicidal behavior in young people. *Journal of Youth and Adolescence, 17*(1), 1–24.

Blythe, P. (1973). *Stress and disease.* London: Barker.

Boas, F. (1911). Growth. In H. Kiddle (Ed.), *A cyclopedia of education.* New York: Steiger.

Bohling, H. R. (1991). Communication with Alzheimer's patients: An analysis of caregiver listening patterns. *International Journal of Aging and Human Development, 33*(4), 249–67.

Boivin, M. & Begin, G. (1989). Peer status and self-perception among early elementary school children: The case of the rejected children. *Child Development, 60,* 591–96.

Bollen, K. & Phillips, D. (1982). Imitative suicides: A national study of the effects of television news stories. *American Sociological Review, 47,* 802–9.

Bolton, C. & Camp, D. J. (1989). The post-funeral ritual in bereavement counseling and grief work. *Journal of Gerontological Social Work, 13,* 49–59.

Bornstein, J. M. (1986). Retraining the older worker: Michigan's experience with senior employment services. Special issue: Career counseling of older adults. *Journal of Career Development, 13*(2), 14–22.

Boston Women's Health Book Collective. (1976). *Our bodies, ourselves* (2nd ed.). New York: Simon & Schuster.

Botwinick, J. (1977). Intellectual abilities. In J. E. Birren & K. W. Schaie (Eds.), *Handbook of the psychology of aging.* New York: Van Nostrand Reinhold.

Botwinick, J. (1981). *Aging and behavior* (2nd ed.). New York: Springer.

Botwinick, J. (1984). *Aging and behavior.* New York: Springer.

Bourne, R. (1913). *Youth and life.* Boston: Little, Brown.

Bove, R. (1987). Retraining the older worker. *Training and Development Journal, 41*(3), 77–78.

Bowe, E. (1988). The pregnant body. In D. Tapley and W. Todd (Eds.), *Complete guide to pregnancy.* New York: Crown.

Bowlby, J. (1969). *Attachment.* New York: Basic Books.

Bowlby, J. (1973). *Separation: Anxiety and anger.* New York: Basic Books.

Bowlby, J. (1982). Attachment and loss: Retrospect and prospect. *American Journal of Orthopsychiatry, 52,* 664–78.

Bowman, L. (1959). *The American funeral: A study in guilt, extravagance, and sublimity.* Washington: Public Affairs Press.

Brabeck, M. (1983). Moral judgment: Theory and research on differences between males and females. *Developmental Review, 3,* 274–91.

Brabeck, M. (1984). Longitudinal studies of intellectual development during adulthood. *Journal of Research and Development in Education, 17*(3), 12–25.

Bradway, K., Thompson, C., & Cravens, R. (1958). Preschool IQs after 25 years. *Journal of Educational Psychology, 49,* 278–81.

Brady, B. A. & Gray, D. D. (1988). Employment services for older job seekers. *The Gerontologist, 27,* 565–68.

Brain, J. L., Blake, C. F., Bluebond-Langner, M., Chilungu, S. W., Coelho, V. P., Domotor, T., Gorer, G., LaFontaine, J. S., Levy, S. B., Loukotos, D., Natarajan, N., Raphael, D., Schlegel, A., Stein, H. G., & Wilder, W. D. (1977). Sex, incest, and death: Initiation rites reconsidered. *Current Anthropology, 18*(2), 191–98.

Bransford, J. & Stein, B. (1984). *The IDEAL problem solver.* New York: W. H. Freeman.

Brazelton, T. B. (1981). *On becoming a family: The growth of attachment.* New York: Delacorte.

Brazelton, T. B. (1984). *Neonatal behavioral assessment scale.* London: Heinemann.

Brazelton, T. B. (1987). *Working and caring.* Reading, MA: Addison-Wesley.

Brazelton, T. B. (1990a). Saving the bathwater. *Child Development, 61,* 1661–71.

Brazelton, T. B. & Als, H. (1979). Four early stages in the development of mother-infant interaction. *The Psychoanalytic Study of the Child, 34,* 349–69.

Brazelton, T. B. & Cramer, B. (1990b). *The earliest relationship.* Reading, MA: Addison-Wesley.

Brenner, A. (1984). *Helping children cope with stress.* San Diego, CA: Lexington Books.

Bretschneider, J. G. & McCoy, N. L. (1988). Sexual interest and behavior in healthy 80 to 102 year olds. *Archives of Sexual Behavior, 17*(2), 109–29. New York: Plenum Press.

Bridges, K. (1930). A genetic theory of the emotions. *Journal of Genetic Psychology, 37,* 514–27.

Brim, O. G. & Kagan, J. (Eds.) (1980). *Constancy and change in human development.* Cambridge, MA: Harvard University Press.

Brim, O. S. (1958). Family structure and sex role learning by children: A further analysis of Kler Koch's data. *Sociometry, 21,* 1–16.

Brittain, W. L. (1979). *Creativity, art, and the young child.* New York: Macmillan.

Brody, E. M., Hoffman, C., Kleban, M. H., & Schoonover, C. B. (1986). Caregiving daughters and their local siblings: Perceptions, strains, and interactions. *The Gerontologist, 29*(4), 529–38.

Brody, E. M., Litvin, S. J., & Hoffman, C. (1992). Differential effects of daughters' marital status on their parent care experiences. *The Gerontologist, 32,* 58–67.

Brody, E. M. & Schoonover, C. B. (1986). Patterns of parent-care when adult daughters work and when they do not. *The Gerontologist, 26,* 372–81.

Brody, G., Stoneman, Z., McCoy, J., & Forehand, R. (1992). Contemporaneous and longitudinal associations of sibling conflict with family relationship assessments and family discussions about sibling problems. *Child Development, 63,* 391–400.

Brody, J. E. (1984, January 15). The growing militance of the nation's nonsmokers. *New York Times,* 31.

Bronfenbrenner, U. (1977). Nobody home: The erosion of the American family. *Psychology Today, 10*(12), 40.

Bronfenbrenner, U. (1978). *The ecology of human development.* Cambridge, MA: Harvard University Press.

Bronfenbrenner, U. (1989). Ecological systems theory. *Annals of Child Development, 6,* 187–249.

Bronfenbrenner, U. & Crouter, M. (1983). The evolution of environmental models in developmental research. In P. Mussen (Ed.), *Handbook of child psychology.* New York: Wiley.

Bronson, M., Pierson, D., & Tivnan, T. (1984). The effects of early education on children's competence in elementary school. *Evaluation Review, 8*(5), 615–27.

Brooks, J. & Lewis, M. (1976). Midget, adult and child: Infants' responses to strangers. *Child Development, 47,* 323–32.

Brooks-Gunn, J. (1987). Pubertal processes. In V. B. Van Hasselt & M. Hersen (Eds.), *Handbook of adolescent psychology.* New York: Pergamon.

Brooks-Gunn, J., Peterson, A., & Eichorn, D. (1985). The study of maturational timing effects in adolescence. *Journal of Youth and Adolescence, 14*(3), 149–61.

Brown, B. B. (1990). Peer groups and peer cultures. In S. Feldman and G. Elliot (Eds.), *At the threshold: The developing adolescent.* Cambridge, MA.: Harvard University Press.

Brown, G. W., Bhrolchain, M. N., & Harris, R. (1975). Social class and psychiatric disturbance among women in an urban population. *Sociology, 9,* 225–54.

Brown, J. K. (1982). Cross-cultural perspectives on middle-aged women. *Current Anthropology, 23*(2), 143–56.

Brown, J., Childers, K., & Waszak, C. (1990). Television and adolescent sexuality. *Journal of Adolescent Health Care, 11,* 62–70.

Brown, M. (1979). Teenage prostitution. *Adolescence, 14*(56), 665–80.

Brown, R. (1988). Model youth: Excelling despite the odds. *Ebony, 43,* 40–48.

Bruch, H. (1981). *Eating disorders.* Canada: Basic Books.

Buchanan, C., Maccoby, E., & Dornbusch, S. (1991). Caught between parents: Adolescents' experience in divorced homes. *Child Development, 62,* 1008–29.

Buehler, C. A., Hogan, M. J., Robinson, B. E., & Levy, R. J. (1985–86). The parental divorce transition: Divorce-related stressors and well-being. *Journal of Divorce,9*(2), 61–81.

Buffum, J. (1988). Substance abuse and high-risk sexual behavior: Drugs and sex—the dark side. *Journal of Psychoactive Drugs, 20*(2), 165–68.

Bukatko, D. & Daehler, M. (1992). *Child development.* Boston: Houghton Mifflin.

Burgess, A. W. (1986). *Youth at risk: Understanding runaway and exploited youth.* Washington, DC: National Center for Missing and Exploited Children.

Burke, T. (1990). Home invaders: Gangs of the future. *The Police Chief, 57,* 23.

Burnham, W. (1911). Hygiene and adolescence. In H. Kiddle (Ed.), *A cyclopedia of education.* New York: Steiger.

Burton, C. A. (1978). *Juvenile street gangs: Predators and children.* Unpublished manuscript, Boston College.

Buskirk, E. R. (1985). Health maintenance and longevity: Exercise. In C. E. Finch & E. L. Schneider (Eds.), *Handbook of the biology of aging* (2nd ed.). New York: Van Nostrand Reinhold.

Butler, R. N. & Lewis, M. I. (1977). *Aging and mental health* (2nd ed.). St. Louis: Mosby.

Buunk, B. & vanDriel, B. (1989). *Variant lifestyles and relationships.* Newbury Park, CA: Sage.

Cahill, M. & Salomone, P. R. (1987). Career counseling for work life extension: Integrating the older worker into the labor force. *Career Development Quarterly, 35* (3), 188–96.

Campbell, A. (1975). The American way of mating: Marriage si; children only maybe. *Psychology Today, 8,* 37–43.

Campbell, A. (1987). Self definition by rejection: The case of gang girls. *Social Problems, 34,* 451–66.

Campbell, B. & Gaddy, J. (1987). Rates of aging and dietary restrictions: Sensory and motor function in the Fischer 344 rat. *Journal of Gerontology, 42*(2), 154–59.

Campbell, S. (1979). Delayed mandatory retirement and the working woman. *The Gerontologist, 19,* 257–63.

Campione, W. A. (1987). The married woman's retirement decision: A methodological comparison. *Journal of Gerontology, 42*(4), 381–86.

Campos, J. (1976). Heart rate: A sensitive tool for the study of emotional development in the infant. In Lewis Lipsitt (Ed.), *Developmental Psychology.* New York: Erlbaum.

Camus, A. (1948). *The plague.* New York: The Modern Library.

Canestrari, J. (1963). Paces and self-paced learning in young and elderly adults. *Journal of Gerontology, 18,* 165–68.

Cangemi, J. & Kowalski, C. (1987). Developmental tasks and student behavior: Some comments. *College Student Journal, 21,* 321–29.

Cantwell, D. P. & Carlson, G. A. (1983). *Affective disorders in childhood and adolescence.* New York: Spectrum.

Caplan, T. & Caplan, F. (1984). *The early childhood years.* New York: Bantam Books, Inc.

Cardon, L., Fulker, D., DeFries, J., & Plomin, R. (1992). Continuity and change in general cognitive ability from 1 to 7 years of age. *Developmental Psychology, 28,* 64–73.

Carey, J. (1983, December 16). Weight gained later in life is more risky for the heart. *USA Today,* 3.

Carey, W. (1981). The importance of temperament-environment interaction. In M. Lewis & L. Rosenblum (Eds.), *The uncommon child.* New York: Plenum.

Carlson, G. A. (1983). Depression and suicidal behavior in children and adolescents. In D. P. Cantwell & G. A. Carlson (Eds.), *Affective disorders in childhood and adolescence* (pp. 335–51). New York: Spectrum.

Carnegie Corporation. (1990). Adolescence: Path to a productive life or a diminished future? *Carnegie Quarterly, XXXV* (1, 2).

Carnevali, D. L. & Patrick, M. (Eds.). (1986). *Nursing management for the elderly* (2nd ed.). Philadelphia: J.B. Lippincott.

Cassell, E. (1979, February 18). Medical technology raises moral issues. *Boston Globe,* A12.

Cassem, N. (1975). Bereavement as a relative experience. In B. Schoenberg & others, (Eds.), *Bereavement* (pp. 3–9). New York: Columbia University Press.

Cassidy, J. & Asher, S. (1992). Loneliness and peer relations in young children. *Child Development, 63,* 350–65.

Cassidy, J., Parke, R., Butkovsky, L. & Braungart, J. (1992). Family-peer connections: The role of emotional expressiveness within the family and children's understanding of emotions. *Child Development, 63,* 603–18.

Catania, J. A., Turner, H., Kegeles, S. M., Stall, R., Pollack, L., & Coates, T. J. (1989). Older Americans and AIDS: Transmission risks and primary prevention research needs. *The Gerontologist, 29,* 373–81.

Cath, S. H. (1975). The orchestration of disengagement. *International Journal of Aging and Human Development, 6,* 199–213.

Celotta, B., Jacobs, G., & Keys, S. G. (1987). Searching for suicidal precursors in the elementary school child. Special issue: Identifying children and adolescents in need of mental health services. *American Mental Health Counselors Association Journal, 9*(1), 38–50.

Centers for Disease Control. (1989a). AIDS and human immunodeficiency virus infection in the United States: 1988 update. *Morbidity and Mortality Weekly Report, 38,* 1–35.

Centers for Disease Control. (1989b). First 100,000 cases of Acquired Immunodeficiency Syndrome—United States. *Morbidity and Mortality Weekly Report, 38,* 561–62.

Centers for Disease Control. (1989c). Update: Heterosexual transmission of Acquired Immunodeficiency Syndrome and Human Immunodeficiency virus infection—United States. *Morbidity and Mortality Weekly Report, 38,* 36–40.

Chall, J. S. (1992). *Stages of reading development,* New York: McGraw-Hill.

Chapman, A. B. (1988). Male-female relations: How the past affects the present. In H. P. McAdoo (Ed.), *Black Families* (2nd ed, p. 200). Newbury Park, CA: Sage.

Charlesworth, R. (1987). *Understanding child development.* Albany, NY: Delmar.

Chebator, P. (1992). *The bar exam.* Boston, MA: Boston College.

Chewning, B., Lohr, S., Van Koningsveld, R., Hawkins, R., Bosworth, K., Gustafson, D., & Day, T. (1986, April). *Family communication patterns and adolescent sexual behavior.* Paper presented at the Society for Research on Adolescence, Madison, WI.

Children's Defense Fund (1989). *A day in the lives of some teens.* Washington, DC: Children's Defense Fund.

Chilman, C. (1974). *Adolescent sexuality in a changing American society.* Washington, DC: ERIC document.

Chiriboga, D. A. (1989). Mental health at the midpoint. In S. Hunter & M. Sundel (Eds.), *Midlife myths.* Newbury Park, CA: Sage.

Chisholm, J. S. (1983). *Navajo infancy.* New York: Aldine.

Chomsky, N. (1957). *Syntactic structures.* The Hague: Mouton.

Chomsky, N. (1965). *The development of syntax in children 5 to 10 years.* Cambridge, MA: MIT Press.

Christiansen, B. A., Roehling, P. V., Smith, G. T., & Goldman, M. S. (1989). Using alcohol expectancies to predict adolescent drinking behavior after one year. *Journal of Consulting and Clinical Psychology, 57*(1), 93–99.

Cicirelli, V. G. (1979). *Social services for elderly in relation to the kin network.* Report to the NRTA-AARP Andrus Foundation, Washington, DC.

Cicirelli, V. G. (1980). Sibling relationships in adulthood: A lifespan perspective. In L. W. Poon (Ed.), *Aging in the 1980s: Psychological issues* (pp. 455–62). Washington, DC: American Psychological Association.

Clarke, A. & Clarke, A. D. (1976). *Early experience: Myth and evidence.* New York: The Free Press.

Clarke-Stewart, A. (1982). *Daycare.* Cambridge, MA: Harvard University Press.

Clemens, S. (1907–1919). Pudd'n-head Wilson and those extraordinary twins. In *The Writings of Mark Twain* (Vol. 14). New York: Harper.

Clifford, P., Tan, S., & Gorsuch, R. (1991). Efficacy of a self-directed behavioral health change program: Weight, body, composition, cardiovascular fitness, blood pressure, health risk and psychological mediating variables. *Journal of Behavioral Medicine, 14*(3), 303–23.

Clifton, R., Perris, E., & Bullinger, A., (1991). Infants' perception of auditory space. *Developmental Psychology, 27,* 187–97.

Cohen, L. H., Burt, C. E., & Bjorck, J. P. (1987). Life stress and adjustment: Effects of life events experienced by young adolescents and their parents. *Developmental Psychology, 23*(4), 583–92.

Cohen-Mansfield, J. & Marx, M. S. (1992). The social network of the agitated nursing home resident. *Research on Aging, 14*(1), 110–23.

Cohn, R. (1979). Age and the satisfactions from work. *Journal of Gerontology, 34,* 264–72.

Cole, C. (1986). Developmental tasks affecting the marital relationship in later life. *American Behavioral Scientist, 29,* 389–403.

Cole, E. & Rothblum, E. (1990). Commentary on "Sexuality and the midlife woman." *Psychology of Women Quarterly, 14*(4), 509–12.

Coleman, J. S. (1961). *The adolescent society.* Glencoe, IL: The Free Press.

Coleman, L. M., Antonucci, T. C., & Adelman, P. K. (1987). Social roles in the lives of middle-aged and older black women. *Journal of Marriage and the Family, 49,* 761–71.

Coll, C. T. G. (1990). Developmental outcome of minority infants: A process-oriented look into our beginnings. *Child Development, 61*(2), 270–89.

Colletta, N. (1982). How adolescents cope with problems of early motherhood. *Adolescence, 16*(63), 499–512.

Colligan, J. (1975). Achievement and personality characteristics as predictors of observed tutor behavior. *Dissertation Abstracts International, 35,* 4293–94.

Comfort, A. (1964). *Aging: The biology of senescence.* New York: Holt, Rinehart & Winston.

Compos, B. E., & Williams, R. A. (1990). Stress, coping and adjustment in mother and young adolescents in single- and two-parent families. *American Journal of Community Psychology, 19*(4), 525–45.

Comstock, G. & Paik, H. (1991). *Television and the American child.* New York: Academic Press.

Conrad, R. (1964). Acoustic confusion in immediate memory. *British Journal of Psychology, 55,* 75–84.

Cooney, T. M., Schaie, K. W., & Willis, S. L. (1988). The relationship between prior functioning on cognitive and personality dimensions and subject attrition in longitudinal research. *Journal of Gerontology: Psychological Sciences, 43*(1), 12–17.

Cooper, C. L. (Ed.), (1984). *Psychosocial stress and cancer.* New York: Wiley & Sons.

Cooper, C. R. & Grotevant, H. D. (1987). Gender issues in the interface of family experience and adolescents' friendship and dating identity. *Journal of Youth and Adolescence, 16*(3), 247–65.

Cooper, H. (1989, November). Synthesis of research on homework. *Educational Leadership,* 85–91.

Cooper R. M., Bilash, I., & Zubek, J. P. (1959). The effect of age on taste sensitivity. *Journal of Gerontology, 14,* 56–58.

Coopersmith, S. (1967). *The antecedents of self-esteem.* San Francisco: W. H. Freeman.

Corder, B. G., Shorr, W., & Corder, R. E. (1974). A study of social and psychological characteristics of adolescent suicide attempts in an urban, disadvantaged area. *Adolescence, 9* (33), 1–6.

Corso, J. (1971). Sensory processes and age effects in normal adults. *Journal of Gerontology, 26,* 90.

Costa, P. T., Jr. & McCrae, R. (1980). Still stable after all these years: Personality as a key to some issues in adulthood and old age. In P. B. Baltes (Ed.), *Life-span development and behavior* (Vol. 3, pp. 65–102). New York: Academic Press.

Cote, J. E. & Levine, C. (1988). The relationship between ego identity status and Erikson's notions of institutionalized moratoria, value orientation stage, and ego dominance. *Journal of Youth and Adolescence, 17*(1), 81–100.

Cowan, R. S. (1987). Women's work, housework, and history: The historical roots of inequality in work-force participation. In N. Gerstel & H. E. Gross (Eds.), *Families and work* (pp. 164–77). Philadelphia: Temple University Press.

Cowley, G. (1989, December 18). Medical mystery tour. *Newsweek,* 59.

Cox, H. (1988). *Later life.* Englewood Cliffs, NJ: Prentice-Hall.

Cox, M., Owen, M., Henderson, V. K., & Margand, N. (1992). Prediction of infant-father and infant-mother attachment. *Developmental Psychology, 28,* 474–83.

Craig-Bray, L., Adams, G. R., & Dobson, W. R. (1988). Identity formation and social relations during late adolescence. *Journal of Youth and Adolescence, 17*(2), 173–88.

Craik, F. I. M. (1977). Age differences in human memory. In J. E. Birren & K. W. Schaie (Eds.), *Handbook of the psychology of aging*. New York: Van Nostrand Reinhold.

Craik, F. I. M. & Simon, E. (1980). Age differences in memory: The roles of attention and depth of processing. In L. W. Poon, J. L. Fozard, L. S. Cermak, D. Arenberg, & L. W. Thompson (Eds.), *New directions in memory and aging: Proceedings of the George Talland Memorial Conference*. Hillsdale, NJ: Lawrence Erlbaum.

Cratty, B. (1979). *Perceptual and motor development in infants and children*. Englewood Cliffs, NJ: Prentice-Hall.

Craven, J. & Wald, F. (1975, October). Hospice care for dying patients. *American Journal of Nursing*, 1816–22.

Criqui, M. H. (1990). Comment on Shaper's "Alcohol and mortality." *British Journal of Addiction, 85*(7), 854–57.

Cristante, F. & Lucca, A. (1987). Cognitive functioning as measured by various developmental tasks and social competence in late childhood. *Archivio-di-Psycologia, 48,* 62–74.

Crockenberg, S. (1981). Irritability, mother responsiveness, and social support influences on the security of infant-mother attachment. *Child Development, 52,* 857–65.

Crockett, L., Losaff, M., & Petersen, A. (1986). Perceptions of the peer group and friendship in early adolescence. *Journal of Early Adolescence, 4*(2), 155–81.

Crohan, S. E., Antonucci, T. C., Adelman, P. K., & Coleman, L. M. (1989). Job characteristics and well-being at mid-life: Ethnic and gender comparisons. *Psychology of Women Quarterly, 13,* 223–35.

Cronin, K. & Wald, K. (1979). *Successful dying*. Unpublished manuscript, Boston College, Chestnut Hill, MA.

Crouter, A. C., Perry-Jenkins, M., Huston, T. L., & McHale, S. M. (1987). Processes underlying father involvement in dual-earner and single-earner families. *Developmental Psychology, 23,* 431–40.

Crowell, D. (1987). Childhood aggression and violence. In D. Crowell, I. Evans, & C. O'Donnel (Eds.), *Childhood Aggression and Violence*. New York: Plenum Press.

Crowley, J. E. (1986). Longitudinal effects of retirement on men's well-being & health. *Journal of Business and Psychology, 1*(2), 95–113.

Crumley, F. (1982). The adolescent suicide attempt: A cardinal symptom of a serious psychiatric disorder. *American Journal of Psychotherapy, 36*(2), 158–65.

Csikszentmihalyi, M. (1990). *Flow*. New York: Harper & Row.

Cullinan, D. & Epstein, M. H. (1979). *Special education for adolescents: Issues and perspectives*. Columbus, OH: Merrill.

Curran, D. (1984). Peer attitudes toward attempted suicide in midadolescents. *Dissertation Abstracts International, 44*(12), 3927B.

Curtiss, S. (1977). *Genie: A psycholinguistic study of a modern-day "wild child."* New York: Academic Press.

Cutler, S. J. & Grams, A. E. (1988). Correlates of self-reported everyday memory problems. *Journal of Gerontology: Social Sciences, 43*(3), 82–90.

Cutler, W., Schleidt, W., & Friedmann, E. (1987). Lunar influences on the reproductive cycle in women. *Human Biology, 59,* 959–72.

Dacey, J. S. (1976). *New ways to learn*. Stamford, CT: Greylock.

Dacey, J. S. (1981). *Where the world is*. Glenview, IL: Goodyear.

Dacey, J. S. (1982). *Adult development*. Glenview, IL: Scott, Foresman.

Dacey, J. S. (1986). *Adolescents today* (3rd ed.). Glenview, IL: Scott, Foresman.

Dacey, J. S. (1989a). *Fundamentals of creative thinking*. Lexington, MA: D.C. Heath/Lexington Books.

Dacey, J. S. (1989b). Peak periods of creative growth across the life span. *The Journal of Creative Behavior, 24*(4), 224–47.

Dacey, J. S. (1989c). Discriminating characteristics of the families of highly creative adolescents. *The Journal of Creative Behavior, 24*(4), 263–71.

Dacey, J. S. (in progress). *The psychology of self-control*.

Dacey, J. S. & Gordon, M. (1971, February). *Implications of post-natal cortical development for creativity research*. Paper presented at the American Education Research Association Convention, New York City.

Dacey, J., Madaus, G., & Crellin, D. (1968, November). *Can creativity be facilitated? The critical period hypothesis*. Paper presented at Ninth Annual Convention of the Educational Research Association of New York State, Kiamesho Lake.

Dacey, J. S. & Packer, A. (1992). *The nurturing parent: How to raise a creative, loving, responsible child*. New York: Fireside/Simon & Schuster.

Dacey, J. S & Ripple, R. E. (1967). The facilitation of problem solving and verbal creativity by exposure to programmed instruction. *Psychology in the School, 4*(3), 240–45.

Damon, W. (1983). *Social and personality development*. New York: Norton.

Damon, W. & Hart, D. (1988). *Self understanding in childhood and adolescence*. Cambridge: Cambridge University Press.

Daniels, D. & Moos, R. (1990). Assessing life stressors and social resources among adolescents. *Journal of Adolescent Research, 5,* 268–89.

Darling, C. A. & Hicks, M. W. (1982). Parental influence on adolescent sexuality: Implications for parents as educators. *Journal of Youth and Adolescence, 11,* 231–45.

Darrach, B. (1992). The war on aging. *Life, 15*(10), 32–45.

Datan, N., Rodeheaver, D., & Hughes, F. (1987). Adult development and aging. *Annual Review of Psychology, 38,* 153–80.

Daugherty, L. R. & Burger, J. M. (1984). The influence of parents, church, and peers on the sexual attitudes and behaviors of college students. *Archives of Sexual Behavior, 13,* 351–59.

DeCasper, A. & Fifer, W. (1980). Studying learning in the womb. *Science, 208,* 1174.

DeChardin, T. (1959). *The phenomenon of man* (B. Wall, Trans.). New York: Harper & Row.

deCuevas, J. (1990). "No, she holded them loosely." *Harvard Magazine, 93,* 60–67.

DeGenova, M. K. (1992). If you had your life to live over again, What would you do differently? *International Journal of Aging and Human Development, 34*(2), 135–43.

Dekovic, M. & Janssens, J. (1992). Parents' child-rearing style and child's sociometric status. *Developmental Psychology, 28,* 925–32.

Dellas, M. & Jernigan, L. P. (1987). Occupational identity status development, gender comparisons, and internal-external locus of control in first-year Air Force cadets. *Journal of Youth and Adolescence, 16*(6), 587–600.

Dembo, R. (1981). Examining a causal model of early drug involvement among inner-city junior high school youths. *Human Relations, 34*(3), 169–93.

Demoise, C. & Conrad, R. (1972). Effects of age and radiation exposure on chromosomes in a Marshall Island population. *Journal of Gerontology, 27*(2), 197–201.

Dennis, W. (1966). Creative productivity between 20 and 80 years. *Journal of Gerontology, 21,* 1–8.

Detting, E. R. & Beauvais, F. (1987). Peer cluster theory, socialization characteristics, and adolescent drug use: A path analysis. *Journal of Counseling Psychology, 34*(2), 205–13.

Devilliers, J. & Devilliers, P. (1978). *Language acquisition*. Cambridge, MA: Harvard University Press.

deVries, H. A. (1980). *Physiology of exercise for physical education and athletics* (3rd ed.). Dubuque, IA: Wm. C. Brown.

deVries, H. A. (1981). Physiology of exercise and aging. In D. F. Woodruff & J. E. Birren (Eds.), *Aging: Scientific perspectives and social issues* (pp. 464–65). New York: Van Nostrand Reinhold.

Dewey, J. (1933). *How we think*. Boston: Heath.

Diamond, M. (1982). Sexual identity, monozygotic twins reared in discordant sex roles, and a BBC follow-up. *Archives of Sexual Behavior, 11,* 181–85.

Dinnerstein, M. (1992). *Women between two worlds: Reflections on work and family*. Philadelphia: Temple University Press.

Dotz, W. & German, B. (1983). The facts about treatment of dry skin. *Geriatrics, 38,* 93.

Dranoff, S. M. (1974). Masturbation and the male adolescent. *Adolescence, 9*(34), 16–176.

Draper, P. & Keith, J. (1992). Cultural contexts of care: Family caregiving for elderly in America and Africa. *Journal of Aging Studies, 6*(2), 113–34.

Duncan, C. (1979). *A death curriculum*. Unpublished doctoral dissertation, Boston College, Chestnut Hill, MA.

Dunn, J. (1983). Sibling relationships in early childhood. *Child Development, 54,* 787–811.

Dunn, J. (1985). *Sisters and brothers*. Cambridge, MA: Harvard University Press.

Dunn, J. (1988). Connections between relationships: Implications of research on mothers and siblings. In R. Hinde & J. Stevenson-Hinde (Eds.), *Relations between relationships*. Oxford: Clarendon Press.

Dunn, J. (1988). Sibling influences on childhood development. *Journal of Child Psychology and Child Psychiatry, 29,* 119–29.

Duvall, E. (1971). *Family development*. Philadelphia: Lippincott.

Earle, J. R. & Perricone, P. J. (1986). Premarital sexuality: A ten-year study of attitudes and behavior on a small university campus. *Journal of Sex Research, 22*(3), 304–10.

Eichorn, D. H., Clausen, J. A., Haan, N., Honzik, M. P., & Mussen, P. H. (Eds.). (1981). *Past and present in middle life*. New York: Academic Press.

Eifrig, D. E. & Simons, K. B. (1983). An overview of common geriatric ophthalmologic disorders. *Geriatrics, 38,* 55.

Eisdorfer, C. & Wilkie, F. (1973). Intellectual changes and advancing age. In L. Jarvik (Ed.), *Intellectual functioning in adults.* New York: Springer.

Eisenberg, N. (1992). *The caring child.* Cambridge, MA: Harvard University Press.

Elias, M. & Elias, J. (1976). Matching of successive auditory stimuli as a function of age and year of presentation. *Journal of Gerontology, 31,* 164.

Elias, M., Elias, P., & Elias, J. (1977). *Basic processes in adult developmental psychology.* St. Louis: Mosby.

Elkin, F. & Handel, G. (1989). *The child and society.* New York: Random House.

Elkind, D. (1978). *The child's reality: Three developmental themes.* Hillsdale, NJ: Erlbaum.

Elkind, D. (1981). *The hurried child.* Reading, MA: Addison-Wesley.

Elkind, D. (1987). *Miseducation: Preschoolers at risk.* New York: Knopf.

Elkind, D. (1989, June 30). Under pressure. *The Boston Globe Magazine,* 24 ff.

Elkind, D. & Bowen, R. (1979). Imaginary audience behavior in children and adolescents. *Developmental Psychology, 15,* 38–44.

Elliott, D. S., Huizinga, D., & Menard, S. (1989). *Multiple problem youth.* New York: Springer-Verläg.

Elsayed, M., Ismail, A. H., & Young, J. R. (1980). Intellectual differences of adult men related to age and physical fitness before and after an exercise program. *Journal of Gerontology, 35,* 383–87.

Engstrom, P. F. (1986). Cancer control objectives for the year 2000. In L. E. Mortenseon, P. F. Engstrom, & P. N. Anderson (Eds.), *Advances in cancer control.* New York: Alan R. Liss.

Ennis, R. (1987). A taxonomy of critical thinking dispositions and abilities. In J. Baron & R. Sternberg (Eds.), *Teaching thinking skills.* New York: W. H. Freeman.

Enright, R. D., Levy, V. M., Harris, D., & Lapsley, D. K. (1987). Do economic conditions influence how theorists view adolescents? *Journal of Youth and Adolescence, 16*(6), 541–60.

Epstein, J. L. (1980). *A longitudinal study of school and family effects on student development.* Baltimore, MD: The Johns Hopkins University Press.

Erikson, E. H. (1958). *Young man Luther: A study in psychoanalysis and history.* New York: W. W. Norton.

Erikson, E. H. (1959). Identity and the life cycle. *Psychological Issues, 1,* 18–164.

Erikson, E. H. (1963). *Childhood and society* (2nd ed.). New York: W. W. Norton.

Erikson, E. H. (1968). *Identity: Youth and crisis.* New York: W. W. Norton.

Erikson, E. H. (1969). *Gandhi's truth: On the origins of militant nonviolence.* New York: W. W. Norton.

Erikson, E. H. (1975). *Life, history and the historical movement.* New York: W. W. Norton.

Erikson, E. H. (1978). *Adulthood.* New York: W. W. Norton.

Erickson, M. (1982). *Child psychopathology.* Englewood Cliffs, NJ: Prentice-Hall.

Estrada, A., Rabow, J., & Watts, R. K. (1982). Alcohol use among Hispanic adolescents: A preliminary report. *Hispanic Journal of Behavioral Sciences, 4*(3), 339–51.

Ewing, C. (1990). *When children kill: The dynamics of juvenile homicide.* Lexington, MA: Lexington Books/D. C. Heath.

Fagot, B. (1985a). Changes in thinking about early sex role development. *Developmental Review, 5,* 83–98.

Fagot, B. (1985b). Beyond the reinforcement principle: Another step toward understanding sex role development. *Developmental Psychology, 21,* 1097–104.

Fagot, B. & Hagan, R. (1991). Observations of parent reactions to sex-stereotyped behaviors: Age and sex effects. *Child Development, 62,* 617–28.

Fagot, B. & Kavanagh, K. (1990). The prediction of antisocial behavior from avoidant attachment classifications. *Child Development, 61,* 864–73.

Fagot, B., Leinbach, M. & O'Boyle, C. (1992). Gender labeling, gender stereotyping, and parenting behaviors. *Developmental Psychology, 28,* 225–30.

Faier, J. (1979, August). Sexual harassment on the job. *Harper's Bazaar,* 90–91.

Falco, M. (1988). *Preventing abuse of drugs, alcohol, and tobacco by adolescents.* Washington, DC: Carnegie Council on Adolescent Development.

Falknor, H. P. & Kugler, D. (1981). *JCAH hospice project, interim report: Phase I.* Chicago: Joint Commission on Accreditation of Hospitals.

Fantz, R. (1961). The origin of form perception. *Scientific American, 204,* 66–72.

Farber, E. & Egeland, B. (1987). *The invulnerable child.* New York: Guilford.

Farrell, M. P. & Rosenberg, S. D. (1981). *Men at midlife.* Boston: Auburn House.

Federal Bureau of Investigation. (1992). *Uniform crime reports for the U.S.* Washington, DC: U.S. Government Printing Office.

Feifel, H. (1977). *New meanings of death.* New York: McGraw-Hill.

Feldman, D. (1979). The mysterious case of extreme giftedness. In A. H. Passow (Ed.), *The gifted and the talented: Their education and development.* Chicago: University of Chicago Press (NSSE).

Feldman, D., Rosenberg, M. S., & Peer, G. G. (1984). Educational therapy: A behavior change strategy for predelinquent and delinquent youth. *Journal of Child and Adolescent Psychotherapy, 1*(1), 34–37.

Feldman, H. (1981). A comparison of intentional parents and intentionally childless couples. *Journal of Marriage and the Family, 43,* 593–600.

Fendrich, M. (1984). Wives' employment and husbands' distress: A meta-analysis and a replication. *Journal of Marriage and the Family, 46,* 871–79.

Ferber, R. (1985). *Solve your child's sleep problems.* New York: Simon & Schuster.

Field, D. (1987). A review of preschool conservation training: An analysis of analyses. *Developmental Review, 7,* 210–51.

Field, T. (1990). *Infancy.* Cambridge, MA: Harvard University Press.

Findlay, S. (1983, December 6). Study finds family link to heart ills. *USA Today,* 1.

Findley, P. (1979, December 13). This law is for you. *Parade,* 5–6.

Finian, M. J. & Blanton, L. P. (1987). Stress, burnout, and role problems among teacher trainees and first-year teachers. *Journal of Occupational Behavior, 8*(2), 157–65.

First, J. (1988). Immigrant students in U.S. public schools: Challenges with solutions. *Phi Delta Kappan, 70,* 205–10.

Fischer, K., Shaver, P., & Carnochan, P. (1990). How emotions develop and how they organize behavior. *Cognition and Emotion, 4,* 81–127.

Fischer, P. & Breakey, W. (1991). The epidemiology of alcohol, drug, and mental disorders among homeless persons. *American Psychologist, 46,* 1115–128.

Fisher, T. D. (1986a). An exploratory study of communication about sex and similarity in sexual attitudes in early, middle, and late adolescents and their parents. *Journal of Genetic Psychology, 147,* 543–57.

Fisher, T. D. (1986b). Parent-child communication and adolescents' sexual knowledge and attitudes. *Adolescence, 21,* 517–27.

Fitzpatrick, M. A. (1984). A typological approach to marital interaction: Recent theory and research. In L. Berkowitz (Ed.), *Advances in experimental social psychology* (Vol. 18). New York: Academic Press.

Flavell, J. H. (1963). *The developmental psychology of Jean Piaget.* New York: Van Nostrand.

Flavell, J. H. (1985). *Cognitive development.* Englewood Cliffs, NJ: Prentice-Hall.

Florian, V. & Zernicky-Shurka, E. (1986). Vocational instructor burnout in two national rehabilitation systems in Israel. *Journal of Applied Rehabilitation Counseling, 17*(1), 41–44.

Fodor, I. & Franks, V. (1990). Women in midlife and beyond: The new prime of life? *Psychology of Women Quarterly, 14,* 445–49.

Fonagy, P., Steele, M., & Steele, M. (1991). Maternal representations of attachment during pregnancy predict the organization of infant-mother attachment at one year of age. *Child Development, 62,* 891–905.

Fordham, S. & Ogbu, J. U. (1986). Black students' school success: Coping with the burden of "acting white." *Urban Review, 18,* 176–206.

Forste, R. & Heaton, R. (1988). Initiation of sexual activity among female adolescents. *Youth and Society, 19*(3), 250–68.

Forstein, M. (1989, April 7–9). *Sexuality and AIDS.* Paper presented at the Conference on the Psychiatric Treatment of Adolescents and Young Adults, Harvard Medical School, Boston, MA.

Foster, S. (1988). *The one girl in ten: A self portrait of the teenage mother.* Washington, DC: The Child Welfare League of America.

Fowler, J. (1974). Toward a developmental perspective on faith. *Religious Education, 69,* 207–19.

Fowler, J. (1975a). *Stages in faith: The structural developmental approach.* Harvard Divinity School Research Project on Faith and Moral Development.

Fowler, J. (1975b, October). *Faith development theory and the aims of religious socialization.* Paper presented at annual meeting of the Religious Research Association, Milwaukee, WI.

Fox, L. H. & Washington, J. (1985). Programs for the gifted and talented: Past, present, and future. In F. D. Horowitz & M. O'Brien (Eds.), *The gifted and talented.* Washington, DC: American Psychological Association.

Fox, N., Kimmerly, N., & Schafer, W. (1991). Attachment to mother/attachment to father: A meta-analysis. *Child Development, 62,* 210–25.

Fozard, J. L. (1980). The time for remembering. In L. W. Poon (Ed.), *Aging in the 1980s: Psychological issues* (pp. 273–87). Washington, DC: American Psychological Association.

Fozard, J. L., Wolf, E., Bell, B., McFarland, R. A., & Podolsky. S. (1977). Visual perception and communication. In J. E. Birren & K. W. Schaie (Eds.), *Handbook of the psychology of aging*. New York: Van Nostrand Reinhold.

Fraiberg, S. (1959). *The magic years*. New York: Scribner.

Fraiberg, S. (1980). *Clinical studies in infant mental health: The first year of life*. New York: Basic Books.

Frankl, V. (1967). *Psychotherapy and existentialism*. New York: Simon & Schuster.

Frasier, M. (1989). Poor and minority students can be gifted, too. *Educational Leadership, 46,* 16–18.

Freeman, E. W. (1982). Self-reports of emotional distress in a sample of urban black high school students. *Psychological Medicine, 12*(4), 809–17.

Freud, A. (1968). Adolescence. In A. E. Winder & D. L. Angus (Eds.), *Adolescence: Contemporary studies*. New York: American Book.

Freud, S. (1955). Totem and taboo. In J. Strachey (Ed. and Trans.), *The standard edition of the complete psychological works of Sigmund Freud* (Vol. 13). London: Hogarth Press (Original work published 1914).

Freud, S. (1966). *The complete introductory lectures on psychoanalysis*. Translated and edited by James Strachey. New York: Norton.

Freudenberger, H. (1973). A patient in need of mothering. *Psychoanalytic Review, 60*(1), 7–10.

Frey, B. A. & Noller, R. B. (1983). Mentoring: A legacy of success. *Journal of creative behavior, 17*(1), 60–64.

Frey, K. & Ruble, D. (1992). Gender constancy and the "cost" of sex-typed behavior: A test of the conflict hypotheses. *Developmental Psychology, 28,* 714–21.

Friedan, B. (1963). *The feminine mystique*. New York: W. W. Norton.

Friedenberg, E. (1959). *The vanishing adolescent*. Boston: Beacon Press.

Friedenberg, E. (1967). *Society's children*. New York: Random House.

Friedman, C. J., Mann, F., & Friedman, A. (1976). Juvenile street gangs: The victimization of youth. *Adolescence, 11,* 527.

Frisch, R. E. (1988, March). Fatness and fertility. *Scientific American,* pp. 88–95.

Fromm, E. (1955). *The sane society*. New York: Holt, Rinehart & Winston.

Fromm, E. (1968). *The art of loving*. New York: Harper & Row.

Fuerst, M. L. (1991). An STD primer: Let's be careful out there. *American Health, X*(4), 45–46.

Fulton, R. (1977). General aspects. In N. Linzer (Ed.), *Understanding bereavement and grief*. New York: Yeshiva University Press.

Fulton, R. & Owen, G. (1988). Death and society in twentieth century America. *Omega, 18,* 379–95.

Furman, W. & Buhrmester, D. (1985). Children's perceptions of the qualities of sibling relationships. *Child Development, 56,* 448–61.

Furstenberg, F. F. (1990). The new extended family. In K. Pasley & M. Ihinger-Tallman (Eds.), *Remarriage and stepparenting*. New York: Guilford.

Gabriel, A. & McAnarney, E. R. (1983). Parenthood in two subcultures: White, middle-class couples and black, low-income adolescents in Rochester, New York. *Adolescence, 18*(71), 595–608.

Gagnon, J. H. & Simon, W. (1969), They're going to learn on the street anyway. *Psychology Today, 3*(2), 46 ff.

Gagnon, J. H. & Simon, W. (1987). The sexual scripting of oral genital contacts. *Archives of Sexual Behavior, 16*(1), 1–25.

Galambos, N. L. & Lerner, J. V. (1987). Child characteristics and the employment of mothers with young children: A longitudinal study. *Journal of Child Psychology and Psychiatry and Allied Disciplines, 28*(1), 87–98.

Gallup, G. (1988). *The Gallup poll*. New York: Random House.

Gallup, G. E. (1977). Ninth Annual Gallup poll of the public's attitudes. *Phi Delta Kappan, 59*(1), 34–48.

Garapon, A. (1983, August/September). Place de l'initiation dans la délinquance juvenile. (Initiation role in juvenile delinquency.) *Neuropsychiatrie de l'Enfance et de l'adolescence, 31*(8–9), 390–403.

Garbarino, J. (1992). *Children and families in the social environment*. New York: Aldine.

Garbarino, J. & Abramowitz, R. (1992). The family as a social system. In Garbarino, J. (1992). *Children and families in the social environment*. New York: Aldine.

Garcia, E. E. (1990). Language-minority education litigation policy: "The law of the land." In S. Barona & E. E. Garcia (Eds.),

Children at risk: Poverty, minority status, and other issues in educational equity (pp. 53–63). Washington, DC: National Association of School Psychologists.

Garcia-Coll, C., Hoffman, J., & Oh, W. (1987). The social ecology and early parenting of Caucasian adolescent mothers. *Child Development, 58*(4), 955–63.

Gardner, H. (1980). *Artful scribbles.* New York: Basic Books.

Gardner, H. (1982). *Art, mind, and brain: A cognitive approach to creativity.* New York: Basic Books.

Gardner, H. (1983). *Frames of mind: The theory of multiple intelligences.* New York: Basic Books.

Gardner, H. (1985). *The mind's new science: A history of the cognitive revolution.* New York: Basic Books.

Gardner, H. (1991). *The unschooled mind.* New York: Basic Books.

Gardner, H. & Winner, E. (1982). Children's conceptions (and misconceptions) of the arts. In H. Gardner, *Art, mind, and brain.* New York: Basic Books.

Garmezy, N. (1987). Stress, competence, and development: The search for stress-resistant children. *American Journal of Orthopsychiatry, 57,* 159–74.

Garmezy, N. & Rutter, M. (Eds.). (1988). *Stress, coping, and development in children.* Baltimore, MD: The Johns Hopkins University Press.

Garn, S. M. & Petzold, A. (1983). Characteristics of the mother and child in teenage pregnancy. *American Journal of Diseases of Children, 137*(4), 365–68.

Garner, H. G. (1975). An adolescent suicide, the mass media and the educator. *Adolescence, 17*(2), 23–27.

Garvey, M. S. (1986). The high cost of sexual harassment suits. *Personnel Journal, 65*(1), 75–78, 80.

Gates, D. & Jackson, R. (1990). Gang violence in L.A. *The Police Chief, 57,* 20–21.

Gay, P. (1988). *Freud: A life for our time.* New York: W. W. Norton.

Geer, J., Heiman, J., & Leitenberg, H. (1984). *Human sexuality.* Englewood Cliffs, NJ: Prentice-Hall.

Gelman, D. (1989, December 18). The brain killer, *Newsweek,* 54–83.

Gelman, R. & Baillargeon, R. (1983). A review of some Piagetian concepts. In P. Mussen (Ed.), *Handbook of child psychology.* New York: Wiley.

Genshaft, J. (1991). The gifted adolescent in perspective. In M. Birely & J. Genshaft (Eds.), *Understanding the gifted adolescent: Educational development and multicultural issues* (pp. 259–62). New York: Teachers College Press.

George, L. K. & Landerman, R. (1984). Health and subjective well-being. *International Journal of Aging and Human Development, 19,* 133–56.

Gerson, L. W., Jarjoura, D., & McCord, G. (1987). Factors related to impaired mental health in urban elderly. *Research on Aging, 9*(3), 356–71.

Gesell, A. (1940). *The first five years of life.* New York: Harper.

Gest, T. (1989, December 1). Is there a right to die? *Time,* 35–37.

Gibbons, M. (1974). Walkabout: Searching for the right passage from childhood and school. *Phi Delta Kappan, 55*(9), 596–602.

Gibson, E. & Wolk, R. (1960). The visual cliff. *Scientific American, 202,* 64–71.

Gierz, M., Haris, J., & Lohr, J. B. (1989). Recognition and treatment of depression in Alzheimer's disease. *Geriatrics, 36,* 901–11.

Gilbert, J. (1986). *A cycle of outrage: America's reaction to the juvenile delinquent in the 1950's.* New York: Oxford University Press.

Giles, H., Coupland, N., Coupland, J., Williams, A., & Nussbaum, J. (1992). Intergenerational talk and communication with older people.

Gilligan, C. (1977). In a different voice: Women's conception of self and morality. *Harvard Educational Review, 47,* 481–517.

Gilligan, C. (1982). *In a different voice: Psychological theory and women's development.* Cambridge, MA: Harvard University Press.

Gilligan, C. (1989). Mapping the moral domain: New images of self in relationship. *Cross-currents, 39,* 50–63.

Gilligan, C., Lyons, N., & Hanmer, T. (1990). *Making connections: The relational worlds of adolescent girls at Emma Willard School.* Cambridge, MA: Harvard University Press.

Gilman, L. (1987). *The adoption resource book.* New York: Harper & Row.

Gimby, G. & Saltin, B. (1983). The aging muscle. *Clinical Physiology, 3,* 209–18.

Gitlin, T. (1987). *Years of hope, days of rage.* New York: Bantam Books, Inc.

Glaser, K. (1965). Attempted suicide in children and adolescents. *American Journal of Psychotherapy, 19*(2), 220–27.

Glaser, K. (1978). The treatment of depressed and suicidal adolescents. *American Journal of Psychotherapy, 32*(2), 252–69.

Glenn, E. N. (1978). *Behavior in the human male.* Philadelphia: Saunders.

Goertzel, V. & Goertzel, M. (1962). *Cradles of eminence.* Boston: Little, Brown.

Goethals, G. & Klos, D. (1976). *Experiencing youth*. Boston: Little, Brown.

Goetting, A. (1986). The developmental tasks of siblingship over the life cycle. *Journal of Marriage and the Family, 48,* 703–14.

Goldberg, S. (1983). Parent-infant bonding: Another look. *Child Development, 54,* 1355–82.

Goldberg, S. & DiVitto, B. (1983). *Born too soon*. San Francisco: Freeman, Cooper.

Goldfinger, S. (Ed.). (1991). When to hang up the keys. *Harvard Health Letter, 17*(1), 1–4.

Goldman-Rakic, P., Isseroff, A., Schwartz, M., & Bugbee, N. (1983). The neurobiology of cognitive development. In P. Mussen (Ed.), *Handbook of child psychology*. New York: Wiley.

Goldsmith, H. H. (1983). Genetic influences on personality from infancy to adulthood. *Child Development, 54,* 331–55.

Goldsmith, H. H. & Campos, J. (1990). The structure of temperamental fear and pleasure in infants: A psychometric perspective. *Child Development, 61,* 1944–64.

Goldsmith, M. F. (1988). Anencephalic organ donor program suspended; Loma Linda report expected to detail findings. *Journal of the American Medical Association, 260,* 1671–72.

Goleman, D. (1987, October 13). Thriving despite hardship: Key childhood traits identified. *New York Times, 82,* 104.

Goodenough, F. (1926). *The measurement of intelligence through drawing*. New York: Holt.

Gottman, J. M. & Krokoff, L. J. (1989). Marital interaction and satisfaction: A longitudinal view. *Journal of Consulting and Clinical Psychology, 57,* 47–52.

Gould, R. (1978). *Transformations*. New York: Simon & Schuster.

Gray, R. E. (1987). Adolescent response to the death of a parent. *Journal of Youth and Adolescence, 16*(6). 511–26.

Green, A. (1978). Self-destructive behavior in battered children. *American Journal of Psychiatry, 135*(5), 579–83.

Greenspan, S. & Greenspan, N. T. (1985). *First feelings*. New York: Viking.

Gregory, T. (1978). *Adolescence in literature*. New York: Longman.

Griffin, G. W. (1987). Childhood predictive characteristics of aggressive adolescents. *Exceptional Children, 54*(3), 246–52.

Grobstein, C. (1981). *From chance to purpose*. Reading, MA: Addison-Wesley.

Grobstein, C. (1988). *Science and the unborn*. New York: Basic Books.

Grossman, H. J. (1983). *Classification in mental retardation*. Washington, DC: American Association on Mental Deficiency.

Grossman, H. Y. & Chester, N. L. (1990). *The experience and meaning of work in women's lives*. Hillsdale, NJ: Erlbaum.

Grotevant, H. D., Thorbeck, W., & Meyer, M. L. (1982). An extension of Marcia's identity status interview into the interpersonal domain. *Journal of Youth and Adolescence, 11*(1), 33–47.

Gruber, K., Jones, R. J., & Freeman, M. H. (1982). Youth reactions to sexual assault. *Adolescence, 17*(67), 541–51.

Grzegorczyk, P. B., Jones, S. W., & Mistretta, C. M. (1979). Age-related differences in salt taste acuity. *Journal of Gerontology, 34,* 834–40.

Guelzow, M. G., Bird, G. W., & Koball, E. H. (1991). An exploratory path analysis of the stress process for dual career men and women. *Journal of Marriage and the Family, 53,* 151–64.

Guilford, J. P. (1975). Creativity: A quarter century of progress. In I. A. Taylor & J. W. Getzels (Eds.), *Perspectives in creativity*. Chicago: Aldine.

Gutman, D. (1973, December). Men, women and the parental imperative. *Commentary,* 59–64.

Guttmacher, A. & Kaiser, I. (1986). *Pregnancy, birth and family planning*. New York: Signet.

Haan, N. (1976). ". . . change and sameness . . ." reconsidered. *International Journal of Aging and Human Development, 7,* 59–65.

Haan, N. (1981). Common dimensions of personality development. In D. M. Eichorn (Ed.), *Present and past in middle life*. New York: Academic Press.

Haan, N. (1989). *Personality at midlife*. In S. Hunter & M. Sundel (Eds.), *Midlife myths*. Newbury Park, CA: Sage.

Hajcak, F. & Garwood, P. (1989). Quick-fix sex: Pseudosexuality in adolescents. *Adolescence, 23*(92), 75–76.

Hakim-Larson, J. & Hobart, C. J. (1987). Maternal regulation and adolescent autonomy: Mother-daughter resolution of story conflicts. *Journal of Youth and Adolescence, 16*(2), 153–66.

Halford, G. S. (1989). Reflection on 25 years of Piagetian cognitive psychology, 1963–1988. *Human Development, 32,* 325–57.

Hall, D. A. (1976). *Aging of connective tissue*. London: Academic Press.

Hall, G. S. (1904). *Adolescence*. (2 vols.) New York: Appleton-Century-Crofts.

Hall, G. S. (1905). *Adolescence.* New York: Appleton-Century-Crofts.

Hallman, R. (1967). Techniques for creative teaching. *Journal of Creative Behavior, 1*(3), 325–30.

Hankoff, L. D. (1975). Adolescence and the crisis of dying. *Adolescence, 10*(39), 373–90.

Hansson, R., O'Connor, M., Jones, W., & Blocker, T. (1981). Maternal employment and adolescent sexual behavior. *Journal of Youth and Adolescence, 10*(1), 55–60.

Hardt, D. (1979). *Death: The final frontier.* Englewood Cliffs, NJ: Prentice-Hall.

Harre, R. & Lamb, R. (1983). *The encyclopedic dictionary of psychology.* Cambridge, MA: the MIT Press.

Harris, J. R. & Liebert, R. (1992). *Infant & child.* Englewood Cliffs, NJ: Prentice-Hall.

Harrison, H. & Kositsky. A. (1983). *The premature baby book.* New York: St. Martin's Press.

Harry, J. (1986). Sampling gay men. *Journal of Sex Research, 22,* 21–34.

Harter, S. (1983). Developmental perspectives on the self-system. In P. Mussen (Ed.), *Handbook of child psychology.* New York: Wiley.

Hartup, W. (1983). Peer relations. In P. Mussen (Ed.), *Handbook of child psychology.* New York: Wiley.

Harvard Health Letter (1991). *Elderly drivers.* Cambridge, MA: Harvard University Press.

Hasher, L. & Zacks, R. T. (1979). Automatic and effortful processes in memory. *Journal of Experimental Psychology: General, 108,* 356–88.

Haskell, M. R. & Yablonsky, L. (1982). *Juvenile delinquency.* Chicago: Rand-McNally.

Hatfield, E., Traupmann, J., & Sprecher, S. (1984). Older women's perceptions of their intimate relationships. *Journal of Social and Clinical Psychology, 2*(2), 108–24.

Hauser, S., Book, B., Houlihan, J., & Powers, S. (1987). Sex differences within the family: Studies of adolescent and parent family interactions. Special issue: Sex differences in family relations at adolescence. *Journal of Youth and Adolescence, 16*(3), 199–220.

Hauser, S. T. & Bowlds, M. K. (1990). In S. Feldman and G. Elliot (Eds.), *At the threshold: The developing adolescent.* Cambridge, MA: Harvard University Press.

Havighurst, R. (1972). *Developmental tasks and education.* New York: McKay.

Havighurst, R. (1982). The world of work. In B. Wolman (Ed.), *Handbook of developmental psychology.* Englewood Cliffs, NJ: Prentice-Hall.

Havighurst, R. J. (1951). *Developmental task and education.* New York: Longmans, Green.

Havighurst, R., Neugarten, B., & Tobin, S. (1968). Disengagement and patterns of aging. In B. Neugarten (Ed.), *Middle age and aging.* Chicago: University of Chicago Press.

Hawton, K. (1982a). Motivational aspects of deliberate self-poisoning in adolescents. *British Journal of Psychiatry, 141,* 286–90.

Hawton, K. (1982b). Adolescents who take overdoses: Their characteristics, problems and contacts with helping agencies. *British Journal of Psychiatry, 140,* 118–23.

Hawton, K. (1982c). Classification of adolescents who take overdoses. *British Journal of Psychiatry, 140,* 124–31.

Hay, D. (1986). Infancy. In M. Rosenzweig & L. Porter (Eds.), *Annual review of psychology.* Palo Alto, CA: Annual Reviews.

Hayes, J. (1989). *The complete problem solver.* Philadelphia: The Franklin Institute Press.

Hayward, M. D. (1986). The influence of occupational characteristics on men's early retirement. *Social Forces, 64*(4), 1032–45.

Held, L. (1981). Self-esteem and social network of the young pregnant teenager. *Adolescence, 16*(64), 905–12.

Herold, E., Mantle, D., & Zemitis, O. (1979). A study of sexual offenses against females. *Adolescence, 14*(53), 65–72.

Hersch, P. (1991, April). Teen epidemic. *American Health,* 42–45.

Hersch, P. (1991, May). Sexually transmitted diseases are ravaging our children. *American Health,* 42–52.

Hertz, R. (1989). Dual-career corporate couples: Shaping marriages through work. In Risman, B. & Schwartz, P. (Eds.), *Gender in intimate relationships.* Belmony, CA: Wadsworth.

Hertzog, C. (1989). Influences of cognitive slowing on age differences in intelligence. *Developmental Psychology, 25,* 636–51.

Herz, E. & Reis, J. (1987). Family life education for young inner city teens: Identifying needs. *Journal of Youth and Adolescence, 16*(4), 361–77.

Hetherington, E. M. (1972). Effects of father absence on personality development in adolescent daughters. *Developmental Psychology, 7,* 313–26.

Hetherington, E. M. (1973). Girls without fathers. *Psychology Today, 6*(9), 47–52.

Hetherington, E. M., Cox, M., & Cox, R. (1985). Long-term effects of divorce and remarriage on the adjustment of children. *Journal of the American Academy of Child Psychiatry, 5,* 518–30.

Hetherington, E. M. & Parke, R. (1986). *Child psychology.* New York: McGraw-Hill.

Hetherington, E. M., Staney-Hagan, M., & Anderson, E. R. (1989). Marital transitions: A child's perspective. *American Psychologist, 44*(2), 303–12.

Hill, C. D., Thompson, L. W., & Gallagher, D. (1988). The role of anticipatory bereavement in older women's adjustment to widowhood. *The Gerontologist, 28*(6), 7–12.

Hill, J. P. & Holmbeck, G. N. (1987). Disagreements about rules in families with seventh-grade girls and boys. *Journal of Youth and Adolescence, 16*(3), 221–46.

Hill, P., Jr. (1987, July). *Passage to manhood: Rearing the male African-American child.* Paper presented at the annual conference of the National Black Child Development Institute, Detroit, MI.

Hinde, R. (1970). *Animal behavior* (2nd ed.). New York: McGraw-Hill.

Hinde, R. (1979). *Towards understanding relationships.* New York: Academic Press.

Hinde, R. (1987). *Individuals, relationships and culture.* New York: Cambridge University Press.

Hirshberg, H. (1990). When infants look to their parents: Twelve-months-olds' response to conflicting parental emotional signals. *Child Development, 61,* 1187–91.

Hoberman, H. & Garfinkel, B. (1989). Completed suicide in children and adolescents. *Annual Progress in Child Psychiatry and Child Development,* 429–45.

Hochhaus, C. & Sousa, F. (1988). Why children belong to gangs: A comparison of expectations and reality. *High School Journal, 71,* 74–77.

Hochhauser, M. & Rothenberger, J. (1992). *AIDS education.* Dubuque, IA: Wm. C. Brown.

Hoffman, S. & Duncan, G. (1988). What are the economic consequences of divorce? *Demography, 25,* 641–45.

Holden, C. (1987). Alcoholism and the medical cost crunch. *Science, 235,* 1132–33.

Holinger, P. C., Offer, D., & Ostrov, E. (1987). Suicide and homicide in the United States: An epidemiologic study of violent death, population changes, and the potential for prediction. *American Journal of Psychiatry, 144*(2), 215–19.

Holland, J. L. (1970). *The self-directed search.* Palo Alto, CA: Consulting Psychologists Press.

Holland, J. L. (1973). *Making vocational choices: A theory of careers.* Englewood Cliffs, NJ: Prentice-Hall.

Holmes, C. T. (1990). Grade level retention efforts: A meta-analysis of research studies. In L. A. Shepard & M. L. Smith (Eds.), *Flunking grades: Research and policies on retention.* New York: The Falmer Press.

Holmes, J. & Rahe, S. (1967). A social adjustment scale. *Journal of Psychosomatic Research, 11,* 213–18.

Honig, A. (1986, May). Stress and coping in children. Part 1. *Young Children,* 50–63.

Hood, J. C. (1983). *Becoming a two-job family.* New York: Praeger.

Horn, J. L. (1975). *Psychometric studies of aging and intelligence.* New York: Raven Press.

Horn, J. L. (1978). Human ability systems. In P. B. Baltes (Ed.), *Life-span development and behavior* (Vol. 1). New York: Academic Press.

Horn, J. L. (1985). Remodeling old models of intelligence. In B. B. Wolman (Ed.), *Handbook of intelligence: Theories, measurements, and applications.* New York: Wiley.

Horn, J. L. & Donaldson, G. (1980). Cognitive development in adulthood. In O. G. Brim, Jr. & J. Kagan (Eds.), *Constancy and change in human development* (pp. 445–529). Cambridge, MA: Harvard University Press.

Horn, J. L., Donaldson, G., & Engstrom, R. (1981). Apprehension, memory, and fluid intelligence decline in adulthood. *Research on Aging, 3,* 33–84.

Horney, K. (1967). *Feminine psychology.* New York: Norton.

Horwitz, A. V. & White, H. R. (1987). Gender role orientations and styles of pathology among adolescents. *Journal of Health and Social Behavior, 28*(2), 158–70.

Hoyer, W. J. & Pludes, D. J. (1980, March). *Aging and the attentional components of visual information processing.* Paper presented at the symposium on Aging and Human Visual Function, National Academy of Sciences, Washington, DC.

Huba, G., Wingard, J., & Bentler, P. (1979). Beginning adolescents' drug use and peer and adult interaction patterns. *Journal of Consulting Clinical Psychology, 47,* 830–41.

Huba, G., Wingard, J., & Bentler, P. (1980a). A longitudinal analysis of the role of peer support, adult models, and peer subcultures in beginning adolescent substance use. *Multivariate Behavior Research, 15,* 259–80.

Huba, G., Wingard, J., & Bentler, P. (1980b). Framework for an interactive theory of drug use. In D. Lettieri, M. Sayers, & H. Pearson (Eds.), *Theories on drug abuse.* Rockville, MD: National Institute on Drug Abuse.

Huba, G., Wingard, J., & Bentler, P. (1980c). Applications of a theory of drug use prevention. *Journal of Drug Education, 10,* 25–38.

Hudgens, R. (1975). Suicide communications and attempts. In *Psychiatric disorders in adolescents* (chap. 5). Baltimore, MD: William & Wilkins.

Hulicka, I. (1967). Short-term learning and memory. *Journal of the American Geriatrics Society, 15,* 285–94.

Hunt, T. & Lindley, C. J. (1989). *Testing older adults.* Washington, DC: Center for Psychological Services.

Hunter, S. & Sundel, M. (1989). *Midlife myths.* Newbury Park, CA: Sage.

Huston, A. (1983). Sex-typing. In P. Mussen (Ed.), *Handbook of child psychology.* New York: Wiley.

Huttenlocher, J., Haight, W., Bryk, A., Seltzer, M., & Lyons, T. (1991). Early vocabulary growth: Relation to language input and gender. *Developmental Psychology, 27,* 236–48.

Ingalls, Z. (1983, January 19). Although drinking is widespread, student abuse of alcohol is not rising, new study finds. *The Chronicle of Higher Education,* 9.

Izard, C. E. (1978). On the ontogenesis of emotions and emotion-cognition relationships in infancy. In M. Lewis & L. Rosenblum (Eds.), *The development of affect.* New York: Plenum.

Izard, C., Haynes, O. H., Chisholm, G., & Baak, K. (1991). Emotional determinants of infant-mother attachment. *Child Development, 62,* 906–17.

Izard, I. & Malatesta, C. (1987). Perspectives on emotional development I: Differential emotions theory of early emotional development. In J. Osofsky (Ed.), *Handbook of infant development.* New York: Wiley.

Jackson, J., Antonucci, T., & Gibson, R. (1990). Cultural, racial and ethnic minority influences on aging. In J. Birren & K. Schaie (Eds.), *Handbook of the psychology of aging* (3rd ed.). New York: Van Nostrand.

Jackson, J. S. & Gibson, R. C. (1985). Work and retirement among the black elderly. In Z. S. Blou (Ed.), *Current perspectives on aging and the life cycle: Vol. 1: Work, retirement, and social policy.* Greenwich, CN: JAI Press.

Jacobs, J. (1971). *Adolescent suicide.* New York: Wiley.

Jacobson, S. & Frye, K. (1991). Effect of maternal social support on attachment: Experimental evidence. *Child Development, 62,* 572–82.

Jacobziner, H. (1965). Attempted suicide in adolescence. *Journal of the American Medical Association, 10,* 22–36.

James, M. (1990). Adolescent values clarification: A positive influence on perceived locus of control. *Journal of Alcohol and Drug Education, 35*(2), 75–80.

Janus, S. & Janus, C. (1993). *The Janus report on sexual behavior.* New York: Wiley.

Jaquish, G. A., Block, J., & Block, J. H. (1984). *The comprehension and production of metaphor in early adolescence: A longitudinal study of cognitive childhood antecedents.* Unpublished manuscript.

Jaquish, G. & Ripple, R. E. (1980). Cognitive creative abilities across the adult life span. *Human Development, 34,* 143–52.

Jasmin, S. & Trygstad, L. (1979). *Behavioral concept and the nursing process.* St. Louis: Mosby.

Jerse, F. W. & Fakouri, M. E. (1978). Juvenile delinquency and academic deficiency. *Contemporary Education, 49,* 108–9.

Johnson, B., Shulman, S., & Collins, W. A. (1991). Systemic patterns of parenting as reported by adolescents: Developmental differences and implications for psychosocial outcomes. *Journal of Adolescent Research, 6*(2), 235–52.

Johnson, J., Christie, J., & Yawkey, T. (1987). *Play and early childhood development.* Glenview IL: Scott, Foresman.

Johnson, T. G. & Goldfinger, S. E. (1981). *The Harvard Medical School health letter book.* Cambridge, MA: Harvard University Press.

Johnston, L. D., O'Malley, P. M., & Bachman, J. G. (1987). *National trends in drug use and related factors among American high school students and young adults, 1975–1986.* Rockville, MD: National Institute on Drug Abuse.

Johnston, W. B. (1987). *Workforce 2000: Work and workers for the 21st century.* Indianapolis, IN: Hudson Institute.

Joint Commission on Accreditation of Hospitals. (1984). *JCAH hospice provider profile.* Chicago: Author.

Jonaitis, M. A. (1988). Nutrition during pregnancy. In D. Tapley & W. Todd (Eds.), *Complete guide to pregnancy.* New York: Crown.

Jones, R. (1978, November 2). Margaret Mead. *Time,* 21–22.

Joselow, F. (1979, August). Fiscal fitness: How to manage money. *Harper's Bazaar,* 91.

Jung, C. G. (1933). *Modern man in search of a soul.* New York: Harcourt, Brace & World.

Jung, C. G. (1966). *The spirit in men, art and literature.* New York: Bollingen Foundations.

Jung, C. G. (1971). *The portable Jung.* (Joseph Campbell, Ed.). New York: Viking Press.

Jurich, A. P., Schumm, W. R., & Bollman, S. R. (1987). The degree of family orientation perceived by mothers, fathers, and adolescents. *Adolescence, 22*(85), 119–28.

Kacmar, K. M. & Ferris, G. R. (1989). Theoretical and methodological considerations in the age-job satisfaction relationship. *Journal of Applied Psychology, 74,* 201–7.

Kagan, J. (1984). *The nature of the child.* New York: Basic Books.

Kagan, J. & Moss, H. (1962). *Birth to maturity: A study in psychological development.* New York: Wiley.

Kail, R. (1990). *The development of memory in children.* New York: W. H. Freeman.

Kalab, K. (1992). Playing gateball: A game of the Japanese elderly. *Journal of Aging Studies, 6*(1), 23–40.

Kalinich, L. J. (1992). The biological clock. In J. M. Oldham & R. S. Liebert (Eds.), *The middle years: New psychoanalytic perspectives* (p. 123). New Haven: Yale University Press.

Kallman, F. & Jarvik, L. (1959). Individual differences in constitution and genetic background. In J. Birren (Ed.), *Handbook of aging and the individual.* Chicago: University of Chicago Press.

Kalter, N. (1987). Long-term effects of divorce on children: A developmental vulnerability model. *American Journal of Orthopsychiatry, 57*(4), 587–600.

Kangas, J. & Bradway, K. (1971). Intelligence at midlife: A 38-year follow-up. *Developmental Psychology, 5,* 333–37.

Kantner, J. F. (1982). Sex and pregnancy among American adolescents. *Educational Horizons, 61*(4), 189–94.

Kantrowicz, B. (1989, December 18). Trapped inside her own world. *Newsweek,* 56–58.

Kart, C. S., Metress, E. K., & Metress, S. P. (1988). *Aging, health and society.* Boston: Jones & Bartlett Publishers, Inc.

Kastenbaum, R. (1959). Time and death in adolescence. In H. Feifel (Ed.), *The meaning of death.* New York: McGraw-Hill.

Kastenbaum, R. & Kastenbaum, B. (1989). *The encyclopedia of death.* Phoenix: Oryx Press.

Kauffman, J. M. (1981). *Characteristics of children's behavior disorders* (2nd ed.). Columbus, OH: Charles E. Merrill.

Kavanaugh, R. (1972). *Facing death.* New York: Penguin Books.

Kelley-Buchanan, C. (1988). *Peace of mind during pregnancy.* New York: Dell.

Kellog, J. (1988). Forces of change. *Phi Delta Kappan, 70,* 199–204.

Kendig, H., Hashimoto, A., & Coppard, L. (Eds.). (1992). *Family support for the elderly: The international experience.* Oxford: Oxford University Press.

Kessler, J. (1988). *Psychopathology of childhood.* Englewood Cliffs, NJ: Prentice-Hall.

Kienhorst, C. W., Wolters, W. H., Diekstra, R. F., & Otto, E. (1987). A study of the frequency of suicidal behavior in children aged 5 to 14. *Journal of Child Psychology and Psychiatry and Allied Disciplines, 28* (1), 153–65.

Kilpatrick, W. (1975). *Identity, and intimacy.* New York: Delacorte.

Kimmel, D. (1974). *Adulthood and aging.* New York: Wiley.

Kimmel, D. C. & Weiner, I. B. (1985). *Adolescence: A developmental transition.* Hillsdale, NJ: Erlbaum.

Kinard, E. & Reinherz, H. (1987). School aptitude and achievement in children of adolescent mothers. *Journal of Youth and Adolescence, 16*(1), 69–87.

King, I. (1914). *The high school age.* Indianapolis: Bobbs-Merrill.

Kinnaird, K. & Gerrard, M. (1986). Premarital sexual behavior and attitudes toward marriage and divorce among young women as a function of their mothers' marital status. *Journal of Marriage and the Family, 48*(4), 757–65.

Kinsey, A., Pomeroy, W., Martin, C., & Gebhard, P. (1948). *Sexual behavior in the human male.* Philadelphia: Saunders.

Kinsey, A., Pomeroy, W., Martin, C., & Gebhard, P. (1953). *Sexual behavior in the human female.* Philadelphia: Saunders.

Kintsch, W. (1970). *Learning, memory, and conceptual processes.* New York: Wiley.

Kirkland, M. & Ginther, D. (1988). Acquired Immune Deficiency Syndrome in children: Medical, legal, and school related issues. *School Psychology Review, 17,* 304–5.

Kitahara, M. (1983). Female puberty rites: Reconsideration and speculation. *Adolescents, 18*(72), 957–64.

Kitchener, K. & King, P. (1981). Reflective judgment: Concepts of justification and their relationship to age and education. *Journal of Applied Developmental Psychology, 2,* 89–116.

Klassen, A. D., Williams, C. J., & Levitt, E. E. (1989). *Sex and morality in the U.S.: An empirical enquiry under the auspices of the Kinsey Institute*. Middletown, CT: Wesleyan University Press.

Klaus, M. & Kennell, J. (1976). *Maternal-infant bonding*. St. Louis: Mosby.

Klaus, M. & Kennell, J. (1983). *Bonding*. New York: Mosby.

Klein, H. & Cordell, A. (1987). The adolescent as mother: Early risk identification. *Journal of Youth and Adolescence, 16*(1), 47–58.

Klein, N. (1978, October). Is there a right way to die? *Psychology Today, 12,* 122.

Kliegl, R., Smith, J., & Baltes, P. (1989). Testing-the-limits and the study of adult age differences in cognitive plasticity of a mnemonic skill. *Developmental Psychology, 25*(2), 247–56.

Kliegl, R., Smith, J., & Baltes, P. (1990). On the locus and process of magnification of age differences during mnemonic training. *Developmental Psychology, 26*(6), 894–904.

Kneisl, C. R. & Ames, S. W. (1986). *Adult health nursing: A biopsychosocial approach*. New York: Addison-Wesley.

Knobloch, H. & Pasamanick, B. (1974). *Gesell and Amatruda's developmental diagnosis*. New York: Harper & Row.

Knowles, M. (1984). *The adult learner: A neglected species*. Houston: Gulf.

Knox, J. (1991, June 26). Distribution of condoms in the schools. *The Boston Globe,* 17.

Koch, H. (1960). The relation of certain formal attributes of siblings to attributes held toward each other and toward their parents. *Monographs of the Society for Research in Child Development, 25* (4), 1–124.

Koenig, H. G., Kvale, J. N., & Ferrell, C. (1988). Religion and well-being in later life. *The Gerontologist, 28*(1), 18–20.

Koff, E., Rierdan, J., & Sheingold, K. (1980, April). *Memories of menarche: Age and preparedness as determinants of subjective experience*. Paper presented at the annual meeting of the Eastern Psychological Association, Hartford, CT.

Kogan, H. (1973). Creativity and cognitive style: A life-span perspective. In P. B. Baltes & K. W. Schaie (Eds.), *Life-span developmental psychology*. New York: Academic Press.

Kogan, N. (1983). Stylistic variation in childhood and adolescence: Creativity, metaphor, cognitive styles. In P. H. Mussen (Ed.), *Handbook of child psychology* (Vol. 3). New York: Wiley.

Kohlberg, L. (1966). A cognitive-developmental analysis of children's sex-role concepts and attitudes. In E. Maccoby (Ed.), *The development of sex differences*. Stanford, CA: Stanford University Press.

Kohlberg, L. (1973). The claim to moral adequacy of a highest stage of moral judgment. *Journal of Philosophy, 60,* 630–46.

Kohlberg, L. (1975). The cognitive-developmental approach to moral education. *Phi Delta Kappan, 56,* 670–77.

Kohlberg, L. (1981). *The philosophy of moral development*. New York: Harper & Row.

Kohner, M. (1977, October 11). Adolescent pregnancy. *The New York Times,* 38.

Kolata, G. (1990). *The baby doctors*. New York: Dell.

Konner, M. (1991). *Childhood*. Boston: Little, Brown.

Kopp, C. (1987). Developmental risk: Historical reflections. In J. Osofsky (Ed.), *Handbook of infant development*. New York: Wiley.

Kornzweig, A. L. (1980). New ideas for old eyes. *Journal of the American Geriatrics Society, 28,* 145.

Koyle, P. R., Jensen, L. C., Olsen, J., & Cundick, B. (1989). Comparison of sexual behaviors among adolescents. *Youth and Society, 20*(4), 461–76.

Kraeer, R. J. (1981). The therapeutic value of the funeral in post-funeral counseling. In O. S. Margolis & others (Eds.), *Acute grief*. New York: Columbia University Press.

Kreider, D. G. & Motto, J. A. (1974). Parent-child role reversal and suicidal states in adolescence. *Adolescence, 10*(35), 365–70.

Kroger, J. & Haslett, S. J. (1988). Separation-individuation and ego identity status in late adolescence. *Journal of Youth and Adolescence, 17*(1), 59–80.

Krouse, H. (1986). Use of decision frames by elementary school children. *Perceptual and Motor Skills, 63,* 1107–12.

Kübler-Ross, E. (1969). *On death and dying*. New York: Macmillan.

Kübler-Ross, E. (1975). *Death: The final stage of growth*. Englewood Cliffs, NJ: Prentice-Hall.

Kübler-Ross, E. & Warshaw, W. (1978). *To live until we say goodbye*. Englewood Cliffs, NJ: Prentice-Hall.

Kurz, S. (1977, November 23). Teenage prostitutes. *Equal Times,* 6.

Labouvie-Vief, G. & Lawrence, R. (1985). Object knowledge, personal knowledge, and processes of equilibration in adult cognition. *Human Development, 28,* 25–39.

Lamb, M. E. (1987). *The father's role: Cross-cultural perspectives.* Hillsdale, NJ: Erlbaum.

Lamb, M., Elster, A., & Tavare, J. (1986). Behavioral profiles of adolescent mothers and partners with varying intracouple age differences. *Journal of Adolescent Research, 1*(4), 399–408.

Lamb, M. & Sternberg, K. (1990). Some thoughts about infant daycare. *Research and Clinical Center for Child Development, 12,* 70–77.

Lancaster, J. B. & Hamburg, B. A. (1986). *School-age pregnancy and parenthood.* New York: Aldine de Gruyter.

Langlois, J. & Downs, A. C. (1979). Peer relations as a function of physical attractiveness: The eye of the beholder or behavioral reality. *Child Development, 50,* 409–18.

Langlois, J., Ritter, J., Rogman, L., & Vaughn, L. (1991). Facial diversity and infant preferences for attractive faces. *Developmental Psychology, 27,* 79–84.

Langone, J. (1991). *AIDS: The facts.* Boston: Little, Brown.

Lansdown, R. & Walker, M. (1991). *Your child's development from birth through adolescence.* New York: Knopf.

Larson, R. & Johnson, C. (1981). Anorexia nervosa in the context of daily experience. *Journal of Youth and Adolescence, 10*(6), 455–71.

Lass, N. & Golden, S. (1971). The use of isolated vowels as auditory stimuli in eliciting the verbal transformation effect. *Canadian Journal of Psychology, 25,* 349.

Lawton, M. P. (1984). Health and subjective well-being. *International Journal of Aging and Human Development, 19,* 157–66.

Lazar, I. & Darlington, R. (1982). Lasting effects of early education: A report from the consortium for longitudinal studies. *Monographs of the Society for Research in Child Development, 47*(195).

Leboyer, F. (1975). *Birth without violence.* New York: Knopf.

Lee, P. R., Franks, P., Thomas, G. S., & Paffenberger, R. S. (1981). *Exercise and health: The evidence and its implications.* Cambridge, MA: Oelgeschlager, Gunn, & Hain.

Lehman, H. C. (1953). *Age and achievement.* Princeton, NJ: Princeton University Press.

Lehman, H. C. (1962). The creative production rates of present versus past generations of scientists. *Journal of Gerontology, 17,* 409–17.

Lehman, H. C. (1964). The relationship between chronological age and high level research output in physics and chemistry. *Journal of Gerontology, 19,* 157–64.

Leiblum, S. R. (1990). Sexuality and the midlife woman. *Psychology of Women Quarterly, 14*(4), 495–508.

Leigh, R. J. (1982). The impoverishment of ocular motility in the elderly. In R. Sekuler, D. Kline, & K. Dismukes (Eds.), *Aging and human visual function* (pp. 173–80). New York: Alan R. Liss.

Lenneberg, E. (1967). *Biological foundations of language.* New York: Wiley.

Leventhal, H. & Tomarken, A. (1986). Emotions: Today's problems. In M. Rosenzweig & L. Porter (Eds.), *Annual review of psychology.* Palo Alto, CA: Annual Reviews.

Levin, J. S. & Markides, K. S. (1986). Religious attendance and subjective health. *Journal for the Scientific Study of Religion. 25,* 31–38.

Levinson, D. (1978). *The seasons of a man's life.* New York: Knopf.

Levinson, D. (1986). A conception of adult development. *American Psychologist, 41*(1), 3–13.

Levinson, D. (in press). *Seasons of a woman's life.* New York: Knopf.

Lewis, C. (1963). *A grief observed.* New York: Seabury Press.

Lewis, D. & Greene, J. (1982). *Thinking better.* New York: Rawson, Wade.

Lewis, M. & Brooks-Gunn, J. (1979). *Social cognition and the acquisition of self.* New York: Plenum.

Lickona, T. (1983). *Raising good children.* New York: Bantam Books, Inc.

Lidz, T. & Lidz, R. W. (1984). Oedipus in the Stone Age. *Journal of the American Psychoanalytic Association, 32*(3), 507–27.

Liebert, R. M., Sprafkin, J. N., & Davidson, E. (1988). *The early window.* Elmsford, New York: Pergamon.

Lifson, A., Hessol, N., & Rutherford, G. W. (1989, June). *The natural history of HIV infection in a cohort of homosexual and bisexual men: Clinical manifestations, 1978–1989.* Paper presented at the Fifth International Conference on AIDS, Montreal.

Lindemann, E. (1944). Symptomology and management of acute grief. *American Journal of Psychiatry, 101,* 141–48.

Lindholm, K. J. (1990). Bilingual immersion education: Educational equity for language minority students. In A. Barona & E. E. Garcia (Eds.), *Children at risk: Poverty, minority status, and other issues in educational equity* (pp. 77–89). Washington, DC: National Association of School Psychologists.

Lips, H. (1993). *Sex and gender.* Mountain View, CA: Mayfield.

Lisina, M. I. (1983). The development of interaction in the first seven years. In W. Hartup (Ed.), *Review of child development research.* Chicago: University of Chicago Press.

Lohr, M. J., Essex, M. J., & Klein, M. H. (1988). The relationships of coping responses to physical health status and life satisfaction among older women. *Journal of Gerontology: Psychological Sciences, 43*(2), 54–60.

Lopata, H. Z. (1973). *Widowhood in an American city.* Cambridge, MA: Schenkman.

Lorand, S. & Schneer, H. I. (Eds.). (1961). *Adolescence: Psychoanalytic approaches to problems and theory.* New York: Hoeber.

Lott, B. (1989). *Women's lives.* Monterey, CA: Brooks/Cole.

Lowenkopf, E. (1982, May/June). Anorexia nervosa: Some nosological considerations. *Comprehensive Psychiatry, 23*(3), 233–39.

Maccoby, E. (1988). Social emotional development and response to stressors. In N. Garmezy & M. Rutter (Eds.), *Stress, coping and development in children.* Baltimore: The Johns Hopkins University Press.

Mack, J. E. & Hickler, H. (1982). *Vivienne: The life and suicide of an adolescent girl.* New York: New American Library.

MacRae, H. (1992). Fictive kin as a component of the social networks of older people. *Research on Aging, 14*(2), 226–47.

Mader, S. (1984). Hearing impairment in elderly persons. *Journal of the American Geriatrics Society, 32,* 548.

Maher, E. L. (1983). Burnout and commitment: A theoretical alternative. *Personnel and Guidance Journal, 61*(7), 390–93.

Magid, K. & McKelvey, C. (1987). *High risk: Children without a conscience.* New York: Bantam Books.

Manchester, W. (1983). *The last lion: Winston Spencer Churchill.* Boston: Little, Brown.

Marcia, J. E. (1966). Development and validation of ego identity status. *Journal of Personality and Social Psychology, 3,* 551–58.

Marcia, J. E. (1967). Ego identity status: Relationship to change in self-esteem, general maladjustment and authoritarianism. *Journal of Personality, 35,* 118–33.

Marcia, J. E. (1968). The case history of a construct: Ego identity status. In E. Vinacke (Ed.), *Readings in general psychology.* New York: Van Nostrand Reinhold.

Marcia, J. E. (1980). Identity in adolescence. In J. Adelson (Ed.), *Handbook of adolescent psychology.* New York: Wiley.

Marcia, J. E. (1983). Some directions for the investigation of ego development in early adolescence. *Journal of Early Adolescence, 3*(3), 215–23.

Marshall, L. (1981). Auditory processing in aging listeners. *Journal of Speech and Hearing Disorders, 46,* 226–40.

Marsiglio, W. (1992). Stepfathers with minor children living at home. *Journal of Family Issues, 13*(2), 195–214.

Martin, C. & Little, J. (1990). The relation of gender understanding to children's sex-typed preferences and gender stereotypes. *Child Development, 61,* 1427–39.

Martin, C., Wood, C., & Little, J. (1990). The development of gender stereotype components. *Child Development, 61,* 1891–1904.

Martin, D. H., Mroczkowski, T. F., Dalu, Z. A., McCarty, J., Jones, R. B., Hopkins, S. J., Johnson, R. B., & Azithromycin for Chlamydial Infections Study Group (1992). A controlled trial of a single dose of Azithromycin for the treatment of chlamydial urethritis and cervicitis. *New England Journal of Medicine, 327*(13), 921–25.

Martocchio, J. J. (1989). Age-related differences in employee absenteeism: A meta-analysis. *Psychology and Aging, 4,* 409–14.

Maslow, A. (1987). *Motivation and personality.* (Revised by R. Frager, J. Fadiman, D. McReynolds, & R. Cox.) New York: Harper & Row.

Mason, J. & Au, K. (1990). *Reading instruction for today.* Glenview, IL: Scott, Foresman/Little, Brown.

Masoro, E. J. (1984). Nutrition as a modulator of the aging process. *Physiologist, 27*(2), 98–101.

Massachusetts Department of Public Health (1991). *Adolescents at risk 1991: Sexually transmitted diseases.* Boston, MA: Author.

Masters, W. & Johnson, V. (1966). *Human sexual response.* Boston: Little, Brown.

Masters, W. & Johnson, V. (1970). *Human sexual inadequacy.* Boston: Little, Brown.

Masters, W. H., Johnson, V. E., & Kolodny, R. C. (1986). *Masters and Johnson on sex and human loving.* Boston: Little, Brown.

Matheny, A. P. (1980). Bayley's infant behavior record: Behavioral components and twin analyses. *Child Development, 51,* 466–75.

Matthews, K., Wing, R., Kuller, L., Meilahn, E., Kelsey, S., Costello, E., Caggiula, A. (1990). Influences of natural menopause on psychological characteristics and symptoms of middle-aged healthy women. *Journal of Consulting and Clinical Psychology, 58*(3), 345–51.

Matthews, S. H. (1987). Provision of care to old parents: Division of responsibility among adult children. *Research on Aging, 9,* 45–60.

Matusow, A. J. (1984). *The unraveling of America.* New York: Harper & Row.

Maurer, D. & Maurer, C. (1988). *The world of the newborn.* New York: Basic Books.

May, R. (1975). *The courage to create.* New York: Norton.

Mayer, M. (1978). *The male mid-life crisis—Fresh start after 40.* New York: Doubleday.

Mazess, R. B. (1982). On aging bone loss. *Clinical Orthopaedics and Related Research, 165,* 239–52.

McAnarney, E. (1979). Adolescent and young adult suicide in the U.S.—A reflection of social unrest? *Adolescence, 14*(56), 765–74.

McCann, I. L. & Holmes, D. S. (1984). Influence of aerobic exercise on depression. *Journal of Personality and Social Psychology, 46*(5), 1142–47.

McCarthy, J. (1979, May 5). Jogging. *Time,* 46.

McCary, J. (1978). *McCary's human sexuality.* New York: Van Nostrand Reinhold.

McCoy, K. (1991, May). It's 4:00 P.M., do you know what your teens are doing? *Family Circle,* pp. 17–21.

McCrae, R. & Costa, P. T., Jr. (1984). *Emerging lives, enduring dispositions: Personality in adulthood.* Boston: Little, Brown.

McEvoy, G. M. & Cascio, W. F. (1989). Cumulative evidence of the relationship between employee age and job performance. *Journal of Applied Psychology, 74,* 11–17.

McFarland, R. (1968). The sensory and perceptual processes in aging. In K. Schaie (Ed.), *Theory and methods of research on aging.* Morgantown, WV: West Virginia University Press.

McGhee, P. (1979). *Humor: Its origin and development.* San Francisco: W. H. Freeman.

McGhee, P. (1988). The role of humor in enhancing children's development and adjustment. *Journal of Children in Contemporary Society, 20,* 249–74.

McIntire, J. (1980). Suicide and self-poisoning in pediatrics. *Resident and Staff Physician, 21,* 72–85.

McKenry, D., Tishler, C., & Kelley, C. (1982). Adolescent suicide: A comparison of attempters and non-attempters in an emergency room population. *Clinical Pediatrics, 21*(5), 911–16.

McKenry, P., Walters, L., & Johnson, C. (1979). Adolescent pregnancy: A review of the literature. *Family Coordinator, 33,* 17–29.

McLaughlin, B. (1990). Development of bilingualism: Myth and reality. In A. Barona & E. E. Garcia (Eds.), *Children at risk: Poverty, minority status, and other issues in educational equity* (pp. 77–89). Washington, DC: National Association of School Psychologists.

Mead, M. (1972, April). *Long living in cross-sectional perspective.* Paper presented to the Gerontological Society, San Juan, Puerto Rico.

Medway, F. J. & Rose, J. S. (1986). Grade retention. In T. R. Kratochwill (Ed.), *Advances in school psychology* (Vol. 5, pp. 141–75). Hillsdale, NJ: Erlbaum.

Meeks, J. E. & Cahill, A. J. (1988). In S. C. Feinstein (Ed.), *Adolescent Psychiatry, 15,* 475–86.

Meilman, P. (1979). Cross-sectional age changes in ego identity status during adolescence. *Developmental Psychology, 15*(2), 230–31.

Meltz, B. F. (1988, November 26). Saving the magic moments. *Boston Globe,* 56.

Menyuk, P. (1982). Language development. In C. Kopp & J. Krakow (Eds.), *The child.* Reading, MA: Addison-Wesley.

Miller, J. B. (1976). *Toward a new psychology of women.* Boston: Beacon Press, Inc.

Mintz, B. I. & Betz, N. E. (1988). Prevalence and correlates of eating disordered behaviors among undergraduate women. *Journal of Counseling Psychology, 35,* 463–71.

Minuchin, P. & Shapiro, E. (1983). The school as a context for social development. In P. Mussen (Ed.), *Handbook of child psychology.* New York: Wiley.

Mitchell, J. (1972). Some psychological dimensions of adolescent sexuality. *Adolescence, 7,* 447–58.

Mitford, J. (1963). *The American way of death.* New York: Simon & Schuster.

Moen, P. (1985). Continuities and discontinuities in women's labor force activity. In G. H. Elder (Ed.), *Life course dynamics: Trajectories and transitions, 1968–1980* (pp. 113–55). Ithaca, NY: Cornell University Press.

Moll, L. & Diaz, R. (1984). Teaching writing as communication: The use of ethnographic findings in classroom practice. In D. Bloome (Ed.), *Literacy and Schooling* (pp. 193–221). Norwood, NJ: Ablex.

Molnar, J. & Rubin, D. (1991). *The impact of homelessness on children: Review of prior studies and implications for future research.* Paper presented at the NIMH/NIAA research conference, Cambridge, MA.

Money, J. (1980). *Love and love sickness.* Baltimore: The Johns Hopkins University Press.

Money, J. (1987). Sin, sickness, or status? Homosexual gender identity and psychoneuroendocrinology. *American Psychologist, 42*(4), 384–99.

Montemayor, R. & Brownlee, J. R. (1987). Fathers, mothers, and adolescents: Gender-based differences in parental roles during adolescence. *Journal of Youth and Adolescence, 16*(3), 281–91.

Moore, B. N. & Parker, R. (1986). *Critical thinking*. Mountain View, CA: Mayfield Publishing Co.

Moore, E. W., McCann, H., & McCann, J. (1985). *Creative and critical thinking*. Boston, MA: Houghton Mifflin.

Moore, G. (1983). *Developing and evaluating educational research*. Boston: Little, Brown.

Moore, K., Jofferth, S., & Wertheimer, I. (1979). Teenage motherhood. *Children Today, 6,* 12–16.

Moore, K. A., Peterson, J. L., & Furstenberg, F. F. (1986). Parental attitudes and the occurrence of early sexual activity. *Journal of Marriage and Family, 48,* 777–82.

Morgan, L. A. (1991). *After marriage ends: Economic consequences for midlife women*. London: Sage.

Morinis, A. (1985). The ritual experience: Pain and the transformation of consciousness in ordeals of initiation. *Ethos, 13*(2), 150–74.

Morison, P. & Masten, A. (1991). Peer reputation in middle childhood as a predictor of adaptation in adolescence: A seven-year follow-up. *Child Development, 62,* 991–1007.

Morris, J. (1974, July). Conundrum. *Ms. Magazine,* 57–64.

Morton, T. (1987). Childhood aggression in the context of family interactions. In D. Crowell, I. Evans, & C. O'Donnell (Eds.), *Childhood aggression and violence*. New York: Plenum Press.

Moskowitz, B. (1979). The acquisition of language. *Scientific American, 239,* 92–108.

Moyers, W. & Bly, R. (1989). *A gathering of men*. New York: N.E.T.

Muehlbauer, G. & Dodder L. (1983). *The losers: Gang delinquency in an American suburb*. New York: Praeger.

Muller, C. F. & Boaz, R. F. (1988). Health as a reason or a rationalization for being retired? *Research on Aging, 10*(1), 37–55.

Murphy, C. (1983). Age-related effects on the threshold, psychophysical function, and pleasantness of menthol. *Journal of Gerontology, 38,* 217–22.

Muuss, R. (1982). *Theories of adolescence* (4th ed.). New York: Random House.

Myers, L. S. & Morokoff, P. J. (1986). Physiological and subjective sexual arousal in pre- and postmenopausal women and postmenopausal women taking replacement therapy. *Psychophysiology, 23*(3), 283–92.

Nagin, D. & Farrington, D. (1992). The stability of criminal potential from childhood to adulthood. *Criminology, 30,* 235–60.

Nagy, P. (1982). Limitations of recent research relating Piaget's theory to adolescent thought. *Review of Educational Research, 52*(4), 513–56.

National Association for the Education of Young Children (1986). How to choose a good early childhood program. Washington, DC: Author.

National Association of Secondary School Principals. (1984). *The mood of American youth*. Reston, VA: Author.

National Center for Education Statistics (1992). *American education at a glance*. Washington, DC: Office of Educational Research and Improvement.

National Commission on Children (1991). *Beyond rhetoric: A new American agenda for children and families*. Washington, DC: U.S. Government Printing Office.

National Institute of Allergy and Infectious Diseases. (1987). *STDA*. Atlanta: Center for Disease Control.

National Institute for Health Statistics (1986). *Obesity in the American population*. Washington, DC: U.S. Government Printing Office.

National Institute on Aging. (1979). *Is male menopause a myth?* Washington, DC: U.S. Government Printing Office.

National Research Council. (1986). *Environmental tobacco smoke: Measuring exposures and assessing health effects*. Washington, DC: National Academy Press.

Nelson, D. A. (1980). *Frequently seen stages in adolescent chemical use*. Minneapolis, MN: CompCare Publications.

Nesselroade, J. R., Pedersen, N. L., McClearn, G. E., Plomin, R., & Bergeman, C. S. (1988). Factorial and criterion validities of telephone-assessed cognitive ability measures: Age and gender comparisons in adult twins. *Research on Aging, 10*(2), 220–34.

Neugarten, B. (Ed.). (1968). *Middle age and aging*. Chicago: University of Chicago Press.

Neugarten, B. & Moore, J. (1968). The changing age-status system. In B. Neugarten (Ed.), *Middle age and aging*. Chicago: University of Chicago Press.

Neugarten, B. L. & Weinstein, K. K. (1964). The changing American grandparent. *Journal of Marriage and the Family, 26,* 199–206.

Newman, P. R. & Newman, B. M. (1976). Early adolescence and its conflict: Group identity versus alienation. *Adolescence, 11*(42), 261–73.

Nightingale, E. O. & Goodman, M. (1990). *Before birth*. Cambridge, MA: Harvard University Press.

Nightingale, E. O. & Wolverton, L. (1988). *Adolescent rolelessness in modern society*. Washington, DC: Carnegie Council on Adolescent Development.

Nilsson, L., Furuhjelm, M., Ingleman-Sundberg, A., & Wirsen, C. (1987). *A child is born*. New York: Delacorte.

Noller, R. B. (1983). *Mentoring: An annotated bibliography*. Buffalo, NY: Bearly Limited.

Northcott, H. C. & Lowe, G. S. (1987). Job and gender influences in the subjective experience of work. *Canadian Review of Sociology and Anthropology, 24*(1), 117–31.

Norton, A. J. & Moorman, J. E. (1987). Current trends in marriage and divorce among American women. *Journal of Marriage and the Family, 49*(1), 3–14.

Nuckolls, K. B., Cassell, J., & Kaplan, B. H. (1972). Psychosocial assets, life crisis, and the prognosis of pregnancy. *American Journal of Epidemiology, 95,* 431–41.

Nuttall, R. & Nuttall, E. (1979). *The impact of disaster on coping behaviors of families*. Unpublished manuscript, Boston College, Chestnut Hill, MA.

Office of the Assistant Secretary for Health. (1988). Report of the second Public Health Service AIDS Prevention and Control Conference. *Public Health Report, 3,* 103.

Office of Technology Assessment (1991). *Adolescent health* (Vol. 1–3). Washington, DC: U.S. Government Printing Office.

Office of Technology Assessment (1991). *National household survey on drug abuse: Main findings, 1988,* DHHS Pub. no. (ADM) 90—1682. Rockville, MD: U.S. Department of Health and Human Services.

Olson, G. & Sherman, T. (1983). Attention, learning and memory in infants. In P. Mussen (Ed.), *Handbook of child psychology*. New York: Wiley.

Ornstein, R. & Sobel, D. (1987). *The healing brain*. New York: Simon & Schuster.

Orr, C. A. & Van Zandt, S. (1987). *The role of grandparenting in building family strengths*. Paper presented at the Annual National Symposium on Building Family Strengths.

Osofsky, J. (1976). Neonatal characteristics and mother-infant interaction in two observational situations. *Child Development, 47,* 1138–47.

Osofsky, J. (Ed.) (1987). *Handbook of infant development*. New York: Wiley.

Ossip-Klein, D. J., Doyne, E. J., Bowman, E. D., Osborn, K. M., McDougall-Wilson, I. B., & Neimeyer, R. A. (1989). Effects of running or weight lifting on self-concept in clinically depressed women. *Journal of Consulting and Clinical Psychology, 57,* 158–61.

Otto, V. (1972). Suicidal attempt in childhood and adolescents—today and after ten years: A follow-up study. In A. L. Annell (Ed.), *Depressive states in childhood and adolescence*. New York: Halstead Press.

Outward Bound, U.S.A. (1988). *Outward bound*. Greenwich, CT: Outward Bound National Office.

Owens, W. (1953). Aging and mental abilities. *Genetic Psychology Monographs, 48,* 3–54.

Packer, L. & Rosenblatt, R. (1979). Issues in the study of social behavior in the first week of life. In D. Shaffer and J. Dunn (Eds.). *The First Year of Life*. New York: Wiley.

Palmore, E. B. (1981). The facts on aging quiz: Part two. *The Gerontologist, 21*(4), 431–37.

Parke, C. M. & Weiss, R. S. (1983). *Recovery from bereavement*. New York: Basic Books.

Parke, R. & Slaby, R. (1983). The development of aggression. In P. Mussen (Ed.), *Handbook of child psychology*. New York: Wiley.

Paul, R. W. (1987). Dialogical thinking. In J. B. Baron & R. J. Sternberg (Eds.), *Teaching thinking skills*. New York: W. H. Freeman.

Paulos, J. (1988). *Innumeracy*. New York: Hill & Wang.

Pearlman, C. (1972, November). Frequency of intercourse in males at different ages. *Medical Aspects of Human Sexuality,* 92–113.

Peplau, L. (1981). What homosexuals want. *Psychology Today,* pp. 28–58.

Perkins, U. (1987). *Explosion of Chicago's black street gangs: 1900 to the present*. Chicago: Third World Press.

Perlmutter, B. F. (1987). Delinquency and learning disabilities: Evidence for compensatory behaviors and adaptation. *Journal of Youth and Adolescence, 16*(2), 89–96.

Perris, E., Myers, N., & Clifton, R. (1990). Long-term memory for a single infancy experience. *Child Development, 61,* 1796–1807.

Perry, W. (1968a). *Forms of intellectual and ethical development in the college years*. New York: Holt, Rinehart & Winston.

Perry, W. (1968b, April). *Patterns of development in thought and values of students in a liberal arts college: A validation of a scheme.* Washington, DC: U.S. Department of Health, Education, and Welfare, Office of Education, Bureau of Research. Final report.

Perry, W. (1981). Cognitive and ethical growth. In A. Chickering (Ed.), *The modern American college.* San Francisco: Jossey-Bass.

Petersen, A. (1988). Adolescent development. In M. Rosenzweig & L. Porter (Eds.), *Annual review of psychology.* Palo Alto, CA: Annual Reviews.

Petersen, A. C., Crockett, I., Richards, M., & Boxer, A. (1988). A self-report measure of pubertal status. *Journal of Youth and Adolescence, 17,* 117–34.

Phares, V. (1992). Where's poppa? The relative lack of attention to the role of fathers in child and adolescent psychopathology. *American Psychologist, 47,* 656–64.

Phillips, D. (1979). Suicide, motor vehicle fatalities, and the mass media: Evidence toward a theory of suggestion. *American Journal of Sociology, 84*(5), 1150–74.

Phillips, R. T. & Alcebo, A. M. (1986). The effects of divorce on black children and adolescents. *American Journal of Social Psychiatry, 6*(1), 69–73.

Piaget, J. (1926). *The language and thought of the child.* New York: Harcourt, Brace, & World.

Piaget, J. (1929). *The child's conception of the world.* New York: Harcourt, Brace & World.

Piaget, J. (1932). *The moral judgment of the child.* New York: Macmillan.

Piaget, J. (1952). *The origins of intelligence in children.* New York: International Universities Press.

Piaget, J. (1953). *The origins of intelligence in the child.* New York: Harcourt, Brace, Jovanovich.

Piaget, J. (1966). *Psychology of intelligence.* Totowa, NJ: Littlefield, Adams & Co.

Piaget, J. (1967). *Six psychological studies.* New York: Random House.

Piaget, J. (1973). *The child and reality.* New York: Viking Press.

Piaget, J. & Inhelder, B. (1969). *The psychology of the child.* New York: Basic Books.

Piccigallo, P. (1988, Fall). Preschool: Head start or hard push? *Social Policy,* pp. 45–48.

Pine, V. (1979, March 25). Funerals criticized. *Boston Globe,* C17.

Pinon, M., Huston, A., & Wright, J. (1989). Family ecology and child characteristics that predict young children's educational television viewing. *Child Development, 60,* 846–56.

Pizer, H. (Ed.). (1983). *Over fifty-five, healthy and alive.* New York: Van Nostrand Reinhold.

Pleck, J. H. (1985). *Working wives/Working husbands.* Beverly Hills, CA: Sage.

Polit, D. (1985). *A review of the antecedents of adolescent sexuality, contraception, and pregnancy outcomes.* Jefferson City, MO: Humanalysis.

Postman, N. (1982). *The disappearance of childhood.* New York: Dell Publishing Co., Inc.

Powledge, T. (1983, July). The importance of being twins. *Psychology Today, 17,* 20–27.

Prechtl, H. (1977). *The neurological examination of the full-term newborn infant.* London: Heinemann.

Rafferty, M. & Shinn, M. (1991). The impact of homelessness on children. *American Psychologist, 46,* 1170–79.

Ralph, N., Lochman, J., & Thomas, T. (1984). Psychosocial characteristics of pregnant and multiparous adolescents. *Adolescence, 19,* 283–94.

Ramsey, P. (1987). *Teaching and learning in a diverse world.* New York: Teachers College, Columbia University.

Ramsey, P. W. (1982). Do you know where your children are? *Journal of Psychology and Christianity, 1*(4), 7–15.

Rando, T. A. (1986). Creation of rituals in psychotherapy. *Forum Newsletter, 6,* 8–9.

Raphael, D., Feinberg, R., & Bachor, D. (1987). Student teachers' perceptions of the identity formation process. *Journal of Youth and Adolescence, 16*(4), 331–44.

Regoli, R. & J. Hewitt. (1991). *Delinquency in society.* New York: McGraw-Hill.

Reinke, B., Ellicott, A., Harris, R., & Hancock, E. (1985). Timing of psychosocial change in women's lives. *Human Development, 28,* 259–80.

Remafedi, G. (1988). Homosexual youth. *Journal of the American Medical Association, 258*(2), 222–25.

Resener, M. (1979, August). Burnout: The new stress disease. *Harper's Bazaar,* 92–93.

Rest, J. (1983). Morality. In P. Mussen (Ed.), *Handbook of child psychology.* New York: Wiley.

Restak, R. (1986). *The infant mind.* New York: Doubleday.

Rhodes, S. (1983). Age-related differences in work attitudes and behavior: A review and conceptual analysis. *Psychological Bulletin, 93,* 328–67.

Rhyne, D. (1981). Basis of marital satisfaction among men and women. *Journal of Marriage and the Family, 43,* 941–55.

Ricks, M. (1985). The social transmission of parental behavior: Attachment across generations. In I. Bretherton & E. Waters (Eds.), *Growing points of attachment: Theory and research.* Monographs of the Society for Research in Child Development. Chicago: University of Chicago Press.

Rierdan, J., Koff, E., & Stubbs, M. (1988). A longitudinal analysis of body image as a predictor of the onset and persistence of adolescent girls' depression. *Working Paper No. 188,* Wellesley College Center for Research on Women.

Riese, M. (1990). Neonatal temperament in monozygotic and dizygotic twin pairs. *Child Development, 61,* 1230–37.

Ringel, S. P. & Simon, D. B. (1983). Practical management of neuromuscular diseases in the elderly. *Geriatrics, 38,* 86.

Ritter, J., Casey, R., & Langlois, J. (1991). Adults' responses to infants varying in appearance of age and attractiveness. *Child Development, 62,* 68–82.

Ritter, J. & Langlois, J. (1988). The role of physical attractiveness in the observation of adult-child interactions: Eye of the beholder or behavioral reality? *Developmental Psychology, 24,* 254–63.

Rivchun, S. B. (1980, August). Be a mentor and leave a lasting legacy. *Association Management, 32*(8), 71–74.

Roazen, P. (1976). *Erik H. Erikson: The power and limits of a vision.* New York: The Free Press.

Robertson, M. J. (1989). *Homeless youth: An overview of recent literature.* Paper presented at the National Conference on Homeless Children and Youth, Institute for Policy Studies at the Johns Hopkins University, Washington, DC.

Robins, L. & Rutter, M. (1990). *Straight and devious pathways from childhood to adulthood.* Cambridge: Cambridge University Press.

Robinson, J. K. (1983). Skin problems of aging. *Geriatrics, 38,* 57–65.

Rodman, H. (1989). Controlling adolescent fertility. *Society, 23*(1), 35–37.

Rogers, D. (1979). *Adult psychology.* Englewood Cliffs, NJ: Prentice-Hall.

Rogoff, B. (1990). *Apprenticeship in thinking: Cognitive development in social context.* New York: Oxford University Press.

Rogoff, B. & Morelli, G. (1989). Perspectives and children's development from cultural psychology. *American Psychologist, 44*(2), 343–48.

Rogow, A. M., Marcia, J. E., & Slugoski, B. R. (1983). The relative importance of identity status interview components. *Journal of Youth and Adolescence, 12*(5), 387–400.

Rohn, R. (1977). Adolescents who attempt suicide. *The Journal of Pediatrics, 90,* 636–38.

Roll, S. & Miller, L. (1978, Spring). Adolescent males' feelings of being understood by their fathers as revealed through clinical interviews. *Adolescence,* 83–94.

Rollins, B. & Feldman, H. (1970). Marital satisfaction over the family life cycle. *Journal of Marriage and the Family, 32,* 20–28.

Roper Organization. (1987). *The American Chicle youth poll.* Storrs, CT: University of Connecticut (Roper).

Rose, S., Feldman, J., Wallace, I., & McCarton, C. (1991). Information processing at 1 year: Relation to birth status and developmental outcome during the first 5 years. *Developmental Psychology, 27,* 723–37.

Rosen, B. M., Bahn, A. K., Shellow, R., & Bower, E. M. (1965). Adolescent patients served in outpatient clinics. *American Journal of Public Health, 55,* 1563–77.

Rosen, E. I. (1987). *Bitter choices: Blue-collar women in and out of work.* Chicago: University of Chicago Press.

Rosen, J. C., Gross, J., & Vara, L. (1987). Psychological adjustment of adolescents attempting to lose or gain weight. Special issue: Eating disorders. *Journal of Consulting and Clinical Psychology, 55*(5), 742–47.

Rosen, R. (1980). Adolescent pregnancy decision-making: Are parents important? *Adolescence, 15*(57), 43–54.

Rosenberg, G. S. & Anspach, D. F. (1973). Sibling solidarity in the working class. *Journal of Marriage and the Family, 35,* 108–13.

Rosenthal, D. A., Gurney, R. M., & Moore, S. M. (1981). From trust to intimacy: A new inventory for examining Erikson's stages of psychosocial development. *Journal of Youth and Adolescence, 10*(6), 525–37.

Rosin, H. M. (1990). The effects of dual career participation on men: Some determinants of variation in career and personal satisfaction. *Human Relations, 43*(2), 169–82.

Rosser, P. & Randolph, S. (1989) Black American infants: The Howard University normative study. In J. K. Nugent (Ed.), *The cultural context of infancy* (Vol. 1). Norwood, NJ: Ablex.

Rotheram-Borus, M. J., & Koopman, C. (1991). HIV and adolescents. *Journal of Primary Prevention, 12*(1), 65–82.

Rotter, J. (1971, June). External control and internal control. *Psychology Today, 5,* 37ff.

Rourke, B., Bakker, D., Fisk, J., & Strang, J. (1983). *Child neuropsychology.* New York: Guilford Press.

Rowe, I. & Marcia, J. E. (1980). Ego identity status, formal operations, and moral development. *Journal of Youth and Adolescence, 9*(2), 87–99.

Rowland, D. L., Heiman, J. R., Gladue, B. A., & Hatch, J. P. (1987). Endocrine, psychological and genital response to sexual arousal in men. *Psychoneuroendocrinology, 12*(2), 149–58.

Roy, A. (1990). Family rituals: Functions and significance for clergy and psychotherapists. *Group, 14*(1), 59–64.

Rubin, K., Fein, G., & Vandenberg, B. (1983). Play. In P. Mussen (Ed.), *Handbook of child psychology.* New York: Wiley.

Rubin, L. B. (1979). *Women of a certain age: The midlife search for self.* New York: Harper & Row.

Rubin, R. H. (1981). Attitudes about male-female relations among black adolescents. *Adolescence, 16*(61), 159–74.

Rubin, Z. (1980). *Children's friendships.* Cambridge, MA: Harvard University Press.

Rudman, D. (1992). The fountain of youth. *Milwaukee Magazine,* pp. 12–13.

Russ-Eft, D., Springer, M., & Beever, A. (1979). Antecedents of adolescent parenthood and consequences at age 30. *Family Coordinator, 16,* 173–78.

Rutter, M. (1979). *Fifteen thousand hours.* Cambridge, MA: Harvard University Press.

Rutter, M. (1980). *Changing youth in a changing society.* Cambridge, MA: Harvard University Press.

Rutter, M. (1981). *Maternal deprivation reassessed.* New York: Penguin Bks., Inc.

Rutter, M. (1987, July). Psychosocial resilience and protective mechanisms. *American Journal of Orthopsychiatry,* 316–31.

Rutter, M. (1987). Continuities and discontinuities from infancy. In J. Osofsky (Ed.), *Handbook of infant development.* New York: Wiley.

Rutter, M. (1988). Stress, coping, and development: Some issues and some questions. In N. Garmezy & M. Rutter (Eds.), *Stress, coping and development in children.* Baltimore: The Johns Hopkins University Press.

Rutter, M. (1989). Pathways from childhood to adult life. *Journal of Child Psychology and Child Psychiatry, 30,* 23–50.

Rutter, M. & Garmezy, N. (1983). Developmental psychopathology. In P. Mussen (Ed.), *Handbook of child psychology.* New York: Wiley.

Sabbath, J. (1969). The suicidal adolescent: The expendable child. *Journal of the American Academy of Child Psychiatry, 8*(2), 272–85.

Sachs, J. (1985). Prelinguistic development. In J. B. Gleason (Ed.), *The development of language.* Columbus, OH: Merill.

Sadler, T. W. (1985). *Langman's medical embryology.* Baltimore: Williams & Wilkins.

Sagan, C. (1977). *The dragons of Eden.* New York: Random House.

St. Peters, M., Fitch, M., Huston, A., Wright, J., & Eakins, D. (1991). Television and families: What do young children watch with their parents? *Child Development, 62,* 1409–23.

Sales, E. (1978). Women's adult development. In I. Fieze (Ed.), *Women and sex roles.* New York: W. W. Norton.

Salthouse, T. (1990). Cognitive competence and expertise in aging. In *Handbook of the psychology of aging.* New York: Academic Press.

Saluter, A. F. (1991). Marital status and living arrangements: March 1991. *Current Population Reports,* P-20 (No. 461). Washington, DC: Bureau of the Census.

Sameroff, A. (1975). Early influence on development: Fact or fancy. *Merrill-Palmer Quarterly of Behavior and Development, 21*(4), 23–34, 267–94.

Sameroff, A. (1986). Environmental context of child development. *Journal of Pediatrics, 109,* 192–200.

Santana, G. (1979, September). *Social and familial influences on substance use among youth.* Paper presented to the American Psychological Association, New York.

Santrock, J. W. (1987). The effects of divorce on adolescents: Needed research perspectives. *Family Therapy, 14*(2), 147–59.

Santrock, J. & Yussen, S. (1992). *Child Development.* Dubuque, IA: Wm. C. Brown.

Saudino, K. & Eaton, W. (1991). Infant temperament and genetics: An objective twin study of motor activity level. *Child Development, 62,* 1167–74.

Scarr, S. (1992). Developmental theories for the 1990s: Developmental and individual differences. *Child Development, 63,* 1–19.

Scarr, S., Weinberg, R., & Levine, A. (1986). *Understanding development.* New York: Harcourt, Brace, Jovanovich.

Schaie, K. W. & Hertzog, C. (1983). Fourteen-year cohort-sequential analyses of adult intellectual development. *Developmental Psychology, 19,* 531–43.

Schemeck, H. M. (1983, September 2). Alcoholism tests back disease idea. *New York Times,* A10.

Schibuk, M. (1989). Treating the sibling subsystem: An adjunct of divorce therapy. *American Journal of Orthopsychiatry, 59,* 226–37.

Schiedel, D. S. & Marcia, J. E. (1985). Ego identity, intimacy, sex role orientation, and gender. *Developmental Psychology, 21*(1), 149–60.

Schiffman, S. (1977). Food recognition of the elderly. *Journal of Gerontology, 32,* 586–92.

Schiffman, S. & Pasternak, M. (1979). Decreased discrimination of food odors in the elderly. *Journal of Gerontology, 34,* 73–79.

Schneider, J. (1984). *Stress, loss and grief.* Rockville, MD: Aspen Pubs.

Schofferman, J. (1987). Hospice care of the patient with AIDS. *The Hospice Journal, 3,* 51–84.

Schowalter, J. E. (1978). Parent death and child bereavement. In A. Weiner, I. Gerber, A. Kutscher, & B. Schoenberg (Eds.), *Bereavement.* New York: Columbia University Press.

Schroedel, J. R. (1990). Blue-collar women: Paying the price at home on the job. In H. Y. Grossman and N. L. Chester (Eds.), *The experience and meaning of work in women's lives.* Hillsdale, NJ: Erlbaum.

Schulman, M. (1991). *The passionate mind.* New York: Macmillan.

Schulz, R. (1978). *The psychology of death, dying, and bereavement.* Reading, MA: Addison-Wesley.

Schulz, R. & Ewen, R. B. (1988). *Adult development and aging: Myths and emerging realities.* New York: Macmillan.

Schumm, W. R. & Bugaighis, M. A. (1986). Marital quality over the marital career. *Journal of Marriage and the Family, 48,* 165–68.

Scott, D. (Ed.). (1988). *Anorexia and bulimia.* New York: New York University Press.

Scott, J. P. (1968). *Early experience and the organization of behavior.* Belmont, CA: Brooks/Cole.

Scott-Jones, D. & White, A. (1990). Correlates of sexual activity in early adolescence. *Journal of Early Adolescence, 10*(2), 221–38.

Sears, R. (1975). Your ancients revisited: A history of child development research. In E. M. Hetherington (Ed.), *Review of child development research.* Chicago: University of Chicago Press.

Sears, R. R. & Sears, P. S. (1982). Lives in Berkeley. *Contemporary Psychology, 27*(12), 925–27.

Sebald, H. (1977) *Adolescence: A social psychological analysis* (2nd ed.). Englewood Cliffs, NJ: Prentice-Hall.

Seginer, R. & Flum, H. (1987). Israeli adolescents' self-image profile. *Journal of Youth and Adolescence, 16*(5), 455–72.

Seiden, R. H. & Freitas, R. P. (1980). Shifting patterns of deadly violence. *Suicide and Life Threatening Behavior, 10,* 195–209.

Select Committee on Children, Youth and Families (1985). *Emerging trends in mental health care for adolescents.* Washington, DC: U.S. Government Printing Office.

Seligman, M. (1975). *Helplessness.* San Francisco: W. H. Freeman.

Selman, R. (1980). *The growth of interpersonal understanding.* New York: Academic Press.

Selye, H. (1956). *The stress of life.* New York: McGraw-Hill.

Selye, H. (1974). *Stress without distress.* Philadelphia: Lippincott.

Selye, H. (1975, October). Implications of stress concept. *New York State Journal of Medicine,* 2139–45.

Selye, H. (1982). History and present status of the stress concept. In L. Goldberger & S. Breznitz (Eds.), *Handbook of stress: Theoretical and clinical aspects.* New York: The Free Press.

Serakan, U. (1989). Understanding the dynamics of self-concept of members in dual-career families. *Human Relations, 42*(2), 97–116.

Sessions, W. (1990). Gang violence and organized crime. *The Police Chief, 57,* 17.

Shaffer, D. & Fisher, P. (1981). The epidemiology of suicide in children and young adolescents. *Journal of the American Academy of Child Psychiatry, 20,* 545–65.

Shahtahmasebi, S., Davies, R., & Wenger, G. C. (1992). A longitudinal analysis of factors related to survival in old age. *The Gerontologist, 32*(3), 404–13.

Shatz, C. (1992, September). The developing brain. *Scientific American, 267*(3), 61–67.

Sheehy, G. (1992). *The silent passage: Menopause.* New York: Random House.

Sher, K. (1993). Children of alcoholics and the intergenerational transmission of alcoholism: A biopsychosocial perspective. In J. Baer, G. Marlatt, & R. McMahon (Eds.), *Addictive behaviors across the life span.* Newbury Park, CA: Sage.

Shields, P. & Rovee-Collier, C. (1992). Long-term memory for context-specific category information at six months. *Child Development, 63,* 245–59.

Shirk, S. R. (1987). Self-doubt in late childhood and early adolescence. *Journal of Youth and Adolescence, 16*(1), 59–68.

Siegal, L. J. & Senna, J. J. (1981). *Juvenile delinquency: Theory, practice, and law.* New York: West.

Siegler, I. (1975). The terminal drop hypothesis: Fact or artifact? *Experimental Aging Research, 1,* 169.

Silber, T. (1980). Values relating to abortion as expressed by the inner city adolescent girl—Report of a physician's experience. *Adolescence, 15*(57), 183–89.

Silverberg, S. B. & Steinberg, L. (1987). Influences on marital satisfaction during the middle stages of the family life cycle. *Journal of Marriage and the Family, 49*(4), 751–60.

Silverman, P. (1974). Anticipatory grief from the perspective of widowhood. In B. Schoenberg, A. Carr, A. Kutscher, D. Peretz, & I. Goldberg (Eds.), *Anticipatory grief* (pp. 320–30). New York: Columbia University Press.

Simonton, D. K. (1975). Age and literary creativity. *Journal of Cross-cultural Creativity, 6,* 259–77.

Simonton, D. K. (1976). Biographical determinants of achieved eminence. *Journal of Personality and Social Psychology, 33,* 218–76.

Simonton, D. K. (1977a). Creativity, age and stress. *Journal of Personality and Social Psychology, 35,* 791–804.

Simonton, D. K. (1977b). Eminence, creativity and geographical marginality. *Journal of Personality and Social Psychology, 35,* 805–16.

Singer, S. (1985). *Heredity.* San Francisco: W. H. Freeman.

Skinner, B. F. (1938). *The behavior of organisms.* New York: Appleton-Century-Crofts.

Skinner, B. F. (1948). *Walden two.* New York: Macmillan.

Skinner, B. F. (1953). *Science and human behavior.* New York: Macmillan.

Skinner, B. F. (1957). *Verbal behavior.* New York: Appleton-Century-Crofts.

Skinner, B. F. (1971). *Beyond freedom and dignity.* New York: Knopf.

Skinner, B. F. (1983, September). Origins of a behaviorist. *Psychology Today, 17*(2), 22–33.

Slaby, R. (1990, Spring). Gender concept development legacy. *New Directions for Psychology, 47,* 21–29.

Slugoski, B. R., Marcia, J. E., & Koopman, R. F. (1984). Cognitive and social interactional characteristics of ego identity statuses in college males. *Journal of Personality and Social Psychology, 47*(3), 646–61.

Smart, J. C. & Ethington, C. A. (1987). Occupational sex segregation and job satisfaction of women. *Research in Higher Education, 26*(2), 202–11.

Smith, B. K. (1989). *Grandparenting in today's world.* Austin, TX: Hogg Foundation for Mental Health.

Smith, D. W. E., Seibert, C. S., Jackson, F. W., & Snell, J. (1992). Pet ownership by elderly people: Two new issues. *International Journal of Aging and Human Development, 34*(3), 175–84.

Smith, R. (1990). *A theoretical framework for explaining the abuse of hyperactive children.* Unpublished doctoral dissertation, Boston College, Chestnut Hill, MA.

Smith-Pointer, R. A., Woodward, N. J., Wallston, B. S., Wallston, K. A., Rye, P., & Zylstral, M. (1988). Health care implications of desire and expectancy for control in elderly adults. *Journal of Gerontology: Psychological Sciences, 43*(1), 1–7.

Sommerstein, J. C. (1986). Assessing the older worker: The career counselor's dilemma. Special issue: Career counseling of older adults. *Journal of Career Development, 13*(2), 52–56.

Sonnenstein, F. L., Pleck, J. H., & Ku, L. C. (1990). Sexual activity, condom use and AIDS awareness among adolescent males. *Family Planning Perspectives, 21*(4), 151–58.

South, S. J. & Spitze, G. (1986). Determinants of divorce over the marital life course. *American Sociological Review, 51*(4), 583–90.

Spence, A. (1989). *Biology of human aging.* Englewood Cliffs, NJ: Prentice-hall.

Sprafkin, J. N., Liebert, R. M., & Poulos, R. W. (1975). Effects of a prosocial televised example on children's helping. *Journal of Experimental Child Psychology, 20,* 119–26.

Spreen, O., Tuppet, D., Risser, A., Tuokko, H., & Edgell, D. (1984). *Human developmental neuropsychology.* New York: Oxford University Press.

Sroufe, L. A. & Cooper, R. (1992). *Child development.* New York: Knopf.

Starr, B. & Weiner, M. (1981). *The Starr-Weiner report on sex and sexuality in the mature years.* New York: Stein & Day.

Steinberg, L. (1987). Single parents, step-parents, and the susceptibility of adolescents to antisocial peer pressure. *Child Development, 58*(1), 269–75.

Steinberg, L. (1990). Autonomy, conflict, and harmony in the family relationship. In S. Feldman & G. Elliot (Eds.), *At the threshold: The developing adolescent.* Cambridge, MA.: Harvard University.

Stern, D. (1977). *First relationships.* Cambridge, MA: Harvard University Press.

Stern, D. (1985). *The interpersonal world of the child.* New York: Basic Books.

Stern, D. (1990). *Diary of a baby.* New York: Harper Collins.

Sternberg, R. (Ed.), (1982). *Handbook of human intelligence.* New York: Cambridge University Press.

Sternberg, R. (1986). *Intelligence applied.* New York: Harcourt, Brace, Jovanovich.

Sternberg, R. (1988). *The triarchic mind: A new theory of human intelligence.* New York: Viking Press.

Sternberg, R. J. (1986). The triangular theory of love. *Psychological Review, 93,* 129–35.

Sternberg, R. J. (1990). *Wisdom.* New York: Cambridge University Press.

Sternberg, R. J. & Detterman, D. K. (Eds.), (1979). *Human intelligence: Perspectives on its theory and measurement.* Norwood, NJ: Ablex.

Stevens, J. C. & Cain. W. S. (1987). Old-age deficits in the sense of smell as gauged by thresholds, magnitude matching, and odor identification. *Psychology and Aging, 2,* 36–42.

Stevens-Long, J. (1988). *Adult life* (3rd ed.). Mountain View, California: Mayfield.

Stiffman, A., Earls, F., Robins, L., & Jung, K. (1987). Adolescent sexual activity and pregnancy: Socioenvironmental problems, physical health, and mental health. *Journal of Youth and Adolescence, 16*(5), 497–569.

Stigler, J., Shweder, R., & Herdt, G. (Eds.), (1990). *Cultural psychology.* New York: Cambridge University Press.

Stimmel, B. (1991). *The facts about drug use.* New York: Consumer Report Books.

Stoddard, S. (1977). *The hospice movement: A better way of caring for the dying.* New York: Stein & Day.

Stoller, E. P., Forster, L. E., & Duniho, T. S. (1992). Systems of parent care within sibling networks. *Research on Aging, 14*(1), 28–49.

Stoneman, Z., Brody, G. L., & MacKinnon, C. E. (1986). Same-sex and cross-sex siblings: Activity choices, roles, behavior and gender stereotypes. *Sex Roles, 15,* 495–512.

Story, M. D. (1982). A comparison of university student experience with various sexual outlets in 1974 and 1980. *Adolescence, 17*(68), 737–47.

Strawbridge, W. & Wallhagen, M. (1992). Is all in the family always best? *Journal of Aging Studies, 6*(1), 81–92.

Strobino, D. (1987). Health and medical consequences. In C. Hayes & S. Hofferth (Eds.), *Risking the future: Adolescent sexuality, pregnancy, and child-bearing* (pp. 107–23). Washington, DC: National Academy Press.

Strom, R. & Strom, S. (1982). *Redefining the grandparent role.* Office of Parent Development International Research and Development Reports. Tempe, AZ: Arizona State University, College of Education.

Subcommittee on Health and the Environment. (1987). *Incidence and control of chlamydia.* Washington, DC: U.S. Government Printing Office.

Sudnow, D. (1967). *Passing on.* Englewood Cliffs, NJ: Prentice-Hall.

Sue, S. & Okazaki, S. (1990). Asian-American educational achievements: A phenomenon in search of an explanation. *American Psychologist, 45,* 913–20.

Sugarman, L. (1986). *Lifespan Development: Concepts, Theories and interventions.* New York: Methuen.

Sullivan, T. & Schneider, M. (1987). Development and identity issues in adolescent homosexuality. *Child and Adolescent Social Work Journal, 4*(1), 13–24.

Suomi, S. J., Harlow, H., & Novak, M. A. (1974). Reversal of social deficits produced by isolation rearing in monkeys. *Journal of Human Evolution, 3,* 527–34.

Super, D. E. (1957). *The psychology of careers.* New York: Harper & Row.

Super, D. E. (1983). Assessment in career guidance: Toward truly developmental counseling. *Personnel and Guidance Journal, 61,* 555–62.

Super, D. E. (1990). A life-span, life-space approach to career development. In D. Brown, L. Brooks, & others (Eds.), *Career choice and development.* San Francisco: Jossey-Bass.

Super, D. E. & Thompson, A. S. (1981). *The adult career concerns inventory.* New York: Teachers College, Columbia University.

Sutton-Smith, B. (1988). *Toys as culture.* New York: Gardner Press.

Swanson, E. A. & Bennett, T. F. (1983). Degree of closeness: Does it affect the bereaved's attitudes toward selected funeral practices? *Omega, 13,* 43–50.

Sweet, J. A. & Bumpass, L. L. (1987). *American families and households.* New York: Sage.

Tager, I. B., Weiss, S. T., Munoz, A., Rosner, B., & Speizer, F. E. (1984). Longitudinal study of the effects of maternal smoking on pulmonary function in children. *New England Journal of Medicine, 309,* 699–703.

Takata, S. R., Zevitz, R. G., Berger, R. J., Salem, R. G., Gruberg, M., & Moore, J. (1987). Youth gangs in

Racine: An examination of community perceptions. *Wisconsin-Sociologist, 24,* 132–41.

Tanner, J. M. (1989). *Foetus into man.* Cambridge, MA: Harvard University Press.

Tapley, D. & Todd, W. D. (Eds.), (1988). *Complete guide to pregnancy.* New York: Crown.

Task Force (1976). *The walkabout.* Bloomington, IN: Phi Delta Kappa.

Taub, H. (1975). Effects of coding cues upon short-term memory. *Developmental Psychology, 11,* 254.

Tauber, M. A. (1979). Parental socialization techniques and sex differences in children's play. *Child Development, 50,* 225–34.

Teicher, J. (1973). A solution to the chronic problem of living: Adolescent-attempted suicide. In J. C. Schoolar (Ed.), *Current issues in adolescent psychiatry.* New York: Brunner/Mazel.

Teicher, J. & Jacobs, J. (1966). The physician and the adolescent suicide attempter. *Journal of School Health, 36,* 406.

Television and behavior: Ten years of scientific progress and implications for the eighties (1982). Washington, DC: U.S. Department of Health and Human Services.

Terman, L. M. (1925). *Genetic studies of genius.* Stanford, CA: Stanford University Press.

Tessler, D. J. (1980). *Drugs, kids, and schools: Practical strategies for educators and other concerned adults.* Santa Monica, CA: Goodyear.

Thomas, A. (1981). Current trends in developmental theory. *American Journal of Orthopsychiatry, 51,* 580–609.

Thomas, A. & Chess, S. (1977). *Temperament.* New York: Brunner/Mazel.

Thomas, A., Chess, S., & Birch, H. (1970). The origin of personality. *Scientific American, 223,* 102–9.

Thomas, J. & Datan, N. (1983). *Change and diversity in grandparenting experience.* Paper presented at the Annual Convention of the American Psychological Association.

Timiras, R. S. (1972). *Developmental physiology and aging.* New York: Macmillan.

Tishler, C. (1981). Adolescent suicide attempts: Some significant factors. *Suicide and Life Threatening Behavior, 11*(2), 86–92.

Tishler, C. & McKenry, P. (1983). Intrapsychic symptom dimension of adolescent attempted suicide. *The Journal of Family Practice, 16*(4), 731–34.

Todd, J., Friedman, A., & Karinki, P. (1990). Women growing stronger with age: The effect of status in the U.S. and Kenya. *Psychology of Women Quarterly, 14,* 567–77.

Toffler, A. (1970). *Future shock.* New York: Bantam Books.

Toffler, A. (1984). *The third wave.* New York: Bantam Books.

Tolan, P. H. (1987). Implications of age of onset for delinquency risk. *Journal of Abnormal Child Psychology, 15*(1), 47–65.

Toolan, J. (1975). Depression in adolescents. In J. Howell (Ed.), *Modern perspectives in adolescent psychiatry.* New York: Brunner/Mazel.

Topol, P. & Reznikoff, M. (1984). Locus of control as factors in adolescent suicide attempts. *Suicide and Life Threatening Behavior, 12*(3), 141–50.

Torgersen, A. M. (1982). Genetic factors in temperamental individuality: A longitudinal study of same-sexed twins from two months to six years of age. *Journal of the American Academy of Child Psychiatry, 20,* 702–11.

Torrens, P. R. (1985). Current status of hospice programs. In P. R. Torrens (Ed.), *Hospice programs and public policy* (pp. 35–59). Chicago, IL: American Hospital Publishing.

Traupmann, J., Peterson, R., Utne, M., & Hatfield, E. (1981). Measuring equity in intimate relations. *Applied Psychological Measurement, 5*(4), 467–80.

Travers, J. (1982). *The growing child.* Glenview, IL: Scott, Foresman.

Treffinger, D. J., Isaksen, S. G., & Firestien, R. (1983). Theoretical perspectives on creative learning and its facilitation: An overview. *Journal of Creative Behavior, 17*(1), 9–17.

Troll, L. E. (1975). *Early and middle adulthood.* Monterey, CA: Brooks/Cole.

Tronick, E. (1989). Emotions and emotional communication in infants. *Child Development, 44,* 112–19.

Troumbley, P., Burman, K., Rinke, W., & Lenz, E. (1990). A comparison of the health risk, health status, self motivation, psychological symptomatic distress, and physical fitness of overweight and normal weight soldiers. *Military Medicine, 155*(9), 424–29.

Tudor, C., Petersen, D., & Elifson, K. (1980). An examination of the relationship between peer and parental influences in adolescent drug use. *Adolescence, 60,* 783–98.

Turner, J. S. & Helms, D. B. (1989). *Contemporary adulthood.* New York: Holt, Rinehart & Winston.

Tversky, A. & Kahneman, D. (1981). The framing of decisions and the psychology of choice. *Science, 211,* 453–58.

Twomey, L., Taylor, J., & Furiss, B. (1983). Age changes in the bone density and structure of the lumbar vertebral column. *Journal of Anatomy, 136,* 15–25.

Uhlenberg, P., Cooney, T., & Boyd, R. (1990). Divorce for women after midlife. *Journals of Gerontology, 45,* 3–11.

U.S. Bureau of Census (1986). *Current population reports,* Series P-20 (Nb. 399), 35–46. Washington, DC: U.S. Government Printing Office.

U.S. Bureau of the Census (1986). *Statistical abstract of the United States, 1986.* Washington, DC: U.S. Government Printing Office.

U.S. Bureau of the Census (1991). *Statistical abstract of the United States, 1991.* Washington, DC: U.S. Government Printing Office.

U.S. Bureau of Census (1992). *Vital statistics.* Washington, DC: U.S. Government Printing Office.

U.S. Department of Commerce. (1988). *Household and family characteristics: March, 1987.* Washington, DC: Bureau of the Census, 3.

U.S. Department of Health and Human Services. (1988). *10 Steps to help your child say "NO."* Rockville, MD: Office for Substance Abuse Prevention.

U.S. Department of Health and Human Services, Division of HIV/AIDS (1990, May and September). *HIV/AIDS surveillance.* Atlanta, GA: U.S. Department of Health and Human Services.

U.S. Department of Health and Human Services (1991). Public Health Service, Centers for Disease Control. Premarital sexual intercourse among adolescent women—U.S., 1970–1988. *Morbidity & Mortality Weekly Report, 39*(51/52), 929–32.

U.S. Department of Labor, Bureau of Labor Statistics (1991). *Child care arrangements for children under 5 years old with employed mothers.* Washington, DC: U.S. Government Printing Office.

U.S. Department of Transportation. (1976). *How to talk to your teenager about drinking and driving.* Washington, DC: National Highway Traffic Safety Administration (GPO 1976, 0–625–636).

U.S. National Center for Health Statistics. (1986). *Advance data from vital and health statistics,* No. 125. DHHS Pub. No. (PHS) 86–1250. Hyattsville, MD: Public Health Service.

U.S. National Center for Health Statistics. (1988). *Vital statistics of the United States, 1968–1987. Death rates by age, race, sex—5 and 10 year age groupings.* Washington, DC: U.S. Government Printing Office.

U.S. National Center for Health Statistics. (1989). *Vital statistics of the United States.* Washington, DC: U.S. Government Printing Office.

U.S. National Center for Health Statistics (1992). *Aging in the eighties.* Washington, DC: U.S. Government Printing Office.

U.S. National Center for Health Statistics. (1992). *Vital statistics of the United States.* Washington, DC: U.S. Government Printing Office.

Vaillant, G. (1977). *Adaptation to life.* Boston: Little, Brown.

VanderMay, B. J. & Neff, R. L. (1982). Adult child incest: A review of research and treatment. *Adolescence, 17*(68), 717–35.

VanderZanden, J. (1989). *Human development* (4th ed.). New York: Knopf.

Van Gennep, A. [1909] (1960). *The rites of passage.* (M. Vizedom & G. Caffee, Trans.) Chicago: The University of Chicago Press.

vanWilkinson, W. (1989). The influence of lifestyles on the patterns and practices of alcohol use among South Texas Mexican Americans. *Hispanic Journal of Behavioral Sciences, 11*(4), 354–65.

Vaughan, V. & Litt, I. (1990). *Child and adolescent development: Clinical implications.* Philadelphia, PA: W. B. Saunders.

Vaughn, B., Stevenson-Hinde, J., Waters, E., Kotsaftis, A., Lefever, G., Shouldice, A., Trudel, M., & Belsky, J. (1992). Attachment security and temperament in infancy and early childhood: Some conceptual clarification. *Developmental Psychology, 28,* 463–73.

Veatch, R. M. (1981). *A theory of medical ethics.* New York: Basic Books.

Veatch, R. M. (1984). Brain death. In J. Schneidman (Ed.), *Death: Current perspectives* (3rd ed.). Mountain View, CA: Mayfield.

Verbrugge, L. M. (1979). Marital status and health. *Journal of Marriage and the Family, 41,* 267–85.

Vigil, J. D. (1988). *Barrio gangs: Street life and identity in southern California.* Austin: University of Texas Press.

Vorhees, C. & Mollnow, E. (1987). Behavioral teratogenesis. In J. Osofsky (Ed.), *Handbook of infant development.* New York: Wiley.

Vygotsky, L. S. (1962). *Thought and language.* Cambridge, MA: MIT Press.

Vygotsky, L. S. (1978). *Mind in society: The development of higher psychological processes.* (M. Cole, V. John-Steiner, S. Scribner, & E. Souberman, Eds.). Cambridge, MA: Harvard University Press.

Vygotsky, L. S. (1981). The development of higher forms of attention in childhood. In J. V. Wertsch (Ed.), *The concept of activity in Soviet psychology* (pp. 189–240). Armonk, NY: Sharpe.

Wade, N. L. (1987). Suicide as a resolution of separation-individuation among adolescent girls. *Adolescence, 22*(85), 169–77.

Wagner, C. (1980). Sexuality of American adolescents. *Adolescence, 15*(59), 567–80.

Walch, S. (1976). Adolescent attempted suicide: Analysis of the differences in male and female behavior. *Dissertation Abstracts International, 38,* 2892B.

Waldrop, J. (1990, December). You'll know it's the 21st century when . . . *American Demographics,* pp. 23–27.

Walker, L. S. & Greene, J. W. (1987). Negative life events, psychosocial resources, and psychophysiological symptoms in adolescents. *Journal of Clinical Child Psychology, 16*(1), 29–36.

Walker, L. & Taylor, J. (1991). Family interaction and the development of moral reasoning. *Child Development, 62,* 264–83.

Wallace, M. (1989). Brave new workplace: Technology and work in the new economy. *Work and Occupations, 16*(4), 363–92.

Wallach, M. A. & Kogan, N. (1965). *Modes of thinking in young children.* New York: Holt, Rinehart & Winston.

Wallerstein, J. (1984). Children of divorce: Preliminary report of a ten-year follow-up of young children. *American Journal of Orthopsychiatry, 54,* 444–58.

Wallerstein, J. & Blakeslee, S. (1989). *Second chances: Men, women and children. Decade after divorce.* New York: Ticknor & Fields.

Walsh, M. (1992). *Moving to nowhere.* New York: Auburn House.

Walsh, R. (1989). Premarital sex among teenagers and young adults. In K. McKinney & S. Sprecher (Eds.), *Human sexuality: The societal and interpersonal context.* Norwood, NJ: Ablex.

Walther, R. R. & Harber, L. C. (1984). Expected skin complaints of the geriatric patient. *Geriatrics, 39,* 67.

Waltz, G. & Benjamin, L. (1980). *Adolescent pregnancy and parenthood.* ERIC document 184528.

Warabi, T., Kase, M., & Kato, T. (1984). Effect of aging on the accuracy of visually guided saccadic eye movement. *Annals of Neurology, 16,* 449–54.

Washington, A. E., Arno, P. S., & Brooks, M. A. (1986). The economic cost of pelvic inflammatory disease. *Journal of the American Medical Association, 225*(13), 1021–33.

Wasow, M. & Loeb, M. B. (1979). Sexuality in nursing homes. *Journal of the American Geriatric Society, 28*(2), 73–79.

Watson, J. (1968). *The double helix.* New York: Atheneum Press.

Weber, M. (1904). *The protestant ethic and the spirit of capitalism.* New York: Charles Scribner's Sons.

Wechsler, D. (1955). *Manual for the Wechsler adult intelligence test.* New York: Psychological Corp.

Wechsler, D. (1958). *The measurement and appraisal of adult intelligence.* Baltimore: Williams & Wilkins.

Weeks, D. (1989). Death education for aspiring physicians, teachers, and the funeral directors. *Death Studies, 13,* 17–24.

Weiner, I. B. (1970). *Psychological disturbances in adolescence.* New York: Wiley Interscience.

Weinrich, J. D. (1987). A new sociobiological theory of homosexuality applicable to societies with universal marriage. *Ethology and Sociobiology, 8*(1), 37–47.

Weisman, M. (1974). The epidemiology of suicide attempts, 1960–1971. *Archives of General Psychiatry, 30,* 737–46.

Weitzman, B., Knickman, J., & Shinn, M. (1990). Pathways to homelessness among New York City families. *Journal of Social Issues, 46*(4), 125–40.

Weitzman, L. J. (1985). *The divorce revolution: The unexpected social and economic consequences for women and children in America.* New York: Macmillan.

Welch, C. (1989). *In the matter of Claire C. Conroy: Pulling the plug on precedent.* Unpublished manuscript, Boston, MA.

Wenz, F. (1979). Economic status, family anomie, and adolescent suicide potential. *Journal of Psychology, 98*(1), 45–47.

Werner, E. (1991). Children of the Garden Island. In N. Lauter-Klatell (Ed.), *Readings in child development.* Mountain View, CA: Mayfield.

Wertheimer, M. (1962). Psychomotor coordination of auditory-visual space at birth. *Science,* 134.

Wertsch, J. V. & Tulviste, P. (1992). L. S. Vygotsky and contemporary developmental psychology. *Developmental Psychology, 28*(4), 548–57.

Westoff, C. F., Calot, G., & Foster, A. D. (1983). Teenage fertility in developed nations. *Family Planning Perspectives, 15,* 105.

Whelan, R. (1982). Presidential message. *Society for Learning Disabilities and Remedial Education Newsletter, 2,* 1–3.

Whisett, D. & Land, H. (1992). Role strain, coping and marital satisfaction of stepparents. *Families in Society: The Journal of Contemporary Human Services,* 79–91.

White, C. B. (1981, September). *Sexual interests, attitudes and knowledge and sexual history.* Paper presented at the Annual Meeting of the American Psychological Association, Los Angeles, CA.

White, H. C. (1974). Self-poisoning in adolescents. *British Journal of Psychiatry, 124,* 24–35.

White, L. (1992). The effect of parental divorce and remarriage on parental support for adult children. *Journal of Family Issues, 13*(2), 234–50.

White, N. & Cunningham, W. R. (1988). Is terminal drop pervasive or specific? *Journals of Gerontology, 43*(6), 141–44.

Whitehead, A. & Mathews, A. (1986). Factors related to successful outcome in the treatment of sexually unresponsive women. *Psychological Medicine, 16*(2), 373–78.

Wiland, L. J. (1986). Vision quest: Rites of adolescent passage. *Camping Magazine, 58*(7), 30–33.

Wilkinson, K. & Isreal, G. (1984). Suicide and rurality in urban society. *Suicide and Life Threatening Behavior, 14*(3), 187–200.

Wilks, J. (1986). The relative importance of parents and friends in adolescent decision making. *Journal of Youth and Adolescence, 15,* 323–34.

Will, J., Self, P., & Datan, N. (1976). Maternal behavior and perceived sex of infant. *American Journal of Orthopsychiatry, 46,* 135–39.

Williams, G. (1979, March 11). How to explain death to a child. *Boston Globe,* A12.

Williams, K. & Bird, M. (1992). The aging mover: A preliminary report on constraints to action. *International Journal of Aging and Human Development, 34*(4), 271–97.

Williams, L. (1989, July 21). Teens feel having sex is their own right. *The New York Times,* 13.

Williamson, J. B., Munley, A., & Evans L. (1980). *Aging and society: An introduction to social gerontology.* New York: Holt, Rinehart & Winston.

Willis, S. & Nesselroade, C. (1990). Long-term effects of fluid ability training in old-old age. *Developmental Psychology, 26*(6), 905–10.

Willis, S. & Schaie, K. W. (1986). Training the elderly on the ability factors of spatial orientation and inductive reasoning. *Psychology and Aging, 1*(3), 239–47.

Wilson, B. F. & Clarke, S. C. (1992). Remarriages: A demographic profile. *Journal of Family Issues, 13*(2), 123–41.

Wilson, E. O. (1978). *Sociobiology.* Cambridge, MA: Harvard University Press.

Wilson, H. S. & Kneisl, C. (1979). *Psychiatric nursing.* Menlo Park, CA: Addison-Wesley.

Wilson, J. & Herrnstein, R. (1985). *Crime and human nature.* New York: Simon & Schuster.

Wilson, P. (1987). Psychoanalytic therapy and the young adolescent. *Bulletin of the Anna Freud Centre, 10*(1), 51–79.

Wing, S. (1992). The challenge of multiculturalism. *American Counselor,* 6–14.

Wingerson, L. *Mapping our genes.* (1990). New York: Plume.

Wittrock, M. (1986). Students' thought processes. In M. Wittrock (Ed.), *Handbook of research on teaching.* New York: Macmillan.

Woititz, J. (1990). Adult Children of Alcoholics. Lexington, MA: Health Communications, Inc.

Wolf, M. & Dickinson, D. (1985). From oral to written language: Transitions in the school years. In J. B. Gleason (Ed.), *The development of language*. Columbus, OH: Merrill.

Wolfson, M. (1989). *A review of the literature on feminist psychology*. Unpublished manuscript, Boston College, Chestnut Hill, MA.

Wolpert, L. (1991). *The triumph of the embryo*. New York: Oxford.

Woodruff-Pak, D. (1988). *Psychology and aging*. Englewood Cliffs, NJ: Prentice-Hall.

Woodward, B. (1991). *The commanders*. New York: Simon & Schuster.

Wyatt, G. (1989). Reexamining factors predicting Afro-American and White American women's age at first coitus. *Archives of Sexual Behavior, 18*(4), 271–97.

Wyers, N. (1987). Homosexuality in the family: Lesbian and gay spouses. *Social Work, 32,* 143–48.

Wynne, E. (1978). Beyond the discipline problem: Youth suicide as a measure of alienation. *Phi Delta Kappan, 59*(5), 307–15.

Wynne, E. (1988, February). Balancing character development and academics in the elementary school. *Phi Delta Kappan,* 424–26.

Yao, E. L. (1988, November). Working effectively with Asian immigrant parents. *Phi Delta Kappan,* 223–25.

Yogman, M. (1982). Development of the father-infant relationship. In H. Fitzgerald, B. Lester, & M. Yogman (Eds.), *Theory and research in behavioral pediatrics*. New York: Plenum.

Young, T. J. (1987). PCP use among adolescents. *Child Study Journal, 17*(1), 55–86.

Youniss, J. & Smollar, J. (1985). Parent-adolescent relations in adolescents whose parents are divorced. *Journal of Early Adolescence, 5*(1), 129–44.

Zepeda, M. (1986). *Early caregiving in a Mexican origin population*. Paper presented at the International Conference on Infant Studies, Los Angeles.

Zietlow, P. H. & VanLear, C. A. (1991). Marriage duration and relational control: A study of developmental patterns. *Journal of Marriage and the Family, 53,* 773–85.

Zigler, E. & Lang, M. (1991). *Child care choices: Balancing the needs of children, family, and society*. New York: The Free Press.

Zoja, L. (1984, June). Sucht als unbewusster Versuch zur initiation, I. Teil I./Addiction as an unconscious attempt toward initiation: I. *Analyische Psychologie, 15*(2), 110–25.

Credits

Text Art and Tables

Chapter 1

Figure 1.1 From John F. Travers, *The Growing Child*. Copyright © 1982 Scott, Foresman and Company, Glenview, IL. Reprinted by permission of the author.
Figure 1.4 From John F. Travers, *The Growing Child*. Copyright © 1982 Scott, Foresman and Company, Glenview, IL. Reprinted by permission of the author.
Table 1.6 From John Dacey, *Adolescents Today,* 3d ed. Copyright © 1986 Scott, Foresman and Company, Glenview, IL. Reprinted by permission of the author.

Chapter 3

Table 3.1 From John F. Travers, *The Growing Child*. Copyright © 1982 Scott, Foresman and Company, Glenview, IL. Reprinted by permission of the author.
Figure 3.1 From John F. Travers, *The Growing Child*. Copyright © 1982 Scott, Foresman and Company, Glenview, IL. Reprinted by permission of the author.
Figure 3.2 From John F. Travers, *The Growing Child*. Copyright © 1982 Scott, Foresman and Company, Glenview, IL. Reprinted by permission of the author.
Figure 3.3 From John F. Travers, *The Growing Child*. Copyright © 1982 Scott, Foresman and Company, Glenview, IL. Reprinted by permission of the author.
Figure 3.4 From Larue Allen and John W. Santrock, *Psychology, The Contexts of Behavior*. Copyright © 1993 Wm. C. Brown Communications, Inc., Dubuque, Iowa. All Rights Reserved. Reprinted by permission.
Figure 3.5 From John F. Travers, *The Growing Child*. Copyright © 1982 Scott, Foresman and Company, Glenview, IL. Reprinted by permission of the author.
Figure 3.8 From John F. Travers, *The Growing Child*. Copyright © 1982 Scott, Foresman and Company, Glenview, IL. Reprinted by permission of the author.
Table 3.3 From John F. Travers, *The Growing Child*. Copyright © 1982 Scott, Foresman and Company, Glenview, IL. Reprinted by permission of the author.
Table 3.4 From John F. Travers, *The Growing Child*. Copyright © 1982 Scott, Foresman and Company, Glenview, IL. Reprinted by permission of the author.

Chapter 4

Figure 4.5 pub perm Modified from K. L. Moore and T. V. N. Persaud, *The Developing Human: Clinically Oriented Embryology,* 5th ed. Copyright © 1993 W.B. Saunders Company, Philadelphia, PA. Reprinted by permission.

Chapter 5

Table 5.2 From John F. Travers, *The Growing Child*. Copyright © 1982 Scott, Foresman and Company, Glenview, IL. Reprinted by permission of the author.
Figure 5.1 From *The Conscious Brain* by Steven Rose. Copyright © 1973 by Steven Rose. Reprinted by permission of Alfred A. Knopf, Inc.
Table 5.4 From John F. Travers, *The Growing Child*. Copyright © 1982 Scott, Foresman and Company, Glenview, IL. Reprinted by permission of the author.
Figure 5.2 From Richard Ferber, *Solve Your Child's Sleep Problems*. Copyright © 1985 by Richard Ferber, M.D. Reprinted by permission of Simon & Schuster, Inc.
Table 5.5 From John F. Travers, *The Growing Child*. Copyright © 1982 Scott, Foresman and Company, Glenview, IL. Reprinted by permission of the author.

Chapter 7

Table 7.2 From John F. Travers, *The Growing Child*. Copyright © 1982 Scott, Foresman and Company, Glenview, IL. Reprinted by permission of the author.
Figure 7.1 From John F. Travers, *The Growing Child*. Copyright © 1982 Scott, Foresman and Company, Glenview, IL. Reprinted by permission of the author.
Figure 7.3 From John F. Travers, *The Growing Child*. Copyright © 1982 Scott, Foresman and Company, Glenview, IL. Reprinted by permission of the author.
Figure 7.4 From John W. Santrock and Steve R. Yussen, *Child Development: An Introduction,* 4th ed. Copyright © 1989 Wm. C. Brown Communications, Inc., Dubuque, Iowa. All Rights Reserved. Reprinted by permission.

Chapter 9

Table 9.2 From John F. Travers, *The Growing Child*. Copyright © 1982 Scott, Foresman and Company, Glenview, IL. Reprinted by permission of the author.
Figure 9.1 From John F. Travers, *The Growing Child*. Copyright © 1982 Scott, Foresman and Company, Glenview, IL. Reprinted by permission of the author.
Figure 9.2 From John F. Travers, *The Growing Child*. Copyright © 1982 Scott, Foresman and Company, Glenview, IL. Reprinted by permission of the author.
Table 9.4 From John F. Travers, *The Growing Child*. Copyright © 1982 Scott, Foresman and Company, Glenview, IL. Reprinted by permission of the author.

Chapter 14

Figure 14.3 Reprinted courtesy of *The Boston Globe*.

Chapter 17

Table 17.3 From John Dacey, *Adult Development*. Copyright © 1982 Scott, Foresman and Company, Glenview, IL. Reprinted by permission of the author.
Figure 17.1 From *The Seasons of a Man's Life* by Daniel J. Levinson, et al. Copyright © 1978 by Daniel J. Levinson. Reprinted by permission of Alfred A. Knopf, Inc.

Chapter 18

Figure 18.1 From L. L. Langley, *Physiology of Man,* 1971, Wadsworth, Belmont, CA. Reprinted with permission.
Figure 18.2 From *From Now to Zero: Fertility, Contraception and Abortion in America*. Copyright © 1968, 1971 by Charles F. Westoff and Leslie Aldridge Westoff. Reprinted by permission of Little, Brown and Company, Boston, MA.

Chapter 19

Figure 19.2 From *The Seasons of a Man's Life* by Daniel J. Levinson, et al. Copyright © 1978 by Daniel J. Levinson. Reprinted by permission of Alfred A. Knopf, Inc.
Figure 19.4 From R. McCrae and P. Costa, Jr., *Emerging Lives, Enduring Disposition: Personality in Adulthood*. Copyright © 1984 Little, Brown and Company, Boston, MA. Reprinted by permission of the author.

Chapter 20

Figure 20.2 Reprinted with the permission of Macmillan Publishing Company from Developmental Physiology and Aging by Paola S. Timiras. Copyright © 1972 P. S. Timiras.
Figure 20.6 From Elias, J. W., and Elias, M. F. Journal of Gerontology, 1976, 31, 164. Reprinted by permission.

Chapter 21

Figure 21.1 From Robert Havighurst, 1963. Copyright by S. Karger AG, Basel, Switzerland. Reprinted by permission of the publisher.

Illustrations

Illustrious, Inc.: 1.4, 1.5, 2.1, 12.1, 12.2, 12.4, 13.1, 13.2, 14.1, 14.2A-B, 14.3, 14.4, 15.2
Rolin Graphics: 1.1, 2.2, 3.1, 3.2, 3.3, 3.5, 3.6, 3.7, 3.8, 4.1, 4.2, 4.6, 5.1, 5.2, 7.1, 7.2, 7.3, 7.4, 9.2, 17.1, 18.2, 18.3, 18.4, 18.5, 19.1, 19.2, 20.6, 22.1

Part Openers

Part 1: © H. Armstrong Roberts; **Part 2, Part 3:** © Laura Dwight; **Part 4:** © H. Armstrong Roberts; **Part 5:** © Erika Stone/Photo Researchers, Inc.; **Part 6:** © Cleo/PhotoEdit; **Part 7:** © Erika Stone/Photo Researchers, Inc.; **Part 8:** © H. Armstrong Roberts; **Part 9:** © D&I MacDonald/PhotoEdit

Chapter 1

Page 4: The Bettmann Archive; **p. 14:** © Omikron/Photo Researchers, Inc.; **p. 16 top:** © David Hurn/Magnum Photos; **p. 16 bottom:** © W. Marc Bernsau/The Image Works; **p. 18:** © Alan Carey/The Image Works

Chapter 2

Page 32: The Bettmann Archive; **p. 34:** © Elizabeth Crews/The Image Works; **p. 36:** Courtesy of Dr. Andrew Schwebel; **p. 39:** © Harriet Gans/The Image Works; **p. 41:** © Christopher Johnson/Stock Boston; **p. 45:** The Bettmann Archives; **p. 48:** © Ulrike Welsch

Chapter 3

Page 56: © Will & Deni McIntyre/Science Source/Photo Researchers, Inc.; **p. 57:** © Spencer Grant/The Picture Cube; **p. 58 top:** © Walter Dawn/Photo Researchers, Inc.; **p. 58 bottom:** © SIU/Photo Researchers, Inc.; **p. 73:** The Granger Collection; **p. 76:** © Alan Carey/The Image Works

Chapter 4

Page 81: © Nancy Durrell McKenna/Photo Researchers, Inc.; **p. 84 top:** © D. W. Fawcett/D. Phillips/Photo Researchers, Inc.; **p. 84 bottom:** © Omikron/Photo Researchers, Inc.; **p. 85:** © Dr. C. Reather/Photo Researchers, Inc.; **p. 86:** © Petit Format/Nestle/Science Source/Photo Researchers, Inc.; **p. 94 top:** © Michael J. Okoniewski/The Image Works; **p. 94 bottom:** Photo courtesy of A. P. Streissguth, H. M. Barr, and D. C. Martin (1984); **p. 95:** © John Griffin/The Image Works; **p. 99:** © Nancy Durrell McKenna/Photo Researchers, Inc.; **p. 100:** © H. Armstrong Roberts, Inc.; **p. 106:** © Eric Roth/The Picture Cube

Chapter 5

Page 112: © Tim Davis/Photo Researchers, Inc.; **p. 117:** Figure 2 from "Discrimination and Imitation of Facial Expressions by Neonates," I. Field et al., Vol. 218, #4568 pp. 179–181, 8 October, 1982 Metzoff et al. © 1982 by AAAS; **p. 123:** © Alan Carey/The Image Works; **5.3:** © William Vandivert and Scientific American, April 1960; **p. 130:** © Barbara Rios/Photo Researchers, Inc.

Chapter 6

Page 142: © Suzanne Szasz/Photo Researchers, Inc.; **p. 144:** © Ulrike Welsch; **p. 149:** © Elizabeth Crews/The Image Works; **p. 154:** © Janice Fullman/The Picture Cube; **p. 156:** © Sandra Johnson/The Picture Cube; **p. 158:** © Majorie Nichols/The Picture Cube; **p. 163:** © Alan Carey/The Image Works; **p. 164:** © Julie O'Neil/The Picture Cube

Chapter 7

Page 170: © Myrleen Ferguson/Photo Edit; **p. 172:** © Elizabeth Crews/The Image Works; **p. 180 right:** © Henry Horenstein/The Picture Cube; **p. 180 left:** © Tony Freeman/PhotoEdit; **p. 181:** © Sharon L. Fox/The Picture Cube; **p. 183:** © Carol Palmer/The Picture Cube; **p. 189:** © Rick Freidman/The Picture Cube; **p. 190 left:** Photo Researchers, Inc.; **p. 190 middle, right:** © Topham/The Image Works; **p. 193:** © Lynn McLaren/Photo Researchers, Inc.

Chapter 8

Page 199: © Pedro Coll/The Stock Market; **p. 201:** © Carol Palmer/The Picture Cube; **p. 202:** © Steve Takatsuno/The Picture Cube; **p. 209:** © Erika Stone/Photo Researchers, Inc.; **p. 211 left:** © Nancy Lutz/The Picture Cube; **p. 211 right:** © Carol Palmer/The Picture Cube; **p. 217:** © Ulrike Welsh/Photo Researchers, Inc.; **p. 220:** © Elizabeth Crews/The Image Works; **p. 221 top:** © Linda Benedict-Jones/The Picture Cube; **p. 221 bottom:** © Erika Stone/Photo Researchers, Inc.; **p. 225:** © Laura Dwight

Chapter 9

Page 230: © Myrleen Ferguson/PhotoEdit; **p. 234:** © Alan Carey/The Image Works; **p. 235:** © Mimi Forsyth/Monkmeyer Press; **p. 238:** © Michael Siluk; **p. 242:** © James Carroll; **p. 246:** © Michael Siluk; **p. 248:** © Myrleen Ferguson/PhotoEdit; **p. 252:** © Roberta Hershenson/Photo Researchers, Inc.

Chapter 10

Page 262: © Robert Brenner/PhotoEdit; **p. 264:** © Felicia Martinez/PhotoEdit; **p. 266:** © Ulrike Welsch; **p. 269:** © Meri Houtchens-Kitchens/The Picture Cube; **p. 272:** © Elizabeth Crews/The Image Works; **p. 273:** © Jack Spratt/The Image Works; **p. 276:** © Michael Heron Jr./The Stock Market; **p. 279:** © Michael Siluk; **p. 291:** © Bob Kalman/The Image Works

Chapter 11

Page 299: © Elizabeth Crews/The Image Works; **p. 301:** Giraudon/Art Resource, N.Y.; **p. 303:** Archives of the History of American Psychology, University of Akron, Akron, Ohio; **p. 306:** University of Chicago; **p. 308:** © Archive/Photo Researchers, Inc.; **p. 309:** © Mark Antman/The Image Works; **p. 313:** © Ulrike Welsch; **p. 314:** © Thelma Shumsky/The Image Works; **p. 315:** © Margaret Thompson/The Picture Cube; **p. 318:** © Barbara Rios/Photo Researchers, Inc.

Chapter 12

Page 328: © Toni Michaels; **p. 334:** © James Carroll; **p. 338:** © Spencer Grant/Photo Researchers, Inc.; **p. 339:** © Jean Claude Lejeune; **p. 341:** The Granger Collection

Chapter 13

Page 347: © Michael Weisbrot/Stock Boston; **p. 352:** © Toni Michaels; **p. 355:** © Jim L. Shaffer; **p. 357:** © Susan Rosenberg/Photo Researchers, Inc.; **p. 360:** © Ulrike Welsch; **p. 364:** © Robert Brenner/PhotoEdit; **p. 367:** © James

L. Shaffer; **p. 370:** © John Maher/The Picture Cube; **p. 371:** © Alan Carey/The Image Works; **p. 373:** © Phil McCarten/PhotoEdit

Chapter 14

Page 379: © Bob Daemmrich/The Image Works; **p. 380 left:** AP/Wide World Photos; **p. 380 right:** © Jaye R. Phillips/The Picture Cube; **p. 381, 382:** © Alan Carey/The Image Works; **p. 385:** © Robert Eckert/The Picture Cube; **p. 391:** © Rohn Engh/The Image Works; **p. 392:** © John Griffin/The Image Works; **p. 396:** © Susan Kuklin/Photo Researchers, Inc.

Chapter 15

Page 410: © Martin Etter/Anthro Photo; **p. 413, 415:** © Ellis Herwig/The Picture Cube; **p. 416:** © Mark Antman/The Image Works; **15.5:** © Irven DeVore/Anthro Photo

Chapter 16

Page 434: © Toni Michaels; **p. 436 top:** © Eric Breitenbach/The Picture Cube; **p. 436 bottom:** © Thelma Shumsky/The Image Works; **p. 438:** © Therese Frare/The Picture Cube; **p. 439 top:** © Bob Kalman/The Image Works; **p. 439 bottom:** © Alan Carey/The Image Works; **p. 445:** © Michael Siluk; **p. 449:** © Toni Michaels

Chapter 17

Page 457: © Bobbi Carrey Collection/The Picture Cube; **p. 459:** AP/Wide World Photos; **p. 461:** © Dion Ogust/The Image Works; **p. 463:** © Toni Michaels; **p. 466:** © Ulrike Welsch; **p. 472:** © Toni Michaels; **p. 474:** © Charles Getewood/The Image Works

Chapter 18

Page 480: © Ellis Herwig/The Picture Cube; **p. 481:** © Toni Michaels; **p. 484:** © Mikki Ansin/The Picture Cube; **p. 486:** Archives of the History of American Psychology, University of Akron, Akron, OH; **p. 489:** The Bettmann Archive; **p. 494:** © Ellis Herwig/The Picture Cube

Chapter 19

Page 499: © Ulrike Welsch; **p. 500:** © Elizabeth Crews/The Image Works; **p. 507:** © Toni Michaels; **p. 512:** © Toni Michaels/The Image Works; **p. 514:** © Nita Winter/The Image Works; **p. 519:** © Mark Antman/The Image Works

Chapter 20

Page 525: © Topham/The Image Works; **p. 526:** © F. B. Grunzweig/Photo Researchers, Inc.; **20.3, 20.4:** The University of Chicago Press; **p. 531:** © Toni Michaels; **p. 535:** © Alan Carey/The Image Works; **p. 539, p. 542, p. 545:** © Toni Michaels

Chapter 21

Page 550: © Toni Michaels; **p. 553:** © Frank Siteman/The Picture Cube; **p. 556, p. 560:** © Toni Michaels; **p. 564 top:** © Frank Siteman/The Picture Cube; **p. 564 bottom:** © Toni Michaels; **p. 565:** The Museum of Modern Art Film Stills Archives

Chapter 22

Page 569: The Granger Collection; **p. 570:** © Toni Michaels; **p. 571:** © William Thompson/The Picture Cube; **p. 574:** © Michael J. Okoniewski/The Image Works; **p. 580:** © Alan Carey/The Image Works; **p. 582:** © Toni Michaels/The Image Works; **p. 584:** © Alan Carey/The Image Works; **p. 586:** © Keystone/The Image Works

Color Plates

Page 1, top left: © David Woods/The Stock Market; **top right:** © David M. Grossman/Photo Researchers, Inc.; **bottom left:** © Elizabeth Hathon/The Stock Market; **bottom right:** © Laura Dwight
Page 2, top left: © Suzanne Szasz/Photo Researchers, Inc.; **top right:** © Michael Newman/Photo Edit; **bottom left:** © Michael Newman/Photo Edit; **bottom right:** © Laura Dwight
Page 3, top left: © Laura Dwight; **top right:** © Michael Siluk; **bottom left and right:** © Laura Dwight
Page 4, top left: © Margaret Miller/Photo Researchers, Inc.; **top right:** © Ariel Skelley/The Stock Market; **bottom left:** © Michael Siluk; **bottom right:** © Bryan Peterson/The Stock Market
Page 5, top left: © Tony Freeman/PhotoEdit; **top right:** © Bill Aron/PhotoEdit; **bottom left:** © Mark Richards/PhotoEdit; **bottom right:** © Jose Carrillo/PhotoEdit
Page 6, top left: © Elena Rooraid/PhotoEdit; **top right:** © Bill Bachman/Photo Researchers, Inc.; **bottom left:** © George Holton/Photo Researchers, Inc.; **bottom right:** © Paul Conklin/PhotoEdit
Page 7, top left: © Diane Rawson/Photo Researchers, Inc.; **top right:** © Tony Freeman/PhotoEdit; **bottom left:** © M. Roessler/H. Armstrong Roberts, Inc.; **bottom right:** © Gary A. Conner/PhotoEdit
Page 8, top left: © Tony Freeman/PhotoEdit; **top right:** © Ulrike Welsch/PhotoEdit; **bottom:** © Amy C. Etra/PhotoEdit

NAME INDEX

A

Abel, E. L., 377
Abeshouse, R. P., 494
Abramowitz, R., 202
Ackerman, J. W., 355
Adalbjarnardottir, S., 273
Adams, G. R., 310
Adams, J., 491, 501
Adelman, P. K., 502, 517
Agee, J., 592
Ainsworth, M. D. S., 158, 159, 163
Alarcon, J., 580
Alban, D. F., 583
Alcebo, A. M., 455
Alexander, A., 391
Alexander, T., 63
Allan, G., 501
Alpaugh, P., 543
Als, H., 147
Amato, P. R., 517
Ambrose, S., 227
American Association of Retired Persons (AARP), 479, 480, 567
American Heart Association, 434
American Psychiatric Association (APA), 357, 358, 384, 385
Ames, S. W., 423
Anderson, A. M., 410
Anderson, E. R., 206, 207, 314
Angelou, M., 405
Angle, C., 366, 367
Anselmo, S., 195, 224
Anspach, D. F., 372, 501
Antonucci, T. C., 502, 517
Arendell, T. J., 502
Asher, S., 269
Ashton, J., 368
Asian-American Drug Abuse Program, 349
Aslin, R., 126
Atherton, C., 567
Au, K., 257
Auel, J., 321, 405
Avery, M. C., 109
Avery, M. E., 105, 106, 107

B

Bachman, J. G., 559
Bachor, D., 310
Baier, J. L., 409
Baillargeon, R., 133, 187, 235, 237
Baker, S., 97
Baker, S. W., 380
Baltes, P. B., 488, 542
Bancroft, L., 363
Bandura, A., 43, 44, 281
Banks, S., 221, 264, 265
Barfield, R., 560
Barnett, R., 517
Barnicle, M., 374
Baruch, G., 517
Baumrind, D., 203, 204, 315
Beaconsfield, P., 83
Beaconsfield, R., 83
Beauvais, F., 352
Beck, A. T., 359
Becker, E., 592
Beever, A., 398
Begin, G., 272
Belbin, E., 492
Belbin, R. M., 492
Belenky, M., 446, 447
Bell, B., 492
Bellah, R., 563
Belsky, J., 14, 212
Bem, A., 579
Bem, S. L., 381
Beneke, W., 479
Bengston, V. L., 558
Benjamin, L., 400
Bennett, T. F., 568
Benson, H., 430
Berg, K., 115
Berg, W. K., 115
Bergman, M., 481
Berk, L., 216
Berk, S. F., 460, 461
Berman, M., 308
Berman, P. L., 524, 568
Berndt, T., 74
Bettelheim, B., 410
Betz, N. E., 357
Bhrolchain, M. N., 418
Bilash, I., 482
Binet, A., 238
Bing, L., 374
Birch, H., 151
Bird, G. W., 460
Bird, M., 536
Birdwood, G., 83
Birren, J., 527, 542
Birtchnell, J., 580
Bishop, J., 80
Bishop, J. E., 57, 69, 438
Black, A., 412
Blackman, J., 92
Blakeslee, S., 208, 227
Blanton, L. P., 425
Block, 1*, 209
Block, 2*, 209
Block, J., 220
Bloom, B., 22, 190, 191, 234, 244
Blume, J., 345, 430
Blumenthal, S. J., 368
Bly, R., 430, 469, 499
Blythe, P., 424
Boas, F., 323
Boaz, R. F., 539
Bohling, H. R., 538
Boivin, M., 272
Bollen, K., 368
Bollman, S. R., 313
Bolton, C., 574
Borgman, *, 425
Bornstein, J. M., 559
Boston Women's Health Book Collective, 383, 497
Botwinick, J., 485, 492, 493–94, 545
Bourne, R., 323
Bove, R., 559
Bowe, E., 98
Bowen, R., 336
Bowlby, J., 156–58, 163, 168, 424
Bowlds, M. K., 426
Bowman, L., 574
Boyd, R., 502
Brabeck, M., 253, 446
Bradway, K., 486
Brady, B. A., 559
Brain, J. L., 410
Bransford, J., 248
Braungart, J., 14
Brazelton, T. B., 7, 115, 117, 118, 120, 132, 145, 147, 153, 155, 159, 168, 227
Breakey, W., 205
Brenner, A., 283, 286
Breslin, J., 522
Bretschneider, J. G., 484, 553, 554
Bridges, K., 165
Brim, O. G., 221, 504
Brittain, W. L., 181
Brody, E. M., 500, 556
Brody, G., 267
Brody, G. L., 221
Brody, J. E., 437
Bronfenbrenner, U., 8, 313
Bronson, M., 27
Brookbank, J., 548
Brooks, J., 125
Brooks-Gunn, J., 215, 331
Brown, *, 425
Brown, B. B., 316
Brown, G. W., 418
Brown, J. K., 483
Brown, M., 370–71
Brown, R., 191
Browne, *, 285
Brownlee, J. R., 315
Bruch, H., 357, 358
Buchanan, C., 208
Buehler, C. A., 455
Buffum, *, 347
Buffum, J., 347
Bukatko, D., 3
Bullinger, A., 125
Bumpass, L. L., 454, 455, 499, 500
Burger, J. M., 387
Burgess, A. W., 370
Burke, T., 373
Burnham, W., 323
Burton, C. A., 373, 374
Buskirk, E. R., 480
Butler, R. N., 552
Buunk, B., 385

C

Cahill, A. J., 369
Cahill, M., 559
Cain, W. S., 482
Caine, L., 592
Calderone, M. S., 405
Camara, *, 314
Camp, D. J., 574
Campbell, A., 374, 499
Campbell, B., 435
Campbell, S., 560
Campione, W. A., 560
Campos, J., 127, 151
Camus, A., 583
Canestrari, J., 486
Cangemi, J., 10
Cantwell, D. P., 359
Caplan, F., 120, 224
Caplan, T., 120, 224
Capote, T., 405
Cardon, L., 235
Carey, J., 438
Carey, W., 151, 152
Carlson, G. A., 359
Carnegie Corporation, 298, 319
Carnevali, D. L., 535
Carnochan, P., 224
Cascio, W. F., 559
Cassell, E., 568

Cassell, J., 425
Cassem, N., 421
Cassidy, J., 269, 270
Catania, J. A., 392
Cath, S. H., 564
Celotta, B., 367
Centers for Disease Control, 392
Chall, J. S., 258
Chapman, A. B., 455
Charlesworth, R., 120
Chess, *, 151, 152, 153
Chess, S., 151
Chester, N. L., 517
Chewning, B., 387
Children's Defense Fund, 399
Chilman, C., 439
Chiriboga, D. A., 502
Chisholm, J. S., 146
Chomsky, N., 138
Christiansen, B. A., 353
Christie, J., 227
Churchill, W. S., 190, 524
Cicirelli, V. G., 501
Clark, Ronald W., 54
Clarke, S. C., 456
Clarke-Stewart, A., 210, 212
Clavell, J., 345
Clifford, P., 479
Clifton, R., 125
Cohen, D., 377
Cohen, L. H., 371
Cohen, S., 377
Cohen-Mansfield, J., 530
Cohn, R., 558
Cole, C., 10
Cole, E., 484
Coleman, J. S., 312
Coleman, L. M., 502, 517
Coll, C. T. G., 146
Colletta, N., 398
Colligan, J., 426
Collins, W. A., 538
Comfort, A., 527
Compos, B. E., 425
Comstock, G., 278, 280
Conrad, R., 530
Cooney, T., 502
Cooney, T. M., 485
Cooper, C. L., 425
Cooper, C. R., 315
Cooper, H., 275
Cooper, R., 20, 99, 163, 266
Cooper, R. M., 482
Coopersmith, S., 216
Coppard, L., 500
Cordell, A., 398
Corder, B. G., 366
Corso, J., 534
Costa, P. T., Jr., 505, 508, 515, 516
Cote, J. E., 310
Cowan, R. S., 461
Cowley, G., 537
Cowley, M., 567
Cox, H., 555
Cox, M., 163, 206
Cox, R., 206
Craig-Bray, L., 310
Craik, F. I. M., 492
Cramer, B., 155, 168
Cratty, B., 179
Craven, J., 582
Cravens, R., 486
Criqui, M. H., 437
Cristante, F., 10
Crockenberg, S., 159
Crockett, L., 401
Crohan, S. E., 502, 517
Cronin, K., 583
Crouter, A. C., 462
Crowell, D., 289
Crowley, J. E., 560
Crumley, F., 359, 364, 366
Csikszentmihalyi, M., 422, 430
Cullinan, D., 372
Cunningham, W., 548
Cunningham, W. R., 542–43
Curran, D., 359, 363, 368, 580
Curtis, R. H., 377
Curtiss, S., 196
Cutler, W., 484

D

Dacey, J. S., 17, 315, 342, 366, 414, 421, 440, 488, 513, 544, 573
Daehler, M., 3
Damon, W., 215
Daniels, D., 425
Darling, C. A., 387
Darlington, R., 213
Darrach, B., 526
Darwin, C., 190
Datan, N., 220, 553, 558
Daugherty, L. R., 387
Davidson, E., 296
Davies, R., 530
de Beauvoir, S., 567
DeCasper, A., 82, 133
DeChardin, T., 586
deCuevas, J., 192
DeGenova, M. K., 584
Dehart, *, 163
Dekovic, M., 202
Dellas, M., 310
Dembo, R., 351
Demoise, C., 530
Dennis, W., 489–90
Detterman, D. K., 492
Detting, E. R., 352
Devilliers, J., 135
Devilliers, P., 135
deVries, H. A., 432
Dewey, J., 243
Diamond, M., 380
Diaz, R., 337
Dickinson, D., 258
DiClemente, *, 393
Dinnerstein, M., 517
DiVitto, B., 106, 162
Dobson, W. R., 310
Dodder, L., 374
Donaldson, G., 487, 488
Donnan, S., 368
Dornbusch, S., 208
Dotz, W., 535
Down, L., 75
Downs, A. C., 150, 221
Dranoff, S. M., 384
Draper, P., 557
Duncan, C., 568
Duncan, G., 502
Duniho, T. S., 556
Dunn, J., 220, 221, 264, 267, 268, 277
Duvall, E., 454

E

Earle, J. R., 443
Eaton, W., 118, 152
Edison, T., 190, 341
Egeland, B., 293
Ehrhardt, *, 217, 380
Eichorn, D. H., 516
Eifrig, D. E., 481, 534
Eisdorfer, C., 493
Eisenberg, N., 266, 267, 281, 282
Elias, J., 532
Elias, M., 532, 534
Elkin, F., 217, 218
Elkind, D., 200, 214, 224, 227, 234, 336, 337, 418
Elliott, D. S., 16
Elsayed, M., 438
Engstrom, P. F., 436
Ennis, R., 245
Epstein, J. L., 401
Epstein, M. H., 372
Erickson, M., 371
Erikson, E. H., 45–50, 54, 276, 307–11, 321, 345, 356, 380, 430, 470–73, 472, 505, 514, 558, 565–66, 578
Estrada, A., 351
Ethington, C. A., 517
Evans, L., 551, 555, 560, 561
Ewen, R. B., 480, 481
Ewing, C., 374

F

Fagot, B., 219, 290
Faier, J., 517
Fakouri, M. E., 372
Falco, M., 353
Falknor, H. P., 582
Fantz, R., 126
Farber, E., 293
Farrell, M. P., 500, 510–11
Farrington, D., 288
Federal Bureau of Investigation, 352, 369, 371
Feifel, H., 576–77
Fein, G., 223, 225
Feinberg, R., 310
Feldman, D., 340, 341, 372
Fendrich, M., 461
Ferber, R., 122, 123
Ferrell, C., 584
Ferris, G. R., 559
Field, D., 237
Field, T., 106, 114, 141, 146
Fifer, W., 82, 133
Findlay, P., 558
Findlay, S., 438
Finian, M. J., 425
Finkelhor, *, 285
Firestien, R., 491
Fischer, K., 224
Fischer, P., 205
Fisher, M. F. K., 524
Fisher, P., 359
Fisher, T. D., 387
Fitzpatrick, M. A., 499
Flavell, J. H., 36, 333–36
Flum, H., 308
Fodor, I., 483
Fonagy, P., 164
Fordham, S., 318
Forste, R., 387, 389
Forstein, M., 392
Forster, L. E., 556
Fowler, J., 588–89
Fowles, J., 475
Fox, L. H., 190
Fox, N., 162
Fozard, J. L., 492, 493
Fraiberg, S., 132, 198
Frankl, V., 585–86
Franks, P., 438
Franks, V., 483
Frasier, M., 191
Freeman, E. W., 359
Freeman, M. H., 391
Freitas, R. P., 359
Freud, A., 33, 305–6, 471
Freud, S., 15, 32–35, 112, 410, 592
Freudenberger, H., 371
Frey, B., 469
Frey, K., 221
Friedan, B., 378, 471, 472
Friedenberg, E., 304, 382
Friedland, R., 475
Friedman, A., 374, 483
Friedman, C. J., 374
Friedmann, E., 484
Frisch, R. E., 331
Fromm, E., 405, 449, 452, 472
Frye, K., 160
Fulton, R., 571, 573, 574
Furiss, B., 535
Furstenberg, *, 399
Furstenberg, F. F., 387

G

Gabriel, A., 399
Gaddy, J., 435
Gagnon, J. H., 383, 439
Galambos, N. L., 517
Gallup, G., 584
Galton, F., 340
Garapon, A., 409
Garbarino, J., 202
Garcia, E. E., 256
Garcia-Coll, C., 398
Gardner, H., 19, 138, 182, 187, 222, 224, 241, 242, 243, 261, 339, 340, 487, 497
Garland, *, 369
Garmezy, N., 290
Garn, S. M., 397
Garner, H. G., 366
Garvey, M. S., 517
Garwood, P., 389
Gates, D., 374

Gay, P., 33
Geer, J., 452
Gelman, D., 537
Gelman, R., 187, 235, 237
Genshaft, J., 191
George, L. K., 584
German, B., 535
Gerrard, M., 443
Gerson, L. W., 538
Gesell, A., 15, 175, 179
Gest, T., 582
Gibbons, M., 417
Gibson, E., 127
Gibson, M., 377
Gibson, R. C., 484
Gierz, M., 536
Gilbert, J., 316
Giles, H., 533
Gilligan, C., 251, 252, 253, 311, 336, 446, 464, 471, 472
Gilman, L., 68
Gimby, G., 480
Ginther, D., 392
Gitlin, T., 472
Gjerde, *, 209
Glaser, K., 365
Gleason, *, 192, 193
Glenn, E. N., 551
Goertzel, M., 189
Goertzel, V., 189
Goethals, G., 308
Goetting, A., 10
Goldberg, S., 106, 162
Goldfinger, S. E., 481
Goldman, C., 524, 568
Goldman, W., 321
Goldman-Rakic, P., 19
Goldsmith, H. H., 151, 152
Goldsmith, M. F., 570
Goleman, D., 292
Goodenough, F., 181
Goodman, M., 99
Gordon, M., 475
Gorsuch, R., 479
Gottman, J. M., 499
Gould, R., 462–66, 470, 505
Gray, D. D., 559
Gray, R. E., 362
Green, A., 366
Greene, J., 248
Greene, J. W., 356
Greenspan, S., 165, 166
Gregory, T., 413
Grobstein, C., 58, 66, 70
Grossman, H. J., 188
Grossman, H. Y., 517
Grotevant, H. D., 311, 315
Gruber, K., 391
Grzegorczyk, P. B., 535
Guelzoe, M. G., 460
Guest, J., 522
Guilford, J. P., 337
Gutman, D., 550
Guttmacher, A., 59, 85, 95, 99, 101, 109

H

Haan, N., 550
Hajcak, F., 389
Hakim-Larson, J., 315
Halford, G. S., 237
Hall, *, 213
Hall, D. A., 535
Hall, G. S., 15, 16, 302, 303–5
Hallman, R., 491
Hamburg, B. A., 398
Hammond, *, 372
Handel, G., 217, 218
Hankoff, L. D., 362
Hanmer, T., 253
Hansberry, L., 522
Hansson, R., 386–87
Harber, L. C., 535
Hardt, D., 569
Haris, J., 536
Harre, R., 6, 124
Harris, J., 377
Harris, J. R., 114, 141
Harris, R., 418
Harrison, H., 106
Harry, J., 386
Harter, S., 215
Hartmann, F., 569
Hartup, W., 214, 268
Harvard Health Letter, 534
Hasher, L., 492
Hashimoto, A., 500
Haskell, M. R., 369, 371
Haslett, S. J., 310
Hatfield, E., 449
Hauser, S., 315
Hauser, S. T., 426
Havighurst, R., 9, 10, 29, 303–4, 306, 562, 564
Hawton, K., 363, 366, 367
Hayes, J., 247, 248
Hayward, M. D., 559
Heaton, R., 387, 389
Heiman, J., 452
Held, L., 399
Helms, D. B., 535, 556, 575
Henry, R., 97
Herdt, G., 239
Herold, E., 391
Herrnstein, R., 288
Hersch, P., 394, 395
Hertz, R., 461
Hertzog, C., 487
Herz, E., 398, 401
Hetherington, E. M., 10, 206, 207, 209, 274, 314
Hewitt, J., 288
Hickler, H., 21
Hicks, M. W., 387
Hill, C. D., 571
Hill, J. P., 315
Hill, P., Jr., 410
Hinde, R., 116, 146, 148, 149, 264, 587
Hirshberg, H., 126
Hobart, C.J., 315
Hochhaus, C., 374
Hochhauser, M., 92, 94
Hoffman, C., 556
Hoffman, J., 398
Hoffman, S., 502
Holden, C., 436
Holinger, P. C., 579
Holland, J. L., 459–60
Holmbeck, G. N., 315
Holmes, C. T., 275
Holmes, D. S., 438
Holmes, J., 426, 573
Honig, A., 285
Hood, J. C., 461
Horn, J. L., 486, 487, 488
Horney, K., 35
Horwitz, A. V., 357
Hoyer, W. J., 492
Huba, G., 353
Hudgens, R., 364
Hughes, F., 553
Huizinga, D., 16
Hulicka, I., 493
Hunt, T., 481
Hunter, S., 484
Huston, A., 222, 278
Huston, T. L., 462
Huttenlocher, J., 139

I

Ingalls, Z., 436
Inhelder, B., 129, 182, 184, 185
Isaksen, S. G., 491
Ismail, A. H., 438
Isreal, G., 580
Isseroff, A., 19
Izard, C. E., 165

J

Jackson, J. S., 484
Jackson, R., 374
Jacobs, J., 365, 366, 367
Jacobson, S., 160
Jacoby, A., 405
Jacobziner, H., 364, 580
James, M., 500
Janssens, J., 202
Janus, C., 390, 405, 442, 443, 484, 485, 551
Janus, S., 390, 405, 442, 443, 484, 485, 551
Jaquish, G. A., 23, 339, 543–44
Jarjoura, D., 538
Jarvik, L., 529
Jasmin, S., 424
Jernigan, L. P., 310
Jerse, F. W., 372
Jofferth, S., 398
Johnson, *, 538
Johnson, A., 2
Johnson, B., 538
Johnson, C., 357, 398, 400, 401
Johnson, J., 227
Johnson, T. G., 481
Johnson, V., 383, 443, 439, 482, 484
Johnston, L. D., 559
Joint Commission on Accreditation of Hospitals, 582
Jonaitis, M. A., 97
Jones, R. J., 391
Jones, S. W., 535
Joselow, F., 518
Journal of Creative Behavior, 497
Jung, C. G., 507, 586
Jurich, A. P., 313

K

Kacmar, K. M., 559
Kagan, J., 23, 164, 165, 504
Kahn, M., 221, 264, 265
Kahneman, D., 245, 246
Kaiser, I., 59, 85, 95, 99, 101, 109
Kalinich, L. J., 498
Kallman, F., 529
Kalter, N., 455
Kangas, J., 486
Kantner, *, 387
Kantrowicz, B., 537
Kaplan, B. H., 425
Karinki, P., 483
Kart, C. S., 479, 537
Kase, M., 481
Kastenbaum, B., 578
Kastenbaum, R., 362, 578
Kato, T., 481
Kauffman, J. M., 372
Kavanagh, K., 290
Kavanaugh, R., 578
Keith, J., 557
Kelley, *, 285
Kelley-Buchanan, C., 90, 92
Kellog, J., 18
Kempe, R. S., 391
Kendig, H., 500
Kennedy, W., 377
Kennell, J., 160–62
Kenny, *, 17
Kessler, J., 188
Kienhorst, C. W., 579
Kilpatrick, W., 412, 414
Kilpatrick, W. K., 475
Kimmel, D., 356, 493
Kimmerly, N., 162
Kinard, E., 398
King, I., 322, 323
King, P., 446
Kinnaird, K., 443
Kinsey, A., 443
Kintsch, W., 492
Kirkland, M., 392
Kitahara, M., 410
Kitchener, K., 446
Klassen, A. D., 394
Klaus, M., 160–62
Klein, H., 398
Klein, N., 574
Kliegl, R., 542
Klos, D., 308
Kneisl, C. R., 423
Knickman, J., 205
Knobloch, H., 175, 176
Knowles, M., 431, 493, 494
Koball, E. H., 460
Koch, H., 267
Koenig, H. G., 584
Koff, E., 326
Kogan, H., 339
Kogan, N., 339
Kohlberg, L., 222, 250–51, 590
Kohner, M., 397
Kolata, G., 88, 109
Konner, M., 14, 15
Koopman, R. F., 310
Kopp, C., 89

Kornzweig, A. L., 534
Kort, C., 475
Kositsky, A., 106
Kowalski, C., 10
Koyle, P. R., 383
Kraeer, R. J., 574
Kreider, D. G., 366
Kroger, J., 310
Krokoff, L. J., 499
Krouse, H., 245
Ku, L. C., 382
Kübler-Ross, E., 362, 571, 576–78, 592
Kugler, D., 582
Kunjifu, *, 44
Kupfer, D. J., 368
Kurz, S., 371
Kvale, J. N., 584

L

Labouvie-Vief, G., 486
Lamaze, F., 104
Lamb, M. E., 213, 462
Lamb, R., 6, 124
Lancaster, J. B., 398
Land, H., 457
Landerman, R., 584
Lang, M., 209, 210
Langlois, J., 150, 221
Langone, J., 93, 94
Lansdown, R., 233, 265
Larson, R., 357
Lawrence, D. H., 452
Lawrence, R., 486
Lawton, M. P., 584
Lazar, I., 213
Leboyer, F., 104
Lee, P. R., 438
Lehman, H. C., 487, 489, 545
Leiblum, S. R., 484
Leigh, R. J., 481
Leinbach, M., 219
Leitenberg, H., 452
Lenneberg, E., 137
Lerner, J. V., 517
Leventhal, H., 164
Levin, J. S., 584
Levine, *, 316
Levine, A., 130
Levine, C., 310
Levinson, D., 466–70, 505, 507, 508, 512, 519, 520
Levitt, E. E., 394
Lewis, C., 571
Lewis, D., 248
Lewis, M., 125, 215
Lewis, M. I., 552
Lickona, T., 251, 261
Lidz, R. W., 410
Lidz, T., 410
Liebert, R., 114, 141, 296
Liebert, R. M., 279, 280, 281, 282
Lindberg, J., 497
Lindemann, E., 572
Lindholm, K. J., 255
Lindley, C. J., 481
Lips, H., 220
Lisina, M. I., 153, 154
Litt, I., 331
Little, J., 219
Litvin, S. J., 556
Litwack, G., 105, 106, 107, 109
Lochman, J., 397
Loeb, M. B., 553
Lohr, J. B., 536
Lohr, M. J., 538
Lopata, H. Z., 556
Lorand, S., 365
Losaff, M., 401
Lott, B., 219
Lowe, G. S., 517
Lowenkopf, E., 358
Lucca, A., 10
Lyons, N., 253

M

McAnarney, E., 368
McAnarney, E. R., 399
McCann, I. L., 438
McCarthy, J., 571
McCary, J., 443, 552
Maccoby, E., 204, 208, 287
McCord, G., 538
McCoy, K., 345
McCoy, N. L., 484, 553, 554
McCrae, R., 505, 508, 515, 516
McCullers, C., 321
McEvoy, G. M., 559
McFarland, R. A., 492
McGhee, P., 187
McHale, S. M., 462
McIntire, J., 363, 364, 580
Mack, D., 425
Mack, J. E., 21
McKelvey, C., 287, 289, 290, 296
McKenry, D., 366, 367
McKenry, P., 359, 360, 364, 398, 400, 401
MacKinnon, C. E., 221
McLaughlin, B., 256
MacRae, H., 557
Magid, K., 287, 289, 290, 296
Malatesta, C., 165
Manchester, W., 15, 168
Mann, F., 374
Mantle, D., 391
Marcia, J. E., 310–11
Markides, K. S., 584
Marshall, L., 481, 534
Marsiglio, W., 457
Martin, *, 204
Martin, C., 219
Martin, D. H., 394, 538
Martocchio, J. J., 559
Marx, M. S., 530
Maslow, A., 10, 12, 13, 29
Mason, J., 257
Masoro, E. J., 435
Massachusetts Department of Public Health, 394, 395
Massachusetts Women's Health Study, 500
Masten, A., 270
Masters, W., 383, 439, 443, 482, 484
Matheny, A. P., 152
Mathews, A., 443
Matthews, K., 483
Matthews, S. H., 500
Matusow, A. J., 472
Maurer, C., 98, 100, 128, 168
Maurer, D., 98, 100, 128, 141, 168
May, R., 452
Mayer, M., 519
Mazess, R. B., 535
Mead, M., 526
Medway, F. J., 275
Meeks, J. E., 369
Meilman, P., 310
Menard, S., 16
Mendel, G., 190
Mendes, *, 382
Menuhin, Y., 241
Menyuk, P., 254, 255
Metress, E. K., 479
Metress, S. P., 479
Meyer, M. L., 311
Miller, J. B., 472, 475
Milling, *, 282
Mintz, B. I., 357
Minuchin, P., 273, 274
Mistretta, C. M., 535
Mitchell, J., 440
Mitford, J., 574
Moen, P., 460
Moll, L., 337
Mollnow, E., 105
Molnar, J., 205
Money, *, 217, 380
Money, J., 217, 218, 385
Montemayor, R., 315
Moore, B. N., 337
Moore, E. W., 338
Moore, G., 27
Moore, J., 454
Moore, K., 398
Moore, K. A., 387
Moorman, J. E., 455
Moos, R., 425
Moravia, A., 321
Morgan, J., 560
Morgan, L. A., 502
Morinis, A., 410
Morison, P., 270
Morokoff, P. J., 483
Morris, J., 379
Morton, T., 289
Moser, *, 347
Moskowitz, B., 192, 198
Moss, H., 23
Motto, J. A., 366
Mowrer, C., 141
Moyers, W., 469, 499
Mozart, W. A., 341
Muehlbauer, G., 374
Muller, C. F., 539
Munley, A., 551, 555, 560, 561
Murdock, J., 583
Murphy, C., 482
Myers, *, 282
Myers, L. S., 483

N

Nagin, D., 288
Nagy, P., 361
National Association for the Education of Young Children, 212
National Association of Secondary School Principals, 371
National Center for Education Statistics, 447
National Center for Health Statistics, 435, 437, 551
National Commission on Children, 202
National Institute of Allergy and Infectious Diseases, 394, 396
National Institute on Aging, 483
National Survey of Family Growth, 382
Neff, R. L., 391
Nesselroade, J. R., 485, 488, 542
Neugarten, B., 454, 550, 557, 562–63, 564
New York Times, 373
Newman, B. M., 318
Newman, P. R., 318
Newton, I., 190
Nightingale, E. O., 99, 316, 319
Nilsson, L., 86, 99, 497
Noesjirwan, J. A., 410
Noller, R. B., 469
Northcott, H. C., 517
Norton, A. J., 455
Nuckolls, K. B., 425

O

O'Boyle, C., 219
Offer, D., 579
Office of Technology Assessment, 347
Ogbu, J. U., 318
Oh, W., 398
Ohio Cancer Information Services *, 427
Okazaki, S., 191
Olsen, T., 548
Olson, G., 133
O'Malley, P. M., 559
Ornstein, R., 434
Osofsky, J., 141, 148, 151
Ossip-Klein, D. J., 438
Ostrov, E., 579
Otto, V., 580
Outward Bound, U.S.A., 416
Owens, W., 486

P

Packer, A., 315
Packer, L., 162
Paffenberger, R. S., 438
Paik, H., 278, 280
Parke, C. M., 571, 572
Parke, R., 10, 207, 274, 280, 281
Parker, R., 337
Parmet, H., 296
Pasamanick, B., 175, 176
Pasternak, M., 482
Patrick, M., 535
Paul, R. W., 338

Paulos, J., 274
Pearlman, C., 550
Peck, S. M., 452
Peplau, L., 484
Perkins, U., 13
Perlmutter, B. F., 372
Perricone, P. J., 443
Perris, E., 125, 133
Perry, W., 444, 445–46, 447
Perry-Jenkins, M., 462
Perutz, K., 454
Petersen, A., 16, 401
Peterson, *, 358
Peterson, J. L., 387
Petzold, A., 397
Pfeffer, *, 282
Phares, V., 163
Phi Delta Kappa Task Force, 417
Phillips, D., 368
Phillips, R. T., 455
Piaget, J., 36–39, 113, 129–31, 137, 182–87, 183, 184, 235–37, 248–50, 332, 333, 336
Piccigallo, P., 213
Pierson, D., 27
Pine, V., 568
Pinon, M., 278
Pizer, H., 535
Plath, S., 377
Pleck, J. H., 382, 461
Pludes, D. J., 492
Podolsky, S., 492
Polit, D., 398
Postman, N., 200
Potok, C., 345
Poulos, R. W., 281
Powell, C., 2
Powledge, T., 63
Prechtl, H., 117
Proctor, W., 430

R

Rafferty, M., 14, 205, 206
Ralph, N., 397
Ramsey, J., 405
Ramsey, P., 216, 276
Ramsey, P. W., 410
Rando, T. A., 572, 574
Randolph, S., 119
Raphael, D., 310
Rebeta-Burditt, J., 377
Reese, H. W., 488
Regoli, R., 288
Reinherz, H., 398
Reinke, B., 512
Reis, J., 398, 401
Remafedi, G., 385–86
Rest, J., 130
Restak, R., 141
Reznikoff, M., 363, 365
Rhodes, S., 559
Rhyne, D., 499
Ricks, M., 163, 164
Riese, M., 63, 152
Ringel, S. P., 531
Ripple, R. E., 23, 543–44
Ritter, J., 150
Rivchun, S. B., 469
Rivers, C., 517
Roazen, P., 308, 514
Robards, J., 568
Robertson, J. L., 558
Robertson, M. J., 370
Robins, L., 19
Robinson, J. K., 535
Rodeheaver, D., 553
Roff, L., 567
Rogers, D., 535
Rogoff, B., 336
Rogow, A. M., 310
Rohn, R., 367
Roper Organization, 346
Rose, J. S., 275
Rose, S., 129
Rosen, B. M., 356
Rosen, E. I., 461
Rosen, J. C., 357
Rosen, R., 398
Rosenberg, G. S., 372, 501
Rosenberg, S. D., 500, 510–11
Rosenblatt, R., 162
Rosin, H. M., 461, 462
Ross, D., 44, 281
Ross, S., 44, 281
Rosser, P., 119
Rothblum, E., 484
Rothenberger, J., 92, 94
Rotheram-Borus, M. J., 370
Rotter, J., 366
Rourke, B., 177
Rousseau, J., 302
Rovee-Collier, C., 133
Rovine, M., 212
Rowe, I., 310
Rowland, D. L., 483
Roy, A., 415
Rubin, D., 205
Rubin, K., 223, 225
Rubin, L. B., 552
Rubin, R. H., 504
Rubin, Z., 269, 270
Ruble, D., 221
Rudman, D., 483
Russ-Eft, D., 398
Rutter, M., 3, 18, 19, 20, 162, 163, 286, 291, 357

S

Sabbath, J., 365
Sachs, J., 134, 135
Sadler, T. W., 85, 92, 100
Sagan, C., 72, 80
St. Peters, M., 278
Sales, E., 550
Salk, J., 17
Salomone, P. R., 559
Salthouse, T., 541
Saltin, B., 480
Saluter, A. F., 455
Sameroff, A., 7, 8, 103, 145, 147
Santana, G., 352
Santrock, J. W., 204, 215, 222, 313
Sarton, M., 548
Saudino, K., 118, 152
Scarr, S., 14, 130, 201
Schafer, W., 162
Schaie, K. W., 485, 487, 488, 542
Schemeck, H. M., 436
Schibuk, M., 265
Schiedel, D. S., 310
Schiffman, S., 482
Schleidt, W., 484
Schneer, H. I., 365
Schneider, J., 568
Schneider, M., 386
Schofferman, J., 583
Schoonover, C. B., 500
Schowalter, J. E., 364
Schroedel, J. R., 518
Schulman, M., 128
Schulz, R., 480, 481, 569, 570
Schumm, W. R., 313
Scott, D., 358
Scott, J. P., 265
Scott-Jones, D., 387, 388
Sears, P. S., 516
Sears, R. R., 516
Sebald, H., 312, 317, 440
Sechehaye, M., 377
Seginer, R., 308
Seiden, R. H., 359
Select Committee on Children, Youth and Families, 356
Self, P., 220
Seligman, M., 571
Selman, R., 270, 271, 273
Selye, H., 421, 422, 423
Senna, J. J., 372
Serakan, U., 460, 462
Sessions, W., 374
Shaffer, D., 359
Shahtahmasebi, S., 530
Shakespeare, W., 302
Shapiro, E., 273, 274
Shaver, P., 224
Shaw, G. B., 454
Sheehy, G., 483
Sherman, T., 133
Shields, P., 133
Shinn, M., 205
Shirk, S. R., 308
Shulman, S., 538
Shweder, R., 239
Siegal, L. J., 372
Siegler, I., 542
Silber, T., 398
Silverberg, S. B., 315
Silverman, P., 572
Simon, E., 492
Simon, T. R., 238
Simon, W., 383, 439
Simons, K. B., 481, 534
Simonton, D. K., 492
Singer, S., 61, 70, 71, 74
Skinner, B. F., 41–43, 54, 138, 545, 548
Slaby, R., 280, 281, 381
Slogoski, B. R., 310
Slugoski, B. R., 310
Smart, J. C., 517
Smith, B. K., 557
Smith, D. W. E., 562
Smith, R., 418
Smollar, J., 315
Sobel, D., 434
Sommerstein, J. C., 558
Sonnenstein, F. L., 382
Sousa, F., 374
South, S. J., 455
Sowarka, D., 542
Spence, A., 433, 480, 482, 526, 529, 530, 548
Spitze, G., 455
Sprafkin, J., 296
Sprafkin, J. N., 281
Sprecher, S., 449
Spreen, O., 105
Springer, M., 398
Sroufe, L. A., 20, 99, 163, 266
Stair, N., 584
Staney-Hagan, M., 206, 207, 314
Starr, B., 552
Steele, M., 164
Stein, B., 248
Steinberg, L., 312, 315, 318
Stern, D., 132, 155
Sternberg, K., 213
Sternberg, R., 238, 341, 448
Sternberg, R. J., 239, 240, 241, 452, 487, 492, 546
Stevens, J. C., 482
Stiffman, A., 398
Stigler, J., 239
Stimmel, B., 94
Stoddard, S., 582
Stoller, E. P., 556
Stoneman, Z., 221
Story, M. D., 443
Strawbridge, W., 557
Strobino, D., 397
Subcommittee on Health and the Environment, 394
Sudnow, D., 570
Sue, S., 191
Sugarman, L., 9
Sullivan, T., 386
Sundel, M., 484
Suomi, S. J., 116
Super, D. E., 460
Sutton-Smith, B., 224
Swanson, E. A., 568
Sweet, J. A., 454, 455, 499, 500
Szinovacz, *, 460

T

Takata, S. R., 374
Tan, S., 479
Tannahill, R., 405
Tanner, J. M., 174, 177, 233
Tapley, D., 58, 66, 75, 89, 95, 104
Taub, H., 493
Tauber, M. A., 221
Taylor, J., 248, 535
Teicher, J., 365, 367, 580
Terman, L. M., 192, 340, 485–86
Tessler, D. J., 351, 353
Thomas, *, 151, 152, 153
Thomas, A., 151
Thomas, G. S., 438
Thomas, J., 558
Thomas, T., 397
Thompson, C., 486
Thorbeck, W., 311

Timson, B., 479
Tishler, C., 359, 360, 364, 367, 580
Tivnan, T., 27
Tobin, S., 562, 564
Todd, J., 483
Todd, W. D., 58, 66, 75, 89, 95, 104
Toffler, A., 419
Tolan, P. H., 371
Tolstoy, L., 592
Tomarken, A., 164
Toolan, J., 360, 580
Topol, P., 363, 365
Torgersen, A. M., 152
Torrens, P. R., 583
Traupmann, J., 449
Travers, J., 58, 59, 85, 98, 120, 194, 273
Treffinger, D. J., 491
Trelease, J., 198
Troll, L. E., 531
Tronick, R., 3, 164, 165
Troumbley, P., 479
Trygstad, L. N., 424
Tudor, C., 352
Tulviste, P., 40
Turner, J. S., 535, 556, 575
Tversky, A., 245, 246
Twomey, L., 535
Tyler, A., 54

U

Uhlenberg, P., 502
Updike, J., 475, 522
U.S. Bureau of Census, 202, 313, 432, 433, 580
U.S. Department of Commerce, 370
U.S. Department of Health and Human Services, 17, 278, 280
U.S. Department of Labor, 210
U.S. National Center for Health Statistics, 454, 459, 479, 579, 580

V

Vaillant, G., 505, 512, 513
Van Gennep, A., 410
Vandenberg, B., 223, 225
VanderMay, B. J., 391
VanderZanden, J., 438
vanDriel, B., 385
VanLear, C. A., 456
vanWilkinson, W., 437
Vaughan, M. E., 548
Vaughn, B., 153
Vaughn, V., 331
Veatch, R. M., 569
Verbrugge, L. M., 438
Vigil, J. D., 373
Vorhees, C., 105
Vygotsky, L. S., 40, 138, 336

W

Wade, N. L., 366
Wagner, C., 387
Walch, S., 365, 367
Wald, F., 582
Wald, K., 583
Waldholz, M., 57, 69, 80
Waldrop, J., 453
Walker, A., 405
Walker, L., 248
Walker, L. S., 356
Walker, M., 233, 265
Wallace, M., 494
Wallach, M. A., 339
Wallerstein, J., 208, 227
Wallhagen, M., 557
Walsh, M., 204, 206
Walsh, R., 387, 388
Walters, L., 398, 400, 401
Walters, R., 43, 44
Walther, R. R., 535
Waltz, G., 400
Warabi, T., 481
Ward, G., 296
Washington, J., 190
Wasow, M., 553
Watson, J., 80, 113
Weber, M., 458
Wechsler, D., 238, 485
Weeks, D., 574
Weinberg, R., 130
Weiner, I. B., 356, 580
Weiner, M., 552
Weinrich, J. D., 442, 587
Weinstein, K. K., 557
Weisman, M., 580
Weiss, R. S., 571, 572
Weitzman, B., 205
Weitzman, L. J., 503
Wenger, G. C., 530
Wenz, F., 363, 367
Werner, E., 60
Wertheimer, I., 398
Wertheimer, M., 115
Wertsch, J. V., 40
Whelan, R., 372
Whisett, D., 457
White, A., 387, 388
White, C. B., 554
White, H. C., 363, 366, 367, 580
White, H. R., 357
White, L., 503
White, N., 542–43
Whitehead, A., 443
Wibbelsman, C., 345
Wilkie, F., 493
Wilkinson, K., 580
Wilks, J., 387
Will, J., 220
Williams, C. J., 394
Williams, G., 568
Williams, K., 536
Williams, L., 382
Williams, P. S., 409
Williams, R. A., 425
Williamson, J. B., 551, 555, 560, 561
Willis, S., 542
Willis, S. L., 485
Wilson, B. F., 456
Wilson, E. O., 441, 442, 587
Wilson, J., 288
Wilson, P., 356
Wing, S., 18
Winner, E., 138, 339, 340
Wittrock, M., 274
Woititz, J., 435, 452
Wold, M., 258
Wolf, E., 492
Wolfe, T., 497
Wolfson, M., 472
Wolk, R., 127
Wolverton, L., 316, 319
Wood, C., 219
Woodriff-Pak, D., 579, 580
Woodward, B., 29
Wright, J., 278
Wright, R., 321
Wyatt, G., 387, 389
Wyers, N., 457–58
Wynne, E., 252, 367

Y

Yablonsky, L., 369, 371
Yao, E. L., 216
Yawkey, T., 227
Yeats, W. B., 485
Yogman, M., 163
Young, J., 261
Young, J. R., 438
Youngs, *, 217
Youniss, J., 315
Yung, *, 372
Yussen, S., 204, 215, 222

Z

Zacks, R. T., 492
Zelnick, *, 387
Zemitis, O., 391
Zepeda, M., 146
Zietlow, P. H., 456
Zigler, *, 213, 369
Zigler, E., 209, 210
Zoja, L., 409
Zubek, J. P., 482

Subject Index

A

Abuse
nature of problem, 284–85
sexual
adolescents, 391
children, 285
stress, 283–85
Acceptance of death, 576, 578
Accommodation
Piaget's theory, 37
visual, 126
Accumulation of errors, aging and, 528
Accumulation of metabolic waste, aging and, 527
Achieved identity, 415
Achievement, academic, homeless children, 206
Achievement orientation, adolescent suicide and, 368
Acquired immunodeficiency syndrome. *See* AIDS-related complex (ARC)
Acrosome, 58, 59
Acting out
coping with stress, 292
depression and, 359
sexual, depression and, 361
Action theory, 14
Active processors of information, in relationships, 146
Activity theory, 563
Adaptation
gender-role, 379
intelligence model, 240
Piaget's theory, 37–38
visual, infancy, 127–28
Adaptation to Life (Vaillant), 512–14
Adaptive behavior, early childhood, 175
Adhesion, in implantation, 83
Adolescence, 298–402
beginning of, 299–301
cognitive development, 332–37
egocentrism, 336–37
Flavell's theory, 333–36
Piaget's theory, 332
communication in, 388
creative thinking, 338–42
critical thinking, 337–38
delinquent behavior, 369–74
family and, 312–16
divorce, 313–14
function loss, 312–13
gender effects, 315
nurturing parents, 315–16
gender roles, 379–81
giftedness, 340–42
historical views, 301–2
identity. *See* Identity; *headings beginning with term* Identity
mental health issues, 355–69
anorexia nervosa, 357–58
bulimia nervosa, 358
death, 361–62
depression, 358–61
suicide, 363–69
moratorium of youth, 308–9, 413–14
parenthood during, 396–402
peers and subcultures, 316–18
physical development. *See also* Puberty
early studies, 322–23
reproductive system, 323–26
pregnancy during. *See* Teenage pregnancy
research areas, 15–17
sexual behavior, 382–91. *See also* Teenage pregnancy
autosexuality, 383–84
first coitus, 387–89
heterosexuality, 383, 386–87
homosexuality, 383, 384–86
Janus Report, 390, 551
masturbation, 383–84
motives for, 389–90
sexual abuse, 391
sexual revolution, 382–83
stages, 383
sexual identity, 379
sexually transmitted diseases, 392–96
substance abuse, 347–55
theories of
Erikson, 307–11
Freud, 305–6
Hall, 303–5
Havighurst, 306
Adolescence (Hall), 16, 303–4
Adolescent egocentrism, 336–37
Adolescent moratorium, 308–9, 413–14
Adoption, 68
Adult children of alcoholics (ACoA), 435
Adult life cycle, personal development stages, 466–67
Adulthood. *See also* Early adulthood; Late adulthood; Middle adulthood
anxieties concerning adolescent subcultures, 316–17
focus of study of, 17
life cycle, 466–68
response needs in, first relationships, 154
seasons of life
females, 511–12
males, 468–70, 506–11
stresses of. *See* Stress, adulthood
transformations, 462–66, 505–6
transition to. *See* Initiation rites
AFP (alpha-fetoprotein), 99
Afterbirth, in birth process, 101
Age, vulnerability to stress and, 286
Age cohorts, 24
Age of Enlightenment, adolescence in, 302
Age-status, aging and, 563
Ageism, 533
Aggression. *See* Crime; Juvenile delinquency; Violence
Aggressive gangs, 369
Aging, 526–30. *See also* Late adulthood
accumulation of errors and, 528
accumulation of metabolic waste and, 527
aging by program theory, 526–27
autoimmunity and, 527–28
cross-linkage theory, 527
environmental effects on, 529–30
gene theory, 529
genetic aspects of, 529
homeostatic imbalance theory, 527
modifiers of ability, 530
pet ownership, 562
physiological aspects of, 526–28
stress and, 286
wear and tear theory, 526
Aging parents, relationship and care of, 500, 556–57
AID (artificial insemination by donor), 66
AIDS (acquired immunodeficiency syndrome), 92–94, 392–93
prenatal development and, 91, 92
AIDS-related complex (ARC), 93
Alarm reaction, general adaptation syndrome, 422
Alcohol use
determining if you have a drinking problem, 352
health and, 435–36
middle adulthood, 480
prenatal development, 91, 95, 105–6
prevalence in adolescence, 348
Allele, 59
Alpha-fetoprotein (AFP), 99
Altruism, coping with stress, 292
Alzheimer's disease, 536–38
Ambivalent attachment, 159
America 2000, 272–73
Amniocentesis, 99
Amniotic sac, 84
Amphetamines
effects, 350
prenatal development, 95
Anal stage, 35
Analgesics, prevalence of use in adolescence, 348
Ancient times
adolescence in, 301
funerals in, 575

Androgyny, 381, 494
Anemia, sickle-cell, 76, 78
Anger, dying and, 577
Anglo-Saxon England, funerals in, 575
Anima, 586
Animism, 184–85
Animus, 586
Anorexia nervosa, 357–58
Anoxia (lack of oxygen), in birth process, 103–4
Anticipation, coping with stress, 292
Anticipatory images, 184
Antigens, 528
Anxieties
  adult, adolescent subcultures and, 316–17
  reaction time in late adulthood, 533
Apgar scale, 117
Apnea, 114–15
Appearance. *See also* Physical appearance
  of not being needed, 367
Apposition, in implantation, 83
Appropriate for gestational age (AGA) infants, 105
ARC (AIDS-related complex), 93
Art, Mind, and Brain: A Cognitive Approach to Creativity (Gardner), 339
Art of Loving, The (Fromm), 449
Artificial insemination by donor (AID), 66
Artificialism, 185
Asceticism, 33, 306
Ascribed identity, 415
Aspirin, as teratogen, 91
Assimilation, Piaget's theory, 37
Athletics. *See* Sports
Attachment, 156–64
  ambivalent, 159
  avoidant, 159
  behavioral theory, 158
  Bowlby's theory, 156–58
  chronology of development, 157
  cognitive theory, 158
  fathers and, 162–63
  intergenerational continuity, 163–64
  irritable infants and, 159–60
  Klaus and Kennell studies, 160–62
  mid-life transition, 508
  psychoanalytic theory, 158
  psychosocial development of, 156–63
  secure, 159
  strange situation technique, 158–59
Attachment (Bowlby), 156
Attention
  automatic, 492
  effortful, 492
Authoritarian parents, 203, 315
Authoritative parents, 203, 315
Autoimmunity, aging and, 527–28
Automatic attention, 492
Autonomous morality, 250
Autonomy, mother-infant interaction, 155, 173
Autonomy versus shame and doubt, 45, 46
Autosexual behavior, 383–84
Autosome, 59
Avoidant attachment, 159

B

Babbling, language development, 135
Babinski reflex, 116
Balancing, early childhood, 180
Bandura's theory, 43–44
Barbiturates
  effects, 350
  prenatal development, 95
Bargaining, dying and, 577
Bartholin's glands, 324
Basal metabolism rate (BMR), 479
Behavioral assessment methods, neonates, 117–18
Behavioral theory, 41–44
  attachment, 158
  of Bandura, 43–44
  language development, 138–39
  of Skinner, 41–43
  subcultures, 317
Belonging, sense of, self-esteem and, 217
Belongingness needs, Maslow's theory, 12
Benzodiazepines, effects, 350
Berkeley Growth Study, 22
Bilingual education, 277–78
Bilingualism, 255–56
Binet intelligence tests, 238
Binocular coordination, 126
Biological death, 570
Biological development. *See individual topics, i.e.,* Birth process
Biological language development theories, 137
Biopsychosocial forces
  behaviorist view, 44
  causes of change, 32
  cognitive structures view, 41
  eras, 467
  Freud's theory, 35
  model, 3, 5–6
    relationships and, 145–46
  psychosocial view, 49
Birth control, 64
Birth process, 100–7
  childbirth strategies, 104
  complications of
    anoxia, 103–4
    breech birth, 103
    Cesarean section, 103
    forceps delivery, 103
    prematurity, 103
    Rh factor, 104
  labor, signs of, 100
  myths/facts about, 101
  premature infants, 103, 104–7
  stages in, 100–1, 102
    afterbirth, 101
    dilation, 100, 102
    expulsion, 101
Blastocyst, 61, 83–84, 87
Blindness, color, 74
BMR (Basal metabolism rate), 479
Bodily-kinesthetic intelligence, 243
Bonding. *See also* Attachment
  lack of, violence and, 290
  siblings, 264–65
Boredom, depression and, 361
Bowlby's theory, 156–58
Brain
  cerebral hemispheres, 177
  development
    early childhood, 176–77
    infancy, 118
Brain death, 569
Brazelton Neonatal Behavioral Assessment Scale, 117–18
Breech birth, 103
Bulimia nervosa, 358

C

Cancer patients, 427
Cannabis
  effects, 350
  prenatal development, 95
Cardiopulmonary resuscitation (CPR), 569
Careers. *See also* Work
  dual-career families, 460–62
  goals, 518–19
  Holland's theory, 459–60
  Super's theory, 460
  working women, 517–18
Case studies, 21
Cataracts, 534
Causality, 130
Centration, preoperational stage, 185
Cerebral hemispheres, 177
Cervix, 324
Cesarean section, 103
Change
  biopsychosocial causes of, 32
  as source of stress, 419–21
Chemicals, prenatal development, 94–95
Child abuse. *See* Abuse
Child development, major theorists, 15
Child prodigies, 340–41
Childbirth. *See also* Birth process
  strategies for, 104
Childhood and Society (Erikson), 45
Childhood research areas, 14–15
Children. *See* Abuse; *specific periods of childhood, i.e.,* Early childhood
Children's Friendships (Rubin), 269
Chlamydia, 394
Chloral hydrate, effects, 350
Cholesterol, 434
Chorionic villi sampling (CVS), 99
Chromosomal disorders, 75–76, 78
Chromosomal sex, 217
Chromosome(s), 59, 62, 63, 69–70
  sex, 59, 69–70, 217–18
    chromosomal disorders, 75–76, 78
  sex-linked inheritance, 74
Chromosome failure, 380
Cigarette smoking. *See* Smoking
Classification, concrete operational stage, 235
Climacteric, 482–84
Climacterium, 482
Clinical death, 569
Clitoris, 324
Closed adoption, 68
CMV. *See* Cytomegalovirus (CMV)
Cocaine
  effects, 350
  prenatal development, 91, 95
  prevalence of use in adolescence, 348
Codeine, effects, 350
Codification of rules stage, 249
Cognitive development. *See* Intelligence; *specific developmental periods*
Cognitive-developmental theory
  gender development, 222
  language, 137–38
Cognitive factors, teenage pregnancy, 401
Cognitive manifestations of depression, 359
Cognitive structures theory. *See* Piaget's theory; Vygotsky's theory
Coitus. *See* Sexual intercourse
Collagen, aging and skin, 527
Colonial New England, funerals in, 575
Color
  of eyes, 74
  infant perception of, 126
Color blindness, 74
Combinations, thinking, 334–35
Commitment, 445, 448, 588
Committee on Human Development, 562–65
Communication. *See also* Language development

with adolescents, 388
about teenage pregnancy, 400
adolescent suicide and, 365, 367
culture, 337
Compartmentalization, 577
Compensation
defense mechanism, 33
in thinking, 335
Competence, sense of, self-esteem and, 217
Competition, 234
Comprehension, reading, 258
Concentration difficulty, depression and, 360
Concrete operational stage, Piaget's theory, 38–39, 235–37
Conditioning
inconsistent, 318
operant, 42
Conservation
concrete operational stage, 235, 236
preoperational stage, 186
Consolidation, thinking, 335–36
Construct(s), 31
Constructed knowledge, 447
Contextual intelligence, 240
Contingency, mother-infant interaction, 155
Continuity, 148
intergenerational, 163–64
Continuity versus discontinuity, 18–19, 35
behaviorist view, 44
cognitive structures view, 41
Freud's theory, 35
personality in middle adulthood, 504–5
psychosocial view, 49
Continuous traits theory, personality development, 515–16
Controlled scribbling, 181
Conventional level of moral development, 250, 251
Convergent thinking, 337
Coordination of secondary schemes, 129
Coping, invulnerable children, 292–93
Core self, 132
Cowper's glands, 325
CPR (Cardiopulmonary resuscitation), 569
Crack cocaine, prevalence of use in adolescence, 348
Crawling, infants, 120
Creative thinking
adolescence, 338–42
metaphor, use of, 339–40
Creativity
adolescence, 338–42
guidelines for improving, 342
late adulthood, 543–46
middle adulthood, 488–92
obstacles and aids, 491
psychohistorical studies of, 489–92
traits, 488–89
Creeping, infants, 120
Crime. *See also* Juvenile delinquency
substance abuse and, 352
Critical thinking, adolescence, 337–38
Cross-linkage theory, aging, 527
Cross-sectional studies, 20, 23–24
creative productivity, 543–44
Crossover, sex roles, 550
Crystallized intelligence, 486
Culture. *See also* Multicultural view; Racial factors
Asian-American culture clashes, 351
development and, 18
multicultural children and tests, 239, 255
multicultural schools, 276–77
multicultural view of teenage pregnancy, 399
subcultures in adolescence, 316–18
Culture transmission models, subcultures, 317
CVS (chorionic villi sampling), 99
Cystic fibrosis, 69, 77, 78
Cytomegalovirus (CMV)
in AIDS, 93
teratogenic effects of, 91, 92

D

Data collection methods, studies, 20–22, 25
Day care
desirable qualities for, 212
developmental outcomes, 212–13
facts about, 209–10
as national concern, 211
types of centers, 210–11
Death, 568–83
acceptance of own death, 576, 578
adolescence and, 361–62
biological, 570
brain death, 569
clinical, 569
death with dignity law, 581–82
defense mechanisms, 576–77
defined, 569
funerals, 574–75
grief, 571–74, 578
pathological grieving, 572–73
role of, 573–74
young children, 286
hospice, 582–83
legal definition, 570
of others, 571
of parent
adolescent suicide and, 366
vulnerability and invulnerability, 291
social, 570
stages of dying, 576–79
suicide. *See* Suicide
Death with dignity law, 581–82
Decentering, 235
Decision-making and reasoning, middle childhood, 245–46
Decline, late adulthood, 541–42
Decoding, reading, 257
Deductive reasoning, 334
Defense mechanisms, 33, 34
dying, 576–77
Deferred imitation, 183
Delayed puberty, 323
Delayed reaction, grief, 572
Delinquent behavior. *See* Juvenile delinquency
Democracy, defining, 333
Denial
coping with stress, 292
depression and, 359
dying and, 576–77
Deoxyribonucleic acid. *See* DNA (deoxyribonucleic acid)
Dependence, depression and, 359
Depression, 364
causes of, 359
death and, 578
depressive equivalents, 360–61
masked depression, 360
mental disorders, 358–61
postnatal, 101
reaction time in late adulthood, 533
suicide and, 364. *See also* Suicide
symptoms of, 359
Deprivation dwarfism, 178
DES (diethylstilbestrol)
in prenatal development, 91, 95
Descriptive studies, 20, 21
Despair
integrity versus, 49, 565–66
separation of infant from mother, 157
Destruction/creation, mid-life crisis, 507
Detachment, separation of infant from mother, 157
Development. *See also specific types and periods*
culture and, 18
day care and, 212
homelessness and, 205
models of, 7–8
normal range of, 327
siblings' effect on, 267–68
stress and, 287
Developmental research
adolescence, 15–17
childhood, 14–15
cross-sectional studies, 20, 23–24
descriptive studies, 20, 21
longitudinal studies, 20, 22–23
manipulative experiments, 20, 21–22
naturalistic experiments, 20, 22
one-time, one-group studies, 20, 22
sequential studies, 20, 24–25
Developmental risk, prenatal, 89–90
Developmental stage theories, 50
Developmental tasks, 306
example of, 10–11
meaning of, 9
sources of, 9–10
Developmentally delayed children, 174, 188–89
homelessness, 205
Dialect differences, 277
Diaphragmatic hernia, 88
Diet. *See* Nutrition
Diethylstilbestrol (DES)
in prenatal development, 91, 95
Differential opportunity structure, 551
Differentiation, sibling relationship, 265–66
Difficult children, 152
Dilation, in birth process, 100, 102
Discontinuity. *See* Continuity versus discontinuity
Discrimination, older workers, 558–59
Disease. *See also* Health; Mental health; *specific disorders*
early childhood development, 178
infectious, as teratogens, 90–94
stress and, 424–25, 426
Disengagement theory, 563
Displacement, defense mechanism, 33
Distantiation, 471
Distorted reactions, grief, 572
Divergent thinking, 337, 544
Diversity, 421
Divorce, 206–9
effects on children, 207–8, 209
emotional, 499
middle adulthood, 502–4
no-fault, 503
parents of adolescents, 313–14
reconciliation fantasies, 208
Dizygotic twins, 63
DNA (deoxyribonucleic acid), 59, 70–72

information contained in, 72
structure and function, 71–72
Dominance (genetic), 59
Dominant traits, 73, 74
Donor insemination (DI), 66
Double standard, 388
Doubt, autonomy versus, 45, 46
Down syndrome, 75–76, 78
Drawing, early childhood, 181–82, 183, 184
Drinking. *See* Alcohol use
Drive-by shootings, 374
Drug(s), prenatal development, 94–95
Drug abuse. *See* Substance abuse
Dual-career families, 460–62
Dual-process model of intelligence, 488
Dualism, 444
Dwarfism, deprivation, 178
Dying. *See* Death

E

Early adulthood, 431–74
adolescent moratorium, 413–14
cognitive development, 444–47
intellectual/ethical development, 444–46
women's ways of knowing, 446–47
initiation rites, 409–11
lack of initiation ceremony, 414–15
love, 448–50
Outward Bound program, 416
personality development, 462–73
Erikson's theory, 470–71
Gould's theory, 462–66
Levinson's theory, 466–67, 468–70
male versus female identity, 471–73
physical development, 432–39
effect of life-style on health, 433–39
organ reserve, 433
psychosocial development, 454–74
marriage and family, 454–58
personal development stages. *See* Personal development, early adulthood
work patterns, 458–62
sexuality, 439–43
Freud's theory, 439
marital intercourse, 443
motivations, 440
premarital experiences, 442–43
sexual scripts, 439–40
sociobiological view, 441–42
stress, 418–27
transition to. *See* Initiation rites
walkabout approach, 417
work patterns, 458–62
Early childhood, 170–225
characteristics, 171, 173
cognitive development, 182–92
intellectual differences, 188–92
preoperational stage, 182–87
humor and, 187
key ages, 175–76
language development, 173, 175, 192–96
patterns, 194
rule learning, 193
speech irregularities, 194–96
motor skills, 173, 175, 179–81
drawing, 181–82
physical development, 174–79
brain, 176–77
psychosocial development, 175, 201–22
day care, 209–12
education, 213–14
family, 201–213
play, 222–25
self, 215–22
Easy children, 151
Ectoderm, 84, 118
Ectopic pregnancy, 101
Education. *See also* Learning; Schools
early childhood, 213–14
middle adulthood, 494–95
middle childhood, 272–78
miseducation and, 200, 214
underachievement, homeless children, 206
EEG (electroencephalogram), 570
Effortful attention, 492
Egg (ovum), 58, 59–60, 59–61, 324
Ego, 33, 34
Egocentric speech, 137
Egocentric stage, 249
Egocentrism, 129
adolescence, 336–37
preoperational stage, 185
Electra conflict, 35
Electroencephalogram (EEG), 570
Embryo, 87
Embryonic period, prenatal development, 84–85
Emergent self, 132
Emile (Rousseau), 302
Emotion(s)
depression and, 359
maternal, prenatal development and, 98–99
Emotional development, infancy. *See* Infancy, emotional development
Emotional disorders. *See* Mental health
Emotional divorce, 499
Emotional energy, late adulthood, 563
Emotional release, play and, 224
Emotional security, sense of, self-esteem and, 217
Empirical research, 31
Empirico-inductive reasoning, 334
Employment. *See* Work
Empty nest syndrome, 499
Encoding, language, 254
Endoderm, 84, 118
Endometriosis, 64
English as second language program (ESL), 277
Entrainment, mother-infant interaction, 155
Environment
effects on aging, 529–30
lack of control over, 366
adolescent suicide and, 366
Epididymis, 325
EPSI (Erikson Psychosocial Stage Inventory), 307
Equal pay for work, 518
Equal potential, sensitive periods versus, 20, 35
Erikson Psychosocial Stage Inventory (EPSI), 307
Erikson's theory, 45–50
adolescence, 307–11
autonomy versus shame and doubt, 45, 46
gender roles, 380
generativity versus stagnation, 45, 49, 514
identity confusion, 47
identity crisis, 47
industry versus inferiority, 45, 47
initiative versus guilt, 45, 47
integrity versus despair, 45, 49
intimacy versus isolation, 45, 48, 470–71
personal development in late adulthood, 565–66
trust versus mistrust, 45, 46
ERT (estrogen replacement therapy), 483
Escape, 445
ESL (English as second language) program, 277
Esteem needs, Maslow's theory, 12
Estrogen replacement therapy (ERT), 483
Ethical development, 444–46. *See also* Moral development
Ethnic group. *See also* Culture; Multicultural view; Racial factors
substance abuse and, 349–51
Exhaustion stage, 422
Existential love, 553
Exosystem, 8
Experience, openness to, 516
Exploration stage, 460
Exploratory play, 222
Expressive language, 193
Expulsion, in birth process, 101
External fertilization, 58, 65–66
Extinction, 42
Extroversion, 515
Eyes
color, 74
color blindness, 74
disease, in middle adulthood, 481

F

Facial expressions, mother-infant interaction, 155
Failure, fear of, intelligence and, 241
Failure-to-thrive (FTT), 121–22
Faith. *See* Spirituality
Fallopian tubes, 61, 324
False assumptions, major, 462
Family. *See also* Fathers; Mothers; Parent(s); Parenting behavior; *specific developmental periods*
adolescent suicide and, 366–67
changing, 202–3
day care, 209–13
divorce, 206–9, 313–14, 502–4
grandparents, 557–58
homeless children, 204–6
late adulthood, 555–58
marriage, 454–58
parenting behavior, 203–4
role in teenage pregnancy, 398–400
status, adolescent suicide and, 368
FAS (fetal alcohol syndrome), 91, 105–6
Fathers. *See also* Divorce
attachment and, 162–63
Fear of failure, intelligence and, 241
Feelings. *See* Emotion(s); Infancy, emotional development
Females. *See also* Fertilization; Mothers; Pregnancy; Puberty; Sex; Sexuality; *headings beginning with terms* Gender *and* Maternal
adolescent, one hundred years ago, 329
cognitive development, 446–47
divorced, middle-aged, 502–4
early-maturing, 328
identity, 471–73

late-maturing, 328–29
menopause, 482
menstrual cycle, 60–64
reproductive system, 324–25
seasons of life theory, 511–12
sex chromosomes. *See* Chromosome(s), sex
sexual harassment on the job, 517–18
women's ways of knowing, 446–47
work, 517–18
Feminine Mystique, The (Friedan), 378, 472
Fertilization, 58–70
external, 58, 65–66
implantation, 61, 83
internal, 58
ovulation, 60–64
ovum, 58, 59–60
sperm, 58–59
zygote, 61–64
Fetal alcohol syndrome (FAS), 91, 105–6
Fetal hydrocephalus, 88
Fetal hydronephrosis, 88
Fetal period, prenatal development, 85–86
Fetal surgery, 88
Fetoscopy, 99
Fetus, 87. *See also* Prenatal development
diagnosis and counseling for problems, 99–100
surgery on, 88
Fimbriae, 324
First relationships, 153–55, 156
adult response needs, 154
bodily needs, 154
developmental sequence, 154–55
mother-infant interaction, 151, 155
psychological needs, 154
First words, 136
Fiscal fitness, 518
Flavell's theory, 333–36
Fluid intelligence, 486
For Character program, 252
Forceps delivery, 103
Foreskin, 325
Formal operational stage, 39, 332
Foster Grandparents Program, 564
Fowler's theory, 588–90
Fragile X syndrome, 74
Frankl's theory, 585–86
Fraternal twins, 63
Free will, 41
Freudian theory. *See* Psychoanalytic theory
Friendships. *See also* Peers
interpersonal understanding, 271
middle childhood, 269–70
FTT (failure-to-thrive), 121–22
Funerals, 574–75
Future shock, 419

G

Galloping, early childhood, 180
Gamete(s), 59
Gamete intrafallopian transfer (GIFT), 68
Gangs, 372–74
characteristics of joiners, 374
Maslow's theory and, 13
social class and, 373–74
Gaze, mother-infant interaction, 155
Gender, adolescent view of, 381
Gender-atypical behavior, 386
Gender constancy, 221
Gender development theories, 222
Gender identity, 217–22
biology and, 217–18
gender constancy and, 221
learning about gender, 219
parents and, 219–20
peers and, 221
siblings and, 220–21
theories of gender development, 222
Gender revolution, 511, 512
Gender role(s), 217
androgyny, 381
aspects of, 379–80
Erikson's studies, 380
late adulthood, 550
reversal, 563
Gender-role adaptation, 379
Gender-role orientation, 379
Gender-role preference, 379
Gender-role reversals, 563
Gender splitting, 511
Gender stability, 221
Gender stereotypes, 217
Gene(s), 59, 69–70
heterozygous, 59
homozygous, 59
recessive, 59
Gene locus, 59, 70
Gene theory of aging, 529
General adaptation syndrome, 422–23
Generative love, 553
Generativity crisis, 519
Generativity versus stagnation, 45, 49, 514
Genetic disorders, 76–77, 78
Genetic theory of homosexuality, 385
Genetics. *See* Heredity
Genital herpes, 395
Genital sex, 218
Genital stage, 35
Genotype, 57, 59, 70
German measles (rubella), 91–92
Germinal period, prenatal development, 83–84
GIFT (gamete intrafallopian transfer), 68
Giftedness, 189–92, 340–42
Gilligan's theory, 252–53
Glans penis, 325
Glaucoma, 481
Glutethimide, effects, 350
Gonadal sex, 217
Gonorrhea, 394
Goodness of fit, 152–153
Gould's theory, 462–66, 505–6
Grade retention, 275
Grandparents, 557–58
Grasping reflex, 116
Grief. *See* Death, grief
Group marriage, 457
Guidance Study, 23
Guilt, initiative versus, 45, 47

H

Habituation, 125
Hair, late adulthood, 535
Hall's theory, 303–5
Hallucinogens
effects, 350
prevalence of use in adolescence, 348
as teratogens, 91
Hashish
effects, 350
prevalence of use in adolescence, 348
Havighurst's theory, 306
Hazing practices, 409
Head control, infants, 119
Healing Brain, The (Ornstein and Sobel), 434
Health. *See also* Adolescence, mental health issues; Mental health; *specific disorders*
evaluating, 540
homelessness, 205
late adulthood, 536–39
life-style and, 433–39
alcohol use, 105–6, 435–36, 480
marital status, 438
nutrition, 434–35
physical fitness, 438
smoking, 436–37
middle adulthood, 479–80
physical and mental, relationship between, 538
stress and, 424–25, 426
Hearing
evaluating, 540
infancy, 115
late adulthood, 534
middle adulthood, 481–82
prenatal development of, 87
Height, early childhood, 176
Helplessness, depression and, 359
Hemophilia, 74
Hepatitis B, 395
Heredity, 70–77
chromosomal disorders, 75–76
DNA, 71–72
early childhood development, 177
genetic disorders, 76–77
Sagan's view, 72
trait transmission, 73–74
Hernia, diaphragmatic, 88
Heroin
effects, 350
prenatal development, 95
prevalence of use in adolescence, 348
Herpes simplex virus, 92
genital, 395
teratogenic effects of, 91
Heteronomous morality, 250
Heterosexual behavior, 383, 386–87
Heterozygous gene pairs, 59
HIV (human immunodeficiency virus), 92–93. *See also* AIDS
Holland's theory, 459–60
Holophrases, 135
Holophrastic speech, 135
Homeless children, 204–6
Homeostatic imbalance theory of aging, 527
Homework, 275
Homosexual behavior, 383, 384–86
causes of, 385
onset of, 385–86
Homosexual marriages, 457–58
Homozygous gene pairs, 59
Hopping, early childhood, 180
Hormonal balance, 327
Hormonal sex, 218
Hormone(s), adolescence, 300
Hospice, 582–83
How We Think (Dewey), 243–44
Human Genome Project, 69
Human immunodeficiency virus (HIV), 92–93
Humor
coping with stress, 292
early childhood, 187
Hunger, homeless children, 205
Hyaline membrane disease, 123
Hydrocephalus, 88
Hydromorphone, effects, 350
Hydronephrosis, 88
Hymen, 324
Hypersensitive youth, 365
Hypothetico-deductive reasoning, 334

I

Id, 33–34
Identical twins, 63
Identification, sibling relationship, 265–66
Identity, 307–11
achieved, 415
achievement, 310
ascribed, 415
confusion, 310
Erikson's theory, 47
foreclosure, 310
gender. *See* Gender identity
moratorium of youth and, 308–09, 310
negative, 309
search for, 308

sense of, self-esteem and, 217
state of, 307
Identity and repudiation versus identity confusion, 45, 47–48
Identity crisis, 47, 307
Identity status, 310–11
Illness. *See* Disease; Health; *specific disorders*
Imaginary audience, 337
Imitation
adolescent suicide and, 368
deferred, 183
infancy, 115
Immanent justice, 250
Immigrant children
language and thought, 255
tests and, 239
Implantation, fertilization, 61, 83
Impotency, 551–52
Impulsivity, coping with stress, 292
In vitro fertilization (IVF), 58, 67–70
Incipient cooperation stage, 249
Inconsistent conditioning, 318
Individuating-reflective faith, 589
Individuation, 468, 471–72, 473
Industry versus inferiority, 45, 47
Infancy, 112–39. *See also* Neonate; Premature infants
cognitive development, 128–34, 139
memory, 133–34
sensorimotor stage, 38, 129–31, 332
emotional development, 164–66
milestones, 165–66
signs, 164
theories, 165
language development, 134–39
babbling, 135
egocentric speech, 137
first words, 136
holophrastic speech, 135
socialized speech, 137
telegraphic speech, 136
two-word stage, 135
vocabulary growth, 136
motor development, 118–21, 139
head control, 119
locomotion, 120–21
toilet training, 120
perceptual development, 123–28
meaning of perception, 124–25
visual perception, 126–28
physical development, 114–18, 139
assessment, 117–18
brain, 118
newborn abilities, 115–17
reflexes, 114–15
psychosocial development. *See* First relationships; Infancy, emotional development; Relationships, infancy
Infectious diseases, 90
Inferiority, industry versus, 45, 47
Infertility, 64–67
Information processing
active processors of information in relationships, 146
adolescence, 335
middle adulthood, 492–93
Inhalants, prevalence of use in adolescence, 348
Inheritance. *See* Heredity
Initiation rites, 409–418
adolescent moratorium, 413–14
analysis of, 410–11
lack of, 414–15
Western countries, 412–418
Initiative, 173
Initiative versus guilt, 45, 47
Inner dialogue, 463
Instincts, 33
Integrity, 427, 578
Integrity versus despair, 49, 565–66
Intellectual/ethical development, cognitive development, 444–46
Intellectualization, 33, 306
Intelligence
components, 239–41
early childhood, 188–92
fluid and crystallized, 486–87
giftedness, 189–92, 340–42
mental retardation, 188–89
multiple, 241–43
new views of, 487–88
tests, 238–39
theories of, 485–87
of Horn, 486–87
of Piaget, 36–38
of Terman, 485–86
of Wechsler, 485
Inter-propositional thinking, 334
Interaction model, 7
Interactions
categories, 149
reciprocal, 7–8, 116, 145–47
Intercourse. *See* Sexual intercourse; Sexuality
Interest in work, 558
Intergenerational continuity, 163–64
Internal fertilization, 58
Internal representation, 130
Interpersonal intelligence, 243
Intimacy, 448. *See also* Love
Intimacy versus isolation, 48, 470–71
Intra-propositional thinking, 334
Intrapersonal intelligences, 243
Intrauterine devices (IUDs), 64
Introjection, defense mechanism, 33
Intuitive-projective faith, 589
Invasion, in implantation, 83
Inversion, in thinking, 335
Invulnerable children, 290–93
characteristics, 293
coping, 292–93
identifying, 290–92
IQ. *See* Intelligence
Irreversibility, preoperational stage, 186
Irritable infants, attachment and, 159–60
Isolation, 471
intimacy versus, 48, 470–71
IUDs (intrauterine devices), 64
IVF (in vitro fertilization), 58, 67–70

J

Janus Report, 390, 551
Jumping, early childhood, 179
Jung's theory, 586
Juvenile delinquency, 287–90
adolescence, 369–74
gangs, 372–74
learning and, 372
nonaggressive offenders, 369–71
causes of, 288–90
depression and, 361
theories of early criminal behavior, 288–90

K

Kaposi's sarcoma, 93
Klinefelter syndrome, 76, 78
Knowledge, women's ways of knowing, 446–47
Knowledge-acquisition components, 240
Kohlberg's theory, moral development, 250–51

L

Labia majora, 324
Labia minora, 324
Labor, signs of, 100
Lamaze method, 104
Language, 184
bilingual education, 277–78
bilingualism, 255–56
dialect differences, 277
expressive, 193
immigrant children, 255
whole, 274
Language acquisition device (LAD), 138
Language and Thought of the Child, The (Piaget), 138
Language development. *See also specific developmental periods*
sequence of, 195
theories of, 136–39
behavioral, 138–39
biological, 137
cognitive, 137–38
nativist, 137
psycholinguistic, 138
Late adulthood, 524–66. *See also* Aging
Alzheimer's disease, 536–37
appearance, 539
cognitive development, 541–46
creativity, 543–46
decline in, 541–42
terminal drop theory, 542–43
creativity, 543–46
decline, 541–42
personal development, 562–66
Erikson's theory, 565–66
physical development, 531–40
appearance, 539
health, 536–39
reaction time, 531–33
sensory abilities, 534–35
social development, 550–62
family relationships, 555–58
gender roles, 550
retirement, 559–61
sexuality, 550–55
work, 558–59
suicide, 579–81
widowhood, 556
Latency stage, 35
Lateralization, cerebral, 176
LCU (life change unit), 426
Lead, as teratogen, 91
Learning. *See also* Education; Schools
about gender identity, 219
juvenile delinquents and, 372
lifelong, 494
middle adulthood, 493–95
observational, 43–44, 281
slow learners, 188–89
styles of, 276
Learning theory of homosexuality, 385
Leboyer method, 104
Legal death, 570
Leisure, retirement, 560
LEP (Limited English Proficiency), 277
Levinson's theory, 466–67, 468–70, 506–12
Libido, 33
Life change unit (LCU), 426
Life course, 467
Life cycle, adult, personal development stages, 466–67
Life stages for creativity, 544–46
Life structure, 467
Life-style, health and. *See* Health, life-style and
Lifespan
charting, 9
divisions of, 6

Lifespan psychology, 2–27. *See also* Developmental research
adolescence, 15–17
adulthood, 17
childhood, 14–15
continuity versus discontinuity, 18–19, 35, 504–5
culture and development, 18
developmental tasks, 9–11
meaning of, 6
models of development, 7–8
needs across lifespan, 10, 12–14
reasons for study of, 6
sensitive periods versus equal potential, 20
systems analysis, 8
Limited English Proficiency (LEP), 277
Linguistic intelligence, 241
Locomotion
infancy, 120–21
late adulthood, 536
Logical-mathematical intelligence, 242
Longitudinal studies, 20, 22–23
Love, 448–50
belongingness needs, Maslow's theory, 12
forms of, 448–49
generative and existential, 553
validation, 449–50
LSD
effects, 350
prevalence of use in adolescence, 348
as teratogen, 91

M

Macrosystem, 8
Magical thinking, adolescent suicide and, 366
Main-effects model, 7
Major false assumptions, 462
Males. *See also* Puberty; Sex; Sexuality; *headings beginning with term* Gender
adolescent, one hundred years ago, 330–31
average-maturing, 329–30
change of life, 482
early-maturing, 329
fathers, attachment and, 162–63
identity, 471–73
late-maturing, 330
reproductive system, 325–26
seasons of life theory, 468–70, 506–11
sex chromosomes. *See* Chromosome(s), sex
Manipulative experiments, 20, 21–22
Marijuana
effects, 350
prenatal development, 91, 95
prevalence of use in adolescence, 348
Marital status, health and, 438
Marriage. *See also* Divorce
changing, 454–57
family and, 454–58
middle adulthood, 499–500
sexual activity before, 442–43. *See also* Adolescence, sexual behavior
traditional, 511, 512
types, 457–58
Masculinity/femininity, 508
Masked depression, 360
Maslow's theory
gangs and, 13
need satisfaction, 12–14
self-actualization, 10, 11–12
Masturbation, 383–84
Maternal age, Down syndrome, 75–76
Maternal emotions, prenatal development and, 98–99
Maternal nutrition, prenatal development and, 96–98
Maturation, early and late, 328–31
Mature Temps, 564
Maturity, components of, 414
Maximum growth spurt, 327
Meiosis, 59, 70
Memory
infancy, 133–34
middle childhood, 249
types, 493
Men. *See* Males; Puberty; Sex; Sexuality; *headings beginning with term* Gender
Menarche, 300, 327
Menopause, 482
Menstrual cycle, 60–64, 300
Mental health
adolescence. *See* Adolescence, mental health issues
depression due to death and. *See* Death, grief
early childhood development, 178
elderly, suicide, 579–81
homeless children, 206
physical health related to, 538
Mental images, 183, 184
Mental retardation, 188–89
Mental structures, 36
Mentoring, 469–70
Meperidine (pethidine), effects, 350
Mescaline, effects, 350
Mesoderm, 84, 118
Mesosystem, 8
Metabolism
accumulation of metabolic waste, aging and, 527
middle adulthood, 479
Metacomponents, of intelligence, 240
Metalinguistics awareness, 254
Metaphor, use of, creative thinking, 339–40
Methadone
effects, 350
prenatal development, 95
Methaqualone, effects, 350
Methylphenidate, effects, 350
Microsystem, 8
Mid-life crisis, 509
work, 518–20
suggestions for dealing with, 520
Mid-life transition, 507–8
Middle adulthood, 478–520
cognitive development, 485–95
creativity, 488–92
information processing, 492–93
intelligence theories, 485–87
learning ability, 493–95
health, 479–80
mid-life crisis, 509
work, 518–20
personality development, 504–16
continuity versus change, 504–5
continuous trait theory, 515–16
Erikson's theory, 514
Gould's theory, 505–6
Levinson's theory, 506–12
Vaillant's theory, 512–14
physical development, 478–84
climacteric, 482–84
muscular ability, 480
sensory abilities, 480–82
weight and metabolism, 479
psychosocial development, 498–520
aging parent relationships, 500
divorce, 502–4
marriage, 499–500
sibling relationships, 500–1
work, 516–20
sex in, 484
unmarried adults in, 500
Middle Ages, adolescence in, 301
Middle childhood, 230–94
characteristics of, 232
cognitive development, 235–43
concrete operational stage, 38–39, 235–37, 332
memory, 249
psychometric view, 237–39
triarchic model of intelligence, 239–41
decision making and reasoning, 245–46
language development, 253–59
bilingualism, 255–56
language use, 254–55
reading, 256–59
moral development, 248–53
motor skills, 233–34
physical development, 232–34
problem-solving skills, 247–48
psychosocial development, 262–94
peers in, 268–72
schools and, 272–78
siblings, 264–68
stress and, 282–93
television and, 278–82
violence and, 287–90
sports, 234
thinking skills, 243–45
Midwives, 104
Mild mental retardation, 188
Mind, structures of, 33–34
Miscarriage, 101
Miseducation, 200, 214
Mistrust, trust versus, 45, 46
Mitosis, 59, 70
Mobility, adolescent suicide and, 368
Modeling, 43–44, 281
Models of development, lifespan psychology, 7–8
Moderate mental retardation, 188
Modifiers of ability, 530
Monogamy, 457
Monozygotic twins, 63
Mons pubis, 324
Mons veneris, 324
Moral development, middle childhood, 248–53
Gilligan's theory, 252–53
Kohlberg's theory, 250–51
Piaget's theory, 248–50
Moral dilemma, 250
Moral reasoning ability, teenage pregnancy, 401
Moratorium of youth, 308–9, 413–14
Moro reflex, 116
Morphine, effects, 350
Mothers. *See also* Divorce; Parent(s); *headings beginning with term* Maternal
with AIDS, 91, 92
interaction with infants, 151, 155. *See also* Attachment
separation of infant from, 156–57
Motivation
depression and, 359
intelligence and, 241
reaction time in late adulthood, 533
sexual
adolescence, 389–90
early adulthood, 440

Motor development
early childhood, 179–81
infancy, 118–21
middle childhood, 233–34
Motor rules stage, 249
Mourning. *See* Death, grief
Multicultural children and tests, 191, 239, 255
Multicultural schools, 276–77
Multicultural view. *See also* Culture; Racial factors
adolescence, 302
communication, 337
drug use, 349
elder care, 557
enrichment through diversity, 51
funerals, 575
immigrant children and tests, 239
initiation rites, 411–12
interactions, 146
language, thought, and immigrants, 255
minority group membership, 425
modeling, 44
racial influences on peer groups, 318
self and others, 216
sports, 539
teenage pregnancy, 399
work, 517
Multiple births, 63, 105
Multiple intelligence theory, 241–43
Muscular ability, middle adulthood, 480
Muscular strength, evaluating, 540
Musical intelligence, 241–42
Mutation, 59
Mutual regulation, 266
Mythical-literal faith, 589
Mythification, 588

N

Narcotics, effects, 350
National Council of Senior Citizens, 564
Nativist view, 137
language development, 137
Naturalistic experiments, 20, 22
Needs
first relationships, 154
Maslow's theory, 12–14
Negative identity, 309
Negative reinforcement, 42
NEO model of personality development, 515
Neonate. *See also* Birth process; Infancy; Premature infants
appropriate for gestational age, 105
assessment
Apgar scale, 117
behavioral, 117–18
Brazelton Neonatal Behavioral Assessment Scale, 117–18
neurological assessment, 117
failure-to-thrive, 121–22
newborn abilities, 115–17
reflexes, 114, 115, 116
respiratory distress syndrome, 123
sleeping disorders, 122–23
small for gestational age, 105
sudden infant death syndrome, 122
very low birth weight infants, 105
Neurological assessment, infancy, 117
Neuroticism, 515
New York Longitudinal Study, 151
Newborn. *See* Birth process; Neonate; Premature infants
Nicotine. *See* Smoking
No-fault divorce, 503
Noetic dimension, 585
Non-identical twins, 63
Nonaggressive offenders, 369–71
Nonorganic FTT, 122
Normal range of development, 327
Novelty, 421
Novice phase of personality development, 468
Nucleotides, 72
Numeration, 237
Nursing homes, sex in, 553
Nurturing parents, 315–16
Nurturist view, 7
Nutrition
early childhood development, 177
health and, 434–35
homelessness, 205
prenatal development and, 96–98

O

Oakland Growth Study, 23
Object permanence, 128, 130
Objectification, 588
Observational learning, 43–44
television and, 281
Occupation. *See* Work
Oedipal conflict, 35
of Horn's theory, 486–87
Old-old, 553
Olfactory sense, 482. *See also* Smell sense
On Death and Dying (Kübler-Ross), 362
One-time, one-group studies, 20, 22
Open adoption, 68
Openness to experience, 516
Operant conditioning, 42
Operations, 37
Opium, effects, 350
Optimal potential, 432–33
Optimum drive level, 421
stimulus reduction versus, 421–22
Oral stage, 35
Organ reserve, 433
Organization, Piaget's theory, 37
Organogenesis, 85
Outward Bound program, 416
Ovaries, 325
Overextensions, speech, 194
Overregularities, speech, 195
Overreliance on limited support, 365
Ovulation, 60
Ovum (egg), 58, 59–61, 324

P

Paradoxical-consolidation faith, 589
Parent(s). *See also* Divorce
adolescents as, 396–402
aging, relationships and care of, 500, 556–57
child abuse and. *See* Abuse
as child abusers, 284
death of
adolescent suicide and, 366
vulnerability and invulnerability, 291
educational pressure and, 214
fantasies about infants, 132
father, attachment and, 162–63
gender differences and, 219–20
guidelines for, premature infants and, 106
loss of, 366
sibling underworld, 266
single. *See* Divorce
Parent-child relationship, vulnerability and invulnerability, 291
Parent-child role reversals, adolescent suicide and, 366–67
Parenting behavior
authoritarian, 203
authoritative, 203
nurturing, 314–15
permissive, 203–4
Passion, 448
Pathological grieving, 572–73
PC pneumonia, 93
PCP, prevalence of use in adolescence, 348
Pedagogy, 494
Peers
adolescence, 316–18
problems with, suicide and, 367–68
gender development and, 221
middle childhood, 268–72
friendships and, 269–70
influence of, 268–272
problems with, 367–69
racial influences, 318
Pelvic inflammatory disease (PID), 64, 395
Penis, 325
Perception, meaning of, 124–25
Perceptual development, infancy, 123–28
Performance, older workers, 559
Performance components, intelligence, 240
Permissive parents, 203–4, 315
Permutations, thinking, 334–35
Personal development. *See also* Personality development
Erikson's theory, 565–66
stages of, during late adulthood, 562–66
Personal intelligences, 243
Personal relationships, substance abuse, 352–53
Personality development
adaptation to life, 512–14
adult life cycle, 466–67
continuity versus change, 504–5
continuous traits theory, 515–16
Erikson's theory, 470–71, 514
Freud's theory, 34–35
generativity versus stagnation, 514
Holland's theory, 459–60
intimacy versus isolation, 470–71, 514
male versus female identity, 471–73
mid-life transition, 507–8
middle adulthood, 504–16
NEO model, 515
novice phase, 468
seasons of man's life, 468–70
seasons of woman's life, 511–12
settling down, 506
transformations, 462–66, 505–6
Personality problems, as cause of suicide, 364–66
Pet ownership, aged persons, 562
Pethidine (meperidine), effects, 350
Peyote, effects, 350
Phallic stage, 35
Phencyclidine
effects, 350
prenatal development, 95
Phenmetrazine, effects, 350
Phenotype, 57, 59, 70
Phenylketonuria (PKU), 77, 78
Phonology, 193
Physical appearance
infants' preferences, 150
late adulthood, 539

Physical development. *See specific periods*
Physical fitness, health and, 438
Physical manifestations of depression, 359
Physiological needs, Maslow's theory, 12
Piaget's theory, 36–39, 41
accommodation in, 37
adaptation in, 37–38
adolescent cognitive development, 332
assimilation in, 37
cognitive development, 38–39
concrete operational stage, 38–39, 235–37, 332
criticisms of, 130, 133
formal operational stage, 39, 332
intelligence role, 36–38
moral development, 248–50
organization in, 37
preoperational stage, 38, 173, 182–87, 332
sensorimotor stage, 38, 129–31, 332
PID (pelvic inflammatory disease), 64, 395
Pituitary gland, 325, 326
PKU (phenylketonuria), 77, 78
Placenta, 83–84
Plantar grasp reflex, 116
Plasticity, 19
Play, 183, 222–25
cognitive development and, 223
development of, 224–25
early childhood, 222–25
emotional release and, 224
meaning of, 222–24
social development and, 224
Poetic-conventional faith, 589
Polyandry, 457
Polygamy, 457
Positive reinforcement, 42
Possible, real versus, 333
Postconventional level of moral development, 250, 251
Postnatal depression, 101
Pragmatics, 193
Precociousness, 340
Preconventional level of moral development, 250, 251
Preference, visual, infancy, 126–27, 150
Pregnancy. *See also* Birth process; Prenatal development; prenatal diagnosis; Teenage pregnancy
ectopic, 101
weight gain in, 96–97
Premarital sex, 442–43. *See also* Adolescence, sexual behavior
Premature foreclosure, 309
Premature infants, 101, 103–7
birth process, 103, 104–7
causes of, 105–6
definition of, 105
outlook for, 106–7
Prenatal development, 82–100
developmental risk, 89–90
diethylstilbestrol, 95
embryonic period, 84–85
fetal period, 85–86
germinal period, 83–84
implantation, 83
maternal emotions, 98–99
maternal nutrition, 96–98
sensory development, 86–89
teratogens, 90–95
thalidomide, 94
Prenatal diagnosis
amniocentesis, 99
chorionic villi sampling, 99
fetoscopy, 99
ultrasound, 100
Preoperational stage, 38, 173, 183–87, 332
centration, 185
conservation, 186
drawing, 184
egocentrism, 185
imitation, 183
irreversibility, 186
language, 184
limits of, 185–87
mental images, 184
play, 183
states and transformations, 186
Preparatory depression, 578
Prepared childbirth, 104
Preseminal fluid, 325
Pretend play, 224
Preterm AGA (appropriate for gestational age), 105
Preterm SGA (small for gestational age), 105
Primary circular reactions, 129, 131
Problem-solving skills, middle childhood, 247–48
Procedural knowledge, 446–47
Prodigiousness, 340–41
Profound mental retardation, 188
Prosocial behavior, television and, 281–82
Prostate glands, 326
prostatectomy and, 552
Prostitution, 370–71
Protective factors, 426
Protest, separation of infant from mother, 156
Psyche, structures of, 33–34
Psychiatric disorders. *See* Mental health
Psychic censor, 33
Psychoanalytic theory, 32–35
anal stage, 35
attachment, 158
defense mechanisms, 33
ego, 33–34
Electra conflict, 35
genital stage, 35
homosexuality, 385
id, 33–34
latency stage, 35
Oedipal conflict, 35
oral stage, 35
phallic stage, 35
superego, 33, 34
Psychogenic model, subcultures, 317
Psychohistorical studies of creativity, 489–92
Psycholinguistic theory, language development, 138
Psychological dimension, 585
Psychological needs, first relationships, 154
Psychometric view, cognitive development, 237–39
Psychopathology, 356. *See also* Mental health
Psychosocial development
attachment, 156–63
early adulthood. *See* Early adulthood, psychosocial development; Personal development stages, early adulthood
early childhood. *See* Early childhood, psychosocial development
Erikson's theory, 45–50, 470–471
infancy. *See* First relationships; Infancy, emotional development; Relationships, infancy
late adulthood. *See* Late adulthood, personal development; Late adulthood, social development
middle adulthood. *See* Middle adulthood, psychosocial development
middle childhood. *See* Middle childhood, psychosocial development
Puberty, 16, 322–32
delayed, 323
early studies, 322–23
rites in, 410
sequence of changes in, 327
timing of, 323, 326–32
Punishment, 42

Q

Quiescence, 421

R

Racial factors. *See also* Culture; Ethnic group; Multicultural view
adolescent sexual behavior, 389
pregnancy and, 399
peer groups, 318
suicide, 580
work, 517
Random scribbling, 181
Rationalization, defense mechanism, 33
RDS (respiratory distress syndrome), 123
Reaction time
evaluating, 540
late adulthood, 531–33
Reading
comprehension, 258
importance of, 256–59
vocabulary, 257
Real versus possible, 333
Realism, 184
Reality testing, adolescent suicide and, 365
Reasoning
empirico-inductive, 334
hypothetico-deductive, 334
middle childhood, 245–46
moral, 401
Recapitulation theory, 303–4
Received knowledge, 446
Recessive gene, 59
Recessive traits, 73, 74
Recidivism rates, 416
Reciprocal interactions, 7–8, 145–47
infancy, 116
Reconciliation fantasies, 208
Reflexes, 114
neonatal, 115, 116
Regression
coping with stress, 292
defense mechanism, 33
Reinforcement, 42
gender identity and, 220
total range of, 158
Relationships
aging parents, middle adulthood, 500, 556–57
infancy, 144–64. *See also* Attachment
biopsychosocial model, 145–46
characteristics of, 149
first relationships, development of, 153–55, 156
information processing, 146
interactions in, 148
personal characteristics, 150–51
reciprocal interactions, 145–47, 147
research background, 144–45
sensitive responsiveness, 147
temperament and, 151–53
love. *See* Love
peers and. *See* peers

siblings. *See* Siblings
sources, 144, 145
substance abuse, 352–53
Relativism, 444
Religion. *See* Spirituality
Religious fanaticism, adolescent suicide and, 366
Representation, 182
Repression, defense mechanism, 33
Reproductive images, 184
Reproductive system, 323–26
female, 324–25
male, 325–26
Repudiation, 307
Repudiation versus identity confusion, 45, 47–48
Research. *See also* Developmental research
empirical, 31
Research articles, understanding, 27
Resilience, 426
Resistance stage, 422
Respiratory distress syndrome (RDS), 123
Response, 42
Restlessness, depression and, 361
Retired Executives' Service Corps, 564
Retirement, 559–61
leisure and, 560
phases of, 560–61
programs for, 564
Retreat, escape method, 445
Rh factor, 104
Risk and resilience, stress, 426
Risk factors, 426
Rivalry, siblings, 266–67
Role discontinuity, 550
Roman Empire, funerals, 575
Rooting reflex, 115, 116
Rubella (German measles), 91–92
Rule(s), moral development, 248–50
Rule-governed play, 222
Rule learning, language development, 193
Running, early childhood, 179
Running away, 370
depression and, 360

S

Safety
sense of, self-esteem and, 217
travel, working women, 518
Safety needs, Maslow's theory, 12
SART (Society for Assisted Reproductive Technology), 68–69
Schemata, 37
Schools. *See also* Education
character development and, 252
combating substance abuse, 353–55
middle childhood, 272–78
bilingual education, 277–78
educational change, 274–76
multicultural, 276–77
social development and, 273–74
vulnerability and invulnerability, 291
Scribbling, early childhood, 181
Scrotum, 326
Seasons of life theories
men, 468–70, 506–11
women, 511–12
Secondary circular reactions, 129, 131
Secular trend, 331
Secure attachment, 159
Sedatives
effects, 350
prevalence of use in adolescence, 348
Self
core, 132
development of, early childhood, 215–22
emergent, 132
responsibility for, 466
sense of, 216
subjective, 132
verbal, 132
Self-actualization, 10, 11–12
Self-esteem, 215–17
low, teenage pregnancy, 401
Self-fulfilling prophecy, 479
Semantics, 193
Senile macular degeneration, 481
Senior Worker Action Program, 564
Sense(s)
infancy, 123–28
late adulthood, 534–35
middle adulthood, 480–82
prenatal development of, 86–89
Sense of space, 130
Sensitive periods versus equal potential, 20, 35
behaviorist view, 44
cognitive structures view, 41
Freud's theory, 35
psychosocial view, 49
Sensitive responsiveness, relationships, 145, 147
Sensorimotor stage, Piaget's theory, 38, 129–31, 332
Separation
of infant from mother, 156–57
mid-life transition, 508
Sequential studies, 20, 24–25
Seriation, 235
Settling down, personality development, 506
Sever mental retardation, 188
Sex. *See also* Females; Males; Sexuality; Sexually-transmitted diseases; *headings beginning with terms* Gender *and* Sexual
chromosomal, 217
genital, 218
gonadal, 217–18
hormonal, 218
middle adulthood, 484
in nursing homes, 553
stress and, 286
suicide and, 580
vulnerability and invulnerability, 291
vulnerability to stress and, 286
Sex chromosomes. *See* Chromosome(s), sex
Sex cleavage, 221
Sex-linked inheritance, 59, 74
Sexual abuse
of adolescents, 391
of children, 285
Sexual harassment on job, 517–18
Sexual identity, 379
Sexual intercourse, 443
first, 387–89
late adulthood, 551–55
premarital, 442–43. *See also* Adolescence, sexual behavior
Sexuality
adolescence. *See* Adolescence, sexual behavior
communication about, 388
early adulthood. *See* Early adulthood, sexuality
first coitus, 387–89
late adulthood, 550–55
Sexually transmitted diseases (STDs), 64, 392–96
AIDS, 392–93
chlamydia, 394
genital herpes, 395
gonorrhea, 394
Hepatitis B, 395
pelvic inflammatory disease, 395
syphilis, 395
Shame, autonomy versus, 45, 46
Shock and denial of death, 576–77
Shootings, drive-by, 374
Sibling(s)
developing bonds, 264–65
development and, 267–68
gender differences and, 220–21
middle adulthood, 500–1
positive aspects, 265–67
Sibling underworld, 266
Sickle-cell anemia, 76, 78
SIDS (sudden infant death syndrome), 122
Silence, 446
Single parents. *See* Divorce
Single persons, middle adulthood, 500
Skeletal system, late adulthood, 535
Skin, late adulthood, 535
Skinner's theory, 41–43
Sleep requirements, childhood, 124
Sleeping disorders, infant, 122–23
Slow learners, 188–89
Slow-to-warm-up children, 152
Small for gestational age (SGA) infants, 105
Smell sense
evaluating, 540
late adulthood, 534
middle adulthood, 482
prenatal development of, 87
Smiling, infancy, 164
Smoking
health and, 436–37
prematurity, 105
prenatal development, 95
prevalence in adolescence, 348
teratogenic effects of, 91
Social class. *See* Socioeconomic status
Social death, 570
Social development. *See also* Relationships; *specific developmental periods*
play and, 224
schools and, 273–74
Social learning theory, of gender development, 222
Social perspective-taking, 270
Social Readjustment Rating Scale, 426
Socialization, 200
Socialized speech, 137
Society for Assisted Reproductive Technology (SART), 68–69
Sociobiological view, sexuality, 441–42
Socioeconomic status
early childhood development, 178
gangs and, 373–74
prematurity and, 105
teenage pregnancy, 401
Solidarity, intimate partners, 471
Solidarity versus isolation, 45, 48
Solidification, thinking, 335–36
Somatic dimension, spirituality, 585
Space, sense of, 130
Spatial intelligence, 242–43
Speech. *See also* Language; Language development
holophrastic, 135
irregularities of, language development, 194–96
socialized, 137
telegraphic, 136

Sperm, 58–59, 326
Spina bifida, 77, 78
Spirituality, 583–90
  adolescent suicide and, 368
  fanaticism, 366
  Fowler's theory, 588–90
  Frankl's theory, 585–86
  Jung's theory, 586
  religious participation, 584–85
  Wilson's theory, 587–88
Sports
  gateball in Japan, 539
  as initiation rite, 415
  in middle childhood, 234
Stability and Change in Human Characteristics (Bloom), 22
Stability versus change, 19, 35
  behaviorist view, 44
  cognitive structures view, 41
  Freud's theory, 35
  psychosocial view, 49
Stage theorists, 505
Stages of dying, 576–79
Stagnation, 514
  generativity versus, 45, 49, 514
Standing, infancy, 120–21
Stanford-Binet Intelligence Test, 238
State of identity, 307
States and transformations, preoperational stage, 186
Stepping reflex, 116
Stereotypes, 298
  gender, 217
Stillbirth, 101
Stimulants
  effects, 350
  prevalence of use in adolescence, 348
Stimulus, 42
  habituation, 125
Stimulus reduction versus optimum drive level, 421–22
STORCH diseases, 90–92
Storm and stress, 303
Strange situation technique, 158–59
Stress
  adulthood, 418–27
    change as source of, 419–21
    disease as result of, 424–25
    general adaptation syndrome, 422–23
    mid-life crisis, 507–11, 520
    relationship between illness and, 426
    risk and resilience, 426
    stimulus reduction versus optimum drive level, 421–22
  maternal, prenatal development, 98–99
  middle childhood, 282–93
    age and, 286
    child abuse, 283–85
    coping patterns, 292–93
    developmental effects, 287
    invulnerable child and, 290–93
    sex and, 286
    temperament and, 287
    types, 285–87
    violence and, 287–90
  storm and stress, 303
Structure building, 467
Structure changing, 467
Subcultures, in adolescence, 316–18
  adult anxieties about, 316–17
  origins, 317–18
Subjective knowledge, 446
Subjective self, 132
Substance abuse
  adolescence, 347–55
  combating, 353–55
  crime and, 352
  ethnic group and, 349–51
  personal relationships and, 352–53
  prevalence of, 347–49
  prevalence of use in adolescence, 348
Sucking reflex, 116
Sudden infant death syndrome (SIDS), 122
Suggestibility, adolescent suicide and, 365
Suicide
  adolescence, 363–69
    attempts, meaning of, 363–64
    family problems and, 366–67
    mental disorders, 363–69
    personality problems as cause, 364–66
    societal factors in, 367–69
  attempts, 363–64, 581
  elderly, 579–81
  gender influence, 580
  racial influence, 580
  stress in childhood, 282
  warning signs of, 369
Superego, 33, 34
Super's theory, 460
Support, overreliance on, adolescent suicide and, 365
Suppression, coping with stress, 292
Symbolic play, 183
Symmetry, 155
  mother-infant interaction, 155
Synchrony, mother-infant interaction, 155
Syntax, 193
Syphilis, 90, 395
  teratogenic effects of, 91
System kids, 370
Systems analysis, 8

T

Talented children, 189–90, 340–42
Taste sense
  evaluating, 540
  late adulthood, 534–35
  middle adulthood, 482
  prenatal development of, 87
Taxonomy of Educational Objectives (Bloom), 244–45
Tay-Sachs disease, 76, 78
Teenage pregnancy
  causes of, 400–402
  cognitive ability and, 401
  moral reasoning ability and, 401
  multicultural view, 399
  physiological causes, 400
  psychological causes, 401
  role of family, 398–400
  social causes, 401
  trends in behavior, 397–98
Teeth, late adulthood, 535
Telegraphic speech, 136
Television
  cognitive development and, 278–80
  extent of watching, 279
  middle childhood and, 278–82
  prosocial behavior and, 281–82
  violence and, 280–81
Temperament, 151–53
  stress effects, 287
  vulnerability and invulnerability, 291
  vulnerability to stress and, 287
Temporizing, 445
Teratogens, prenatal development and, 90–95
Terminal drop, late adulthood, 542–43
Tertiary circular reactions, 130, 131
Testes, 326
Tests, intelligence, 238–39
Tetrahydrocannabinol, effects, 350
Thalidomide
  prenatal development, 94
  as teratogen, 91
Theories. *See also specific theoretical approaches*
  importance of, 30–31
Thinking skills
  adolescence
    combinations, 334–35
    compensation, 335
    consolidation, 335–36
    convergent thinking, 337
    creative thinking, 338–42
    critical thinking, 337–38
    divergent thinking, 337
    empirico-inductive thinking, 334
    hypothetico-deductive, 334
    inter-propositional thinking, 334
    intra-propositional thinking, 334
    inversion, 335
    permutations, 334–35
    solidification, 335–36
  divergent, 337, 544
  immigrant children, 255
  infancy, 132
  middle childhood, 243–45
Time sequences, 130
Time-variable designs, 20, 22–25
Timing of puberty, 326–32
Tobacco. *See* Smoking
Toilet training, 120
Tonic neck reflex, 116
Total range of reinforcement, 158
Touch sense
  evaluating, 540
  late adulthood, 535
  prenatal development of, 86
Toxoplasmosis, 90–91
Traditional marriage enterprise, 511, 512
Trait(s), inheritance, 73–74
Trait theorists, 504
Tranquilizers, prevalence of use in adolescence, 348
Transactional analysis, 147
Transactional model, 7–8
  development, 145
Transformations
  personality development, 462–66, 505–6
  preoperational stage, 186
Transformations (Gould), 462–63
Transience, 419
Transition, 467
  adolescent suicide and, 368
  to adulthood, 412–13. *See also* Initiation rites
Transsexual operation, 379
Travel safety, working women, 518
Triarchic model of intelligence, 239–41
Trisomy, 59
Trophoblast, 83
Trust versus mistrust, 45, 46
L-Tryptophan, 479
Tuberculosis, 93
Turner syndrome, 76, 78
Twins, 63, 105
Two-word stage, language development, 135

U

Ultrasound, prenatal diagnosis, 100
Umbilical cord, 84
Unconscious, Freud's theory, 32, 33
Universalizing faith, 589
Unmarried persons, middle adulthood, 500

Ureter, 325, 326
Urethra, 325, 326
Uterus, 325

Vaillant's theory, 512–14
Valachi Papers, The (Maas), 289
Validation, love and, 449–50
Variable accommodation, 126
Vas deferens, 326
Verbal evocation, 183
Verbal self, 132
Very low birth weight (VLBW), 105
Violence, 287–90. *See also* Crime; Juvenile delinquency
  adolescent suicide and, 368
  depression and, 361
  gangs, 369
  television and, 280–81
  theories of early criminal behavior, 288–90
Viruses, 94. *See also* Cytomegalovirus (CMV); German measles (rubella); Herpes simplex virus; Human immunodeficiency virus (HIV)
Vision
  evaluating, 540
  infancy, 115
    adaptation, 127–28
    perception, 126–28
    preference, 126–27
    variable accommodation, 126
  late adulthood, 534
  middle adulthood, 481
  prenatal development of, 87
Visual cliff experiment, 127–28
Visual tracking, infancy, 126
Vivienne: The Life and Suicide of an Adolescent Girl (Mack and Hickler), 21
Vocables, 135
Vocabulary, 257
  growth of, 136
Vocalizations, mother-infant interaction, 155
Vulva, 325
Vygotsky's theory, 40–41
  language development, 138

W

Walkabout approach, as initiation rite, 417
Walking. *See* Locomotion
We Are Aging (Botwinick), 545
Wear and tear theory of aging, 526
Wechsler intelligence scales, 238
Wechsler intelligence theories, 485
Weight
  early childhood, 176
  gain in pregnancy, 96–97
  middle adulthood, 479
Whole language, 274
Widowhood, 551, 556
Wild Strawberries (film), 565
Wilson's theory, 587–88
Withdrawal, coping with stress, 292
Women. *See* Females; Fertilization; Mothers; Pregnancy; Puberty; Sex; Sexuality; *headings beginning with terms* Gender *and* Maternal
Words, first, 136
Work
  career choices, 459–60
  dual-career family, 460–62
  early adulthood, 458–62
  employment patterns, 459
  late adulthood, 558–59
  middle adulthood, 516–20
    mid-life crisis, 518–20
    women, 517–18

X chromosome. *See* Chromosome(s), sex
XXY syndrome, 76
XYY syndrome, 76, 78

Y chromosome. *See* Chromosome(s), sex
Young/old, mid-life transition, 507

Zone of proximal development, 40
Zygote, 59, 61–64, 84, 87
Zygote intrafallopian transfer (ZIFT), 68